Composite Materials

Kamal K. Kar
Editor

Composite Materials

Processing, Applications, Characterizations

Springer

Editor
Kamal K. Kar
Advanced Nanoengineering Materials Laboratory
Material Science Programme
Indian Institute of Technology Kanpur
Kanpur, Uttar Pradesh, India

Advanced Nanoengineering Materials Laboratory
Department of Mechanical Engineering
Indian Institute of Technology Kanpur
Kanpur, Uttar Pradesh, India

ISBN 978-3-662-49512-4 ISBN 978-3-662-49514-8 (eBook)
DOI 10.1007/978-3-662-49514-8

Library of Congress Control Number: 2016950571

Printed on acid-free paper

This Springer imprint is published by Springer Nature
The registered company is Springer-Verlag GmbH Germany
The registered company address is: Heidelberger Platz 3, 14197 Berlin, Germany

Dedicated to my wife, Sutapa, and my little daughter, Sristisudha, for their loving support and patience, and my mother, late Manjubala, and my father, Khagendranath

Preface

Composite materials have emerged as a major class of advanced elements and are either used or being considered as substitutions of metals/traditional materials in aerospace, automotive, civil, mechanical, and other industries. The outstanding features of composites are their high specific stiffness, high specific strength, and controlled anisotropy, which make them very attractive materials. A unique feature of composites is that the characteristics of the finished product can be tailored to a specific engineering requirement by a careful selection of matrix and filler. Due to these reasons, high performance advanced composites are widely used in several applications. As a result, a number of edited books have been published in the last decade outlining a variety of topics of current interest related to composite research. This is also an edited book for graduate students and researchers from various fields of science and technology, who wish to learn about the recent development of composites.

In ▶ Chap. 1, focus is directed towards the processing and applications of fiber/filler reinforced polymer based composite materials, as they constitute an important class of advanced materials and share a significant part of engineering material market. In the last few decades, several processes, i.e., hand lay-up, vacuum bag, pressure bag, filament winding, pultrusion, resin transfer molding, vacuum assisted resin transfer molding, compression molding, structural reaction injection molding, structural foam reaction injection molding, sandwich molding, etc., are developed by composite engineers around the globe. These processes with application of composites are discussed in this chapter.

The characterization of a material is one of the essential/critical tasks for developing a product in the advanced applications. Chapter 2 addresses the major characterization techniques, i.e., volume fractions of constituents, voids in composites, surface roughness, tensile strength, flexural strength, interlaminar shear strength, impact strength, notch strength, fracture toughness, modulus, creep, stress relaxation, thermal properties, glass transition temperature, electrical properties, magnetic properties, piezoelectric properties, tribological properties, rheological properties, biological properties, etc. These characterization studies are also addressed in the adverse environments.

Initially, the composite manufacturing processes developed for aerospace industries have been given too much emphasis on the quality, and as a result the production volume is less. In ▶ Chap. 3, a new manufacturing method, i.e., rubber pressure molding technique for long fiber/fabric reinforced polymer composites is discussed. A steel die and a rubber punch are used to produce the component in this process. This type of combination gives a nearly uniform hydrostatic pressure distribution over the surface of the composite part and hence produces a good quality product with uniform distribution of fibers and resin with lesser voids.

As composite materials are markedly superior to the conventional materials, they are better and acceptable over the traditional metallic/ceramic/polymer. But the functionally graded materials are another class of composite materials, wherein the compositions of two or more components vary spatially or in a stepwise manner. This leads to innumerable advantages and a component can be used as a multifunctional product. It is also possible to achieve the contradictory properties within a component using the concept of functionally graded material. ▶ Chapter 4 defines the concept of functionally graded material, and various manufacturing techniques involved in making the graded composites.

Electrodeposition is one of the most successful industrial processes for the deposition of mainly metallic coatings over the traditional products to improve the performance of material in the aggressive environment. The versatility of technique also allows the deposition of composites with special features at the micrometer and nanometer levels. Although the process is most suitable for the metal matrix composites, recent study reveals the potential in allowing the deposition of ceramic and polymer based composites as well. In ▶ Chap. 5, an overview of the fundamentals of electrodeposition and processing of composite materials are discussed.

The polycarbonate, owing to its unique combination of properties such as high toughness and excellent compatibility with several polymers, proves to be a suitable matrix for structural applications. Whereas, carbon fiber, due to its superior tensile strength as high as 230 GPa and being lightweight, finds vast applications in aerospace, military, and civil industries. ▶ Chapter 6 discusses the preparation and detailed characterization of short carbon fiber reinforced polycarbonate composites.

Ionic polymer metal composites are technologically matured electro-active polymers. These are frequently known as “soft actuators-sensors” or “artificial muscles” because of their electromechanical coupling behavior. The various factors, which determine the efficiency of such devices, are discussed in ▶ Chap. 7 along with the working principle of ionic polymer metal composites including fabrications and applications.

Carbon based polymer nanocomposites (PNCs) are of recent research interest and cover numerous applications including structural, drug delivery, shape memory polymers, etc. In Chap. 8, different allotropes of carbon viz. carbon nanotubes, graphene, graphene oxide, fullerenes, metallofullerenes, carbon nanohorn, carbon nanodiamond, etc., are discussed in brief. Different types of nonconducting polymeric materials used for the fabrication of PNCs are also introduced. Several types of carbon-based PNCs and their fabrication methodologies have been emphasized to represent a broad overview on carbon based PNCs using nonconducting polymer matrices.

Solid foams form an important class of lightweight cellular engineering materials that are used in engineering design mostly as core materials in the sandwich composite structures. These foams are formed by trapping of gas bubbles in a solid matrix thereby resulting in light-weight materials. The fundamentals, current state of research, and applications of thermosetting polymer based syntactic foams are critically addressed in ▸ Chap. 9.

Carbon-carbon composites (C/Cs) became of interest in 1970s when the light-weight high temperature resistant material was required for the space vehicles. The two critical applications most often mentioned are re-entry vehicles and rocket nozzles. Other few commercial applications are the requirements for high performance lightweight brakes for aircraft and other high performance braking applications. Since there are not many materials that can withstand the temperatures up to 3000 °C, carbon fiber-carbon matrix composite has become a material of interest. ▸ Chapter 10 summarizes various aspects of C/Cs in terms of matrix and reinforcement precursors, fabrication and effects of various processing parameters on their mechanical and thermal properties, and their applications.

The metal matrix composites have various advantages over the traditional composites. But one of the most important disadvantages is that the metal matrices are poor in chemical and mechanical compatibility with the reinforcements. In ▸ Chap. 11, authors have discussed various reinforcing fiber, matrixes, processing, characterizations, and applications of metal matrix composites.

Functionally graded materials (FGMs), due to their characteristic spatial gradient of structural/compositional features, exhibit unique properties that are not possessed by their constituent components. Applications of FGMs cover diverse areas including biomedicine, tissue engineering, structural composites for advanced applications, etc. ▸ Chapter 12 deals with a case study of synthesis and characterization of FGMs consisting of soft magnetic iron/nickel particles as filler in styrene butadiene rubber matrix.

Copper-graphite composite is a typical electrical sliding contact material. Graphite has unique antifriction properties but also possess some disadvantages, i.e., low strength and current-carrying capacity. Addition of graphite particles to copper matrix reduces density, increases stiffness, raises the service temperature, and provides a mechanism for tailoring the coefficient of thermal expansion. ▸ Chapter 13 addresses the various types of copper-graphite composite, processing, performance of composites, and applications.

Ceramic materials have excellent strength and thermal stability. However, the brittleness is a major barrier to these monolithic ceramics. Therefore, various particulates, i.e., whisker, fibers, or fabric materials with different forms and orientations are used to enhance the toughness of ceramic matrix composites (CMCs) at elevated temperature in several advanced applications. ▸ Chapter 14 focuses on the processing, properties, and applications of CMCs

Although traditional ceramics are manufactured by ceramic industries (in terms of volume as well as the selling price), a variety of new ceramics has been immersing progressively with better properties and performances. ▸ Chapter 15 addresses zirconia and its nanocomposites, and their performances in thermal and chemical barrier coating, buffer layer for high temperature superconducting films, ball heads for hip replacements, automobiles, and series of optical and/or biological sensors.

Superionic conductors, fast-ionic conductors or solid electrolytes are special class of materials, in which the ionic conductivity is very high and in some cases it can be comparable to that of molten salts and ionic solutions. ▶ Chapter 16 briefly discusses the conducting properties of fast ion conductors based on phosphate glass, which includes recent development in understanding of the nearly constant loss phenomenon, scaling behavior of AC conductivity, few models in the field of ion transport in glass electrolytes, etc.

Carbon nanotube and graphene are allotrope of carbon having hexagonal arrangement of sp2-hybridized carbon. They exhibit excellent mechanical, thermal, and electronic properties. In ▶ Chap. 17, the authors focus on ceramic composites reinforced by carbon nanotubes and graphene.

Bamboo is a potential natural fiber with its good stiffness, rigidity, easy availability, and hydrophobic nature. This hydrophobicity renders improper interfacial adhesion during polymer matrix based composite fabrication. This adhesion can be improved by suitable surface treatment of bamboo fiber, which is an important research topic to many scientists and is covered to some extent in ▶ Chap. 18.

Naturally occurring superhydrophobic/superoleophobic surfaces have received tremendous attention owing to their unique multilevel surface structure. ▶ Chapter 19 reviews the basic physical understanding and the structure-property correlations of such liquid repelling surfaces. The chapter further explores the recent progress achieved towards the development of artificial liquid repellent surfaces by mimicking the natural ones.

The editor and authors hope that readers from materials science, engineering, and technology will be benefited by reading of these high-quality review articles related to composite materials and their processing and applications. This book is not intended to be a collection of all research activities on composites worldwide, as it would be rather challenging to keep up with the pace of progress in this field. The editor would like to acknowledge many composite researchers, who have contributed to the contents of the book. The editor would also like to thank all the publishers and authors for giving permission to use their published images and original work.

There were lean patches when I felt that 1 would not be able to take time out and complete the book, but my wife, Sutapa, and little daughter, Srishtisudha, played a crucial role to inspire me to complete it. I hope that this book will attract more researchers to this field, and that it will form a networking nucleus for the composite community. Please enjoy the book and communicate to the editor/authors any comments that you might have about its content.

Advanced Nanoengineering Materials Laboratory
Department of Mechanical Engineering
Indian Institute of Technology Kanpur
Kanpur, Uttar Pradesh, India

Kamal K. Kar

Contents

About the Editor

Prof. Kamal K. Kar, Department of Mechanical Engineering and Materials Science, is Umang Gupta Chair Professor from July 2014 to June 2017. He pursued higher studies from Indian Institute of Technology Kharagpur, India, and Iowa State University, USA, before joining as a Lecturer in the Department of Mechanical Engineering and Materials science at IIT Kanpur in 2001. He was a BOYSCAST Fellow in the Department of Mechanical Engineering, Massachusetts Institute of Technology, USA, in 2003 (May-December). He was a Head of Interdisciplinary Programme in Materials Science.

Prof. Kar is an active researcher in the area of nanostructured carbon materials, nanocomposites, functionally graded materials, nanopolymers, and smart materials for structural, energy, and biomedical applications. His research works have been recognized through the office of Department of Science and Technology, Ministry of Human Resource and Development, National Leather Development Programme, Indian Institute of Technology Kanpur, Defence Research and Development Organisation, Indian Space Research Organization, Department of Atomic Energy, Department of Biotechnology, Council of Scientific and Industrial Research, Aeronautical Development Establishment, Aeronautics Research and Development Board, Defence Materials and Stores Research and Development Establishment, Hindustan Aeronautics Limited Kanpur, Danone research and development department of beverages division France, Indian Science Congress Association, Indian National Academy of Engineering, etc.

Prof. Kar has been invited to serve as an associate editor of *Materials Express* published by American Scientific Publishers, and member of the editorial boards of *Advanced Manufacturing: Polymer and Composites Science* published by Maney Publishing, *International Journal of Plastic Technology* published by Springer, *Recent Patents on Nanotechnology* by Bentham Science, etc.

Prof. Kar has published more than 165 papers in international referred journals, 114 conference papers, 5 books on nanomaterials and their nanocomposites, 30 review articles/book chapters, and more than 55 national and international patents. He has guided 15 Ph.D. students and 75 M.Tech. students. Currently 15 Ph.D. students, 10 M.Tech. students, and few visitors are working in his group.

Contributors

Sandeep S. Ahankari Department of Mechanical Engineering, Annasaheb Dange College of Engineering and Technology, Sangli, Maharashtra, India

Syed Nadeem Akhtar Advanced Nanoengineering Materials Laboratory, Department of Mechanical Engineering, Indian Institute of Technology Kanpur, Kanpur, Uttar Pradesh, India

S. Aravindan Department of Mechanical Engineering, Indian Institute of Technology Delhi, New Delhi, India

Navajit Singh Baban Advanced Nanoengineering Materials Laboratory, Materials Science Programme, Indian Institute of Technology Kanpur, Kanpur, Uttar Pradesh, India

Soma Banerjee Advanced Nanoengineering Materials Laboratory, Materials Science Programme, Indian Institute of Technology Kanpur, Kanpur, Uttar Pradesh, India

J. K. Chakravartty Materials Group, Bhabha Atomic Research Centre, Mumbai, Maharashtra, India

Jayesh Cherusseri Advanced Nanoengineering Materials Laboratory, Materials Science Programme, Indian Institute of Technology Kanpur, Kanpur, Uttar Pradesh, India

Sarit K. Das Department of Mechanical Engineering, Indian Institute of Technology Madras, Chennai, Tamil Nadu, India

Malay K. Das Advanced Nanoengineering Materials Laboratory, Department of Mechanical Engineering, Indian Institute of Technology Kanpur, Kanpur, Uttar Pradesh, India

Mahuya Das JIS College of Engineering, Kalyani, Nadia, West Bengal, India

Samudra Dasgupta Aeronautical Development Establishment (DRDO), Bangalore, Karnataka, India

Kinshuk Dasgupta Materials Group, Bhabha Atomic Research Centre, Mumbai, Maharashtra, India

Gaurav K. Gupta Advanced Nanoengineering Materials Laboratory, Materials Science Programme, Indian Institute of Technology Kanpur, Kanpur, Uttar Pradesh, India

A. S. M. A. Haseeb Department of Mechanical Engineering, University of Malaya, Kuala Lumpur, Malaysia

M. N. Jagdish Kumar Aeronautical Development Establishment (DRDO), Bangalore, Karnataka, India

Kamal K. Kar Advanced Nanoengineering Materials Laboratory, Materials Science Programme, Indian Institute of Technology Kanpur, Kanpur, Uttar Pradesh, India

Advanced Nanoengineering Materials Laboratory, Department of Mechanical Engineering, Indian Institute of Technology Kanpur, Kanpur, Uttar Pradesh, India

Sudhir Kumar Advanced Nanoengineering Materials Laboratory, Materials Science Programme, Indian Institute of Technology Kanpur, Kanpur, Uttar Pradesh, India

Oindrila Manna Department of Mechanical Engineering, Indian Institute of Technology Madras, Chennai, Tamil Nadu, India

Ayan Manna Department of Biomedical Engineering, Faculty of Engineering, University of Malaya, Kuala Lumpur, Malaysia

Deepak Pandey Advanced Nanoengineering Materials Laboratory, Materials Science Programme, Indian Institute of Technology Kanpur, Kanpur, Uttar Pradesh, India

Sumit Pramanik Advanced Nanoengineering Materials Laboratory, Materials Science Programme, Indian Institute of Technology Kanpur, Kanpur, Uttar Pradesh, India

Department of Biomedical Engineering, Faculty of Engineering, University of Malaya, Kuala Lumpur, Malaysia

K. Rajkumar Department of Mechanical Engineering, SSN College of Engineering, Kalavakkam, Tamil Nadu, India

S. Ram Materials Science Centre, Indian Institute of Technology Kharagpur, Kharagpur, West Bengal, India

J. Ramkumar Advanced Nanoengineering Materials Laboratory, Department of Mechanical Engineering, Indian Institute of Technology Kanpur, Kanpur, Uttar Pradesh, India

K. Ravi Sekhar Aeronautical Development Establishment (DRDO), Bangalore, Karnataka, India

N. L. Ravikumar Advanced Nanoengineering Materials Laboratory, Department of Mechanical Engineering, Indian Institute of Technology Kanpur, Kanpur, Uttar Pradesh, India

B. N. Ravishankar Aeronautical Development Establishment (DRDO), Bangalore, Karnataka, India

S. Sankaran Directorate of Extramural Research & Intellectual Property Rights, DRDO HQ, DRDO Bhawan, New Delhi, India

K. Shahi Department of Physics, Indian Institute of Technology Kanpur, Kanpur, Uttar Pradesh, India

S. D. Sharma Advanced Nanoengineering Materials Laboratory, Materials Science Programme, Indian Institute of Technology Kanpur, Kanpur, Uttar Pradesh, India

Department of Mechanical Engineering, Rewa Engineering College, Rewa, India

Raghunandan Sharma Advanced Nanoengineering Materials Laboratory, Materials Science Programme, Indian Institute of Technology Kanpur, Kanpur, Uttar Pradesh, India

G. P. Singh Centre for Nanotechnology, Central University of Jharkhand, Ranchi, Jharkhand, India

D. P. Singh Solid State Ionics Laboratory, Materials Science Programme, Indian Institute of Technology Kanpur, Kanpur, Uttar Pradesh, India

L. Sowntharya Advanced Nanoengineering Materials Laboratory, Materials Science Programme, Indian Institute of Technology Kanpur, Kanpur, Uttar Pradesh, India

Ashis Tripathy Department of Biomedical Engineering, Faculty of Engineering, University of Malaya, Kuala Lumpur, Malaysia

Polymer-Based Composite Structures: Processing and Applications

1

S. D. Sharma, L. Sowntharya, and Kamal K. Kar

Contents

S.D. Sharma
Advanced Nanoengineering Materials Laboratory, Materials Science Programme, Indian Institute of Technology Kanpur, Kanpur, Uttar Pradesh, India

Department of Mechanical Engineering, Rewa Engineering College, Rewa, India
e-mail: sds1970@gmail.com

L. Sowntharya
Advanced Nanoengineering Materials Laboratory, Materials Science Programme, Indian Institute of Technology Kanpur, Kanpur, Uttar Pradesh, India
e-mail: sowsindhu@gmail.com

K.K. Kar (✉)
Advanced Nanoengineering Materials Laboratory, Materials Science Programme, Indian Institute of Technology Kanpur, Kanpur, Uttar Pradesh, India

Advanced Nanoengineering Materials Laboratory, Department of Mechanical Engineering, Indian Institute of Technology Kanpur, Kanpur, Uttar Pradesh, India
e-mail: kamalkk@iitk.ac.in

K.K. Kar (ed.), *Composite Materials*,
DOI 10.1007/978-3-662-49514-8_1

Abstract

The present chapter focuses on the processing and applications of polymer-based composite structures, as they constitute an important class of materials and share a significant part of engineering material market. These processing techniques are hand layup, vacuum bag, pressure bag, filament winding, pultrusion, resin transfer molding, vacuum-assisted resin transfer molding, compression molding, structural reaction injection molding, structural foam reaction injection molding, sandwich molding, etc.

Keywords

Polymer • Composite • Hand layup • Vacuum bag • Pressure bag • Filament winding • Pultrusion • Resin transfer molding • Vacuum-assisted resin transfer molding • Compression molding • Structural reaction injection molding • Structural foam reaction injection molding • Sandwich molding

Introduction

Composite materials have emerged as a major class of structural elements. These materials are light weight, flexible, and also have high corrosion resistance, impact strength, fatigue strength, etc. Because of these properties, composite materials are being considered as a replacement of traditional materials used in the aerospace, automotive, and other industries. The specialty of composites is that the engineering properties, which are required in the end product, can be achieved by a careful selection of matrix and reinforcement. The outstanding features of fiber-reinforced polymer composites (FRPs) are their high specific stiffness, high specific strength, and controlled anisotropy, which make them very attractive structural materials. Due to these reasons, the high-performance advanced composites, i.e., FRPs are widely used in the pulp and paper, semiconductor, metal refining, power, waste treatment, petrochemical, pharmaceutical and other industries. The following few examples are the products made of FRPs: pressure vessels, scrubbers, hoppers, hoods, ducts, fans, stacks, pipes, pumps, pump bases, valve bodies, elevator buckets, heat-exchanger

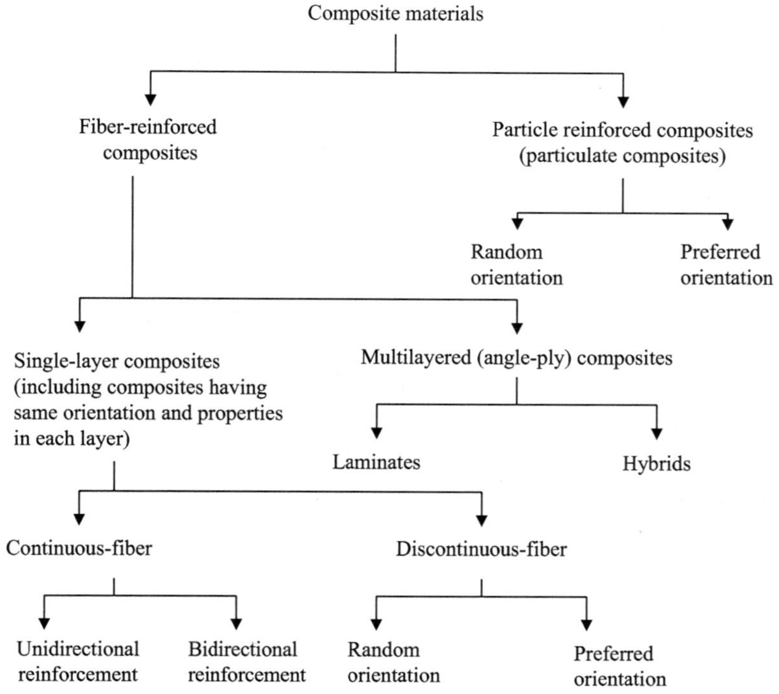

Fig. 1.1 Classification of composite materials

shells and tube sheets, mist-eliminator blades, grating, floor coatings, tank lining systems, etc.

It consists of two or more chemically distinct constituents having a distinct interface, which separates them as reinforcing particle and matrix. Also, it has a unique combination of properties that are noticeably different from the constituent properties. Generally, a discontinuous phase (called reinforcement) is embedded into a continuous phase (called matrix). These materials are generally classified on the basis of geometry of the reinforcement used. The reinforcement may be in the form of fibers or particles. Accordingly, two broad classifications are fibrous composites and particulate composites. A more detailed classification is given in Fig. 1.1 [1].

In the present chapter, focus is directed toward the FRPs as they constitute an important class of materials and share a significant part of engineering material market. A market statistics of polymer matrix composites by end-use application/ market percentage based on the US glass fiber composite shipments in 2014 (totaling 8.2 billion lbs) is given in Fig. 1.2, which includes both types of matrix materials, i.e., thermosets and thermoplastics [2]. A market forecast for US composite shipment by application segments is also shown for 2020.

FRPs offer many attractive properties but their manufacturing process presents new challenges day by day. Most of these raw materials are easily available in abundance. Cost of carbon fiber, which is used as reinforcement, is also considerably

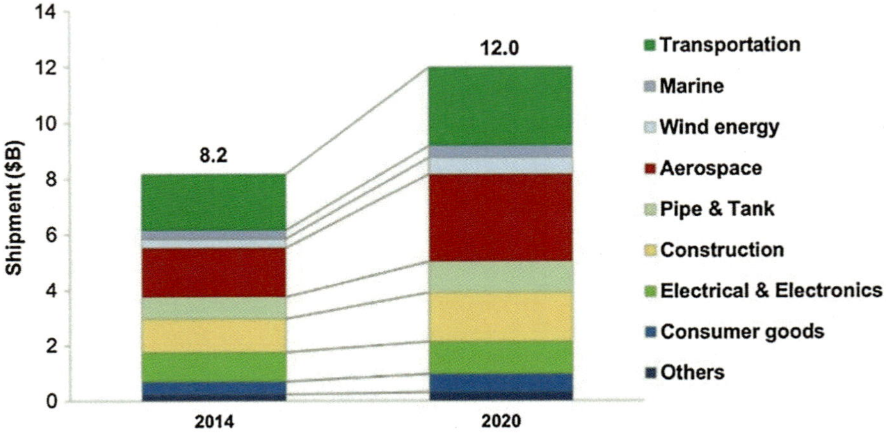

Fig. 1.2 US composite material market forecast by application segments (Courtesy: American composites manufacturers association)

reduced in the last decades, and cost of materials per unit volume is competitive to the traditional materials. Still, the percentage use of composites is less. Major bottleneck is the manufacturing process. Initially, the processes developed for aerospace industry gave too much emphasis on the quality and as a result, the production volume was less [2].

There are several methods to manufacture FRPs, such as vacuum bagging, autoclave, filament winding, pultrusion, matching die set compression molding, resin transfer molding (RTM), resin infusion, and other liquid composite molding (LCM). Among these methods, autoclave is the best method for manufacturing of some aeronautic parts. Recently, LCM (such as RTM, infusion, bladder molding, etc.) is becoming more and more popular in the aeronautic industry. But a major cost issue for manufacturing of FRPs and parts using autoclave is the requirement of expensive tooling and disposable bagging materials. Other disadvantages are long cure times, high energy consumption, volatile toxic byproducts, creation of residual stress and voids in the materials, and the use of expensive tooling that are capable of withstanding high autoclave temperature. Therefore, a key step to further increase the overall usage of FRP products is to decrease the manufacturing cost [2].

Polymer Matrix Composites

Polymer matrix composites (PMCs) constitute a major category of composite materials with a wide range of applications. In PMCs, fibers or any other reinforcing agent is embedded in the polymer matrix. They offer very attractive properties which can be tailored to the specific requirements by proper selection of fiber, matrix, fiber configuration (short, long, straight, woven, braided, laminated, etc.) and fiber surface treatment. PMCs are lightweight material coupled with high stiffness and strength along the direction of reinforcing fiber. They also exhibit dimensional stability, temperature and chemical resistance, and relatively easy processing. Advanced PMCs are replacing metal components in many uses, particularly in aerospace and automotive industries [2].

Matrix Materials

The polymer matrix binds the fibers together so as to transfer the load to and between them and protects them from environments and handling at work. Thermosets and thermoplastics are the two basic types of resin used in the manufacturing of advanced PMCs [2].

Thermoset Resins

Nowadays, in the advanced composite industry, thermoset resins dominate over the thermoplastic resins. In order to produce a product using thermoset resins, addition of a curing agent or hardener and impregnation onto a reinforcing material followed by curing are key steps [2]. Some properties of important thermoset resins are given in Table 1.1.

Epoxies The epoxy resins have relatively low molecular weight with low shrinkage (1–5 %) during cure. They can be cured partially and stored in that state. The cured epoxy resins exhibit good mechanical and thermal properties, high resistance to chemicals and corrosion. Their glass transition temperature varies from 100 °C to 270 °C. However, they are more expensive compared to the polyester resin. These resins vary from low-viscosity liquids to high-molecular-weight solids. The curing agent or hardener is one of the essential ingredients for epoxy. These compounds control the reaction rate and determine the performance characteristics of the finished part. The aromatic amines like 4,4′-methylene-dianiline (MDA) and 4,4′-sulfonyldianiline (DDS) are the most commonly used curing agents in the advanced composite industry [2].

Table 1.1 Typical properties of some thermoset resins

Resin	Density (g/cm^3)	Tensile strength (MPa)	Tensile modulus (GPa)	HDT (°C)	Cure shrinkage (%)	Glass transition temp. (°C)
Epoxies	1.2–1.3	55–130	2.75–4.10	48	1–5	100–270
Polyester	1.1–1.43	34.5–103.5	2.1–3.45	60–205	5–12	70–120
Vinyl ester	1.12–1.32	73–81	3.0–3.5	93–135	5.4–10.3	102–150
Phenolic	1.00–1.25	30–50	3.6	165–175	0–0.01	260
Polyimide (PMR-15)	1.32	38.6	3.9	–	0–0.006	320–330
Polyurethane	1.1–1.5	1–69	0.069–0.69	50–205	0.02	135

Polyesters Polyester resins are widely used due to their good mechanical properties, corrosion resistance, low weight, and low cost. Their tensile strength varies from 34.5 to 103.5 MPa. Glass transition temperature of polyester resins varies from 70 °C to 120 °C, which is lower than epoxy resin. They consist of linear long-chain polymers dissolved in the vinyl monomer (usually styrene). Unsaturated polyesters are combinations of anhydrous forms of saturated and unsaturated dicarboxylic acids (e.g., phthalic anhydride and maleic anhydride, respectively). These acids react with diols to produce the ester group in the polymer backbone. The unsaturation sites in the polymer chains are reactive, and they cross-link with the styrene monomer via a free radical reaction. A peroxide catalyst, such as methyl ethyl ketone peroxide (MEKP), is used as an initiator. The speed of reaction can be increased at a given temperature by using an accelerator in the resin, such as cobalt octoate. There should be a reasonable working time before the resin changes from liquid to solid state. This can be achieved by the presence of inhibitors in the resin, such as quinines or catechols. The addition of heat to the system accelerates this cross-linking reaction significantly [2].

Polyurethanes Polyurethanes are another group of resins used in the advanced composites. These compounds are formed as a result of reaction between the polyol component and an isocyanate compound, typically toluene diisocyanate (TDI). Other examples are methylene diisocyanate (MDI) and hexamethylene diisocyanate (HDI). Polyurethanes are versatile polymers and perform well for tensile strength (1–69 MPa), abrasion resistance (abrasion index, NBS ranging from 110 to 435), and tear strength (12–192 kN/m). These resins are used to manufacture fiber-reinforced structural foams [2].

Phenolic and Amino Resins Phenolic and amino resins are another group of thermosetting resins. These are low in cost and have good electrical and heat-insulating properties along with good mechanical properties. Dielectric strength of mica-filled phenolic resins varies from 13,790 to 15,760 V/m, whereas their tensile strength varies from 38 to 48 MPa. Amino resins are formed by controlled reaction of formaldehyde with various compounds that contain the amine group – NH_2. Urea-formaldehyde and melamine-formaldehyde are common examples in this category. They have low smoke and heat release properties in the event of a fire. Hence, aircraft interiors are made by these materials [2].

Polyimides Polyimides are relatively new class of materials used for high-temperature applications in the composite industry. The glass transition temperature of polyimides ranges from 320 °C to 330 °C. The thermal stability and flame resistance of these resins are better than the epoxy resins. Polyimide-based composites have excellent retention of strength in the adverse environmental conditions, but they have a very low elongation at break and are brittle [2].

Table 1.2 Typical properties of some thermoplastic resins

Resin	Density (g/cm^3)	Tensile strength (MPa)	Tensile modulus (GPa)	HDT (°C)	Cure shrinkage (%)	Glass transition temp. (°C)
Acrylonitrile-Butadiene-Styrene, molded	0.88–3.5	24.1–73.1	0.78–6.1	65–220	–	105–109
Polystyrene	1.02–1.18	17.9–60.7	0.3–3.35	62–98	0.002–0.008	83–100
Nylon (PA6)	1.12–1.14	41–166	2.6–3.2	68–85	0.003–0.015	47
Polycarbonate (molded)	0.95–1.51	46.1–93.1	1.8–3.0	78–187	–	143–152
Polysulfone	1.37–1.48	60–131	–	172–213	–	–
Polyethylene, HDPE (glass filled)	0.94–1.53	11–113	0.7–13.6	51.7–127	0.001–0.003	–

Thermoplastic Resins

Currently, thermoplastic resins represent a relatively small part of the PMC industry. These resins are nonreactive solids, i.e., no chemical reaction occurs during processing. The final product can be made by application of heat and pressure. Unlike thermoset resins, the thermoplastic resins can usually be reheated and reformed into another shape, if desired. Common examples of thermoplastic resins are polyethylene, polystyrene, nylon, polycarbonate, polysulfone, polyphenylene sulfide, acrylonitrile-butadiene-styrene (ABS), etc. [2]. Some properties of important thermoplastic resins are given in Table 1.2.

Polyethylene Polyethylene (PE) is a translucent thermoplastic material with sufficient strength for many product applications, good flexibility over a wide range of temperatures, and excellent corrosion resistance. There are two types of polyethylene: low-density (LDPE) and high-density (HDPE). LDPE has a branched-chain structure and HDPE has a straight-chain structure. LDPE has a tensile strength from 6.2 to 17.2 MPa, whereas HDPE is stronger with tensile strength varying from 20 to 37 MPa. Polyethylenes are mainly used in electrical insulation, chemical tubing, housewares, and packaging [2].

Polystyrene Polystyrene is a clear, odorless, and tasteless plastic material, which is relatively brittle unless modified. The impact properties of polystyrene can be improved by copolymerization with the polybutadiene rubber. It has good dimensional stability and low cure shrinkage (0.002–0.008 %). But they are susceptible to chemical attack by organic solvents and oils. They have adequate tensile strength (17.9–60.7 MPa) within operating temperature limits [2].

Polyamides (Nylons) Polyamides or nylons are thermoplastics characterized by the presence of a repeating amide group in their main-chain structure. They have sufficient tensile strength (41–166 MPa), good toughness (107 J/m), and good chemical resistance to make them popular in many industries. They are typically used for unlubricated gears, speedometer and windshield wiper gears, antenna mounts, packaging, and for many general-purpose applications [2].

Polycarbonate Polycarbonates form another group of thermoplastics having carbonate group in their structure. The tensile strength of polycarbonates at room temperature is relatively high (about 62 MPa), and their impact strength is also very high (640–854 J/m) as measured by Izod test. They have high heat deflection temperature, good electrical insulating properties, good corrosion resistance, and high dimensional stability. Typical applications for polycarbonates include safety shields, cams and gears, aircraft components, helmets, electrical relay covers, and computer terminals [2].

Polysulfone Polysulfones are characterized by the presence of a sulfone group in their structure and these polymers are known for their toughness and stability at high temperatures. They have high heat deflection temperature of 174 °C at 1.68 MPa and have a high tensile strength (for thermoplastics) of 70 MPa. Polysulfones resist hydrolysis in aqueous acid and alkaline environments. These polymers are typically used for connectors, TV components, capacitor films, medical instruments, and corrosion-resistant piping [2].

Reinforcing Materials

Fibers as reinforcing materials offer two advantages. Firstly, the bulk material is always stronger when produced as small-diameter fibers due to the natural exclusion of large-scale defects. Secondly, the fiber configuration allows the tailoring of properties in the specific directions. Fibers added to the resin system provide strength to the finished part. According to the desired properties required in the finished product, reinforcing material is selected. Various forms of fibers such as yarns, rovings, chopped strands, woven fabric, and mats are used in the advanced composite. Each of these has its own special application. Yarns and rovings are used in processes such as filament winding or pultrusion. When preforms are used in parts manufacture, woven fabric or mats are required. Four basic types of fiber as reinforcing materials are used in the advanced composite industries [2]. These are glass, carbon, aramid, and boron. Typical properties of these fibers are mentioned in Table 1.3.

Glass Fibers

Glass fibers consist of $(-SiO4-)_n$ tetrahedra units in the backbone and are amorphous in nature. It is the most common reinforcing material for PMCs, as they offer high strength at relatively low cost. They are available as continuous and short fibers. However, they have poor abrasion resistance, poor adhesion to polymer matrix, and tendency to absorb moisture. In order to overcome these drawbacks, coupling agents

Table 1.3 Typical properties of some fibers

Fiber	Density (g/cm^3)	Tensile strength (MPa)	Tensile modulus (GPa)	Elongation at break (%)
E-Glass	2.54	3450	72	4.5
S-Glass	2.49	4300	87	5.3
Carbon (HT)	1.8	5000	250	1.6
Aramid (Kevlar 49)	1.44	3600	131	2.8
Boron	2.6	3500	400	0.8

(e.g., silanes) are used. The glass used for making fibers is classified into five major types. The letter designation is based on the characteristic property of the glass:

(i) A-glass is a high-alkali glass; it has very good resistance to chemicals, but lower electrical properties.
(ii) C-glass is a chemical grade, which offers extremely high chemical resistance.
(iii) E-glass has low alkali content and it is electrical grade. It provides good insulation property and strong resistance to water.
(iv) S-glass has 33 % higher tensile strength than E-glass.
(v) D-glass has superior electrical properties with low dielectric constant.

E-glass fiber dominates the current PMC industry and has very good mechanical and electrical characteristics at reasonable cost. The mechanical properties like strength and stiffness of E-glass fiber are 3450 MPa and 72 GPa, respectively. S-glass contains a higher percentage of alumina and silica as compared to E-glass. Typical values of strength and stiffness for S-glass fibers are 4300 MPa and 87 GPa, respectively [2, 3].

Carbon/Graphite Fibers

Carbon fiber has high strength and high stiffness and used in a variety of structural and electrical applications. It is also called as graphite fiber. Typical strength and stiffness of carbon fiber (HT Type) are 5000 MPa and 250 GPa, respectively. But, it may vary depending on the precursor used [4]. It is manufactured by heating, oxidizing, and carbonizing of polyacrylonitrile polymer fibers (PAN) or other precursors, like rayon and petroleum pitch. The PAN-based fiber is the most commonly used fiber in the advanced composite industry today.

Aramid Fibers

Aramid fibers (Kevlar) are made from aromatic polyamides and possess unique properties like high tensile strength and modulus, temperature stability, dimensional stability, flex performance, textile processability, and resistance to chemical. Typical values for strength and stiffness of aramid fibers (Kevlar 49) are 3600 MPa and 131 GPa, respectively [3]. The disadvantage of aramid fibers is that they are difficult to machine.

Boron Fibers

Boron fibers are manufactured by chemical vapor deposition in which a fine tungsten wire or graphite filament is used as a core and boron trichloride gas as boron source. The fiber diameters vary from 0.1 to 0.2 mm. They have low density, high tensile strength, and high modulus of elasticity. Typical values for strength and stiffness are 3500 MPa and 400 GPa, respectively [5]. Boron fibers are difficult to weave, braid, or twist because of their high stiffness. They can be formed into resin-impregnated tapes for hand layup and filament winding. These fibers have limited applications in the aerospace industry because of their high cost.

Current Manufacturing Methods of Polymer Matrix Composites

The manufacturing of polymers and development of new technologies for their composites have contributed a lot in the composite manufacturing science. In this section, a review of currently available manufacturing methods for FRPs has been presented. The recent developments in each technology area are surveyed and various new applications and advantages/disadvantages have been discussed.

Hand Layup

Hand layup is an oldest open-mold process used for the composite manufacturing. This process is simple, and it is a low-volume and labor-intensive process. Large components, such as boat hulls, can be prepared by this technique. Reinforcing mat or woven fabric or roving is placed manually in the open mold, and resin is poured, brushed, or sprayed over and into the glass plies. Squeegees or rollers are used to remove the entrapped air manually to complete the laminated structure as shown in Fig. 1.3. The most commonly used matrixes are polyesters and epoxies that can be cured at room temperature. The time of curing depends on the type of polymer used for composite processing. For example, for epoxy-based system, normal curing time at room temperature is 24–48 h. A catalyst and accelerator are added to the resin, which enables room-temperature curing of the resin. In order to get high-quality part surface, a pigmented gel layer is first applied on the mold surface. Hand layup is the most commonly preferred process for the manufacture of polymeric composites.

Avila and Morais have investigated the effect of curing process and stacking sequence on the mechanical properties of composites prepared by hand layup [6]. After completion of the stacking sequence, three sets of plates are selected,

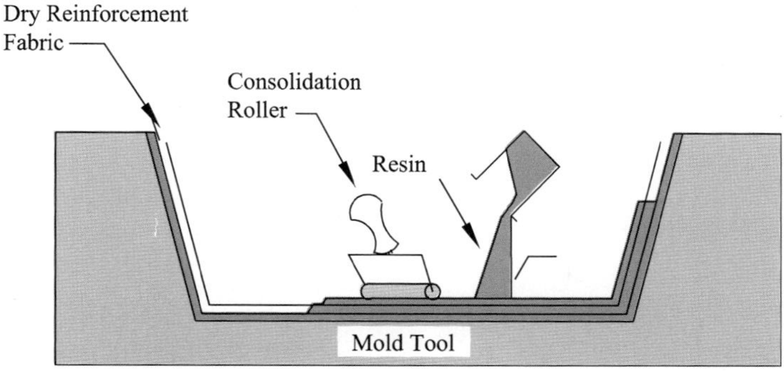

Fig. 1.3 Schematic of hand layup

and each one is cured by different processes, such as cure in air, cure through vacuum assisted, and cure under compression. The void formation rate has been identified by a microscopic analysis. A hand layup molding process that involves a forming of reinforcing fiber preform on the screen frames has been developed by Otaka [7]. The preforms are made from the cut fibers and adhesives under reduced pressure (e.g., by connecting to the vacuum pumps). Finally, the preforms are set on the hand layup molds, and resins are coated on the preforms. The method is demonstrated for glass, carbon, and ceramic fibers. Mariatti and Chum have studied the consequence of laminate configuration on the properties of glass fiber-reinforced plastics prepared by hand layup [8]. After examining the effect of water absorption on the flexural properties of GFRP composites, it is found that the configurations of the laminate system govern the properties of GFRPs. The occupational exposure to styrene vapor during the hand layup molding of glass fiber-reinforced polyester composite has been analyzed by Andersson and Alenius [9]. Styrene concentrations and working temperatures are measured continuously during the study. Air flow, temperature, and styrene concentration are also calculated using a CFD code, which is found to be in fairly close agreement with the styrene concentration measured at the same point. This process is extremely flexible, and wide variety of shapes can be made. The process does not need large capital investment, but the production rate is very slow due to the manual operation and large number of steps involved. Due to this, the results may be variable and subject to operator differences. Since the external pressure is not applied, quality of the product is poor with low fiber volume percent and increased void content. Also, the surface of the product is rugged due to the resin shrinkage. Rahamani et al. have investigated the effect of fiber orientations, resin types, and number of laminates on mechanical properties of laminated composites [10]. Composites were manufactured by hand layup process, using a fiber-to-resin ratio of 40:60 (w:w). To investigate the effect of fiber orientation, angles of 0, 35, 45, and 90 were selected. Results show that the mechanical properties, in terms of tensile, flexural, and impact strengths, are mainly dependent on the fiber orientations followed by the number of laminates. An automatic layup process has been developed by Chaple et al. for producing FRP roof sheets of uniform thickness and providing safe environment for workers [11]. Vasanthanathan et al. have developed a new method for fabrication of composite shell structures without any joints and wrinkling using carbon fibers and epoxy resin [12].

Vacuum Bag Molding

In vacuum bag molding, the entrapped air and excess resin are removed using vacuum. After fabrication of the layup, a perforated release film or peel ply is placed over the laminate. The bleeder ply, which is placed above the peel ply, is made of fiber glass cloth, nonwoven nylon, polyester cloth, or other material that absorbs excess resin from the laminate, followed by a breather ply of a nonwoven fabric. The vacuum bag is placed over the entire assembly and sealed at the mold flange as shown in Fig. 1.4.

A vacuum is created under the bag, and thus the laminate can be merged by applying a pressure up to one atmosphere. This process provides a high reinforcement, improved adhesion between layers, and great control of fiber volume percent compared to the hand layup. Chapman et al. have developed composite sections for aircraft fuselages and other structures [13]. A novel vacuum bag process for molding a laminated board using high-energy focused ultrasound and high-pressure water tank has been developed by Liu et al. [14]. In this process, a pre-impregnated material is loaded onto a mold and sealed in a silicone rubber vacuum bag. The bag is placed in a water tank and connected to the vacuum system. Water is filled in the tank till the required water pressure is obtained, and the resin is cured by using high-energy focused ultrasound. After curing, water is discharged from the tank, and the laminated board is taken out. Caron has developed a reusable vacuum bagging technique for forming polymeric materials [15]. It includes a fabric layer bearing reinforcing fibers and a release surface disposed on at least one side of the fabric layer. Multiple mold cycles of the vacuum less than ambient pressure can be

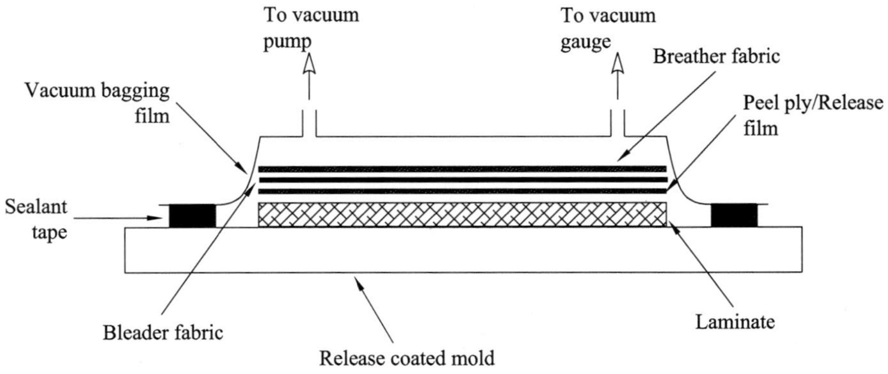

Fig. 1.4 Schematic of vacuum bag molding

subjected to the vacuum bag without a significant leakage. A double-vacuum-bag process for the manufacture of composite has been developed by Hou and Jensen [16]. The process includes a heating chamber with double-vacuum environment around the preform provided by inner and outer vacuum bags. Vacuum is drawn in the first environment, then the second one, wherein the two vacuum levels are different. Curing is done in more than one stage. Zhou and Yang have developed another vacuum bag pressing for reinforcing structure with external bonding of fiber using prepreg vacuum bagging additionally [17]. A special carbon fiber/epoxy prepregs with multifilaments of nylon fibers bonded on both surfaces and their laminates manufactured by vacuum bag molding with decreased void content, and good mechanical strength has been reported by Kaneko et al. [18]. Johnson and Lewit have developed a composite manufacturing method in which the curable liquid sealant acts as vacuum bag [19]. In this method, the reinforcement is kept on the open mold, and a liquid-form sealant layer is applied over the reinforcement to create an airtight chamber to entrap the reinforcing fiber material between the mold and the sealant layer. After curing the sealant layer, a vacuum pressure is created in the airtight chamber to pass resin through the reinforcing fiber. Another method for vacuum molding of wet resin glass fiber-reinforced plastics has been developed by Kook [20]. Major advantages of vacuum bag molding are higher fiber content in the laminate, lower void content, better fiber wet-out, and reduced volatile emissions as compared to the hand layup. Large cruising boats and racing car components can be manufactured by vacuum bag molding. Disadvantages of vacuum bag molding include expensive and disposable bagging materials, labor intensive, inconsistent performance, trapped air/volatiles, wrinkles, loss of seal, and requirement of higher level of operator skills.

Alms et al. have developed a new method called vacuum-induced preform relaxation (VIPR) in which the permeability of the reinforcing fabric can be influenced during infusion and hence the resin flow can be manipulated locally [21]. In this process, an external vacuum chamber is used that seals against the flexible molding surface of a VIP mold, and a secondary vacuum is applied to locally reduce the compaction pressure on the fibers, which increases the permeability of that region steering resin flow to that region [21]. Harper has discussed the use of reusable vacuum bags made of silicone and their advantages over conventional consumables in vacuum bag molding [22].

Pressure Bag Molding

Pressure bag molding is identical to the vacuum bag molding except that the pressure, usually provided by compressed air or water, is applied to the flexible bag that covers the prepreg composite. The application of pressure forces out the entrapped air, vapors, and excess resin. It also facilitates better wetting of fibers. Shima has developed an internal-pressure bag molding for manufacturing of hollow FRPs efficiently [23]. A production method for FRPs has been proposed by Horibe et al. [24]. This method is used to manufacture a boat hull. The reinforcement (glass chopped strand mat/carbon fiber cloth) impregnated with resin (vinyl ester) is put in a vacuum bag, which is then pressure cured by compressing at ≥ atmospheric pressure from the outside of the bag film. Dai et al. have developed a mandrel for pressure bag molding used to manufacture a shaft [25]. The mandrel comprises a cylinder having an open base part at one end and an open clamping part for a tube at the other end. Coiled prepreg sheets are mounted on this mandrel and inserted into a pressure bag mold. Hot fluid is passed through the mandrel tube, which assists in curing of the prepreg sheets thermally. Another method for manufacturing of FRPs with complex structures by internal-pressure molding using a flexible bag has been proposed by Kanemaru and Sanae [26]. The reinforcing fabric is put in the mold and a bag is prepared. The polymer containing a curing agent is forced into the mold with simultaneous application of pressure to the bag. Asano and Nojima have manufactured a full-sized FRP pipe by this bag molding [27]. The thermosetting resin-impregnated molding material is put around a core mold and covered with a bag. Heat-shrinkable belts are used to wrap the film bag and further pressurize the molding material while heating. Another modified method for preparing FRP has been proposed by Howard [28]. Mold parts defining the mold cavity are movable and are pressed by a gas bag during the curing process, against the polymerizable composition, which is turned and subjected to a mold pressure as it cures within the mold cavity. A glass fiber-reinforced stadium seat is manufactured by this process. The pressure bag molding produces quality composites with high fiber content and lower void content. The method requires disposable bagging material and high skill of the operators. An improvement of pressure bag molding is the use of autoclaves for curing of the composite. An autoclave enables the curing of composite at higher temperature so that the curing reaction is more efficient and viscosity goes down enhancing wetting of fibers.

Autoclaves are basically heated pressure vessels. These are usually provided with the vacuum systems. The bagged layup is cured inside the autoclave as shown in Fig. 1.5. The pressures required for curing are typically in the range of one to six bars and takes several hours to complete the curing. This method accommodates higher temperature matrix resins having properties higher than the conventional resins, such as epoxies. Component size is limited by the autoclave size. It is mostly used in the aerospace industry to manufacture high-strength/weight ratio parts from pre-impregnated high-strength fibers for aircraft, spacecraft, and missiles.

Antonucci et al. have analyzed the heat transfer in the autoclave technology [29]. The heat transferred between the energy-carrying fluid and the bag-composite-tool element triggers exothermic curing reactions, leading to compaction of composite. Simplified energy balance equations are used assuming a negligible thermal gradient throughout the thickness. The relation between interlaminar fracture toughness and pressure condition in the autoclave molding of composite laminates has been proposed by Jiang et al. [30]. A change in curing pressure affects the morphology of matrix and fibers, and the interfacial strength of fiber and matrix. Conventionally, a high pressure has been achieved by pressurized air systems, which have a risk of fire occurring within the autoclave. The use of inert nitrogen instead of air eliminates the risk of fire. Chen et al. have studied the dimensional control of

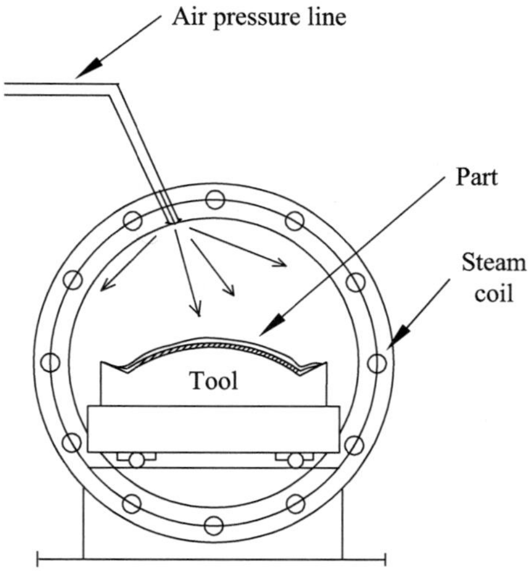

Fig. 1.5 Schematic of autoclave

laminates using Taguchi method [31]. The optimum curing conditions for stabilizing the laminate thickness are obtained from dynamic characteristic analysis, and an equation predicting the laminate thickness is postulated. An autoclave curing scheme for FRPs has been developed by Boey [32]. The system can operate at $\leq$20 bar pressure while achieving a 0.1-bar vacuum within the layup film bagging. Autoclaves are widely used to cure high-performance FRPs at high temperature and pressure. This process involves application of higher heat and uniform pressure on the component during curing, which results in a denser and low void percentage product. The autoclave equipment and tooling are expensive and it is only suitable for high-end applications.

Filament Winding

This process consists of a rotating mandrel on which pre-impregnated fibers or reinforcement is wound in the preset patterns. The method provides the best control of fiber placement. The wet method is shown in Fig. 1.6. Here, the fiber is allowed to pass through a bath containing low-viscosity resin. In the dry method, the pre-impregnated reinforcing layers are wound on the mandrel, and then the component is removed and postcured. Conventionally, this process is used to make pressure vessels, rocket motor cases, tanks, ducting, golf club shafts, and fishing rods. Recently, non-cylindrical and nonspherical composite parts are also produced by filament winding technology. Polyesters, vinyl esters, epoxies, and phenolics are the typical thermoset resins used in the filament wound parts. Beckwith has discussed a filament winding technology for manufacture of composites made of continuous fibers impregnated with resins [33]. Materials of fabrication, resins, winding patterns, and filament winding applications are discussed briefly [33]. A quality comparison of the conventional and robotic winding of fiber composite components has

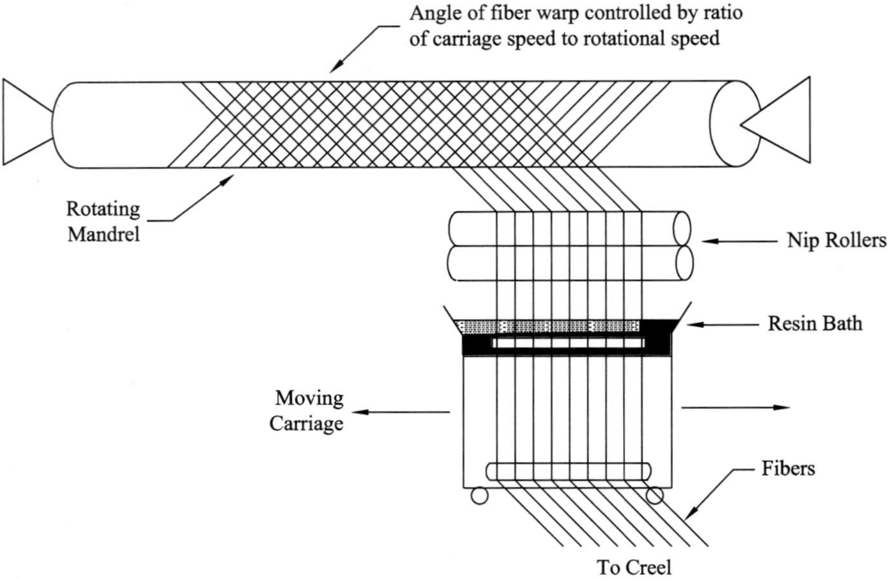

Fig. 1.6 Schematic of filament winding

been proposed by Munro [34]. The concepts of hybrid filament winding processes, lightweight winding systems, and advanced filament winding techniques are also discussed here. Polini and Sorrentino have presented a study on filament winding of asymmetric shape parts by robotic system [35]. In the robotized filament winding process, a roving is deposited on a die according to a designed robot trajectory, and the wound parts are polymerized in a furnace for thermoset matrix. It is possible to wind asymmetrical or complex geometries using this technique. The critical aspect of the robotized filament winding process is the winding trajectory, which influences the mechanical performance and the winding time of the composite part. The work discusses the most important criteria for a proper winding trajectory, its influence on the winding time, and helps to optimize the designing of the winding trajectory. Materials, process, machines, and tools needed for filament winding of composites has been presented by Parnas and Ardic [36]. They have also discussed the design and analysis methods, test methods, current situation, and future perspectives of the filament winding process. Sohl has discussed the materials and equipment used in the filament winding process together with the specialized topics such as residual stresses and strains, macrobuckling, void content, and quality assurance aspects, related to the filament winding process [37]. This process is best suited for parts with rotational symmetry, but it is possible to wind odd-shaped parts using a robotized winding. It requires special equipment and may result in variation in the part thickness in case of tapered parts. The tooling and setup cost is high and it is only suitable for a limited variety of components.

Martinec et al. have developed a method to calculate and determine the accurate 3-D trajectory of the end effector of an industrial robot during the manufacture of composite profiles using a dry fiber winding technology on a polyurethane core with a circular cross section. The algorithm can be applied to any manufacturing process, where it is necessary to determine the 3-D trajectory of a robot end effector [38]. Zhang et al. have developed smart mandrels using shape memory polymers and analyzed their behavior using simulation [39]. The simple and efficient demolding of SMP mandrels is verified by filament winding experiments on bottle-shaped and air duct-shaped mandrels. Rojas et al. have developed a unified approach and a generic mathematical model that leads to harness the benefits of filament winding using complex shape mandrels [40]. A numerical tool is also developed and applied to predict the evolution of the filament winding angle of the fiber tow placed over the surface of two axisymmetric geometries: a convex and a concave one. Experimental validation is carried out by manufacturing these geometries using a four-axis filament winding machine.

Pultrusion

Pultrusion process generally involves pulling of continuous fibers through a bath containing resin, mixed with a catalyst, and then it is passed through a preforming fixture followed by a heated die. The excess resin is removed and the component is partially pre-shaped while passing through the performing fixtures. The sectional geometry and finish of the product are determined by a heated die. Further curing is often required for a good quality product. Figure 1.7 shows a schematic of pultrusion process. The strength and weight of the profiles produced by this process compete with the traditional metal profiles made of steel and aluminum. The composite sandwich panels are designed, pultruded, tested, and evaluated for their mechanical performance in relation to the panels, which are manufactured from vacuum-assisted resin transfer molding (VARTM). Pultruded panels show significant advantages over VARTM panels, including increased tensile strength and stiffness, increased bending stiffness at panel level, and reduction in material and production costs. Lam et al. have studied the pultrusion process and developed a numerical procedure to

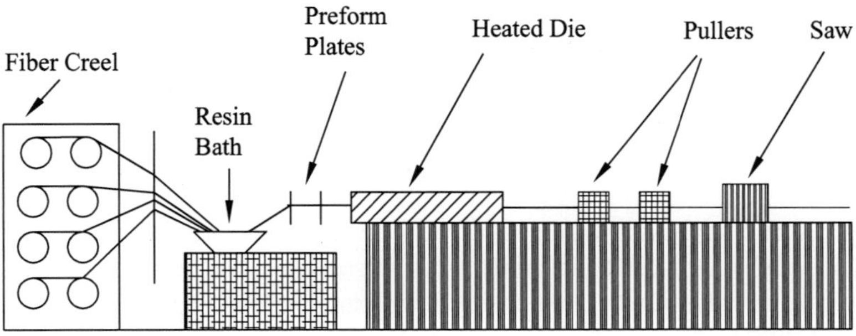

Fig. 1.7 Schematic of pultrusion process

reach an optimum combination of pull-speed and temperatures in dies for producing a uniformly cured composite pultrudate [41]. The important parameters in pultrusion that affect the degree and uniformity of the curing of thermosetting composites are the pull-speed and die temperature. A required degree of cure having maximum uniformity can be achieved by optimizing the above parameters using an algorithm. This algorithm is based on the objective function, representing the effects of both parameters on the distribution of degree of cure across the cross section of a pultruded part. Trivisano et al. have developed a mathematical model of the pultrusion process for epoxy-based composites [42]. The dimensionless mathematical model is applied to the description and optimization of the process variables like die temperature, pulling speed, and composite thickness. Esfandeh et al. have investigated the factors affecting the surface quality of pultruded profiles [43]. For a rod made of unsaturated polyester resin and E-glass roving, increasing material packing in the die, employing release agents compatible with resin, and also type and level of curing agent have significant influence on the surface quality of the product. The pulling speed and heating profile of the die are also important factors that must be selected according to the resin formulation used. A method for continuous manufacture of fiber-reinforced plastic foam composites has been developed by Murata and Muranaka [44]. The setup consists of a means for conveying long-board cores in the longitudinal direction, plural molds for impregnating filament bundles with foamable thermosetting resins, and a mold for integrating the impregnated bundles with the cores and foaming the resins. Methven et al. have studied the possibilities for the manufacture of fiber-reinforced composites using microwave heating, with emphasis on microwave-assisted pultrusion (MAP) [45]. The interlaminar shear strength and the ultimate tensile strength of the profiles increase after postcuring, indicating limited cross-linking in the MAP. Pultruded profiles with specific geometries are only suitable for a continuous process.

Simacek and Advani have modeled the resin infiltration process during thermoset tape pultrusion, by modifying the preexisting simulation tool for liquid molding processes [46]. The formulated capability can be used not only to optimize the impregnation dynamics within the pressure chamber but can also be used to predict the required forces for the selected pulling rate. The proposed model does allow one to handle a variety of tape cross-sections, not just rectangular prisms. Boyd et al. have presented an experimental methodology for determining the through thickness properties of pultruded GRP materials and their application in finite element analysis (FEA) of adhesively bonded joints [47]. The finite element analysis is validated using an infrared thermography-based experimental mechanics technique known as thermoelastic stress analysis. Belingardi et al. have proposed alternative lightweight materials (GMT, GMTex, and GMT-UD) for automobile bumper subsystems using pultrusion process [48]. Major parameters, such as impact energy, peak load, crash resistance, energy absorption, and stiffness, have been taken as evaluation criteria to compare the proposed material solutions with pultruded and steel solutions. The results predicted by the finite element analysis have been evaluated and interpreted in comparison with other existing solutions to put in evidence the effectiveness of the proposed innovative materials and design concept solutions.

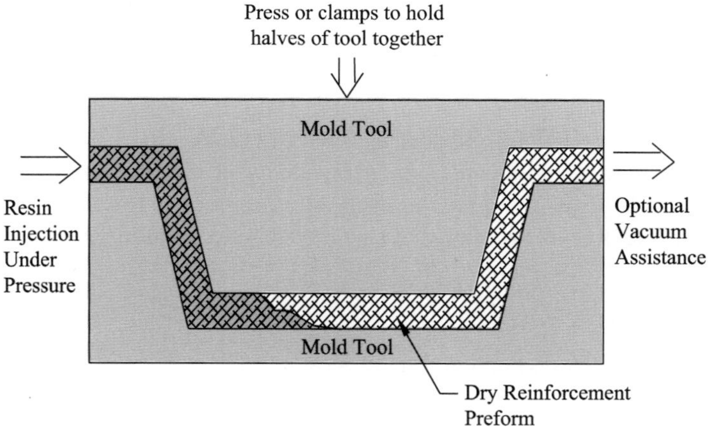

Fig. 1.8 Schematic of resin transfer molding

Resin Transfer Molding

Resin transfer molding (RTM) is a low-pressure closed molding process for moderate- and high-volume production. This process basically involves placement of the dry stack of reinforcement in the bottom part of the mold, and then the other half is clamped over the bottom mold. For complex shapes, preforms are used. After closing the mold, a low-viscosity resin containing catalyst is pumped in, which displace the air through strategically located vents. The resin/catalyst ratios are controlled by metered mixing equipment and injected into the mold port as shown in Fig. 1.8. The commonly used matrix resins include polyester, vinyl ester, epoxy, and phenolics. Both injection and curing can take place at either ambient or elevated temperature. In order to have optimum surface finish, a gel coat is applied to the mold surface prior to molding. High-quality parts such as automotive body parts, bathtubs, and containers are produced by this method. Otten and Siebrecht have developed a device for manufacturing a composite component by RTM [49]. It comprises (a) a trough, into which starting material for manufacturing the component can be introduced, (b) a cover that can be attached to the trough to enclose the starting material, and (c) a heating device for manufacturing the component by heating the starting material. The trough is manufactured from a material with a coefficient of expansion that is essentially identical to the coefficient of expansion of the starting material for manufacturing the component.

A new injection approach for resin transfer molding has been designed and tested by Lawrence et al. [50]. The resin does not follow the same path every time to impregnate the preform, and a micron-size gap between the mold edge and the preform may produce flow deviations and change the flow pattern causing premature arrival of the resin at the vent locations. Hence, to distribute the resin throughout the mold effectively and efficiently, multiple gates and vent locations are required. An automated RTM intelligent workstation with scattered injection gates and vents that will be activated by the control methodology selected based on the online flow pattern of the resin during impregnation has been developed recently. The online flow pattern is noticed by the point sensors placed in the mold walls. This method assists in distribution of resin throughout the mold and also controls the opening and closing of the individual gate during mold filling. Devillard et al. have characterized the race-tracking and preform permeability at the time of resin filling in RTM process [51]. In the first step, the simulations are generated offline that represent the possible interruption that may probably occur during injection process. In the second step, the sensors, which are placed in the mold, identify the model within the database that corresponds to the experiment in terms of the filling characteristics. The final step uses the available information to demonstrate bulk permeability, race-tracking strength, and location. The accuracy and reliability of this method has been determined by numerical case studies and experiments. A system to control the flow of resin using real-time feedback in RTM process during mold filling has been developed by Nielsen and Pitchumani [52]. This assists in eliminating the defects related to filling process in the composite products. A real-time response of the system can be forecasted by using numerical simulations. This helps to guide the flow through a scheduled target by finding the proper combination of flow rates at the injection port. The performance of the controller, which has been implemented on a laboratory-scale RTM process, is evidenced on various configurations of preform and required fill patterns. Mogavero et al. have proposed a way to maintain the constant flow rate in RTM process using a nonlinear control method [53]. This simulation also analyzes the impact of various parameters on the ability to provide a constant flow rate. The theoretical model evaluated using flow experiments is in correspondence with the computer models. A similar method for low-cost estimation of permeability in RTM has been proposed by Ferland et al. [54]. A rectangular mold is used for the numerical observation of the principal permeability. The experimental data has an inbuilt correlation with Darcy's law and allows an approximate calculation of both numerical and experimental errors. Kranbuehl et al. have developed another intelligent closed-loop skillful automated method for controlling the RTM process of an epoxy composite with graphite fiber preform [55]. During the impregnation and curing in RTM process, the processing properties of the epoxy resin are monitored and controlled by the resin transfer model using an in situ frequency-dependent electromagnetic sensor (FDEMS). Once it is correlated with degree of cure and viscosity, the FDEMS sensor monitors the reaction progress of the resin, viscosity, and fabric impregnation that are predicted by the RTM processing model. This eases the skillful control of the RTM process in the mold through a careful in situ monitoring and evaluation. The consequence of process variables on the quality of RTM moldings has been studied by Hayward and Harris [56]. Carbon fiber-reinforced epoxy and glass fiber-reinforced polyester resins gave a high level of reproducibility under identical processing conditions. The variation in injection pressure has no effect on the quality of moldings. A wide range of resin viscosities has been successfully molded. RTM can produce laminates having high fiber volume with very low void contents. It is safe for the health and environment due to the enveloping of resin. Component prepared by RTM has molded surface on both sides. The disadvantages of RTM process are need of heavy and expensive tooling to withstand pressures, limitation in size of the components, and very expensive scrap parts due to un-impregnated areas.

Han et al. have proposed a new high-speed RTM process in which resin flow direction is changed from in-plane to through-thickness for the preform with an aim to reduce the impregnation time [57]. Computational fluid dynamic (CFD) simulation is introduced to investigate the impregnation state of preform with respect to the number and size of vacuum gates. Impregnation tests, under conditions similar to those in the CFD simulation, are carried out, and the impregnation behaviors of the resin are compared with the simulation results. Finally, an optimization process is carried out to find the number of vacuum gates and permeability ratio that result in the minimum impregnation time. Yang et al. have developed a mathematical model to describe the formation of micro- and mesoscale voids with particular attention on the influence of flow direction and fabric shear on the impregnation of the unit cell [58]. Experimental validation has been conducted by measuring the formation and size of voids; a good agreement between the model prediction and experimental results has been found. Eck et al. have presented multi-objective optimization of a fiber-reinforced composite part, and its manufacturability with the RTM process is considered during **the optimization** [59]. A new CPU time-efficient method, called process estimator, is also developed based on the local material parameters that allow to calculate the filling times of complex and fine meshed parts with varying permeabilities and porosities in some seconds to minutes.

Vacuum-Assisted Resin Transfer Molding (VARTM)

RTM process is modified by using vacuum for efficient resin transfer and impregnation. In vacuum-assisted resin transfer molding (VARTM), the mold consists of a single die and preform is placed over it. Above the preform, a flexible transparent sheet made of nylon or mylar plastic is placed and sealed. Now, the entrapped air is removed by application of vacuum between the preform and the sheet. Finally, the part is kept inside the oven and cured. The VARTM process is shown in Fig. 1.9. The application of pressure enhances the concentration of the fiber and also renders improved adhesion between sandwich layers. Tanaka and Yotsukura have developed a method of manufacturing FRPs by VARTM [60]. A process model for the VARTM simulation by the response surface method (RSM) has been developed by Dong [61]. The design of experiments (DOE) assists in identifying the substantial process variables. A quadratic regression model formulated by the response surface method is validated against the 3-D VARTM model and experiments.

Kelkar et al. have evaluated the performance of woven fabric composites manufactured by VARTM process used in defense and aerospace applications [62]. For applications in composite armored vehicle (CAV), tension-compression fatigue behavior of plain-woven S2 glass/epoxy and twill-woven S2 glass/vinyl ester composites have been evaluated. Low-velocity impact loading tests have been

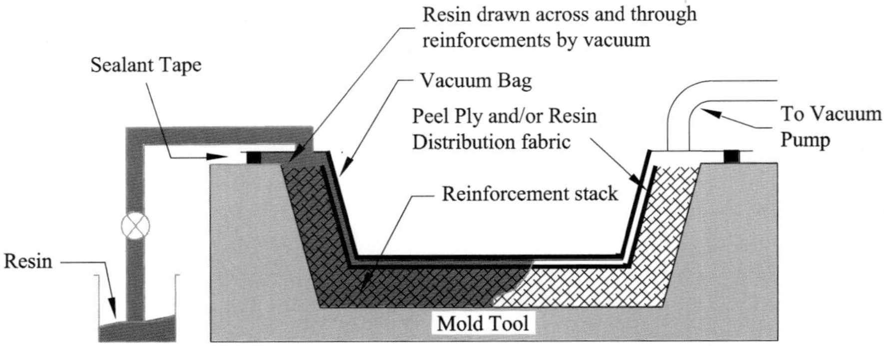

Fig. 1.9 Schematic of vacuum-assisted resin transfer molding

analyzed on the stitch-bonded E-glass/vinyl ester and woven roving E-glass/vinyl ester composites for naval application. The system for controlling the flow rate in VARTM has been investigated by Bender et al. [63]. In this system, the vacuum pressure applied on the injection barrel generates a computer-controlled difference in vacuum between the injection and vent gate. A fuzzy-logic controller has been enforced in Lab VIEW and the optimization of the system has been done using virtual model. A parametric study explains the effect of process and material parameters on the performance of the control system. Tackitt and Walsh have conducted another experimental analysis on variation in thickness in the VARTM process [64]. This thickness variation results in change in volume percent of the fiber and the related mechanical properties. The thickness gradient is studied using a range of linear variable differential transformers (LVDTs) to observe displacements of the surface during infusion. Surface displacements are displayed as a 3-D plot in order to visualize the entire part surface at the time of infusion. An artificial neural network model to forecast the compressive strength of VARTM-processed polymer composites has been proposed by Seyhan et al. [65]. The ply layup compressive strength of E-glass/polyester composites by VARTM processes is analyzed by a three-layer feed-forward artificial network (ANN) model, which has one-output and three-input neurons and two hidden neurons. The back-propagation algorithm fulfills the understanding of ANN. The average error is 3.28 %. The predicted values and measured values are in good correspondence. Tuccillo et al. have analyzed and characterized the non-isothermal resin infiltration during the VARTM process [66]. Embedded optic fiber sensors assist in real-time monitoring of the resin flow during the experimental infusion tests and to demonstrate the capability and the reliability of the simulation. Mechanical properties of the honeycomb sandwich composites made by VARTM have been studied by Eum et al. using static tests [67]. VARTM is a very effective method to manufacture the honeycomb sandwich composites, and mechanical properties of composite structures made by VARTM are satisfactory for real-life applications. Chen et al. have proposed a new method to simulate the flow in the VARTM process [68]. This method considers the equivalent permeability approach to substantially minimize the time required for computation of VARTM flow model of large parts. The computational model of the 3-D control volume/FEM simulation for VARTM process is simplified by the equivalent permeability model of flow channel. The results obtained from conventional 3-D computational simulations and experiments validate this new model. VARTM process provides several benefits compared to the traditional RTM process as follows: lower tooling cost, processing at room temperature, shorter mold filling time, and better scalability for large structures. In this process, the efficiency of the process and quality of the product depend on resin level in the mold and sufficient wet-out of the fibrous preform.

Kamar et al. have investigated the ability of graphene nanoplatelets (GnPs) to improve the interlaminar mechanical properties of glass-reinforced multilayer composites [69]. A novel method was developed for the inclusion of GnPs into the interlaminar regions of plain-weave, glass fabric fiber-reinforced/epoxy polymer composites processed with VARTM. Flexural tests showed a 29 % improvement in flexural strength with the addition of only 0.25 wt% GnP. At the same concentration, mode-I fracture toughness testing revealed a 25 % improvement. Matsuzaki et al. have proposed an active flow control scheme by forecasting resin flow from the monitored time to the filling ends using numerical flow simulation and taking corrective action using dielectric heating at a specific targeted location to decrease the viscosity of the resin [70]. Because dry-spot configuration can be forecasted early, the flow can be actively controlled before the occurrence of an adverse flow front. Hsiao and Heider have described a membrane-based VARTM process variation used in aerospace applications [71]. The additional membrane, which is permeable to gas and impermeable to the resin, facilitates the improved means to avoid dry-spot formation and to continuously degas the resin during the infusion and curing stages compared to conventional VARTM.

Compression Molding

Compression molding utilizes coupled metal molds employing bulk-molding compound (BMC), sheet molding compound (SMC), or preform mat. The preform of required dimensions is placed on a press, and pressure and temperature are applied for curing as shown in Fig. 1.10. Cycles can range from few minutes to few hours. Some of the common thermoset resins used in compression-molded parts are polyesters, vinyl esters, phenolics, rubbers, etc.

Dumont et al. have studied the compression molding of SMC and developed a model to explain the effect of various processing parameters such as the axial punch velocity, initial SMC temperature, and geometry of the mold on local normal stress levels [72]. The model is coupled into a finite element code especially developed for the modeling of compression molding of composite materials. Though the model is simple, good comparisons are obtained between simulated and experimental results. Another method for preferential curing in compression molding of FRPs has been developed by Yen and Lukas [73]. The molding process involves applying heat to a mold billet located in a mold cavity defined by steel inserts. A layer of material

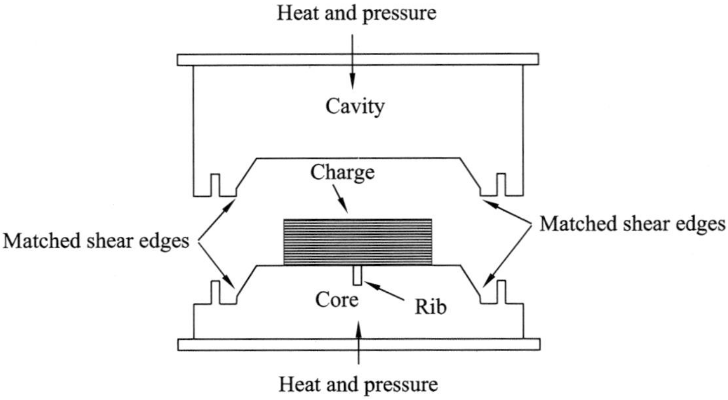

Fig. 1.10 Schematic of compression molding

having a low thermal conductivity is positioned contiguously to some of the inserts so that a portion of the mold billet in the vicinity of the low thermal conductivity material layer remains cooler than portions of the mold billet that are remote from the layer of material. Odenberger et al. have worked on the experimental visualization of the flow at the time of mold closure in compression molding of sheet molding compound (SMC) [74]. A continuous high-resolution close-up monitoring is utilized to capture the advancing flow front, and its three phases are defined, namely, flow, squish, and boiling. The actual flow is very complex and air is likely to be entrapped during the initial phase. The controlling process parameters during first phase are mold closing speed, mold temperature, and amount of preheating in the mold. The flow is stable and seemingly viscous during the second phase. During the boiling phase, bubbles are noticed in the low-pressure region at the flow front, which favors the void formation. The flow behavior of glass mat-reinforced polypropylene in compression molding has been investigated by Zhang et al. [75]. The effects of blank plies and process parameters on the flow are considered. The flow area increases with the number of blank plies. A high molding pressure ($\geq$15 MPa) and temperature (80–90 °C) are in favor of the flow. Jo and Kim have studied the effect of product geometry on fiber orientation of compression-molded rib-type products [76]. The image processing method is used to analyze the distribution of fiber orientation. The effects of product size and fiber content on the orientation state are discussed. A method for calculating the distribution of fiber content in compression-molded long-fiber-reinforced thermoplastic products has been developed by Hojo et al. [77]. The effects of molding variables on the fiber distribution are also studied. Lee et al. have presented a model for the flow, heat transfer, and curing that occurs during compression molding of FRPs [78]. This model applies to thin parts with slowly varying thickness and arbitrary planar geometry. The model takes into consideration the temperature-dependent viscosity, non-Newtonian flow behavior, and the heat loss during curing. This molding offers good part design flexibility and produces fast molding cycles. Complex features such as inserts, ribs, bosses, and attachments can be molded. Good surface finishes are obtainable in this process, which contributes to lower part finishing cost. Subsequent trimming and machining operations are minimized and labor costs are low. Compression molding is not suitable for fragile mold features. Dimensions across the parting line are difficult to maintain. Uneven parting lines, which are present in a mold design, are problem. Flash removal is difficult in high-impact composites.

Goodship et al. have developed a novel process called gas-assisted compression molding (GasComp) based on the injection of nitrogen gas during a conventional compression molding cycle [79]. Injection parameter studies are carried out on polypropylene-based recycled glass mat-reinforced thermoplastic (GMT). The parameters investigated are gas ramp rate, gas injection delay time, and peak pressure. Taketa et al. have proposed a new compression molding approach using unidirectionally arrayed chopped strands (UACS) [80]. UACS is a sheet made by introducing slits into a conventional prepreg. It is demonstrated that complexly shaped components, such as rib structures, can be fabricated by stacking UACS plies and curing them by hot pressing. Moreover, the layer structure is maintained after the processing, and there are few resin-rich regions in it. The UACS laminates show superior mechanical properties compared to those of sheet molding compound (SMC). Levy and Hubert have investigated the interstrand void content (ISVC) reduction in compression molding of randomly oriented strands (ROS) of thermoplastic composite [81]. The compaction is ruled by an ordinary differential equation that is solved numerically. The model is validated experimentally using an instrumented hot press with Carbon-PEEK prepreg strands.

Structural Reaction Injection Molding

Structural reaction injection molding (SRIM) involves the manual placement of a glass fiber mat into the production mold and injection of the resin into the mold, where it permeates and surrounds the glass fiber mat to form the part as shown in Fig. 1.11. Donatti has developed another SRIM process having void reduction [82]. The glass mat is placed in the mold at a location, where bubbles and voids are likely to occur. This reduces the formation of bubbles and stabilizes the cell structure of the completed molded component. Polushkin et al. have analyzed the development of the impregnation front and the pressure profile in SRIM process [83]. There is a good agreement between the calculated and experimentally noticed flow front position and pressure profile. A numerical method for cure kinetics in SRIM process has been proposed by Duh et al. [84]. The method uses either axial flow or radial flow of reactive resins through a fiber preform inside a mold cavity. Castro et al. have developed an innovative approach to manufacture composite parts with complex geometry [85]. The first step is to separate the complex parts into several simple substructures that are separately produced by an optimal process. This is followed by assembly of these substructures on a specially designed tool and shifted into a mold cavity for resin impregnation by SRIM process. A composite link is manufactured by this method. The mold filling is simulated by a 3-D computer

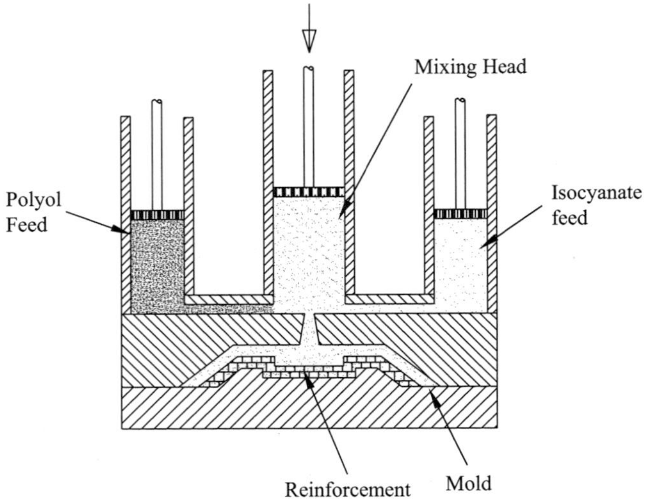

Fig. 1.11 Schematic of structural reaction injection molding

model, and the modeled results are in correspondence with the experimental results. The reason for void content and its effect on the material properties of SRIM composites have been investigated by McGeehan et al. [86]. Void content can be reduced by evacuating the mold cavity. Changing the injection rate and shot size has limited effect on the void content. The tensile strength decreases with an increase in void content. Environmental conditioning shows that the moisture absorption is proportional to the void content. Young et al. have developed the mold filling processes for SRIM and RTM under isothermal conditions using 2-D and 3-D computer simulations [87]. The critical review on SRIM has been done by Tang and Wang [88]. Generally this process is used to produce corrosion-resistant load-bearing parts, such as automotive body parts and structural beams. This process can produce a wide variety of structural composite parts. When compared to other composite materials, SRIM process improves the impact and abrasion properties and lowers the part weight. But it requires expensive setup and robotic-controlled preform equipment for making quality products.

Deng et al. have developed a computer simulation model to analyze the curing and cooling of reaction injection molding process of nylon 6 based on theory of balance equation of energy, reaction kinetics, and crystallization kinetics [89]. The model is applied to predict the conversion and temperature distribution within a plate-type mold. The effects of the feed temperature, wall temperature, and reaction rate constant on the temperature and conversion have been discussed for searching the optimum processing conditions. Kapila et al. have studied the effects of different gates and their sizes on the component and molding parameters in injection molding process [90]. The influence of different gate sizes of film gate and submarine gate to check air traps and various process parameters such as clamping force and fill time is described.

Structural Foam Reaction Injection Molding

Polyurethanes can be processed into either a solid or foam. These are formed as a result of reaction between the polyol and isocyanate component. A blowing agent added to the polyurethane formulation activates the curing reaction and results in a cellular structure. Polyurethane foams having density of 240 kg/m^3 and greater are usually called structural foams. Significant increase in the flexural modulus is noticed after the addition of long glass fibers to the structural polyurethane foam. Processing method is the same as structural RIM except the addition of blowing agent. Hashisaka et al. have designed special molds and molding of polyurethane foams with reduced voids and pinholes [91]. The runners with mixing units are designed with structures so as not to allow materials to remain inside. A method of making unreinforced reaction injection molded structural foams using a polyisocyanate component, an organic isocyanate-reactive component of an organic polyol, and an isocyanate foaming agent has been proposed by Shidaker and Bareis [92]. Modesti et al. have conducted a comparative study of physical and chemical blowing agents used in the structural polyurethane foams [93]. The component equations for the vaporization rate of the two blowing agents and the polymerization kinetics data are reported. The temperature and density profile from experimental results are in good agreement compared with the simplified theoretical model predictions. The skin thickness of polyurethane foam, using a numerical algorithm, which modeled the foaming behavior in reaction injection molding, has been predicted by Tighe and Manzione [94]. The predicted value is in good agreement with the observed values from the actual structural foams. Foaming as a function of cavity thickness is also treated in the proposed model. Structural foam RIM process has many advantages such as 20–40 % weight savings compared to the traditional composites, high-impact properties, less labor, and styrene-free processing. It needs high initial investment but the running cost is low.

Heim and Tromm have used a special mold technology to produce injection molded components with locally graded foam structures and locally different densities [95]. The characteristics of the density in the different foamed areas and the mechanical properties depend on the process-specific process parameters. The density reduction in the highly foamed areas is directly set by the process, and the density in adjacent thin-walled areas results as a function of process control. Li et al. have introduced a reaction injection molding process of phenolic foam sandwich plate in order to resolve the difficulties in repetitive continuous manufacturing process of such product [96]. The phenolic foam produced here has uniform pore diameter and good fireproof property, and there need not any glue between phenolic foam and color steel plate, which could make it a potential building material instead of polysterol sandwich plate. Samkhaniani et al. have presented a numerical approach for simulation of mold filling in reaction injection molding with polyurethane foam [97]. Polyurethane foam formation includes several complex phenomena such as chemical reactions, heat generation, and blowing-agent evaporation. Foam properties are variable during formation, foam viscosity increases, and conductivity reduces. Foam phase is considered compressible and two phases are immiscible. This study prepares a numerical model to reduce several experimental runs with expensive prototypes for mold design.

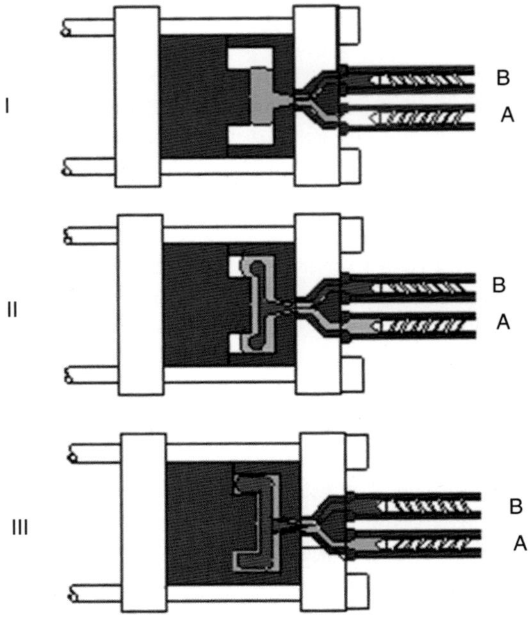

Fig. 1.12 Schematic of sandwich molding

Sandwich Molding

Sandwich molding involves injection of two different materials simultaneously into a mold cavity to make products having required features of both materials. It may consist of foamed, fiber-reinforced, or recycled material. The skin may be selected for its visual impression. Various stages of the sandwich molding process are shown in Fig. 1.12. In stage I, the skin material is shot into the mold cavity followed by core material injection in stage II. The injection of core material is continued further in stage III, where the product takes the final shape. Kohlitz and Krawinkel have developed a procedure for the production of a multipart sandwich structure suitable for automobile chassis [98]. A sealing material is applied in between the foam and skin layers to prevent the penetration of foamed or injected material to the visible side of the product. Resin-made boxes having flat bottoms with less warpage have been manufactured by sandwich molding, as reported by Kato and Kanemitsu [99]. Mitamura et al. have made laminates consisting of synthetic resin matrix and reinforcing fiber layers and have a surface with concave and convex patterns [100]. A method for fabrication of a lightweight composite part by sandwich molding used for automobiles and buildings has been proposed by Zanella [101]. Giehl and Mitschang have designed an integrated process for producing fiber-reinforced sandwich and profile structures in one step [102]. The residual stresses and warpage in sandwich injection molding have been modeled by Lundstrom and Toll [103]. The derived models compare well to the experiments made with a two-layered structure. This process offers the possibility of combining materials with different properties and hiding a recycled constituent within a fresh skin material. But it involves high investment cost and the method cannot produce thin sandwich products of thickness less than 2 mm.

Modeling of the interface distribution and flow-induced residual stresses and birefringence in the sequential co-injection molding (CIM) of a center-gated disk have been carried out by Kim and Isayev using a numerical scheme based on a hybrid finite element/finite difference/control volume method [104]. The influence of the processing variables including melt and mold temperatures and volume of skin melt on the birefringence and interface distribution was analyzed for multilayered PS-PC-PS, PS-PMMA-PS, and PMMA–PC–PMMA molded disks obtained by CIM. Zaverl et al. have studied the effect of processing parameters on mechanical properties and material distribution of co-injected polymer blends within a complex mold shape [105]. A partially bio-sourced blend of poly(butylene terephthalate) and poly(trimethylene terephthalate) PTT/PBT is used for the core, with a tough biodegradable blend of poly(butylene succinate) and poly(butylene adipate-co-terephthalate) PBS/PBAT for the skin. A ½ factorial design of experiments is used to identify significant processing parameters from the skin and core melt temperatures, injection speed and pressure, and mold temperature. Interactions between the processing effects are considered, and the resulting statistical data produced accurate linear models indicating that the co-injection of the two blends can be controlled. Impact strength of the normally brittle PTT/PBT blend is shown to increase significantly with co-injection and variations in core to skin volume ratios to have a determining role in the overall impact strength.

Challenges in Manufacturing of Polymer Matrix Composites

The manufacture of advanced composite parts involves the preparation of a network of fibers according to the desired shape of the component, impregnating this fiber network with the resin, and curing of the resin while the part is being supported. The major challenges are efficient handling of the fibers and application of uniform pressure during the curing of resin. The fibers are stiff in nature and do not flow. For this reason, the continuous fibers are stacked in layers, which are successively impregnated with the resin or formed into a preform and kept in a mold. The resin is supplied into the mold under pressure to make the fiber wet. In addition to these problems, getting high volume percent of fibers in the composites is another major problem. Appropriate curing of the matrix is required to achieve required properties in the composites. The pressure, temperature, component geometry, and distribution of pressure over the surface of component are some of the deciding factors of quality product. Due to the constraints imposed by fibers and resin systems, the manufacturing technologies developed so far are centered on a specific application area, and the processing parameters are optimized as per specific need.

Applications of Polymer Matrix Composites

It is highly impossible to provide a complete list of PMC applications. However, some applications classified according to major market segments are indicated here.

Agriculture In agriculture sector, PMCs are used in the production of feed troughs, containers and enclosures, fencing, equipment components, partitions, staging, flooring, silos and tanks, etc.

Aerospace and Aircraft PMCs has wide applications in aerospace industry such as construction of containers, gliders, control surfaces, and light aircraft, internal fittings, window masks, partitions and floors, galley units and trolleys, satellite components, aerials and associated enclosures, structural members, ground support equipment components and enclosures, etc.

Business and Appliance Equipment Molded items like enclosures, frameworks, covers, and fittings and assemblies for internal use, switchgear bodies and associated electrical and insulation components, etc.

Building and Construction External and internal cladding, permanent and temporary formwork and shuttering, partitions, polymer concrete, prefabricated buildings, kiosks, cabins and housing, structural and decorative building elements, bridge elements and sections, quay facings, signposts and street furniture, staging, fencing and walkways, etc.

Consumer Product Components For domestic and industrial furniture, sanitary ware, sporting goods, caravan components, archery and playground equipment, garden furniture, notice boards, theme park requirements, swimming pools, aqua tubes, diving boards, seating and benches, simulated marble components, skis and snowboards, etc.

Corrosion-Resistant Equipment Chemical plant, linings, oil industry components, pipes and ducts, chimneys, grid flooring, staging and walkways, pressure vessels, processing tanks and vessels, fume hoods, scrubbers and cooling tower components, assemblies and enclosures, etc.

Defense Aircraft vehicle, aerospace and satellite components, enclosures and containers, personnel armor, rocketry and ballistic items, shipping and transit containers, simulators and allied, etc.

Electrical and Electronic Internal and external aerial components and fittings, circuit boards, generation and transmission components, insulators, switch boxes and cabinets, booms, distribution posts and pylons, telegraph poles, fuse tubes, transformer elements, ladders and cableways, etc.

Engineering and Industrial Assemblies and fittings, sundry enclosures, pallets, safety helmets, trays, bins, profiles and medical items, assemblies and equipment components, etc.

Marine Applications PMCs are used in the manufacture of canoes and boats, yachts, surf and sailboards, lifeboats and rescue vessels, buoys, boat accessories and subassemblies, window masks and internal moldings and fittings for ferries and cruise liners, work boats and trawlers, etc.

Transportation Automotive, bus, camper and vehicle components generally, both underbody, engine and body panels, truck, rail and other vehicle components and fittings, land and sea containers, railway track and signaling components, traffic signs, seating, window masks and partitions, etc.

Water Control Engineering and Sewage PMCs are used in the manufacture of tanks, pipes, process and storage vessels, staging, pump components, partitions, walkways, scrubbers, and weirs.

Concluding Remarks

The composite materials have many advantages and can be tailored to meet the specific functional requirements, yet they are not popular in everyday products. This is because of the manufacturing difficulties and costs associated with the production of consumer goods at competitive pricing. Traditional metallic materials still dominate at many places, where a composite material might be more appropriate with respect to material economy and function. The major obstacle in this direction is not the material cost; it is the manufacturing cost of a composite item. The existing manufacturing methods are costly thereby increasing the product cost. No doubt, the technology for manufacturing of high-quality composite products (like in aerospace and automotive sector) is available, but the equipment, tooling, and running costs are so high that the production of general-purpose items is not economical. Composite materials cannot be utilized to their full potential unless we have the technology for competitive production of composite items. This is a major issue today and there is a need for innovation in manufacturing technology of composites. In addition to these problems, application of high pressure to minimize the void content in the FRP product having a complex geometry and getting high volume percent of fiber is not feasible. Again there is a major problem if the product has a complex geometry like flat, flat with curve, and flat with curve to cone.

Acknowledgment The authors acknowledge the financial support provided by the Indian Space Research Organization, India, for carrying out this work.

References

1. Agarwal BD, Broutman LJ (1990) Analysis and performance of fiber composites. Wiley, New York
2. Peters ST (ed) (1997) Handbook of composites. Chapman & Hall, London
3. Mallick PK (1993) Fibre-reinforced composites: materials, manufacturing, and design. Marcel Dekker, New York
4. Chung DDL (1994) Carbon fiber composites. Butterworth-Heinemann, Boston
5. Mazumdar SK (2002) Composites manufacturing: materials, product and process engineering. CRC Press, Boca Raton
6. Avila AF, Morais DTS (2005) A multiscale investigation based on variance analysis for hand lay-up composite manufacturing. Compos Sci Technol 65(6):827–838
7. Otaka M (1997) Hand lay-up molding process. Japanese Patent JP 09,314,686
8. Mariatti M, Chum PK (2005) Effect of laminate configuration on the properties of glass fiber-reinforced plastics (GFRPs) mixed composites. J Reinf Plast Compos 24(16):1713–1721
9. Andersson IM, Alenius S (1996) A comparison between measured and numerically calculated styrene concentrations in hand lay-up molding. Ann Occup Hyg 3(6):399–415
10. Rahamani H, Heyder S, Najafi M et al (2014) Mechanical performance of epoxy/carbon fiber laminated composites. J Reinf Plast Compos 33(8):733–740
11. Chaple AR, Khedakar SS, Dharmadhikari SR et al (2013) Newly developed automatic lay-up process for manufacturing of FRP sheets. Int J Comput Eng Res 3(3):92–97
12. Vasanthanathan A, Nagaraj P, Muruganantham B (2013) On the fabrication of carbon fabric reinforced epoxy composite shell without joints and wrinkling. Steel Compos Struct 15(3):267–279
13. Chapman MR, Watson RM, Anderson DA et al (2006) Composite sections for aircraft fuselages and other structures, and methods and systems for manufacturing such sections. US Patent 2,006,108,058
14. Liu L, Wang W, Liu Z (2006) Process for molding laminated board using high-energy focused ultrasound and high-pressure water tank. Chinese Patent 1,775,498
15. Caron MC (2005) Reusable vacuum bag for forming polymeric materials against the mold. US Patent 2,005,086,916
16. Hou TH, Jensen BJ (2005) Double vacuum bag process for resin matrix composite manufacturing. US Patent 2,005,253,309
17. Zhou C, Yang Q (2005) Vacuum bag pressing method for reinforcing structure with external bonding of fibre. Chinese Patent 1,641,130
18. Kaneko T, Mitani K, Tomoo S et al (2004) Prepregs and their fibre-reinforced plastics with decreased void and good mechanical strength manufactured by vacuum bag molding. Japanese Patent 2,004,346,190
19. Johnson K, Lewit S (2004) Curable liquid sealant used as vacuum bag in composite manufacturing. US Patent 2,004,046,291
20. Kook JS (2000) Method for vacuum molding of wet resin glass fibre-reinforced plastics. Korean Patent 2,000,067,592
21. Alms JB, Advani SG, Glencey JL (2011) Liquid composite molding control methodologies using Vacuum Induced Preform Relaxation. Compos Part A 42(1):57–65
22. Harper A (2013) Why consider reusable vacuum bags. Reinf Plast 57(3):40–42
23. Shima K (2006) Internal-pressure bag molding method for manufacturing hollow FRP moldings efficiently. Japanese Patent 2,006,130,875
24. Horibe I, Nishimura A, Honma K (2002) Production method of fibre-reinforced plastics. Japanese Patent 2,002,307,463
25. Dai F, Aota Y, Watanabe T (2001) Mandrels for pressure-bag molding and manufacture of shafts using them. Japanese Patent 2,001,030,278
26. Kanemaru R, Sanae T (2000) Manufacture of fibre-reinforced plastics (FRP) with complex structures by internal pressure molding using a flexible bag. Japanese Patent 2,000,334,851

27. Asano K, Nojima K (1997) Manufacture of full-sized FRP pipes by bag molding. Japanese Patent 09,123,196
28. Howard RD (1993) A method for preparing a fibre reinforced polymer composite. UK Patent 9,305,941
29. Antonucci V, Giordano M, Inserraimparato S et al (2001) Analysis of heat transfer in autoclave technology. Polym Compos 22(5):613–620
30. Jiang J, Motogi S, Fukuda T (1998) Relation between interlaminar fracture toughness and pressure condition in autoclave molding process of GFRP composite laminates. J Soc Mater Sci Jpn 47(6):606–611
31. Chen MY, Chen SM, Zeng LZ et al (1992) Dimensional control of polymer composite laminate. MRL Bull Res Dev 6(1):23–27
32. Boey FYC (1989) Development of an autoclave curing system for fiber reinforced polymer composites. Polym Test 8(6):375–384
33. Beckwith SW (1998) Filament winding the string and the glue. Compos Fab 14(1):8–12
34. Munro M (1988) Review of manufacturing of fiber composite components by filament winding. Polym Compos 9(5):352–359
35. Polini W, Sorrentino L (2005) Winding trajectory and winding time in robotized filament winding of asymmetric shape parts. J Compos Mater 39(15):1391–1411
36. Parnas L, Ardic S (2001) Filament winding. In: Akovali G (ed) Handbook of composite fabrication, 1st edn. Rapra, Shropshire, pp 103–126
37. Sohl C (2001) Continuous molding of thermoset composites. In: Anthony K, Carl Z (eds) Comprehensive composite materials, vol 2, Polymer matrix composites. Elsevier, Oxford, pp 845–852
38. Martinec T, Mlynec J, Petru M (2015) Calculation of the robot trajectory for the optimum directional orientation of fibre placement in the manufacture of composite profile frames. Rob Comput Integr Manuf 35:42–54
39. Zhang L, Du H, Liu L et al (2014) Analysis and design of smart mandrels using shape memory polymers. Compos Part B 59:230–237
40. Rojas EV, Chapelle D, Perreux D et al (2014) Unified approach of filament winding applied to complex shape mandrels. Compos Struct 116:805–813
41. Lam YC, Li J, Joshi SC (2003) Simultaneous optimization of die-heating and pull-speed in pultrusion of thermosetting composites. Polym Compos 24(1):199–209
42. Trivisano A, Maffezzoli A, Kenny JM et al (1990) Mathematical modeling of the pultrusion of epoxy based composites. Adv Polym Technol 10(4):251–264
43. Esfandeh M, Reza DAM, SABET SAR et al (2002) Investigation of factors affecting the surface quality of pultruded profiles. Iran J Polym Sci Technol 15(4):229–235
44. Murata T, Muranaka T (2002) Apparatus and method for continuous manufacture of fibre-reinforced plastic foam composites. Japan Patent 2,002,331,593
45. Methven JM, Ghaffariyan SR, Abidin AZ (2000) Manufacture of fiber-reinforced composites by microwave assisted pultrusion. Polym Compos 21(4):586–594
46. Simacek P, Advani SG (2015) Simulating tape resin infiltration during thermoset pultrusion process. Compos Part A 72:115–126
47. Boyd SW, Barton JMD, Thomsen OT et al (2010) Through thickness stress distributions in pultruded GRP materials. Compos Struct 92(3):662–668
48. Belingardi G, Bevene AT, Koricho EG et al (2015) Alternative lightweight materials and component manufacturing technologies for vehicle frontal bumper beam. Compos Struct 120:483–495
49. Otten V, Siebrecht D (2006) Resin-transfer-molding device arrangement and method for manufacturing a component. US Patent 2,006,186,580
50. Lawrence JM, Devillard M, Advani SG (2004) Design and testing of a new injection approach for liquid composite molding. J Reinf Plast Compos 23(15):1625–1638
51. Devillard M, Hsiao KT, Gokce A et al (2003) On-line characterization of bulk permeability and race-tracking during the filling stage in resin transfer molding process. J Compos Mater 37(17):1525–1541

52. Nielsen DR, Pitchumani R (2002) Closed-loop flow control in resin transfer molding using real-time numerical process simulations. Compos Sci Technol 62(2):283–298
53. Mogavero J, Sun JQ, Advani SG (1997) A nonlinear control method for resin transfer molding. Polym Compos 18(3):412–417
54. Ferland P, Guittard D, Trochu F (1996) Concurrent methods for permeability measurement in resin transfer molding. Polym Compos 17(1):149–158
55. Kranbuehl DE, Kingsley P, Hart S et al (1994) In situ sensor monitoring and intelligent control of the resin transfer molding process. Polym Compos 15(4):299–305
56. Hayward JS, Harris B (1990) Effect of process variables on the quality of RTM moldings. SAMPE J 26(3):39–46
57. Han SH, Cho FJ, Lee HC et al (2015) Study on high-speed RTM to reduce the impregnation time of carbon/epoxy composites. Compos Struct 119:50–58
58. Yang B, Jin T, Bi F et al (2015) Influence of fabric shear and flow direction on void formation during resin transfer molding. Compos Part A 68:10–18
59. Eck B, Cardona SC, Binetruy C et al (2015) Multi-objective composite part mechanical optimization enhanced by a Process Estimator. Compos Struct 119:620–629
60. Tanaka Y, Yotsukura R (2006) Manufacture of fibre-reinforced resin articles by VARTM (vacuum-assisted resin transfer molding) and apparatus therefore. Japanese Patent 2,006,167,933
61. Dong C (2006) Development of a process model for the vacuum assisted resin transfer molding simulation by the response surface method. Compos Part A 37(9):1316–1324
62. Kelkar AD, Tate JS, Chaphalkar P (2006) Performance evaluation of VARTM manufactured textile composites for the aerospace and defense applications. Mater Sci Eng B Solid 132 (1–2):126–128
63. Bender D, Schuster J, Heider D (2006) Flow rate control during vacuum-assisted resin transfer molding (VARTM) processing. Compos Sci Technol 66(13):2265–2271
64. Tackitt KD, Walsh SM (2005) Experimental study of thickness gradient formation in the VARTM process. Mater Manuf Processes 20(4):607–62765
65. Seyhan AT, Tayfur G, Karakurt M et al (2005) Artificial neural network (ANN) prediction of compressive strength of VARTM processed polymer composites. Comput Mater Sci 34 (1):99–105
66. Tuccillo F, Antonucci V, Calabro AM et al (2005) Practical and theoretic analysis of resin flow in vacuum assisted resin transfer molding processes. Macromol Symp 228(1):201–218
67. Eum SH, Kim YH, Han JW et al (2005) A study on the mechanical properties of the honeycomb sandwich composites made by VARTM. Key Eng Mater 300(4):2746–2751
68. Chen R, Dong C, Liang Z et al (2004) Flow modeling and simulation for vacuum assisted resin transfer molding process with the equivalent permeability method. Polym Compos 25(2):146–164
69. Kamar NT, Hossain MM, Khomenko A et al (2015) Interlaminar reinforcement of glass fiber/ epoxy composites with graphene nanoplatelets. Compos Part A 70:82–92
70. Matsuzaki R, Kobayashi S, Todoroki A et al (2013) Flow control by progressive forecasting using numerical simulation during vacuum-assisted resin transfer molding. Compos Part A 45:79–87
71. Hsiao KT, Heider D (2012) Vacuum assisted resin transfer molding (VARTM) in polymer matrix composites. In: Advani SG, Hsiao KT (eds) Manufacturing techniques for polymer matrix composites (PMCs), 1st edn. Woodhead, Cambridge, pp 310–347
72. Dumont P, Orgeas L, Favier D et al (2007) Compression moulding of SMC: in situ experiments, modelling and simulation. Compos Part A 38(2):353–368
73. Yen CLE, Lukas KS (2006) Preferential curing technique in compression molding of fibre-reinforced composites. US Patent 2,006,091,583
74. Odenberger PT, Andersson HM, Lundstroem TS (2004) Experimental flow-front visualization in compression molding of SMC. Compos Part A 35(10):1125–1134
75. Zhang GP, Shen CY, Dai GC (2005) Flow behavior of glass mat reinforced polypropylene in compression molding. Boligang Fuhe Cailiao (Fiber Reinforced Plastics/Composite) 3:38–40

76. Jo SH, Kim EG (2002) Effect of product geometry on fiber orientation of compression-molded rib type products. J Mater Process Technol 130:156–160
77. Hojo H, Kim EG, Tamakawa K (1987) The content distribution of compression-molded long fiber-reinforced thermoplastic products. Int Polym Proc 1(2):60–65
78. Lee CC, Folgar F, Tucker CL (1984) Simulation of compression molding for fiber-reinforced thermosetting polymers. J Eng Ind Trans ASME 106(2):114–125
79. Goodship V, Brzeski I, Wood BM et al (2014) Gas-assisted compression moulding of recycled GMT: effect of gas injection parameters. J Mater Process Technol 214(3):515–523
80. Taketa I, Okabe T, Kitano A (2008) A new compression-molding approach using unidirectionally arrayed chopped strands. Compos Part A 39(12):1884–1890
81. Levy A, Hubert P (2015) Interstrand void content evolution in compression moulding of randomly oriented strands (ROS) of thermoplastic composites. Compos Part A 70:121–131
82. Donatti J (2003) The structural reaction injection molding process having void reduction. US Patent 2,003,155,687
83. Polushkin EY, Polushnkina OM, Malkin AY et al (2002) Modeling of structural reaction injection molding. Part II: comparison with experimental data. Polym Eng Sci 42(4):846–858
84. Duh RJ, Mantell S, Vogel JH et al (2001) Optimization of cure kinetics parameter estimation for structural reaction injection molding/resin transfer molding. Polym Compos 22(6):730–741
85. Castro J, Conover S, Wilkes C et al (1997) Manufacturing of composite links by structural reaction injection molding. Polym Compos 18(5):585–594
86. McGeehan JA, Gillespie JW Jr, Hulway JA (1993) Processing/performance relationships considering voids in structural reaction injection molding. Polym Eng Sci 33(24):1627–1633
87. Young WB, Fong KHLH, Lee LJ (1991) Flow simulation in molds with preplaced fiber mats. Polym Compos 12(6):391403
88. Tang H, Wang J (2006) Summarization and research progress on structural reaction injection molding. Gongcheng Suliao Yingyong (Eng Plast Appl) 34(5):72–74
89. Deng X, Li DX, Chen JS et al (2012) Numerical simulation of the curing and cooling in reaction injection molding process of nylon 6. Appl Mech Mater 161:205–211. doi:10.4028/www.scientific.net/AMM.161.205
90. Kapila A, Singh K, Arora G et al (2015) Effect of varying gate size on the air traps in injection molding. Int J Curr Eng Technol 5(1):161–166
91. Hashisaka K, Ue M, Tabata K (2006) Molds and molding of polyurethane foams with reduced voids and pinholes by using them. Japanese Patent 2,006,224,371
92. Shidaker TA, Bareis DW (2004) Unreinforced reaction injection molded structural foams. US Patent 2,004,192,800
93. Modesti M, Adriani V, Simioni F (2000) Chemical and physical blowing agents in structural polyurethane foams: simulation and characterization. Polym Eng Sci 40(9):2046–2057
94. Tighe SC, Manzione LT (1988) Simulation of foaming in reaction injection molding. Polym Eng Sci 28(15):949–954
95. Heim HP, Tromm M (2015) General aspects of foam injection molding using local precision mold opening technology. Polymer 56:111–118
96. Li XF, Lau KT, Yin YS et al (2011) Reaction injection molding process and fireproof property of phenolic foam sandwich panel. Adv Mater Res 410:341–344
97. Samkhaniani N, Gharehbaghi A, Ahmadi Z (2013) Numerical simulation of reaction injection molding with polyurethane foam. J Cell Plast 49(5):405–421
98. Kohlitz S, Krawinkel S (2006) Procedure for the production of a multi-part sandwich structure. German Patent 102,004,052,852
99. Kato N, Kanemitsu Y (2005) Resin-made boxes having flat bottoms with less warpage manufactured by sandwich molding. Japanese Patent 2,005,075,422
100. Mitamura S, Aoki S, Seki T et al (2005) Fibre-reinforced laminates and their manufacture. Japanese Patent 2,005,305,673
101. Zanella G (2005) Process for fabrication of a lightweight composite part by molding. French Patent 2,863,933

102. Giehl S, Mitschang P (2005) Process integration: fibre-reinforced sandwich and profile structures in one step. Kunststoffe 2005(11):76–78
103. Lundstrom TS, Toll S (2003) Modelling of residual stresses and warpage in sandwich injection moulding. Int Polym Proc 18(1):95–106
104. Kim NH, Isayev AI (2015) Birefringence and interface in sequential co-injection molding of amorphous polymers: simulation and experiment. Polym Eng Sci 55(1):88–106
105. Zaverl M, Valerio O, Misra M et al (2014) Study of the effect of processing conditions on the co-injection of PBS/PBAT and PTT/PBT blends for parts with increased bio-content. J Appl Polym Sci 132(2):41278

Polymer-Based Composite Materials: Characterizations

2

Jayesh Cherusseri, Sumit Pramanik, L. Sowntharya, Deepak Pandey, Kamal K. Kar, and S. D. Sharma

Contents

J. Cherusseri • L. Sowntharya • D. Pandey
Advanced Nanoengineering Materials Laboratory, Materials Science Programme, Indian Institute of Technology Kanpur, Kanpur, Uttar Pradesh, India
e-mail: jayeshpuli@gmail.com; sowsindhu@gmail.com; dpandey@oceaneering.com

S. Pramanik
Advanced Nanoengineering Materials Laboratory, Materials Science Programme, Indian Institute of Technology Kanpur, Kanpur, Uttar Pradesh, India

Department of Biomedical Engineering, Faculty of Engineering, University of Malaya, Kuala Lumpur, Malaysia
e-mail: prsumit@gmail.com

K.K. Kar (✉)
Advanced Nanoengineering Materials Laboratory, Materials Science Programme, Indian Institute of Technology Kanpur, Kanpur, Uttar Pradesh, India

Advanced Nanoengineering Materials Laboratory, Department of Mechanical Engineering, Indian Institute of Technology Kanpur, Kanpur, Uttar Pradesh, India
e-mail: kamalkk@iitk.ac.in

S.D. Sharma
Advanced Nanoengineering Materials Laboratory, Materials Science Programme, Indian Institute of Technology Kanpur, Kanpur, Uttar Pradesh, India

Department of Mechanical Engineering, Rewa Engineering College, Rewa, India
e-mail: sds1970@gmail.com

K.K. Kar (ed.), *Composite Materials*,
DOI 10.1007/978-3-662-49514-8_2

Abstract

Various types of composite materials are becoming an inevitable part of our day-to-day life since these are used for a variety of applications. A better understanding on the various properties of the composites is very helpful in their targeted applications, and hence characterizing the composite materials by different techniques play a major role in the development of long-life, high-quality composite products. The polymer-based composite materials provide large amount of flexibility and lightweight to the final product. The selection of various reinforcements and polymer matrices is very critical in designing a desired product. In this chapter, various techniques used for characterizing the polymer-based composite materials in order to examine their mechanical, thermal, electrical, magnetic, piezoelectric, tribological, rheological, and biological properties are discussed.

Keywords

Creep • Fatigue • Flexural strength • Glass transition temperature • Piezoelectric • Polymer composite • Scanning electron microscopy • Stiffness • Stress relaxation • Voids

Introduction

Composite materials have been in use for thousands of years. For example, mud bricks, which are composed of mud (matrix) and straw (reinforcing fiber), can resist both tensile and compressive loads, thus resulting in an excellent building material. It is a combination of at least two materials, one being a binding material (or matrix) and the other one is a reinforcing material. The matrix holds the reinforcing materials together. However, many composite products are made with more than two raw materials, but those materials are not miscible together and have different nature. Composites exist in nature and can be made by humans also. A good example for a natural composite is wood. It consists of long fibers of cellulose bound together by a much weaker substance called lignin. There are several kinds of man-made composites like structural beam, which is an inhomogeneous mixture of cement (matrix), stone (reinforcing particle), and steel rod (reinforcing fiber). The selection of materials for the matrix and reinforcement, the form of reinforcement, and the manufacturing process are very critical for obtaining a composite product with desired properties. The matrix material can be a metal, ceramic, glass, polymer, or organic material in solid (plate or powder) or semisolid forms. The reinforcing material can be a metal, ceramic, glass, textile, polymer, or organic material in the form of lamina, fiber (short and long), particle, whisker, etc. The main advantage of composite material is their high strength and stiffness combined with lightweight feature. The composites have major applications in the advanced fields such as structure, thermal engine and blade, automobile, aerospace, rocket, missile, satellite, military, biomedical, electronics, packaging, etc. The characterization of a composite is one of the essential tasks for developing the desired composite products for particular applications. The major characterization studies used for composites for the evaluation of performance in the targeted areas are mechanical, thermal, electrical, magnetic, piezoelectric, tribological, rheological, and biological.

Volume Fraction

In a composite material, the parameter "volume fraction" plays a major role in characterizing its various properties such as mechanical, thermal, electrical, etc. The evaluation of this parameter is of great importance for material scientists. For example, in a fiber-reinforced composite, the fiber volume fraction significantly determines its properties. A composite can't be prepared with 100 % of fibers. There is a maximum limit of fiber content in a composite structure. And this maximum volume fraction is able to achieve only when the unidirectional fibers are hexagonally close packed. For a composite material, let the total volume be "V" and masses of its constituent materials be M_1, M_2, M_3, etc. Then the composite density (ρ_{com}) can be expressed (Eq. 2.1) as

$$\rho_{com} = \frac{M_1}{V} + \frac{M_2}{V} + \frac{M_3}{V} + \ldots \tag{2.1}$$

In terms of the densities and volumes of the constituents, Eq. 2.1 can be written as Eq. 2.2 based on the rule of mixtures.

$$\rho_{com} = \frac{V_1 \rho_1}{V} + \frac{V_2 \rho_2}{V} + \frac{V_3 \rho_3}{V} + \ldots \tag{2.2}$$

where V_i and ρ_i are the volume fractions and densities of the ith phase, respectively.

The fiber volume fraction (V_f) can be written in terms of fiber weight fraction (W_f) [1] as

$$V_f = \frac{W_f \rho_m}{W_f \rho_m + W_m \rho_f} \tag{2.3}$$

where ρ_m, ρ_f, W_f, and W_m are the density of the matrix, the density of the fiber, the weight fraction of the fiber, and the weight fraction of the matrix, respectively. Correspondingly, the fiber weight fraction (W_f) can be expressed in terms of fiber volume fraction (V_f) [1] as

$$W_f = \frac{V_f \rho_f}{V_f \rho_f + V_m \rho_m} \tag{2.4}$$

The fiber volume fraction determines the strength of the composites. Ravi et al. have studied the effect of fiber volume percent on the dynamic damage behavior of woven glass fiber-reinforced polymer (FRP) composites [2]. Delamination is the major damage noticed for specimens with higher fiber volume percent, while the matrix cracking and interface debonding occurs for materials with low fiber volume percent. Fiber volume percent in combination with fiber distribution is the main parameter affecting the interlaminar fracture toughness (ILFT) [3]. It has a significant influence on the longitudinal tensile and compressive strengths and the in-plane shear strength of FRPs. The strength parameters vary in direct proportion to the fiber volume percent up to 50 % [4]. The fiber volume percent of a composite are determined by chemical matrix digestion method as described in ASTM D3171 [5], the burn test as described in ASTM standard D2584 [6], or by photomicrographic techniques. For glass fiber composites, burn test is used because the glass shows good oxidation resistance (500–600 °C), which is much more than the complete burning of matrix.

Voids

Voids form at the interface of composite structures. These are generally formed as gas bubbles trapped inside the cured composite materials. The main sources of voids include the material constituents and the synthesis processes. The presence of voids in a composite structure is found to affect its structural and mechanical properties. It has been predicted that the interlayer and intra-layer reactions caused by the stress and chemically driven diffusion are responsible for the usual void behavior in composites. Voids present in the composite material can be evaluated by matrix digestion and matrix burn-off techniques [1].

The volume fraction of void (V_{void}) can be calculated if the composite density and the density of the matrix and fibers are known. The equation for determining the volume fraction of void is given by Eq. 2.5 [1] as

$$V_{void} = 1 - \rho_{com}^{exp} \left(\frac{W_m}{\rho_m} + \frac{W_f}{\rho_f} \right) \tag{2.5}$$

where ρ_{com}^{exp} is the experimentally determined composite density. Void is undesirable and has adverse effect on the mechanical properties of composites. Wu et al. have developed an analytical model to estimate the elastic properties of polymer composites with voids of various sizes and locations [7]. These voids have significant influence on the interlaminar shear strength (ILSS) of the composite materials

[8]. Bowles and Frimpong have studied the influence of voids on the ILSS of graphite fiber-reinforced PMR-15 composites [9]. As the void content increases, an increase in scatter in the strength data is observed. The role of voids in the initiation and propagation of cracks in composites reinforced with unidirectional carbon fiber (CF) has been investigated by Chambers et al. [10]. Void area greater than 0.03 mm^2 results in deterioration of mechanical properties. Voids assist in the propagation of crack in the resin-rich interply regions. An increased void content increases composite's susceptibility to water penetration and weathering. The procedure of measuring the voids in composites, required for the calculation of density, is given in ASTM D 1505 [11] and ASTM D 3800 [12].

Surface Roughness

Surface roughness is an important parameter for ascertaining the surface quality and aesthetic value. The average surface roughness (Ra value) is one of the most frequently used parameters for surface roughness, which describes the height of irregularities and gives an indirect indication of the sharpness and depth of surface notches [13]. It is an absolute average roughness over one sampling length [14]. Researchers have used this Ra value for studying the impacts of various process parameters on the surface quality of FRP composites. Davim and Reis have analyzed the effects of machining parameters on the surface quality of plastics reinforced with CFs [15].

Surface Topography

Scanning electron microscope (SEM) is also used for studying the topography of solids [16]. In the field of FRPs, it is used to reveal the actual distribution of fibers and matrix in the composite. It is also used for the analysis of fractured surfaces [17] and helps to examine the crack propagation in fibrous composite materials in order to gain an insight about composite strength, the adhesion between the phases, and the mode of failure [18]. It serves as an effective means to confirm the uniformity of fiber distribution across the cross section and presence of internal flaws (like voids, broken fibers, foreign inclusions, etc.) at microscopic level, which explain the physical and mechanical behavior of composites and the discrepancies associated with their properties.

Mechanical

Measurement of mechanical properties of composites involves testing on the suitably designed specimens. The interpretation of material response over the entire range of loads is necessary, if the advanced design procedures are to be used for the efficient material utilization. The macromechanical properties of composites may be computed through micromechanical analysis, and the design of composite materials may be started with constituent material properties. However, the validity of

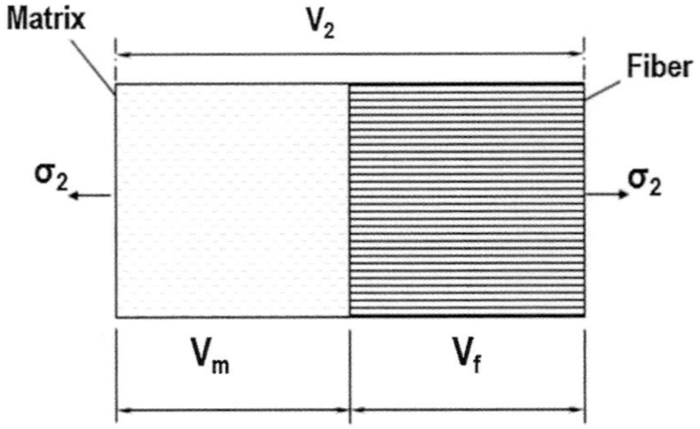

Fig. 2.1 Schematic of slab model showing transverse stiffness of a unidirectional composite

micromechanical analysis is to be checked through experiments. Another purpose of mechanical characterization is to check the adequacy of new fabrication procedure and to assure the material uniformity. The mechanical properties of composites include (i) strain and yield strength in tension, compression, shear, and torsion, (ii) ILSS between the matrix and fiber, (iii) flexural fatigue strength, (iv) impact strength, (v) stress relaxation, and (vi) creep.

Strength

A high strength-to-weight ratio is required for a composite in its applications. Hence, it is very important to determine its strength. Thus, characterizing the strength of a composite is very critical before its application. This section briefly discusses about the various mechanical characterization tools used for finding out the strength of the composite materials.

Tensile Strength The ASTM standard D-3552-96 defines the test method for determining the tensile properties of a fiber-reinforced metal matrix composite (MMC). However, the tensile properties of advanced composite materials and polymer matrix composites can be tested by ASTM standard D-3039, and the tensile properties of MMCs can be characterized by ASTM standard D-3518 specimen. Strength characteristics of composite materials generally follow the mixture rules. These are mathematical expressions giving the property of the composite in terms of the quantity and arrangement of its constituting materials. Let us assume a part of the unidirectional composite material having unit dimensions with the fibers having volume fraction, V_f. Then, the volume fraction of the matrix, V_m (Eq. 2.6), is given [19] as

$$V_m = 1 - V_f \tag{2.6}$$

The slab model is shown in Fig. 2.1.

If a stress σ_1 is applied in a direction of the fiber orientation, then both the fiber and matrix act parallel to resist the load. In this case, strain in both fiber (ε_f) and matrix (ε_m) must be the same (let it be ε_l) in fiber direction and can be written (Eq. 2.7) as

$$\varepsilon_f = \varepsilon_m = \varepsilon_l \tag{2.7}$$

It is clear that the load on the composite must be equal to the sum of the total load of the fiber phase and the matrix phase. Load on each phase is the phase area times the stress in each phase, but phase area is numerically equal to its own volume fraction. Hence, the composite strength (σ_{com}, Eq. 2.8) can be written [19] as

$$\sigma_{com} = \sigma_f V_f + \sigma_m V_m \tag{2.8}$$

where, σ_f, σ_m, V_f, and V_m are the total stress on the fiber phase, the total stress on the matrix phase, the volume fraction of the fiber phase, and the volume fraction of the matrix phase, respectively.

Consider the ultimate strained condition of the fibers in the unidirectional composite at which the fiber begins to break. The breaking strength of the composite (σ_b) can be written from the rule of mixtures [19] as

$$\sigma_b = \sigma_{fb} V_f + \sigma_{mb}(1 - V_f) \tag{2.9}$$

where σ_{fb} and σ_{mb} are the breaking strength of the fiber and the breaking strength of the matrix, respectively. At the breaking strain of the fiber, the load carried by the fiber becomes zero, which results in transferring the entire load to the matrix. The breaking strength of the composite is now reduced according to the fiber fraction as

$$\sigma_b = \sigma_{mb}(1 - V_f) \tag{2.10}$$

The maximum of these two strengths gives the breaking strength of the whole composite. For a certain range of fiber fraction in the composite, the strength actually reduces on the addition of fibers. These relations are presented in Fig. 2.2.

The tensile properties of the composites can also be determined by bending a sandwich beam specimen containing a thin layer of composite material, which is bonded to the top and bottom of a thick substrate such as an Al honeycomb [20, 21]. This method can also be used for determining off-axis properties and transverse properties as reported by Lantz [21]. Richards et al. have showed that the satisfactory results can be achieved for 45^0 specimens having a value of 12 for the ratio of the specimen length between the grips to its width [22]. Pagano and Halpin have found that a uniform state of strain and stress exists at the center of an off-axis tensile coupon if the length-to-width ratio is sufficiently large [23]. A standard test method to determine the tensile properties of polymer matrix composite materials has been described in ASTM D3039 [24].

The strength and toughness are the properties that are strongly enhanced by the addition of nanoparticles in the matrix. This is not only observed in the polymer matrix composites but also in other composites. For example, certain nanoconstituents like silicon carbide (SiC), silicon nitride (Si_3N_4), etc. are added

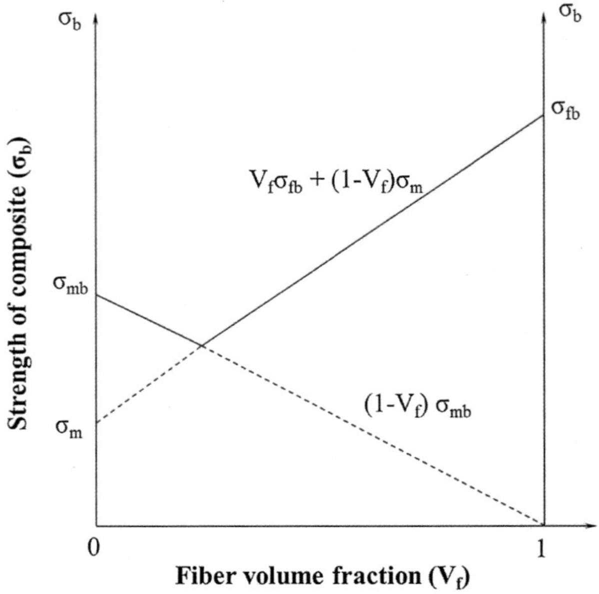

Fig. 2.2 Schematic showing the strength of a unidirectional composite in fiber direction

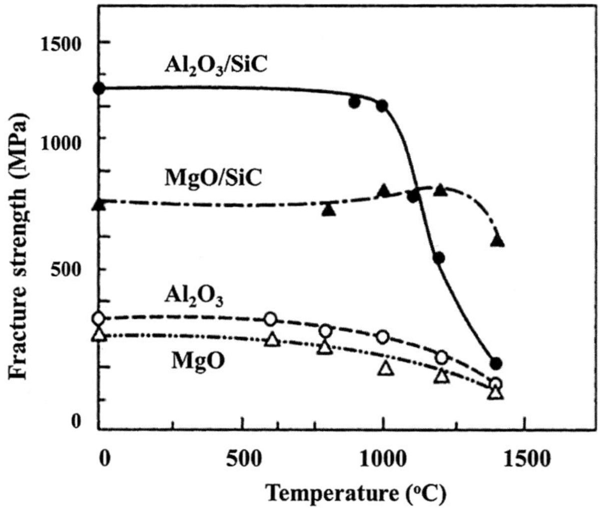

Fig. 2.3 Temperature dependence fracture strength for ceramic oxides and ceramic nanocomposites [25]

into the non-polymer matrix like alumina to form Al_2O_3/SiC, Al_2O_3/Si_3N_4, MgO/SiC, and SiN_4/SiC nanocomposites. The strength of alumina is almost three times augmented with the help of the dispersion of 5 vol.% SiC nanoparticles in the alumina matrix. It has been observed that the annealing also helps to improve the fracture strength by the development of sub-grain boundaries. For example, in the case of MgO/SiC nanocomposite, the annealing enhances its fracture strength over a wide temperature range [25]. A significant improvement of high-temperature

Fig. 2.4 Experimental setup to measure the interlaminar shear strength

strengths for Al_2O_3/SiC and MgO/SiC nanocomposites with respect to Al_2O_3 and MgO matrices is shown in Fig. 2.3.

Flexural Strength In a bend test of a specimen, bending strength of the outermost fiber at the instant of failure is defined as flexural strength. The nominal flexural strength (σ_N, Eq. 2.11) can be written [26] as:

$$\sigma_N = f_r^o\left(1 - \frac{D_b}{D}\right)^{-1} \tag{2.11}$$

where f_r^o is the average tensile strength of the boundary layer and is given by

$$f_r^o = \frac{M_o(D - D_B)}{2I} \tag{2.12}$$

where M_o, D, D_b, and I are the bending moment, the beam depth, the half value of boundary layer thickness, and the moment of inertia of cross section, respectively. The expression for moment of inertia (I) can be written as

$$I = \frac{D^3}{12} \tag{2.13}$$

Plastics have higher flexural strength as compared to their straight tensile. The breaking force required for a specimen of specified width and thickness is known as "modulus of rupture."

Interlaminar Shear Strength Evaluation of ILSS can be done by a similar test procedure adopted for the flexural strength. The experimental setup is shown in

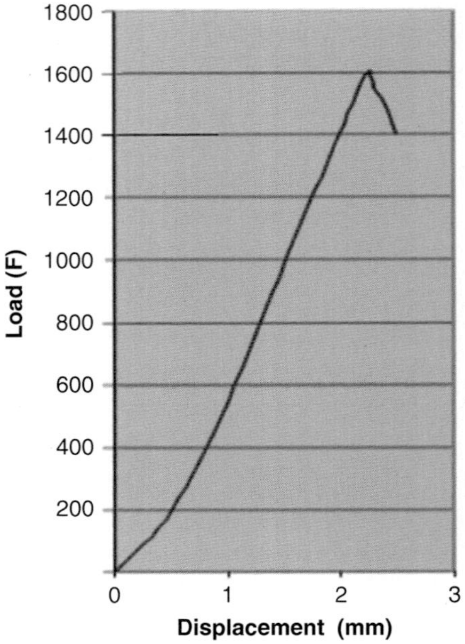

Fig. 2.5 Load versus displacement curve during a test for determining ILSS

Fig. 2.4. The composite specimen dimensions and geometry are recommended in ASTM D-3039-79 and ASTM D-3410 in tension and compression mode, respectively, and ASTM D-695 for advanced composite materials in compression mode. The ILSS (τ, Eq. 2.14) can be expressed [27] as follows:

$$\tau = \frac{3F}{4bd} \tag{2.14}$$

where F, b, and d are the load at the moment of break, the width of the specimen, and the thickness of the specimen, respectively. A representative curve of load versus displacement for a specimen under testing is shown in Fig. 2.5.

Hsu and Herakovich have examined the problem by separating each layer into an outer and an inner region [28]. The inner region is analyzed using classical laminate theory, while the region near the free edge is analyzed by employing the perturbation technique and concluded that their analysis renders better results than the finite difference solution by Pipes and Pagano [29]. Wang and Dickson have developed a method to calculate the interlaminar stresses in symmetric composite laminates [30]. They have used the extended Galerkin method with the interlaminar stresses, and the displacements are represented by complete sets of Legendre polynomials. This method provides systematic steps for direct computation of interlaminar stresses. A substantial study has been reported on the ILSS of composite materials at room temperature. ILSS depends on the resin formulation [31–35]. A group of researchers has altered the resin to improve the ILSS of composites [36–41]. Another group of researchers has altered the fiber surface using elastomer [42], silane [43, 44], styrene

[45], styrene–isoprene [46], polybutadiene [47], methacryloxypropyltrimethoxysilane [46], methacryloxypropyltrimethoxysilane-modified polybutadiene [46], and oxidation to improve the ILSS of composites made of epoxy [42, 45], phenolic [43], nylon 6 [44], and vinyl ester [47]. The dependence of ILSS on fiber orientations of the neighboring plies around the interfaces has been studied extensively by Zhang et al. [48], Okada and Nishijima [49], Vishwanath et al. [50], and Costa et al. [51]. Ray has studied the ILSS of epoxy composites having glass fiber loading of 55, 60 and 65 % by weight [52, 53]. A rapid decrease in ILSS with increasing void content in FRP composites is reported by Costa et al. [51] and Mouritz [54]. Ray [52, 53] has studied the impact of strain on ILSS.

Impact Strength The impact strength of a composite is a very important design parameter in most of the applications, e.g., aeronautics (bird impact on the engine), and personal vehicle armors. An analytical solution of the transversal impact on a yarn has been developed by Smith in 1958. In a yarn point impact for an elastic material, the transverse and longitudinal waves propagate with respective velocity (U, Eq. 2.15) written [55] as

$$U = C\left[\sqrt{\varepsilon(1+\varepsilon)} - \varepsilon\right] \tag{2.15}$$

where C and ε are the strain–wave velocity and strain of the yarn, respectively. The strain–wave velocity (C) can be determined from Eq. 2.16 [55]:

$$C = \frac{2L(1+3\varepsilon)+d_2}{t_2(1+2\varepsilon)} \tag{2.16}$$

where L, d_2 and t_2 are the strain–wave front traveling distance after reflection at the clamp, the distance traveled by the transverse wave after reflection, and the transverse wave overtaking time after reflection.

Chocron-Benloulo et al. have shown the trend of residual velocity with striking velocity analytically, for 8 ply Kevlar 29 panel shot by a 64 grain (4.147 g) fragment simulating projectile, where the striking velocity was 300–1500 m/s [56]. This analytical model is shown in Fig. 2.6.

An experiment has been conducted on a mechanically needle-punched nonwoven fabric, which is made of natural fiber [57]. Polypropylene is used as a polymer binder fiber, and the natural fibers used are flax, hemp, and kenaf. The blending ratios of natural/polymeric fiber are varied from 50/50 to 70/30. The fiber or polypropylene mats are used as a reference material. The prepregs weight ranges from 1600 to 2400 g/m^2. The objective of this experiment is to measure the dynamic stress using a reflection polariscope with single-flash technique. The impact test apparatus consists of a loading apparatus and a photoelastic element setup as illustrated in Fig. 2.7. In this process, pressurized air is used to accelerate the steel projectile and an anvil is used to transmit the impulse. It uses a piezoelectric force sensor and avoids damaging the composite at the contact point. The boundary conditions are different for

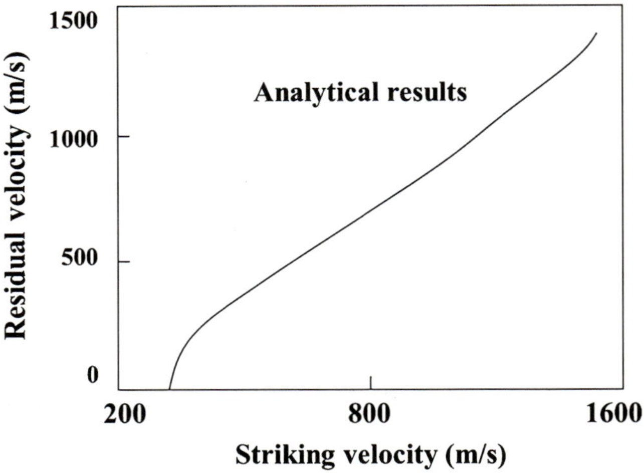

Fig. 2.6 Analytical model of residual velocity versus striking velocity for 8 ply Kevlar 29 panel [56]

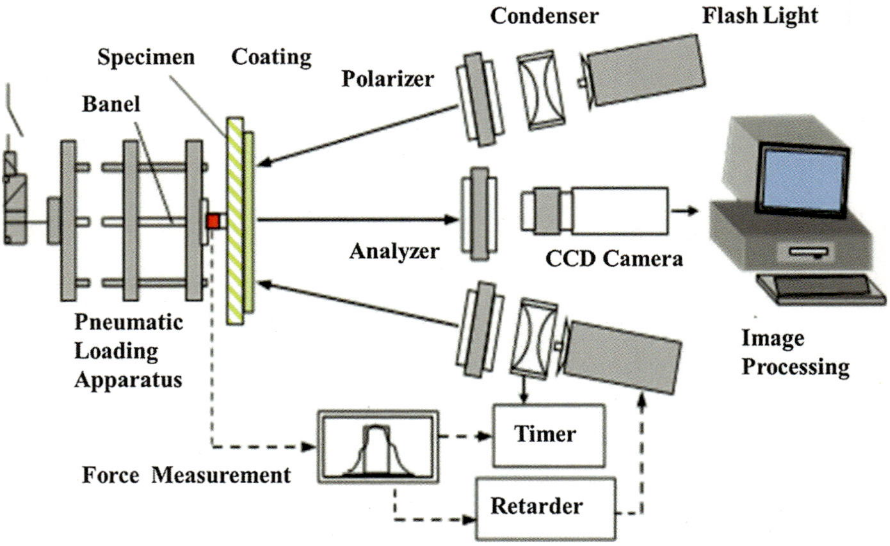

Fig. 2.7 Schematic of impact testing device [57]

various specimens like disk, beams, or plates, which are in the loading apparatus. Now, the wave propagation can be very easily detected in its early stages (provided that the reflection should occur at the point of interest of the specimen). It is possible to investigate the effect of superposition and reflection of propagating waves at different profile and/or boundary conditions. It also helps in understanding the fracture nature of FRP composites after the impact loading. The applications of the test are (i) locating the artificial cracks by analyzing impact strains in the specimen, (ii) determining the viscoelastic properties of composites and photoelastic coating material due to high strain rates, (iii) analyzing the strain–wave propagation in FRP composite disks, (iv) having the ability to trace stress concentration in the impact-loaded orthotropic composites, etc.

Another procedure to determine the impact toughness is notched Izod test. The hammer has been dropped from a height of 0.6 m with a velocity of 3.4 m/s. This test is used to determine the impact behavior of composites with end-shaped fibers. The impact toughness depends on the various parameters like behavior of the matrix, fiber end, and fiber length.

Notch Strength Geometry effects, locally concentrated damages, and notches influence the local stress concentration. The presence of notch is the main reason for the failure of composite materials. This is due to the square planar array of broken fibers. The extents of fiber pullout and the stress concentration around the notch lead to failure. The extent of fiber pullout mainly depends on the Weibull modulus. Normalized notch strength (σ, Eq. 2.17) can be written as follows [58]:

$$\sigma = \frac{\mu*}{c_i} + \sigma_p \tag{2.17}$$

where $\mu*$, c_i, and σ_p are the strength of the critical link at the notch, the notch size, and the normalized pullout stress, respectively. The expression for normalized pullout stress (σ_p) is given by Eq. 2.18 [58].

$$\sigma_p = \left(\frac{1}{m+1}\right)^{\frac{m}{m+1}} \Gamma\left(\frac{m+2}{m+1}\right) \tag{2.18}$$

where m is the Weibull modulus.

The notch strength of composite (σ_f^*) is given by Eq. 2.19 [58].

$$\sigma_f^* = \left[\frac{(c_{i_n} - 1)(m+1)}{n_{i_n}\left(c_{i_n}^{m+1} - 1\right)}\right]^{\frac{1}{m+1}} \tag{2.19}$$

where i_n and n_i are the initial notch size and the number of nearest neighbors, respectively.

The normalized notch strength predicted by an analytical local load-sharing model and by the Batdorf-type model is compared for better understanding. The normalized fiber bundle notch strength (σ_f^*/σ_c) versus initial notch size (i_n) for fiber Weibull moduli (m) of 5 and 10 predicted by the above models are shown in Fig. 2.8. The notch strength predicted by both the models are nearly parallel for m = 10, but there is a huge difference in the notch strength obtained by two methods for m = 5. According to the Batdorf model, the fibers with m = 5 are weaker than m = 10, which is in contradiction to the local load-sharing analytical model.

Fracture Toughness Fracture toughness is the energy absorbed by the composite during loading or impact in order to propagate a crack up to the failure. Actual

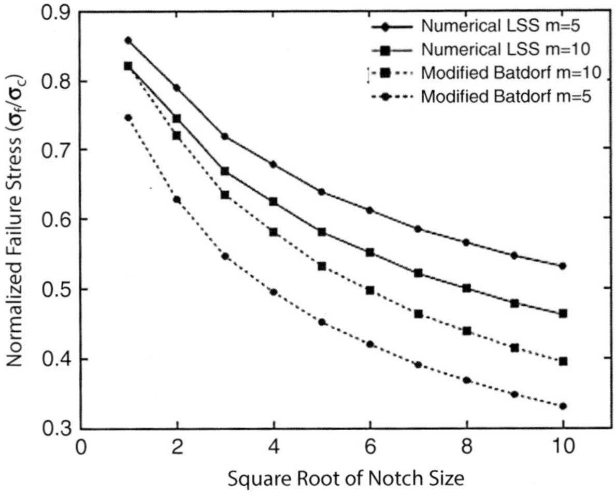

Fig. 2.8 Graph showing notch strength of Ti MMCs [58]

fracture toughness is predicted by a single pullout fiber testing. Fiber properties such as fiber end shape, fiber length, friction at interface, and matrix properties have an influence on the fracture and impact toughness. ASTM D-5528:01 defines the standard test method to determine the mode-I interlaminar fracture toughness of unidirectional FRP matrix composite. However, for the mixed mode-I and mode-II, it is based on the standard number ASTM D-6671/D 6671 M-04e1. Estimation of the fracture load of notched plates is a tedious job but it has been established more or less. The influence of parameters like the notch geometry, notch root radius, notch angle, notch depth, and relative width and thickness of the specimen have been extensively studied. Dash et al. have studied how the critical stress intensity factor K_C is affected by the crack size and environmental effects on fracture toughness [59]. These new experimental procedures are more reliable and have been used for studying the fracture toughness. The total fracture toughness (R_t) is given by Eqs. 2.20, 2.21, and 2.22 [59, 60]:

$$R_t = R_r + R_{po} + R_s \tag{2.20}$$

$$\approx \frac{V_f \sigma_f^*}{\tau_f}\left[\frac{\sigma_f^*.d}{6}\left(\frac{1}{4}+\frac{\sigma_f^*}{E_f}\right)+\frac{R_m}{2}\right]+(1-V_f)R_m \tag{2.21}$$

$$R_s = V_f R_f + (1-V_f)R_m + V_f\frac{l_d}{d}R_i \approx V_f\left(\frac{l_d}{d}-1\right)R_m \tag{2.22}$$

where R_r, R_{po}, R_s, R_f, R_m, R_i, d, τ_f, l_d, V_f, and E_f are the toughness due to stress redistribution, toughness due to the fiber pullout, toughness due to a creation of new surface in the fiber, matrix and interface, toughness at the fiber, toughness at the matrix, toughness at the interface, fiber diameter, frictional shear stress, de-bond length, volume fraction of the fiber, and Young's modulus of the fiber, respectively.

For most of the fiber composites with thermoset matrices, particularly with CFs, fiber pullout is the primary source of fracture toughness and is represented by Eq. 2.23 [60].

$$R_{po} = \frac{2V_f \tau_f l_{po}^2}{d} \tag{2.23}$$

where l_{po} = pullout distance. When the critical transfer length (l_c) is more than the fiber length (l), all the fibers are pulled out assuming l_{po} varies from 0 to 0.5 within the mean value of 0.25; the R_{po} is given by Eq. 2.24 [60].

$$R_{po} = \frac{V_f \tau_f l^2}{6d} l \leq l_c \tag{2.24}$$

R_{po} is maximum for $l = l_c$.

For the composite having fiber length (l) greater than the critical fiber length (l_c), the fraction of fiber can be pulled out up to $\frac{l_c}{l}$ and the l_{po} varies from 0 to 0.5.

Therefore, R_{po} is written by Eq. 2.25 [60].

$$R_{po} = \frac{V_f \tau_f l^3}{6d\, l} l > l_c \tag{2.25}$$

Recently, Dash et al. have used an epoxy resin (Araldite–LY5052 and a hardener–HY5052) reinforced by a plain-woven carbon fabric (INDCARF–30 of 3 K-filament counts: a product of IPCL) [59]. A specimen fabricated of CF-reinforced composite with ASTM standard D-3039 dimensions is shown in Fig. 2.9. A thick diamond-edged wheel with thickness 0.03 mm has been used to cut a sharp notch of varying sizes in different specimens.

For a fracture toughness test, a specimen with single-edge sharp notch is used. This specimen is then subjected to a tensile loading on a servo-hydraulic tensile testing machine. It has been clamped initially by using pin-loaded grips. The specimen experiences a monotonic uniaxial tension at a certain displacement rate. Ample precaution has to be taken to maintain a uniform temperature around the test jig. Rate of crack displacement is closely monitored, and the critical load is calculated at the complete fracture load. The fracture equation is given by Eq. 2.26 [61].

$$K_c = \alpha \sigma_{nc} \sqrt{\pi a} \tag{2.26}$$

where K_C, α, σ_{nc}, and a are the notch toughness, the boundary modification factor, the critical nominal stress, and the crack length, respectively.

The delamination results in complete fracture and also reduction in stiffness, which is a very important parameter for designers. The current challenge is to increase the lifetime and load-bearing capacity of the composites by controlling the delamination behavior. The crack plane due to delamination in composites is well defined, and the elasticity of the material is retained in the vicinity of the crack tip except in the very thin layer of interface. Hence, examining the energy release rate assists in analyzing the crack formed. Wilkins et al. have found the critical strain

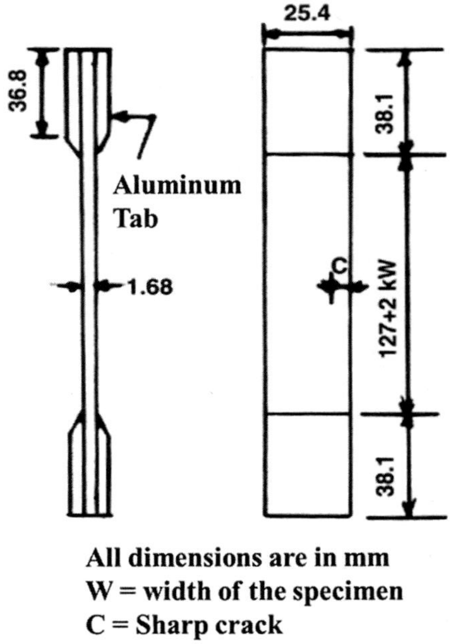

Fig. 2.9 A specimen with ASTM standard D-3039 [59]

energy release rate for mode-I and mode-II delamination [62]. Han and Koutsky have found the interlaminar fracture energy of polyester composites reinforced with glass using width-tapered double-cantilever beam (WTDCB) specimens [63]. A nonlinear theory for the energy release rate using double-cantilever beam (DCB) has been developed by Devitt et al. [64]. Mall et al. have used the DCB specimen to investigate the effect of loading rate on ILFT [65]. A substantial study of ILFT has been reported at room temperature on composite materials comprising glass fiber and polyester resin [66–73]. The ILFT of composite is influenced by the sizing agent [66, 67], materials used as matrix [68, 69], loading of particles [70], geometry of the fabric [71], fiber volume percent [74], fiber orientation of the fiber [72, 75], processing pressure [76], aging condition [68], humidity [69], and temperature [77]. It is clear from the literatures that for structural application of composite materials, ILFT is an important material property.

Modulus or Stiffness

The reinforcements such as fibers in a matrix material may have random or preferred orientations. If the fibers are aligned in the same direction, then they suffer with high stress. Hence, random orientation of material (known as anisotropic, which means that it is having different properties in different directions) is preferred. The most important factor of controlling the anisotropy is the optimization of properties for

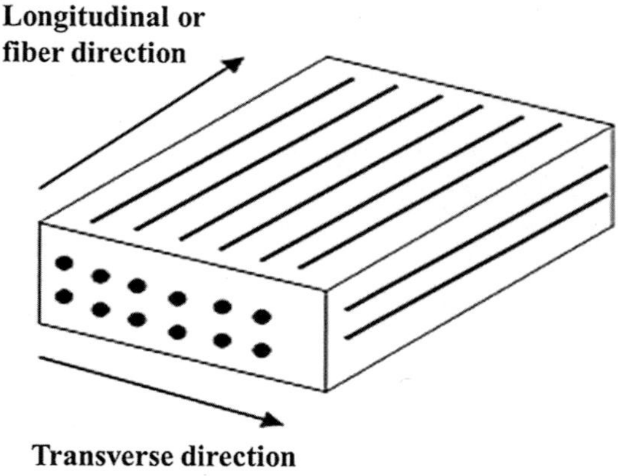

Fig. 2.10 Longitudinal and transverse direction of fiber-reinforced composite

specific applications. At a microscopic level, properties of composites depend on the individual properties of the fiber and matrix as well as the orientation of the fiber in the matrix. The unidirectional fibers are the simplest arrangement of fibers in a fiber-reinforced composite and are shown in Fig. 2.10. In these composites, minimum properties are given along the transverse direction, whereas the maximum properties are given along the fiber-oriented direction.

Unidirectional Fiber–Matrix Composites Unidirectional composites possess different tensile moduli in various directions and are presented in Fig. 2.11a-b.

Longitudinal Tensile Modulus The following assumptions have been used for determining the longitudinal tensile modulus:

(i) Longitudinal load produces equal strain in fiber and matrix.
(ii) Fibers are uniform, parallel, and continuous.
(iii) Perfect bonding between fiber and matrix. The net load applied (F_l) on the oriented direction of the fibers is shared by fibers (F_f) and matrix (F_m), which can be written by Eq. 2.27.

$$F_l = F_f + F_m \tag{2.27}$$

The stresses depend on the cross-sectional area of fiber (A_f) and matrix (A_m). Then, Eq. 2.27 is written as Eq. 2.28.

$$\sigma_l A = \sigma_f A_f + \sigma_m A_m \tag{2.28}$$

where A, σ_f, and σ_m are the total cross-sectional area of the composite, the fiber stress, and the matrix stress, respectively. The total cross-sectional area of the composite is given by

$$A = A_f + A_m \tag{2.29}$$

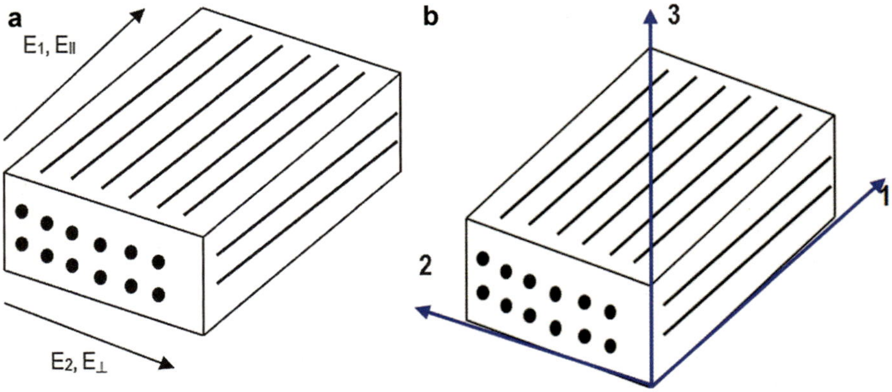

Fig. 2.11 Different tensile moduli labeled (**a**) in various ways and (**b**) in different directions

From Hooke's law, Eq. 2.28 is written as Eq. 2.30

$$E_1\varepsilon_1 A = E_f\varepsilon_f A_f + E_m\varepsilon_m A_m \tag{2.30}$$

where E_1, E_f, E_m, ε_1, ε_m, and ε_f are the axial Young's modulus of the composite, the axial modulus of the fiber, the axial modulus of the matrix, the strain on the composite, the strain on the fiber, and the strain on the matrix, respectively.

However, the strain is same on the fibers (ε_f), the matrix (ε_m), and the composite (ε_1) and can be written as

$$\varepsilon_1 = \varepsilon_f = \varepsilon_m \tag{2.31}$$

Then the Eq. 2.30 is reduced to Eq. 2.32.

$$E_1 A = E_f A_f + E_m A_m \tag{2.32}$$

Dividing both sides by the area "A," then Eq. 2.32 changes to Eq. 2.33.

$$E_1 = E_f \frac{A_f}{A} + E_m \frac{A_m}{A} \tag{2.33}$$

For a unidirectional composite,

$$\frac{A_f}{A} = V_f \tag{2.34}$$

$$\frac{A_m}{A} = V_m \tag{2.35}$$

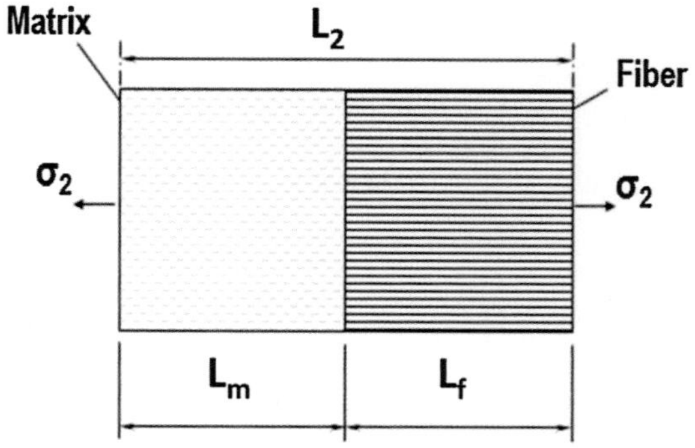

Fig. 2.12 Model showing transverse stiffness of a unidirectional composite

where V_f and V_m are the volume fraction of the fiber and volume fraction of the matrix, respectively. Then the tensile modulus of the composite in the longitudinal direction can be represented by Eq. 2.36 [78].

$$E_l = E_f V_f + E_m V_m \tag{2.36}$$

For density, this is similar to the rule of mixtures. For polymer composites, $E_f >> E_m$. Therefore, $E_l \approx E_f V_f$.

Transverse Tensile Modulus For predicting the transverse tensile property, plenty of mechanical models have been used. One of such a simple model is Jones model [79]. In this model, a load, which is applied in a direction normal to the fibers, contributes for the transverse stiffness. The fibers are stacked together and the model is being simplified. It is assumed that the stresses are equal in each of the components [i.e., composite stress in the transverse direction (σ_2) = fiber stress (σ_f) = matrix stress (σ_m)], and also the Poisson contraction effects are ignored. A model showing transverse stiffness of a unidirectional composite is depicted in Fig. 2.12.

The total extension (δ_2) = $\delta_f + \delta_m$. If the total elongation in the load direction is L_2, and the elongation on the fibers and matrix are L_f and L_m, respectively, then length and strains are written by Eqs. 2.37 and 2.38, respectively.

$$\varepsilon_2 L_2 = \varepsilon_f L_f + \varepsilon_m L_m \tag{2.37}$$

where ε_2, ε_f, and ε_m are the composite strain, fiber strain, and matrix strain, respectively.

Dividing Eq. 2.37 both sides by L_2, then

$$\varepsilon_2 = \varepsilon_f \frac{L_f}{L_2} + \varepsilon_m \frac{L_m}{L_2} \tag{2.38}$$

However, $L_f/L_2 = V_f$ and $L_m/L_2 = V_m = (1-V_f)$. Therefore, strain becomes (Eq. 2.39),

$$\varepsilon_2 = \varepsilon_f V_f + \varepsilon_m (1 - V_f) \tag{2.39}$$

where V_f and V_m are the volume fraction of the fiber and volume fraction of the matrix, respectively.

By using Hooke's law, Eq. 2.39 can be written as Eq. 2.40:

$$\frac{\sigma_2}{E_2} = \frac{\sigma_f V_f}{E_f} + \frac{\sigma_m (1 - V_f)}{E_m} \tag{2.40}$$

However, $\sigma_2 = \sigma_f = \sigma_m$; hence, Eq. 2.40 becomes Eq. 2.41 [79].

$$E_2 = \frac{E_m E_f}{E_m V_f + E_f (1 - V_f)} \tag{2.41}$$

where E_2, E_m, and E_f are the modulus of composite in the transverse direction, modulus of the matrix, and modulus of the fiber, respectively. It is observed that E_2 is not particularly sensitive to V_f. If $E_f >> E_m$, E_2 becomes independent of the property of the fiber. The matrix dominates the transverse modulus and is virtually independent of the reinforcement. Jones model is not relevant to most of the practical situations. Hence, improved mechanical models have been developed by using the theoretical studies such as Halpin–Tsai model [80]. The semiempirical Halpin–Tsai equations [81, 82] correct the Eq. 2.41 as

$$E_2 = \frac{E'_m E_f}{E'_m V_f + E_f (1 - V_f)} \tag{2.42}$$

where $E'_m = [E_m/(1 - \nu^2_m)]$ and ν_m is the Poisson's ratio of the matrix.

Hull et al. in 1996 have predicted the rule of mixtures for longitudinal (E_1) and transverse (E_2) modulus of glass/polyester composite ($E_f = 73.7$ MPa, $E_m = 4$GPa) as shown in Fig. 2.13 [83].

Stiffness of Short Fiber Composites For short fiber (aligned) composites, which are usually not found in polymers, the rule of mixtures for modulus [84] in the fiber direction is given as:

$$E_c = \eta_L E_f V_f + E_m (1 - V_f) \tag{2.43}$$

where E_c, η_L, E_f, E_m, and V_f are the modulus of the short fiber composite in the fiber direction, the length correction factor, the modulus of the fiber, the modulus of the matrix, and the volume fraction of the fiber, respectively. The length correction factor (η_L) is obtained theoretically (for $L > 1$ mm, $\eta_L > 0.9$).

The rule of mixture expression is brought into picture for composites in which fibers are not perfectly aligned by incorporating both η_L and η_o, where "η_o" is the correction for the non-unidirectional reinforcement. It is 1.0 for unidirectional,

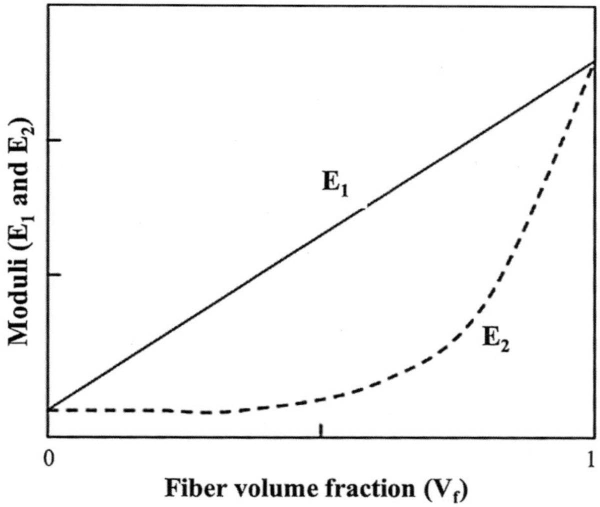

Fig. 2.13 Schematic showing rule of mixture prediction for longitudinal (E_1) and transverse (E_2) modulus [83]

0.5 for biaxial, 0.25 for biaxial at 45°, 0.375 for random in plane, and 0.2 for random in three dimensions.

In short fiber-reinforced thermoset polymer composites, the fibers are much longer than their critical length, and orientation effects determine the elastic properties. For the randomly oriented fiber, the following equations (Eqs. 2.44, 2.45, and 2.46) give sufficiently accurate estimates [85].

$$E_r = \frac{3}{8}E_{11} + \frac{5}{8}E_{22} \tag{2.44}$$

$$G_r = \frac{1}{8}E_{11} + \frac{1}{4}E_{22} \tag{2.45}$$

$$\nu_r = \frac{E_r}{2G_r} - 1 \tag{2.46}$$

where E_r, E_{11}, E_{22}, G_r, and ν_r are the modulus of randomly oriented fiber composite, longitudinal modulus of the composite, transverse modulus of the composite, shear modulus of the randomly oriented fiber composite, and Poisson's ratio of the randomly oriented fiber composite, respectively.

Fatigue

Fiber-reinforced composites have a good rating regarding fatigue life, which can be evaluated by S-N curve. Tension–tension fatigue of advanced composites can be tested with ASTM standard D-3479 specimen. Most of the composite materials are inhomogeneous and anisotropic, and the fatigue behavior of inhomogeneous composites is much more complicated. A model proposed by Epparachchi et al. [86] is applicable to the fatigue data for glass/epoxy and glass/polypropylene specimens. The empirical fatigue law used is given below.

$$\frac{d\sigma}{dt} = -a\sigma_{max}(1 - R)^{\gamma}t^{-k} \quad (2.47)$$

where σ, t, σ_{max}, R, and γ are the applied stress, time, maximum stress applied, stress ratio, and the set fixed to 1.6 (derived from assumption on fatigue crack propagation rate), respectively, and a and k are constants.

The above model can be arranged as follows [86]:

$$\left(\frac{\sigma_{ult}}{\sigma_{max}} - 1\right)\frac{1}{(1 - R)^{\gamma}}f^{\beta} = \alpha\left(N^{\beta} - 1\right) \quad (2.48)$$

where σ_{ult}, σ_{max}, f, and N are the ultimate stress, the maximum stress applied, the loading frequency, and the number of loading and unloading cycles till failure, respectively, and α and β are constants (can be determined experimentally).

A model has also been developed for the fatigue testing of E-glass fiber/epoxy composite, and the model equations [87] are given as follows:

$$\frac{d\sigma}{dt} = -C_2 t^{-m_1} \quad (2.49)$$

$$C_2 = A.f(R, \sigma_u, \sigma_{max}) \quad (2.50)$$

where σ, t, m_1, A, R, σ_u, and σ_{max} are the residual strength after n cycles, time, a material constant, a constant, stress ratio, the initial residual strength, and the maximum applied stress, respectively.

Creep

Creep can be defined as high-temperature progressive deformation at constant stress. It is a performance-based behavior since it is not an intrinsic material response. It is highly dependent on environmental conditions such as temperature. A creep test involves a tensile specimen under a constant load maintained at a constant temperature, and strain measurements are recorded over a period of time. A standard method of testing the creep and creep rupture of a continuous fiber-reinforced ceramic composites subjected to tensile loading at elevated temperature is according to ASTM C-1337:96. The creep resistance of MgO/SiC nanocomposite is found to be 2–3 times greater than that of pure MgO matrix at a temperature of 1300 °C and for the Al_2O_3/SiC nanocomposite, it is about 3–4 times greater than that of a pure Al_2O_3 matrix at a temperature of 1400 °C. In these two cases, a tremendous improvement in creep resistance is observed. This is attributed to the suppression of grain boundary slidings by the reinforced intergranular nano-sized SiC particles.

Stress Relaxation

Glass fiber-reinforced polymer (GFRP) composites used for structural applications require good, long-term mechanical properties. This behavior can be predicted by accelerated mechanical testing such as creep and stress relaxation. The dimensional stability of load-bearing structures and the retention of clamping force for bolts fastened to composites are determined by stress relaxation behavior. The stress relaxation studies of pineapple leaf fiber-reinforced polyethylene composites and banana fiber-reinforced polyester composites have been studied [88, 89]. Barpanda and Mantena have studied the effects of hybridization on the stress relaxation and creep characteristics of pultruded glass–epoxy–graphite hybrid composites [90]. The dynamic mechanical analysis is used for the stimulated stress relaxation and creep testing. Geethamma et al. have analyzed the tensile stress relaxation of natural rubber composites reinforced with short-coir fiber [91]. The strain rate, fiber orientation, and loading affect the relaxation mechanisms of both filled and unfilled composite. Kawai et al. have studied stress relaxation of unidirectional T800H/3631 epoxy/carbon composite at comparatively high tensile strain levels and at elevated temperature [92]. This test has been carried out at 100 °C on plain coupon samples having different fiber orientations, $\theta = 0$, 10, 30, 45, and 90°. After maintaining the constant strain, relaxation of the axial stress with time takes place in a short period. In composite materials, the stress relaxation rate inclines to become zero, regardless of the orientation of the fiber. The relaxation modulus associated with stress relaxation depends on the strain level. The creep and stress relaxation studies to determine the time-dependent longitudinal strength of epoxy and polyester composites reinforced with unidirectional glass fiber have been reported by Lifshitz and Rotem [93]. The risk of delayed fracture can be avoided for stress values lower than 0.4 and 0.65 times the ultimate failure stress for the glass–polyester and glass–epoxy composites, respectively. Gutman and Soncino have analyzed the stress relaxation in polyester–fiber–glass composite under different temperature and chemical environments [94]. Wortmann and Schulz have studied the stress relaxation properties of polypropylene fibers at different temperatures (−50 to 30 °C) [95]. The moduli and relaxation time show noticeable temperature dependencies. Ikeda has assessed the chemical resistance of fiber-reinforced plastics using stress relaxation analysis [96].

Performance Under Adverse Conditions

Polymer matrix composites are sensitive to the changes in environmental conditions such as extreme temperatures, moisture, etc. The effects of temperature and moisture on the parameters such as tensile strength, ILFT, ILSS, elastic modulus, fatigue, etc. have to be examined for better understanding. Hashemi et al. have determined the influence of temperature and geometry on the mode-I, mode-II, and mixed mode-I/mode-II ILFT of the CF/PEEK composites [97]. The study is done within the temperature range 20–130 °C. An increase in ILFT with increase in temperature is observed. However, the ILFT at sub-zero temperatures has not been determined. Sloan and Seymour have investigated the impact on mode-I interlaminar fracture and crack growth in graphite/epoxy as a result of exposing to seawater [98]. Takeda et al. have studied the ILFT of epoxy matrix CFRP and GFRP composites irradiated by high-energy electrons [99]. In case of GFRP, a significant decrease in the fracture toughness is noticed. The effect of moisture on the ILFT of particle-filled glass fiber-reinforced polyester resin composites has been investigated by Srivastava and Hogg [70]. The samples are exposed to water at 20 and 40 °C. Mode-I and mode-II ILFT are measured at an interval of 2 months. It is reported that mode-I toughness is increased for all composites with an increasing moisture uptake. Watt et al. have studied the changes to the mode-I ILFT of stitched fiberglass/vinyl ester composites after exposure to elevated temperatures between 100 and 300 °C [100]. The ILFT of the stitched composites degrades significantly by heating. Pavlidou and Papaspyrides have analyzed the effect of hygrothermal history on the ILSS of glass/polyester composites [101]. The experimentation is carried out at a temperature of 35 °C. Davies et al. have studied the effect of seawater aging on the shear behavior of glass-reinforced composites at 20 and 50 °C [102]. Apicella et al. have studied polyesters and their glass mat-reinforced composites at 20 °C and 90 °C for 50 days [103]. Choqueuse et al. have done a detailed research in a 2-year period on distilled water aging at temperatures 5, 20, 40, and 60 °C [104].

Thermal

Thermal characterization is very important to achieve precise measurement before the application. The thermal behavior of composite materials is evaluated based on the coefficient of thermal expansions (CTEs). Since CTEs vary from material to material, the thermal properties of the composite may be altered. The CTEs are given by Eqs. 2.51 and 2.52 [1].

$$\alpha_l = \frac{V_f \alpha_{fl} E_{fl} + V_m \alpha_m E_m}{V_f E_{fl} + V_m E_m} \tag{2.51}$$

$$\alpha_t = V_f \alpha_{ft} \left(1 + \nu_{flt} \frac{\alpha_{fl}}{\alpha_{ft}}\right) + V_m \alpha_m (1 + \nu_m) - (V_f \nu_{flt} + V_m \nu_m) \alpha_l \tag{2.52}$$

where α_l, α_t, α_{fl}, α_{ft}, E_{fl}, α_m, E_m, ν_{flt}, and ν_m are the coefficient of thermal expansion in the longitudinal direction, coefficient of thermal expansion in the transverse direction, coefficient of thermal expansion of fibers in the longitudinal direction, coefficient of thermal expansion of the fibers in the transverse direction, the modulus of fibers in the longitudinal direction, coefficient of thermal expansion of the matrix, the modulus of matrix, the in-plane Poisson's ratio of fibers, and Poisson's ratio of the matrix, respectively, and the other variables are as discussed above.

Similarly, the longitudinal (Eq. 2.53) and transverse (Eq. 2.54) thermal conductivities of the composite can be written [1] as

$$K_l = V_f K_{fl} + V_m K_m \tag{2.53}$$

$$K_t = \frac{K_{fl} K_m}{V_f K_m + V_m K_{ft}} \tag{2.54}$$

where K_l, K_{fl}, K_m, K_t, and K_{ft} are the thermal conductivity of the composite in the longitudinal direction, the thermal conductivity of fibers in the longitudinal direction, the thermal conductivity of the matrix, the thermal conductivity of the composite in the transverse direction, and the thermal conductivity of fibers in the transverse direction, respectively, and the other variables are as discussed above.

Ceramic matrix composites are the materials for high-temperature applications. Various composites have been prepared with SiC matrices, SiC fibers, and boron nitride (BN) fiber coatings [105]. High deposition temperature CVD is the most desirable technique for BN fiber coatings; however, it may lead to problems in the composite fabrication. In this composite, there appears to be three important mechanisms for BN oxidation corrosion: (i) borosilicate glass formation, (ii) gettering (is the process of removing device-degrading impurities from the active regions) of oxygen by SiC, and (iii) volatilization [105]. Rochais et al. have performed initial tests on known materials and a carbon–carbon composite with the help of different analysis methods [106]. These methods explain the modifications in the experimental setup as indicated in Fig. 2.14, a procedure used to determine the temperature diffusivity. The experimental setup has three main parts: (i) an optical setup for positioning and focusing the probe beams and the pump used, (ii) device used for measuring the intensity of the probe beam, and (iii) various electronic devices, which are used for detecting the signal and carrying out the experiment. Maximum power of the pump beam, which is a continuous Ar laser, is 2 W. Its frequency is modulated by a frequency generator-driven acousto-optic modulator operated at frequencies up to 2 MHz. Then, the pump beam is directed with the help of dichroic mirror, which is finally focused with the help of microscope onto the sample surface in the heating stage. The probe beam, which is a laser diode (wavelength 780 nm, power 20 mW), passes through a quarter-wave plate and the dichroic mirror, which is then focused with the same microscope on to the sample surface. After reflecting back, it retracts through the quarter-wave plate again, and from there it is sent to the photodiode by a beam splitter cube. A pump beam photon is protected from reaching the detector by using an optical filter. The amplitude and phase of the periodic signal collected by the photodiode is extracted by a lock-in amplifier. The dichroic mirror orientation is controlled by a PC controller, which governs the distance "r" between the probe and pump beam locations. Heating stage window, which is made of fused silica, transmits 93 % of the intensity of the two beams. Specifically designed objective (40 × magnification) with a large working distance, which is capable of correcting the spherical aberrations produced due to the heating stage window, has been used. Heating stage could attain a highest temperature of 1500 °C under a vacuum level better than 10^{-3}mbar. Sample polishing is necessary in order to ensure good

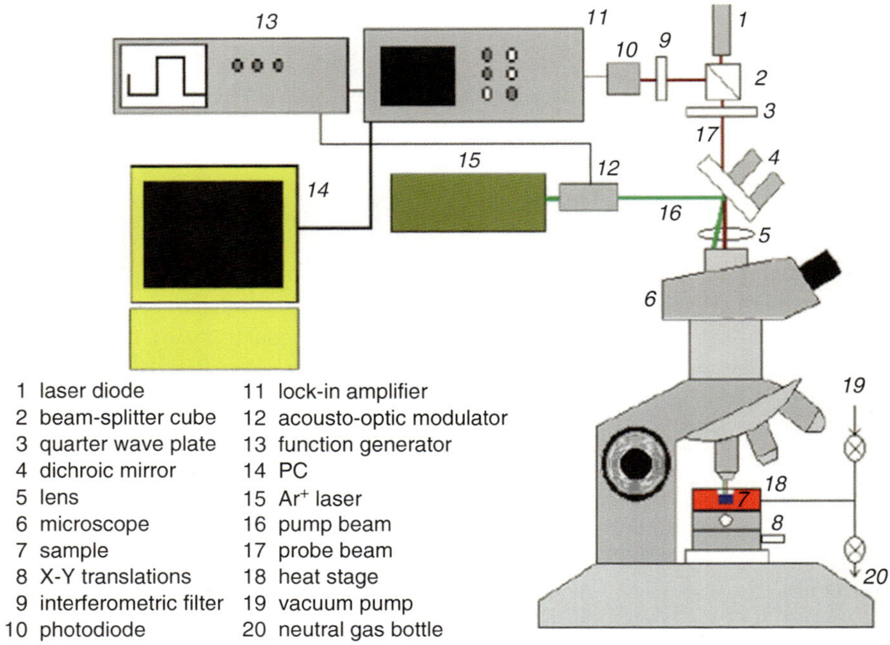

Fig. 2.14 Schematic of a thermal microscope setup [106]

reflection. The sample has a diameter less than 7 mm and thickness c.a. 1 mm. Homogeneous temperature gradient is maintained by keeping the thickness of sample as small as possible.

Glass Transition Temperature

Glass transition temperature (denoted by "Tg") is the temperature at which a material experiences a significant change in properties from hard and brittle to soft and pliable. Typically, Young's modulus changes sharply. At glass transition temperature, the polymeric structure turns "rubbery" upon heating and "glassy" upon cooling. The glass transition temperature of composites can be determined with the help of differential scanning calorimetry (DSC) technique. It is to be noted that in the case of low Tg materials, in order to achieve the benefit of orientation enhancement effect, the Tg should be in the vicinity of or lower than the measuring temperature.

Electrical

The uses of electrically conductive polymers (ECPs) have been limited in the past due to their drawbacks such as poor mechanical properties, poor environmental and thermal stabilities, etc. It has been shown that the composite structure using an insulating polymer as the host matrix improves the physical and chemical properties of ECPs. These composites can have practical advantages over the homogeneous materials. In the high-temperature region, the electrical conductivity (σ) of the films is found to follow the equation given below [107]:

$$\sigma = \sigma_0 \exp\left(-\frac{E_a}{kT}\right) \tag{2.55}$$

where E_a, T, and k are the activation energy, absolute temperature, and Boltzmann constant, respectively. It has been reported that the mechanical properties of the conductive composites can be improved with a decrement in the electrical conductivity [107]. Lee and Chen have proposed an idea for improving the mechanical strength of conventional ceramic superconductors, where they have found applications in magnetic levitation and shielding [108]. A series of composites have been prepared, which are flexible even at high temperatures (e.g., $(Bi,Pb)_2Sr_2Ca_2Cu_3O_x$ (2223)). A low-melting-point thermoplastic polymer, nylon-6,6, is used as structural matrix for the composite preparation. The superconducting $(Bi,Pb)_2Sr_2Ca_2Cu_3O_x$, (2223) powder with critical temperature $T_c \sim 104$ K has been prepared by conventional solid state reactions having a nominal composition of $Bi_{1.6}Pb_{0.4}Sr_2Ca_2Cu_3O_y$ [108].

Magnetic

Magnetic materials in the field of micromachining have been an area of interest for applications such as electromagnetic interference shielding, integrated inductive components, and magnetic microactuators and sensors. However, those materials possessing desirable hard magnetic properties might not be compatible with that of the standard microelectronic processes. For example, in preparation of magnets demanding high temperature, sintering, and pressing, such as in neodymium–iron–boron and samarium–cobalt magnets [109]. Magnetic polymer composites are made of strontium ferrite powder embedded in the polyimide organic matrix. These are considered to be compatible with the micromachining and microelectronic processes and they exhibited good magnetic properties, such as a high coercivity (c.a. 320 kA/m), square magnetization curve, a residual induction approaching 0.3 T, and a maximum energy product of 11,900 T (A/m) and are comparable to the bulk ferrite [109]. Nanostructured magnetic materials exhibit properties stemming from the intrinsic character of the particles and interactions between the particles. These magnetic properties often differ from both the individual constituent atoms and bulk crystalline counterparts. For soft magnetic properties, exchange coupling between the magnetic nanocrystalline grains is required. Therefore, freestanding nanoparticles must be consolidated to allow for this exchange coupling. A ferromagnetic nanocomposite having magnetic nanocrystalline particles in the polymeric matrix can replace conventional ferrites for various applications such as chokes, high-frequency inductors, filters, sensors, core-shaped planar transformers, etc. The increase in resistivity leads to a major reduction in the eddy-current losses in the magnetic/polymeric nanocomposite solid, while the coupling between adjacent magnetic nanoparticles produces remarkably better soft magnetic properties at larger frequencies compared with ferrites or conventional bulk materials.

Piezoelectric

Piezoelectric and electrostrictive materials are found applications in transducers and sensors. Their role in a sensing device is to detect displacement, acceleration, force, and sound. Hence, they are widely used in applications, such as strain and vibration measurement; medical, structural, and health monitoring; underwater imaging; etc. In the last two decades, the ceramics (e.g., lead titanate), polymers (e.g., polyvinylidene difluoride and its copolymers), and ceramic polymer composites (e.g., piezoelectric ceramic particles LaRC Si polyamide binder matrix composite) have been used as piezoelectric materials. Some of the electrostrictive materials have been found to produce a relatively high strain (e.g., lead magnesium niobate) with less hysteresis than that of piezoelectric materials. The piezoelectricity is commonly defined by the following equations [110]:

$$\varepsilon_i = S_{ij}^E \sigma_j + d_{mi} E_m \tag{2.56}$$

$$D_m = d_{mi} \sigma_i + \zeta_{ij}^{\sigma} E_k \tag{2.57}$$

where ε, S, σ, d, E, D, and ζ are the strain vector (in m/m), matrix of compliance coefficients (in m^2/N), stress vector (in N/m^2), matrix of piezoelectric strain constants (in m/V), vector of applied electric field (in V/m), vector of electric displacement (in C/m^2), and permittivity (in F/m), respectively. Here, i and j varies from 1 to 6, and m and k varies from 1 to 3. Also, the superscripts E and σ represent that the measurements are taken at constant electric field and constant stress, respectively.

In polycrystalline ceramics, the poled axis is usually denoted as the 3-axis, and with the ceramic having C_α symmetry, the number of nonzero matrix elements is 10 (like s_{11}^E, s_{12}^E, s_{13}^E, s_{33}^E, s_{44}^E, d_{31}, d_{33}, d_{15}, ε_{11}^T, and ε_{33}^T). Piezoelectric materials exhibit dispersion and the various material constants are function of frequencies. Most of the constants can be calculated by using resonance measurements. The resonance measurements are carried out by using small signals on the specimens. A test methodology has been developed to characterize the piezoelectric fiber composite for structural actuations as an alternative to monolithic piezoceramic wafers. By using interdigitated electrode pattern, the performance of the abovementioned actuator system has been improved, which helps in the orientation of the electric field's primary component in the plane of the structure, thus permitting the use of piezoelectric effect along the active fibers. This technology can be applied in an integral twist actuation of helicopter rotor blades for large harmonic control. The actuators

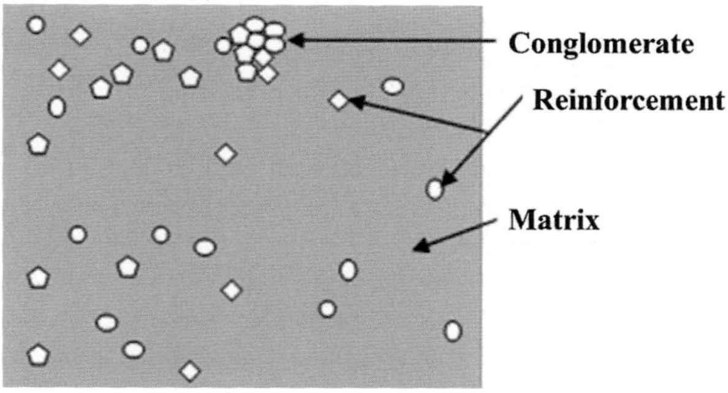

Fig. 2.15 Schematic of piezoelectric ceramic particles reinforced LaRC Si polyamide composite

required in this application must be able to counter the effect of harsh operating conditions for rotor blades of the helicopter. Thus, actuators that can withstand a huge number of electrical and mechanical cycles with appreciable centripetal and bending loads are desired. The performance of several actuators including S2 glass-reinforced and E-glass-laminated actuators have been studied. It has been observed that the abovementioned interdigitated electrode piezoelectric fiber composite actuator has a wide range of ability for the heavy loading requirements. The micromechanical piezoelectric composites consisting of LaRC Si polyamide binder and piezoelectric ceramic particles, to serve as sensors and actuators for space applications, have been developed. A schematic of this composite is shown in Fig. 2.15. In this schematic, piezoelectric ceramic particles (reinforcement) are indicated by lighter in color, and the darker region is an indicative of LaRC Si polyamide binder (matrix).

It has been shown that the ceramic particles are bound together in a conglomerate configuration. This typical feature plays a significant role to determine the thermomechanical response of the composite. It is known that a combination of negative coefficient of thermal expansion for ceramic phase and a positive coefficient of thermal expansion for LaRC Si phase results in a nonzero coefficient of thermal expansion for a piezoelectric composite. An application of thermal loading will result in internal stresses due to the thermomechanical and the piezoelectric response. The theoretical models have been developed for studying the piezoelectric behavior of lead titanate (up to 70 vol.%) and polyvinylidene difluoride composites. It has been observed that the data are matched with the experimental results. The piezoelectric polymer (PEP) composites have shown different piezoelectric relaxations. The dispersion has been prepared by adding 3–0 PEP composite piezoelectric spheres in a continuous polymer medium. When an electric field is applied to such a composite, each piezoelectric spheres experiences polarization and can be represented as a dipole. The dipole moment locally modifies the applied field and the surrounding medium. When only a small volume fraction of piezoelectric spheres are present, the influence of this dipole field on the neighboring field will be negligible.

Tribological

Tribological properties of composites are highly significant in long-term application point of views. The erosive wear behavior of Al_2O_3-fiber MMCs has been widely studied [111]. The erosion testing apparatus consists of a stainless steel T-tube in the slurry tank above an erodent reservoir, as shown in Fig. 2.16. The T-tube acts as a vacuum pump by circulation mechanism of water via. a loop arrangement, and the SiC particles (size ~ 600–800 μm) are driven up for impinging the slurry jet onto the specimen. The system gives a certain concentration (e.g., 5.6 wt%) of SiC particles, a jet velocity (e.g., 7.3 m/s) with a flow rate of water (e.g., 4.6 lit/min). The impact angle can be easily adjusted in this arrangement. The distance between the targets can be altered as per our requirements, and the ejector nozzle as well as their relative positions can be easily shuffled and can be monitored. The samples of dimension 25 mm × 25 mm × 5 mm are cut from an Al-MMC cast in two directions – one parallel to the fiber (Al-MMC-P) and the other vertical to the fiber (Al-MMC-V). An Al-alloy (K8510) matrix blank sample without fiber has been used. It has been observed that in the initial stages, the erosion rate (E) rises with the impact angle to

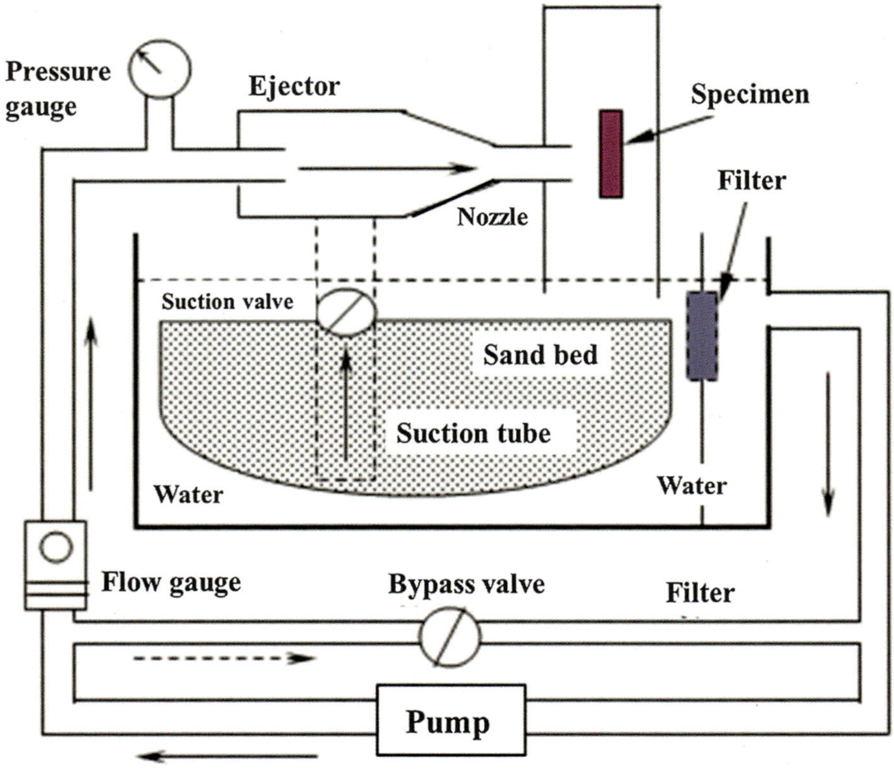

Fig. 2.16 Schematic of erosion test apparatus of sand/water slurry [111]

the maximum and decreased very lethargically then after, with further increase in the erosion angle (α).

Bearing Strength

For an aircraft, bearing strength greatly influences the composite component designs. When applied through a cylindrical fastener surface, the maximum stress that a composite can withstand is known as the bearing strength. The bearing stress is an applied comprehensive load on an interface divided by the bearing area. Bearing strength of composite materials can be tested with standard specimen ASTM.00008.10. Furthermore, bearing strength of advanced composite materials can be tested with standard specimen SACMA SRM 9–88. Due to high dependence of notch sensitivity on laminate configuration and a lower shear and bearing strength, having a satisfactory efficiency for a structural coupling with mechanical fastening is very difficult for the composite materials when compared with metals. These properties actually restrict the structural performance of the composite structure and require special reinforcement techniques. Fink and Kolesnikov have observed a high bearing strength for CFRP/Ti hybrid composites while performing bearing tests with specimens having different Ti content unidirectional prepreg plies [112]. In this study, carbon specimens with different laminate configurations have also been tested for comparison. A 6 mm diameter bolt is used and has given a clearance fit with the clamping torque of 0–0.6 mm. The tested specimens featured an edge distance to diameter ratio of 3 and a width to diameter ratio of 4 and 5. The characteristics of static tests conducted up to the failure in on-axis and 90°-off-axis loading directions are illustrated in Fig. 2.17.

Here, the maximum specific on-axis bearing strength of the hybrid material is noticed at a Ti content of 55 %, which is slightly lower in comparison to that of 50/40/10 carbon laminate, while 90°-off-axis bearing strength is about 40 % higher.

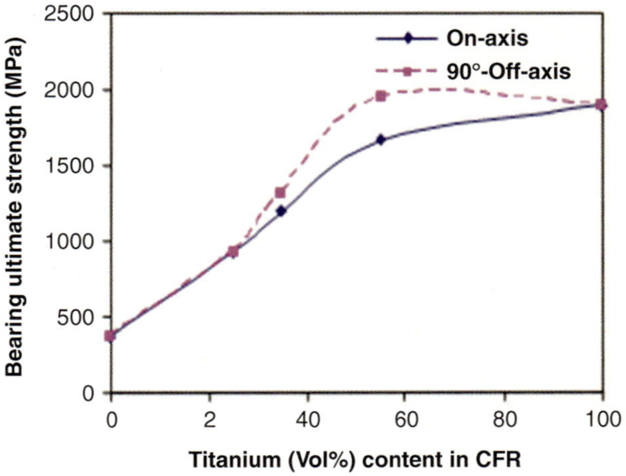

Fig. 2.17 Dependency of bearing strength of a hybrid material on Ti content [112]

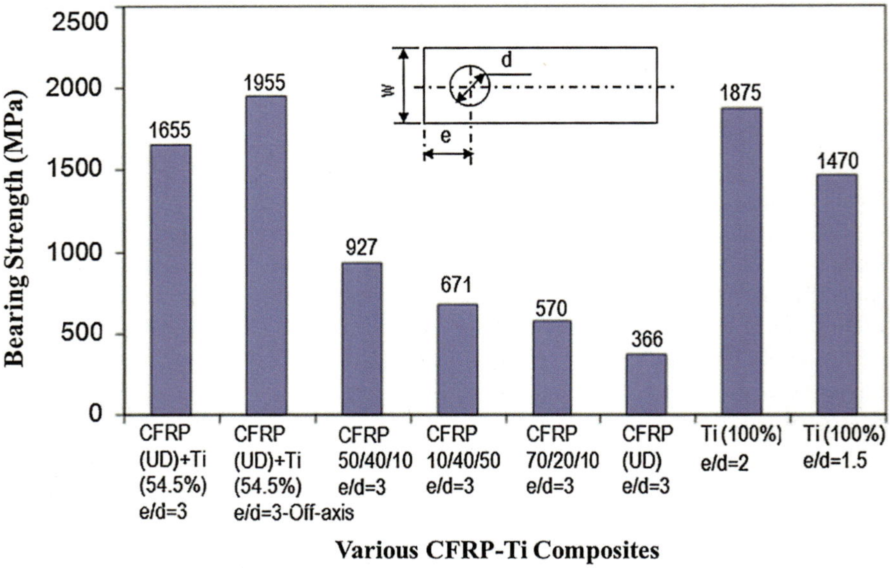

Fig. 2.18 Bearing strength of various CFRP-Ti composites [112]

However, the laminate-specific strength doesn't reflect the absolute weight efficiency of the joint, since the laminate represents only a fraction of the whole local joint design of the structure. The realization of high absolute joint strengths contributes to the reduced complexity and weight of secondary joint components such as bolts, fittings, and connecting parts to a high extent, thus reducing the overall joint weight. The test results have shown the advantage of Ti-reinforced composite material, which offers an increase in strength about 80 % under on-axis loading and 180 % under 90°-off-axis loading compared to the bearing properties of a 50/40/10 CF laminate and is shown in Fig. 2.18. Higher strength increase rates result when comparing with laminates having a higher degree of anisotropy. Hybrid material containing 54 % Ti offers a bearing strength similar to that of a Ti-alloy.

A bearing test apparatus for an aerospace grade unidirectional CF prepreg tape is shown in Fig. 2.19. The materials such as standard moduli CFs and 180 °C cure modified, toughened epoxy resin have been used for the bearing test experiment [113]. In this experiment, the fasteners with free running nut, 6.35 mm diameter steel hex head bolts, and plain steel washers having 12.7 mm diameter are used. The experimental matrix is comprised of six coupons with the following local constraint conditions: (i) a torque of 0.6, 1.3, 2.4, 3.5, or 7.0 Nm is given to the fastener and (ii) 0.5 or 0.1 mm gap is given on each side of the composite coupon. A mode of failure has been observed and is concluded that the level of local constraints (applied torque or gap size) affects the bearing strength. The gap values between 0.5 and 1.0 mm do not influence the initial failure load. Although, if failure is initiated, debris filling the gap will increase that leads the bearing strength values that do not actually shows the pin-bearing behavior. For an increase in torque values above 1.3 Nm,

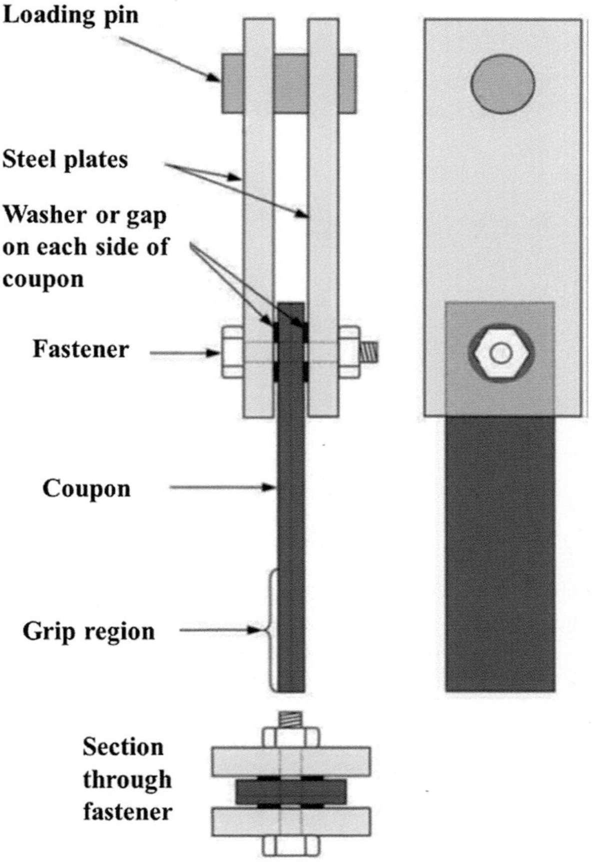

Fig. 2.19 Bearing test apparatus for aerospace grade unidirectional carbon fiber prepreg tape [113]

there is a slight increase in bearing strength, which is steady in nature, and with the increase in values of torque between 0 and 1.3 Nm, the bearing strength increases rapidly. A typical bearing stress–strain for each constraint condition is shown in Fig. 2.20.

Rheological

Rheology is the study of the flow of liquids but also solids under conditions in which they flow rather than deform elastically. It is applicable to substances, which have a complex structure, including muds, sludges, suspensions, polymers, many foods, bodily fluids, biological materials, and composites. The rheological performance of long fiber-reinforced thermoplastic composite has been developed, and its relation to structural aspects, which stem from different processing routes used, and correlation of observed facts with available theories have been focused by Cervenka and Alan [114]. The materials for the study are selected in such a way that the role of all

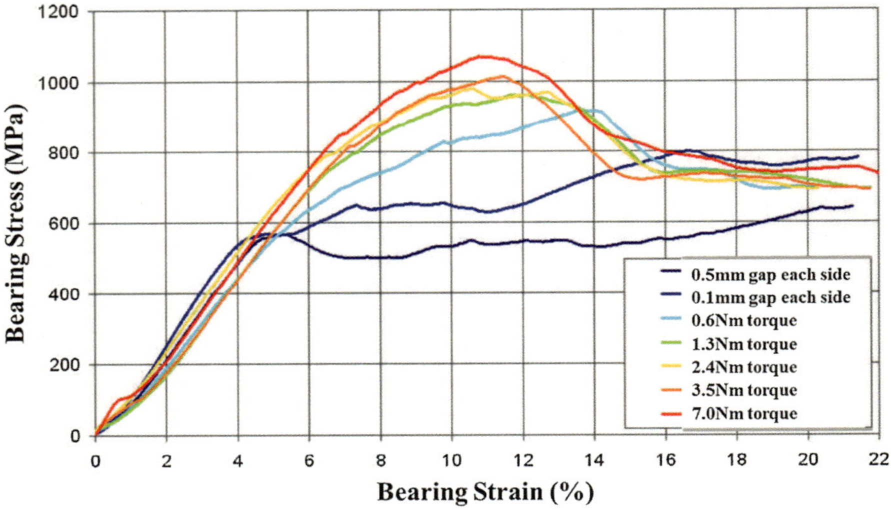

Fig. 2.20 A typical bearing stress–strain for each constraint condition [113]

important variables known to control the composite performance can be assessed. The important variables are as follows:

1. The type of matrix: polypropylene and polyamide
2. The nature of the fiber: glass, carbon, and Kevlar
3. Processing methods: classical injection molding and multiple life-feed techniques
4. Initial fiber length: 5 and 10 mm long pellets used as the feedstock for both the processing methods.

In this study, measurements of rheological parameters and quantification of relevant structural features together with generation of data for composites have been conducted simultaneously. The different structures of thermoplastic matrix/fiber constituents of different mechanical properties are related to the rheological behavior of heterogeneous melts and the interfacial behavior between the matrix and the reinforcement. Shi et al. have studied the rheological properties with good amount of dispersion of single-walled nanotubes (SWNTs) at low concentrations (0.05 wt%) in uncross-linked poly(propylene fumarate) [115]. It has been observed that the SWNTs aggregate at higher concentrations. This aggregation ratifying the mechanical and rheological properties has been revealed with the help of scanning electron microscopic study of fractured nanocomposite surfaces. The oscillatory shear viscoelastic (dynamic and linear) response for an uncross-linked nanocomposite and an uncross-linked polymer with varying concentrations of SWNTs, which are made compatible with the polymer using 12-aminododecanoic acid surfactant, is shown in Fig. 2.21 [115]. The complex viscosity magnitude ($|\eta^*|$) and the elastic modulus (G') both transforms abruptly from liquid-like behavior of the matrix polymer ($G' \propto \omega^2$, $|\eta^*| \propto \omega^0$) to solid-like behavior of the nanocomposites having 0.05 wt% and higher SWNT ($G' \propto \omega^0$, $|\eta^*| \propto \omega^{-1}$).

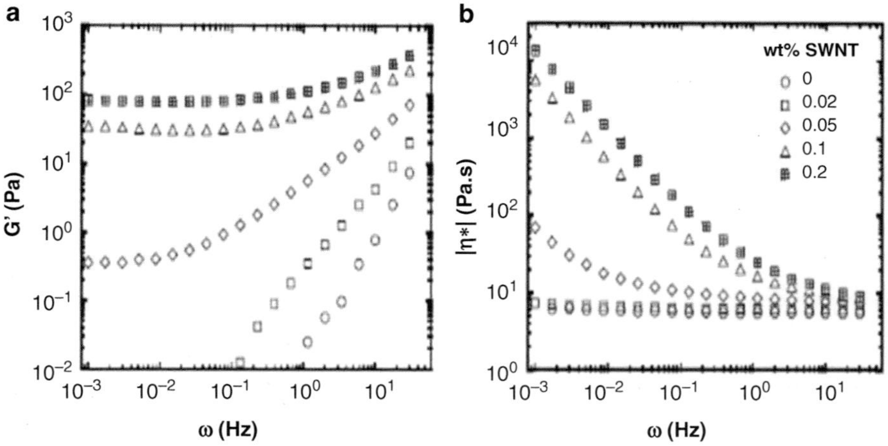

Fig. 2.21 Schematic showing dependency of (**a**) elastic modulus and (**b**) complex viscosity magnitude on frequency [115]

Biological

Biological characterizations of composite materials include toxicity and degradation testing procedures, which are most important in biological applications [116–119]. The interphases of such composites are particularly important in the case of bio-related applications; hence, their properties in terms of biocompatibility, biodegradability, and bioactivity have to be well understood before practical applications [120, 121]. For example, the dental patients have routinely demanded high aesthetic restorations since a long time ago. The major problems related to the use of resin composites for large restorations in posterior teeth are the side effects and environmental issues related with dental amalgams. The current developments in dental researches have evoked the interest on the use of ceramic composite materials for restoration application [122]. Research on novel biodegradable, biocompatible, and bioactive composite materials for various biological applications such as bone replacement [123–127], tissue replacements [128–132], etc., are on the way to success.

Concluding Remarks

In this chapter, the various experimental characterizations of composites are discussed briefly. Since these characterizations are not adequate for determining complete properties of the composite systems, advanced techniques have to be developed in order to obtain a broad picture. The need of highly performing advanced composites is increasing day by day due to the developments on their processability, designing flexibility, etc. The advanced composites are often presenting greatest challenges in their experimental characterizations. Among these, biological characterization is also very important since the toxicity of composites will directly influence the human life. Hence, characterizing composite materials for the better understanding of their real behaviors, in all possible directions, is very critical for their practical applicability and acceptance.

Acknowledgment The authors acknowledge the financial support provided by the Indian Space Research Organization, India, for carrying out this work.

References

1. Astrom BT (2002) Manufacturing of polymer composites. Nelson Thornes, Cheltenham
2. Ravi S, Iyengar NGR, Kishore NN, Shukla A (2000) Influence of fiber volume fraction on dynamic damage in woven glass fabric composites: an experimental study. Adv Compos Mater 9:319
3. Gdoutos EE, Pilakoutas K, Rodopoulos CA (2000) Failure analysis of industrial composite materials. McGraw-Hill, New York
4. Kelly A (1990) Concise encyclopedia of composite materials. MIT Press, Cambridge
5. ASTM D 3171–06 (2006) Standard test methods for constituent content of composite materials. Annual Book of ASTM Standards 15.03
6. ASTM D 2584–02 (2006) Standard test method for ignition loss of cured reinforced resins. Annual Book of ASTM Standards 8.01
7. Wu Y, Shivpuri R, Lee LJ (1998) Effect of macro and micro voids on elastic properties of polymer composites. J Reinf Plast Compos 17:1391
8. Summerscales J (1987) Non-destructive testing of fibre-reinforced plastics composites. Elsevier Science Publishers, London
9. Bowles KJ, Frimpong S (1992) Void effects on the interlaminar shear strength of unidirectional graphite fiber-reinforced composites. J Compos Mater 26:1487
10. Chambers AR, Earl JS, Squires CA, Suhot MA (2006) The effect of voids on the flexural fatigue performance of unidirectional carbon fibre composites developed for wind turbine applications. Int J Fatigue 28:1389
11. ASTM D 1505–03 (2006) Standard test method for density of plastics by the density-gradient technique. Annual Book of ASTM Standards 8.01
12. ASTM D 3800–99 (2006) Standard test method for density of high-modulus fibres. Annual book of ASTM standards 15.03
13. Zahavi E, Torbilo V (1996) Fatigue design: life expectancy of machine parts. CRC Press, Boca Raton
14. Graham TS (2002) Industrial metrology: surfaces and roughness. Springer, Berlin
15. Davim JP, Reis P (2005) Damage and dimensional precision on milling carbon fiber-reinforced plastics using design experiments. J Mater Process Technol 160:160
16. Savage G (1993) Carbon-carbon composites. Chapman and Hall, London
17. Kessler MR (2004) Advanced topics in characterization of composites. Trafford Publishing, Victoria
18. Theocaris PS, Stassinakis CA (1981) Crack propagation in fibrous composite materials studied by SEM. J Compos Mater 15:133
19. Kreider KG (ed) (1974) Metallic matrix composites. Academic, London
20. Hill RG (1968) Evaluation of elastic moduli of bilaminate filament-wound composites – crazed and uncrazed coupons from a rocket case and laboratory panel are subjected to static and dynamic loading. Exp Mech 8:75
21. Lantz RB (1969) Boron epoxy laminate test method. J Compos Mater 3:642

22. Richards GL, Airhart TP, Ashton JE (1969) Off-axis tensile coupon testing. J Compos Mater 3:586
23. Pagano NJ, Halpin JC (1968) Influence of end constraint in the testing of anisotropic bodies. J Compos Mater 2:18
24. ASTM D 3039 (2006) Standard test method for tensile properties of polymer matrix composite materials. Annual Book of ASTM Standards 15.03
25. Niihara K et al (1996) High temperature strength and creep behavior of ceramic based nanocomposites roles of intergranular nano-sized particulates. In: Hui D (ed) Proceedings of the 3rd international conference on composite engineering (ICCE/3), New Orleans, 1996
26. Daniel IM, Gdoutos EE, Rajapakse YDS (eds) (2009) Major accomplishments in composite materials and sandwich structures. Springer, London
27. Thomas S, Joseph K, Malhotra SK, Goda K, Sreekala MS (eds) (2012) Polymer composites, volume 1, macro- and microcomposites. Wiley-VCH, Weinheim
28. Hsu PW, Herakovich CT (1977) Edge effects in angle-ply composite laminates. J Compos Mater 11:422
29. Pipes RB, Pagano NJ (1970) Interlaminar stresses in composite laminates under uniform axial extension. J Compos Mater 4:538
30. Wang JTS, Dickson JN (1978) Interlaminar stress in symmetric composite laminates. J Compos Mater 12:390
31. Gu Y, Liang G (2006) Self-reinforced interlayer shearing intensity resin base fibre reinforced composite material preparation method. Chinese Patent 1,762,686
32. Sato H, Suzuki Y (2005) Polyacrylonitrile-based carbon fibre and method for producing the same. JP Patent 2,005,314,830
33. Barelko VV et al (2005) Fibre-glass reinforcing weave filler for polymeric glass composites. RU Patent 2,245,477
34. Goto K et al (2004) Epoxy resin composition prepreg and fibre-reinforced composite material. JP Patent 2,004,292,594
35. Muraki T, Nishiyama S (2003) Method for producing high-density fibrous structure and the resultant high-density fibrous structure. JP Patent 2,003,073,968
36. Mortaigne B, Regnier N (2000) Study of epoxy and epoxy–cyanate networks thermal degradation to predict materials lifetime in use conditions. J Appl Polym Sci 77:3142
37. Wu SJ, Lin TK, Zhang JX, Shyu SS (2000) Properties of cyanate ester-cured epoxy/polyphenylene oxide blends as a matrix material for Kevlar fiber composites. J Adhes Sci Technol 14:1423
38. Ratna D, Chongdar TK, Chakraborty BC (2004) Mechanical characterization of new glass fiber reinforced epoxy composites. Polym Compos 25:165
39. DeCarli M, Kozielski K, Tian W, Varley R (2005) Toughening of a carbon fibre reinforced epoxy anhydride composite using an epoxy terminated hyperbranched modifier. Compos Sci Technol 65:2156
40. Matheswaran M, Padmanabhan K, Kishore (1995) Static and impact behavior of thermoplastic modified glass fabric/epoxy composites. J Mater Sci Lett 14:951
41. Sainathan N, Padmanabhan K, Sashidhara S, Rao RMVGK, Kishore (1995) Influence of particulate graphite additions on the shear related behavior of glass fabric reinforced epoxy composites. J Reinf Plast Compos 14:445
42. Podgaiz RH, Williams RJJ (1997) Effects of fiber coatings on mechanical properties of unidirectional glass-reinforced composites. Compos Sci Technol 57:1071
43. John NA, Brown JR (1998) Flexural and interlaminar shear properties of glass-reinforced phenolic composites. Compos Part A 29A:939
44. Cho D, Yun SH, Kim J, Lim S, Park M, Lee S-S, Lee G-W (2004) Influence of silane coupling agents on the interlaminar and thermal properties of woven glass fabric/nylon 6 composites. Macromol Res 12:119
45. Barraza HJ, Aktas L, Hamidi YK, Long J Jr, Orear EA, Altan MC (2003) Moisture absorption and wet-adhesion properties of resin transfer molded (RTM) composites containing elastomer-coated glass fibers. J Adhes Sci Technol 17:217

46. Park R, Jang J (2004) Effect of surface treatment on the mechanical properties of glass fiber/vinylester composites. J Appl Polym Sci 91:3730
47. Li Y, Mai Y-W, Ye L (2005) Effects of fibre surface treatment on fracture-mechanical properties of sisal-fibre composites. Compos Interfaces 12:141
48. Zhang C, Hoa SV, Ganesan R (2002) Experimental characterization of interlaminar shear strengths of graphite/epoxy laminated composites. J Compos Mater 36:1615
49. Okada T, Nishijima S (1990) Investigation of interlaminar shear behavior of organic composites at low temperatures. Adv Cryog Eng 36B:811
50. Vishwanath B, Verma AP, Rao CVSK (1991) Effect of fabric geometry on friction and wear of glass fiber-reinforced composites. Wear 145:315
51. Costa ML, Almeida SFMD, Rezende MC (2005) Critical void content for polymer composite laminates. AIAA J 43:1336
52. Ray BC (2006) Loading rate sensitivity of glass fiber-epoxy composite at ambient and sub-ambient temperatures. J Reinf Plast Compos 25:329
53. Ray BC (2006) Loading rate effects on mechanical properties of polymer composites at ultralow temperatures. J Appl Polym Sci 100:2289
54. Mouritz AP (2000) Ultrasonic and interlaminar properties of highly porous composites. J Compos Mater 34:218
55. Smith JC, Blandford JM, Schiefer HF (1960) Stress–strain relationships. In yarns subjected to rapid impact loading. Text Res J 30:752
56. Chocron-Benloulo IS, Sanchez-Galvez V (1996) Impact resistance of polymeric matrix composites. In: Hui D (ed) Proceedings of the third international conference on composite engineering (ICCE/3), New Orleans, 1996
57. Mueller DH (2004) Improving the impact strength of natural fiber reinforced composites by specifically designed material and process parameters. INJ Winter 31:28
58. Foster GC (1998) Tensile and flexure strength of unidirectional fiber reinforced composites: direct numerical simulations and analytic models. Dissertation, Virginia Polytechnic Institute and State University
59. Dash PK, Chatterjee AK (2004) Effects of environment on fracture toughness of woven carbon/epoxy composite. J Inst Eng 85:1
60. Kim J-K, Mai Y-W (1993) Interfaces in composites. In: Chou TW (ed) Structure and properties of fibre composites, vol 13. VCH Publishers, Weinheim, pp 239–289
61. Metcalfe AG (ed) (1974) Interfaces in metal matrix composites. Academic, London
62. Wilkins DJ, Eisenmann JR, Camin RA, Margolis WS, Benson RA (1982) Characterizing delamination growth in graphite-epoxy. In: Reifsnider KL (ed) Damage in composite materials. ASTM International, Philadelphia, p 168
63. Han KS, Koutsky J (1981) The interlaminar fracture energy of glass fiber reinforced polyester composites. J Compos Mater 15:371
64. Devitt DF, Schapery RA, Bradley WL (1980) A method for determining the mode I delamination fracture toughness of elastic and viscoelastic composite materials. J Compos Mater 14:270
65. Mall S, Law GE, Katouzian M (1987) Loading rate effect on interlaminar fracture toughness of a thermoplastic composite. J Compos Mater 21:569
66. Feih S, Wei J, Kingshott P, Sorensen BF (2005) The influence of fibre sizing on the strength and fracture toughness of glass fibre composites. Compos Part A Appl S 36:245
67. Tanoglu M, Seyhan AT (2003) Investigating the effects of a polyester preforming binder on the mechanical and ballistic performance of E-glass fiber reinforced polyester composites. Int J Adhes Adhes 23:1
68. Compston P, Jar P-YB, Davies P (1998) Matrix effect on the static and dynamic interlaminar fracture toughness of glass-fiber marine composites. Compos Part B Eng 29:505
69. Kuboki T, Jar PYB, Forest TW (2003) Influence of interlaminar fracture toughness on impact resistance of glass fiber reinforced polymers. Compos Sci Technol 63:943

70. Srivastava VK, Hogg PJ (1998) Moisture effects on the toughness, mode-I and mode-II of particles filled quasi-isotropic glass-fiber reinforced polyester resin composites. J Mater Res 33:1129
71. Kang TJ, Lee SH (1997) Mechanical properties of nonwoven glass fiber composites. Polym Polym Compos 5:29
72. Shetty MR, Kumar KRV, Sudhir S, Raghu P, Madhuranath AD, Rao RMVGK (2000) Effect of fiber orientation on mode-I interlaminar fracture toughness of glass epoxy composites. J Reinf Plast Compos 19:606
73. Sela N, Ishai O, Banks-Sills L (1989) The effect of adhesive thickness on interlaminar fracture toughness of interleaved CFRP specimens. Composites 20:257
74. Yang Z, Sun CT (2000) Interlaminar fracture toughness of a graphite/epoxy multidirectional composite. J Eng Mater Technol 122:428
75. Yuan Q, Karger-Kocsis J (1996) On the efficiency of interleaves in carbon fiber/epoxy composite laminates by the fractal approach. J Mater Sci Lett 15:842
76. Barikani M, Saidpour H, Sezen M (2002) Mode-I interlaminar fracture toughness in unidirectional carbon-fiber/epoxy composites. Iran Polym J 11:413
77. Shindo Y, Shinohe D, Kumagai S, Horiguchi K (2005) Analysis and testing of mixed-mode interlaminar fracture behavior of glass-cloth/epoxy laminates at cryogenic temperatures. J Eng Mater Technol 127:468
78. Stinchcomb WW, Ashbaugh NE (eds) (1993) Composite materials fatigue and fracture. ASTM International, Philadelphia
79. Wang R-M, Zheng S-R, Zheng Y (2011) Polymer matrix composites and technology. Woodhead Publishing, Cambridge
80. Halpin Affdl JC, Kardos JL (1976) The Halpin-Tsai equations: a review. Polym Eng Sci 16:344
81. Halpin JC, Tsai SW (1967) Environmental factors in composite design. In: Airforce materials laboratory technical report, AFML-TR-67-423
82. Clyne TW, Withers PJ (1993) An introduction to metal matrix composites. Cambridge University Press, Cambridge
83. Hull D, Clyne TW (1996) An introduction to composite materials. Cambridge University Press, Cambridge
84. Lu L, Fuh J (2001) Laser-induced materials and processes for rapid prototyping. Kluwer Academic Publishers, Boston
85. Campbell FC (2010) Structural composite materials. ASM International, Materials Park
86. Epparachchi JA, Clausen PD (2000) A new approach to a fatigue damage model for glass-fiber reinforced plastic composites. In: Hui D (ed) Proceedings of the seventh international conference on composites engineering, ICCE/7, Denver 2000
87. Toumi RB, Renard J, Monin M, Nimdum P (2013) Fatigue damage modelling of continuous E-glass fibre/epoxy composite. Procedia Eng 66:723
88. Sreekala MS, Kumaran MG, Joseph R, Thomas S (2001) Stress-relaxation behaviour in composites based on short oil-palm fibres and phenol formaldehyde resin. Compos Sci Technol 61:1175
89. Pothan LA, Neelakantan NR, Rao B, Thomas S (2004) Stress relaxation behavior of banana fiber-reinforced polyester composites. J Reinf Plast Compos 23:153
90. Barpanda D, Mantena PR (1998) Effect of hybridization on the creep and stress relaxation characteristics of pultruded composites. J Reinf Plast Compos 17:234
91. Geethamma VG, Pothen LA, Rhao B, Neelakantan NR, Thomas S (2004) Tensile stress relaxation of short-coir-fiber-reinforced natural rubber composites. J Appl Polym Sci 94:96
92. Kawai M, Kazama T, Masuko Y, Tsuda H, Takahashi J, Kemmochi K (2004) Stress relaxation behavior of unidirectional carbon/epoxy composites at elevated temperature and analysis using visco-plasticity mode. JSME Int J A-Solid M 47:8

93. Lifshitz JM, Rotem A (1970) Time-dependent longitudinal strength of unidirectional fibrous composites. Fibre Sci Technol 3:1
94. Gutman EM, Soncino R (1995) Environmental effect on stress relaxation in polyester-fiberglass composite. Polym Compos 16:518
95. Wortmann F-J, Schulz KV (1995) Stress relaxation and time/temperature superposition of polypropylene fibers. Polymer 36:315
96. Ikeda S (1987) Evaluation of the chemical resistance of fiber-reinforced plastics by stress relaxation method. Nippon Zairyo Gakkai 26:127
97. Hashemi S, Kinloch AJ, Williams JG (1990) The effects of geometry, rate and temperature on the mode I, mode II and mixed-mode I/II interlaminar fracture of carbon fiber/poly(ether ether ketone) composites. J Compos Mater 24:918
98. Sloan FE, Seymour RJ (1992) The effect of seawater exposure on Mode I interlaminar fracture and crack growth in graphite/epoxy. J Compos Mater 26:2655
99. Takeda N, Tohdoh M, Takahashi K (1995) Interlaminar fracture toughness degradation of radiation-damaged GFRP and CFRP composites. Adv Compos Mater 4:343
100. Watt A, Goodwin AA, Mouritz AP (1998) Thermal degradation of the Mode I interlaminar fracture properties of stitched glass fiber/vinyl ester composites. J Mater Sci 33:2629
101. Pavlidou S, Papaspyrides CD (2003) The effect of hygrothermal history on water sorption and interlaminar shear strength of glass/polyester composites with different interfacial strength. Compos Part A 34:1117
102. Davies P, Mazeas F, Casari P (2001) Sea water aging of glass reinforced composites: shear behaviour and damage modelling. J Compos Mater 35:1343
103. Apicella A, Migliaresi C, Nicolais L, Iaccarino L, Roccotelli S (1983) The water ageing of unsaturated polyester-based composites: influence of resin chemical structure. Composites 14:387
104. Choqueuse D, Davies P, Mazeas F, Baizeau R (1997) Aging of composites in water: comparison of five materials in terms of absorption kinetics and evolution of mechanical properties. In: Gates TS, Zureick A-H (eds) High temperature and environmental effects on polymeric composites, vol 2. ASTM International, Philadelphia, p 73
105. Jacobson NS, Fox DS, Opila EJ (1998) High temperature oxidation of ceramic matrix composites. Pure Appl Chem 70:493
106. Rochais D, Houedec HL, Enguehard F, Jumel J, Lepoutre F (2005) Microscale thermal characterization at temperatures up to 1000°C by photoreflectance microscopy: application to the characterization of carbon fibers. J Phys D Appl Phys 38:1498
107. Kiralp S, Kucukyavuz Z, Qasrawi AF (2003) Preparation and characterization of conducting polybutadiene/polythiophene composites. Turk J Chem 27:417
108. Lee SL, Chen TM (1993) Preparation and characterization of a ceramic superconductor/nylon 6,6 composite. Chin J Phys 31:1175
109. Lagorce LK, Allen MG (1997) Magnetic and mechanical properties of micromachined strontium ferrite/polyimide composites. J Microelectromech Syst 6:307
110. Moheimani SOR, Fleming AJ (2006) Piezoelectric transducers for vibration control and damping. Springer, London
111. Fang Q et al (1996) Erosive wear behaviour of Al_2O_3-fiber reinforced aluminium based metal matrix composites (MMCs). In: Hui D (ed) Proceedings of the third international conference on composite engineering, ICCE/3, New Orleans, 1996
112. Fink A, Kolesnikov B (2005) Hybrid titanium composite material improving composite structure coupling. Accessed 09 July 2015. http://www.dlr.de/fa/Portaldata/17/Resources/dokumente/publikationen/2005/08_fink.pdf
113. Ferguson R (2004) Effect of local constraint on measured bearing stress in carbon/epoxy laminate. In: Proceedings of the 2nd international conference on composites testing and model identification, Bristol, 21–23 Sept 2004
114. Cervenka A, Allan PS (1997) Characterization of finite length fiber composites: Part I. Introductory paper (technical report). Pure Appl Chem 69:1693

115. Shi X, Hudson JL, Spicer PP, Tour JM, Krishnamoorti R, Mikos AG (2005) Rheological behavior and mechanical characterization of injectable poly(propylenefumarate)/single-walled carbon nanotube composites for bone tissue engineering. Nanotechnology 16:S531
116. Fung YC (1993) Biomechanics: mechanical properties of living tissues. Springer, New York
117. Wang M, Porter D, Bonfield W (1994) Processing, characterisation, and evaluation of hydroxyapatite reinforced polyethylene composites. Br Ceram Trans 93:91
118. Wang M, Bonfield W (2001) Chemically coupled hydroxyapatitepolyethylene composites: structure and properties. Biomaterials 22:1311
119. Wang M, Kokubo T, Bonfield W (1996) A-W glass–ceramic reinforced polyethylene composite for medical applications. Bioceramics 9:387
120. Wang M, Bonfield W, Li M, Guiu F (1996) Interphase in composite materials. Key Eng Mater 127:583
121. Guild FJ, Bonfield W (1993) Predictive modeling of hydroxyapatitepolyethylene composite. Biomaterials 14:985
122. Bergman MA (1999) The clinical performance of ceramic inlays: a review. Aust Dent J 44:157
123. Bonfield W (1987) Materials for the replacement of osteoarthritic hip joints. Met Mater 3:712
124. Kikuchi M, Itoh S, Ichinose S, Shinomiya K, Tanaka J (2001) Self-organization mechanism in a bone-like hydroxyapatite/collagen nanocomposites synthesized in vitro and its biological reaction in vivo. Biomaterials 22:1705
125. Reis RL, Cunha AM, Fernandes MH, Correia RN (1997) Bioinert and biodegradable polymeric matrix composites filled with bioactive SiO_2-3CaO.P_2O_5-MgO glasses and glass–ceramics. In: Bioceramics Vol.10. Elsevier Science, pp 415–418 Paris, France
126. Sousa RA, Reis RL, Cunha AM, Bevis MJ (2000) Structure and properties of hydroxylapatite reinforced starch bone-analogue composites. Key Eng Mater 192–195:669
127. Ladizesky NH, Pirhonen EM, Appleyard DB, Ward IM, Bonfield W (1998) Fibre reinforcement of ceramic/polymer composites for a major load-bearing bone substitute material. Compos Sci Technol 58:419
128. Wang M (2003) Developing bioactive composite materials for tissue replacement. Biomaterials 24:2133
129. Wang M, Bonfield W, Hench LL (1995) Bioglass/high density polyethylene composite as a new soft tissue bonding material. In: Wilson J, Hench LL, Greenspan D (eds) Bioceramics 10. Elsevier Science, Oxford
130. Wang M, Yue CY, Chua B, Kan LC (1999) Hydroxyapatite reinforced polysulfone as a new biomaterial for tissue replacement. Bioceramics 12:401
131. Wang M, Yue CY, Chua B (2001) Production and evaluation of hydroxyapatite reinforced polysulfone for tissue replacement. J Mater Sci Mater Med 12:821
132. Wang M, Chua B (2002) Fatigue performance of a bioactive composite developed for hard tissue replacement. Bioceramics 15:935

Newly Developed Rubber Pressure Molding Technique for Fabrication of Composites

3

S. D. Sharma and Kamal K. Kar

Contents

S.D. Sharma (✉)
Advanced Nanoengineering Materials Laboratory, Materials Science Programme, Indian Institute of Technology Kanpur, Kanpur, Uttar Pradesh, India

Department of Mechanical Engineering, Rewa Engineering College, Rewa, India
e-mail: sds1970@gmail.com

K.K. Kar
Advanced Nanoengineering Materials Laboratory, Materials Science Programme, Indian Institute of Technology Kanpur, Kanpur, Uttar Pradesh, India

Advanced Nanoengineering Materials Laboratory, Department of Mechanical Engineering, Indian Institute of Technology Kanpur, Kanpur, Uttar Pradesh, India
e-mail: kamalkk@iitk.ac.in

K.K. Kar (ed.), *Composite Materials*,
DOI 10.1007/978-3-662-49514-8_3

Abstract

A new manufacturing method, i.e., rubber pressure molding (RPM) for processing of long fiber/fabric-reinforced polymer composites having complex geometry, is developed. Studies are conducted for performance evaluation of the newly proposed RPM using silicone, natural, polybutadiene, and styrene–butadiene rubbers. Polyester and epoxy resin are used with glass fiber fabric to manufacture the fiber-reinforced plastic (FRP) products.

Keywords

Rubber pressure molding • Composites manufacturing • FRP composites manufacturing • Compression molding • GFRP composites manufacturing

Introduction

Fiber-reinforced polymer (FRP) composites offer a combination of low specific gravity, high strength-to-weight ratio, and specific modulus that are either comparable to or better than many traditional metallic materials. As a result, the FRP composites have emerged as a major class of structural material and are either used or being considered as substitutions for metals in many weight-critical applications. Several methods, i.e., filament winding process, pultrusion method, vacuum bagging technique, autoclave technique, matching die set compression molding, resin transfer molding, etc., have been developed to manufacture FRP products. Major challenges in manufacturing of composites are efficient handling of fibers and application of uniform pressure during curing of the resin. Due to this, manufacturing cost of the FRP increases. In order to reduce the manufacturing cost of good quality FRP composites, a new manufacturing technique known as **Rubber Pressure Molding** (RPM) has been developed recently.

The RPM is based on the matching die set, where the die is made from mild steel and the punch from flexible rubberlike material, i.e., natural rubber, butyl rubber, silicone rubber, polybutadiene rubber, chloroprene rubber, styrene–butadiene rubber, etc., to apply hydrostatic pressure on the various surfaces of a complex-shaped product.

This chapter presents the performance evaluation of RPM using different rubbers like silicone rubber, natural rubber (NR), polybutadiene rubber (PBR), and butyl rubber (BR) and resin systems like epoxy and polyester. Burn test, scanning electron microscopy (SEM), surface roughness, and mechanical tests, like interlaminar fracture toughness (ILFT), interlaminar shear strength (ILSS), Young's modulus, tensile strength, percentage elongation at break, etc., are carried out to know the fiber content, void content, presence of delamination, bonding between fiber and resin, microstructure, and mechanical properties of the composite materials. These properties are also compared with FRP components made by the conventional technique, i.e., regular compression molding, where both parts of the mold are made of steel, to evaluate its performances in the structural applications.

Performance of RPM is also evaluated by using different hardness values of the rubber punch. For this, natural rubber is used and polyester resin with glass fiber is selected for experimentation as this is commonly used for general-purpose FRP products. Since the hardness of rubber changes with the percentage of filler used in the rubber, the amount of filler material (i.e., carbon black, N 330) in the raw natural rubber is varied from 0 to 75 phr with an increment of 15 phr. FRP composites are made using RPM and characterizations of composite laminates are carried out by burn test, tensile test, ILSS/crushed strength, and ILFT tests. The results are compared with the corresponding values for the composites made by conventional method to evaluate the performance of RPM. In addition to this, performance of RPM is also evaluated when the same mold is used repeatedly to manufacture FRP composite. It is found that surface roughness of rubber mold does not deteriorate when multiple laminates are prepared from a single rubber mold.

Manufacturing of Composites using Rubber Pressure Molding

Rubber Pressure Molding Setup for Cooler Pump Cap

To prepare the cooler pump cap (Fig. 3.1) from FRP, a split steel die, steel cone, and flexible rubber punch are used. A schematic diagram of the overall setup is shown in Fig. 3.2. Split steel die has an interior conforming to the external shape of the

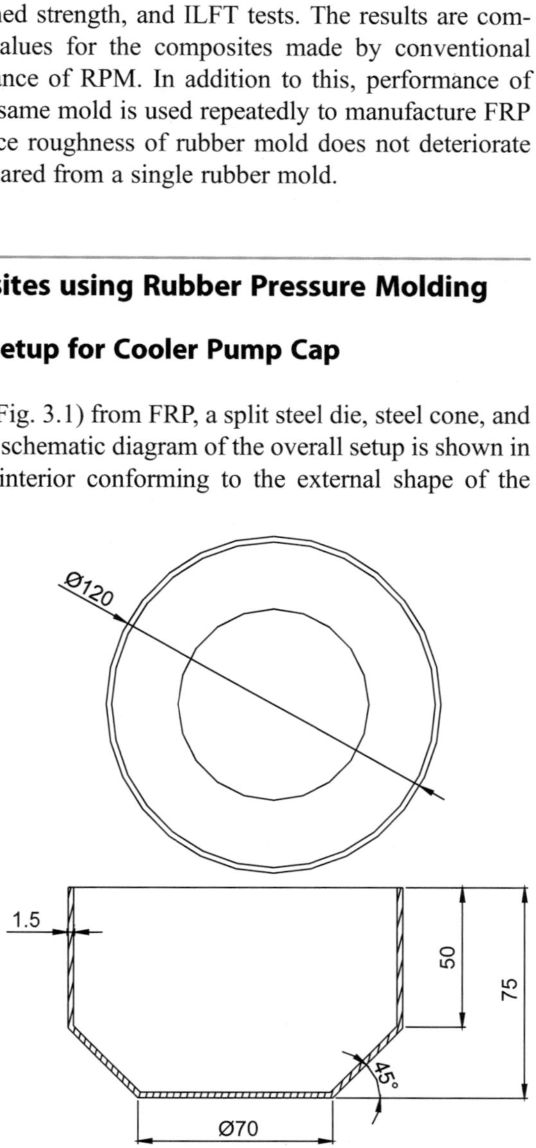

Fig. 3.1 Cooler pump cap [1–4]

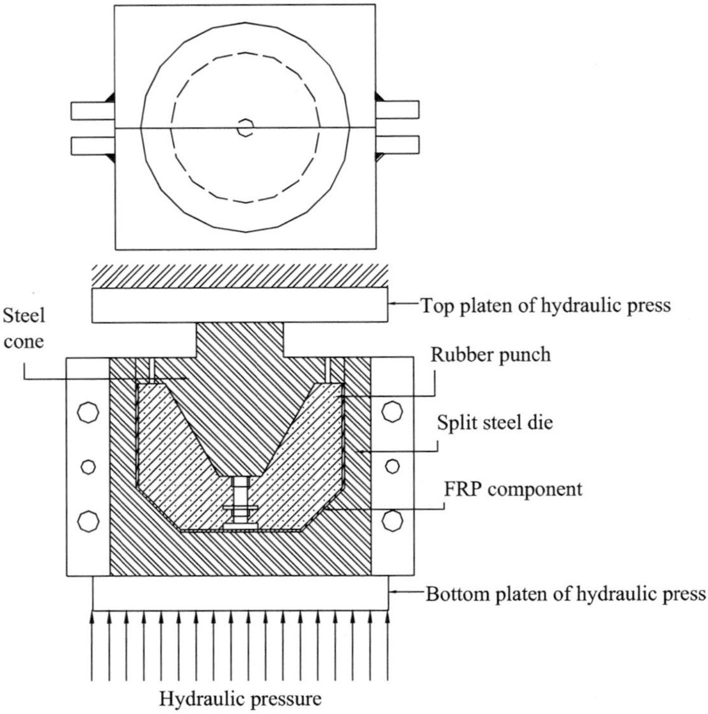

Fig. 3.2 A schematic diagram of RPM setup for cooler pump cap [1–4]

component. The die is made in two parts as shown in Fig. 3.3 for easy removal of component once it is cured. A preform of glass fiber fabric wetted with epoxy/polyester resin is placed in between split steel die and rubber punch as shown in Fig. 3.2. The rubber punch is pressed by a suitably designed steel cone. The force on the steel cone is applied by the hydraulic press. The product is cured in 16 h at room temperature, i.e., 25 °C and pressure of 0.65 MPa.

Preparation of Split Steel Die

The mild steel die is made in two parts (split type). Its assembly drawing is shown in Fig. 3.3. Two parts of the die are symmetric to each other. To make the die, a rectangular steel block is cut into two pieces and their surfaces are polished to get smooth surface. Then two parts are tack welded and machining is done on the welded assembly to obtain the exact internal dimensions of die. The inner surface of die is polished to obtain very smooth surface as the shape of rubber punch and preform depends on the surface finish of die. Four steel flat bars (125 × 25 × 10 mm), with a dowel hole (Φ8) and two M12 holes on each flat bar (shown in Fig. 3.3), are welded. To sustain high pressure, supporting webs are provided on the steel flat bar. Then the

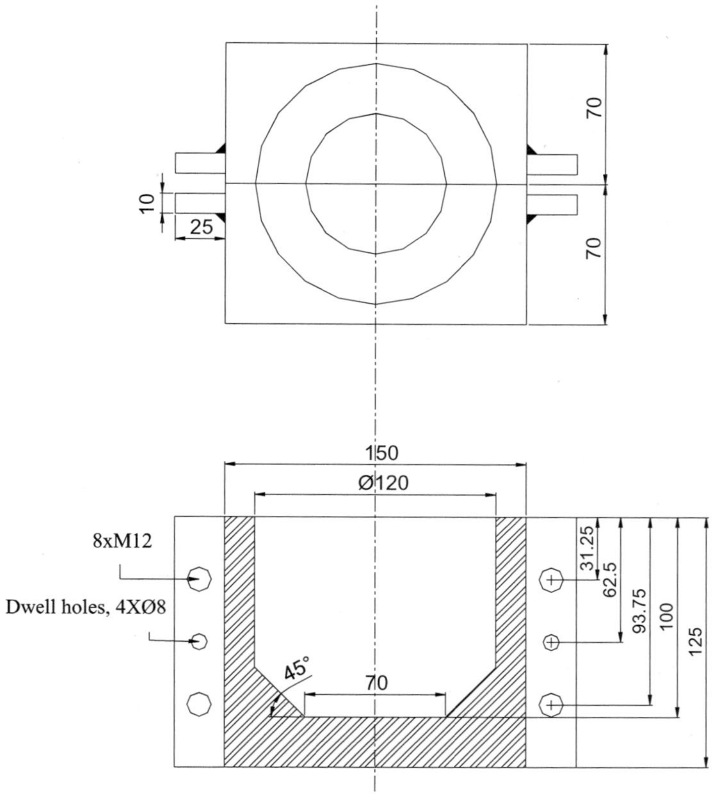

Fig. 3.3 Split steel die [1–4]

halves of the welded assembly are detached by breaking the welded points. Different allowances are taken into consideration to prepare actual size of FRP product.

As the component prepared is of only 1.5 mm thickness, a single die set is used for preparing rubber punch as well as FRP components. However, to prepare thicker component, two die sets will be required, one for preparing the rubber punch and another one for preparing the component. In that case the internal dimension of the die to prepare rubber punch should be less and offset by thickness of the component.

Design and Fabrication of Rubber Punch

Six types of rubber (natural, butyl, silicone, polybutadiene, styrene–butadiene and chloroprene rubber) are used to cast six different rubber punches. The rubber punches have the same inner shape as that of the FRP component. The die used to cast the rubber punch is same that is used to prepare the component, i.e., split steel die. This section describes design and fabrication of rubber punch and procedure to prepare rubber punch. The split steel die (Fig. 3.3) and the steel cone are used for

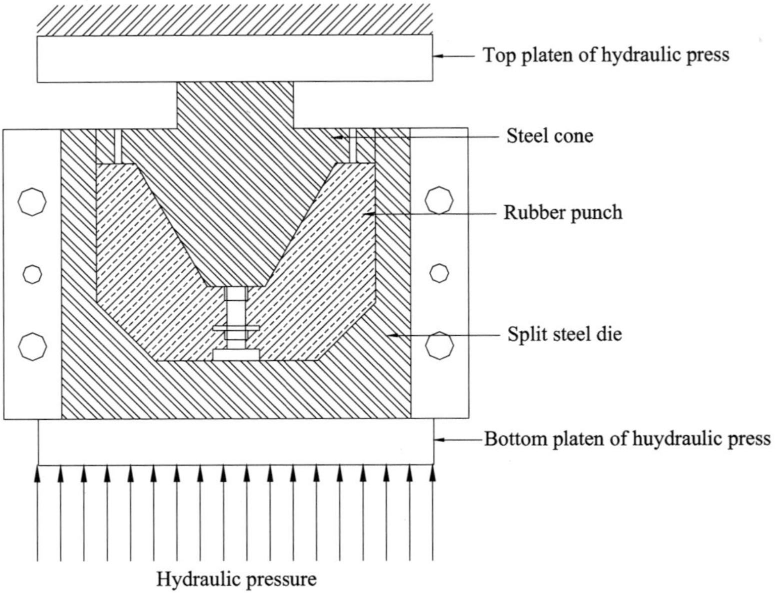

Fig. 3.4 A schematic diagram of setup used for casting of rubber punch [1–4]

Table 3.1 Composition of ingredients in phr (per hundred rubber) [1–4]

Rubber (100 phr)	Zinc oxide (in phr)	Stearic acid (in phr)	Carbon black (in phr)	Sulfur (in phr)	Accelerator (TMTD) (in phr)
Natural	5.0	3.0	45.0	5.0	2.5
Butyl	5.0	3.0	45.0	5.0	2.5
Silicone	5.0	3.0	45.0	5.0	2.5
Polybutadiene	5.0	3.0	45.0	5.0	2.5

casting the rubber punch by pressing uncured rubber between split steel die and steel cone.

The material for steel cone chosen is mild steel. It consists of a circular base with a conical projection. The circular base has a dimension of $\Phi 119.8 \times 10$ mm thickness. The conical portion has a diameter of 85 mm, 54° half cone angle, and height of 75 mm. Two holes of diameter 3 mm each are provided on the circular base to bleed off extra rubber material during curing of rubber punch. The whole assembly is shown in Fig. 3.4.

Preparation of Uncured Rubber The uncured rubber is prepared in "two roll mixing mill." The formulations used to mix these rubber ingredients with different rubbers are given in Table 3.1.

Molding and Curing of Rubber Punch The split steel die is preheated before putting rubber in it. The preheating is done in a hydraulic press to a temperature of

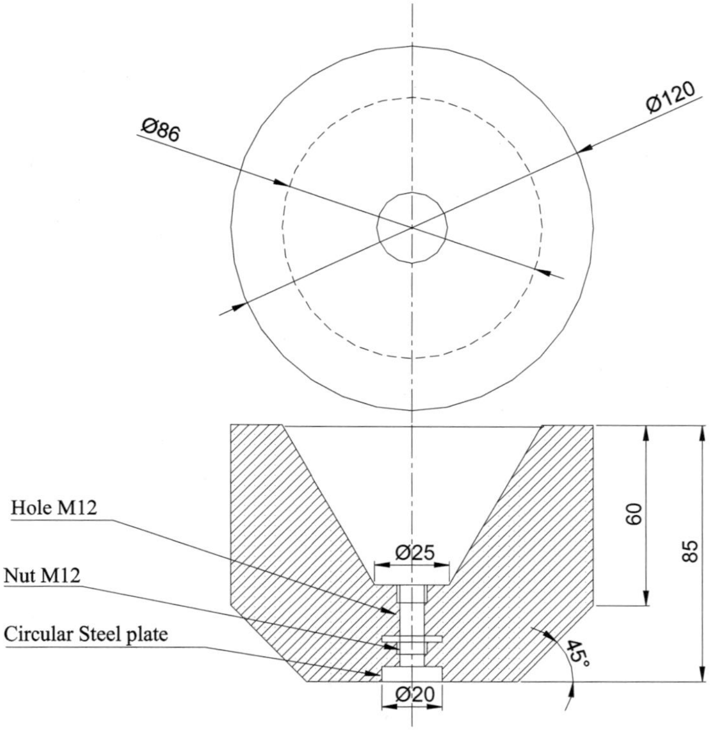

Fig. 3.5 Schematic diagram of rubber punch [1–4]

150 °C. When temperature reaches close to 150 °C, uncured rubber is filled in the die and the steel cone is placed over it. A nut and bolt system is placed inside the rubber punch (Fig. 3.5) for easy removal of the component from the rubber punch after curing of preform. The pressure of 6 MPa is applied on the die set. The temperature and pressure are maintained for 30 min. The cured rubber punch is taken out by removing bolts on the metal flats of die. The photograph of the rubber punch is shown in Fig. 3.6.

Fabrication of Fiber-Reinforced Plastic Product

This section describes the fabrication of FRP product (cooler pump cap) using RPM. The FRP products were made using six varieties of rubber punches. These are (i) natural, (ii) butyl, (iii) silicone, (iv) styrene–butadiene, (v) neoprene, and (vi) polybutadiene rubber punch. Pump cap is prepared using above six varieties of rubber to check the performance of different rubber punches. The tooling required for molding a pump cap consists of the split steel die, rubber punch, and steel cone. The photographs of split steel die, rubber punch, and steel cone together are shown in Fig. 3.7. The cooler pump cap is fabricated using glass fiber fabric and epoxy/polyester resin. Major steps of the fabrication process are described here.

Fig. 3.6 Photograph of rubber punch [1–4]

Fig. 3.7 Split steel die with rubber punch and steel cone [1–4]

Preparation of Die and Punch The oil and grease present on the surface of the split steel die and the rubber punch are removed by using acetone. Then a thin layer of mold release agent is applied on the surface of the die and rubber punch to prevent adhesion of resin to steel die/rubber punch and easy removal of the FRP product from the metal die/rubber punch after curing. After applying a thin coating of release agent, it is kept at room temperature for an hour to let the solvent evaporate out to leave a thin layer of coating agent on the surface of the steel die and rubber punch.

Making of Preform The hand lay-up technique is used to prepare the preform (uncured form of product). The glass fabric is cut in right shape using template

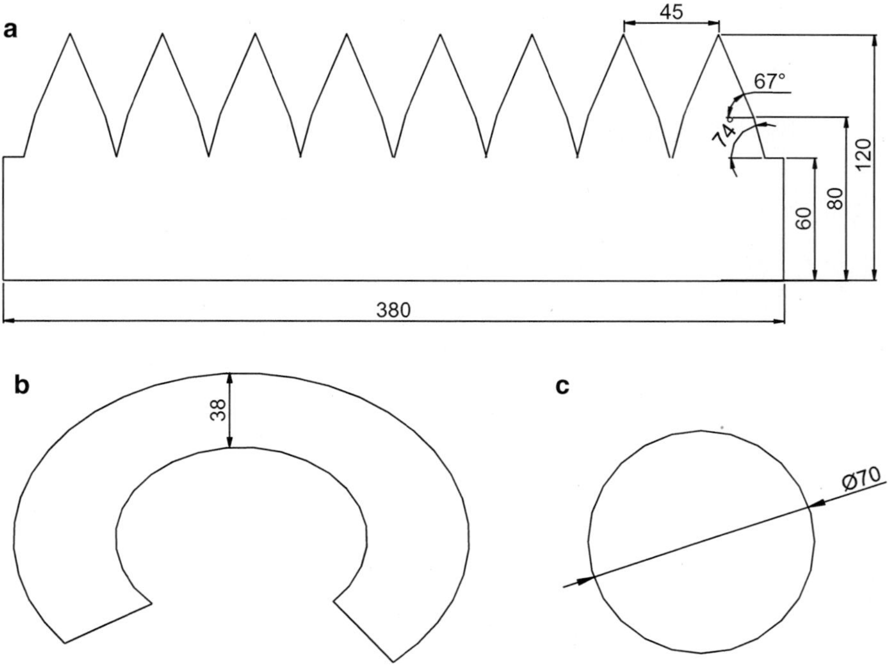

Fig. 3.8 (**a**) Template 1, (**b**) template 2, and (**c**) template 3 [1–4]

1 (Fig. 3.8a) from a woven glass fiber sheet. The template is designed in such a way that the uncut portion covers cylindrical portion of the product and the cut portion covers the conical and base portion of the product. The conical and base of the product do not give continuous laying, so template 2 and template 3 (Figs. 3.8b and c) are inserted in conical and base portion after every two layers of template 1. The matrix used to prepare the product is epoxy or polyester resin. The epoxy resin is prepared by adding 10 % hardener (HY951) by weight, whereas in case of polyester resin, the unsaturated resin is mixed with 1.2 % catalyst (methyl ethyl ketone peroxide) and 1.2 % accelerator (cobalt octoate) by weight. Total numbers of template 1, template 2 and template 3 used to prepare the preform are 4, 2 and 2 respectively. Using hand lay-up technique, the glass fabric and the epoxy/polyester resin are placed on the rubber punch; then the rubber punch with preform is placed inside the split steel die, and four nuts of the die set are tightened till the two portions of the split die touch each other. Then, steel cone is placed over the rubber punch.

Curing of Preform The steel die and the rubber punch with preform are loaded in a hydraulic press. The pressure used in the hydraulic press is 0.65 MPa and the preform is allowed to cure at room temperature, i.e., 25 °C for 16 h.

Removal of the Product from Die The steel die is separated into two parts by removing the bolts after curing of preform. To take out the product from the rubber

Fig. 3.9 FRP components made by RPM [1–4]

Fig. 3.10 Products made of FRP using RPM [1–4]

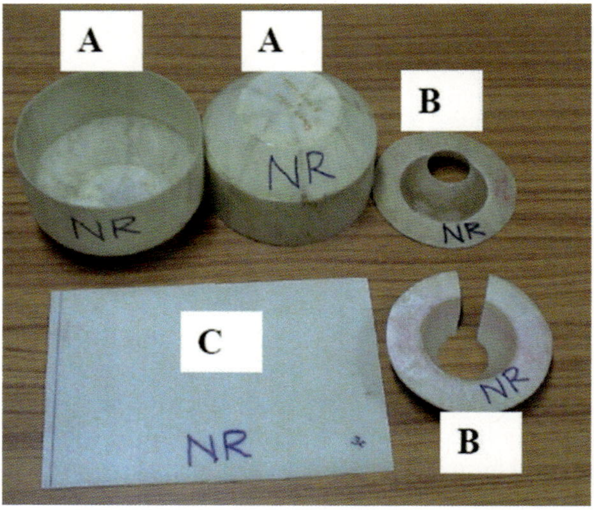

punch, a M12 bolt is placed on the M12 nut inside the rubber punch and screwed slowly to take out the product from the rubber punch. The purpose of placing the nut and a steel plate inside the rubber punch is to take out the product easily without hammering the rubber and the FRP product. The removal of product from the die and punch is done very carefully so as not to damage the surface of the die, punch, and the product. The product is then smoothened with the help of an emery paper and shown in Figs. 3.9 and 3.10.

Theoretical Analysis

This technique is based on the use of a flexible rubber punch that applies hydrostatic pressure on the surface of the product. This overcomes many difficulties of prevailing fabrication methods of complex-shaped FRP components. In order to verify the uniform pressure distribution over the complex-shaped product and to find out other parameters like most suitable strain energy density function for the rubber, optimum pressure during fabrication, and rubber hardness, a finite element analysis is carried out using ANSYS (version 7.0). There are three different material types – steel, FRP, and rubber. The properties of the above materials are defined as follows:

Steel: it is a linear, elastic, and isotropic material with Young's modulus of 210 GPa and Poisson's ratio of 0.3.

FRP: it is a linear, elastic, and orthotropic material having nine elastic constants. These are as follows: $E_{11} = 16.2$, $E_{22} = 15.3$, and $E_{33} = 7.3$ GPa; Poisson's coefficients $\nu_{12} = 0.115$, $\nu_{13} = 0.115$, and $\nu_{23} = 0.3$; and shear moduli $G_{12} = 3.9$, $G_{23} = 2.8$, and $G_{13} = 3.9$ GPa [5].

Rubber: it is a nonlinear and hyperelastic material. For defining hyperelastic material in the finite element analysis, it is necessary to know the strain energy density function. It is represented by 2-, 3-, 5-, 7-, and 9-parameter Mooney–Rivlin [6]; 1st-, 2nd-, and 3rd-order Ogden [7]; neo-Hookean [8]; 1st-, 2nd-, and 3rd-order polynomial [7]; Arruda–Boyce [9]; Gent [10]; and Yeoh [11].

The stress strain data are calculated using these strain energy density functions. These data are compared with the experimental stress strain data for natural rubber containing 45 phr carbon black (A_{45}) and shown in Fig. 3.11. The formulation of rubber is given in Table 3.2. Mooney–Rivlin 3 parameters, 5 parameters,

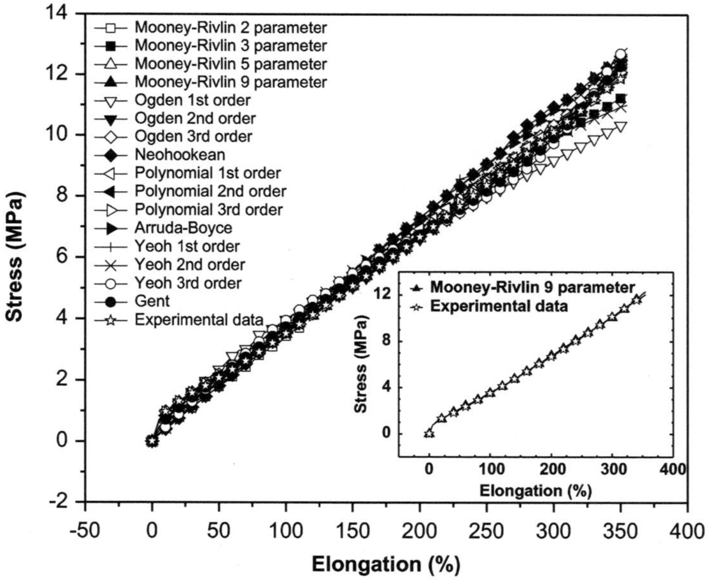

Fig. 3.11 Experimental stress–elongation curve of NR vulcanizates containing 45 phr carbon black and its comparison with 2-, 3-, 5-, and 9-parameter Mooney–Rivlin; 1st-, 2nd-, and 3rd-order Ogden; neo-Hookean; 1st-, 2nd-, and 3rd-order polynomial; Arruda–Boyce; 1st-, 2nd-, and 3rd-order Yeoh; and Gent. Inset figure shows the comparison between experimental data and Mooney–Rivlin nine parameters [12, 13]

Table 3.2 Formulations of natural rubber vulcanizates, hardness values, and pressure distribution on various rubber molds for an applied pressure of 0.65 MPa [12, 13]

Raw materials		A_0	A_{15}	A_{30}	A_{45}	A_{60}	A_{75}
Natural rubber (ISNR 10)		100	100	100	100	100	100
Zinc oxide		5.00	5.00	5.00	5.00	5.00	5.00
Stearic acid		3.00	3.00	3.00	3.00	3.00	3.00
Carbon black (N330)		0.0	15.0	30.0	45.0	60.0	75.0
Sulfur		5.00	5.00	5.00	5.00	5.00	5.00
TMTD		2.50	2.50	2.50	2.50	2.50	2.50
Hardness of rubber (Shore A)		50 ± 1	61 ± 1	69 ± 1	72 ± 1	79 ± 1	85 ± 2
Pressure distribution (MPa)	Variation	0.42–0.51	0.41–0.50	0.39–0.48	0.62	0.37–0.45	0.42–0.49
	Average	0.46	0.45	0.44	0.62	0.41	0.46

A_0, A_{15}, A_{30}, A_{45}, A_{60}, A_{75}: suffix indicates loading of filler (phr: per hundred rubber)

7 parameters, and 9 parameters; Ogden 2nd and 3rd orders; polynomial 1st, 2nd, and 3rd orders; Yeoh 2nd and 3rd orders; neo-Hookean; and Arruda–Boyce fitted well with experimental data. However, the Mooney–Rivlin 2 parameters, Ogden 1 order, and Yeoh 1st order moderately fitted with experimental data. The material constants of the above models are calculated and used in the finite element analysis. These are summarized in Table 3.3.

The analysis is carried out for cooler pump cap at the pressure of 1 MPa. The pressure profile for RPM is shown in Fig. 3.12. It shows that pressure is uniformly distributed over the complex-shaped geometry. It varies from 0.71 to 0.92 with an average of 0.81 MPa. The same analysis is carried out for the conventional technique and shown in Fig. 3.13. The pressure distribution is uneven and widely varies from 0.18 to 0.62 with an average of 0.40 MPa (Fig. 3.13). The maximum pressure is concentrated at the corner point, which is also a stress concentration zone. Also high pressure point is observed on the periphery of FRP component. This may be basically due to the heterogeneous structure of FRP. The soft rubber mold evenly distributes this high pressure, which is not possible by the steel mold due to the rigid structure. It is worth to mention here that the pressure transferred to the product is less in the case of conventional technique. At the processing pressure of 1 MPa, RPM transfers pressure ~81 % to the component, whereas in case of conventional technique it is ~40 %. The excess pressure compared to the conventional technique, which is shown in RPM, is due to the hydrostatic pressure of rubber mold developed during the compression of rubber. Now the same analysis is carried out again at a pressure of 10 MPa for both the techniques. The pressure transferred to FRP component in RPM is 8.25 MPa, whereas in case of conventional technique it is 4.06 MPa. The pressure transferred in RPM is ~82.5 %, whereas in case of conventional technique it is ~40.6 %. The variation of pressure in conventional technique is ~ 1.88–6.25 MPa, whereas in case of RPM, it varies from 7.07–9.23 MPa.

Effect of Processing Pressure

The above analysis is carried out at a pressure of 1 and 10 MPa for natural rubber vulcanizates containing 45 phr of carbon black loading. The hardness of this rubber compound is 72 shore A. Now the analysis is extended over a wide range of pressure, i.e., 0.5–50 MPa, to see the pressure distribution over the product for both techniques. In all cases of RPM, pressure is uniformly distributed over the surface of product, but the variation is observed in conventional technique. The results are shown in Fig. 3.14. The transmitted pressure remains constant within 81–82 % with increasing applied pressure from 1 to 50 MPa in the case of RPM, whereas in the case of conventional technique the transmitted pressure is ~40–41 %. This proves that the RPM is a better technique to fabricate complex-shaped FRP component. This says that without changing the cross-sectional area, RPM produces more pressure to the FRP component compared to the conventional technique. In addition to this, the pressure is uniformly distributed over the complex-shaped product in the case of RPM. As there is no improvement of transmitted pressure beyond 82 %, there is no need to use high processing pressure.

Table 3.3 Model parameters for various strain energy density functions obtained from uniaxial stress–strain data of NR vulcanizates containing 45 phr carbon black loading, A_{45} [12, 13]

Mooney–Rivlin	C10	C01	C11	C20	C02	C30	C21	C12	C03
2 Parameter	5.78	−5.86	–	–	–	–	–	–	–
3 Parameter	4.80	−4.86	0.12	–	–	–	–	–	–
5 Parameter	8.76	−8.89	2.17	−0.32	−0.50	–	–	–	–
9 Parameter	17.19	−17.44	1730	−826.8	−922.9	−0.05	0.70	203.6	−113.7
Polynomial									
1st order	5.78	−5.86	–	–	–	–	–	–	–
2nd order	8.76	−8.89	2.17	−0.32	−4.50	–	–	–	–
3rd order	17.18	−17.44	1730	−827	−923	−0.05	0.703	203.06	−113.7
Yeoh									
1st order	0.05	–	–	–	–	–	–	–	–
2nd order	0.04	0.17	–	–	–	–	–	–	–
3rd order	0.03	0.38	−0.009	–	–	–	–	–	–
Ogden	**μ1**	**a1**	**μ2**	**a2**	**μ3**	**a3**	**d1**	**d2**	**d3**
1st order	0.05	5.56	–	–	–	–			
2nd order	0.02	5.57	0.03	5.57	–	–			
3rd order	0.02	5.57	0.02	5.57	0.02	5.57	–	–	–
Neo-Hookean	0.047	–	–	–	–	–	–	–	–
Arruda–Boyce	0.07	1.44							
Gent	0.05	0.97							

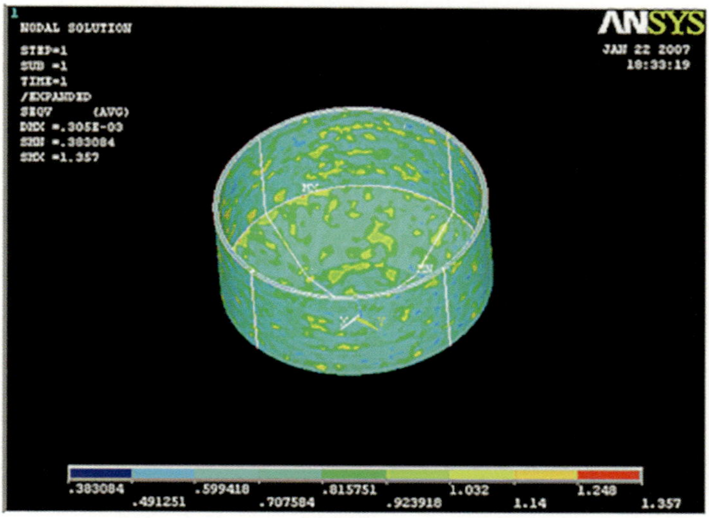

Fig. 3.12 Pressure distribution on FRP component made by RPM where the rubber mold contains 45 phr carbon black [13]

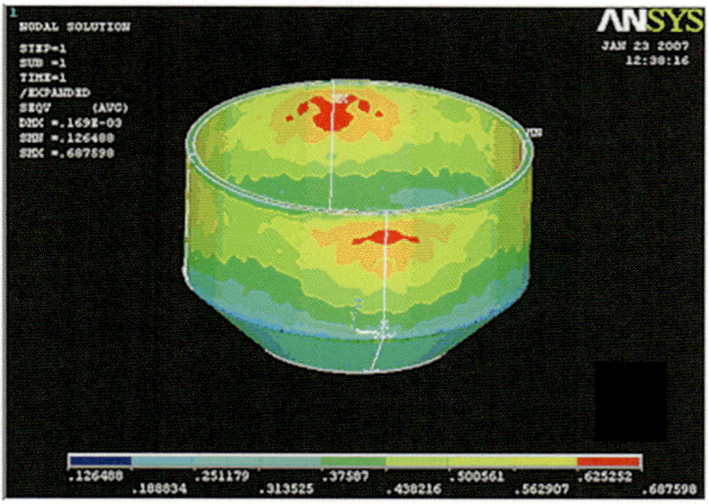

Fig. 3.13 Pressure distribution on FRP component made by conventional technique [13]

Effect of Strain Energy Density Function

The above analysis is carried out using nine-parameter Mooney–Rivlin strain energy density function for natural rubber vulcanizates containing 45 phr of carbon black loading. To see the effect of other forms of strain energy density function on pressure distribution over the FRP surface, 2-, 3-, 5-, and 7-parameter Mooney–Rivlin; 1st-,

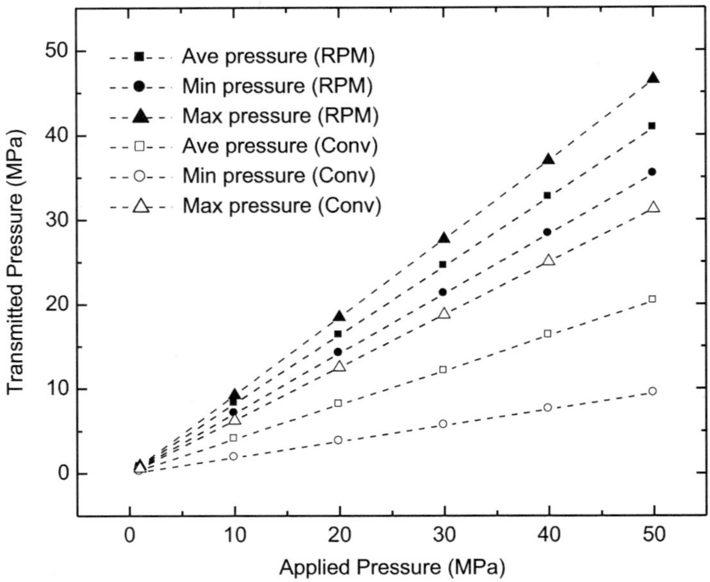

Fig. 3.14 Applied pressure versus transmitted pressure for RPM and conventional technique [13]

2nd-, and 3rd-order Ogden; neo-Hookean; 1st-, 2nd-, and 3rd-order polynomial; Arruda–Boyce; and 1st-, 2nd-, and 3rd-order Yeoh models are used. The pressure is found to be uniformly distributed over the surface. No significant difference is observed.

Effect of Poisson's Ratio

Poisson's ratio used for rubber in the above analysis is 0.499. Now to see the effect of Poisson's ratio on the pressure distribution of FRP component, the analysis for nine-parameter Mooney–Rivlin strain energy density function is extended to 0.45 and 0.4 Poisson's ratio. The results are shown in Fig. 3.15. The pressure is uniformly distributed over the complex-shaped FRP component. As Poisson's ratio is reduced from 0.499 to 0.45 and 0.40, the average pressure on the FRP component changes from 0.52 to 0.50 and 0.42 at the applied pressure of 0.65 MPa. As Poisson's ratio is reduced, the transmitted pressure on the FRP component is also decreased. In addition to this, the pressure is not uniformly distributed on the curved surface of FRP component. As for example, at Poisson's ratio of 0.499, the variation of pressure is 0.45–0.60 MPa, whereas in case of 0.45 and 0.40 Poisson's ratio, the variations of pressure are 0.05–0.73 and 0.01–0.72 MPa, respectively. Now the analysis is further extended to the zero Poisson's ratio. The system is unstable for rubber. But for the understanding purpose, material properties of steel (Young's modulus of 210 GPa) are used. All other conditions are the same. Only Poisson's

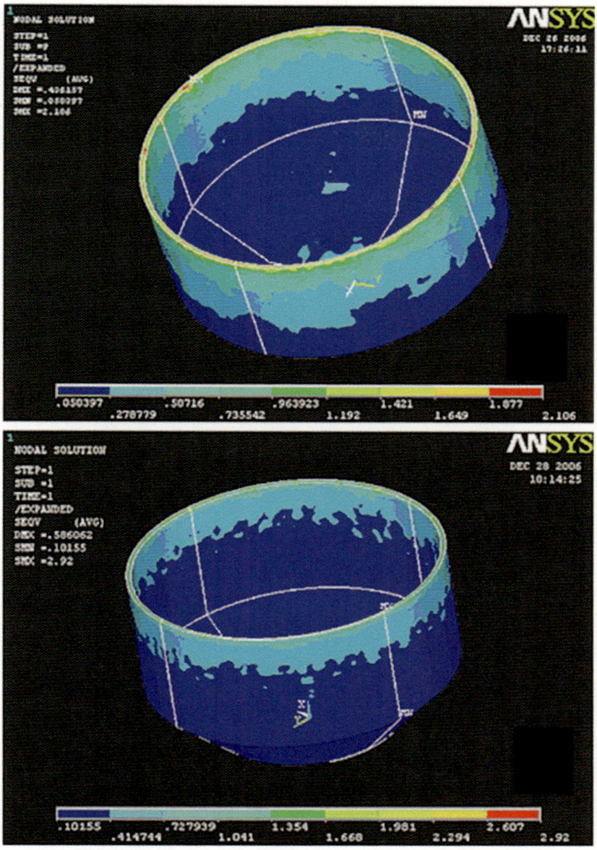

Fig. 3.15 Distribution of pressure on FRP component made by RPM. A Poisson's ratio, 0.45; B Poisson's ratio, 0.40. Applied pressure on the top of rubber mold is 0.65 MPa [13]

ratio of 0 instead of 0.3 is used here. In addition to this, top, middle, and bottom parts of the processing setup are steel, FRP, and steel, respectively. The variations of pressure at the processing pressure of 0.65 MPa are 0.22–0.39 and 0.23–0.45 MPa for Poisson's ratio of 0.3 and 0, respectively. As such, there is no difference of pressure distribution on the FRP component.

Effect of FRP Thickness

The thickness of FRP component used in earlier analysis is 1.5 mm. In order to see the effect of thickness of FRP component on the pressure distribution, the analysis is carried out for 1, 2, 3, and 4 mm thickness, and the results are shown in Figs. 3.16 and 3.17, when the component is made by RPM. The volume percent of fiber in composite is constant. Only variable is the thickness of product. The dimensions of bottom mold, which is made of steel are same as those of earlier analysis. As the thickness of FRP component increases, to compensate this increased thickness, the dimensions of rubber mold are reduced accordingly. But the applied pressure is

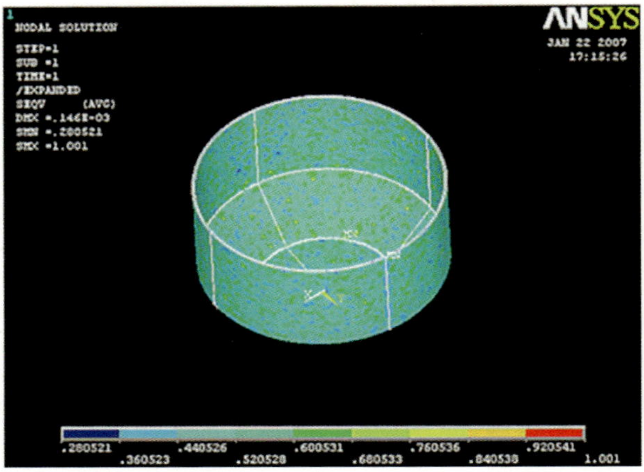

Fig. 3.16 Effect of thickness of FRP components made by conventional technique on pressure distribution. A: 1 mm thickness [13]

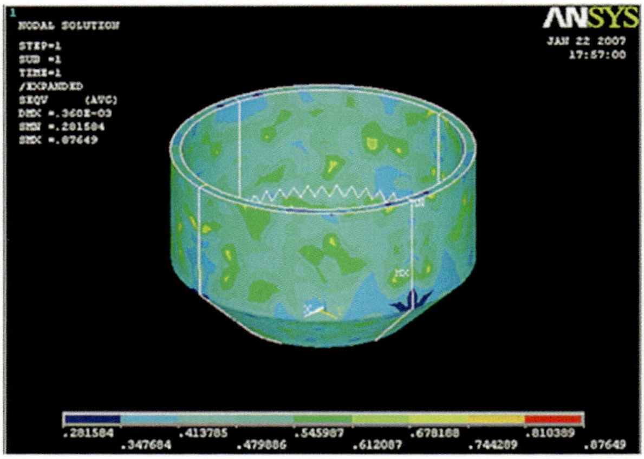

Fig. 3.17 Effect of thickness of FRP components made by conventional technique on pressure distribution. D: 4 mm thickness [13]

0.65 MPa in all cases. All other boundary conditions are similar to that of earlier analysis. The pressure is uniformly distributed on the FRP component. The variations of pressure are 0.44–0.68, 0.48–0.62, 0.47–0.59, and 0.48–0.61 MPa for 1, 2, 3, and 4 mm thickness of FRP component, respectively. The variation is very small. As such there is no remarkable effect of thickness of FRP component on pressure distribution during processing of FRP component through RPM. This indicates that FRP component of any thickness can be successfully made through the newly proposed RPM without losing its quality. The same analysis is also carried out for conventional technique (pressure diagrams are not shown here). The results are similar to that of earlier one, i.e., pressure is not uniform throughout the surface of FRP component. In addition to this, as such there is no effect of thickness on pressure distribution on FRP component made by conventional technique.

Effect of Fiber Volume Percent

Now another analysis is carried out to see the effect of fiber vol% on pressure distribution of FRP component. The thickness of FRP is constant. As FRP composites (45 vol%) are orthotropic materials having elasticity constants of $E_{11} = 16.2$, $E_{22} = 15.3$, and $E_{33} = 7.3$ GPa, Poisson's coefficients are $\nu_{12} = 0.115$, $\nu_{13} = 0.115$, and $\nu_{23} = 0.3$ and shear moduli $G_{12} = 3.9$, $G_{23} = 2.8$, and $G_{13} = 3.9$ GPa [5]. These values are used in earlier studies. Now to address the above question, i.e., the effect of vol% of fiber in FRP components on pressure distribution, there is a need of above nine materials constant as a function of vol% of fiber. Unfortunately, systematic literature data is not available. In order to achieve a better understanding, E_{11} values are varied from 16 to 18 to 20 to 22 GPa. All other parameters, i.e., E_{22}, E_{33}, ν_{12}, ν_{13}, ν_{23}, G_{12}, G_{23}, and G_{13} are the same as the earlier one. It varies from 0.45 to 0.60, 0.44 to 0.63, 0.44 to 0.66, and 0.42 to 0.69 MPa when the E_{11} is varied from 16.2 to 18, 20, and 22 GPa, respectively. In all cases the pressure is uniformly distributed on the FRP component. In addition to this, there is no remarkable effect of vol% of fiber in FRP component on pressure distribution. This suggests that FRP component of any volume percent of fiber can be successfully made through the newly proposed RPM without losing any product quality.

Effect of Gap Between Fiber-Reinforced Plastic and Rubber Mold

For a better understanding, an analysis is carried out in the absence of FRP component to see the pressure distribution. The gap between bottom steel and top rubber mold is equal to that of the FRP component, which is 1.5 mm and used in earlier investigation. Now the gap between bottom steel mold and top rubber mold is varied from 1 to 2 and 3 mm. It is filled by air. There is no displacement of air during compression when the processing pressure of 0.65 MPa is applied on the top of rubber mold. The results are shown in Fig. 3.18 for 2 mm gap. The stress components are summarized in Table 3.4. Up to 2 mm gap, there is no remarkable difference on pressure distribution on the rubber mold. The pressure is uniformly distributed throughout the surface. But at the gap of 3 mm, the peculiar pressure distribution is observed, which is not included here (need more detailed investigation).

Effect of Rubber Hardness

RPM shows that the pressure is uniformly distributed over the complex-shaped product, and ~82 % of the applied pressure is transferred to the FRP component during processing of FRP components. As the RPM uses a rubber mold, now the question arises about the optimum hardness of rubber mold, where the pressure distribution should be uniform and the transmitted pressure should be as maximum as possible. At the same time processing pressure should be minimum. To find out

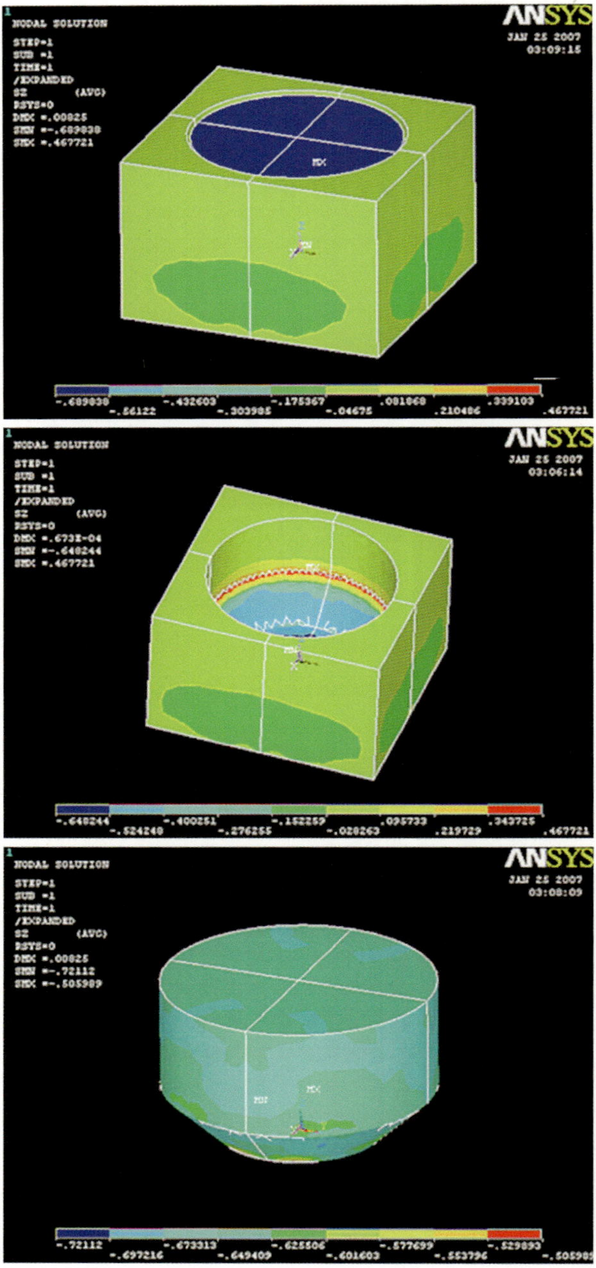

Fig. 3.18 Pressure distribution in the absence of FRP component [13]

the answer, an in-depth analysis is carried out for RPM, where the hardness of rubber mold varies from 50 (0 phr carbon black loading) to 85 shore A (75 phr carbon black loading) and the applied pressure is varied from 0.5 to 50 MPa. The formulation used to get hardness of 50 to 85 shore A is given in Table 3.2. The hardness of the

Table 3.4 Stress components on rubber and steel molds in absence of FRP component. The gap is varied from 1 to 2 mm [13]

	On steel mold (MPa)	On rubber mold (MPa)	
Stress component	Variation	Average	Variation
Gap between steel and rubber mold: 1 mm			
$\mathbf{S_z}$	0.65–0.45	0.65	-
$\mathbf{S_x}$	0.44–0.03	0.65	0.67–0.62
$\mathbf{S_y}$	0.43–0.03	0.64	0.67–0.62
$\mathbf{S_{eqv}}$ (equivalent stress)	0.38–0.53		0.93E-4 -0.1E-3
Gap between steel and rubber mold: 2 mm			
$\mathbf{S_z}$	0.65–0.52	0.65	-
$\mathbf{S_x}$	0.43–0.02	0.65	0.67–0.62
$\mathbf{S_y}$	0.46–0.08	0.64	0.67–0.62
$\mathbf{S_{eqv}}$ (equivalent stress)	0.23–0.57	-	0.82 E-4 -0.11E-3

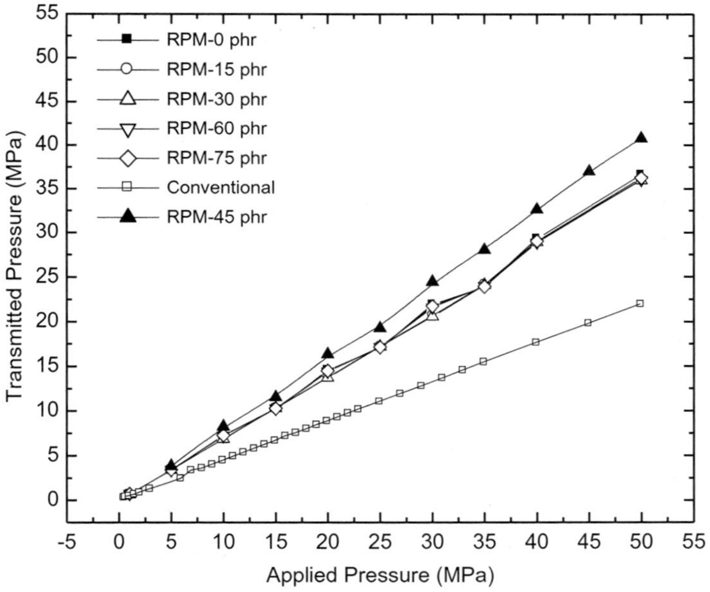

Fig. 3.19 Transmitted pressure versus applied pressure for RPM where the loading of carbon black is varied from 0 to 75 phr and conventional technique [13]

corresponding vulcanizates is also summarized in the same table. It is found that the pressure is uniformly distributed over the surface over the entire range of hardness, i.e., 50 to 85 shore A. Now to find out the optimum hardness of rubber mold, the transmitted pressure is plotted against applied pressure for various hardness of rubber mold and shown in Fig. 3.19. It shows that 72 shore A hardness of the rubber mold (45 phr carbon black loading) is very good for RPM, where the distribution of pressure is excellent and transmitted pressure is ~82 %. No significant difference is observed for other vulcanizates, i.e., A_0, A_{15}, A_{30}, A_{60}, and A_{70}. Why all these rubbers show same transmitted pressure except 72 shore A hardness of the rubber mold (45 phr carbon black loading) is not resolved at this stage.

Performance of RPM to Fabricate FRP Components Based on Glass Fiber and Epoxy Resin

The silicone rubber mold is known for the last few decades and used to make a nonstructural item like toys. Its use in structural application is not known. Keeping this in mind, to develop a new process for fabrication of FRP components, a RPM based on silicone rubber is also developed. In addition to this, natural rubber, polybutadiene, and butyl rubbers are also used to make the rubber punch. This section evaluates the performance of FRP products made by the newly proposed RPM using these four types of rubber. As 45 phr carbon black loading in natural rubber shows the best performance with respect to the uniform pressure distribution over the complex-shaped product and mechanical properties, the rubber punch is made with the same carbon black loading. Again 0.65 MPa processing pressure is used to fabricate the FRP components having three important geometry elements, i.e., (i) cylindrical, (ii) conical, and (iii) flat surface. Figure 3.9 shows such component made by RPM. The mechanical characterization for performance evaluation in structural application is carried out on flat specimen.

Burn test and microstructure studies are performed to check the product qualities like volume percent of fiber/matrix, void content, presence of delamination, interaction between fiber and matrix, etc. It is observed from the burn test that the void content in all three parts of FRP component made by the RPM is within the range of 3.0 %. The differences in volume percent of fiber and void content within the component are also very small. Uniformity is maintained in all parts as the vol% of fiber is in agreement in all parts of the product. The volume percent of fiber and void content in the FRP component made by the RPM are summarized in Table 3.5. To evaluate its performance, these complicated FRP components having three geometry elements are also made by the conventional method at the same applied processing pressure of 0.65 MPa, where both parts of the mold are made of mild steel. A comparison is made between the products made by the RPM and conventional methods. The fiber and void percents of the FRP components are measured and included in Table 3.5. It shows that fiber volume percent in all these three parts varies from 38 to 43 (~5 %, cylindrical to flat) and void percent within 2.7–4.5, whereas in the RPM the deviation from cylindrical to flat portion is within 1.5 %. The higher volume percent of fiber in RPM is due to the hydrostatic pressure developed by the rubber mold during fabrication of FRP component. The small deviation of both fiber volume percent and void content in RPM compared to the conventional process is attributed to the uniform pressure distribution in FRP component during fabrication.

Electron microscopic studies for all portions of FRP product show good interaction between the fiber and matrix in the RPM (Fig. 3.20a) compared to conventional

Table 3.5 Burn test data of FRP components made of glass fiber and epoxy resin, standard deviation 0.1 in all cases [1–4]

Process name	Part name	Volume percent (%)	
		Fiber	Void
RPM using silicone rubber	Cylindrical	45.2	2.6
	Conical	44.8	2.4
	Flat	45.5	2.5
RPM using natural rubber	Cylindrical	45.4	3.0
	Conical	44.6	2.7
	Flat	45.3	2.9
RPM using polybutadiene rubber	Cylindrical	45.7	2.9
	Conical	45.3	2.6
	Flat	47.1	2.7
RPM using butyl rubber	Cylindrical	46.3	2.8
	Conical	44.9	2.5
	Flat	45.3	2.6
Conventional Method	Cylindrical	38.3	3.9
	Conical	42.9	4.5
	Flat	43.1	2.7

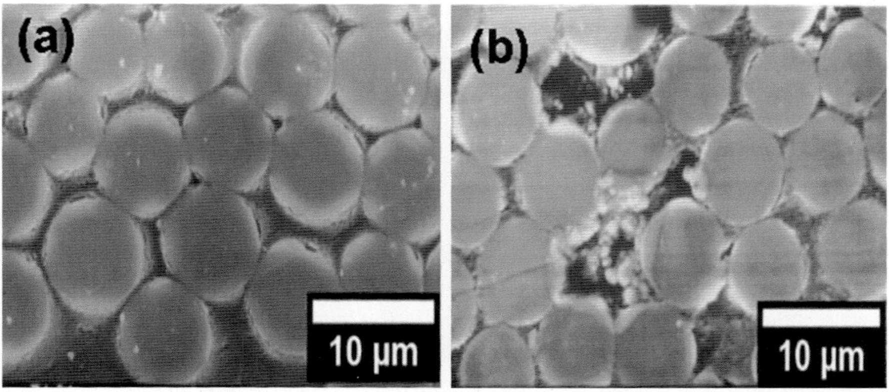

Fig. 3.20 SEM micrographs of FRP component made by (**a**) RPM and (**b**) conventional techniques [1–4]

technique (Fig. 3.20b). It also demonstrates that the void percent of FRP component is more in the case of conventional process. This information is also observed in the burn test (Table 3.5). Thus, the FRP products prepared by rubber punch give better uniformity throughout the surface.

The aerospace, automobile and other industries have used FRP materials for the last four decades. However, the biggest drawback, which has been noticed, is its low resistance to delamination. The delamination in laminates not only leads to complete

Table 3.6 Mechanical properties of FRP components [1–4]

Method of preparation	Interlaminar fracture toughness G_{IC} (J/m^2)	Interlaminar shear strength (MPa)	Tensile strength (MPa)	Modulus of elasticity (GPa)	Elongation (%)
RPM using silicone rubber	214 ± 3	57 ± 2.0	353 ± 10	21.0 ± 0.6	2.2 ± 0.1
RPM using natural rubber	166 ± 16	25 ± 5.0	309 ± 28	20.0 ± 0.2	2.0 ± 0.1
RPM using polybutadiene rubber	178 ± 3	45 ± 5.0	338 ± 6	20.0 ± 0.8	2.1 ± 0.1
RPM using butyl rubber	218 ± 7	60 ± 2.5	353 ± 10	21.0 ± 0.6	2.2 ± 0.1
Conventional method	210 ± 19	57 ± 8.0	312 ± 35	19.0 ± 1.0	2.0 ± 0.1

fracture but also decreases its stiffness, which is a very important design parameter for the designers. In the present scenario, it is a challenge for the researchers to reduce this delaminating behavior of composites in order to increase its life and load-bearing capacity. The parameter "energy release rate, G_{IC}" suits for studying the crack due to the delamination in composites, because the crack plane is well defined and the material remains elastic in the vicinity of the crack tip except in the very thin layer of the interface. In the load–displacement curve, a hysteresis loop is found in each cycle. The first cycle is excluded for the calculation of G_{IC}. The first loading cycle in all experiments is observed to be nonlinear because of some disturbances like mylar films placed (to create crack) in the midplane of the specimen stick to both cantilevers or, during the cutting of specimen to the specified size, some filler materials stick to the precrack surface. When the machine is switched off after the crack propagation, one can observe the drop of load with time, which indicates that the crack still grows after stopping the machine till self-arrest. When the specimen is unloaded to zero load, a small permanent deflection is observed. However, the permanent deflection at zero load is much smaller than the displacement in the loaded condition, and its effect is neglected in this analysis. The results of ILFT are tabulated in Table 3.6.

The average value of ILFT for the specimen made by the RPM using silicone and butyl rubber is greater than that for conventional method, whereas corresponding values of ILFT for natural and polybutadiene rubbers are marginally lower (within 20 %) than those for conventional method. But the variation of results is within the experimental error band. The lower ILFT of FRP components made by the RPM may be due to the higher volume percent of fibers, which decreases the adhesion between fiber and matrix material. It is worth to mention that the standard deviation of ILFT for FRP component made by RPM is very small compared to conventional technique as seen from Table 3.6. This suggests that the newly proposed RPM is a better technique to apply a uniform pressure on the curved surface of FRP component, which is not possible in the conventional technique.

ILSS is also another important material property for the design of laminated composite structures subjected to the transverse loads. The delamination in FRP products can be caused because of the shear stress, as the laminated composites are made of several plies and bonded by the polymeric materials. To find out the suitability of this RPM, the ILSS is carried out on the specimen with 0^0 fiber orientation (warp direction) made by (i) conventional method and (ii) RPM using various rubber punches. The results are included in Table 3.6. From this table it is noticed that the ILSS of the specimen made by the conventional method (fiber volume 54 %) is 57 ± 8.0 MPa and in the RPM using silicone rubber (fiber volume 51 %) and butyl rubber (fiber volume 52 %) is 57 ± 2.0 MPa and 60 ± 2.5 MPa, respectively. It is clear from the above values that specimens prepared by both techniques have same ILSS. In case of RPM using natural rubber (fiber volume 48 %), the value of ILSS is 25 ± 5.0 MPa which is substantially lower (~55 %) than that for the conventional method. Similarly, RPM using polybutadiene rubber gives lower ILSS value by 21 % than that for the specimens prepared by the conventional method. The decrease of ILSS may be due to the decrease of stiffness of FRP components made by RPM. The decrease of stiffness is attributed to the presence of lower volume percent of fibers, i.e., 48 % for natural rubber and 50 % for polybutadiene rubber, compared to the 54 % in conventional process. Though same processing pressure of 0.65 MPa is used here, why is higher volume percent of fibers observed in conventional technique? This is basically due to the lower thickness of FRP specimen used in ILSS, which is 2 mm, whereas the thickness in ILFT and tensile test is 4 and 2.5 mm, respectively. In addition to this the FRP component is made by hand lay-up technique. The volume percent of fibers depends on the quantity of resin used to make this component. The standard deviation of ILSS in RPM using various rubbers is less (2–5 %) compared to the conventional technique (8 %). This again proves that the pressure is uniformly distributed throughout the curved surface of FRP component in case of RPM.

Tension test is also conducted on the FRP laminates to evaluate the mechanical properties like tensile strength, elastic modulus percentage of elongation, etc. These properties are very useful for the design and analysis of structure made by composite materials. The results of mechanical properties using both techniques are also included in Table 3.6. The specimens are found to fail at the central portion, and the fracture line makes 45^0 to the line of loading. It is linear at the low load but nonlinear at higher load due to breaking of fibers. It is observed from the measurement of tensile test that the tensile strength, elastic modulus, and elongation of FRP specimens made by the RPM method using various rubbers are equal to or slightly better (~10–13 %) than those of the specimens prepared by conventional method. The variations of percentages of elongation of the specimens made by the RPM and conventional method are negligible, and average values are within the experimental error bands. This again proves that the pressure is uniformly distributed throughout the curved surface of FRP component and possible only in RPM.

Performance of RPM to Fabricate FRP Components Based on Glass Fiber and Polyester Resin

The previous section concludes that the newly proposed RPM to fabricate complex-shaped FRP components is a new development in manufacturing area of composite materials. The technique uses a matching die set, where the die is made of hard metal like steel and punch from flexible rubber like material. Butyl rubber shows best performance compared to natural rubber and polybutadiene rubber (Section "Performance of RPM to Fabricate FRP Components Based on Glass Fiber and Epoxy Resin"). In ILFT, the FRP specimens of glass fiber and epoxy resin prepared by RPM using butyl rubber have equal ILFT and marginally higher (5 %) ILSS compared to the FRP specimens prepared by the conventional method. In tension test, the specimens made by the RPM using butyl rubber have slightly higher value (~10–13 %) of tensile strength, modulus of elasticity, and strain% compared to the specimens prepared by the conventional method. The standard deviation of ILFT, ILSS and tensile strength for FRP component made by RPM is very small within (6 %), whereas in conventional process, it is 35 %. This suggests that the pressure is uniformly distributed throughout the curved surface.

It is well known that the polyester resin is one of the least expensive resins available to the FRP products when compared with other thermoset resins, i.e., vinylesters and epoxies. Again this polyester resin is best suited for applications insensitive to weight and which do not require high adhesion or fracture toughness. For instance, a simple inexpensive solid fiberglass part is to be fabricated in open tooling in one operation and requires no secondary bonding. If shape accuracy is not critical, resistance to water is of no concern, and ventilation of the workspace is excellent, and then polyester is a great candidate. In spite of these advantages, it historically exhibited poor performance in the areas of adhesion and elongation, rendering the finished part prone to microcracking and secondary bond failures. Keeping this in mind, a study has been conducted to evaluate the performance of RPM to fabricate FRP components from glass fiber and polyester resin. As butyl rubber shows good result compared to natural rubber and polybutadiene rubber, it is used in the newly proposed RPM. The FRP components are also made by conventional technique to evaluate its performance in the general applications. Six FRP components are made from each technique (RPM and conventional processes).

Burn test and microstructure studies are performed to check the qualities of products like volume percent of fiber/matrix, void content, presence of delamination, interaction between fiber and matrix, etc. It is observed from the burn test that the void content in all parts of FRP component made by the RPM is within the range of 3.0 %. The differences (i.e., standard deviation) of volume percent of fiber/matrix and void content within the component are also very small. Uniformity is maintained in all parts as the vol% of fiber is in agreement with the product.

The vol% of fiber and void content in the FRP component made by the RPM are summarized in Table 3.7. To evaluate its performance, this FRP component having two geometry elements is also made by the conventional method, where both parts of the mold are made of mild steel. A comparison is made between the products made

Table 3.7 Burn test data of FRP components made of glass fiber and polyester resin, standard deviation 0.1 in all cases [14]

Process name	Part name	Volume percent (%)	
		Fiber	Void
RPM using butyl rubber	Conical	46.3	2.8
	Flat	45.3	2.6
Conventional	Conical	38.3	3.9
	Flat	43.1	2.7

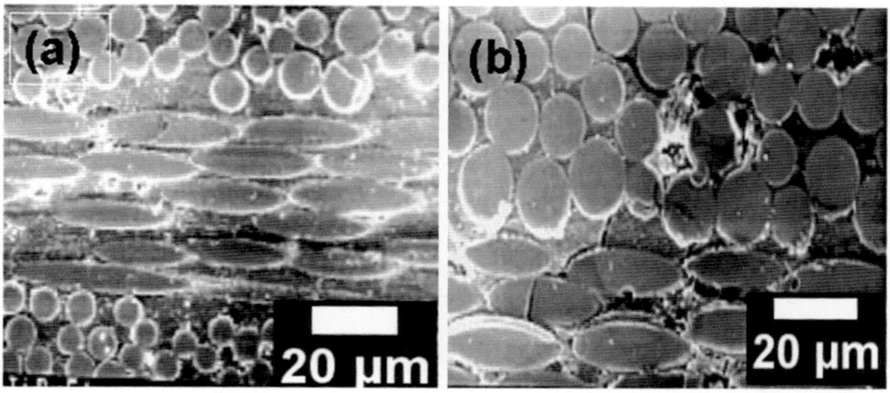

Fig. 3.21 SEM micrographs of FRP component made by (**a**) RPM and (**b**) conventional method [1–4]

by the RPM and conventional methods. The fiber, matrix, and void percents of the FRP components are measured and included in Table 3.7.

It shows that fiber volume percent in all these two parts varies from 38 to 43 (~5 % deviation, conical to flat) and void percent within 2.7–3.9, whereas in the RPM the deviation from conical to flat portion is within 0.7 %. The higher volume percent of fiber in the FRP components made by RPM is due to the hydrostatic pressure developed by the rubber mold during fabrication of products. The small deviation of both fiber volume percent and void content in the RPM compared to the conventional process is attributed to the uniform pressure distribution on the FRP component during fabrication. Electron microscopic studies for all portions of FRP product show good interaction between the fiber and matrix in the RPM (Fig. 3.21a) compared to conventional method (Fig. 3.21b). It also demonstrates that the void percent of FRP component is more in the case of conventional process. This information is also observed in the burn test (Table 3.7) discussed in the earlier section. Thus, the FRP products prepared by the RPM using butyl rubber punch give better uniformity throughout the surface.

The average value of ILFT for the specimen made by the RPM using butyl rubber is 477 ± 21 J/m^2. The fiber volume percent of the specimen made by the RPM using butyl rubber punch is 50 %, whereas the average value of ILFT for the specimen made by the conventional method with volume percent of fiber 45 % is 436 ± 48 J/m^2. Specimens prepared by the RPM using butyl rubber have slightly higher (10 %) ILFT

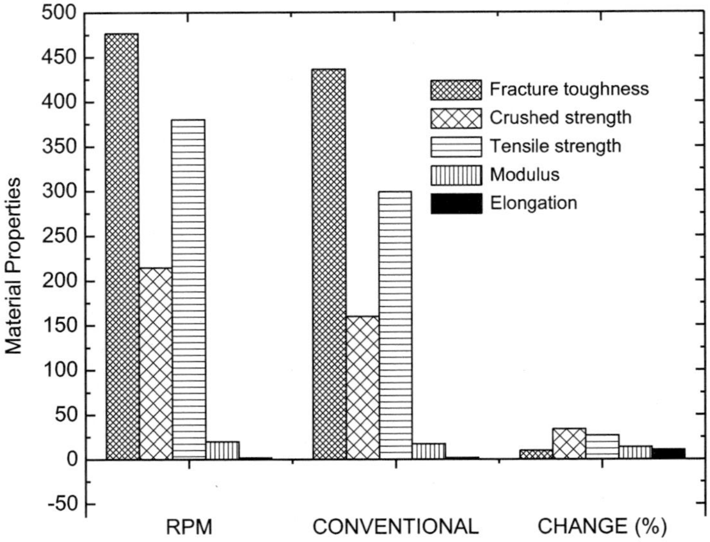

Fig. 3.22 Interlaminar fracture toughness, crushed strength, tensile strength, modulus, and elongation of FRP component made by RPM and conventional process [14]

compared to that for specimens prepared by the conventional method as shown in Fig. 3.22. It is worth to mention that the standard deviation of ILFT for FRP component made by RPM is small within ±21 %, whereas for conventional technique it is ±48 %. This suggests that RPM is a best technique to apply uniform pressure on the curved surface of FRP component, which is not possible in the conventional technique.

Composite laminates are made of several plies, which are bonded together by the matrix. Since there is no reinforcement between the two plies, delamination failure is likely to occur under shear stress. Therefore, there is considerable interest to determine ILSS through shear mode. To find out the suitability of the RPM, the ILSS is carried out on the specimen with 0^0 fiber orientation (warp direction) made by (i) conventional and (ii) RPM using butyl rubber punch. The failure mode of composite is shown in Fig. 3.23a. It is a brittle failure rather than delamination (Fig. 3.23b).

Tension test is conducted on the composite laminates to evaluate the material properties like elastic constants, strength, and Poisson's ratio that can be used for design and analysis of structures.

The specimens are found to fail at the central portion and the fracture line makes 45^0 to the line of loading. It is linear at the low load but nonlinear at the higher load due to the breaking of fibers. The tensile strength of the FRP specimens with 45 % fiber volume prepared by the RPM using butyl rubber is 380 ± 35 MPa, whereas the tensile strength of the FRP specimens with 40 % fiber volume made by the conventional method is 299 ± 52 MPa. But the tensile elastic modulus of the specimen made by the conventional method is 17.5 ± 2.3 GPa, whereas the elastic modulus of

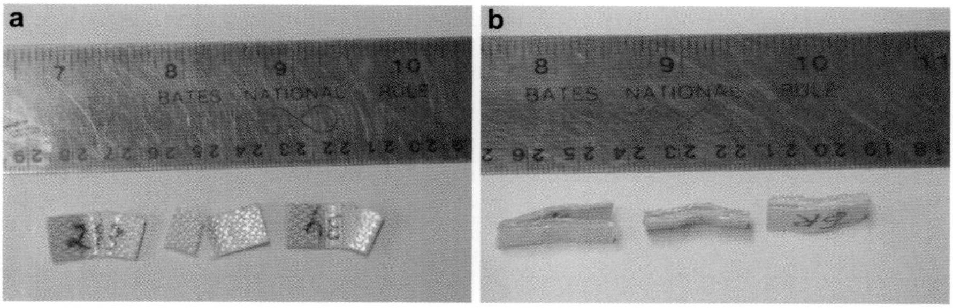

Fig. 3.23 (**a**) Brittle and (**b**) delamination failures of FRP composites [14]

the specimen prepared by the RPM using butyl rubber is 20 ± 1.0 GPa. Similarly the percentage elongations of the specimens made by the conventional method and RPM are 1.8 ± 0.1 % and 2.0 ± 0.5 %, respectively. The average Poisson's ratio of the specimens made by conventional method is 0.19 ± 0.009 and that of specimens prepared by RPM is 0.19 ± 0.005. The tensile strength, tensile elastic modulus and percentage of elongation/strain of FRP components made by both techniques are also shown in the same Fig. 3.22. It is observed from the measurement of tensile test that the tensile strength, elastic modulus and elongation of FRP specimens made by the RPM method using butyl rubber are better (~27 %, 14 %, and 11 %, respectively) than those of the specimens prepared by the conventional method. This again proves that the pressure is uniformly distributed throughout the curved surface of FRP component and possible only in RPM.

Effect of Rubber Hardness on the Properties of FRP Composites Made by RPM

The FRP specimens of glass fiber and epoxy resin prepared by RPM using silicone rubber with 45 phr loading of carbon black (N 330) have equal ILFT, same or marginally higher (5 %) ILSS, and slightly higher value (~10–13 %) of tensile strength, modulus of elasticity, and strain% compared to the FRP specimen prepared by conventional method (Section "Performance of RPM to Fabricate FRP Components Based on Glass Fiber and Epoxy Resin"). When silicone rubber is replaced by natural rubber and polybutadiene rubber, ILFT and ILSS of FRP laminates decrease compared to the FRP specimen prepared by conventional method without any change of tensile strength, whereas the FRP specimens prepared by RPM using butyl rubber have equal ILFT and ILSS compared to the specimens prepared by conventional method. In tension test, the FRP specimens have slightly higher value (~10–13 %) of tensile strength, modulus of elasticity, and strain%. In another experiment, epoxy resin is replaced by polyester resin to see the performance of FRP laminates made by RPM. The results are better than conventional process with respect to the ILFT, Young's modulus, tensile strength, etc.

Table 3.8 Mechanical properties of FRP laminate made by conventional method and by RPM using natural rubber with variation in carbon black content [19]

Manufacturing method	Carbon black loading in natural rubber (phr)	Hardness (shore A)	Average tensile strength (MPa)	Average tensile modulus of elasticity (GPa)	Average fiber volume percent V_f (%)	Average void content (%)
Conventional method	NA	–	300 ± 34.0	14.2 ± 0.5	49	2.7
RPM	0	50	294 ± 18.0	15.3 ± 1.0	52	1.7
RPM	15	61	336 ± 14.0	16.5 ± 0.3	54	1.8
RPM	30	69	352 ± 9.0	16.0 ± 1.0	55	1.6
RPM	45	72	348 ± 10.0	16.5 ± 0.8	55	1.7
RPM	60	79	350 ± 15.0	16.8 ± 1.1	56	1.8
RPM	75	85	353 ± 14.0	18.1 ± 0.8	57	1.6

Since the hardness of rubber changes with the percentage of filler used in the rubber, it is important to know the effect of rubber hardness on the properties of the FRP composites made by RPM. This will be helpful to understand the optimum range of rubber composition to produce good quality FRP products by RPM. In the present study, the amount of filler material (i.e., carbon black, N 330) in the raw natural rubber is varied from 0 to 75 phr (Table 3.2) with an increment of 15 phr. These natural rubber molds of varying hardness are used to produce FRP composites. Though the ILFT and ILSS of FRP components made by the RPM using natural rubber are not good compared to the conventional compression molding process, an attempt is made here to improve the mechanical properties of FRP components by changing the hardness of rubber mold. In addition to this, epoxy resin is replaced by polyester resin, as it is mostly used in general-purpose applications. Now the question arises for performance of composites made by RPM using various loading of carbon black through mechanical properties. To find out an answer, this investigation is carried out for natural rubber over a range of filler loading, i.e., 0 to 75 phr. Characterizations of composite laminates are carried out by burn test, tensile test, ILSS/crushed strength, and ILFT tests. The results are compared with the corresponding values for the composites made by conventional method to evaluate the performance of RPM.

It is observed from the burn test that the void content in FRP component made by the RPM is within the range of 2.0 %. The differences (i.e., standard deviation) of volume percent of fiber/matrix and void content within the component are also very small. Uniformity is maintained in the composites. The volume percent of fiber and void content in FRP component made by the RPM are summarized in Tables 3.8, 3.9, and 3.10. It shows that fiber volume percent varies from 52 % to 55 % (~2 % deviation) and void percent within 1.6–1.7 %, when the rubber mold is loaded with 0 phr carbon black. To evaluate its performance, this FRP component is also made by the conventional method, where both parts of the mold are made of mild steel. A comparison is made between the products made by RPM and conventional methods and shown in Tables 3.8, 3.9, and 3.10. It shows that fiber volume percent varies from 49 % to 51 % (~3 % deviation) and void percent within 2.7–2.8 % in conventional technique. The higher volume percent of fiber in FRP components made by the RPM is due to the hydrostatic pressure developed by the rubber mold during fabrication of products. The small deviation of both fiber volume percent and void content in the RPM compared to the conventional process is attributed to the uniform pressure distribution on the FRP component. To understand the effect of RPM on the volume percent of fibers and voids in FRP components and to get its best performance, composites are fabricated using different loading of carbon black in the rubber mold. The loading of carbon black varies from 0 to 75 phr with an increment of 15 phr. Volume percent of fibers and void content are measured for FRP components made by the RPM using rubber mold of various compositions, i.e., A_0, A_{15}, A_{30}, A_{45}, A_{60}, and A_{75} (given in Table 3.2). The values are included in Tables 3.8, 3.9, and 3.10. It shows that the fiber volume percent increases (from 52 % to 59 %) with an increase in hardness of rubber mold. When the loading of carbon black increases in rubber mold, the hardness increases from 50 shore A for A_0 vulcanizate to 85 shore A for A_{75} vulcanizate (Table 3.8). The change in hardness with carbon black loading in rubber vulcanizates can be explained as follows: Medalia [15] and Kraus [16] suggested that the occluded rubber, bound rubber, and immobilized rubber shell around the carbon black surface overlap each other and form a complicated interlinked three-dimensional network structure. This adsorbed hard immobilized rubber percent increases with carbon black loading. Since adsorbed rubber perturbs the relaxation process of a matrix, an increase in hardness value (which increases with carbon black loading) is observed [17, 18]. The increased hardness may also be attributed to the superposition of different relaxation processes, hydrodynamic effect of the carbon black particles imbedded in the elastomer continuum, stronger rubber–carbon black interaction and tightening of the network associated with a shell of hindered rubber at the carbon black surface which increases hardness of the elastomer [17, 18]. Though hardness increases from 50 to 85 shore A, it maintains its elasticity to generate hydrostatic pressure during compression. That is not possible in the case of conventional steel mold.

The tension test is performed on FRP specimens prepared by the RPM using natural rubber containing carbon black as a filler material in the range of 0–75 phr with an increment of 15 phr (formulation no: A_0, A_{15}, A_{30}, A_{45}, A_{60}, A_{75}, given in Table 3.2). The results of tension test are shown in Table 3.8. The average value of tensile strength of the FRP specimens prepared by the RPM using natural rubber is 294, 336, 352, 348, 350, and 353 MPa for 0, 15, 30, 45, 60, and 75 phr carbon black loading (formulation no A_0, A_{15}, A_{30}, A_{45}, A_{60}, A_{75}, given in Table 3.2), respectively.

The variation of tensile strength with hardness of natural rubber mold is shown in Fig. 3.24. Tensile strength increases with loading of carbon black in the mold, i.e., hardness of rubber mold in the initial stage and then reaches an almost constant value of 350 MPa. It is noticed from this data that the increment in the average value of

Table 3.9 Interlaminar fracture toughness of composite materials made by conventional ($V_f = 0.49$) and RPM using natural rubber. Material constants (A_1 and A_2) are used to calculate interlaminar fracture toughness [19]

Manufacturing method	Carbon black loading (phr)	Average interlaminar fracture toughness (J/m^2)	Material constant, A_1 (N^{-1}) X 10^6	Material constant, A_2 (N)	Average fiber volume percent V_f (%)	Average Void content (%)
Conventional method	NA	1089 ± 100.0	3.26 ± 1.99	2586 ± 310	49	2.8
RPM	0	1098 ± 45	3.73 ± 0.81	2464 ± 199	54	1.6
RPM	15	1067 ± 40	4.51 ± 0.61	2294 ± 158	55	1.5
RPM	30	1276 ± 45	4.66 ± 0.74	2356 ± 182	56	1.8
RPM	45	1296 ± 35	3.57 ± 0.84	2687 ± 174	57	1.5
RPM	60	1265 ± 52	3.90 ± 1.04	2607 ± 181	57	1.7
RPM	75	1034 ± 40	4.44 ± 0.81	2185 ± 161	58	1.7

Table 3.10 Crushed strength of composite materials made by conventional ($V_f = 0.51$) and RPM using natural rubber [19]

Manufacturing method	Carbon black loading in natural rubber (phr)	Average crushed strength (MPa)	Average fiber volume percent V_f (%)	Average void content (%)
Conventional method	NA	256 ± 20.0	51	2.7
RPM	0	149 ± 5.3	55	1.0
RPM	15	150 ± 5.7	56	1.2
RPM	30	230 ± 5.5	58	1.4
RPM	45	242 ± 5.4	58	1.5
RPM	60	235 ± 5.9	59	1.3
RPM	75	182 ± 5.5	59	1.6

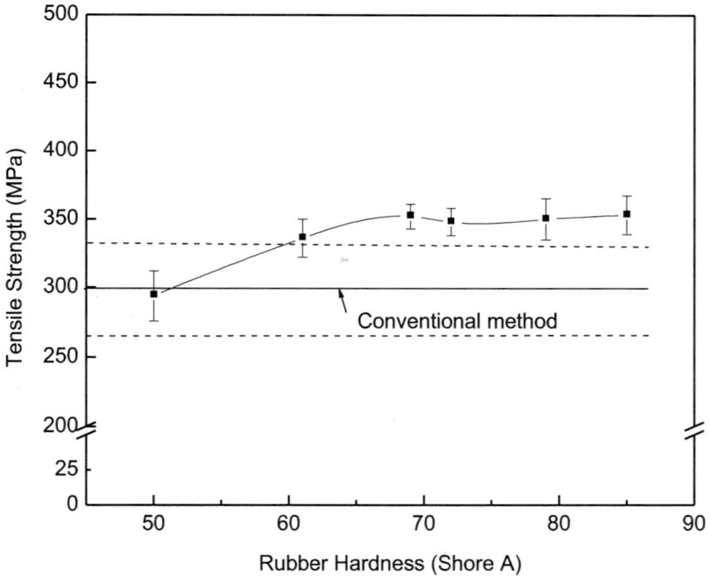

Fig. 3.24 Variation of tensile strength of FRP composites with hardness of rubber mold used in RPM [19]

tensile strength due to varying carbon black content in the rubber mold is within a range of 53 MPa and the optimum loading of carbon black in the rubber mold to get FRP composites with improved tensile strength should be greater than 30 phr. To evaluate the performance of RPM, the tensile strength of FRP component made by the conventional process is also measured. The tensile strength of FRP component is 300 MPa. The higher tensile strength of FRP component in RPM is attributed to the higher volume percent of fiber (Table 3.8) and less voids in the component.

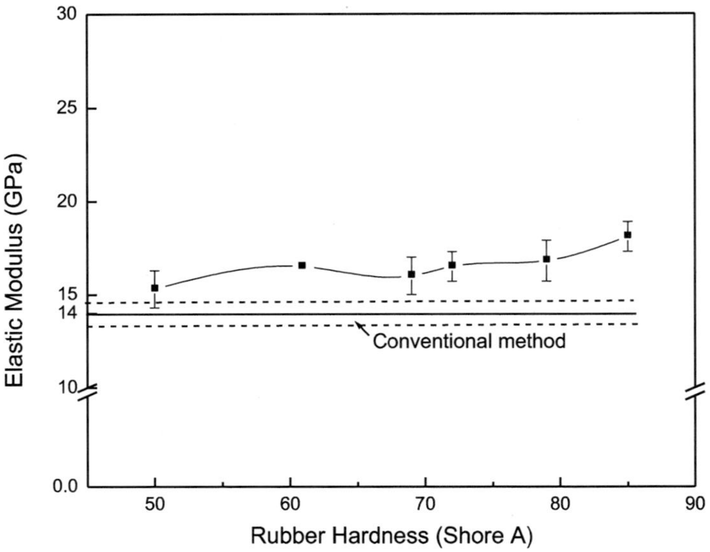

Fig. 3.25 Variation of tensile elastic modulus of FRP composites with hardness of rubber mold used in RPM [19]

Similarly, the average value of tensile elastic modulus of FRP specimens made by conventional method is 14.2 GPa, whereas the average value of tensile elastic modulus of FRP specimens prepared by RPM using natural rubber with different carbon black content is 15.3, 16.5, 16.0, 16.5, 16.8, and 18.1 GPa for 0, 15, 30, 45, 60, and 75 phr carbon black loading, respectively. It is noticed from Table 3.8 and Fig. 3.25 that the elastic tensile modulus of specimens prepared by RPM using natural rubber also gives marginally higher value than that of the specimens prepared by conventional method. Similarly the higher tensile modulus of FRP component in RPM is due to the higher volume percent of fiber (Table 3.8) and less voids in the component.

ILFT is an important design parameter, which should be considered when characterizing a composite laminate. The laminated composites are prone to delamination when subjected to impact loading. The delamination occurs due to low ILFT. Also, the defects like microcracks, inclusions, voids, etc. tend to grow in an interlaminar mode [20]. The average value of G_{IC} (ILFT) for the specimen made by conventional method with volume percent of fiber 49 % is 1090 J/m^2. The average value of G_{IC} (ILFT) for the specimen made by RPM using natural rubber with different carbon black content is 1132, 1187, 1293, 1289, 1325, and 1060 J/m^2 for carbon black content of 0, 15, 30, 45, 60, and 75 phr, respectively (Table 3.9).

A graph showing the variation of ILFT of composite laminate made by RPM with an increase of hardness of rubber mold is shown in Fig. 3.26. It increases with increasing carbon black loading of 30 phr (69 shore A hardness) and then attains an almost constant value of 1293 to 1325 J/m^2 in the loading of 30–60 phr (69–79 shore A hardness). After that it again decreases to 1060 J/m^2 for the vulcanizate of 75 phr

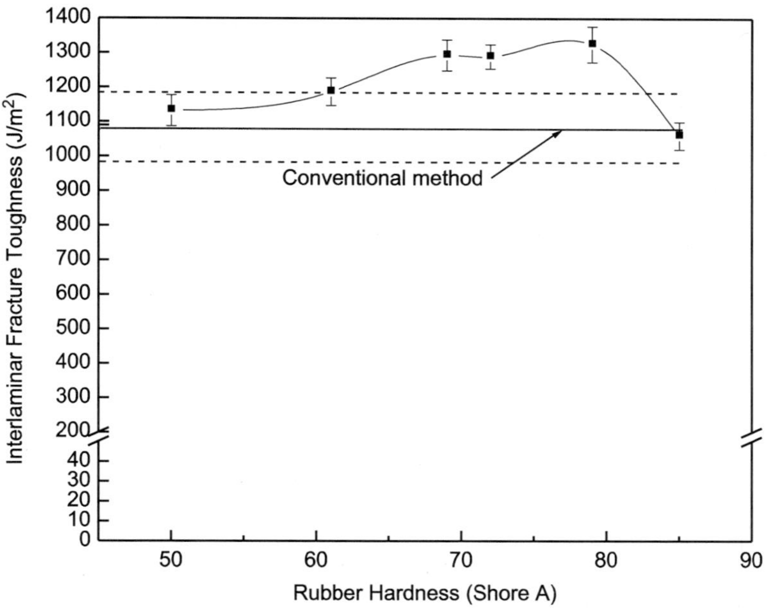

Fig. 3.26 Variation of interlaminar fracture toughness of FRP composites with hardness of rubber mold used in RPM [19]

loading (85 shore A). ILFT also confirms that the optimum level of carbon black loading in the rubber mold should be within the range of 30–60 phr (69–79, i.e., 70–80 shore A hardness), and the FRP component made by RPM is better than that made by the conventional process.

ILSS is an important material property for the design of laminated composite structures subjected to transverse loading, which are likely to fail by delamination under shear stress conditions [21]. The failure mode of composite is shown in Fig. 3.23a. It is a brittle failure rather than delamination (Fig. 3.23b). Similarly the failure strength is denoted by crushed strength and the value is given in Table 3.10. Again the author would like to mention that this does not represent the crushing strength of the material in its true sense, but is used here to get an idea of the material strength in the particular mode of failure (i.e., crushing) in the short beam test specimen. The variation of crushed strength of composites made by RPM with hardness of rubber mold used in RPM is also shown in Fig. 3.27. A similar trend as observed in ILFT is also observed here, i.e., it increases with increasing hardness of rubber mold, i.e., carbon black loading in the rubber mold and attains a plateau value in the range of 30–60 phr (69–79 shore A hardness) and after that decreases with loading of filler or hardness of rubber mold. Crushed strength also suggests the optimum loading of carbon black for the rubber mold in the range of 30–60 phr (or hardness of rubber mold in between 70 and 80 shore A). FRP components made by RPM using 30–60 phr loading of carbon black show good tensile strength, tensile elastic modulus, ILFT, and volume percent of fiber and with less void and standard deviation. However, it shows lower

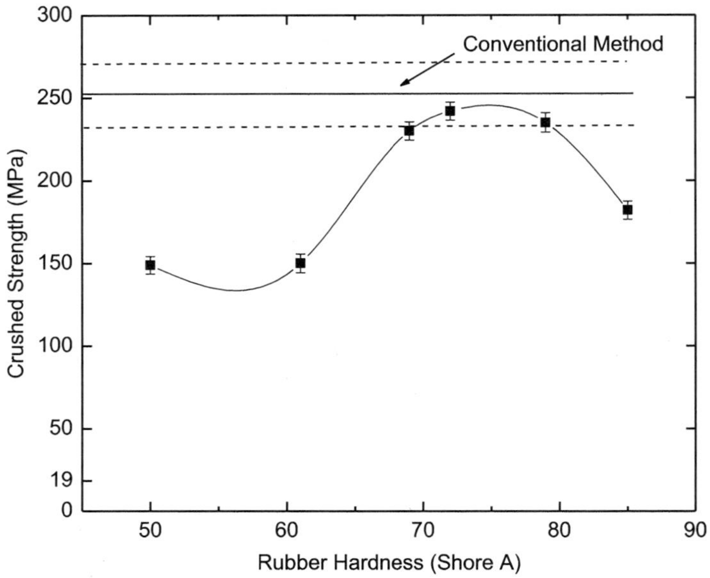

Fig. 3.27 Variation of crushed strength of FRP composites with hardness of rubber mold used in RPM [19]

crushed strength. This may be due to the higher volume percent of fiber, which is responsible for higher shear stress concentration in the FRP laminates.

Successive Laminate Preparation from the Same Rubber Mold

In RPM, the same set of rubber mold and hard mold is to be used to prepare many components. It is of interest to study the variation of surface roughness if the same rubber mold is used successively. The roughness of the rubber mold surface and FRP laminates is determined after each laminate preparation. The variation of Ra values of rubber molds after making successive laminates is plotted in Fig. 3.28a for 0, 15, 30, 45, and 60 phr carbon black loading rubber molds, respectively (formulation no.: A_0, A_{15}, A_{30}, A_{45}, A_{60}). Figure 3.28b shows the variation of Ra values of rubber molds after making successive laminates for 75, 90, 105, and 120 phr carbon black loading (formulation no.: A_{75}, A_{90}, A_{105}, and A_{120}).

Hundred specimens are made in each case to see the effect of multiple laminate preparations from the same rubber mold. The virgin rubber surface is having a small value of Ra and varies from 0.45 to 0.64 μm, but after 3–4 laminate preparations, the values are stabilized. However, the roughness values are so small and the variation of roughness is confined mostly in the range 0.45 and 1.17 μm.

It is thus concluded that surface roughness of rubber mold does not deteriorate when multiple laminates are prepared from a single rubber mold. This is true for all combinations of carbon black and natural rubber, which are used in this

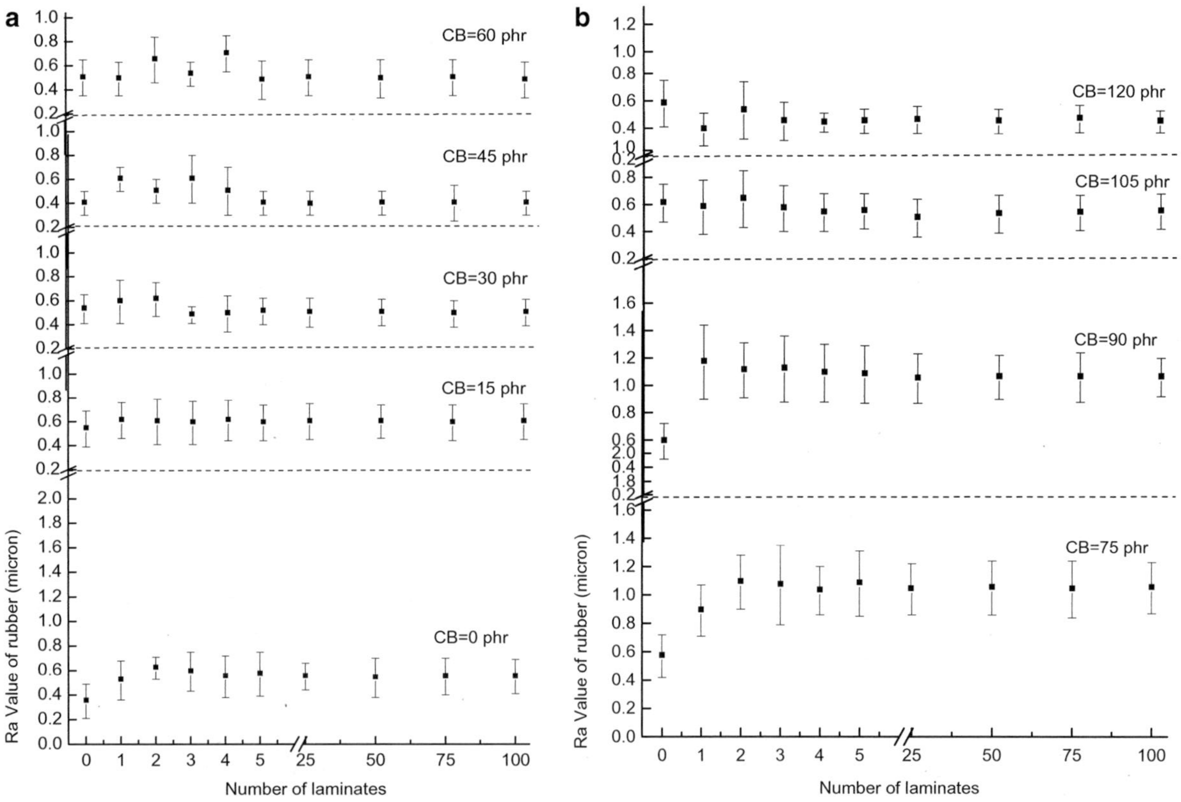

Fig. 3.28 (**a**) Variation of Ra value of rubber mold after making successive laminates for 0, 15, 30, 45, and 60 phr loading of carbon black in the rubber mold [22]. (**b**) Variation of Ra value of rubber mold after making successive laminates for 75, 90, 105, and 120 phr loading of carbon black in the rubber mold [22]

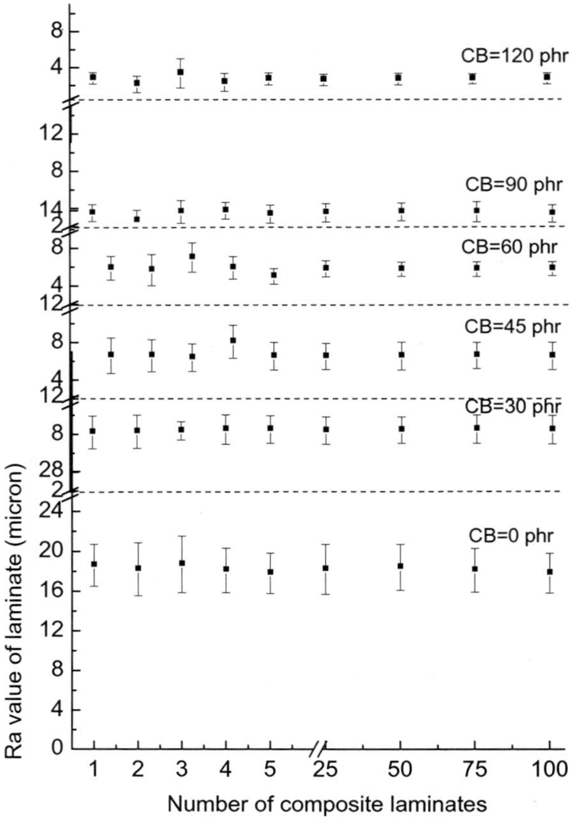

Fig. 3.29 Variation of Ra value of FRP laminates prepared with same rubber mold [22]

experimentation. Even after making of 100 FRP laminates, the change of surface roughness of rubber mold is less than 25 % with respect to the virgin rubber mold.

Figure 3.29 shows the variation of surface roughness of FRP laminate from the same rubber mold. The study is carried out for various loading of carbon black. Representative plots for formulation no.: A_0, A_{30}, A_{45}, A_{60}, A_{90}, and A_{120} are shown in Fig. 3.29. It is clear that there is not much variation in the surface roughness of laminates when these are prepared from the same rubber mold. Values are stabilized after 3rd to 4th laminate. The surface roughness of both rubber mold and the laminates did not deteriorate when multiple laminates are prepared from a single rubber mold. The results are encouraging for repeated use of the rubber mold in the RPM.

Concluding Remarks

The newly developed FRP manufacturing method known as rubber pressure molding (RPM) is evaluated based on physical and mechanical characterization of FRP components made by it. Theoretical analysis using ANSYS software indicates that there is uniform pressure distribution over the surface of FRP component as compared to conventional compression molding technique. The burn test and SEM results show that there is uniform fiber distribution and less voids are present in case of FRP products made by RPM. Mechanical properties like ILFT, ILSS, Young's modulus, tensile strength, and percentage elongation at break also indicate the improvement due to better pressure transmission and uniform distribution over the area of FRP component. Four types of rubber (silicone, natural, polybutadiene, and butyl) were tried for making rubber punch, and two resin systems (epoxy and polyester) were used for manufacturing composites. Hardness of rubber punch also plays an important role on the quality of FRP, and hence experimentation was done to find optimum value for rubber hardness in case of natural rubber. Rubber hardness was varied by varying the filler content, and it was found that FRP components made by RPM using 30–60 phr loading of carbon black show good tensile strength, tensile elastic modulus, interlaminar fracture toughness, volume percent of fiber, and less void content. It was also found that surface roughness of rubber mold does not deteriorate when multiple laminates are prepared from a single rubber mold. This proves that newly developed RPM has great potential for low cost manufacturing of good quality FRP products.

Acknowledgment The authors acknowledge the financial support provided by the Indian Space Research Organization, India, for carrying out this work.

References

1. Kar KK, Sharma SD, Behera SK et al. (2006) Development of a rubber pressure molding technique for fiber reinforced plastics. Kautsch Gummi Kunstst 59(4):169–173
2. Kar KK, Sharma SD, Behera SK et al. (2006) Development of rubber pressure molding technique using butyl rubber to fabricate fiber reinforced plastic components based on glass fiber and epoxy resin. J Appl Polym Sci 101(2):1095–1102
3. Kar KK, Sharma SD, Behera SK et al (2007) Development of rubber pressure molding technique using silicone rubber to fabricate fiber reinforced plastic components based on glass fiber and epoxy resin. J Elastomers Plast 39(2):117–131
4. Kar KK, Sharma SD, Behera SK et al (2006) Development of rubber pressure moulding technique using polybutadiene rubber to fabricate fibre reinforced plastic components based on glass fibre and epoxy resin. Curr Sci 90(11):1492–1499
5. Shukla M (2006) Experimental and numerical investigation of induced distortions and stresses of angles composites during autoclave manufacturing. Dissertation, Indian Institute of Technology Kanpur
6. Yeoh OH (1993) Some forms of the strain energy function for rubber. Rubber Chem Technol 66 (5):754–771
7. Bradley GL, Chang PC, Mckenna GB (2001) Rubber modeling using uniaxial test data. J Appl Polym Sci 81(4):837–848
8. Horgan CO, Schwartz JG (2005) Constitutive modeling and the trousers test for fracture of rubber-like materials. J Mech Phys Solids 53(3):545–564
9. Boyce MC, Arruda EM (2000) Constitutive models of rubber elasticity: a review. Rubber Chem Technol 73(3):504–523
10. Gent AN (1996) A new constitutive relation for rubber. Rubber Chem Technol 69(1):59–61
11. Yeoh OH (1990) Characterization of elastic properties of carbon-black-filled rubber vulcanizates. Rubber Chem Technol 63(5):792–805

12. Kar KK, Sharma SD, Kumar P et al (2007) Analysis of rubber pressure molding technique to fabricate fiber reinforced plastic components. Polym Compos 28(5):637–649
13. Kar KK, Sharma SD, Kumar P et al (2007) Pressure distribution analysis of fiber reinforced plastic components made by rubber pressure molding technique. J Appl Polym Sci 105 (6):3333–3354
14. Kar KK, Sharma SD, Sah TK et al (2007) Development of rubber pressure molding technique using butyl rubber to fabricate fiber reinforced plastic components based on glass fiber and polyester resin. J Reinf Plast Compos 26(3):269–283
15. Medalia AI (1974) Filler aggregates and their effect on reinforcement. Rubber Chem Technol 47 (2):411–433
16. Kraus G (1970) Structure-concentration equivalence principle in carbon black reinforcement of elastomers. J Polym Sci 8(9):601–606
17. Kar KK, Bhowmick AK (1998) Analysis of high strain hysteresis loss of nonlinear rubbery polymer. Polym Eng Sci 38(1):38–48
18. Kar KK, Bhowmick AK (1998) Effect of holding time on high strain hysteresis loss of carbon black filled rubber vulcanizates. Polym Eng Sci 38(12):1927–1945
19. Kar KK, Sharma SD, Kumar P (2007) Effect of rubber hardness on the properties of fiber reinforced plastic composites made by the newly proposed rubber pressure molding technique. Polym Compos 28(5):618–630
20. Moore DR, Pavan A, Williams JG (eds) (2001) Fracture mechanics testing methods for polymers, adhesives and composites. ESIS Publication, New York
21. Pagano NJ, Pipes RB (1973) Some observations on the interlaminar strength of composite laminates. Int J Mech Sci 15(8):679–688
22. Sharma SD, Kar KK, Kumar P (2006) Surface roughness of fiber reinforced plastic laminates fabricated using rubber pressure molding technique. Polym Compos 27(5):504–512

Functionally Graded Composites: Processing and Applications

4

Sandeep S. Ahankari and Kamal K. Kar

Contents

Abstract

The new generation demands advanced materials that fulfills today's commercial applications. Functionally graded materials (FGMs) are the modern, emerging materials that meet the current needs of our society. This chapter presents the classification of FGMs based on various domains. It also highlights the present

S.S. Ahankari (✉)
Department of Mechanical Engineering, Annasaheb Dange College of Engineering and Technology, Sangli, Maharashtra, India
e-mail: asandeeps@gmail.com

K.K. Kar
Advanced Nanoengineering Materials Laboratory, Materials Science Programme, Indian Institute of Technology Kanpur, Kanpur, Uttar Pradesh, India

Advanced Nanoengineering Materials Laboratory, Department of Mechanical Engineering, Indian Institute of Technology Kanpur, Kanpur, Uttar Pradesh, India
e-mail: kamalkk@iitk.ac.in

K.K. Kar (ed.), *Composite Materials*,
DOI 10.1007/978-3-662-49514-8_4

status of various processing methods adopted to fabricate these FGMs and their characterizations. The possible areas of applications of FGMs are also briefed.

Keywords
Functionally graded material • Processing • Composites • Classification • Characterization • Thin films • Applications

Introduction

Functionally graded materials (FGMs) are the composites wherein the compositions of two or more components vary continuously or in a stepwise manner to optimize the performance of material for a specific application. In 1972 only, the importance of functionally graded composites (FGCs) has been accredited, but the lack of availability of processing techniques for these FGMs could not turn out into a reality [1, 2]. It took nearly 30 years for the development of different processing methods for these materials [3, 4].

Consider an example of metal–ceramic functionally graded material (FGM). Herein, the toughness, electrical conductivity, and machinability of a metal are combined with the low density, high strength, high stiffness, and high-temperature resistance of ceramics. When considered either of metal or ceramic separately, these contradictory properties cannot be brought into a single material.

But, if the concentration of ceramic is allowed to change in the metal matrix along one direction, say thickness – composition, microstructure, and hence properties will vary continuously with position (Fig. 4.1) [4]. The brittleness of ceramic will be removed and the metal becomes lighter and stiffer. This will make the material genuinely useful. Now, this material is hard but would not break; it will conduct electricity but can withstand high temperatures. The demand for such materials comes from the automotive (lightweight and strong materials would increase fuel efficiency), electronics, telecommunications, aerospace, and defense industries. FGMs are a category of relatively recent and promising materials, where the engineer gradually varies

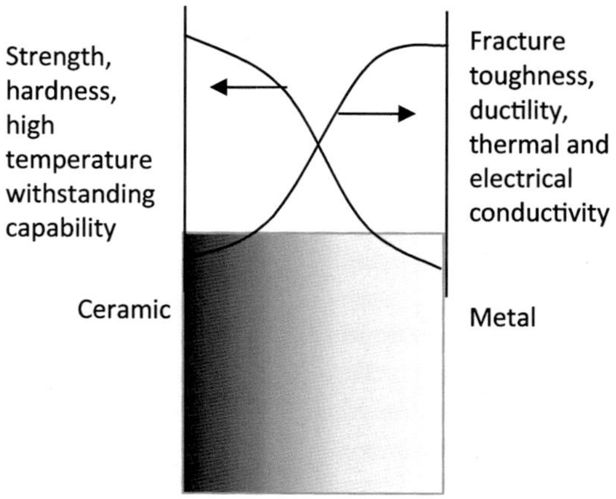

Fig. 4.1 Metal–ceramic-based functionally graded composite

Fig. 4.2 Gradation in a cross section of an animal bone [7]

the material functionality throughout the structure. Compositional gradient brings in the improvement of properties of the isotropic materials when compared with the same average composition of the components [5, 6]. FGMs can be either artificial or natural. Human teeth, animal bones, and bamboo are the natural FGMs. In a typical bone section, one can observe the reduction in the volume fraction and the size of porosity from inner to the outer surface (Fig. 4.2) [7]. Similarly, the volume fraction of vascular bundles of cellulose increases and the size decreases from inner to the outer surface, respectively (Fig. 4.3). The effective elastic modulus varies continuously in the radial direction due to the graded distribution of the vascular bundles of cellulose.

The parameters that can be varied are volume fractions, shape and size of the reinforcing fillers, angle of lamination, diameter, orientation, distribution of coating or chemical composition of fibers, porosity, etc. (Fig. 4.4). Gradations can be made in several directions, i.e., in-plane and/or out-of-plane grading of facings of sandwich panels [8].

Classification of Functionally Graded Materials

General Classification

Depending on the composition/microstructure that varies either continuously or in a stepwise manner over a definable geometric orientation, FGMs can be classified into (a) continuously graded and (b) stepwise-graded materials as shown in Fig. 4.5. In case of continuously graded structure, the change in composition, microstructure, and hence in the properties occurs with position. On the other hand, stepwise-graded materials give rise to a multilayered structure with an interface present between the

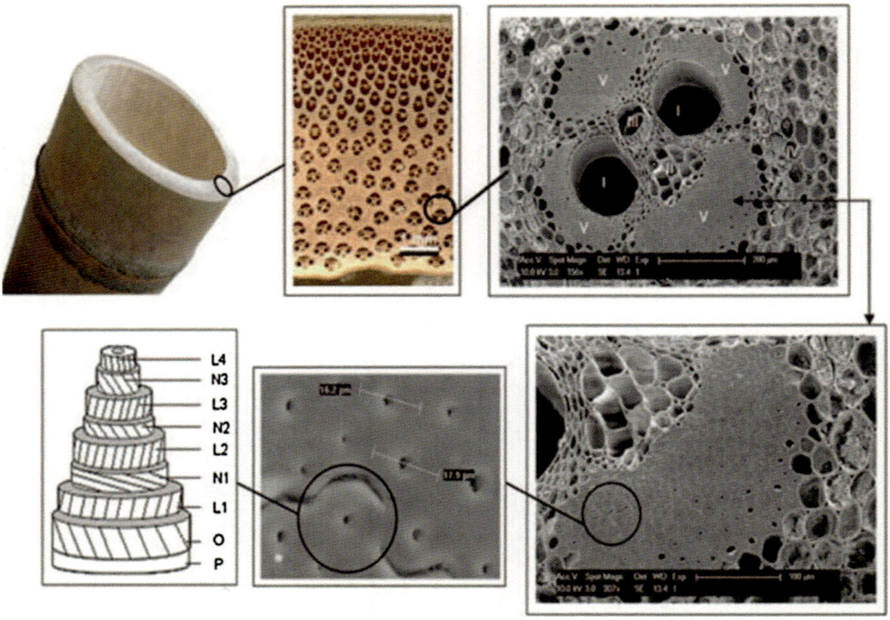

Fig. 4.3 Microstructure of functionally graded bamboo fiber [9]

adjacent discrete layers. Various processes have been explored that produce these types of FGMs. These can also be categorized into two domains – thin FGMs and volume FGMs. In general, thin-sectioned or surface-coated FGMs are developed using physical vapor deposition (PVD), chemical vapor deposition (CVD), or plasma spraying technique. The volume FGMs produce three-dimensional-graded component, and the processing techniques used are powder metallurgy, centrifugal casting, solid freeform fabrication methods, etc. [11, 12]. The vapor deposition techniques can process the graded material having composition gradients at the nanometer level.

Classification Based on Processing Routes

The preparation of FGMs can be divided into two main steps: (a) formation of gradation (processing of FGMs) and (b) consolidation into a bulk, monolithic material without disruption of gradation (Fig. 4.6). In metal–ceramic FGMs, sintering or solidification follows the gradation step, whereas in thermoset polymeric FGMs, curing of the polymer matrix (thermoset) is done during/after the formation of gradation. The processing techniques are chosen in such a way that the formed gradient should not destroy or alter during consolidation. The processing techniques that make the gradation possible are further classified into three different categories, namely, constitutive, homogenizing, and segregating processes. Constitutive process

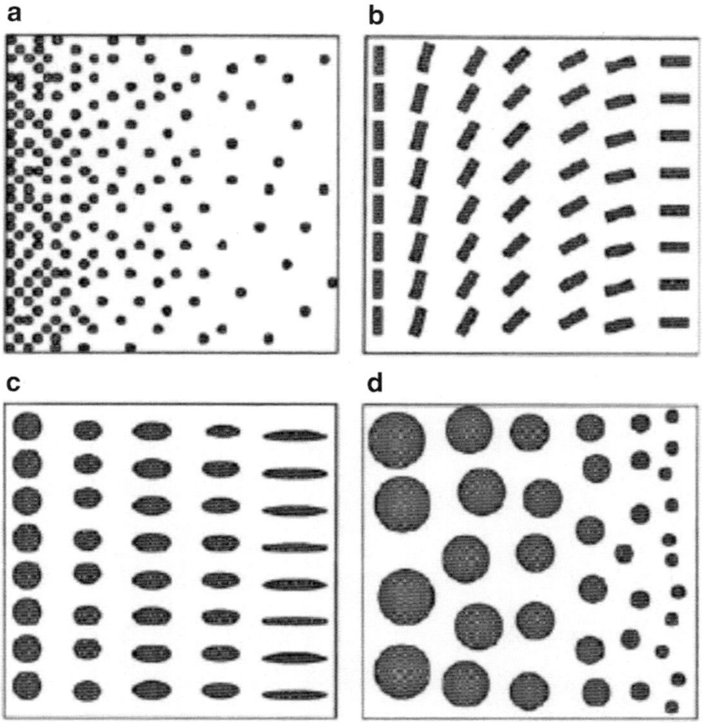

Fig. 4.4 Gradation of (**a**) volume fraction, (**b**) shape, (**c**) orientation, and (**d**) size [10]

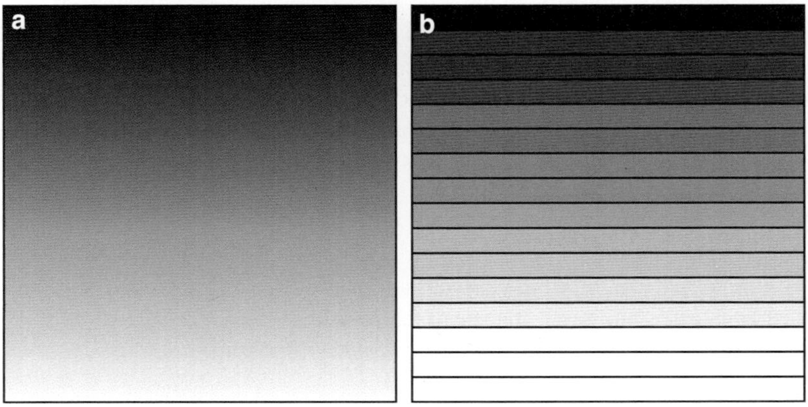

Fig. 4.5 (**a**) Continuously graded and (**b**) stepwise-graded materials

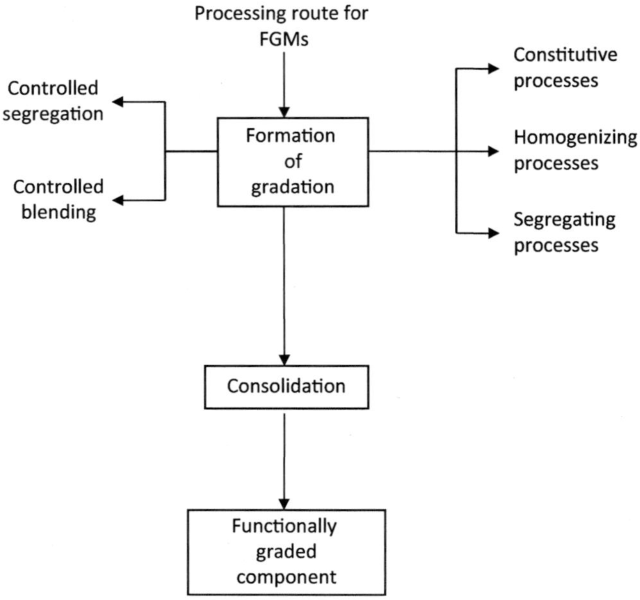

Fig. 4.6 A block diagram showing a general route for preparation of FGMs

forms the stepwise buildup of the graded component, while the other two processes make the continuous gradation in the structure by material transport. In constitutive process, the FGM is constructed with "layer-by-layer" technique. These techniques are also called constructive processes as gradients are literally constructed in space.

In homogenizing and segregation processes, the gradients within a component are dependent on natural transport phenomena such as the flow of fluid, diffusion of atomic species, or conduction of heat. Homogenizing process converts the sharp interface between the phases into a gradient by mass transfer, whereas the segregating process requires an external field (e.g., electric, magnetic, gravitational, etc.) to convert the macroscopically homogeneous composite into a graded composite. Homogenizing and segregating processes produce continuous gradients.

Powder metallurgy, a constitutive process (stepwise variation in composition), is a well-accomplished route to produce graded materials. Formation of green components and sintering with or without pressure, which assist hot consolidation, are the two important steps in this route. This route adopts the following processing sequence:

(a) Powder preparation
(b) Weighing and mixing of the powder as per the spatial distribution requirement
(c) Stacking and ramming of the premixed powders
(d) Sintering (pressure-assisted hot consolidation, if possible)

Powder preparation is possible through chemical reactions, electrolytic deposition, grinding, or pulverization (comminution). These techniques allow the mass production of powder with controlled size [13]. Herein, the gradient of the volume fraction of phases of a multiphase material can be carried out. A combination of phases can be metal–metal, metal–ceramic, or ceramic–ceramic. Even the gradients of pore size can be obtained by employing the variation of particle size in the compact direction [14]. With powder metallurgy route, the powder components can be prepared with continuous as well as stepwise changes in the mixture.

Sintering is a processing technique used, where appreciable diffusion mass transport is required. It is possible to produce density-controlled materials and components from metal–ceramic powders by applying thermal energy, usually executed at elevated temperature of $T > 0.5T_m$. Successful sintering usually results in a dense polycrystalline solid. During the sintering process, densification (the process of removing porosity) and grain growth (the average grain size increases) are two basic phenomena that occurred at the microstructure level. Sintering states are classified as:

(a) Solid-state sintering: Powder compact is densified wholly in solid state at sintering temperature. Herein, material transport mechanism is diffusion.
(b) Liquid state sintering: A liquid phase is present in the powder compact during sintering. Herein, material transport mechanism is "liquid flow."

In powder processing step, focus is given on the precision in weighing amounts and its dispersion. During sintering, the atmospheric condition is maintained appropriately as this high-temperature process is more sensitive to its surrounding. Sintering (consolidation) behavior depends on particle size and shape, composition of the powder, porosity, etc. Hence, problems must be handled based on the existing knowledge of the sintering mechanisms. If different temperature regions are required for densification, then the processing through liquid phase sintering, spark plasma sintering, and laser assisted sintering are suggested [15–17].

Powder Metallurgy technique gives the stepwise variation in the composition of the structure [18]. For continuous gradation of the composition, centrifugal methods (segregating processes) are employed. Through powder metallurgy route, porosity gradient, gradients of the concentration of one of the phases, and the gradient in the chemical composition of the single phase material are possible [19]. Porosity gradients can be made by depositing the powder mixture of different particle shapes. Pore size gradation can also be made by varying the particle size along the thickness [20]. Most of the published papers in the literature describe the gradient of volume phases in the material.

Powder can be deposited in the die, and preparation of graded powder compacts with stepwise as well as continuous changes in the mixture is possible. The processing methods making stepwise changes are constitutive processes of die compaction method, sheet lamination, slurry dipping, and solid freeform fabrication processes. Continuous gradation can be made (segregating processes) through centrifugal sedimentation, gravity sedimentation, centrifugal powder forming, eletrophoretic deposition, slip casting, etc. [21–25].

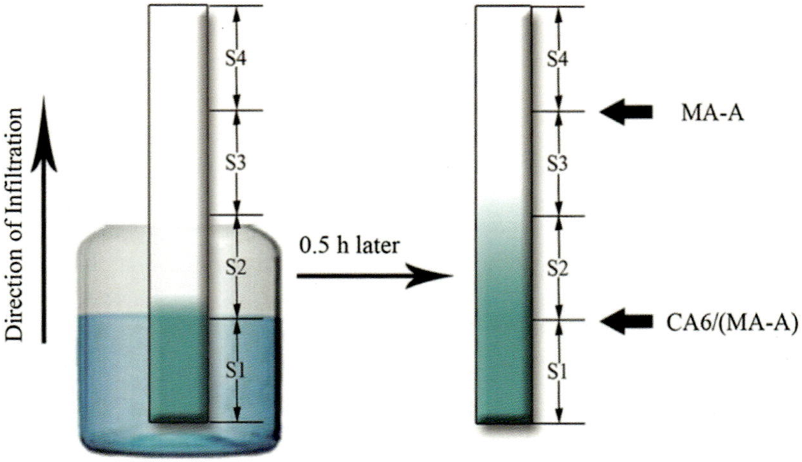

Fig. 4.7 Schematic of infiltration process [28]

Formation of gradation becomes simpler with melt-processing techniques as the transport of the filler in melt is easy. Consolidation subsequently happens followed by gradation. The processes like centrifugal casting, sedimentation casting, directional solidification, and infiltration processing are the popular processing methods employed in general, where mostly metal is used as one of the constituents in making the gradation [26–28]. Yi et al. have manufactured calcium hexaluminate (CA6)/spinel–alumina (MA–A)-graded composites via infiltration technique (Fig. 4.7) [28].

An example of natural transport phenomena is centrifugal method. FGMs are fabricated under a centrifugal force to produce continuous gradation in the component. With the application of centrifugal force, processing methods are classified into three classes as shown in Fig. 4.8 [29]. These are – centrifugal method (application of centrifugal casting), centrifugal slurry method (centrifugal sedimentation), and centrifugal pressurization method [30, 31]. In case of centrifugal method, a centrifugal force is applied to molten metal dispersed with hard ceramic particles, forming the desired gradation. Difference in the centrifugal force on molten metal and solid particles due to their density difference forms the gradation in the composition. In contrast, a centrifugal force is applied on the slurry containing high-velocity particles and low-velocity particles during the fabrication of FGMs by the centrifugal slurry method [30]. After complete sedimentation, the liquid part of the slurry will be removed, and therefore, it will no longer be the part of FGM. Centrifugal force is only used for simple pressurization in case of centrifugal pressurization method. In this method, compositional gradation should be formed prior to the application of centrifugal force [32, 33].

Watanabe et al. have fabricated a nanoparticle distributed FGM (Cu/SiC and Al/TiO_2) employing centrifugal mixed-powder processing technique as shown in Fig. 4.9. One can see the stepwise procedure followed during processing: a powdered mixture is first inserted into a spinning mold (a), metal matrix ingot melted and

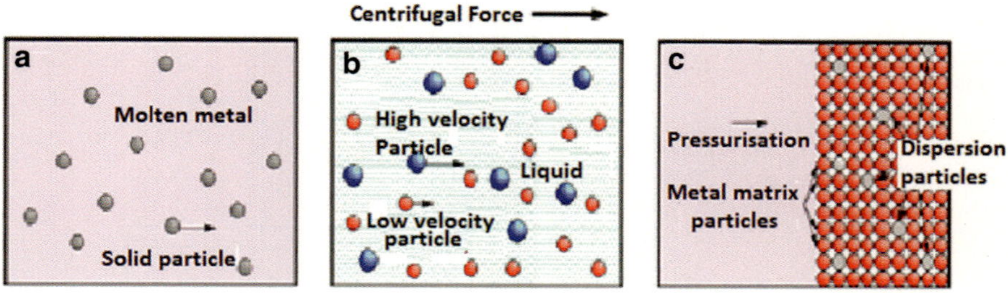

Fig. 4.8 FGMs with three types of fabrication methods under centrifugal force. (**a**) Centrifugal method, (**b**) centrifugal slurry method, and (**c**) centrifugal pressurization method [29]

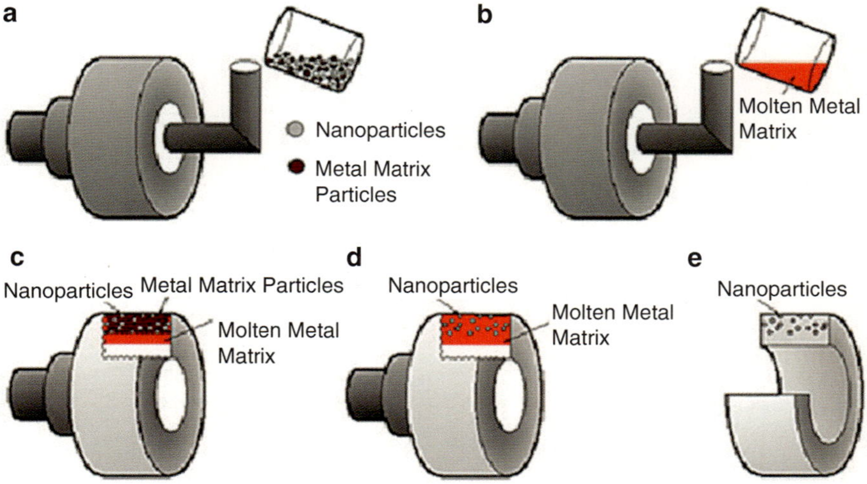

Fig. 4.9 Al/TiO_2 functionally graded material made by centrifugal mixed-powder processing method [26]

that molten metal is poured with the powder mixture into the spinning mold (b), molten metal matrix getting penetrated into the space between the particles due to the centrifugal force (c), and at the same time, the metal matrix powder is getting melted by the heat from molten matrix poured from the container, and finally an FGM ring with functional nanoparticle distributed on its surface can be obtained (e).

Gravity sedimentation, electrophoretic deposition, pressure infiltration, etc. are the few processes that make the continuous changes in the mixture (either homogenizing or segregating processes), whereas the processes like die compaction, sheet lamination, and various computer-controlled solid freeform fabrication processes are the examples that produce the graded components with discrete/stepwise changes in the mixture (constitutive processes). In electrophoretic deposition, the colloidal particles suspended in a liquid migrate under the influence of an electric field and get deposited onto an electrode (Fig. 4.10). This technique includes nearly all material classes such as polymers, pigments, ceramics, and metals too.

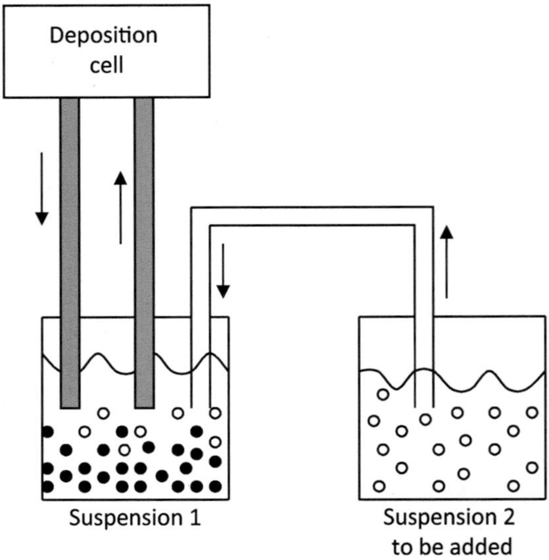

Fig. 4.10 Elecrophoretic deposition setup to prepare FGMs

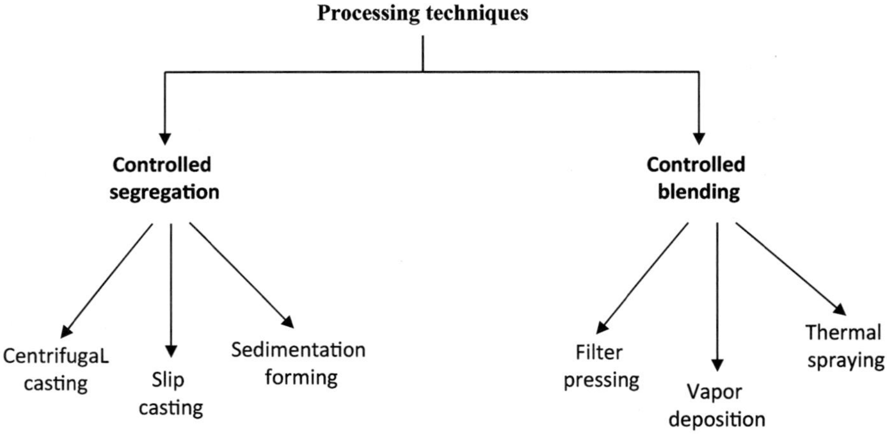

Fig. 4.11 Comparison of controlled segregation and controlled blending

Based on the Quality of Gradation Produced

Depending on the processing techniques that produce the functionally graded parts (Fig. 4.11) and the quality of gradation produced, they can also be classified into controlled segregation and controlled blending techniques.

Controlled Segregation

In controlled segregation, the action of gravity on the density difference of the powder components is the main driving force for gradation. As the segregation rates strongly depend on the particle size and morphology of the powder, the processes involved under this category are a bit of time consuming with poor gradient control. Most of the bulk FGMs employ segregation approach, e.g., sedimentation forming, slip casting, centrifugal casting, and thixotropic casting.

Controlled Blending

In controlled blending, the mixing components are blended either before or during forming, and the mixing ratio is continuously varied from 100% component A through to 100 % component B.

This approach avoids the earlier system-dependent issues of powder density and gravitational settling mechanisms and offers a unique advantage of being able to produce precisely controllable functionally graded components. Controlled blending also enables very rapid processing rates. Because of precise control over gradation and rapid processing rates, controlled blending is widely used in making functionally graded thin films. The processing techniques that are generally used for producing functionally graded thin films are thermal spraying (blended powder feed), vapor deposition (CVD/PVD blended gas feed), electrophoretic deposition (blended slurries), filter pressing (blended slurries), blended spray drying, etc. [34–38].

Classification Based on the Size of Gradation

Based on the size/thickness of gradation, FGMs can be classified into the following domains.

Functionally Graded Thin Films

Functionally graded coating and developed interface assist in reducing thermal and residual stresses as well as inhibit crack propagation. The graded coats are used to connect two different materials with eliminated stress singularity.

Sputter deposition and physical and chemical vapor deposition techniques are being explored to process functionally graded films [39]. These films have already captured energy, heavy machinery, and aircraft sectors, which demand for high-performance materials that can withstand severe conditions [40–42]. These surface deposition methods are used to deposit functionally graded thin surface coatings only. They are energy intensive and produce poisonous gases as by-products [43]. Other methods used in producing the functionally graded thin films include plasma spraying, electrode deposition, electrophoretic, ion beam-assisted deposition, etc. Plasma spray processing offers a flexible and economic route for producing FGMs. In plasma spraying, a deposited layer is formed through the sequential buildup of splats. The processing employs either a single torch system – utilizing multiple feeders and blended or multiple torch system – with independent feeding systems for each component. Plasma spraying is used for many years to apply layered and graded deposits to enhance the durability of thick ceramic thermal barrier coatings (TBCs) used in gas turbines/diesel engines [41, 42]. General thermal barrier coating of ceramic layer on the metal matrix cannot serve the purpose. In

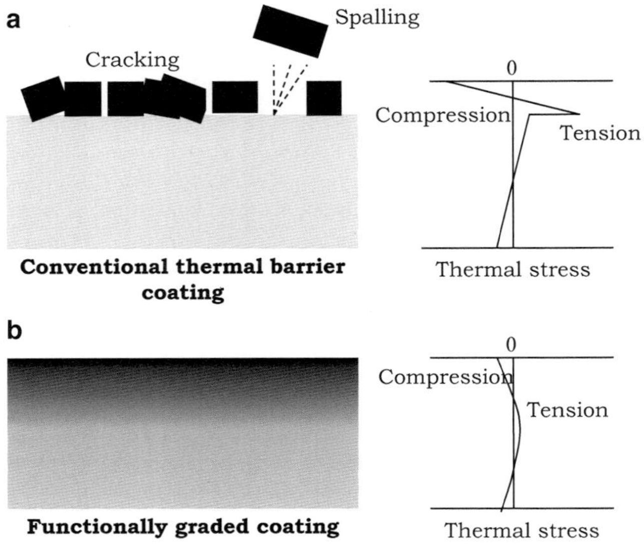

Fig. 4.12 Stress variation along the cross section of (**a**) conventional thermal barrier coating and (**b**) functionally graded coating material

severe conditions, at the interface of the phases, crack initiates and the outer ceramic layer chips off (spalling) exposing the metal to high-temperature atmosphere (Fig. 4.12). High stresses develop at the interface due to mismatch of thermal expansion coefficients of the metal and ceramic components leading to the failure of the composite. These functionally graded coatings minimize the mismatch of coefficients of thermal expansion of the base (metal) and coat (ceramic) material and prolong the service life of the coat. These graded coatings are also effectively used on afterburners, combustors, stationary vanes, etc. In the recent years, the concept of FGM is galloping to other research areas like thermoelectric, photoelectric, and nuclear energy conversion materials [44–49]. All these aforementioned processes are not used for producing volume FGMs as they are energy intensive and uneconomical.

Three-Dimensional Functionally Graded Components

Recent advancements in fabrication processes of metal-/ceramic-based FGMs are now making it possible to develop new technologies that have been previously relegated to the province of our imagination. In ceramic- and metal-based FGMs, lots of processing methods that developed earlier are available with slight modifications. Few examples include thermal spraying, powder metallurgy, coating process, melt processing, etc. with certain advantages and disadvantages of each technique. One of the major drawbacks associated with experimental studies is the preparation of FGMs having large-scale gradation.

Presently, there is hardly any reliable and inexpensive processing technique for producing FGMs that allows bulk production of large parts. Traditional processing routes like powder metallurgy, vapor deposition, solidification processing, etc. are costly. Of these, gravity and centrifugal casting look more economic and attractive [45]. In centrifugal casting, the gradation in the material composition is produced due to the difference in material densities and the spinning of mold [26]. But these techniques are segregation controlled and hence time consuming with poor gradation control. Another disadvantage is that it can manufacture axisymmetrical components only. To overcome these manufacturing problems, researchers started looking for other manufacturing methods.

Solid freeform fabrication (SFF) is a designation for a group of techniques that manufactures a three-dimensional layered (functionally graded) component employing a computer-aided design (CAD) data. These processes come under the controlled blending category. It has also been referred to as rapid prototyping (RP), computer-automated manufacturing, or additive manufacturing. A unique characteristic of solid freeform fabrication is its direct manufacturing capability. It does not require tooling, fixturing, and other unnecessary peripheral activities that are generally associated with conventional processes [46]. It offers many advantages like higher speed of production, maximum material usage, ability to produce intricate-shaped components, less energy intensive, and design freedom as parts are manufactured directly from CAD data. The basic operation of any SFF system consists of slicing a three-dimensional computer model into thin cross sections, translating the result into two-dimensional position information, and using this data to control the placement of solid material. Few important processing methods are - stereolithography, three-dimensional printing, selective laser sintering, selective laser melting, laminated object manufacturing, fused deposition modeling, etc. [50, 51].

Stereolithography uses an UV laser to build up the successive layer of a three-dimensional object in a vat of photopolymer. At every buildup of the layers, the platform lowers down and rewets the surface of the object (Fig. 4.13). This rewetted surface is cured by the UV laser that moves and cures the portion that is the part of the object. This process continues till the object is manufactured completely. The material used for stereolithography is photopolymers. This is one of the disadvantages of this process.

Similarly, direct laser melting (DLM) is a process that manufactures three-dimensional component layer wise by melting the metal powder with laser scanning (Fig. 4.14). DLM process mainly consists of a laser beam, powder delivery system, building platform, and process-control computer system. It adopts the following procedure:

(i) A layer of metal powder placed on the platform (using a layering bar).
(ii) The powder layer is melted (with the laser beam) to bond it with the previously solidified part present underneath.
(iii) The platform is lowered by one layer thickness for the next powder layering and laser melting (in the same way as in stereolithography). Han et al. have manufactured gear having functionally graded properties using the same DLM method (Fig. 4.15) [51].

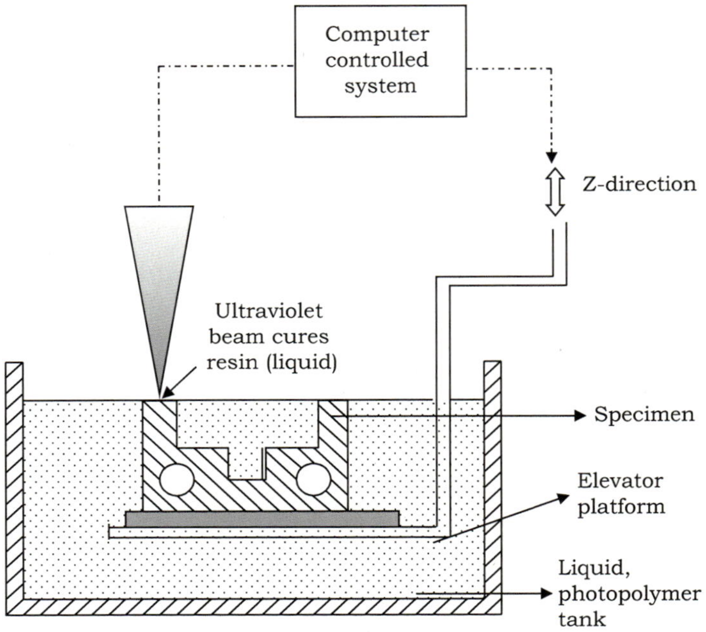

Fig. 4.13 Schematic of stereolithography technique

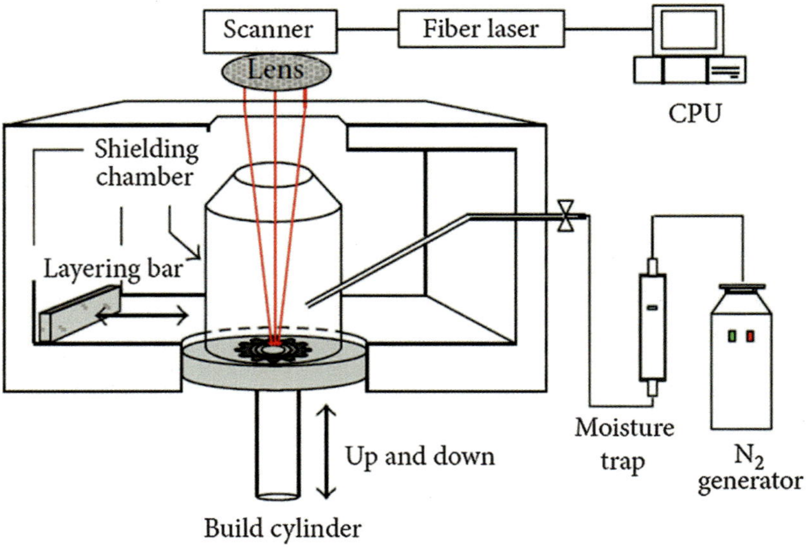

Fig. 4.14 Schematic drawing of direct laser melting system [51]

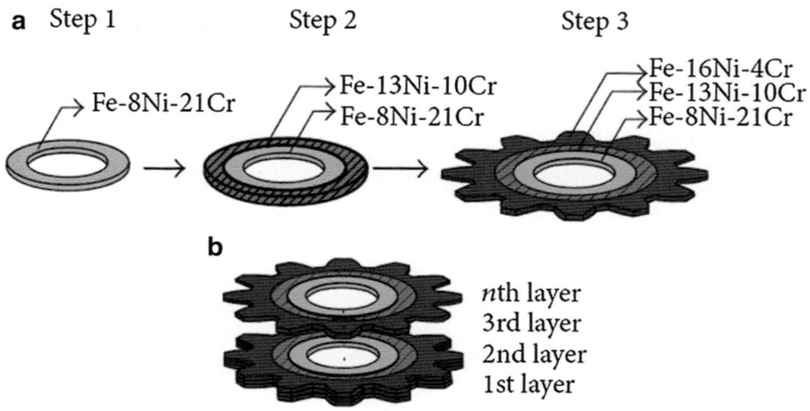

Fig. 4.15 Schematic of gear fabrication technique. (**a**) Layer fabrication, (**b**) layer-by-layer fabrication [51]

The processing techniques developed for making three-dimensional products (solid freeform fabrication techniques) are not yet competent with the traditional composite materials. Mass production of these components by economic means veritably requires an expedient alternative. Another concern is a poor surface finish that makes compulsory to carry out the secondary finishing operation.

Based on the Phases Involved in the Processing

Based on the phases involved in the making of FGMs, they can be classified into:

(i) Metal-based FGMs
(ii) Ceramic-based FGM
(iii) Polymer–metal FGM

As discussed earlier, many processing methods developed earlier are used for manufacturing of ceramic- and metal-based FGMs with slight modifications (in the existing method). Most of the methods that produce ceramic and metal FGMs are discussed in earlier sections. Constructive/constitutive manufacturing processes like powder metallurgy, lamination, coating, etc. as well as the transport-based processes like thermal diffusion, melt infiltration, centrifugal separation, etc. are mostly used to metal and ceramic FGMs. Furthermore, most of the constructive manufacturing processes mentioned earlier are the batch processes that generally result in discrete interfaces. These discrete interfaces act as stress concentrators and are often weaker than the matrix and can lead to the structural failure of the graded component [47].

Polymer-Based Functionally Graded Materials

Unfortunately, there were hardly any methods available till the late 1990s that were capable of producing polymer-based FGMs. Polymers are soft with poor wear and impact resistance and poor stiffness (maximum Young's modulus ~5 GPa). The gradations in these polymers are intended to allow enhanced properties of either of these. In polymer-based FGMs, the cognizance of different processing methods for producing FGMs is very limited as compared to ceramic- and metal-based systems [48]. Nonavailability of standard processing techniques is one of the main hurdles to explore the applicability of polymeric FGMs in diversified areas. Polymers with porosity gradient in the polyurethane foams show an increased impact strength [52]. By now, i.e., 2015, the polymers have been graded in the form of graded interpenetrating polymer networks, graded biodegradable polyesters, graded index polymer thin films and fibers and micro-lenses, centrifuging prior to polymerization, graded fiber with lamination technique, etc. [53–59]. There is scarcely any work sufficed except those related to biomedical and optical fields [49, 60–63]. In the recent past, researchers have attempted to make the gradation of properties in polymers like wear resistance, toughness, hardness, impact resistance, etc. employing either lamination or centrifuging techniques [35, 47, 64–67].

Kikutani has taken the concept of structure from natural bamboo to prepare the liquid crystalline polymeric fibers, the diameter and stacking density of which have been gradually increased along one direction in a thermoplastic matrix [61]. The design of the bone that changes from stiff, dense external part to a porous internal structure shows the adaptation of gradation in the living tissue. Pompe et al. have presented different routes for producing graded materials for orthopedic implants of the shoulder and knee joint replacement [49]. Using the concept of metal–ceramic FGMs, Liu et al. have prepared epoxy–polyurethane FGM for thermal protection system like hot substance transport pipes that operate in nonuniform temperature fields [62]. Variation in the molecular weight, i.e., polydispersity index, alters the properties of the polymers. For multiphase polymers like multiblock copolymers, blends, etc., morphology can be altered by varying the processing conditions. Xie et al. have made the gradation in the phase structure of polypropylene/EVA (ethylene-(vinyl acetate)) blend during annealing process [63]. The effect of gradation of crystalline phase on the semicrystalline polymers has been analyzed by Akiyama [68]. Butcher et al. have created functionally graded epoxy with a gradation of soda–lime glass microspheres employing a gravity casting technique to study the failure of the material at the graded interface [69]. Functionally graded polymer electrolytes have been produced by Ogumi et al. by using plasma polymerization technique [70]. The microstructure of sintered (liquid crystalline copolyesters) Vectran fibers has been varied by changing the processing temperature along one direction to improve the ability to absorb impact energy [71]. Homogeneous, non-isotropic microstructural end, resulting from the lesser sintering temperature, has shown much higher absorption of impact energy when put at the impact side. Parmeswaran and Shukla have developed a continuous FGM using epoxy resin and cenospheres employing buoyancy-assisted casting process [64]. They have observed the enhancement in the quasi-static and dynamic modulus and the drop in the material density in the direction of increasing cenosphere volume fraction. The elastic and fracture properties of cenosphere-graded epoxy resin are studied by Shukla and Jain using dynamic photoelasticity in conjunction with high-speed photography [65]. The use of thermally sprayed functionally graded coatings of WC–Co on the polyimide matrix to improve the erosion and oxidation resistance of polymer matrix composites is studied by Ivosevic [66]. Fukui et al. have carried out a simulative experiment to see the effects of gradient distribution of the components on the fracture behavior in a corundum/plaster composite system [67]. The general fracture behavior of a stepped and continuously graded material under flexural loading has been investigated by Tilbrook et al. via finite element analysis [72].

Very few methods have been suggested for forming functionally graded polymer composites (FGPCs). One of the techniques employs stacking of the layers by lamination technique. Uhlig et al. have modified brittle polyisocyanurate resins by a rubber modifier and have made them useful in structural applications [73]. Higher modifier contents improve the toughness, but the high-temperature withstanding ability and resistance to water absorption decrease. They have graded the modifier content along with the gradients in the fiber density and orientation by lamination of a stack glass fiber prepregs, each lamina containing differently modified resin. Polymers can be reinforced with long fibers and gradation can be made by varying the orientation as well as by changing the composition of the fibers [74]. Polypropylene composites with continuous gradation of glass–fiber–mat content have been manufactured by Lee and Jang [75]. They have used lamination technique with each layer containing different amount of glass fiber in the mat. Higher impact energy and flexural strength have been reported when the face of the sample with high fiber content is loaded in comparison to their isotropic composite counterparts. On the other hand, they also have observed the drop in these properties when low fiber content side took the load. Ikeda has prepared the functionally graded styrene–butadiene rubber (SBR) vulcanizates with gradation of network chain density along the thickness direction by lamination technique and compared them with homogeneous compounding sheets [76]. To control the bending actuation for smart structures, the graded distribution of SMA (shape memory alloy) wires in polyurethanes has been made by Bruck et al. using a lamination technique [77]. The other technique avails centrifugal force as a fundamental variable parameter in gradation of one of the phases. In polymer composites, Lee et al. have developed a continuous gradient of fiber distribution in epoxy–carbon fiber system with the application of the centrifugal force [78]. It is reported that for a given amount of fiber, the flexural strength of the functionally graded composite with carbon fiber-rich surface taking up the load is higher than the conventional isotropic material. The gradation effect employing SiC particles and glass fibers as the reinforcing materials has been studied by Klingshirn et al. by centrifuging prior to polymerization [79]. They have observed that the filler concentration after centrifugation in the outer portions of the specimen increases up to 27 % for the glass fibers and up to 45 % for SiC particles, for an average filler content of 20 %. By controlling the

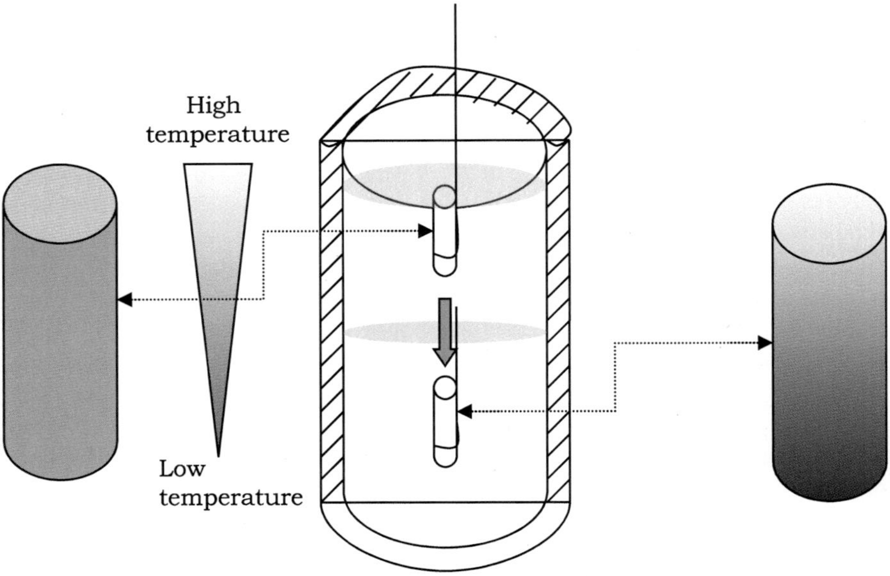

Fig. 4.16 Schematic of compositionally graded polymer blends using uniaxial thermal gradient (Bridgman method)

centrifuging parameters, Krumova et al. have successfully varied the hardness value from 210 to 640 MPa over the sample length by continuous gradation of SiC in epoxy resin matrix [35]. Ohgaki and Yamashita have successfully reinforced hydroxyapatite (HAp) by sedimentary HAp distributions on a PMMA matrix using a centrifuge to achieve both the bending strength and flexibility of PMMA and biocompatibility of HAp to avert stress convergence at the interface [65]. Tsotra and Friedrich have used centrifugation technique to make the graded distributions of short carbon fibers in epoxy resin/polyaniline blends to enhance the electrical conductivity and mechanical properties [80]. The gradation can be made smooth by adjusting the speed of rotation and viscosity of the resin.

Several researchers have tried to implement different methods to produce FGPCs with new ideas. These methods include frontal polymerization, dissolution diffusion, curing of the polymer matrix with UV irradiation, emulsion blending, coating, etc. [81–84]. Recently Koide et al. have prepared compositionally graded polymeric blends of PCL/PEO (poly(ϵ-caprolactone)/polyethylene oxide) system by using nonequilibrium self-organization processes with a uniaxial thermal gradient (Bridgman method) (Fig. 4.16) [82]. A gradient in the solidification direction has maintained, and PCL molecules have diffused from the solidification interface to the liquid phase (liquid phase has PEO due to its higher crystallization temperature).

Liquid crystal devices with a refractive index distribution property are prepared by Sato et al. They have cured the crystals with UV irradiation by maintaining a nonuniform electric field [83].

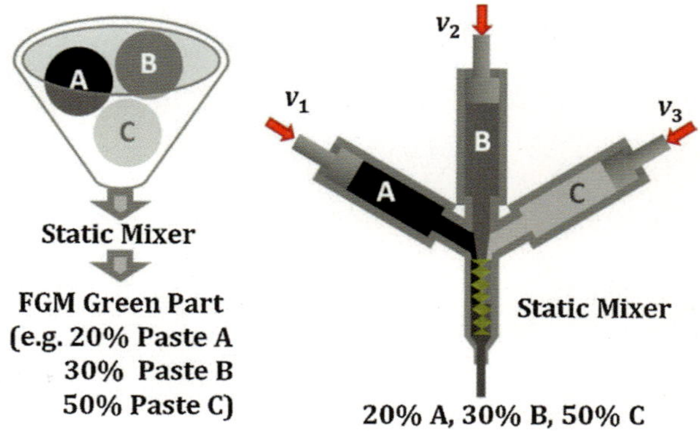

Fig. 4.17 Triple extruder mechanism design [98]

Optical polymer devices such as microlens array and optical waveguides can be prepared with gradation of refractive index. Lambros et al. have chosen poly(ethylene-co-(carbon monoxide)) to employ photodegradation technique (this polymer easily degrades with UV irradiation) and controlled the gradation by varying the UV irradiation time [84]. Abanto-Bueno and Lambros have manufactured a functionally graded photosensitive polyethylene copolymer using UV irradiation and studied its crack growth resistance behavior [85].

Various extrusion-based fabrication methods have been demonstrated for the manufacture of graded extrudates [86, 87]. The principal methods are:

- Extrusion using co-extrusion dies [88–91]
- Hybrid extrusion and spiral winding method [92, 93]
- Hybrid twin-screw extrusion and electrospinning method [94–97]

All three methods are amenable to industrial scale-up and generate scaffolds, which are reproducible in any geometry and properties. All three methods can be used with and without solvents (dry versus wet extrusion methods) and used in the area of interface tissue engineering, targeting regenerative medicine for the bone and cartilage repair and regeneration. These methods possess the ability to introduce various ingredients of the formulations in a time-dependent fashion into the extruder for the manufacture of spatially (radially and axially) graded scaffolds [91, 93, 95, 97]. Leu et al. have employed triple extruder for fabricating FGM. It involves the computer control of flows of multiple aqueous pastes as shown in Fig. 4.17, mixing and extrusion of the (mixed) paste to fabricate a three-dimensional component (layer-by-layer) according to a CAD model (See Fig. 4.18)

Pojman has invented a new technique to prepare functionally graded polymers by frontal polymerization [99]. It is a method for converting a monomer into a polymer via localized reaction zone that propagates through the coupling of the heat released by the polymerization reaction and thermal diffusion. An example of a material making with distinct utility is an optical limiter wherein a gradient of nonlinear optical dye is dissolved in a polymer matrix [100]. An optical limiter is a device that strongly attenuates intense optical beams but allows high transmittance at low-level

Fig. 4.18 Extrusion of pink- and green-colored $CaCO_3$ pastes for tests with and without mixing [98]

light. Such a device would be very useful for protecting human eyes from intense laser pulses. The process is tardy and can produce gradation not more than about 1 cm.

Additionally, several researchers have done work on graded materials having interpenetrating polymer networks (IPNs) [73, 101, 102]. Danielsson et al. have fused different core materials with interpenetrating polymer networks based on PVC and polyurea in a precursor stage [103]. The cellular structure of foam has been kept intact. They have shown that the layered foam–core materials improve sandwich manufacturing process. Most of the work developed by these individuals involves producing a gradient by diffusing one component into another pregelled component followed by curing or producing a gradient in the polymer using a gradient of illumination [104]. The diffusing method can require as much as 280 h for producing a gradient over 10 μ. Using the absorption of light to produce a gradient is limited to polymers with a thickness less than 1 mm. Hardly any of these techniques can be used to produce gradients in polymers which are several centimeters in thickness.

Few techniques process the individual layers to make a graded stack. Solid freeform fabrication (SFF) or ink-jet deposition (IJD) process has the advantage of incorporating nano-reinforcements into a low viscosity resin matrix. This process speedily photo-cures each successive layer to develop FGPCs. Making of graded refractive index lenses (GRIN) with ceramic nano-phase dispersion is an example to produce flat lenses instead of the traditional spherical lens geometry [105]. Conventional manufacturing begins with a billet of material and machines away the material not part of the desired object. Contrarily, the layered manufacturing system starts with no material and produces only required material within the boundary of the object. Typically the object is constructed, one planar layer at a time, with each consecutive layer bonded to the preceding layers, starting from the bottom layer and proceeding to the top layer. The thickness of each layer is a function of process capability, the material, and the requirements of the final object. A major advantage of this type of process is that it breaks down the object into thin layers, each of which can be processed individually. The changes can be made from layer to layer and even within each layer. The compositional and structural gradients can be introduced. SFF is an ideal technique for meeting the needs of GRIN lens fabrication. The disadvantage with this technique is the requirement of less viscous resin for uniform incorporation of the filler and is not applicable to the highly viscous systems like elastomers.

Most of the processing techniques used to manufacture FGPCs are scarce of either of the following points:

- Extending the gradation of phase/reinforcing particles to several centimeters.
- Competing with the uniformly dispersed composites when it comes to the mass production of the graded components.
- Most of the processing techniques require specially designed equipments that unfortunately raise the cost of fabrication of such graded materials.

A need for developing a simple, inexpensive technique that generally facilitates polymer composite processing is highly demanded.

When it comes to the gradation of the filler in an elastomer matrix, one more problem adds to the above listed difficulties, i.e., too high viscosity of the elastomers. Actually, there are hardly few papers that have reported the gradation effects employing an elastomer as a matrix medium even though automobiles, trains, and aircraft rely on this charming material for safety and comfort. One really can find enormous applications of rubber. Gallant et al. have manufactured the functionally graded polymers with continuous gradation of reinforcing particles employing twin-screw extrusion process [47]. They have developed this continuous processing technique for viscoelastic energy absorption, control applications, sensing applications, etc.

Characterization of Functionally Graded Composites

This section focuses more on the characterization techniques employed for (i) metal–ceramic FGMs and (ii) polymer-based FGM.

Characterization of Metal–Ceramic Functionally Graded Materials

Metallic elements or alloy systems are, in general, reinforced with ceramic phase (incorporated in the form of particles, flakes, or fibers) to enhance the stiffness, hardness, wear resistance, creep and abrasion resistance, temperature withstanding ability, and strength to density ratio compared with the base material properties. If the reinforcements are coherent with the metal matrix, they will develop local strain fields upon loading, which can impede the movement of dislocations, which in turn enhance the aforementioned properties.

Microstructural analysis is required to understand the effect of graded dispersion of ceramic phase into the metallic phase. Variation in the composition and structure

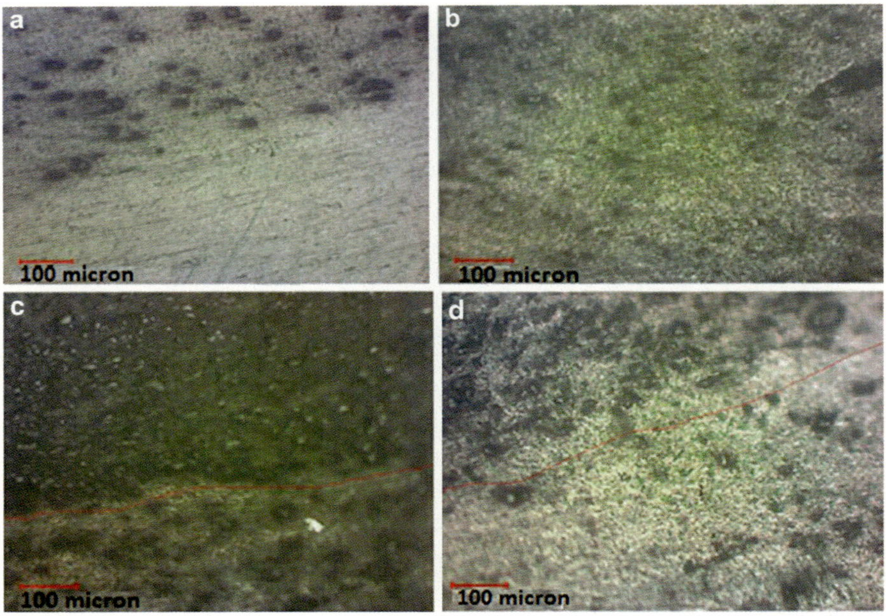

Fig. 4.19 Optical micrograph of (**a**) 0–5 % SiC, (**b**) 5–10 % SiC, (**c**) 10–20 % SiC, (**d**) 20–40 % SiC [106]

over volume results in corresponding change in the properties of FGM. The density of functionally graded component can be carried out using ISO standard 2738. The porosity of the sintered and unsintered components can be determined by Archimedes principle. The compacts are first weighed in air and then tied with string and weighed while hanging in water. It is calculated by Eq. 4.1:

$$\rho_s = \frac{m_a + \rho_w}{m_a + m_w} \tag{4.1}$$

where ρ_s is the density of the sintered specimen (kg/m^3), ρ_w is the density of water, m_a is the mass of the sample in air, and m_w is the mass of the sample in water, respectively. The porosity can be determined using Eq. 4.2:

$$E = 1 - \frac{\rho_s}{\rho_t} \tag{4.2}$$

where E = porosity (%) and ρ_t is the theoretical density (kg/m^3), which is calculated by measuring mass and volume (accurate dimensions) of the specimen.

Optical micrographs of the functionally graded composites can display the variation in the dispersion of the ceramic phase. Kumar et al. have varied the SiC content (from 0 to 40 wt%) in the Al matrix [106]. Figure 4.19a shows uninterrupted grains with lesser porosity in between the grains as this layer is hardly containing

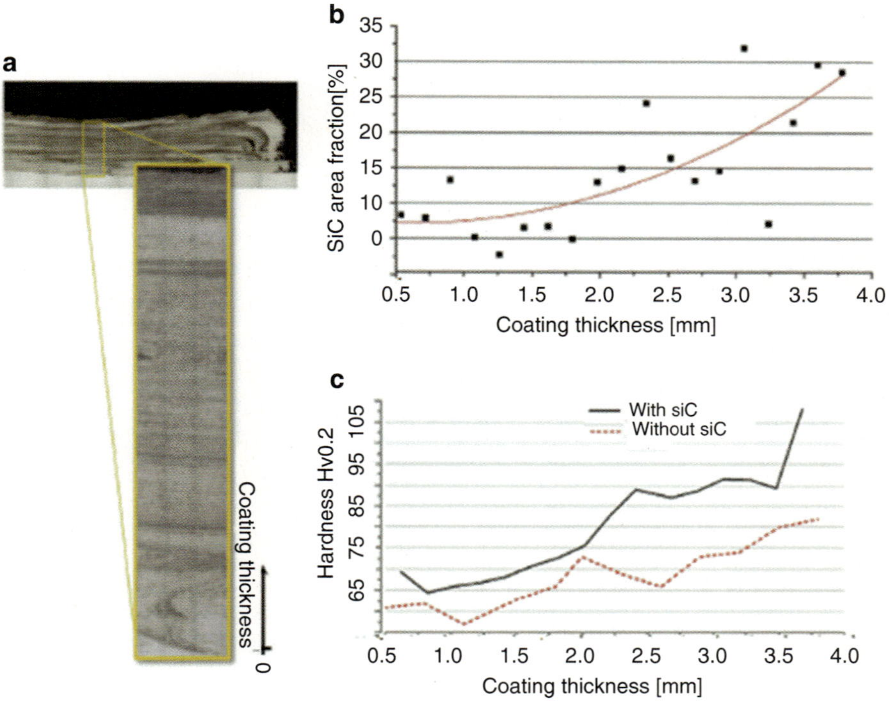

Fig. 4.20 Hardness profile and SiC area fraction analysis along coating thickness. (**a**) Reinforcement distribution, (**b**) SiC area fraction determined by image processing, and (**c**) Vickers hardness profile [107]

SiC particles. The layer (Fig. 4.19b) having SiC up to 10 wt% shows uniform distribution of the hard phase SiC particles between the grains of Al. With further increase in the wt. fraction of SiC to 40 % (Fig. 4.19d), one can see the clustering of the carbide phase at the grain boundaries that is hindering the interparticle contacts and densification.

Gandra et al. have prepared Al–SiC-graded composite coating [107]. Stacking depositions with increasing SiC particle concentration is produced along the thickness as shown in Fig. 4.20a. They have used image processing technique to analyze the percentage of SiC fraction area along the coating thickness (Fig. 4.20b). It is evidenced in the increase in the hardness profile (increasing from 70 HV for depositions without particles to 110 HV at the coating surface) along the same direction. Wear tests have revealed that SiC particles strengthen the coating resistance to wear by impeding the dislocation movement within the plastic deformation region of the Al alloy matrix.

Based on the frictional force and coefficient of friction, they are able to distinguish two wear stages. The stage of run-in period (lower friction forces) represents abrasive wear and the other, steady-state wear stage (higher friction forces), represents adhesive wear mode. SEM micrographs show homogeneous wear track of a

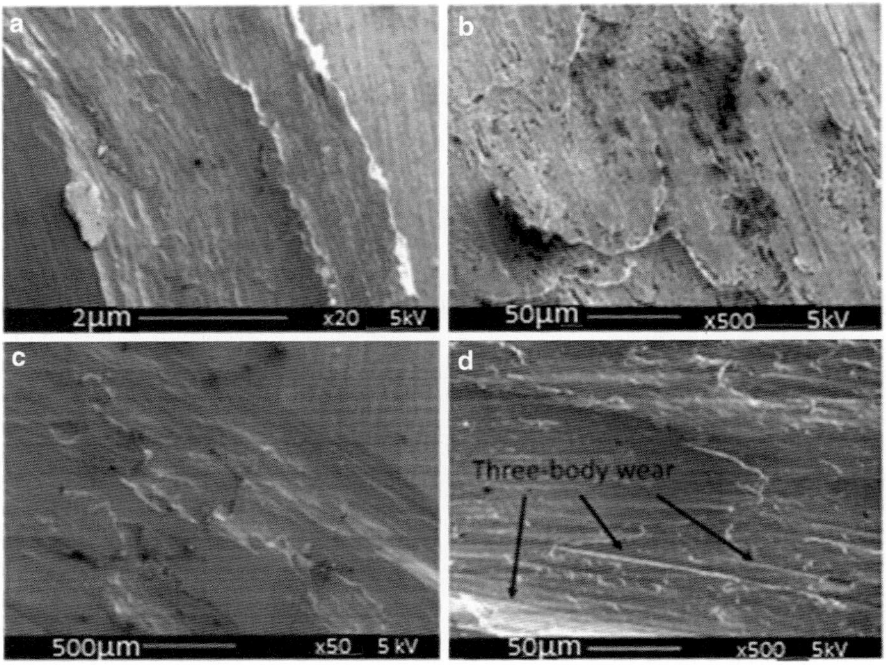

Fig. 4.21 Scanning electron microscopy of a coating wear track. (**a**) Friction Surfacing (FS) coating (without SiC reinforcement), (**b**) evidence of delamination on the FS coating wear track (without SiC reinforcement), (**c**) SiC-reinforced coating wear track, (**d**) detail of SiC-reinforced coating with evidence of three-body abrasive wear [107]

coating produced without hard phase (Fig. 4.21a, b with some signs of delamination). Figure 4.21c, d represents wear track. The use of SiC reinforcement results in improved wear resistance.

Tian et al. have fabricated Ni–Ti multilayer thin films (representing unique properties of shape memory effect and superplasticity) using Denton d.c. magnetron sputtering system [108]. Transmission electron microscopy is used to study the microstructure of the cross section of the film. Figure 4.22 shows the bright field image consisting of seven distinct layers along thickness. They have studied the SAD (selective area diffraction) patterns displaying the grains of BCC lattice. The compositional analysis is carried out by electron energy loss spectroscopy (EELS).

Tuskamoto has carried out nano- and micro-indentation studies of graded ZrO_2/Ti that can be applicable for biomedical applications [109]. The compositional variation is studied employing XRD pattern as shown in Fig. 4.23. Figure 4.23 a and b shows the XRD patterns for the ZrO_2 surface layer and Ti layer in the FGM parts, respectively. XRD pattern for ZrO_2 surface layer shows peaks for tetragonal crystal structure and for Ti layer; most peaks for hexagonal Ti can be seen without any peak for TiO_2. It is observed that for ZrO_2(0 mol% Y_2O_3)/Ti FGMs, most of the zirconia is

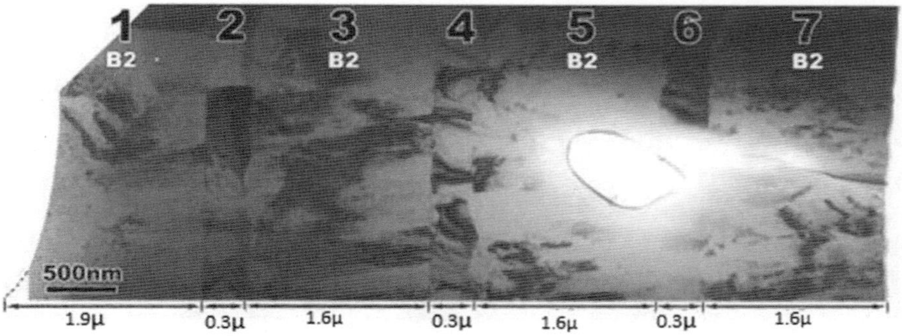

Fig. 4.22 Bright field image of a typical cross section of the film, the layer numbering starts from the substrate (left side) [108]

Fig. 4.23 X-ray diffraction pattern for ZrO_2 surface and Ti layer in FGM part. (**a**) ZrO_2 surface, (**b**) Ti layer in the FGM part [109]

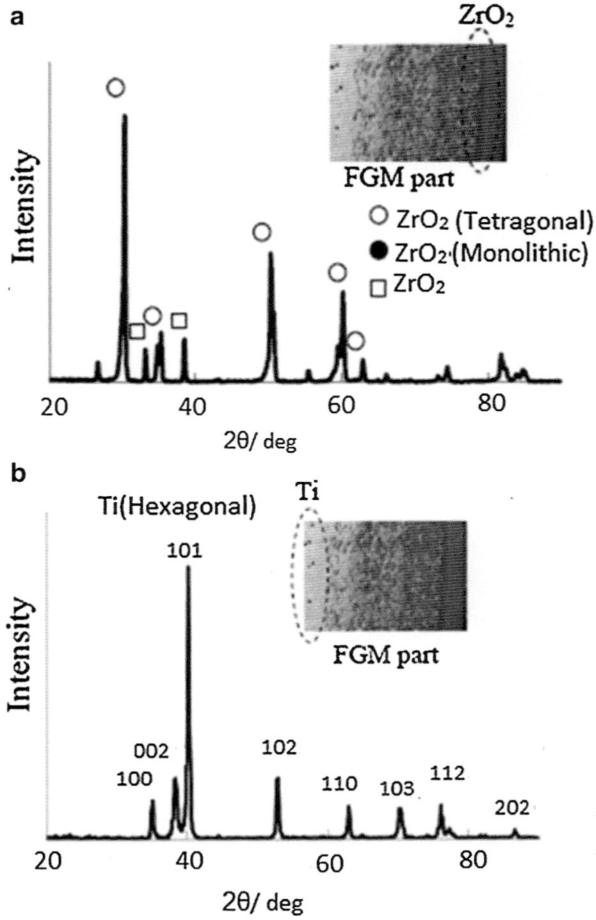

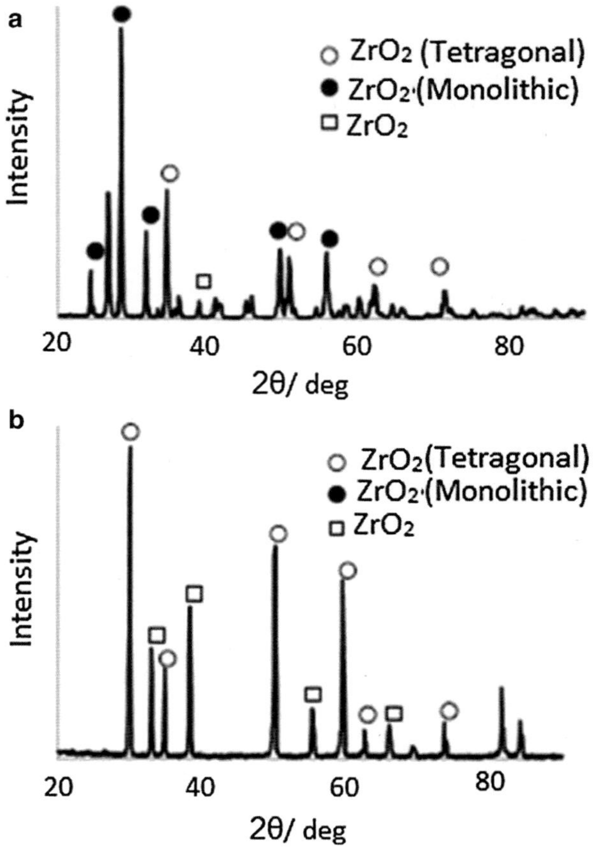

Fig. 4.24 X-ray diffraction pattern for ZrO_2 surface in the ZrO_2 (0 and 8 mol% Y_2O_3)/Ti FGMs. (**a**) ZrO_2(0 mol% Y_2O_3)/Ti FGMs, (**b**) ZrO_2(8 mol% Y_2O_3)/Ti FGMs [109]

getting transformed during cooling process. Figure 4.24a displays the XRD pattern showing tetragonal zirconia transforms to its monoclinic crystal structure. With incorporation of yttria, zirconia gets stabilized and does not transform to its monoclinic structure as shown in Fig. 4.24b.

Characterization of Polymer-Based Functionally Graded Materials

Yuan et al. have prepared PMMA-graded microporous foams and used SEM to characterize the morphology of the foam sample (Fig. 4.25) [110]. The foamed specimens are freeze fractured in liquid nitrogen and the fracture surface is sputter coated with gold.

Krumova et al. have carried out SEM and microhardness studies on SiC-graded epoxy composites [6]. The gradient of SiC particles is obtained by centrifuging at 300 rpm for 10 min. The SEM images of the sample (parallel to the centrifugal force direction, Fig. 4.26) help to reveal the particle volume content.

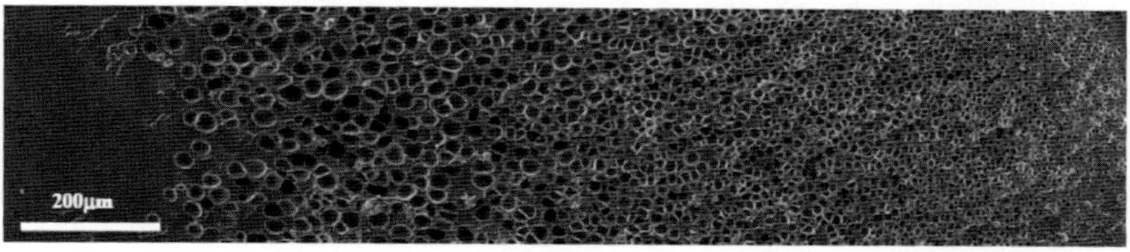

Fig. 4.25 Morphology of graded microporous foams after saturating at 28 MPa and 50°C for 1 h [110]

Fig. 4.26 SEM micrograph showing the gradation of SiC particles in epoxy matrix [6]

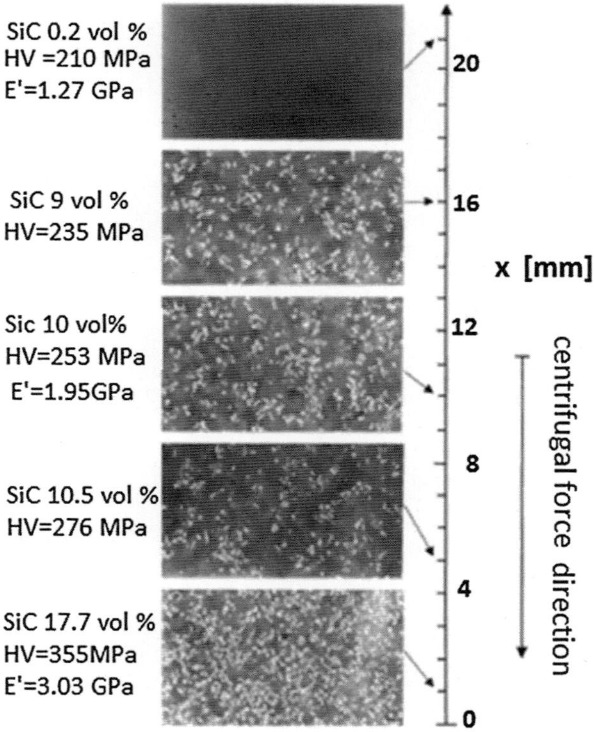

They have measured the microhardness at every 0.5–2 mm over the length of the sample with a Vickers square-based diamond pyramid indenter. After the removal of the test load, the Vickers microhardness *HV*, which shows the irreversible plastic deformation only, is derived from the projected area of indentation as (Eq. 4.3):

$$HV = 1.854\,{}^{P}\!/_{d^2} MPa \tag{4.3}$$

where *P* is the maximum load applied (N) and *d* (m) is the length of the impression diagonal. The corresponding variation of Vickers hardness over the length of the sample x is shown in Fig. 4.27.

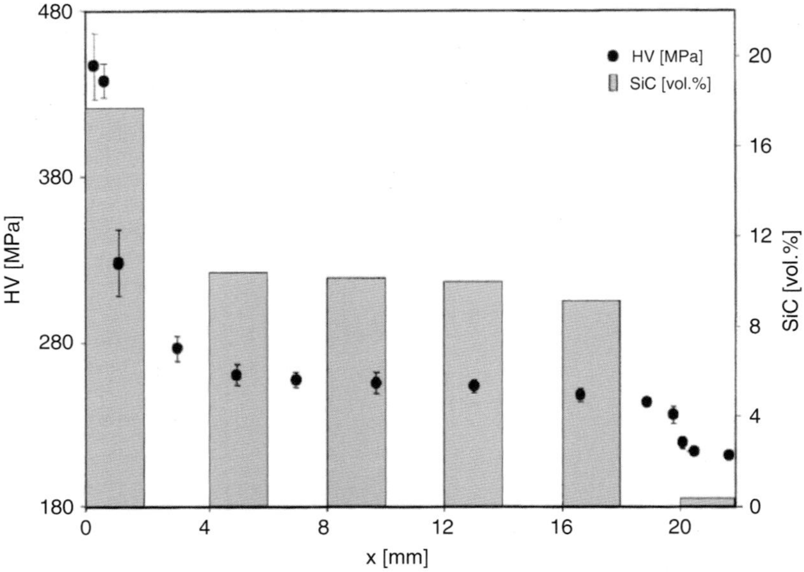

Fig. 4.27 Variation of Vickers hardness and SiC volume content along the centrifugal direction (ω = 300 rpm) [6]

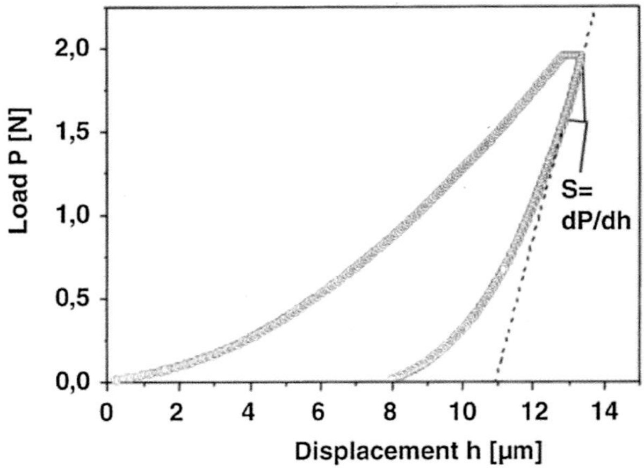

Fig. 4.28 Load displacement curve registered during the universal hardness measurements on epoxy resin/SiC FGM [6]

The slope of the unloading curves during universal hardness measurements is used as a measure of the elastic properties of the material. The load–displacement curve of the microhardness test is conducted on an area with 23 vol.% SiC particles (Fig. 4.28). The elastic modulus is derived using Eq. 4.4:

$$E = \left(\frac{\Pi}{A}\right)^{1/2} \cdot \frac{S}{2}\left(1 - \nu^2\right) \tag{4.4}$$

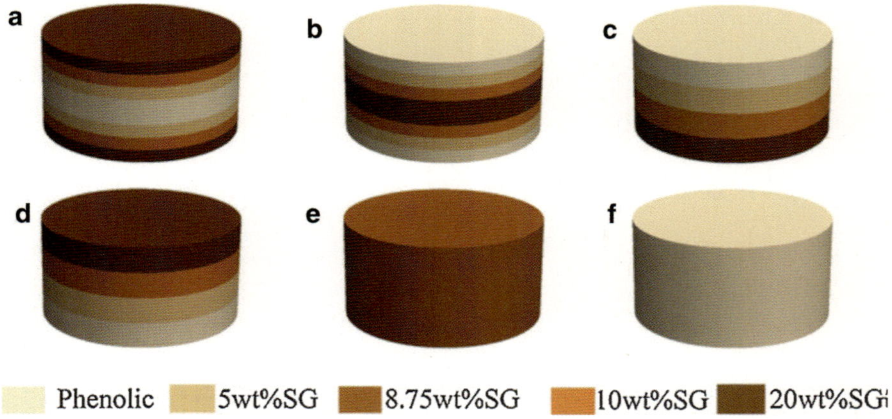

Fig. 4.29 Schematic illustration of the samples – (**a**) FGN-1, (**b**) FGN-2, (**c**) FGN-3, (**d**) FGN-4, (**e**) NGN, and (**f**) phenolic [111]

where $S = \delta P/\delta h$ is the initial unloading stiffness, $A = 24.5\ h^2$ is the projected area of contact between indenter and sample, and ν is the Poisson's ratio of the material.

Koide et al. have fabricated functionally graded polymer blends by uniaxial thermal gradient (Bridgman) method [82]. One of the two different kinds of polymeric compositionally graded materials (PCL/TDP) system, the concentration of TDP (4,4′-thiodiphenol), decreases with solidification direction. From variable temperature FT-IR measurements, it is observed that the hydrogen bonding between the PCL (polycaprolactone) and TDP increases with decreasing temperature. The % crystallinity of PCL (using differential scanning calorimetry, DSC) in the Bridgman sample increases in the solidification direction.

Bafekrpour et al. have fabricated another functionally graded graphite/phenolic nanocomposites using combined powder stacking and compression molding techniques [111]. It is analyzed that the electrical and thermal properties of such nanocomposites can be manipulated by changing the gradient patterns. They have prepared four different graded patterns as shown in Fig. 4.29 and compared them with nongraded nanocomposites. The graded nanocomposite with highest synthetic graphite at the outermost layers and the lowest at the center displayed 97% improvement in thermomechanical properties and the best creep recovery of 34.7% among all graded patterns (Fig. 4.30).

The creep response of the graded nanocomposites is divided into four regions: (i) sudden increase in the strain related to the elastic response of the samples, (ii) nonlinear increase in the strain value in the transition region, (iii) linear increase in the strain due to viscous flow, and (iv) recovery region. The neat phenolic material, FGN-2, shows highest creep and the FGN-1 shows the lowest value of creep.

The variation of CTE (coefficient of thermal expansion) across the thickness is shown for all four graded patterns using measured CTE of individual layers (Fig. 4.31). The variation in CTE values over the thickness induces thermal stresses

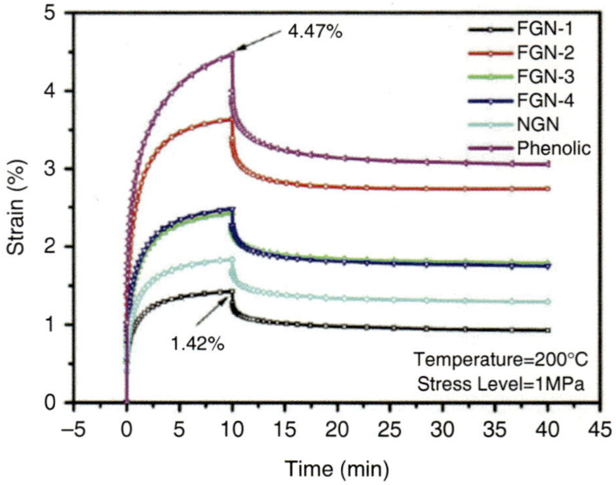

Fig. 4.30 Creep recovery response from different configurations of functionally graded nanocomposites (FGNs), nongraded nanocomposites (NGNs), and phenolic [111]

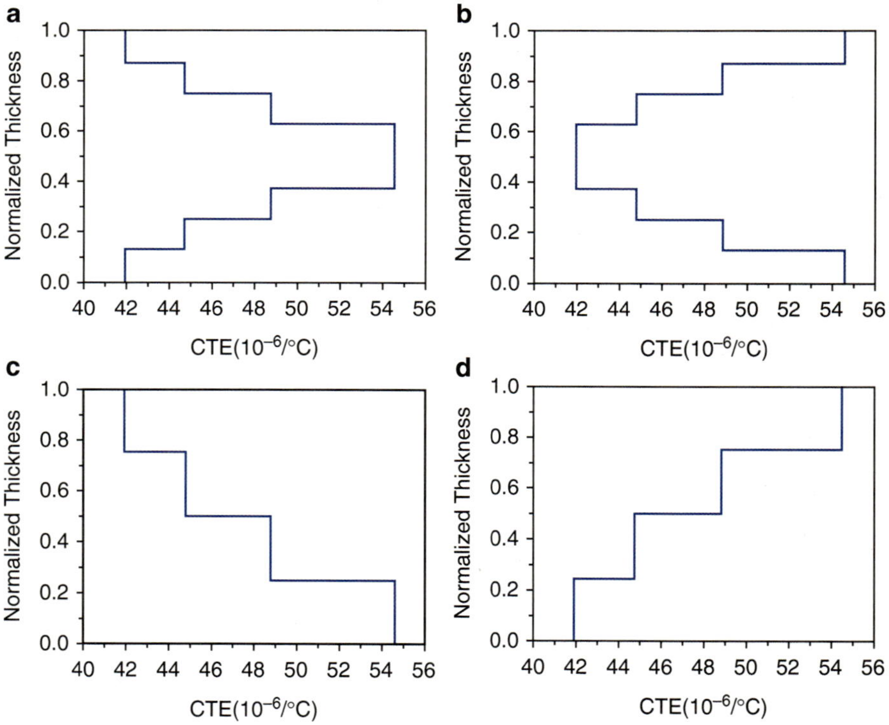

Fig. 4.31 Variation of CTE (coefficient of thermal expansion) across the thickness. (**a**) FGN-1, (**b**) FGN-2, (**c**) FGN-3, (**d**) FGN-4 [111]

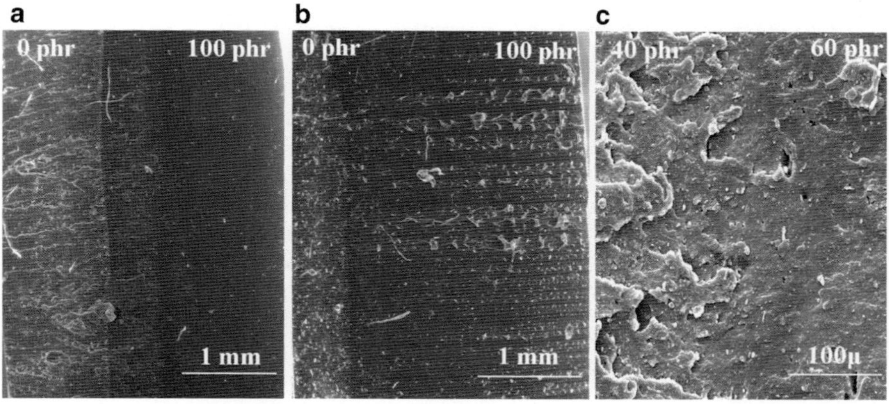

Fig. 4.32 SEM micrograph showing (**a**) fractured surface of 0-20-40-60-80-100 carbon black-graded NR FGPNC, (**b**) micrograph at the interface of 40–60 layer, and (**c**) side surface of the specimen elongated to 100 % of 0-20-40-60-80-100 carbon black-graded NR FGPNC [112]

during cooling. In case of FGN-1 (Fig. 4.31a), CTE values increase as we go to the core part of the component that results in developing thermal residual compressive stresses. This prevents the delamination of the inner layers during tensile stresses and increases the load-bearing capacity of the structure.

Ahankari et al. have prepared various stacks of carbon black (CB)-reinforced natural rubber (NR) functionally graded polymer nanocomposites (FGPNCs) by construction layer method [112]. Figure 4.32 shows SEM fractograph of 0-20-40-60-80-100-graded structure (the numbers stand for phr (parts per hundred) values of the CB and the sequence shows the stacking order of the layers). One can see, at every layer interface, the fracture plane has changed. Stretching of the ligament ends at the interface of 40 and 60 layers as shown in Fig. 4.32b. Stacking of the layers might be leaving small tiny air pores, which are formed at the interface, are stretched, and can be easily seen in Fig. 4.32b. Figure 4.32c shows the SEM micrograph of side surface after 100 % elongation of the sample. One can observe the formation of corrugations/serrations (very large at the 100 phr end, and close/zero completely at 0 phr side).

The tensile stress–elongation curves corresponding to UDPNC and various FGPNCs, employing an average 50 phr CB, are shown in Fig. 4.33 [112]. It reveals that the roughness in the variation of CB (in 0-20-40-60-80-100-graded structure, the step variation is of 20, and in 100-50-0-0-50-100, the step variation is of 50) in the adjacent layers of the FGPNCs increases the modulus. Depending on the strength, modulus, and surface properties required, design can be optimized and layers can be stacked accordingly.

The enhancement of modulus in all FGPNCs is credited to the spatial variation of filler in the matrix. It is shown that for any possible combination of stacking sequences that govern the spatial or stepwise variation of filler demonstrates an enhancement in the modulus. At 100 % elongation, modulus increment in FGPNCs

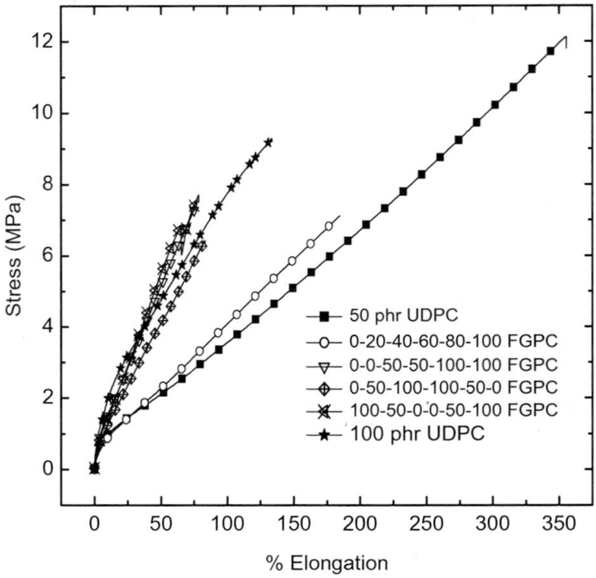

Fig. 4.33 Comparison of stress–elongation curves of uniformly dispersed polymer nanocomposites (UDPNCs) and various FGPNCs employing an average amount of 50 phr CB [112]

employing average 30 %, 40 %, and 50 phr is 35 %, 57 %, and 52 %, respectively (Fig. 4.34).

Ahankari et al. also have made carbon black-reinforced natural rubber FGPNCs (Fig. 4.35) [113]. Dynamic mechanical characterization has revealed that FGPNCs show much higher storage modulus than the corresponding UDPNCs for any given combination of stacking sequence (see Fig. 4.36). Loss tangent of FGPNCs is also observed to be lesser when compared to UDPNCs leading to less hysteretic losses followed by lesser heat buildup in the composite (not shown here). Hysteresis measurements are also accorded with the results of dynamic mechanical characterization.

Ahankari et al. have varied the concentrations of sulfur and accelerator in the nanocomposites of carbon black (CB)-filled SBR matrix to introduce the gradation of the cross-link density [114]. These curatives are varied from 1 to 11 phr (per hundred rubber) along the span of 3 mm thick sheet using the construction-based layering method. Variation in the cross-link density is calculated using Flory–Rehner equation (Eq. 4.5):

$$-ln(1 - V_r) - V_r - \chi V_r^2 = 2V_s \eta_{swell} \left(V_r^{1/3} - \frac{2V_r}{f} \right) \tag{4.5}$$

where V_r is the volume fraction of rubber in swollen gel, V_s is the molar volume fraction of toluene, χ is the rubber–solvent interaction parameter (0.413), η_{swell} is the cross-link density of rubber, and f is the functionality of the cross-links, respectively. With increasing cross-link density along thickness (Fig. 4.37), hardness and modulus increase, while the ultimate properties like tensile strength and elongation at break decrease (not shown here).

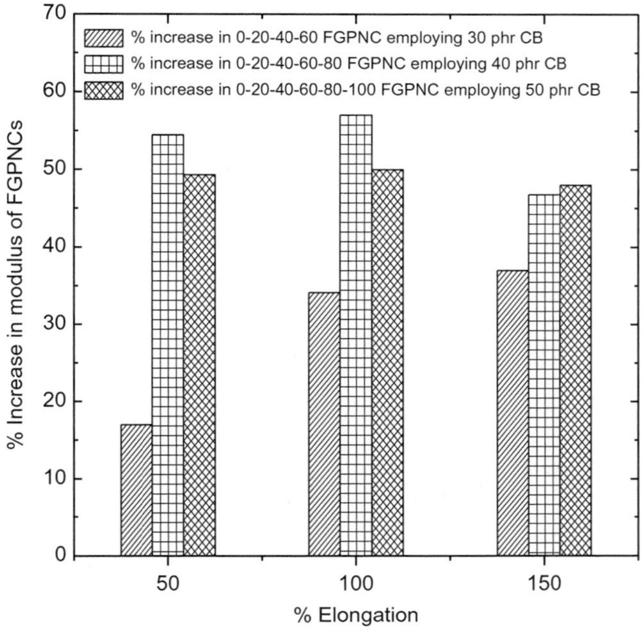

Fig. 4.34 % increase in the modulus of FGPNCs with respect to the corresponding UDPNCs [112]

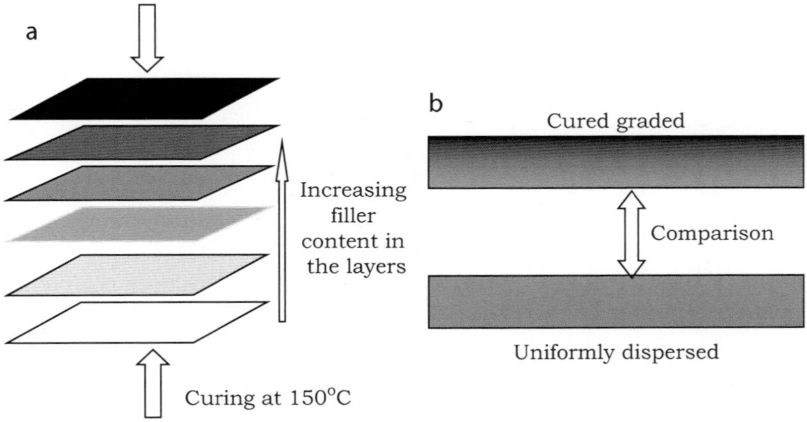

Fig. 4.35 (**a**) Stacking of the layers containing increasing amounts of CB in NR. (**b**) Comparison of FGPCs and UDPCs at a given thickness and filler loading [113]

The dynamic mechanical analysis of CB-graded SBR nanocomposites (FGPNCs) has shown the increment in the storage modulus than the UDPNCs employing the same average amount of curatives (not shown here). The peak position of tan δ_{max} remains at the same temperature while the value mitigates in FGPNCs (Fig. 4.38). In FGPNCs, tan δ peak displays the broadness in the transition region.

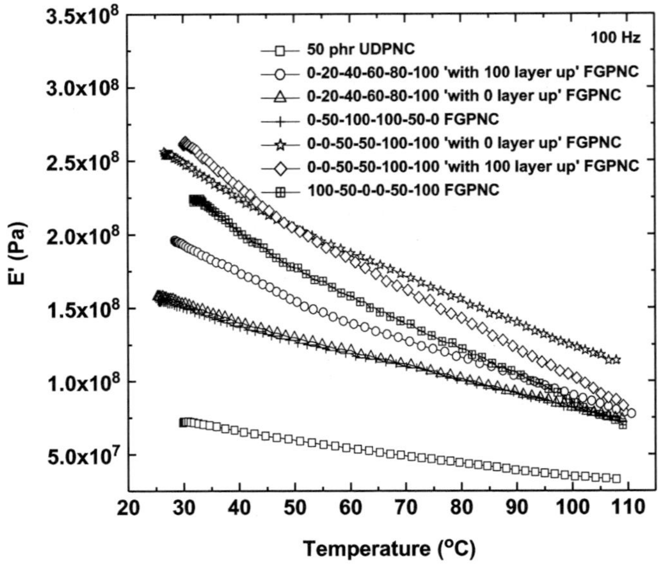

Fig. 4.36 Comparison of storage modulus values of UDPNCs and FGPNCs at 100 Hz [113]

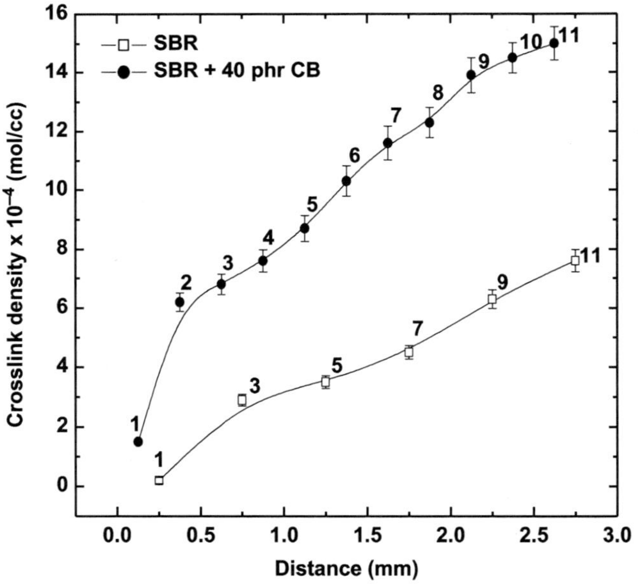

Fig. 4.37 Variation of cross-link density along the thickness of the vulcanizate (one surface of this FGPNCs has a sulfur/accelerator ratio of 1/1 and opposite surface of this FGPNCs has a sulfur/accelerator ratio of 11/11, numbers indicate the corresponding phr values of sulfur and accelerator) [114]

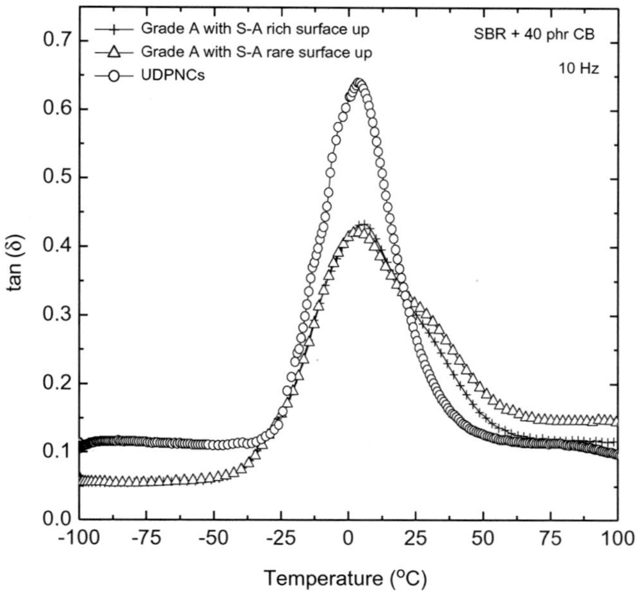

Fig. 4.38 Comparison of damping (tan δ) of the UDPNC and FGPNC employing the same average amount of sulfur and Mercaptobenzothiazole (MBT) in the composite (S–A in figure stands for sulfur–accelerator) [114]

Gradation of the material property "glass transition temperature" is also brought in the carbon black-reinforced SBR nanocomposite by varying the concentration of process oil [115]. Glass transition temperature (T_g) of FGMs is varied from −56 °C to −80 °C along the span of 3 mm thick sheet. Thermogravimetric analysis is carried out to check decomposition of different ingredients in the vulcanizates. Differential thermogravimetric (DTG) curves (Fig. 4.39) for the vulcanizates clearly reveal that there are two different degradation zones. The first mass loss corresponds to the oil and antioxidant decomposition with an onset at 155 °C. The second mass loss corresponds to the decomposition of SBR. The onset of degradation of SBR employing higher amounts of oil has shifted to higher temperature from 401 °C to 416 °C.

As the T_g is sensitive to the molecular interactions, measurement of the transition temperature is used to determine the structural changes occurring in the layers of FGMs. Carbon black hardly affects the T_g. DSC scans have shown the shift in the T_g for different layers of FGMs employing various oil content. With increasing oil concentration, the T_g decreases. With the variation of oil from 0 to 100 phr in the span of 3 mm thickness, the T_g decreases from −56 °C to −8 °C (Fig. 4.40).

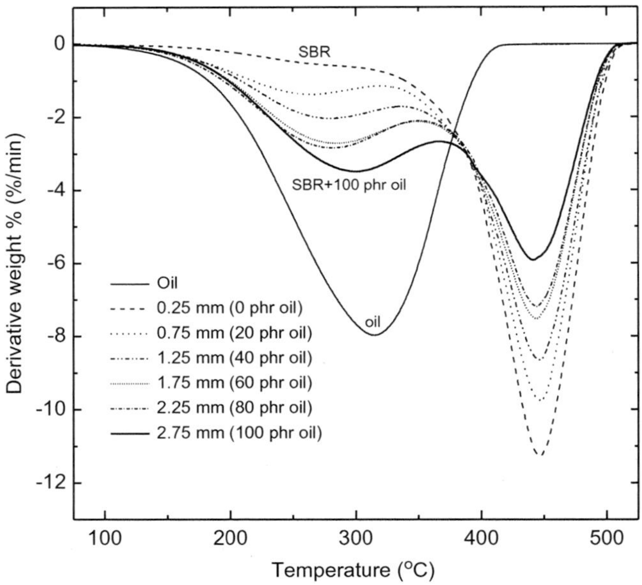

Fig. 4.39 First derivative of the weight curve of different layers of carbon black-filled SBR nanocomposites [115]

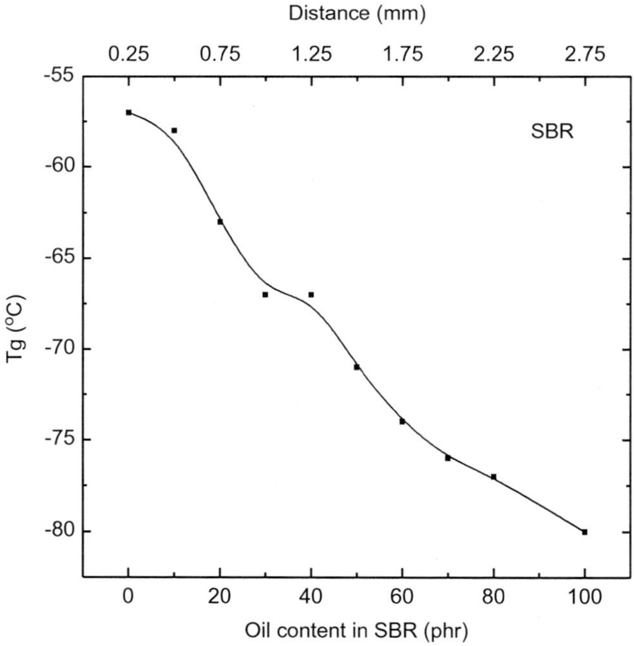

Fig. 4.40 Variation of glass transition temperature of different layers of carbon black-filled SBR nanocomposites [115]

Applications

The importance of FGMs was first realized in thermal applications. The ceramic–metal-graded composites, in the 1980s, were considered in Japan, for thermal barrier coatings protecting the space plane from high surface temperature and withstanding high-temperature gradient. The FGMs have attracted all-encompassing interest since then in diverse areas of mechanical and structural engineering, dentistry, orthopedics, electronics, bioengineering, etc. The applications of FGMs have encompassed the following areas.

Biomedical Living tissues like bones and teeth are characterized as FGMs from nature. To replace these tissues, a compatible material is needed that will serve the purpose of the original bio-tissue. The ideal candidate for this application is functionally graded material/composite. FGM has a wide range of applications in dental and orthopedic implants for teeth and bone replacement [116]. FGMs are the trump for the replacement of bones and joints. They not only provide biological compatibility but also strength and corrosion resistance. For the manufacturing of scaffolds for tissue engineering, the concept of gradation plays a primary role. Hydroxyapatite and collagen are the principal components for such scaffolds [117]. At Center for Nanoscale Materials and Biointegration lab, at the University of Alabama, Birmingham, Vohra and his team have successfully synthesized functionally graded scaffold as shown in Fig. 4.41 [117].

Another application of FGMs is in biological prosthesis owing to the minimized material mismatch, and when graded porosity is created, the bone can grow into it and become an integral part of the prosthesis [118]. The functionally gradient nanohydroxyapatite-reinforced polyvinyl alcohol (nanoHA/PVA) gel biocomposites have been a promising and excellent artificial articular cartilage repair material [119]. Fibrocartilage is a significant complex hierarchical soft tissue that supports

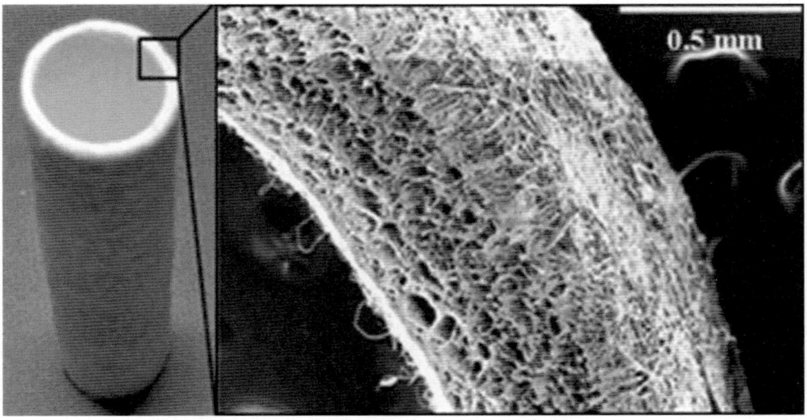

Fig. 4.41 Functionally graded electrospun vascular graft scaffold [117]

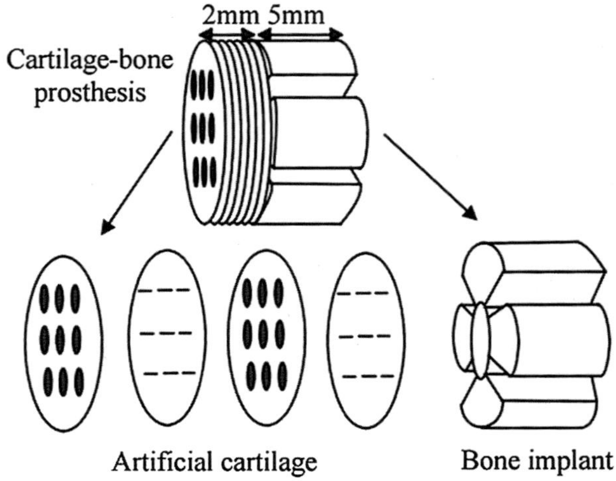

Fig. 4.42 Illustration showing a graded porous implant [14]

and constrains musculoskeletal joints. Sporting injuries, trauma, and degenerative pathologies contribute to loss in fibrocartilage integrity. Functional gradients in modulus and collagen are observed in this type of tissue and should ideally be considered in scaffold design. Graded porous implants can be employed for repairing bone–cartilage tissue (Fig. 4.42) [14].

The attachment of dissimilar materials is often a major engineering challenge because of high levels of localized stresses develop at such interfaces. An example of a biological system, where the properties abruptly change over a short distance, is a tendon-to-bone insertion site. These two materials with huge mismatch in stiffness (modulus of the bone is approximately 20 GPa and that of orthotropic tendon is 450 and 45 MPa, respectively) are connected through a short insertion of 1 mm (in humans). The load is transferred through the insertion by collagen fibers with mineral inclusions whose concentration varies from the mineral bone to tendon where mineral is absent. The transfer of stress without stress concentration is possible with two gradations, i.e., variation of mineral content from the bone to tendon and a variable collagen fiber orientation. ZrO_2/Ti functionally graded material is appealing in both medical and aerospace industries. Having good biocompatibility and non-toxicity, Ti can be of great interest in the medical field. ZrO_2-dispersed Ti matrix can be used for surgical implants [120].

Aerospace Actually, the concept of FGM has come up from aerospace industries. The contradictory properties like thermal conductivity and thermal barrier property in a component can be met with the concept of FGM. It enables the production of lightweight, strong, and durable materials. In the recent past, functionally graded coating is tried on the C–C composites to increase the oxidation resistance of these materials at much higher temperatures [121]. Even the manufacturing of functionally graded carbon–carbon is also attempted [122]. This material finds various applications in aerospace industries such as for rocket nozzles, space shuttles, exit cones, nose-tip of reentry vehicles, etc. [123]. Silicon carbide–carbon (SiC–C) and carbon–carbon

(C–C) composites are widely used as protective shields on the outside of space reentry vehicles and also in the combustion chamber components. It has been shown that applying a graded SiC–C interlayer between a C–C component and a SiC coating improves the lifetime of a space vehicle nose cone exposed to 1900 °C in an oxygen atmosphere. Other examples include heat exchange panels, spacecraft truss structure, reflectors, solar panels, camera housing, turbine wheels (operating above 40,000 rpm), nose caps, and leading edge of missiles and space shuttle [124, 128]. Thermally sprayed-graded coatings can enhance the properties of the coated parts, such as thermal and wear resistance. Yttria-stabilized zirconia sprayed on the top of NiCoCrAlY bond coat employed at various parts of aircraft engines [125]. Similarly, WC–Co wear-resistant cermet-graded coat used on airplane landing gear parts.

Defense One of the most important characteristics of functionally graded material is the ability to inhibit crack propagation. This property makes it useful in defense application, as penetration-resistant materials used for armor plates and bulletproof vests [126]. FGMs have been known to increase the level of ballistic protection, up to 20-folds, at a reduced weight. This can be achieved because the FGMs are made up of an extremely hard surface layer (to absorb the energy of impact), a multilayered-graded interface, and a tough metal backing that accommodates deformation after ballistic impact [127]. Exhaust wash structure that separates exhaust gas from aircraft structure for vehicles, which have internally exhausted engines, i.e., stealth aircraft and UAVs with engines that do not exhaust directly to atmosphere. Hot, high-speed engine exhaust gas flows over the top surface of exhaust wash structures which, in turn, causes large deflections. An FGM patch applied to the underside of the exhaust wash structure can be designed such that thermally induced deflection of the FGM patch is in a direction opposite to the exhaust wash structure deflection. Ceramic–metal FGMs are particularly suited for thermal barriers in space vehicles. FGMs have the added advantage that the metal side can be bolted onto the airframe rather than bonded as are the ceramic tiles used in the orbiter. Other possible uses include combustion chamber insulation in ramjet or scramjet engines [128]. Metal–ceramic FGMs are also employed in armor applications, where the hard ceramic frontal surface blunts the projectile, whereas the metallic back surface precludes penetration [114].

Energy Thermoelectric power generator effectively utilizes garbage as an energy source. Instead of single material to be operated at too high as well as at low temperature, the concept of gradation is utilized. Energy materials are aimed at the efficient conversion of energy. This is applied to any other energy conversion materials like sun energy–laser conversion. They also provide thermal barrier and are used as protective coating on turbine blades in gas turbine engine. Other examples are functionally graded dielectric films for capacitors, functionally graded composite electrodes for solid oxide fuel cells, graded thermoelectric materials for thermogenerators, graded thermal sensors, etc. [129–131].

Yin et al. designed a roofing system such that solar energy can be harvested through photovoltaics, and the heat in the PV module can be utilized for warming

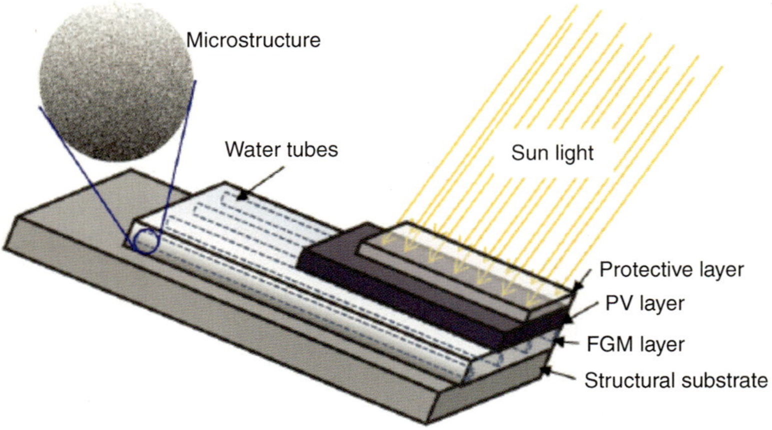

Fig. 4.43 Hybrid solar roofing panel with a functionally graded layer [132]

water that is flowing through the tubes [120]. Thus module temperature can also be controlled and PV efficiency can also be optimized (Fig. 4.43).

Automobile (SiC–SiC) combustion chambers, (AI–SiC) engine cylinder liners, (SiCw/AI alloy) diesel engine pistons, (E-glass/epoxy) leaf springs, (AI–C) drive shafts, (AI–SiC) flywheels, racing car brakes, and (SiCp/AI alloy) shock absorbers are few examples [133].

Optoelectronics Another application area of FGMs is optoelectronics. It has become possible to fabricate the photonic devices that could operate in a wide spectral range. FGMs can be used in the devices such as fibers, GRIN (graded index) lenses, GRINSCH (graded index of refraction separate confinement heterostructure) lasers, peltier junctions, etc. [134]. FGM also finds its application in optoelectronics as graded refractive index materials and in audio–video disks and magnetic storage media [135]. Metal–ceramic-graded composites are considered the best choices for the management of heat transfer and thermal response problems and in electronic packaging [136, 137].

Commercial and Industrial Pressure vessels, fuel tanks, cutting tool inserts, laptop cases, wind turbine blades, firefighting air bottles, MRI scanner cryogenic tubes, eyeglass frames, musical instruments, drilling motor shaft, X-ray tables, and helmets are few examples [121, 138]. Cho et al. have formed a functionally graded layer between the Cr–Mo shank and the ceramic tip and observed the improved thermal strength (Fig. 4.44) [139].

Others FGMs offer great promise, where the conditions are severe, e.g., heat exchanger tubes, rocket heat shields, wear-resistant linings for handling abrasive ore particles, etc. [140]. Carbide tools requiring superior wear resistance and

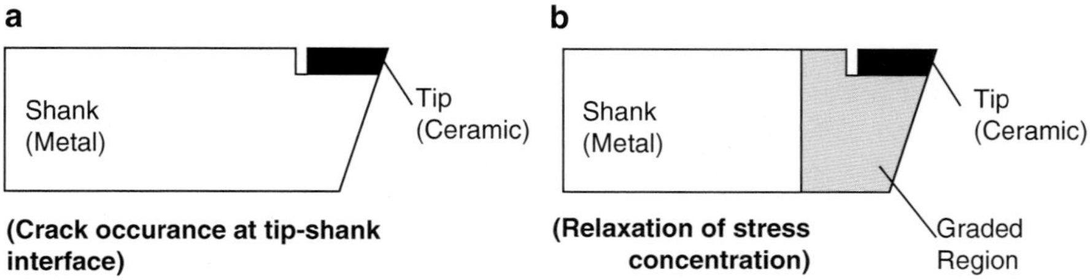

Fig. 4.44 Lathe metal cutting bites: (**a**) Conventional bimaterial type, (**b**) FGM design [139]

toughness avail the FGM concept. Graded WC–Co cutting tools have been used extensively, and it has been shown that resistance to wear may be achieved by appropriately grading the composition of the tool [141]. Chromium nitride films (of varying stoichiometry) produced by cathodic arc deposition are being evaluated for coatings on metal working dies. Functional gradient tool (FGT) is an important developing direction of modern precision mold; it can effectively improve the performance and reduce the cost of die mold [142].

Wear-resistant coatings based on FGMs applied on industrial machinery components (diesel engines, paper and pulp, oil industries) can reduce weight, increase adhesion strength, decrease internal stresses, and improve the resistance against propagation of surface defects. These coatings are usually deposited through high-velocity oxy-fuel or plasma spray processes that are based on cermet compositions such as WC–Co, Cr_3C_2–Mo (W, Ti)-Ni, and Mo_2C–Mo and perform well in abrasive and sliding wear. Mo-based alloy coatings offer excellent scuff resistance in dry unlubricated sliding contact; however, they are prone to brittle breakout of the coating and spallation. The use of WC–Co/SS FGMs for weight reduction and adhesion improvement due to increased material compatibility improves life [143]. The trilayer deposition of 50 nm Cr, 0.4 μm CrN-rich layer, and 3.6 μm Cr_2N-rich layer onto a tool steel substrate can give a combination of adhesion, wear resistance, and intrinsic residual stress. A graded interlayer can be used for joining two dissimilar materials so that the effect of the material mismatch is diffused over a greater distance. In addition, a graded layer may provide additional plastic deformation when metals are bonded to brittle materials.

FGM technology can be used in cellular phones for minimization of size and effective transmission. Other examples include titanium watches, baseball cleats, razor blades, fire retardant doors, etc.

Sensors for measuring the properties of molten glass require protective sensor sheaths in order to shield them from the extremely corrosive molten glass environment. $MoSi_2$ has shown to possess excellent corrosion resistance in molten glass, making it a candidate material for advanced sensor sheath applications [144]. $MoSi_2$-based compositionally graded coatings are used for protecting components against corrosion in molten glass environments [145].

Carbon nanotube-reinforced FGMs find are used in a variety of other applications like piezoelectric actuators, as furnace liners and thermal shielding elements in microelectronics, tools and dies for better thermal management, better wear resistance, reduce scrap, and improved process productivity. They can also be used as gas adsorbents, templates, actuators, catalyst supports, probes, chemical sensors, nanopipes, nano-reactors, etc. [146, 147].

Incorporation of FGM with SMA to form a new class of advanced composites called functionally graded shape memory alloy (FG–SMA) composites, which can withstand higher temperatures and possess better mechanical properties. These composites have great potential to be used in new sensor technology, information technology, and the emerging field of smart material systems [148–151]. These composites have higher corrosion resistance and different shape memory and pseudo-elastic behavior compared to homogenous SMAs [152].

The traditional solid-state welding process employed for joining aluminum and steel at the electric transition joints (ETJs) face the problems of cracking and separation at the interface. These problems are caused by the stress singularities due to mismatch of thermal and mechanical properties. The concept of FGMs overcomes the drawback of the bimetallic plate, such as ETJs due to gradual variation of composition (and hence thermal and mechanical properties) with 100 % of one material to the 100 % of the other material at the other end [153, 154].

Other areas of application of FGM are:

- Functionally graded layer between the shank and tip of a cutting tool enhancing thermal strength.
- Sandwich panels with graded facing–core interfaces.
- Graded fibers or graded particulate medium around the opening reducing stress concentration. Graded fibers/particles (variable geometry, volume fractions, or orientations) in the vicinity of the hole or an opening reduce the stresses [155, 156].
- Functionally graded prosthesis joints.
- Functionally graded piezoelectric actuators [157, 158]. Functionally graded polyester–calcium phosphate materials for bone replacement with a controlled in vitro polyester degradation rate.
- Functionally graded TBCs for combustion chambers.
- FGM metal–ceramic armor.
- Functionally graded heated floor systems.
- Solid oxide fuel cells. FGMs with graded porosity can be used for applications as electrodes of solid oxide fuel cells, etc. It is observed that the maximum power density increased when compared with the conventional design [159].
- Functionally graded dental implants.
- Functionally graded blades.

Concluding Remarks

This chapter presents an overview of the processing techniques that mark the emergence of advanced materials as FGMs, which will revolutionize the materials world. FGMs offer a relative variation of the properties of materials over distance (generally thickness) and direction. But there are many problems that need to be tackled to make this product. The biggest hurdle, the cost of manufacturing, must be made comparable to the traditional manufacturing methods. Solid freeform fabrication seems to be promising manufacturing method, but there are still lots of issues that need to be resolved including the cost of manufacturing, surface finish, etc.

To overcome the shortcomings of the traditional composites, researchers are highlighting on functionally graded materials (and nanoscale composite microstructures also). Such materials enhance the properties of the component in extreme environments. The major hurdle with constructing these structures is controlling microstructural changes and producing the composition gradients within the structure.

The following are the challenges that functionally graded materials need to overcome [111]:

- Development of low cost and high degree of automation for making the gradation
- Development of large size, complex-shaped functionally graded components
- Development of more uniform and accurate control of composition gradient in the structure
- Development of a process for a mass production

The list of applications is endless, and it will increase as the processing technology, cost of production, and properties of FGM improve [160]. The development of FGM needs an integration of multidisciplinary domains to work together in designing, manufacturing, and exploring more areas of applications.

Acknowledgment The authors acknowledge the financial support provided by the Department of Science and Technology, India, for carrying out this work.

References

1. Bever MB, Duwez PE (1972) Gradients in composite materials. Mater Sci Eng 10(1):1–8
2. Shen M, Bever MB (1972) Gradients in polymeric materials. J Mater Sci 7(7):741–746
3. Lee WY, Bae YW, More KL (1995) Synthesis of functionally graded metal-ceramic microstructures by chemical vapor deposition. J Mater Res 10(12):3000–3002
4. Ghosh A (1997) Functionally graded materials: manufacture, properties, and applications. American Ceramic Society, Westerville
5. Wen B, Wu J, Yu J (2004) A flat polymeric gradient material: preparation, structure and property. Polym 45(10):3359–3365
6. Krumova M, Klingshirn C, Haupert F, Friedrich K (2001) Microhardness studies on functionally graded polymer composites. Compos Sci Technol 61(4):557–563
7. Wang Q, Wang C (2012) Preparation and evaluation of a biomimetic scaffold with porosity gradients in vitro. Anis Da Academia Brasileira de Ciencias 84(1):9–16
8. Birman V, Keli T, Hosder S (2012) Functionally graded materials in engineering. In: Structural interfaces and attachments in biology. Springer, New York, pp 19–41
9. Abdul Khalil HPS, Bhat IUH, Jawaid M, Zaidon A, Hermawan D, Hadi YS (2012) Bamboo fibre reinforced biocomposites: a review. Mater Des 42:353–368

10. Neubrand A (2006) Functionally graded materials In: Editors-In-Chief: Robert KHJB, Merton WC, Bernard CF, Edward I Encyclopedia of materials: science and technology. Elsevier, Oxford, pp 3407–3413
11. Kayikci R, Sava S (2014) Fabrication and properties of functionally graded Al/AlB_2 composites. J Compos Mater 49(16):2029–2037
12. Mahamood RM, Akinlabi ET, Shukla M, Pityana S (2012) Functionally graded material: an overview. In: Proceedings of the world congress on engineering, VIII, London
13. Jamaludin SSS, Mustapha F, Nuruzzaman DM, Basri SN (2013) A review on the fabrication techniques of functionally graded ceramic-metallic materials in advanced composites. Acad J 8(21):828–840
14. Miao X, Sun D (2010) Graded/gradient porous biomaterials. Materials 3:26–47
15. Eso O, Fang Z, Griffo A (2005) Liquid phase sintering of functionally graded WC–Co composites. Int J Refract Met H 23(4–6):233–241
16. Tang X, Zhang H, Du D, Qu D, Hu C, Xie R, Feng Y (2014) Fabrication of W–Cu functionally graded material by spark plasma sintering method. Int J Refract Met H 42:193–199
17. Leite LJ, Salmoria GV, Paggi RA, Ahrens CH, Pouzada AS (2012) Microstructural characterization and mechanical properties of functionally graded PA12/HDPE parts by selective laser sintering. Int J Adv Manuf Technol 59(5–8):583–591
18. Neemat-Alla MM, Ata MH, Bayoumi MR, Khair-Eldeen W (2011) Powder metallurgical fabrication and microstructural investigations of Aluminium/Steel functionally graded material. Mater Sci Appl 2:1708–1718
19. Nazari A, Riahi S (2010) Effect of layer angle on tensile behavior of oblique layer functionally graded steels. Turk J Eng Environ Sci 34:17–24
20. Miao X, Sun D (2009) Graded/gradient porous biomaterials. Materials 3:26–44
21. Lee P, ASCE SM, Yin H, ASCE M (2014) Size effect on functionally graded material fabrication by sedimentation. J Nanomech Micromech 5:2153–5434
22. Zarabiana M, Yara AY, Vafaeenezhada S, Sania MAF, Simchia A (2013) Electrophoretic deposition of functionally-graded NiO–YSZ composite films. J Eur Ceram Soc 33(10):1815–1823
23. Hamidreza FH, Mohandesib JA, Çimenoğluc H (2015) Micro-scratch and corrosion behavior of functionally graded HA-TiO2 nanostructured composite coatings fabricated by electrophoretic deposition. J Mech Behav Biomed Mater 46:31–40
24. Katayama T, Sukenaga S, Saito N, Kagata H, Nakashima K (2011) Fabrication of Al_2O_3-W functionally graded materials by slip casting method. IOP Conf Ser Mater Sci Eng 18. Symposium 14:1-4
25. Gerdes T, Park HS, Pontiller P, Saberi A, Willert-Porada M (2014) Functionally graded refractory ceramics with dense surface coating. In: Blucher materials science proceedings, MM& FGM, vol 1. São Paulo: Blucher
26. Watanabe Y, Inaguma Y, Sato H, Fujiwara EM (2009) A novel fabrication method for functionally graded materials under centrifugal force: the centrifugal mixed-powder method. Materials 2(4):2510–2525
27. Mazare L, Miranda G, Soares DF, Silva FS (2010) Influence of solidification rates on a directional solidification process for the production of functionally graded materials. Int J of Mater Prod Technol 39(1/2):44–58
28. Yi S, Huang Z, Huang J, Fang M, Liu Y, Zhang S (2014) Novel calcium hexaluminate/spinel-alumina composites with graded microstructures and mechanical properties. Sci Rep 4:4333
29. Watanbe Y, Sato H (2011) In book: Nanocomposites with unique properties and applications in medicine and industry. Edited by John Cuppoletti, Source: InTech. Rigeka, Croatia
30. Watanabe Y, Eri MF, Sato H (2010) Fabrication of functionally graded materials by centrifugal slurry-pouring method and centrifugal mixed-powder method. J Jpn Soc Powder Metall 57 (5):321–326
31. Jayachandran M, Tsukamoto H, Sato H, Watanbe Y (2013) Formation behavior of continuous graded composition in Ti-ZrO_2 functionally graded materials fabricated by mixed-powder pouring method. J Nanomater, Special issue, Article ID 504631

32. Drenchev L, Sobczak J, Malinov S, Sha W (2003) Numerical simulation of macrostructure formation in centrifugal casting of particle reinforced metal matrix composites Part 2: simulations and practical applications. Modeling Simul Mater Sci Eng 11:651–674
33. Tsukamoto H (2010) Design against fracture of functionally graded thermal barrier coatings using transformation toughening. Mater Sci Eng A 527(13–14):3217–3226
34. Fu Y (2006) Review on current state and development of thermally sprayed functional gradient coatings. Cailiao Baohu 39(6):41–45
35. Schulz U, Peters M, Bach FW, Tegeder G (2003) Graded coatings for thermal, wear and corrosion barriers. Mater Sci Eng A 362(1–2):61–80
36. Nagarajan N, Nicholson PS (2004) Nickel-alumina functionally graded materials by electrophoretic deposition. J Am Ceram Soc 87(11):2053–2057
37. Acikbas NC, Suvaci E, Mandal H (2006) Fabrication of functionally graded SiAlON ceramics by tape casting. J Am Ceram Soc 89(10):3255–3257
38. Prinz D (1999) Functionally graded materials. Trans Tech Publications, Switzerland
39. Padhi D, Park S, Balasubramanian G, Rocha-Alvarez JC, Xia L, Witty DR, M'Saad H (2006) Method to deposit functionally graded dielectric films via chemical vapor deposition using viscous precursors. US Patent 7,802,538
40. Okamura H (1991) State of the art of material design projects for severe service applications. Mater Sci Eng A 143(1–2):3–9
41. Goswami B, Ray AK, Sahay SK (2004) Thermal barrier coating system for gas turbine application: a review. High Temp Mater Processes 23(2):73–92
42. Buyukkaya E (2008) Thermal analysis of functionally graded coating AlSi alloy and steel pistons. Surf Coat Technol 202(16):3856–3865
43. Groves JF, Wadley HNG (1997) Functionally graded materials synthesis via low vacuum directed vapor deposition. Compos Part B 28(1–2):57–69
44. Araki M, Sasaki M, Kim S, Suzuki S, Nakamura K, Akiba M (1994) Thermal response experiments of SiC/C and TiC/C functionally gradient materials as plasma-facing materials for fusion application. J Nucl Mater 212(215):1329–1334
45. Gao JW, Wang CY (2000) Modeling the solidification of functionally graded materials by centrifugal casting. Mater Sci Eng A 292(2):207–215
46. Dutta D, Prinz FB, Rosen D, Lee W (2001) Layered manufacturing: current status and future trends. J Comput Inf Sci Eng 1(1):60–71
47. Gallant FM, Bruck HA, Kota AK (2004) Fabrication of particle-reinforced polymers with continuous gradient architectures using twin screw extrusion process. J Compos Mater 38:1873–1893
48. Kieback B, Neubrand A, Riedel H (2003) Processing techniques for functionally graded materials. Mater Sci Eng A 362(1–2):81–106
49. Pompe W, Worch H, Epple M, Friess W, Gelinsky M, Greil P, Hemple U, Scharnweber D, Schulte K (2003) Functionally graded materials for biomedical applications. Mater Sci Eng A 362(1–2):40–60
50. Kirihara S (2014) Three dimensional printing of metals and ceramics graded dendrites for energy and material flows modulation. In: Blucher materials science proceedings 1(1), MM & FGM
51. Han SW, Ji WJ, Moon YH (2014) Fabrication of gear having functionally graded properties by Direct Laser Melting Process. Adv Mech Eng 6:609–735
52. Bourban P, EBuhler M, Manson JAE, Mathieu LMM, Pioletti D, Stadelmann V (2008) Cellular gradient polymer composites. US Patent 2,008,007,332
53. Li J, Liu T, Xia S, Pan Y, Zheng Z, Ding X, Peng Y (2011) A versatile approach to achieve quintuple-shape memory effect by semi-interpenetrating polymer networks containing broadened glass transition and crystalline segments. J Mater Chem 21:12213–12217
54. Senda K, Matsuda T, Kawanishi T, Tanaka K, Usui H (2013) Preparation of a functionally graded fluoropolymer Thin film and its application to antireflective coating. Jpn J Appl Phys 52(5S1)

55. Mukawa Y, Kondo A, Koike Y (2012) Optimization of the refractive-index distribution of graded-index polymer optical fiber by the Diffusion-Assisted Fabrication Process. Appl Phys Express 5(4) 042501-1/3
56. Arrue J, Jiménez F, Aldabaldetreku G, Durana G, Zubia J, Lomer M, Mateo J (2008) Analysis of the use of tapered graded-index polymer optical fibers for refractive-index sensors. Opt Express 16(21):16616–16631
57. Koerdta M, Kibbenb S, Hesselbachc J, Braunerc C, Herrmannd AS, Vollertsenb F, Krolla L (2014) Fabrication and characterization of bragg gratings in a graded-index perfluorinated polymer optical fiber. Procedia Technol 15:138–146
58. Kusanagi H, Masuda S, Nose T, Sato S (1999) Graded index type liquid crystal microlens using small amount of polymer. Mater Sci Forum 308–311:591–596
59. Li HZL, Scheiding S, Gebhardt A, Risse S, Eberhardt R, Yi AY (2013) Fabrication of three dimensional functional microstructures on curved substrates using three dimensional microlens projection. J Micro Nano Manuf 1(1)
60. Mu ZG, Wang YS (2007) Study on impact property of EP/PU graded interpenetrating polymer networks (IPNs). Reg uxingShuzhi 22(4):18–24
61. Kikutani T (1999) High-speed melt spinning process and control of synthetic fiber structure. Sen-i Gakkaishi 55(11):391–395
62. Liu XQ, Wang YS, Zhu JH (2004) Epoxy resin/polyurethane functionally graded material prepared by microwave irradiation. J Appl Polym Sci 94:994–999
63. Xie XM, Matsuoka M, Takemura K (1992) Formation of gradient phase structure during annealing of a polymer blend. Polym 33(9):1996–1998
64. Parmeswaran V, Shukla A (2000) Processing and characterization of a model functionally gradient material. J Mater Sci 35:21–29
65. Shukla A, Jain N (2004) Dynamic damage growth in particle reinforced graded materials. Int J Impact Eng 30:777–803
66. Ivosevic M, Knight R, Khalidindi R, Palmese GR, Sutter JK (2006) Solid particle erosion resistance of thermally sprayed functionally graded coatings for polymer matrix composites. Surf Coat Technol 200(16):5145–5151
67. Fukui Y, Kinoshita H, Nakanishi K (1992) A simulative experiment for characterizing the strength of functionally gradient material. JSME Int J Ser I 35(1):597–607
68. Akiyama S (1999) Effect of gradient of the crystalline phase on semicrystalline polymers. Polym Prepr Jpn 48(1):94
69. Butcher RJ, Rousseau CE, Tippur HV (1999) A functionally graded particulate composite: preparation, measurements and failure analysis. Acta Mater 47(1):259–268
70. Ogumi Z, Abe T, Nakamura S, Inaba M (1999) Functionally gradient polymer electrolyte prepared by plasma. Solid State Ion 121(1–4):289–293
71. Stellbrink KKU, Hausser G, Steegmuller R (1999) One component composites as functionally gradient materials. J Thermoplast Compos Mater 12(3):188–200
72. Tilbrook MT, Moon RJ, Hoffman M (2005) Finite element simulations of crack propagation in functionally graded materials under flexural loading. Eng Fract Mech 72:2444–2467
73. Uhlig C, Bauer J, Bauer M (1994) Rubber toughening of polycyanurates. Polym Mater Sci Eng 71:748
74. Elghandour EI (1994) Effect of stacking sequence of laminated composites with different cure on the fracture mechanics at elevated temperature. In: 3rd international symposium on structural and functional gradient materials, Laussane
75. Lee NJ, Jang J (2000) The effect of fiber content gradient on the mechanical properties of glass-fiber-mat/polypropylene composites. Compos Sci Technol 60(2):209–217
76. Ikeda Y (2003) Graded styrene-butadiene rubber vulcanizates. J Appl Polym Sci 87(1):61–67
77. Bruck HA, Moore CL, Valentine TM (2003) Bending actuation in polyurethanes with a symmetrically graded distribution of one-way shape memory alloy wires. Exp Mech 44(1):62–70
78. Lee NJ, Jang J, Park M, Choe CR (1997) Characterization of functionally gradient epoxy/carbon fiber composite prepared under centrifugal force. J Mater Sci 32(8):2013–2020

79. Klingshirn C, Koizumi M, Haupert F, Giertzsch H, Friedrich K (2000) Structure and wear of centrifuged epoxy-resin/carbon fiber functionally graded materials. J Mater Sci Lett 19 (3):263–266
80. Tsotra P, Friedrich K (2004) Short carbon fiber reinforced epoxy resin/polyaniline blends: their electrical and mechanical properties. Compos Sci Technol 64:2385–2391
81. Agari Y, Shimada M, Ueda A, Nagai S (1996) Preparation, characterization and properties of gradient polymer blends: discussion of poly (vinyl chloride)/poly(methyl methacrylate) blend films containing a wide compositional gradient phase. Macro Mol Chem Phys 197(6):2017–2033
82. Koide S, Yazawa K, Asakawa N, Inoue Y (2007) Fabrication of functionally graded bulk materials of organic polymer blends by uniaxial thermal gradient. J Mater Chem 17:582–590
83. Sato S, Nose T, Masuda S, Yanase S (1999) Functionally graded optical polymer materials prepared using UV curable liquid-crystals with an electric field. Mater Sci Forum 308:567–572
84. Lambros J, Santare MH, Li H, Sapna GH (1999) A novel technique for the fabrication of laboratory scale model functionally graded materials. Exp Mech 39(3):184–190
85. Abanto-Bueno J, Lambros J (2002) Investigation of crack growth in functionally graded materials using digital image correlation. Eng Fract Mech 69:1695–1711
86. Kalyon DM, Erisken C, Ozkan S, Ergun-Butros A, Yu X, Wang H, Valdevit H, Ritter A (2014) Functionally-graded polymeric graft substitutes and scaffolds for tissue engineering can be fabricated via various extrusion methods. Tissue Sci Eng 5(1)
87. Laia MS, Jorge DR, Neri O (2014) Water-based robotic fabrication: large-scale additive manufacturing of functionally graded hydrogel composites via multichamber extrusion. 3D Print Addict Manuf 1(3):141–151
88. Ergun A, Chung R, Ward D, Valdevit A, Ritter A (2012) Unitary bioresorbable cage/core bone graft substitutes for spinal arthrodesis coextruded from polycaprolactone biocomposites. Ann Biomed Eng 40:1073–1087
89. Ergun A, Yu X, Valdevit A, Ritter A, Kalyon DM (2011) Vitro-analysis and mechanical properties of twin screw extruded single-layered and coextruded multilayered poly (caprolactone) scaffolds seeded with human fetal osteoblasts for bone tissue engineering. J Biomed Mater Res A 99:354–356
90. Ergun A, Yu X, Valdevit A, Ritter A (2011) Compressive fatigue behavior of osteoblast seeded tissue constructs of poly(caprolactone) multilayered scaffolds for bone graft substitute applications. Ortho Res Soc Trans 36:1850
91. Ergun A, Yu X, Valdevit A, Ritter A, Kalyon DM (2012) Radially and axially-graded multizonal scaffolds targeting critical-sized bone defects from polycaprolactone/hydroxyapatite/tricalcium phosphate. Tissue Eng Part A 18:2426–2436
92. Ozkan S, Kalyon DM, Yu X, McKelvey C, Lowinger M (2009) Multifunctional protein-encapsulated polycaprolactone scaffolds: fabrication and in vitro assessment for tissue engineering. Biomaterials 30:4336–4347
93. Ozkan S, Kalyon DM, Yu X (2010) Functionally graded beta-TCP/PCL nanocomposite scaffolds in vitro evaluation with human fetal osteoblast cells. J Biomed Mater Res A 92:1007–1018
94. Erisken C, Kalyon DM, Wang HA (2008) Hybrid twin screw extrusion/electrospinning method to process nanoparticle-incorporated electrospun nanofibres. Nanotechnology 19:165302
95. Erisken C, Kalyon DM, Wang H (2008) Functionally and continuously graded electrospun polycaprolactone and ß-tricalcium phosphate nanocomposites for interface tissue engineering applications. Biomaterials 29:4065–4073
96. Erisken C, Kalyon DM, Wang H (2010) Viscoelastic and biomechanical properties of osteochondral tissue constructs generated from graded polycaprolactone and beta-tricalcium phosphate composites. J Biomeh Eng 132:091013
97. Erisken C, Kalyon DM, Ornek-Ballanco C, Wang H, Xu J (2011) Osteochondral tissue formation through adipose-derived stem cell differentiation using biomimetic tissue scaffolds with graded stimulator concentrations. Tissue Eng A 17:1239–1252

98. Leu MC, Deuser BK, Tang L, Landers RG, Hilmas GE, Watts JL (2012) Freeze-form extrusion fabrication of functionally graded materials. CIRP Ann Manuf Technol 61:223–226
99. Pojman JA, Chekanow YA, Case C, McCardle T (2000) Functionally graded polymeric materials prepared via frontal polymerization. J Appl Polym Sci 78:2398
100. Perry JW, Mansour K, Lee IYS, Wu XL, Bedworth PV, Chen CT, Ng D, Marder SR, Miles P, Wada T, Tian M, Sasabe H (1996) Organic optical limiter with a strong nonlinear absorptive response. Science 273(5281):1533–1536
101. Furukawa M, Okazaki T, Shiiba T (1997) Aggregation structure and mechanical properties of functionally graded polyurethane elastomers. J Nat Rubber Res 41(20):2355–2364
102. Mathew AP, Varghese H, Thomas S (2005) Electrical properties of nanostructured interpenetrating polymer networks based on natural rubber. J Appl Polym Sci 98 (5):2017–2026
103. Danielsson M, Grenestedt JL (1998) Gradient foam core materials for sandwich structures: preparation and characterization. Compos Part A 29(8):981–988
104. Mcdonald PM, Zhu XX, Uemura Y (2003) Diffusion in transient polymer networks: solvent, solute and polymer. Polym Prep 44(1):289
105. Wang HR, Cima MJ, Sachs EM (2004) Alumina-doped silica gradient-index (GRIN) lenses by slurry based three-dimensional printing (S-3DP). J Non-Cryst Solids 349:360–367
106. Kumar R, Chandrappa CN (2014) Synthesis and characterization of Al-SiC functionally graded material composites using powder metallurgy techniques. Int J Innov Res Sci Eng Technol 3(8):2319–8753
107. Gandra J, Vigarinho P, Pereira D, Miranda RM, Velhinho A, Vilakha P (2013) Wear characterization of functionally graded Al–SiC composite coatings produced by Friction Surfacing. Mat Des 52:373–383, Elsevier, Oxford
108. Tian H, Schryvers D, Mohanchandra KP, Carman GP, Humbeeck JV (2009) Fabrication and characterization of functionally graded Ni–Ti multilayer thin films. Funct Mater Lett 2 (2):61–66
109. Tsukamoto H (2015) Nano- and micro-indentation characterization of ZrO2/Ti functionally graded materials fabricated by Spark Plasma Sintering. Mat Sci Eng A 640:338–349, Elsevier, Oxford
110. Yuan H, Li J, Xiong Y, Luo G, Shen Q, Zhang L (2012) Preparation and characterization of PMMA graded microporous foams via one-step supercritical carbon dioxide foaming. J Phys Conf Ser 419. ISSN 1742–6596
111. Bafekrapour E, Simon GP, Habsuda J, Naebe M, Yang C, Bronwyn F (2012) Fabrication and characterization of functionally graded synthetic graphite/phenolic nanocomposites. Mater Sci Eng A 545:123–131, Elsevier, oxford
112. Ahankari SS, Kar KK (2010) Processing and mechanical behavior of carbon black graded rubber compounds. J Appl Polym Sci 115(6):3146–3154
113. Ahankari SS, Kar KK (2010) Hysteresis measurements and dynamic mechanical characterization of functionally graded natural rubber – carbon black nanocomposites. J Appl Poym Sci 50(5):871–877
114. Ahankari SS, Kar KK (2008) Processing of styrene butadiene rubber-carbon black nanocomposites with gradation of crosslink density: static and dynamic mechanical characterization. Mater Sci Engg A 491(1–2):454–460
115. Ahankari SS, Kar KK (2008) Processing and characterization of functionally graded materials through mechanical properties and glass transition temperature. Mater Lett 62(19):3398–3400
116. Al-Sanabani FA, Madfa AA, Al-Qudaimi NH (2014) Alumina ceramic for dental applications: a review article. Amer J Mater Res 1(1):26–34
117. Henriques B (2013) Inhomogeneous materials perform better: functionally graded materials for biomedical application. J Powder Met Min 2(3):1–2
118. Oshkour AA, Osman NAA, Yau YH, Tarlochan F, Abas WABW (2013) Design of new generation femoral prostheses using functionally graded materials: a finite element analysis. Proc Inst Mech Eng J Eng Med 227(1):3–17

119. Abedi G, Sotoudeh A, Soleymani M et al (2011) A collagen-poly(vinyl alcohol) nanofiber scaffold for cartilage repair. J Biomat Sc Polyme 22(18):2445–2455
120. Nugroho AW, Leadbeater G, Davies IJ (2010) Processing of a porous titanium alloy from elemental powders using a solid state isothermal foaming technique. J Mater Sci Mater Med 21:3103–3107
121. Kim JI, Kim WJ, Choi DJ, Park JY, Ryu WS (2005) Design of a C/SiC functionally graded coating for the oxidation protection of c/c composites. Carbon 43(8):1749–1757
122. Forest ML, Parker CA, Dillon F, Siegmund TH, Copra RJ, Fatz AE, Braunisch PF, Cordell T (2006) Manufacture of functionally graded carbon carbon composites US Patent 1,085,293,3
123. Udupa G, Rao SS, Gangadharan KV (2014) A review of carbon nanotube reinforced aluminium composite and functionally graded composites as a future material for aerospace application. Int J Mod Eng Res 4(7):13–22
124. Kumar S, Reddy KVVSM, Kumar A, Rohini Devi G (2013) Development and characterization of polymer–ceramic continuous fiber reinforced functionally graded composites for aerospace application. Aerosp Sci Technol 26(1):185–191
125. Clarke DR. Materials selection guidelines for low thermal conductivity thermal barrier coatings. Materials Department, College of Engineering, University of California, Santa Barbara. 93106–5050
126. Gupta N, Prasad VVB, Madhu V, Basu B (2012) Ballistic studies on TiB_2-Ti functionally graded armor ceramics. Def Sci J 62(6):382–389
127. Aydin M, Apalak MK, Gunes R, Reddy JN (2014) An experimental study on ballistics performance: functionally graded sandwich plate impacted by a 9 mm parabellum projectile. Blucher Mat Sci Proc 1(1):1–4
128. Bohidar SK, Sharma R, Mishra PR (2014) Functionally graded materials: a critical review. Int J Sci Footpr 2(4):18–29
129. Subramanyam G, Cole MW, Sun NX, Kalkur TS, Sbrockey NM, Tompa GS, Guo X, Chen C, Alpay SP, Rossetti GA Jr, Dayal K, Chen L, Schlom DG (2013) Challenges and opportunities for multi-functional oxide thin films for voltage tunable radio frequency/microwave components. Jpn J Appl Phys 114(191301):1–35
130. Woolley RJ, Skinner SJ (2014) Functionally graded composite $La_2NiO_4 + \delta$ and $La_4Ni_3O_{10}$-δ solid oxide fuel cell cathodes. Solid State Ion 255:1–5
131. Dharmin P, Khushbu P, Chetan J (2012) A review on stress analysis of an infinite plate with cut-outs. Int J Sci Res Pub 2(11)
132. Yin HM, Yang DJ, Kelly G, Garant J (2013) Design and performance of a novel building integrated PV/thermal system for energy efficiency of buildings. Sol Energy 87:184–195
133. Miteva AM (2014) An overview of functionally graded materials. Int Virt J Sci Tech Innov Indus 7(3):13–16
134. Sun H, Woodward J, Yin J, Moldawer A, Pecora EF, Nikiforov AY, Negro LD, Paiella R, Ludwig K Jr, Smith DJ, Moustakas TD (2013) Development of AlGaN-based graded-index-separate-confinement heterostructure deep UV emitters by molecular beam epitaxy. J Vac Sci Technol 31(3) 03C117
135. Bulatova R, Bahl C, Andersen K, Kuhn LT, Pryds N (2015) Functionally graded ceramics fabricated with side-by-side tape casting for use in magnetic refrigeration. Int J Appl Ceram Technol 12:891–898
136. Lee HL, Chang WJ, Chen WL, Yang YC (2012) Inverse heat transfer analysis of a functionally graded fin to estimate time-dependent base heat flux and temperature distributions. Energy Convers Manage 57:1–7
137. Das M, Guven I, Madenci E (2006) Coupled BEM and FEM analysis of functionally graded underfill layers in electronic packages. Proc 55th IEEE Electron Compon Technol Conf 995–1005
138. Bahria A, Salehia M, Akhlaghia M (2014) Using a pseudo-functionally graded interlayer in order to improve the static and dynamic behavior of wind turbine blade T-bolt root joints. Compos Interface 21(8):749–770

139. Cho JR, Park HJ (2002) High strength FGM cutting tools: finite element analysis on thermoelastic characteristics. J Mater Process Technol 130–131:351–356
140. Reddy BS, Kumar JS, Reddy CE, Kumar Reddy KV (2013) Buckling analysis of functionally graded material plates using higher order shear deformation theory. J Compos Mater vol. 2013 1–12. Article ID 808764
141. Fang ZZ, Fan P, Guo J (2011) Functionally graded cemented tungsten carbide with engineered hard surface and the method for making the same. EP2350331A2
142. Liu CS, He BH (2014) Research on functional gradient mold coating optimized preparation by electroplating and arc spray. J Chem Pharm Res 6(6):2136–2139
143. Tokita M (2014) Recent progress of spark plasma sintering (sps) method and industrial use of functionally graded materials (FGMs). Blucher Mater Sci Proc 1(1)
144. Petrovic JJ, Castro RG, Vaidya RU, Peters MI, Mendoza D, Hoover RC, Gallegos DE. (2008). Molybdenum Disilicide Materials for Glass Melting Sensor Sheaths. 25th Annual Conference on Composites, Advanced Ceramics, Materials, and Structures: A: Ceramic Engineering and Science. Proceedings, Materials Science and Technology Division Los Alamos National Laboratory Los Alamos, 22(3), 59–64.
145. Vaidya RU. Plasma sprayed functionally graded and layered $MoSi_2$-Al_2O_3 composites for high temperature sensor sheath application. In: International conference on intelligent processing and manufacturing of materials, Vancouver
146. Bharti I, Gupta N, Gupta KM (2013) Novel applications of functionally graded nano, optoelectronic and thermoelectric materials. J Mater Mech Manuf 1(3):221–224
147. Udupa G, Rao SS, Gangadharan KV (2012) Future applications of carbon nanotube reinforced functionally graded composite materials. In: International conference on advances in engineering, science and management, pp 399–404
148. Asadi H, Akbarzadeh AH, Wang Q (2015) Nonlinear thermo-inertial instability of functionally graded shape memory alloy sandwich plates. Compos Struct 120:496–508
149. Asadi H, Akbarzadeh AH, Chen ZT, Aghdam MM (2015) Enhanced thermal stability of functionally graded sandwich cylindrical shells by shape memory alloys. Smart Mater Structs 24(4) (in press)
150. DiOrio AM, Luo X, Lee KM, Mather PT (2011) A functionally graded shape memory polymer. Soft Matter 7:68–74
151. Lu H, Huang WM, Leng J (2014) Functionally graded and self assembled carbon nanofiller and boron nitride in nanopaper in electrical actuation of shape memory nanocomposites. Compos Part B Eng 62:1–4
152. Aysha SCPM, Varghese B, Baby A (2014) A review on functionally graded materials. Int J Eng Sci 3(6):90–101
153. Nemat-Alla M (2009) Reduction of thermal stresses by composition optimization of two-dimensional functionally graded materials. Acta Mech 208(3):147–161
154. Nemat-Alla M, Ata MH, Bayoumi MR, Eldeen WK (2011) Powder metallurgical fabrication and microstructural investigations of Aluminum/Steel functionally graded material. Mat Sci App 2:1708–1718
155. Yang Q, Gao CF, Chen W (2010) Stress analysis of functionally graded material plate with a circular hole. Arch Appl Mech 80:895–907
156. Cho HK, Rowlands RE (2009) Optimizing fiber direction in perforated orthotropic media to reduce stress concentration. J Compos Mater 43:1177–1198
157. Sheng GG, Wang X (2013) Nonlinear vibration control of functionally graded laminated cylindrical shells. Compos Part B Eng 52:1–10
158. Sheng GG, Wang X (2009) Active control of functionally graded laminated cylindrical shells. Compos Struct 90(4):448–457
159. Ni M, Leung MKH, Leung DYC (2007) Micro-scale modeling of solid oxide fuel cells with microstructurally graded electrodes. J Power Sources 168:369–378
160. Woodward B, Kashtalyan M (2012) Performance of functionally graded plates under localised transverse loading. Compos Struct 94(7):2254–2262

Nano-/Microcomposites by Electrodeposition

5

A. S. M. A. Haseeb

Contents

Abstract

Electrodeposition – one of the most commercially successful and inexpensive deposition methods – allows the fabrication of composite materials with feature sizes ranging from nanometers to micrometers. This method continues to attract a great deal of attention for the preparation of composites for structural and functional applications. This chapter presents an overview on the electrodeposition of composites. The fundamentals of electrodeposition are first presented. A concise account on the developments in the processing, properties, and applications of electrodeposited nano- and microcomposites is then given.

A.S.M.A. Haseeb (✉)
Department of Mechanical Engineering, University of Malaya, Kuala Lumpur, Malaysia
e-mail: haseeb@um.edu.my

K.K. Kar (ed.), *Composite Materials*,
DOI 10.1007/978-3-662-49514-8_5

Keywords
Electrodeposition • Electroplating • Codeposition mechanism • Microcomposites • Nanocomposites • Particulate composites • Laminated composites • Metal matrix • Polymer • Ceramics

Introduction

The fact that metal can be deposited from an electrolyte containing its ions by applying a cathodic potential to a substrate in an electrochemical cell is known to us from school days when Faraday's laws of electrolysis are taught. The process of electrodeposition is also known as electroplating, electrochemical deposition, electrochemical synthesis, etc. Electrodeposition is one of the most successful industrial processes for the deposition of mainly metallic coatings. However, the versatility of the technique also allows the deposition of composites with features at the micrometer and nanometer levels.

Electrodeposition has a history that dates back to antiquity. Artifact recovered from Loma Negra on the north coast of Peru suggests [1] that a variation of this technique was used as early as 200 CE to produce a thin gold layer on miniature copper mask. Based on recovered artifact that resembles battery in archaeological sites near the village of Khuyut Rabbou'a (near Baghdad, Iraq), a suggestion has been made that electrodeposition was used in what is today Iraq in circa 200 BC. However, the modern scientific efforts on electrodeposition date back to the days of Alessandro Volta (1745–1827) who discovered the electrochemical effect. It was Luigi Brugnatelli who first reported the electrodeposition of gold in 1805. While British and Russian scientists independently discovered the metal deposition process in 1883, the first patent of electrodeposition was actually issued to George Elkington and Henry Elkington in 1840 in Birmingham. An artist's depiction of an early electrodeposition shop and a modern electrodeposition facility in a clean room in a semiconductor industry is shown in Fig. 5.1

Since then, the process of electrodeposition has been used in wide range of industries including the automotive and aviation industries for the deposition of coatings for decorative, corrosion-resistant, and wear-resistant applications. Recently this process has established itself as an essential step in the semiconductor manufacturing process for copper metallization, where structural features down to nanometer level are routinely deposited.

Electrodeposition is a versatile technique. It can be used for the deposition of a wide range of materials. Particularly, it has been used to deposit composite materials for decades. This method is suitable for the deposition of different types of composites including laminated and particulate composites. One particular advantage this method can offer is that the structural dimensions of composites can be tailored from nanometer to millimeter levels. Although most suitable for metal matrix composites, recent study revealed the potential of this technique in allowing the deposition of ceramic- and polymer-based composites as well.

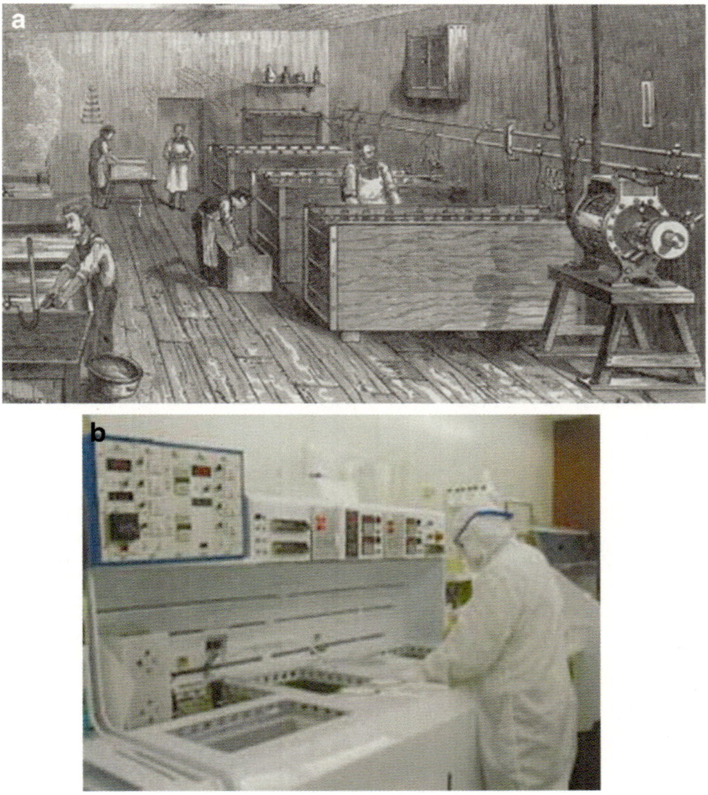

Fig. 5.1 (**a**) Artist's depiction of an early electrodeposition shop involved in nickel electrodeposition by dynamo-electricity and (**b**) a modern electrodeposition facility in a clean room in a semiconductor industry

In this chapter, electrodeposition of composite materials will be discussed. In section "Electrodeposition: An Overview," an overview of the fundamentals of electrodeposition is given. Two main types of composites, namely, particulate composites and laminated composites, will then be presented in sections "Particulate Composites by Electrodeposition" and "Layered Nanocomposites," respectively.

Electrodeposition: An Overview

Fundamentals of Electrodeposition

The principle of electrodeposition is quite straightforward. When two electrodes are immersed in an electrolyte containing a metal ion, the following reaction occurs at the cathode:

$$M^{+n} + ne^{-} = M \tag{5.1}$$

Metal ions contained in the electrolyte are discharged at the cathode and metal deposits on the cathode surface in the form of a thin film or powder/dendrite depending upon the deposition conditions. On the anode, a reverse reaction, i.e., metal dissolution, can take place as follows:

$$M = M^{+n} + ne^{-} \tag{5.2}$$

Other anodic reaction that consumes electron can also occur at the anode. Figure 5.2 presents a schematic of electrodeposition of silver on a spoon. The thickness of the deposited metal, h, can be described by Faraday's law:

$$h = \frac{t.A.I}{n.F.\rho} \tag{5.3}$$

where t is the deposition time, A is the atomic weight, I is the deposition current density, n is the number of electron transferred per atom, F is Faraday's constant, and ρ is the density of the deposited metal.

The above assumes that all the current goes into the metal deposition reaction. However, under some conditions, side reaction(s) can occur at the cathode. For instance, hydrogen evolution which robs away some current can take place at the cathode. In such a case, the thickness given by Faraday's law needs to be multiplied by a factor (less than 1) which is known as current efficiency.

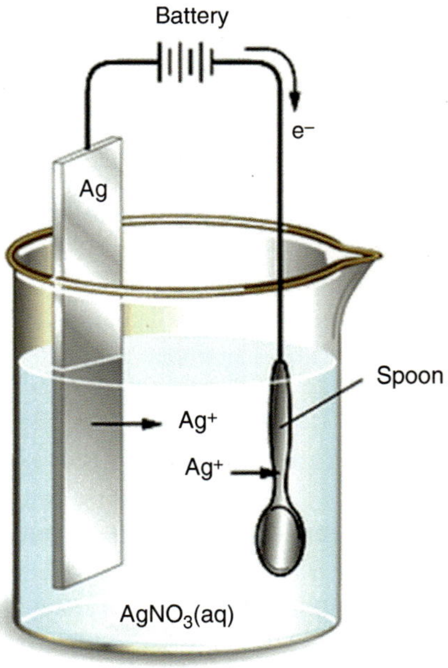

Fig. 5.2 Schematic of electrodeposition of silver spoon

The thermodynamics of electrodeposition is best described by the Nernst equation:

$$E_{\mathrm{eq}} = E_0 + (RT/nF)\ln\left(a_{\mathrm{reac}}/a_{\mathrm{prod}}\right) \tag{5.4}$$

where E_{eq} is the equilibrium potential, E_0 is the standard potential, R is the universal gas constant, and a_{reac} and a_{prod} represent the activities of the reactant and product, respectively.

In order to cause electrodeposition to happen at a perceptible rate, the applied potential should be greater than the equilibrium potential. The difference between the applied and equilibrium potential as given below is called overpotential:

$$\eta = E_{\mathrm{appl}} - E_{\mathrm{eq}} \tag{5.5}$$

Butler-Volmer equation is usually used to describe the kinetics of the electrodeposition reaction

$$i = i_0\left[\exp\left(\frac{(1-\alpha)nF\eta}{RT}\right) - \exp\left(\frac{-\alpha nF\eta}{RT}\right)\right] \tag{5.6}$$

where i is the deposition current density, i_0 is the exchange current density, α is the transfer coefficient, and η is the overpotential.

For large negative overpotential, the above equation can be simplified to

$$i \approx -i_0\exp\left(\frac{-\alpha nF\eta}{RT}\right) \tag{5.7}$$

This equation shows that as the overpotential, η, increases, the deposition current density and therefore deposition rate increase. However, under some conditions, mass transfer at electrodes may be limited. This will lead to a case where an increase in potential no longer results in an increase in current and a limiting current density is achieved. Deposition under such mass transfer-limited condition can lead to course powdery or dendritic deposits.

The following are the key electrodeposition parameters that influence the characteristics of electrodeposits:

- Overpotential/current density
- Bath chemistry: metal ion, complexing agent, surfactant, supporting electrolyte, and other additives
- Temperature
- pH
- Hydrodynamic condition of the bath

Nucleation and Growth

Nucleation and growth of a metal film in electrodeposition are by and large similar to that during vapor deposition except that ions rather than neutral atoms are involved at the initial steps. Ions from the electrolyte are first adsorbed on the substrate surface. The adsorbed ions also known as adions then diffuse on the surface to find a low-energy site. It is then incorporated into the lattice of the existing substrate or can form a nucleus by combining with requisite number of other wandering adions. Compared with vapor deposition, electrodeposition has got some additional complications which include: the presence of a very high electric field at the electrolyte substrate; ions are involved at the initial stage of deposition rather than atoms and the discharge of ions is an additional step; and the presence of other ions and species such as specifically adsorbed anion, solvated cations, water molecules, impurities, reaction products, etc. at the interface. The way these factors influence the nucleation and kinetics of growth is far from being clearly understood.

Overpotential is one of the most important factors that influences the rate and size of nucleation. It has been shown that the critical nucleus size, N_c, in electrodeposition is related to overpotential, η, according to the following:

$$N_c \infty \frac{1}{\eta^2} \tag{5.8}$$

The above equation implies that smaller grain size can be achieved at higher overpotential or in other words higher deposition current density. However, other factors, such as mobility of adions and the presence of other ions (which can be present originally or form simultaneously during electrodeposition) at the interface, can have overriding influence in determining crystal size or structure as they can affect local equilibrium.

Electrodeposition as a Deposition Process

One of the main characteristics of electrodeposition is that it is basically an atomistic deposition process. Thus it allows nanometric control of deposition leading to the creation of structural features a few nanometer in dimension. One the other hand, this process can also be speeded up to produce features in the range of micrometers even millimeters. It can be scaled up or down over a very wide range. Electrodeposition invariably occurs under nonequilibrium conditions and can lead to a wide range of structures. It has been demonstrated that this process can routinely produce amorphous structure, nanocrystalline materials, and metastable phases. It is a room temperature or near-room temperature technology. It is also a low-energy process and is therefore suited to dealing with soft matter like biological structure like DNA. Therefore the latter can be employed as templates to create novel nanostructures. One of the main advantages of electrodeposition is that it is an inexpensive process compared with other rival processes such as vacuum deposition.

In the next section ("Particulate Composites by Electrodeposition"), developments in the area of electrodeposited composites with particle reinforcements will be discussed.

Particulate Composites by Electrodeposition

Preparation of metal matrix composites by electrodeposition has been a topic of research since the late 1930s of the last century (Fink 1928). The early work of Fink and Prince [2] involved copper matrix composites containing graphite particles. The intension was to develop self-lubricating materials for automotive applications. Research on the electrodeposition of composites progressed slowly in the 1950s and 1960s [3]. Interest in electrodeposited MMC became more intense in the 1970s and 1980s. These early efforts were mainly directed toward the development of materials for wear-resistant applications. As a result, commercial MMC coatings have been developed which are currently being used in industries such as aerospace, automotive, manufacturing, chemical processing, and hydraulics (Yeh et al. 1994).

In earlier studies, metal was used exclusively as the matrix for the electrodeposited composites. A typical metal matrix composite prepared by electrodeposition is shown in Fig. 5.3. But in recent years, techniques are being developed to produce polymer matrix composites for novel applications. Ceramic matrix composites are also being developed, although mainly by a slightly different mechanism known as electrophoresis [5]. While most of the earlier studies involved reinforcing particles of micrometer dimensions, nanoparticles are nowadays increasingly being used as reinforcement. While research on the development of improved electrodeposition technique and new composites for wear- and corrosion-resistant applications continues, there is now a new trend toward the development of electrodeposited composites for functional applications like electrocatalysis, photocatalysis, actuators, etc. (see a review by [6]).

In what follows is an overview of the electrodeposited particular composites.

Matrix and Reinforcement Materials

Table 5.1 summarizes the matrix and reinforcement materials used in particulate composites deposited by electrodeposition. As has been mentioned earlier, the majority of the studies used metals as the matrix. This is because metals are the easiest to electrodeposit. Among the metals, nickel is the most widely used matrix as the technology of nickel electrodeposition is well established and it has a fairly good combination of properties like strength, hardness, toughness, corrosion resistance, etc. Other metal matrices investigated include Cu, Co, Au, Cr, Ag, Zn, and Al. Different alloys such as Ni-P, Cu-Zn, Ni-W, Zn-Ni, and Ni-Co have also been employed as matrix in the electrodeposited MMC. Recently polymers have been used as the matrix in a growing number of studies. Conductive polymers such as polypyrrole and polypropylene have been used as the matrix for polymer matrix

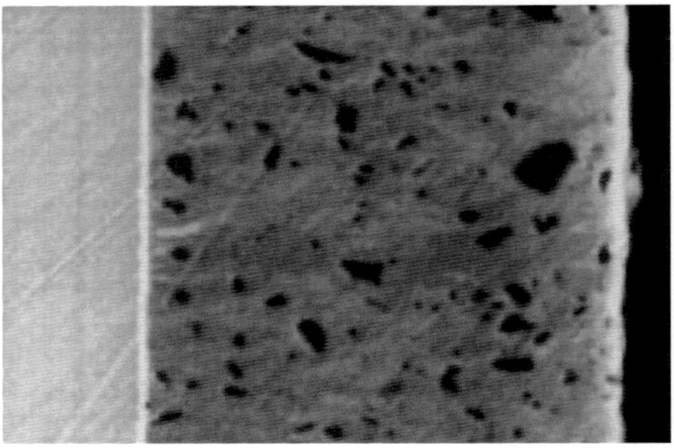

Fig. 5.3 SEM cross section of AuCu/B_4C electrodeposited from cyano-alkaline bath [4]

Table 5.1 Summary of materials used in particulate composites by electrodeposition

Components		Material type	Specific materials
Matrix		Metals	Ni, Cu, Co, Au, Cr, Ag, Zn, Al
		Alloys	Ni-P, Cu-Zn, Ni-W, Zn-Ni, Ni-Co
		Polymers	Polypyrrole, polypropylene
		Ceramics	Hydroxyapatite
Reinforcement	Microparticles	Ceramics	SiC, SiO_2, Al_2O_3, TiO_2, Si_3N_4, ZrO_2, ZrC, WC, CeO, Fe_3O_4, SnO_2, NiO, silicate, diamond
		Metals	Al, Ti, Cr
		Polymer	Polyethylene, polystyrene, PTFE
		Microcapsule/ liquid	Oil, paraffin
	Nanoparticles	Ceramics	SiC, Al_2O_3, Zr_2O, Si_3N_4, TiO_2, SiO_2, CNT, diamond
		Polymer	PTFE

composites by different researchers. A few studies have also used ceramics, e.g., hydroxyapatite as the matrix in electrodeposited/electrophoretically deposited composites.

As for the reinforcement materials, ceramic particles of micrometer dimensions have been the most popular. Ceramic particles such as SiC, SiO_2, and Al_2O_3 have been widely used. Others include TiO_2, Si_3N_4, ZrO_2, WC, ZrO_2, ZrC, CeO, Fe_3O_4, and diamond. In a number of studies, metal particles like Al, Ti, and Cr have been used together with a Ni matrix. The main aim has been to improve oxidation resistance of the composite coatings at high temperature. Polymer particles like polyethylene, polystyrene, and PTFE have been incorporated into metal matrix, most commonly Ni in different studies. Studies have also been conducted on an interesting group of particles which are in the form of microcapsules containing different types of liquid such as oil, paraffin, etc. for novel self-lubricating applications.

Incorporation of nanoparticles into electrodeposited composites has been a recent phenomenon. Increasing number of studies have been using nanoparticles as the reinforcement (see a review by [7]). Nanoparticles used in the electrodeposited MMC mainly include ceramics, carbon nanotube, PTFE, etc.

Mechanism of Codeposition of Particles in Electrodeposition

Codeposition of particles into electrodeposits is a complex process, the exact nature of which is not well understood. The main steps involved in the codeposition of particles into metal deposit during electrodeposition are described in the literature [7–9]. It is generally held that codeposition of particles involves the following main steps:

1. Ionic cloud formation on the suspended particles in the bath: this involves adsorption of ions on the particles. The adsorbed ions impart charges onto the particle surface.
2. Convective transfer of the particles toward the cathode.
3. Diffusion through a hydrodynamic boundary layer.
4. Diffusion through a concentration boundary layer.
5. Adsorption of particles at the cathode and their incorporation into the metal deposit.

The salient features of the mechanism of codeposition are given in Fig. 5.4 [8]. Stage 5 is a critical step in the codeposition of particles and involves complex interactions between the particle and electrode. At the initial stage, the particle is loosely adsorbed through a reversible electrostatic contact. The extent of adsorption depends on the balance between various forces. At a later stage, the particle that manages to stick to the electrode will be trapped into the depositing metal matrix. The exact nature of the particle-electrode is not well understood. It has been suggested [10, 11] that the interaction depends upon the force balance on a particle at the electrode surface (Fig. 5.5). Forces (F_r) that tend to remove the particle from the electrode are counteracted by friction force (F_f). The friction force, in turn, mainly results from forces that encourage particle adhesion (F_a). The removal forces (F_r) are hydrodynamic in nature. The adhesion force (F_a) can be contributed by a number of factors such as London-van der Waals force, electroosmotic force, electrophoretic force, and hydration force. Forces acting on the particles, like gravity, buoyancy, and hydrodynamic forces, may also contribute to adhesion depending upon electrode geometry.

The earliest model for the codeposition is due to Guglielmi [12]. The model of Guglielmi aims at correlating the particle concentration in the deposit (α) to that in the electrodeposition solution (C) and to find a relationship between particle incorporation as a function of current density. It was found experimentally in many systems that α increases with C initially reaching a saturation value – a trend that apparently followed Langmuir adsorption isotherm. Based on this, Guglielmi suggested his model as follows:

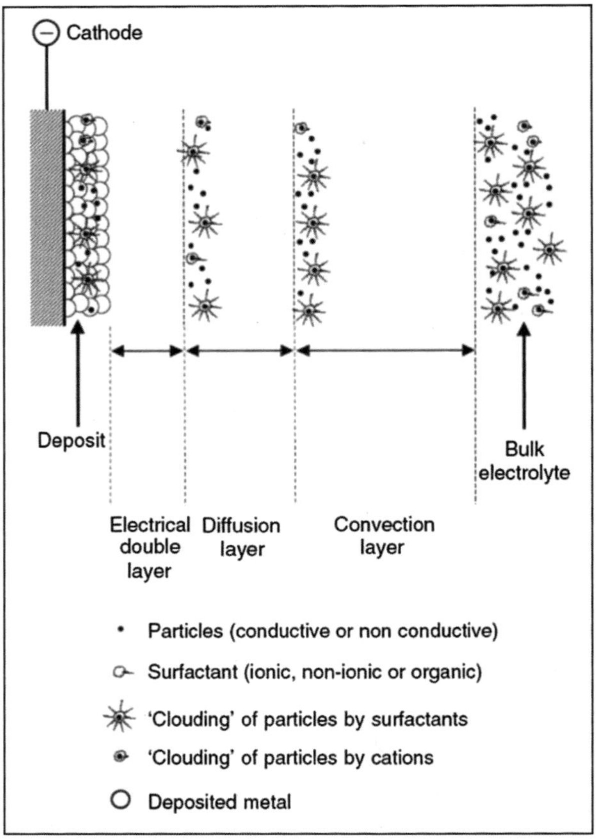

Fig. 5.4 Schematic showing the mechanism of particle codeposition into electrodeposit. The features are formation of ionic clouds around the particles (bulk electrolyte, typical dimension ≈ cm), convective movement toward the cathode (convection layer, typical thickness <1 mm), diffusion through a concentration boundary layer (diffusion layer, typical thickness ≈ hundreds of μm), and electrical double layer (typical thickness ≈ nm) followed by adsorption and entrapment of particles [8]

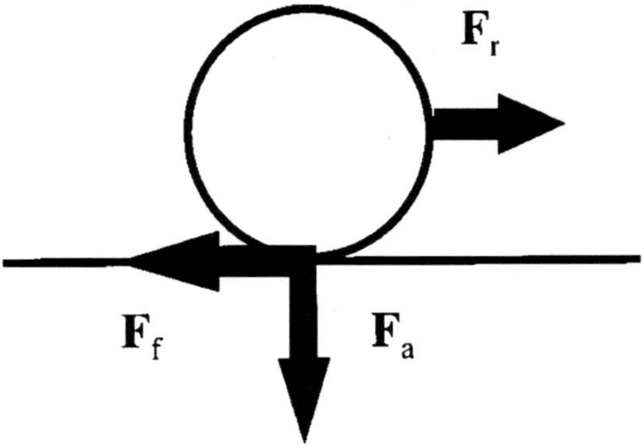

Fig. 5.5 Force balance on a particle adsorbed on an electrode, F_f = friction force, F_a = adhesion force, and F_r = force [10, 11]

$$\frac{C}{\alpha} = \frac{M.J_0}{n.F.d.V_0} \exp\{\eta.(A-B)\}.\left(\frac{1}{k}+C\right)$$

where

α: volume % of particles in suspension in the deposit (vol%)
η: overpotential (V)
A: constant in Tafel equation for metal deposition (V^{-1})
B: constant in Tafel equation for particle deposition (V^{-1})
C: volume % of particles in suspension in the bath ($g\ l^{-1}$ or vol%)
d: density of the electrodeposited metal ($g\ dm^{-3}$)
F: Faraday's constant ($F = 96{,}500$ C) ($C\ mol^{-1}$)
J_0: exchange current density ($A\ dm^{-2}$)
k: coefficient of adsorption ($l\ g^{-1}$ or $vol\%^{-1}$)
M: molecular weight of the electrodeposited metal ($g\ mol^{-1}$)
n: valence of the electrodeposited metal
V_0: constant for particle deposition ($dm\ s^{-1}$)

One of the main drawbacks of Guglielmi's model is that it does not take into account hydrodynamic effects in the bath and particle characteristics. Hwang et al. [104] and later Bercot et al. [13] put forward some modifications of the Guglielmi's model. Bercot suggested a corrective factor to be incorporated into Guglielmi's model to take into account the effects of adsorption and hydrodynamics for their experimental system which consists of 500 nm PTFE nanoparticles in nickel. Other models adopted statistical approach [14] and trajectory approach for non-Brownian particles [10].

Concentration of particles in the bath is a decisive parameter that influences the amount of codeposited particle in the composite. In general, the amount incorporated into the deposit increases with the amount of particles in the bath. However, the amount in the deposit usually reaches a saturation, beyond a certain concentration in the bath.

The type and magnitude of electric current employed also have a great influence on the incorporation. As indicated above, electrodeposition of metals is carried out on the cathode (negative electrode) and involves reduction. On the other hand, electrodeposition of conducing polymer is an oxidation process, and deposition in such a case happens on the anode (positive electrode). Deposition can be controlled by potential or current, the latter being more common in industrial practice. The most commonly used current type is dc. However, pulsed current has been found to offer a number of advantages in composite deposition. These include higher incorporation, finer structure, better dispersion, etc. Pulsed current can take various forms such as positive pulse, pulse reverse, square/rectangular pulse, sinusoidal pulse, etc. The density of deposition current has been found to have great influence on the incorporation of particles. Generally incorporation of particle increases with the increase of current density. However the incorporation can reach a saturation value and then decrease with further increase in current density. For instance, in the case of CTN, a different effect was observed. Although the incorporation increases initially, it decreases after a peak incorporation is achieved. This has been attributed to the fact that CNT, being highly conductive, can coat itself with the electrodeposited metal while still in solution near the electrode. Such coated CTN agglomerates settle down, thereby decreasing effectively the amount of CNT in the bath.

Particulate Composites for Engineering Applications

Wear and Friction

The main application of electrodeposited composites has been in the form of wear-resistant coating. Control of the tribological properties such as wear and friction has traditionally been one of the main driving forces behind the development of composite coating by electrodeposition. Incorporation of ceramic particles has been found to improve the wear resistance of MMC substantially. Mostly carbide particles such as SiC have been used for wear-resistant applications. Studies show that there exists an optimum concentration of particulate which gives the best tribological performance. Lower friction coefficient was reported for Ni- TiO_2 nanocomposites compared with pure Ni [15]. Su et al. [16] observed a decrease in both wear rate and friction coefficient in C-W/MWCNT nanocomposite coatings (Fig. 5.6). An eight-fold increase in diamond tool life was reported in glass machining applications when electrodeposited Ni-CNT was used as coating [17].

Wear studies have also been done on composites containing polymer particles. Ni-PTFE system is among the most widely studied electrodeposited composites. PTFE particles have been used either singly [18] or in combination [19] with hard particles to control friction characteristics. Incorporation of polyethylene particulates has also been found to improve the wear resistance of nickel [20].

The size of particles has been found to influence the tribological performance of electrodeposited composites. Nanosized SiC particles were reported to impart better wear resistance to nickel matrix composites as compared with micrometer-sized particles [21].

Benea et al. [22] reported improved tribo-corrosion resistance as well in Ni-SiC composites (Fig. 5.7).

There has been a search for self-lubricating composites possessing low friction coefficient. Graphite has been the first reinforcement used in electrodeposited Ni to reduce friction as mentioned earlier (Fink 1928). Graphite has also been added to copper to develop high-performance wear-resistant, low friction materials for electrical connector.

A novel approach to develop self-lubricating coating has been to incorporate microcapsules containing lubricating oil into metal matrix. Such composites are discussed further in section "Microcapsule-Containing Composites."

Corrosion Resistance

Studies revealed that the corrosion resistance can be improved by incorporating reinforcing particles into metal matrix. Both ceramic and polymer particles can improve the corrosion resistance of nickel deposit. Garcia et al. [23] have found in electrochemical corrosion tests that the corrosion rate as indicated by corrosion current is much lower in Ni-SiC composites as compared with Ni. The composites also have improved pitting corrosion resistance. Both nano- and micro-sized particles showed improved corrosion resistance.

Corrosion resistance was also found to improve with the concentration of titania (both anatase and rutile) in nickel matrix (Fig. 5.8).

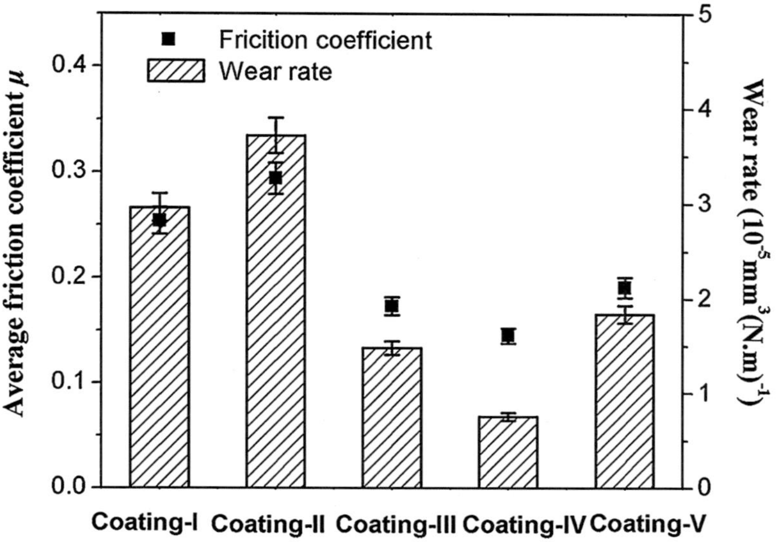

Fig. 5.6 A comparison of friction coefficients and wear rates of Co–W and Co–W/MWCNT composite coatings deposited under different conditions (coating I: Co-11.6% W; coating II: Co-11.5% W-11.5% MWCNT; coating III: Co-10.7% W-8.6% MWCNT; coating IV: Co-10.5% W-16.1% CNT; coating V: Co-12.3% W-10.4% MWCNT) [16]

Fig. 5.7 Tribo-corrosion rate of nickel and nickel – nano-SiC composite in 0.5 M Na_2SO_4 solution at different sliding loads [22]

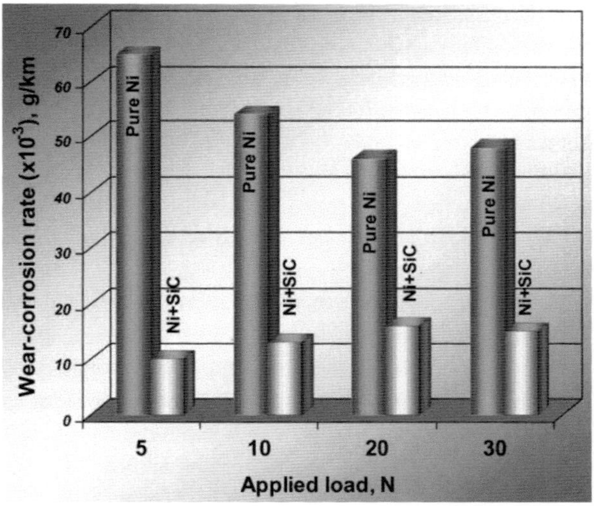

Improved corrosion resistance of SiC nanoparticle reinforced Ni was also reported by others [22, 25]. Corrosion of Ni-Al_2O_3 was found to improve in $NaSO_4$ solution as a result of Al_2O_3 incorporation [26, 27]. Corrosion resistance of Zn improved as a result of incorporation of CNT [28]. Improvement in corrosion

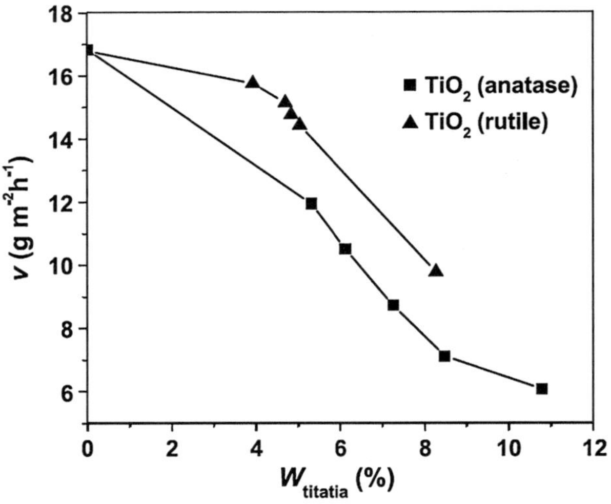

Fig. 5.8 Corrosion rate as a function of the amount of TiO_2 particles in the deposit. ■ anatase (average diameter = 12 nm) and ▲ rutile (average diameter = 1 μm). Electrolyte: 250 g l^{-1} $NiSO_4 \cdot 6H_2O$, 25 g l^{-1} H_3BO_3, 5 g l^{-1} saccharin, pH 3, and 30 °C. Current density: 66.7 mA cm^{-2} [24]

resistance was reported for a number of other electrodeposited systems as well, e.g., Ni/Fe_2O_3 [29], Ni-W/silicate [30], Zn-Al/SnO_2 [31], and Ni-W/SiO_2 [32].

One possible explanation for improved corrosion resistance has been a reduction in area of exposed metals as a result of the presence of ceramic particle in the sample [23].

Although majority of the studies reported improved corrosion resistance in electrodeposited composites, decreased corrosion resistance was also observed by a few researchers [26, 33]. The presence of titania as well as alumina nanoparticles was found to cause increased corrosion both in salt spray and electrochemical tests. Medeliene [26] also observed a decreased corrosion resistance for Ni-B_4C. This was attributed to higher conductivity of the latter.

More studies are necessary to understand the corrosion mechanisms in electrodeposited composites with different constituents.

High-Temperature Oxidation-Resistant Composite Coatings

The aim of achieving high-temperature oxidation resistance has led to many investigations into electrodeposited composite coatings. Electrodeposition can provide super alloy-type coatings at a lower cost. The main idea has been to codeposit particles of beneficial oxide-forming elements like Al [34, 35], Cr, Ti [36, 37], etc. into a suitable metal matrix. Mostly nickel has been utilized as the matrix, although other matrices such as Cu and Co have also been studied [36, 38]. Earlier studies used micrometer-sized metal particles.

It has been observed that the addition of Al particles into Ni matrix can substantially improve the oxidation resistance (Fig. 5.9) as evidenced by lower mass gain.

When aluminum-containing nickel coating is exposed to high temperature, a protective, adherent alumina scale forms which reduces the occurrence of further oxidation. Likewise, chromium particles are protected through the formation of a chromia

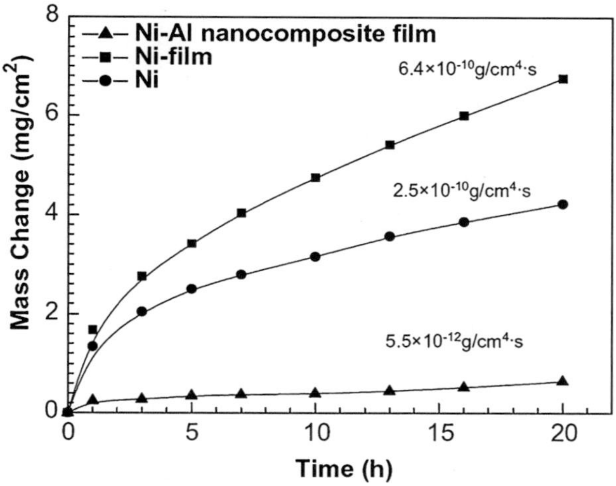

Fig. 5.9 Variation of mass change with time for various samples oxidized at 1050 °C for 20 h. [39]

scale. The size of the particles has been found to have a profound effect on the oxidation resistance. Nanometer-sized aluminum or chromium particles have been found to form the continuous protective layer at a much lower concentration as compared with micrometer-sized particles [35, 39, 40]. It has been found [38, 41, 42] that the addition of rare earth oxide can also substantially improve the oxidation resistance.

Novel Functional Composites

Microcapsule-Containing Composites

Microcapsule provides a novel way to incorporate liquid into composite (Fig. 5.10). Electrodeposition, being a low-temperature deposition process, has a unique capability to incorporate these liquid-containing microcapsules into the deposit. The process involves, as the first step, the preparation of microcapsules containing the desired liquid. The prepared microcapsules are then dispersed in the electrodeposition solution. The microcapsules then codeposit with metal matrix during electrodeposition.

The microcapsules can contain variety of liquids that can impart unique properties to the composites. The microcapsules can contain lubricant which gives the composite self-lubricating property. Microcapsules containing corrosion inhibitor give self-healing corrosion-resistant properties at damaged areas, those containing flux impart self-fluxing property to solder, and those containing phase-change material provide enhanced heat removal capability useful in electronic packaging. [44].

Metal matrix composites with oil-containing microcapsules have been one of the earliest development in this area [45]. Lubricating oil-containing polyterephthalamide microcapsules were prepared using the interfacial polymerization of an oil-soluble monomer (terephthaloyl dichloride) and a mixture of two water-soluble monomers (diethylenetriamine and 1,6-hexamethylenediamine). These were then incorporated into nickel electrodeposits. Liqun et al. [44]

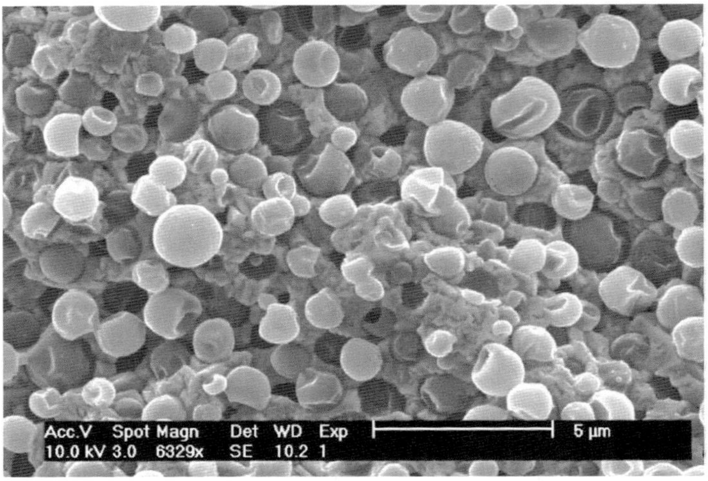

Fig. 5.10 SEM surface morphology of Cu-microcapsule composite containing 40 vol% microcapsules [43]

incorporated microcapsules containing liquid organosilica into a copper matrix. Stappers et al. [46] developed copper matrix composite containing a phase-change material (PCM) such as paraffin. This composite is unique in that it combines the high heat conductivity of copper with the high heat capacity of the PCM. It can therefore act as a high-efficiency heat sink. Since paraffin expands by 10–15 % upon melting, the Cu-PCM composite can also be employed as mechanical sensor (Fig. 5.9) [43].

Itagaki et al. [47] developed an electrodeposition method for the production of self-fluxing electronic solder. They synthesized microcapsules containing abietic acid, which is the main ingredient of the flux, and incorporated them into tin during electrodeposition.

Polymer-Based Functional Composites

Electrodeposition provides a unique way to fabricate conducting polymer-based composites. During such fabrication, electro-polymerization, being an oxidation process, takes place at the anode (rather than at the cathode where metal deposition occurs). Suitable suspended reinforcement particles also codeposit with the polymer under proper deposition conditions. Common conducting polymers that have been used as the matrix include polypyrrole [48–50] and polyaniline ([51]; Gurunathan et al. 2000). The second-phase particles incorporated in conducting polymers have been CNT, carbon black, nano-TiO_2, etc. Jurewicz et al. [49] showed that an electrodeposited polypyrrole matrix composite containing multiwalled carbon nanotube can store a large amount of charge and acts as a supercapacitor. Polyaniline matrix composite containing nanometer-sized carbon black particle acts as a support for Pt and Pt-Ru particles and exhibits excellent electrocatalytic activity for methanol oxidation [51]. Polyaniline-TiO composites are suggested for sensors and electrochromic device applications [104]. Nakayama et al. [50] reported on the development of polypyrrole-mesoporous silica composite that shows enhanced performance as a functional electrode in electrochemical sensing/reaction. Electrodeposition allows the synthesis of polymer matrix composites with better dispersion of different types of particles compared with other synthesis processes.

Biocompatible Hydroxyapatite Composites
Hydroxyapatite (HA), $Ca_{10}(PO_4)_6(OH)_2$, is a very important bioceramic whose composition is similar to that of bone tissue and is ideally suited for biomedical applications. It has been demonstrated that electrodeposition or one of its variation, electrophoretic deposition, can be used to deposit HA. In order to improve its mechanical properties, chitosan – a polymer – has been codeposited together with HA to form composites (Pang and Zhitomirsky 2008; [52]). Chitosan is a natural cationic polysaccharide having properties valuable in biomedical applications such as antimicrobial activity, corrosion resistance, biocompatibility, and good mechanical properties. In order to improve the antimicrobial properties, silver particles have also been incorporated into electrodeposited HA-chitosan composite (Pang and Zhitomirsky 2008). Thus, electrodeposition has opened up unique possibility of development of biocompatible antimicrobial coatings with controlled Ag^+ release rate. Electrodeposition also allows the synthesis of novel HA-alumina composites with improved strength [53]

Composites in Catalyst and Energy Applications
Novel composite coatings with special catalytic, electrocatalytic, and photochemical activity have been deposited by electrodeposition. Composites with such properties are usually based on metal-semiconductor or conducting polymer-semiconductor systems. A review on this topic is available in the literature [6]. Diaz et al. [54] fabricated a nanocrystalline Pt/CeO_2 composite electrode by electrodeposition. They observed that the composite is very effective in oxidizing alcohol. Such material has good potential in direct alcohol fuel cells. Electrodeposited Ni with a small loading of catalytic compounds like $LaNiO_3$ [55, 107], RuO_2 [56, 107] TiO_2 [57], and FeS [58] has been found to be able to catalyze various commercially useful reactions.

Electrodeposited composites are being investigated as electrode materials in energy applications, e.g., fuel cell, supercapacitors, etc. Copper/nickel oxide composites exhibited long life cycle and high retention of specific capacitance (97 % retained after 500 cycles) in supercapacitor [108]. Uysal et al. [59] reported on the application of Sn-Ni/MWCNT nanocomposite as negative electrodes in Li-ion batteries. Harthøj et al. [60] found that Co/CeO_2 composite exhibited improved high-temperature oxidation resistance when used in fuel cell interconnects.

Layered Nanocomposites

In the last section ("Particulate Composites by Electrodeposition"), composites with particle reinforcement were discussed. This section will deal with layered or laminated composites. The history of electrodeposition of laminated composites dates back to 1921 [61] when Dr. W. Blum at the US National Bureau of Standards first deposited Cu/Ni laminated structure from two separate electrodeposition solutions each containing copper ions and nickel ions, respectively. A micrograph of the earliest Cu/Ni laminated composite is reproduced from Blum's work in Fig. 5.11. The dark-etched thicker bands are copper layers and the thinner brighter ones are nickel layers. The scale on the micrograph is worth noting. Nickel layers on the photograph have a thickness of 10 μm. In fact, the thinnest sublayer deposited in that study was about 5 μm in thickness. In contrast, layer thickness <5 nm can nowadays routinely be produced by electrodeposition, thanks to the extensive research work that has been done in the past couple of decades to produce nanometer-thick films by this technique.

Since electrodeposition can easily provide nanometric layers, recently there has been a growing interest in the electrodeposition of laminated composites having nanometer-thick sublayers. These composites are known as laminated nanocomposites, multilayers, compositionally modulated multilayers, etc. These are usually deposited in the form of thin films consisting of a stack of two different materials, A and B, arranged alternately.

Laminated nanocomposites, as a result of layering at the nanometer scale, can exhibit improvement in the conventional properties or possess some new properties not found in normal metallurgical alloys. The following effects can contribute to the unique behavior of multilayer systems [62]:

1. Thin-film effect: this arises from the limited thickness of one or both types of layers.
2. Interface effect: due to the interaction between neighboring layers.
3. Coupling effect: which acts between layers of the same type, acting through the intervening layers.
4. Periodicity effect: as a result of the overall periodicity of the multilayer.

While physical properties are influenced by all the above factors to a variable extent, mechanical properties are likely to be affected more by the first two factors. Properties of laminated nanocomposites that have been widely studied are mainly strength, hardness, corrosion resistance, wear resistance, optical, magnetic, electronic transport and superconducting, elastic properties, etc.

In this review on electrodeposited laminated nanocomposites, first the common electrodeposition techniques used for these materials will be presented. This will be followed by a discussion on their structure. Finally, properties of laminated nanocomposites will be presented.

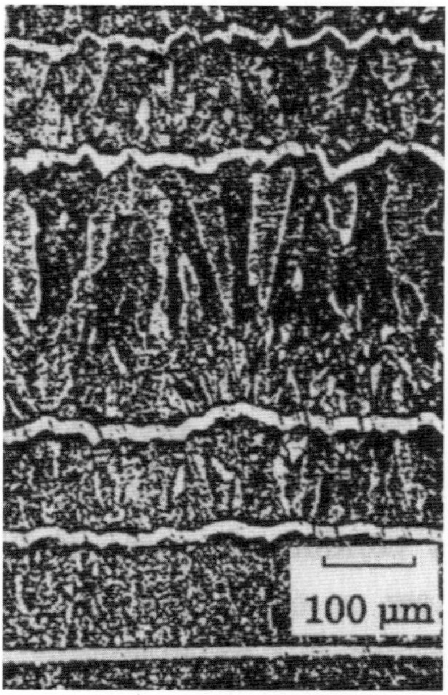

Fig. 5.11 The first Cu/Ni layered structure produced by dual-bath electrodeposition [61, 104]

Electrodeposition Techniques for Laminated Nanocomposites

There are two main ways to deposit laminated nanocomposites: single-bath technique (SBT) and dual-bath technique (DBT). In the single-bath technique, the deposition of two different constituent layers is carried out in the same electrodeposition solution or bath by alternately changing parameters such as current/voltage, sometimes in combination with mass transport. In the dual-bath technique, constituent layers of the composite are deposited alternately in two separate baths.

The single-bath technique has been used by most researchers. This process can be used when the constituent elements of the composite are capable of being electrodeposited from a single electrolyte. Moreover, the deposition potential of the two constituents should be far enough apart to allow the separated deposition of each element. The single-bath technique is in fact a pulsed-plating process in which at higher overpotential or current pulses the less noble constituent and at lower overpotential or current pulses the more noble constituent are deposited. A scheme for the deposition of multilayers from a single bath [63] is shown in Fig. 5.11. "A" represents the more noble and "B" the less noble constituent. In order to avoid the alloying of A layer by B and vice versa, some measures have to be taken during deposition. For instance, the bath should have a very low concentration of the more noble metal ion (A). The limiting current density, i_{LA}, of the more noble metal will be very small in such a case (Fig. 5.12). Under such a condition, the less noble metal will deposit in a relatively pure state at higher overpotentials. Although the more

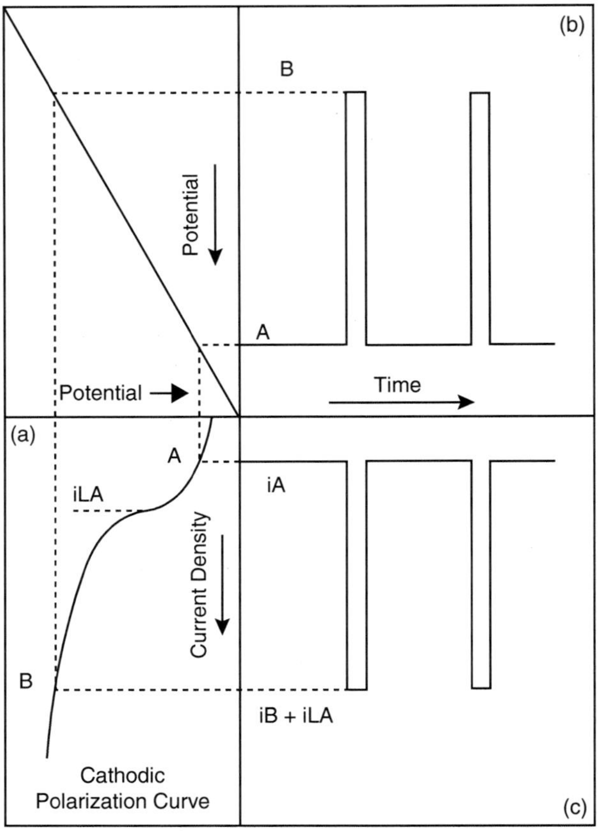

Fig. 5.12 A scheme for the electrode position of multilayers from a single bath. (**a**) Cathodic polarization curve, (**b**) potentiostatic deposition, and (**c**) galvanostatic deposition [63]

noble metal also deposits at higher overpotentials, its presence in the deposit will be negligible as a consequence of its very low concentration. At low overpotential, on the other hand, layers containing only the more noble metal will be deposited. The concentration of the more noble metal ion in the bath was recommended to be only 1 % of the less noble metal ion concentration.

The DBT is conceptually very simple and merely involves deposition of two constituents from two separate solutions in an alternate manner. However, there are some electrochemical requirements that need to be met. Each of the constituents should be chemically and electrochemically compatible in the solution of the other.

Otherwise, dissolution or other surface chemical/electrochemical reaction can occur before deposition which can adversely affect the quality of the interface between the constituent layers. Therefore optimization of the deposition bath may be necessary for some systems. For instance, in Zn-Ni system, direct deposition of Ni on Zn is possible if alkaline electrolytes are used [64, 65]. Moreover, electrolytes containing nickel-complexing agents help eliminate the possibility of the unwanted displacement reaction for zinc. Between the two deposition steps, the substrate should also be cleaned to avoid cross contamination.

Structure of Electrodeposited Nanolaminated Composites

The structure of nanolayered composites has additional features compared with that of normal metallurgical alloys. The presence of a large number of interfaces, their possible intermixing, coherency, or otherwise, makes the structure of layered nanocomposites more complicated. As such, it is difficult to make a complete structural characterization of laminated nanocomposites. Electron microscopic techniques like scanning electron microscopy and transmission electron microscopy have been employed to characterize the layered structure, the latter being more suitable for thinner layers. Auger depth profiling has also been used for such composites [66].

Using high-resolution transmission electron microscopy, Haseeb et al. [67] demonstrated that dual-bath electrodeposition can give rise to Cu/Ni nanolayered composite with individual layer thickness <4 nm and having crystallographic continuity across the interface. Yahalom et al. [68] also revealed the evidence of epitaxial stacking at the Cu/Ni interface by high-resolution TEM. However, the interface was suggested to be not atomically sharp. Undulation was also observed in the layered structure obtained by Yahalom et al.

X-ray diffraction can be used to characterize the nanolayered composites when the constituent layer thickness is a few nanometer. However the diffraction data coming from layering at the nanometer scale are very sensitive to uniformity of layer thickness, interface quality, coherency at the interface, etc. Using low-angle X-ray diffraction, the layer thickness or periodicity (combined thickness of two consecutive layers, A + B) of the laminated nanocomposite can be determined. Valizadeh et al. [69] characterized their Au/Co nanolaminates using the peaks that appear as a result of nanolamination in the low-angle X-ray diffraction pattern (Fig. 5.13). The position of the peaks can provide the thickness, while the intensity can provide information about the uniformity or quality of the nanolaminates. Because of the presence of nonuniformity at the interface, the peak intensity may decrease or in the worst case the peaks may disappear altogether. By comparing with theoretical models, the thickness of the laminates and the nonuniformity at the interface can be estimated.

When the layer thickness goes down to a few atomic diameters and coherency is preserved at the interface, a one-dimensional superlattice, in effect, is formed. This can give rise to additional satellite peaks surrounding a Bragg peak in the X-ray diffraction pattern in the high-angle region. The position and intensity of the satellite peaks can provide information on the thickness and quality of the composites. An XRD pattern showing the satellite peaks in Cu/Ni system is presented in Fig. 5.14 [70].

While electron microscopy provides information on very small scale (μm/nm), information provided by XRD is that averaged over a larger length scale (mm/cm). Thus the presence of additional peaks due to the periodicity in the electrodeposited nanolaminates in a number of work indicates that this process is capable of producing nanolaminates with good quality over a large area.

The versatility of electrodeposition allows it to be used to deposit composite nanowires. Nanolaminated composite nanowire of different types such as CoNi/Cu

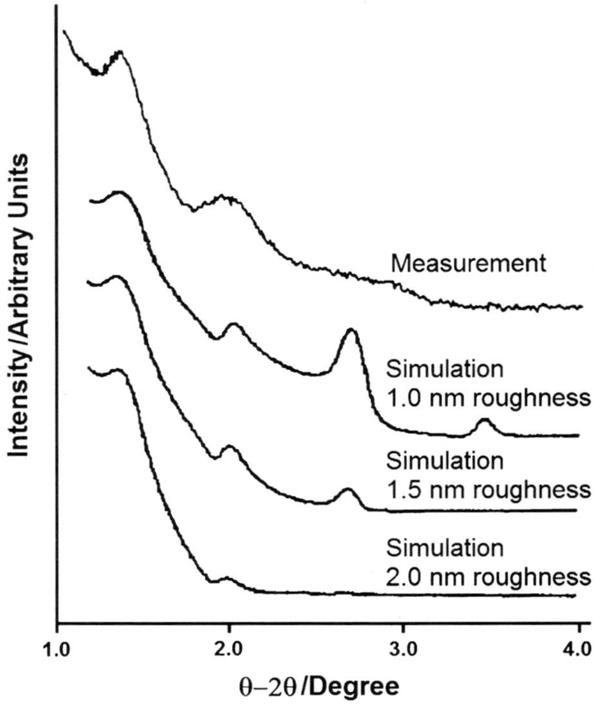

Fig. 5.13 Low-angle XRD pattern for Au/Co nanolaminates (5 nm sublayer) with characteristic peaks from a periodic chemical modulation. Also are shown the simulations of a 5.5 nm sublayer with different interface roughness of 1.0, 1.5, and 2.0 nm [69]

[71], Au/Co [72], and Pb/Cu [73] has been deposited by template-assisted deposition. Nanowire composites are electrodeposited within template containing slender pores of a few nanometer diameter. Such nanowire composite can provide interesting magnetic, superconducting properties. Figure 5.15 shows SEM and TEM images of an example of such nanowires obtained by Guan and Podlaha [72].

Properties of Electrodeposited Nanolaminated Composites

Strength

A great improvement in strength has been observed in laminated nanocomposites. Tench and White [74–76], Menezes and Anderson [77], and Simunovich et al. [78] reported a several hundred percent increase in strength depending upon the sublayer thickness. In general, as the sublayer thickness decreases, the strength of laminated nanocomposites increases. A number of studies [78] also revealed that there exists a threshold thickness, about 1 nm or so, up to which the strength continues to increase as the sublayer thickness decreases. However, as the thickness goes below the threshold value, the strength decreases.

It has been suggested that two factors may lead to strengthening in the laminated nanocomposites:

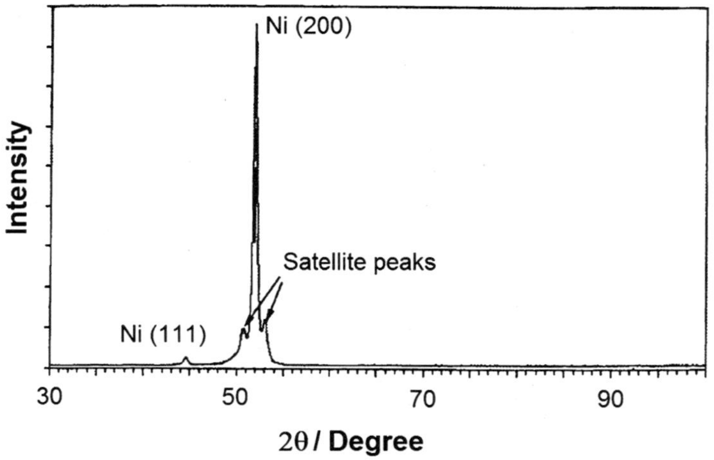

Fig. 5.14 X-ray diffraction pattern from Cu/Ni nanolaminates showing the presence of satellite peaks flanking the main (200) peak [70]

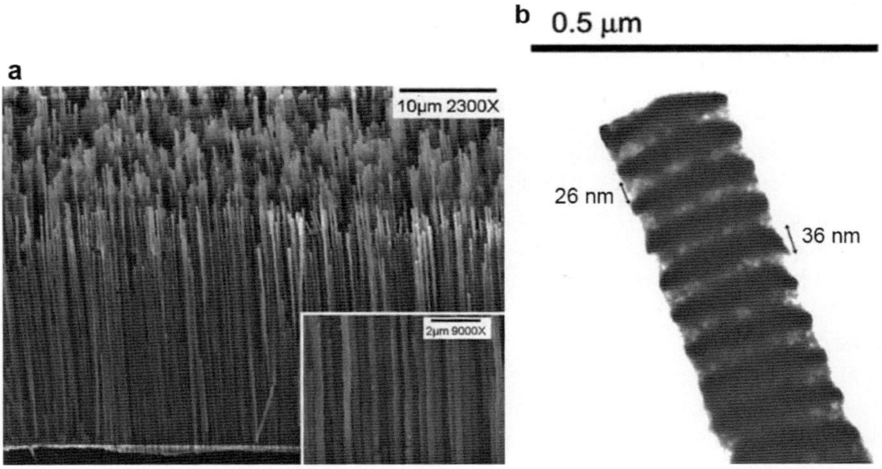

Fig. 5.15 (**a**) SEM micrograph of Au/Co multilayer nanowires, and (**b**) TEM bright-field image of a multilayer nanowire [72]

(a) Strengthening resulting from decreased feature size which is given by the relationship, $\sigma \propto d^a$, where σ = yield strength and d = thickness of the individual layer. The exponent a can assume different values: $a = -0.5$ (Hall-Petch strengthening), $a \neq 0.5$ (modified Hall-Petch strengthening), and $a = -1$ (Orowan-type strengthening) depending upon the sublayer thickness, in-plane grain size of the composite, and the nature of the dislocation pileup [79].

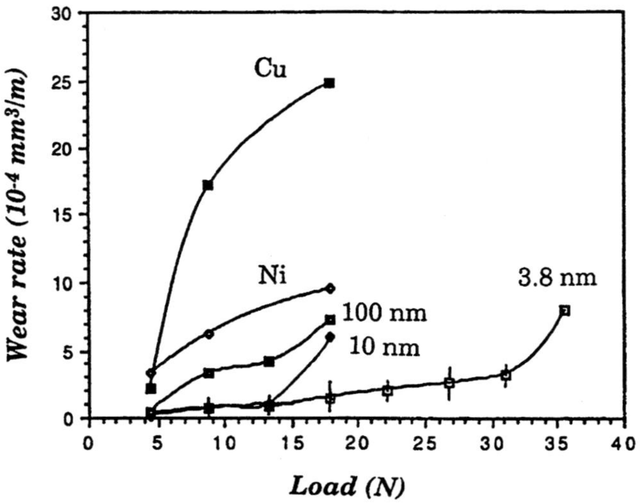

Fig. 5.16 Wear rate as a function of load for copper, nickel, and Cu/Ni nanolaminates [82]

(b) Koehler-type strengthening which results from difficulties in moving dislocations from the layer with low modulus constant to the one with high modulus constant due to image force [80].

The decrease in strength at very small sublayer thickness could be due to the increase in the coherency of the interface [81] which allows easy passage of dislocations through the interface. This could also be due to intermixing at the interface and the consequent loss of distinct nature of each sublayer (Tench et al. 1991; Ebrahimi et al. 1991).

Wear Resistance

A number of studies appeared in the literature on the wear and frictional properties of laminated nanocomposites. Tribological properties of Cu/Ni laminated nanocomposites have received most attention. These studies covered dry and lubricated conditions, as well as sliding wear and fretting wear conditions. Ruff et al. [82–84] reported on the wear properties of Cu/Ni composites with constituent layer thicknesses of 3.8, 10 and 100 nm under sliding wear conditions. In the unlubricated tests, the wear rate of composites was much lower than that of either copper or nickel (Fig. 5.16). Wear rate of composites was found to be dependent strongly upon the sublayer thickness. Decrease in wear rate was observed with a decrease in sublayer thickness. All the laminated composites showed a critical load above which wear rate increased rapidly. The critical load increased as the sublayer thickness was decreased. Lubricated tests also revealed improved wear resistance in composites.

Zhang [85] also found improved wear resistance and frictional properties in Cu/Ni laminated nanocomposites under sliding wear conditions. Under fretting

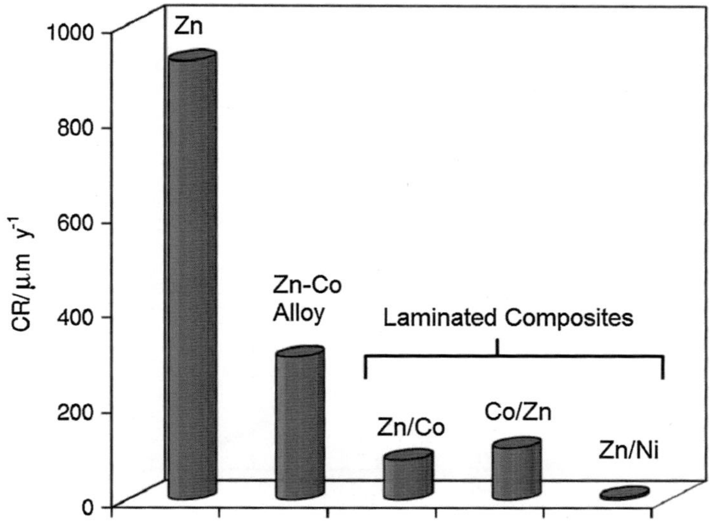

Fig. 5.17 Comparison between the corrosion rates of three laminated composites, pure Zn and Zn-Co alloy [92]

wear conditions, Cu/Ni showed [86] better wear performance as compared with either of the constituents, Cu or Ni. Nanocomposites with thinner sublayers, 5 nm, exhibited better wear resistance compared with that with 10 nm sublayer thickness. Zhang et al. [87] reported that fretting wear resistance of Cu/Ni composites improved dramatically when sublayer thickness was <60 nm.

Georgiou et al. [88] reported that nanostructured gradient Co-Sn layered composites possessed significantly higher wear resistance when compared to pure Sn coatings. Layered nanocomposites containing composition modulated Ni-W multilayer exhibited enhanced hardness without an increase in the internal stress and surface crack formation tendency [89].

Corrosion Resistance

There has been a growing interest in the corrosion properties of laminated nanocomposites. Repeated barriers that such laminated structure provides are believed to lead to improved corrosion resistance. A number of nanolayered systems have been studied for their corrosion resistance such as Zn-Co [90–92], Zn-Ni [93–95], and Cu/Ni [96]. Thangaraj et al. [92] made a comparative study on the corrosion resistance of monolithic Zn-Co alloy and Zn-Co laminated nanocomposite. They used the same total thickness for all coating (15 μm). The nanocomposites contain sublayers down to 25 nm thickness. Thangaraj et al. found a substantial improvement in corrosion resistance in laminated nanocomposites (Fig. 5.17).

A laminated composites containing 600 sublayer (each 25 nm thick) showed about six times better corrosion resistance than Zn-Co alloy deposit. Co-rich outer layer was found to provide enhanced corrosion resistance. Zn-Ni composite was found to have better corrosion resistance than Zn-Co. Based on Mott-Schottky plots

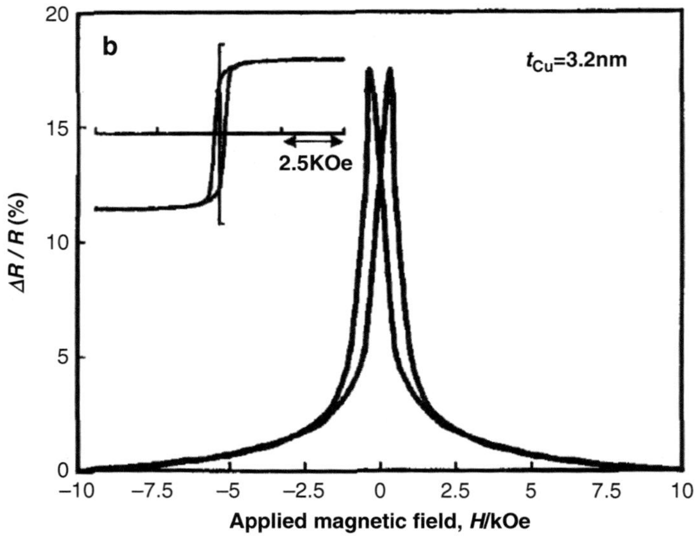

Fig. 5.18 Magnetoresistance curve for Co/Cu nanolaminates [98]

obtained in electrochemical impedance spectroscopy, it was suggested that the increased corrosion resistance of the nanolaminated composites was related to their inherent barrier properties of n-type semiconductor films at the interface between the laminates.

Better corrosion resistance of Zn-Co nanolaminated composites was also reported by others. [90] Fei and Wilcox [94] and Chawa et al. [93] reported that the corrosion resistance of Zn-Ni laminated composite coating was much better than that of a single-layer zinc or nickel deposit. A dependence of corrosion resistance on the sequence of the layers was observed [94]. Zn-Ni coatings having a Ni sublayer adjacent to substrate and Zn sublayer as the top layer exhibited the best protective performance.

Troyon and Wang [96] observed that by adding organic additive in the deposition bath, the structure of Cu/Ni nanolaminates can be improved which resulted in improved corrosion resistance.

Magnetic Properties

Giant magnetoresistance (GMR) is one of the properties of laminated nanocomposites that is of great technological importance and has attracted great interest in recent years. Laminated nanocomposites studied for GMR have been mostly produced by single-bath technique. GMR is characterized by a large increase as the resistance of nanolayered system when the magnetic field is changed. This is a property of great technological importance and can be used in high-density magnetic recording, sensors, etc. A number of researchers [97–99] have reported the attainment of GMR in electrodeposited Co/Cu nanocomposites. Kashiwabara [98] obtained a magnetoresistance value as high as 18 % in their nanolaminates optimized for sublayer thickness (Fig. 5.18).

GMR has also been exhibited by Ni-Co-Cu/Cu layer nanocomposites. [99]

GMR is observed in nanocomposite system consisting of alternate stacks magnetic and nonmagnetic sublayer with only a couple of nanometer sublayer thickness. This is a property which is very sensitive to quality of the nanocomposites with respect to sublayer thickness uniformity, interface quality, crystallinity, etc. Thus, the attainment of GMR in electrodeposited nanocomposites is a testimony to the precise deposition control at the nanometer level that is currently possible in electrodeposition.

Other attractive magnetic properties have been found other than nanolayered composites such as hard magnetic properties in Fe/Pt [100] and high permeability [101].

Recently, Schelhas et al. [102] reported electrodeposition to be a reproducible and tunable deposition process for the fabrication of multilayered Au-Ni composite in the form of a nanowire. This process allowed the authors to create a multistate system with both in-plane and out-of-plane easy axes. This was achieved by balancing magnetic shape anisotropy with dipole coupling between the two layers.

Conclusions

Electrodeposition has a fairly long history with regard to composite preparation. As far as published literature goes, electrodeposited laminated and particulate composites go back to 1921 [61] and 1928 (Fink 1928), respectively. Composites prepared by electrodeposition are already in use in different industry including engineering, automotive, and mining. Electrodeposition, being an atomistic deposition process, has the capability of producing composites, particularly laminated nanocomposites with structural features down to nm. Current availability of nanoparticle has made it possible to produce particulate nanocomposites routinely. This process can also be speeded up to produce mm-scale features. Electrodeposition is particularly suited to deposit composite coatings on large objects with complex shapes.

Although this process is most suitable for metal matrix composites, recent research has proven the versatility of this technique in producing ceramic- and polymer-based composites. Initial interest in electrodeposited composites has been because of their novel engineering properties like wear resistance, corrosion resistance, etc. While these properties continue to attract interest, recent studies have demonstrated that electrodeposited composites can give rise to novel properties like magnetoresistance, catalytic properties, and sensing/actuating properties suitable for advanced devices. Future research and development is expected to take electrodeposition to a new height as a process for the synthesis advanced composites.

References

1. Schwarzacher W (2006) Interface 15:32
2. Fink CG, Prince JD (1928) Trans Am Electrochem Soc 54:315
3. Williams RV (1966) Electroplat Met Finish 19:92
4. Bozzini B, Giovannelli G, Cavallotti PL (1999) J Appl Electrochem 29:685

5. Wang Z, Shemilt J, Xiao P (2002) J Eur Ceram Soc 22:183
6. Musiani M (2000) Electrochim Acta 45:3397
7. Low CTJ, Wills RGA, Walsh FC (2006) Surf Coat Technol 201:371
8. Roos JR, Celis JP, Fransaer J, Buelens C (1990) J Met 42:60
9. Yeh SH, Wan CC (1994) J Appl Electrochem 24:993
10. Fransaer J, Celis JP, Roos JR (1992) J Electrochem Soc 139:413
11. Fransaer J (1994) Ph D thesis, Catholic University of Leuven
12. Guglielmi NJ (1972) Electrochem Soc 119:1009
13. Bercot P, Pena-Munoz E, Pagetti J (2002) Surf Coat Technol 157:282
14. Celis JP, Ross JR, Buelens C (1987) J Electrochem Soc 134:1402
15. Benea L, Danaila E, Celis JP (2014) Mater Sci Eng A 610:106
16. Su F, Liu C, Huang P (2014) Appl Surf Sci 309:200
17. Suzuki T, Konno T (2014) Precis Eng 38:659
18. Li WH, Zhou XY, Xu Z, Yan MJ (2009) Surf Eng 25:353
19. Guo Z, Xu R, Zhu X (2004) Surf Coat Technol 187:141
20. Hamid ZA, Ghayad IM (2002) Mater Lett 53:238
21. Garcia I, Fransaer J, Celis JP (2001) Surf Coat Technol 148:171
22. Benea L, Bonora PL, Borello A, Martelli S (2002) Wear 249:995
23. Garcia I, Conde A, Langelaan G, Fransaer J, Celis JP (2003) Corros Sci 45:1173
24. Li J, Sun Y, Sun X, Qiao J (2005) Surf Coat Technol 192:331
25. Vaezi MR, Sadrnezhaad SK, Nikzad L (2008) Colloids Surf A Physicochem Eng Asp 315:176
26. Medeliene V (2002) Surf Coat Technol 154:104
27. Szczygieł B, Kołodziej M (2005) Electrochim Acta 50:4188
28. Praveen BM, Venkatesh TV, Naik YA, Prashantha K (2007) Surf Coat Technol 201:5836
29. Geng S, Qi S, Zhao Q, Zhu S, Wang F (2012) Int J Hydrog Energy 37:10850
30. Sassi W, Dhouibi L, Bercot P, Rezrazi M, Triki E (2014) Electrochim Acta 117:443
31. Fayomi OSI, Popoola API, Aigbodion VS (2015) J Alloys Comp 623:328
32. Sassi W, Dhouibi L, Bercot P, Rezrazi M (2015) Appl Surf Sci 324:369
33. Erler F, Jakob C, Romanus H, Spiess L, Wielage B, Lampke T, Steinhauser S (2003) Electrochim Acta 48:3063
34. Susan DF, Barmak K, Marder AR (1997) Thin Solid Films 307:133
35. Zhou Y, Peng X, Wang F (2005) Oxid Met 64:169
36. Huang Z, Peng X, Wang F (2006) Oxid Met 65:223
37. Naploszek-Bilnik I, Budniok A, Losiewicz B, Pajak L, Lagiewka E (2005) Thin Solid Films 474:146
38. Shaigan N, Ivey DG, Chen W (2008) J Power Sources 185:331
39. Zhou Y, Peng X, Wang F (2004) Scr Mater 50:1429
40. Zhang Y, Peng X, Wang F (2004) Mater Lett 58:1134
41. Cesuniene NPA, Taicas L (1993) Plat Surf Finish 80:11
42. Zhou Y, Peng X, Wang F (2006) Scr Mater 55:1039
43. Malfliet A, Deferme G, Stappers L, Fransaer J (2007) J Electrochem Soc 154:D50
44. Liqun Z, Wei Z, Feng L (2004) J Mater Sci 39:495
45. Alexandriou S, Kiparissides C, Fransaer J, Celis JP (1995) Surf Coat Technol 71:267
46. Stappers L, Yuan Y, Fransaer J (2005) J Electrochem Soc 152:C457
47. Itagaki M, Shitanda I, Watanabe K, Koyano H, Gijutsu H (2003) J Surf Finish Soc Jpn 54:599
48. Han G, Yuan J, Shi G, Wei F (2005) Thin Solid Films 474:64
49. Jurewicz K, Delpeux S, Bertagna V, Beguin F, Frackowiak E (2001) Chem Phys Lett 347:36
50. Nakayama M, Yano J, Nakaoka K, Ogura K (2002) Synth Met 128:57
51. Wu G, Li L, Li JH, Xu BQ (2005) Carbon 43:2579
52. Pang X, Zhitomirsky I (2005) Mater Chem Phys 94:245
53. Zhen-jun W, Li-ping H, Zong-zhang C (2006) Trans Nonferrous Met Soc China 16:259
54. Diaz DJ, Greenletch N, Solanki A, Karakoti A, Seal S (2007) Catal Lett 119:319
55. Anani A, Mao Z, Srinivasan S, Appleby AJ (1991) J Appl Electrochem 21:683

56. Iwakura C, Furukawa N, Tanaka M (1992) Electrochim Acta 37:757
57. Gierlotka D, Rowinski E, Budniok A, Lagiewka E (1997) J Appl Electrochem 27:1349
58. Assunc NA, de Giz MJ, Tremiliosi-Filho G, Gonzalez ER (1997) J Electrochem Soc 144:2794
59. Uysal M, Gul H, Alp A, Akbulut H (2014) Inter J Hydrog Energy 39:21391
60. Harthøj A, Holt T, Møller P (2015) J Power Sources 281:227
61. Blum W (1921) Trans Am Electrochem Soc 40:307
62. Schuller IK (1988) Mater Res Soc Symp Proc 103:335
63. Yahalom J, Zadoc O (1987) J Mater Res 22:499
64. Fei JY, Hui BP, Liang GZ, Xin WL (2004) Surf Technol 33:37
65. Pourbaix M (1966) Atlas of electrochemical equilibria in aqueous solutions. Pergamon Press, Oxford
66. Celis JP, Roos JR, Blanpain B, Gilles M (1988) In: Proceedings of the 12th World Congress on surface finishing, vol 2. AITE, Paris, p 435
67. Haseeb ASMA, Celis JP, Roos JR (1994) J Electrochem Soc 141:230
68. Yahalom J, Tessier DF, Timsit RS, Rosenfeld AM, Mitchell DF, Robinson PT (1989) J Mater Res 4:755
69. Valizadeh S, Svedberg EB, Liesner P (2002) J Appl Electrochem 32:97
70. Ebrahimi F, Liscano AJ (2001) Mater Sci Eng A301:23
71. Tanga XT, Wanga GC, Shima M (2007) J Magn Magn Mater 309:188
72. Guan M, Podlaha EJ (2007) J Appl Electrochem 37:549
73. de Horne FM, Piraux L, Michotte S (2005) Appl Phys Lett 86:152510
74. Tench D, White J (1984) Metall Trans 15A:2039
75. Tench DM, White JT (1990) J Electrochem Soc 137:3061
76. Tench DM, White JT (1991) J Electrochem Soc 138:3757
77. Menezes S (1989) J Electrochem Soc 137:440
78. Simunovich D, Schlesinger M, Snyder DD (1994) J Electrochem Soc 141:L10
79. Misra A, Verdier M, Kung H, Embury JD, Hirth JP (1999) Scr Mater 41:973
80. Koehler JS (1970) Phys Rev B2:547
81. Rao SI, Hazeldine PM, Dimiduk DM (1995) MRS Symp Proc 362:67
82. Ruff AW, Lashmore DS (1991) Wear of materials. In: Proceedings of the international conference, vol 1. Wear of Materials, Orlando, p 137.
83. Ruff AW, Myshkin NK (1989) Trans ASME 111:156
84. Ruff AW, Wang ZX (1989) Wear 131:59
85. Zhang W, Xue Q, Zhang X (1998) Wear 214:74
86. Haseeb ASMA, Celis JP, Roos JR (2003) Thin Solid Films 444:199
87. Zhang W, Xue Q (1998) Wear 214:23
88. Georgiou EP, Buijnsters JG, Wang H, Drees D, Basak AK, Celis JP (2015) Surf Coat Technol (in press). doi:10.1016/j.surfcoat.2014.12.061
89. Lee S, Choi M, Park S, Jung H, Yoo B (2015) Electrochim Acta 153:225
90. Kirilova I, Ivanov I (1999) J Appl Electrochem 29:1133
91. Kirilova I, Ivanov I, Rashkov S (1998) J Appl Electrochem 28:1359
92. Thangaraj V, Eliaz N, Hegde AC (2009) J Appl Electrochem 39:339
93. Chawa G, Wilcox GD, Gabe DR (1998) Trans Inst Metal Finish 76:117
94. Fei JY, Wilcox GD (2006) Surf Coat Technol 200:3533
95. Kirilova I, Valkova T, Ivanov I (2002) J Appl Electrochem 32:85
96. Troyon M, Wang L (1996) Appl Surf Sci 103:517
97. Gomez E, Labarta A, Llorente A, Valles E (2002) Surf Coat Technol 153:261
98. Kashiwabara S, Jyoko Y, Hayashi Y (1997) Phys B 239:47
99. Nabiyouni G, Schwarzacher W, Rolik Z, Bakonyi I (2002) J Magn Magn Mater 253:77
100. Leistner K, Fahler S, Schlorb H, Schultz L (2006) Electrochem Commun 8:916
101. Perez L, de Abril O, Sanchez MC, Aroca C, Lopez E, Sanchez P (2000) J Magn Magn Mater 215–216:337
102. Schelhas LT, Banholzer MJ, Mirkinb CA, Tolbert SH (2015) J Magn Magn Mater 379:239

103. Ebrahimi F, Kong D (1991) Scripta Mater 40:609
104. Gurunathan K, Trivedi DC (2000) Mater Lett 45:262
105. Hwang BJ, Hwang CS (1993) J Electrochem Soc 140:979
106. Ogden C (1986) Plat Surf Finish 5:130
107. Tavares AC, Trasatti S (1998) In: Sealey SA (ed) Modern Chlor-Alkali technology, vol 7. SCI, London, p 65
108. Yin JL, Park JY (2014) Inter J Hydrog Energy 39:16562

Short Carbon Fiber-Reinforced Polycarbonate Composites

6

Raghunandan Sharma, Kamal K. Kar, Malay K. Das, Gaurav K. Gupta, and Sudhir Kumar

Contents

R. Sharma • G.K. Gupta • S. Kumar
Advanced Nanoengineering Materials Laboratory, Materials Science Programme, Indian Institute of Technology Kanpur, Kanpur, Uttar Pradesh, India
e-mail: raghumsp @iitk.ac.in; ggbhai1982@gmail.com; sudhirdubey1983@gmail.com

K.K. Kar (✉)
Advanced Nanoengineering Materials Laboratory, Materials Science Programme, Indian Institute of Technology Kanpur, Kanpur, Uttar Pradesh, India

Advanced Nanoengineering Materials Laboratory, Department of Mechanical Engineering, Indian Institute of Technology Kanpur, Kanpur, Uttar Pradesh, India
e-mail: raghumsp @iitk.ac.in

M.K. Das
Advanced Nanoengineering Materials Laboratory, Department of Mechanical Engineering, Indian Institute of Technology Kanpur, Kanpur, Uttar Pradesh, India
e-mail: mkdas@iitk.ac.in

K.K. Kar (ed.), *Composite Materials*,
DOI 10.1007/978-3-662-49514-8_6

Abstract

Factors affecting the performance of short fiber-reinforced polymer (SFRP) composites are reviewed with an emphasis on the factors such as fiber orientation and volume fraction, fiber length, and matrix–fiber interfacial properties. The review is followed by a case study on short carbon fiber-reinforced polycarbonate (PC) composites of varying volume fraction (1 %, 3 %, 5 %, 7 %, 10 %, and 15 %). Tensile and three bending tests reveal an increase in the modulus and yield strength of the composites with increasing fiber content. To evaluate the effect of interface in each matrix and fiber system with changing matrix mobility dynamic mechanical analysis is performed, which reveal an increase in the storage modulus of the composites and a shift in the glass transition temperature caused by the introduction of CFs. The surface morphology of the fractured specimens revealed homogenous dispersion of CF in the polymer matrix.

Keywords

Bisphenol • A polycarbonate • CFRP composite • Melt mixing • Impact fracture and mechanical strength

Introduction

Fiber-reinforced polymer (FRP) composites have diverse areas of application including space exploration, civil aviation, military, marine, civil infrastructure, automobiles, medical, and sports. Combination of properties such as low density, high specific strength and stiffness, and excellent corrosion resistance makes FRP composites an appropriate substitute of conventional metals for load-bearing applications. The principal advantage of these composites over metals is their high specific strength/stiffness, which makes them suitable for various applications ranging from space to daily life [1–4]. Lower density of polymer matrix reduces the product weight considerably while high strength fibers carry the load. Apart from this, the FRP composites exhibit a combination of desired properties not otherwise shown by either of the constituent materials. A large variation in properties is possible by changing the polymer matrix, processing conditions, and the reinforcing fiber type, volume fraction (V_f), length, and alignment. This leads to the attainment of desired properties by a suitable choice of the specific matrix, fiber, and other parameters depending on their suitability for specific end-user applications [5]. Based on the length of the reinforcing fiber, the FRP composites are classified as short FRP

(SFRP) and continuous FRP composites. Continuous fibers, being easy to orient and disperse, exhibit high dimensional stability, low shrinkage, and high impact strength. However, for complex shapes, their processing becomes more difficult with increased product cost. On the other hand, the short fiber-reinforced polymer (SFRP) composites, due to their comparatively easy and rapid production, lower fabrication cost, and suitability to form intricate shapes, offer considerable technical advantages over the formers [6]. The SFRP composites fill the mechanical property gap between the continuous fiber laminates used as primary structures by the aircraft and aerospace industries and the unreinforced polymers used in non-load-bearing applications [7].

Commercially used reinforcing fibers for FRP composite fabrication include glass fiber, carbon fiber, and aramid fiber. Among these, the carbon fiber (CF), due to its superior tensile strength as high as 230 GPa and being lightweight, finds vast applications in aerospace and military industries [8, 9]. Studies on both polyacrylonitrile (PAN) and pitch-based CF-reinforced composites exhibit superior properties preferred for the structural applications [10]. Despite being expensive, the CF holds considerable importance in FRP composites as well as carbon–carbon (CC) composites.

Considerable research has been carried out to study the effect of various parameters on the performance of carbon fiber-reinforced polymer composites. This chapter aims to summarize the literature results concerning the effect of these parameters on SFRP composite performance along with a case study of short carbon fiber-reinforced polycarbonate composites. The case study deals with the effect of fiber V_f on the viscoelastic, mechanical, and fracture response of the composites. Short CF-reinforced bisphenol A polycarbonate composites with varying CF content ($V_f = 0.01$–0.15) were synthesized by the melt mixing followed by injection molding. The polycarbonate (PC), owing to its unique combination of properties such as high toughness, excellent compatibility with several polymers, large heat distortion resistance, and lower cost, proves to be a suitable matrix for the structural applications [11].

Factors Affecting SFRP Composite Performance

Effect of Fiber Orientation and Volume Fraction

Factors such as relative orientation, V_f, and dispersion of the fibers have a significant influence on the mechanical strength and other properties of FRP composites [6]. With respect to the orientation, two extremes are possible: (1) parallel alignment of the fibers and (2) random alignment. Alignment of short fibers is more difficult compared to the continuous ones. The aligned FRP composites show anisotropic behavior, while in the second case, it turns out to be isotropic [12]. Preferred alignment of fibers along the loading direction increases the fracture load [13]. Another important factor affecting the composite properties is the dispersion of fiber in the matrix. Inhomogeneous dispersion produces local fiber-deficient regions prone to the matrix failure. Better overall properties are realized when the fiber is uniformly dispersed [14]. High fiber content is required to achieve a high-performance SFRP composite. Therefore, the effect of fiber content on their mechanical properties is of particular interest and significance. Increase in the V_f leads to the increase in the strength and modulus [15, 16]. This follows the rule of mixture for composites [1]. Again, for a matrix with lower toughness, increased V_f improves the toughness of the composite [17]. However, very high fiber loading may affect the fiber wetting efficiency and fiber dispersion adversely.

A study on the penetration energy for CF-reinforced plastics under low-velocity impact conditions by Caprino and Lopresto has shown that for a given fiber type, the penetration energy is substantially influenced by the total fiber volume and diameter [18]. Other secondary parameters affecting the penetration energy include resin type and content, fiber architecture, stacking sequence, and orientations. The work on the fiber-reinforced PC composites by Friedrich reveals that the fracture toughness of the composite, K_{cc}, is affected by V_f, orientation, and distribution of short carbon fibers across the plaque thickness and by the quality of their interfacial bonding to the matrix [17]. An increase in composite toughness is observed with increasing the extent of reinforcement if the matrix is brittle and the fibers are well bonded as well as oriented perpendicular to the crack front. An opposite behavior has been observed for the matrices, which behave in a highly ductile manner even in the presence of fibers.

Effect of Fiber Length

The mechanical properties of the fiber-reinforced composites depend on the degree to which the load transfer between the matrix and fiber takes place. The important factor of this load transfer mechanism is the magnitude of the interfacial adhesion between the fiber and matrix [19]. Under an applied stress, this fiber–matrix bond ceases at the fiber ends, yielding a matrix deformation pattern. In other words, there is no load transfer from the matrix to the fiber at each fiber extremity. Some critical fiber length is necessary for the effective strengthening and stiffening of the composite material [20]. Critical fiber length (l_c) depends on the fiber diameter and its ultimate strength by the relation (Eq. 6.1):

$$l_c = \frac{r_f \sigma_{cu}}{2\tau_i} \tag{6.1}$$

where, τ_i, r_f, and σ_{cu} are the interfacial shear stress between matrix and fiber, radius of fiber, and ultimate strength of composite, respectively. A mathematical study relating the effect of fiber length and orientation distributions on the tensile properties of the SFRP composites by Fu et al. shows that the composite strength increases rapidly with increasing the fiber length for smaller mean lengths ($<l_c$) and saturates for larger mean lengths ($>5l_c$) [21]. Moreover, the smaller value of l_c leads to the higher strength of the composites. In addition, the thermal stability and dynamic mechanical response of the SFRP composites show significant dependence on fiber length. Rezaei et al. have studied the thermal degradation behavior and dynamic mechanical response of CF-reinforced PP composites with varying fiber length from 0.1 to 10 mm. For a constant fiber content ($V_f = 0.1$), the thermal degradation temperature of the composite increases (~10 °C) with increasing fiber length in the above range. On the other hand, contrary to the longer fibers (10 mm), the storage and loss moduli of these composites show no significant change for the smaller fibers (<2 mm) [22].

In SFRP composites, various parameters such as fiber concentration and shear applied during the extrusion change the fiber length and length distribution. With increasing V_f, due to the increased fiber–fiber interaction and fiber-extruder wall contact, the shear stress applied to the fibers during the mixing/extrusion increases, which, in turn, breaks the fibers to reduce the fiber length [23, 24]. Again, this reduction in fiber length lowers the composite strength [25]. This is at least partially the reason why the addition of very short fibers to a polymer matrix does not lead to a significant increase in composite strength [26, 27], modulus [28], and toughness [19].

Matrix–Fiber Interfacial Properties

The performance of fiber-reinforced polymer (FRP) composites depends largely on the quality of the fiber–matrix interface, which determines the way by which the load transfer takes place from polymer to the fiber [29]. The interfacial properties become more important in case of SFRP composites where the interfacial shear strength determines the critical length. Both chemical bonding as well as mechanical locking between rough fiber surface and polymer chains may form the interface. The nature and the strength of bonding at the fiber–matrix interface depend on the fiber surface activity, roughness, the presence of functional groups, compatibility with the polymer matrix, and the thermal properties of the matrix [30]. Various surface modifications are employed to improve the surface activity and roughness. Surface oxidation, grafting of polar groups (–COH, –CO, and –COOH) by chemical/plasma treatment, and the use of sizing agents are some of the ways to improve surface activity for better interfacial strength [31, 32, 33]. The oxygen-containing groups on the CF surface influence the cross-linking reaction of epoxy resin in the composite, which leads to higher degree of cross-linking at the interface [34]. Apart from this, presence of a secondary nanosized material such as carbon nanotubes (CNTs) on the fiber surface improves the interfacial strength by local stiffening of the matrix near the interface [35–37]. Studies by Sager et al. have exhibited a 71 % increase in the interfacial shear strength for randomly oriented multi-walled CNT-coated CF [38]. More enhancements in the interfacial strength can be attained by using coiled CNT coating on the fibers [39]. Recently, coating of graphene oxide on fiber surface has been reported to enhance the mechanical properties of SFRP [40, 41].

Processing of SFRP Composites

Reinforcement of a polymer matrix requires homogeneous dispersion of the reinforcing phase. Dry powder impregnation, solution impregnation, and melt mixing are the generally employed techniques to achieve good dispersion of the short fibers in the thermoplastic polymer matrix [42].

In dry powder impregnation, developed by Price in 1973 [43], strands of reinforcement are pulled continuously through a bed of powdered resin allowing polymer particles to be incorporated into spaces between fibers. In this process the wet-out is obtained by melting the polymer powder in the tow and applying pressure to cause the melt to flow in a direction parallel to the fibers. The uniform dispersion of powders during the dry powder impregnation minimizes the flow length during wet-out [44, 45].

Another technique of impregnation is to dissolve the polymer in a suitable solvent and use that low viscosity medium to wet the fibers. This impregnation operates at low temperatures (~50 °C) followed by drying at ~90 °C to eliminate the unwanted solvent [46–48]. The impregnated fibers are then hot-pressed to prepare the composite. It is relatively easy to operate, provided the availability of a suitable solvent for the polymer matrix. Smith et al. have fabricated repair patches of the CF/PC composite material using the solution impregnation technique. Observations show that PC granules in dichloromethane offer a relatively fast-dissolving solution providing a better impregnation [49]. Homogeneous wetting of long fibers can be achieved using this technique. However, the short fibers may tend to agglomerate in the solution.

For the short fiber reinforcements, melt mixing provides a homogeneous dispersion of the fibers. It is the most frequently used technique to disperse short fibers, particles, and nanoscale reinforcements in the thermoplastic polymer matrix [50, 51]. The procedure involves the uniform dispersion of the reinforcing particles/fibers into the molten polymer using a single or double screw extruder followed by the injection molding to shape a final product. In an extruder, several factors such as mass flow rate, pressure, mixing time, shear rate, zone temperature, and rheological properties control the product quality [52, 53]. For a particular matrix-reinforcement system, optimizations of these parameters are essential to ensure uniform dispersion.

Experimental

The main goal of the present chapter is to analyze the effects of short carbon fiber content on the dynamic mechanical, mechanical, and impact fracture behavior of bisphenol A polycarbonate composite.

Materials

Polycarbonates are polymers having functional groups linked together by carbonate groups (-O-(C=O)-O-) in a long molecular chain (Fig. 6.1). Polycarbonate of bisphenol A, the most common form of polycarbonates, is generally synthesized by a polymerization formed by the reaction of bisphenol A and phosgene. Applications of polycarbonates take advantage of their high impact strength, transparency to visible light, low flammability, and high heat distortion stability.

Fig. 6.1 Repeating unit of polycarbonate made of bisphenol A

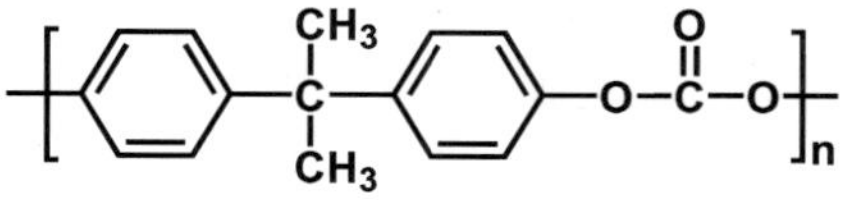

Bisphenol A polycarbonate (Lexan 143R grade) was obtained from M/S SABIC Innovative (General Electric Co.) Plastics India Limited. It is an amorphous solid having high transparency. Its melt flow rate and density as provided by the manufacturer are 11.0 g/10 min and 1.2 g/cm^3, respectively.

PAN-based unidirectional carbon fiber supplied by M/S Fothergill Private Limited (UK) having a density of 1.8 g/cm^3 was used. The diameter of a single filament of CF is 8–10 μm as determined by the scanning electron microscope.

Preparation of Composite and Sampling

Carbon fiber-reinforced polycarbonate composites were prepared by the single-screw extruder (Haake Rheocord – 9000). The PC/CF composites, ranged from 1 % to 15 % of fiber volume, were prepared by diluting the PC/CF master batch with polycarbonate to obtain the required concentration of CF. The temperature of the extruder was set from 270 °C to 300 °C in all four zones. Then samples were molded by injection molding. The set temperatures of all three zones in injection molding were 290 °C, 300 °C, and 315 °C.

Characterization of PC/CF Composites

Various characterization techniques are employed to study the mechanical, thermal, and rheological properties of PC/CF composites. Tensile and impact strength, interfacial shear strength, glass transition temperature, and viscoelastic properties are the most important parameters to characterize PC/CF composites.

Tensile Strength

The tensile and flexural tests were performed on the universal testing machine (UTM Instron 1195) with the maximum load capacity of 100 KN according to the ASTM D-638 at 25 °C. In addition, the flexural tests were carried out on the same UTM machine according to the ASTM D-790.

Fracture and Impact Strength

Impact test was performed by a CEAST impact tester using the ASTM D 256, whereas the abrasive resistance of the composites was measured by a taber abraser using 1000 g load and CS-17 abrasive wheel for 1000 cycles as per ASTM D1044. ASTM D2240 is used for hardness measurements.

Dynamic Mechanical Analysis

Dynamic mechanical analysis (DMA) provides valuable insight into the structure, morphology, and viscoelastic behavior of polymeric materials. It proves an important tool to study the effects of temperature/frequency on the viscoelastic/rheological properties of the polymeric composites under various modes of operation, namely, tensile, bending, and shear. Viscoelastic/rheological properties of FRP composites are influenced significantly by the fiber–matrix mechanical properties, geometric characteristics, fiber length, V_f, and interfacial bonding [16, 54]. Thus, DMA becomes an important tool to probe into the fiber–matrix interaction and interfacial bonding in a particular FRP composite system. In DMA, the response of the material to a dynamic load is monitored. The measured parameters include storage modulus (E'), loss modulus (E''), damping factor (tan δ), dynamic viscosity, etc. E' relates to the amount of energy stored in the composite as elastic energy whereas E'' measures the energy loss during a loading cycle. In the present study, dynamic mechanical analysis over a temperature range of 25–200 °C was performed with Pyris Diamond DMA (Perkin Elmer USA) by using sinusoidal loading of frequency 1Hz in bending mode.

Thermal Analysis (TG/DTA)

The thermal analysis was performed by using a Pyris Diamond Thermogravimetric/Differential Thermal Analyzer (TG/DTA) to study the thermal stability of composites with increasing fiber content. Thermogravimetric analysis (TGA) is used in this study to determine the thermal decomposition and stability of composites with different types of fiber reinforcement. On the other hand, the differential thermal analysis (DTA) provides information about the energy evolution due to various chemical and/or physical changes encountered by the specimen during a thermal cycle. In the present study, the analysis temperature range for the composite as well as the pure PC samples was 35–650 °C. All the samples were analyzed in air atmosphere at a heating rate of 10 °C/min.

Microstructural Analysis

Microstructural analysis of the FRP composites reveals the orientation and dispersion of the reinforcing fibers and the presence of various defects. In addition, study of the microstructure of fractured surface (fractography) reveals the causes and mechanism of failure. For SFRP composites, optical microscopy, and scanning electron microscopy (SEM) are the mostly used techniques to study microstructure. In the present study, SEM (Zeiss EVO MA-15) was used to study the fractured surfaces of Izod and Charpy specimens.

Results and Discussion

Mechanical Analysis

Tensile/Flexural Properties: The tensile curves for composites with varying fiber concentration are shown in Fig. 6.2. With increasing the fiber content, the tensile strength increases consistently. For the composite with 15 vol% CF, the tensile strength shows an enhancement of ~54 % as compared to the pure PC (Fig. 6.3a). Similarly, as the CF concentration increases from 0 to 15 vol%, the tensile modulus of CF-reinforced composites increases significantly (~177 %) as compared to the pure PC. The increase in the tensile modulus follows almost a linear trend as shown in Fig. 6.3a. Additionally, the percentage elongation of the composite shows a substantial decrease for increased fiber concentration, reaching as low as 0.8 % for 15 % CF loading as compared to 7.5 % for pure PC (Fig. 6.3b). This suggests ductile to brittle transition with increasing CF content. Stiff fibers restrict the motion of polymer chains, which, in turn, increases the strength as well as stiffness of the composite.

As shown in Fig. 6.4, the flexural tests of the composites exhibit large improvements in the flexural modulus as well as strength with increasing the fiber content. As compared to the pure PC, the strength and modulus of 15 % CF-reinforced composite show an increment of 50 % and 250 %, respectively. This can be attributed to the presence of randomly oriented fibers, which affect the movement of polymer chains when the composite material is flexed. Both tensile as well as flexural tests show similar improvements in the strength whereas the modulus exhibits larger improvement for flexural test (250 % compared to 177 %). This may be attributed to the presence of both tensile and compressive stresses during the flexural test.

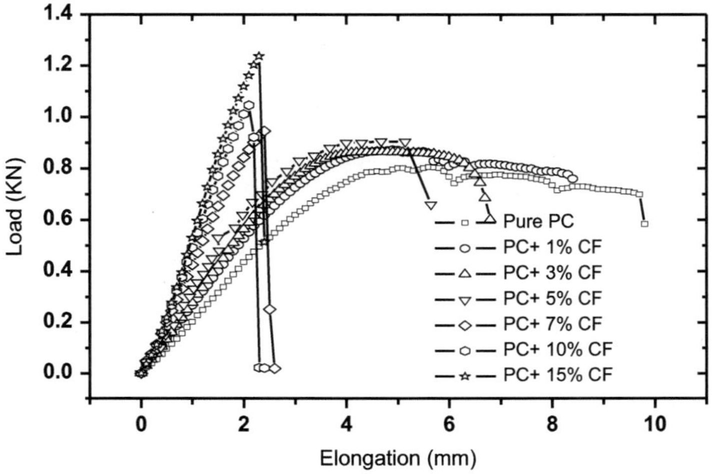

Fig. 6.2 Stress–strain curve under tension loading with varying carbon fiber concentration

Fig. 6.3 (**a**) Tensile strength and tensile modulus, (**b**) elongation at break of PC/CF composites with varying CF concentration

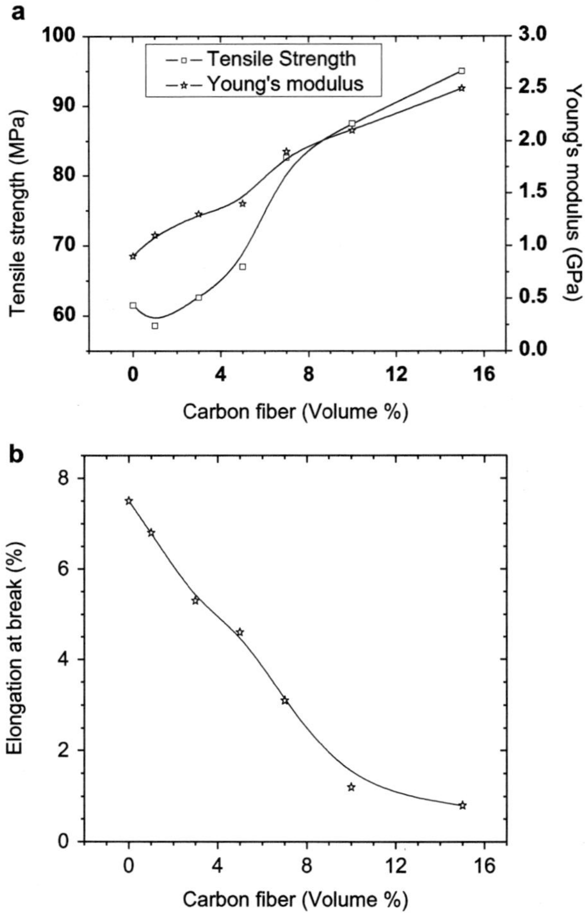

Few factors affect the mechanical performance of FRP composites. These include the moduli and strengths of the fiber and matrix; aspect ratio, length distribution, V_f, and uniformity and orientation of the fibers; integrity of the fiber–matrix interface; interfacial bond strength; etc. In addition, the presence of foreign objects such as voids and inherent defects affects the fiber–matrix interface, which, in turn, alters the mechanical performance [55–57]. The work by Caldeira et al. on vapor-grown carbon fiber (VGCF)-reinforced PC composites also reveals that the good fiber–matrix adhesion is essential for effective improvement in the tensile modulus and strength [58]. Because of their higher intrinsic rigidity, PAN-based carbon fibers give rise to composites with better mechanical properties than the VGCF. Present as well as the previous studies show improvements in tensile strength and toughness of SFRP composites with increasing fiber content.

Impact Strength: Impact tests show that the impact strength of the composite exhibits a drop with increasing the fiber concentration. As shown in Fig. 6.5, the Izod

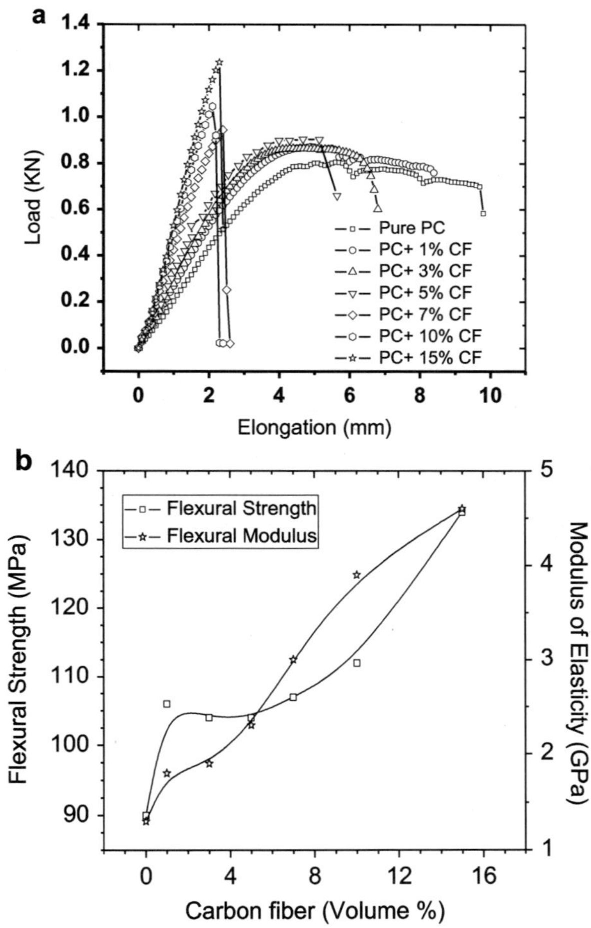

Fig. 6.4 (**a**) Stress–strain curve under three-point bending and (**b**) flexural strength and modulus of elasticity with varying carbon fiber concentration

impact strength for the notched samples decreases from 825 J/m for pure PC to 307 J/m for 5 vol% CF-reinforced PC. However, above the threshold of 5 % CF, there is a marginal increase in the impact strength for both the unnotched as well as the notched specimen. Similarly, the Charpy impact strength decreases from 197 J/m for pure PC to 83 J/m for 5 % CF-reinforced PC, whereas beyond the threshold of 5 % CF content, the impact strength shows a marginal increase. This marginal increase in the impact strength of the composite can be attributed to the phenomenon of fiber pullout where a portion of energy is spent to pull out the fiber from the matrix.

The observed decrease in the impact strength may be a consequence of several factors. Primarily, these factors include the weak fiber–matrix bonding and insufficient fiber concentration, which remains insufficient to provide proper load transfer from the polymer matrix to the fibers. Poor fiber–matrix interface requires very small amount of energy for crack propagation, which, consequently, reduces the impact

Fig. 6.5 (**a**) Notched Charpy and Izod impact strength (**b**) Unnotched impact strength with varying CF content

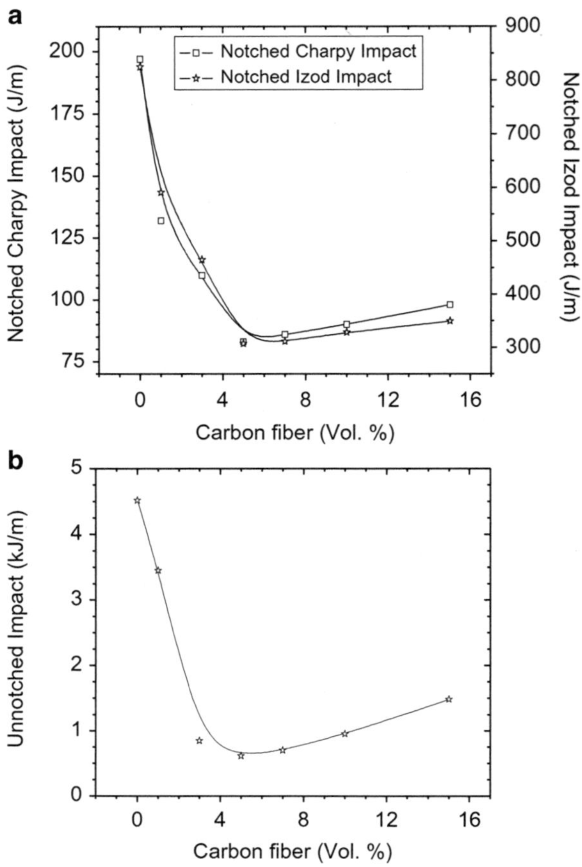

strength. As observed in tensile testing the ductility decreases with increasing CF content in the composites, which reduces the effective crack length and hence lowers the impact energy. Other than this, fiber reinforcement in the polymer matrix creates free volume by breaking the polymer chains, which causes a reduction in the impact strength of the composite.

Durometer Hardness: The hardness of the prepared composites shows an increasing trend with fiber concentration. As shown in Fig. 6.6a, the hardness of 15 % CF-reinforced composite increases by 70 % as compared to pure PC. The increase in the hardness of the composite may be attributed to the presence of CF, which acts as a barrier or obstacle hindering the growth and propagation of the microcracks from the surface to the interior of the material.

Abrasion Resistance: The abrasion resistance results for various compositions of CF in the PC matrix reveal that with an increase of the fiber concentration, the weight loss increases from 18.0 mg for pure PC to 50.0 mg for 15 % reinforced PC (Fig. 6.6b). This clearly shows that the abrasive resistance of the composite material decreases with increase in fiber concentration. This result is in agreement with the

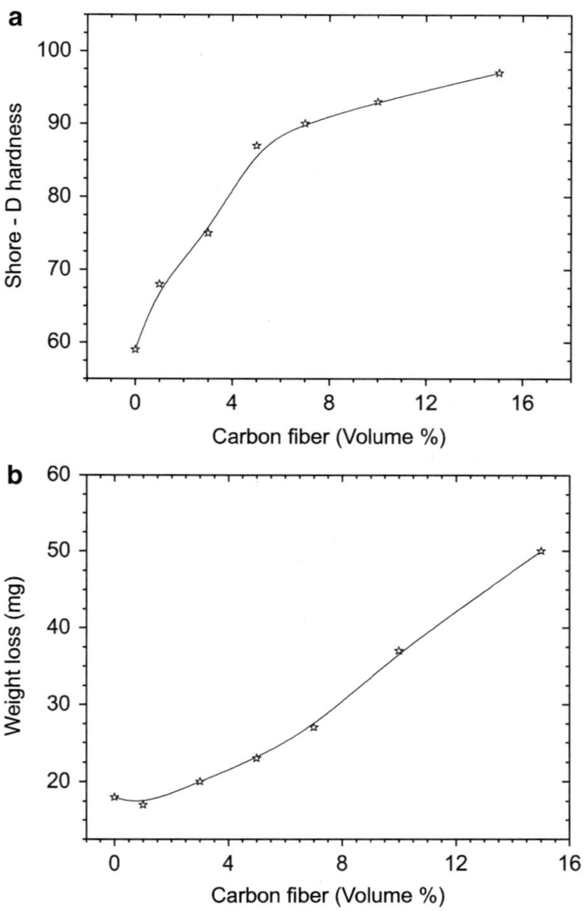

Fig. 6.6 (**a**) Durometer hardness and (**b**) abrasion loss with varying CF content

other studies on FRP composites [59, 60]. Abrasive wear of FRP composites depends on the nature of polymer matrix, fiber content, and their orientation relative to the abraser surface and the sliding direction. Fibers normal to the abraser surface show highest wear resistance [61]. Again, the abrasive wear resistance exhibits significant improvement when particulate fillers like PTFE are used along with the reinforcing fiber [60].

Scanning Electron Microscopy

SEM analysis of the impact-fractured surfaces of PC/CF composite has been carried out to investigate the fracture mechanism (fractography), the dispersion of the CF, and the microstructural changes evolved with the incorporation of the reinforcement into the polymer matrix.

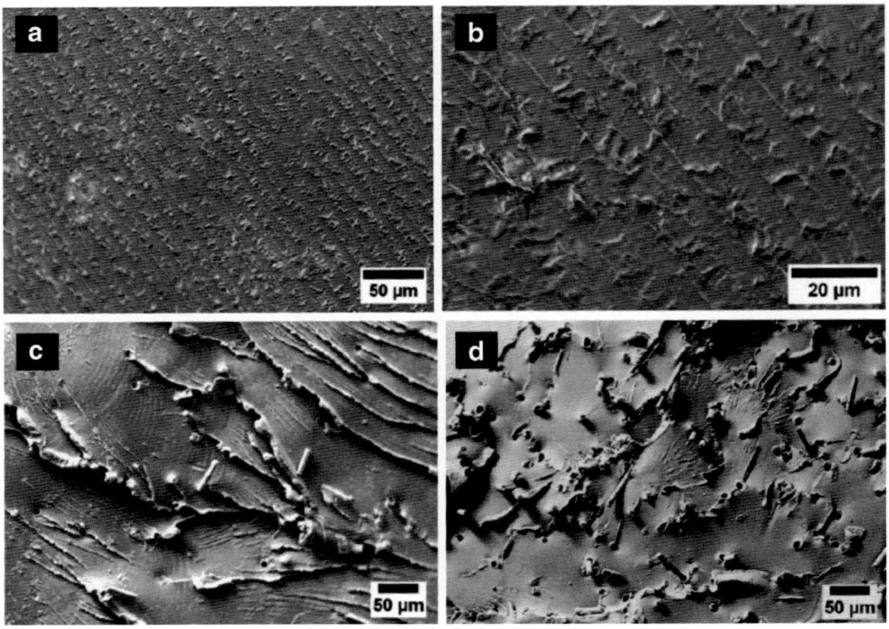

Fig. 6.7 SEMs of PC/CF composite fractured surfaces (**a**, **b**) pure PC (**c**) PC + 1 % CF (**d**) PC + 3 % CF

The average diameter and length of CF, as observed by SEM, lie in between 8–10 μm and 3–4 mm, respectively. The SEM micrographs of the CF-reinforced composites confirm a uniform dispersion of CF in the polymer matrix without any sign of fiber agglomeration for all compositions (Figs. 6.7 and 6.8). From Fig. 6.7a, b, it appears that the fracture is ductile in nature. There is smoothness in the fractured surface with very little ruggedness. At low fiber concentrations, i.e., for 1 vol%, 3 vol%, and 5 vol% CF, a fine terrain like topography in the fractured surface is observed. The smoothness of the fractured surface goes on decreasing with increase in the fiber concentration suggesting a ductile to brittle transformation in the composite material (Fig. 6.8b, c, and d). With increasing the fiber content, there is a remarkable difference in the surface morphology of the fractured surface. The micrographs also reveal the fiber pullout, which is found to be more dominant at higher fiber concentration, i.e., beyond 5 % loading. This can be attributed to poor interfacial bonding between the polymer and CF.

Dynamic Mechanical Analysis

DMA of the composites exhibits significant improvement in the E' with decreased damping (tan δ). However, the E'' also found to increase with increasing fiber concentration, which indicates poor interfacial bonding. This section includes detailed discussion on DMA results.

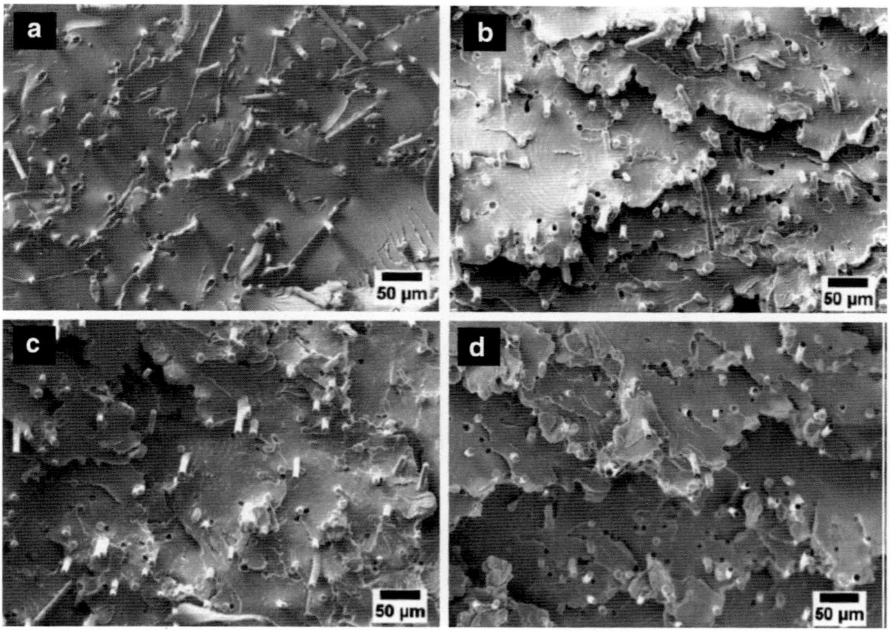

Fig. 6.8 SEMs of PC/CF composite fractured surfaces (**a**) PC + 5 % CF, (**b**) PC + 7 % CF, (**c**) PC + 10 % CF and (**d**) PC + 15 % CF

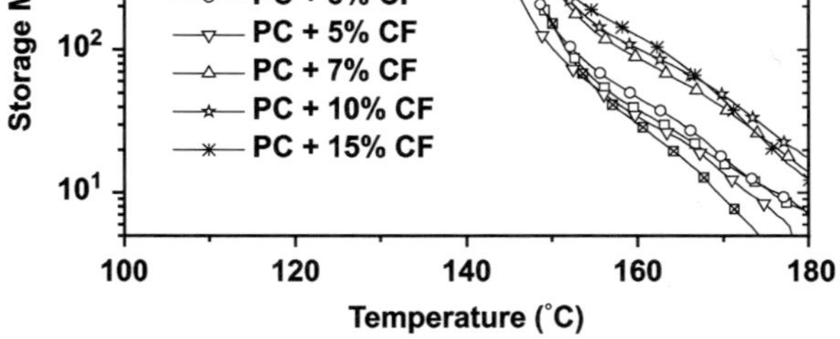

Fig. 6.9 Variation of storage modulus with temperature

Storage Modulus (E′): The E′ curve, Fig. 6.9, shows an increase in modulus in the glassy state region with the increase in fiber content. Inclusion of elastic fibers in the viscoelastic polymer matrix increases the E′ significantly. Below Tg, the E′ increases from ~6.5 GPa for pure PC to ~17.5 GPa for CF/PC composite with 15 vol% CF. Further, the E′ in the rubbery state (above Tg) for 15 vol% CF/PC

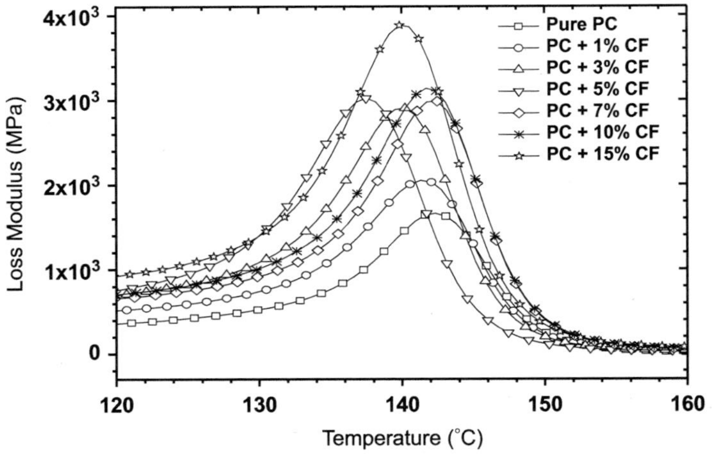

Fig. 6.10 Variation of loss modulus with temperature

composites is an order of magnitude higher than pure PC and remains constantly well above then that of pure PC at all temperatures. The decrease in modulus at higher temperature is due to the decrease in intermolecular force of attraction and softening of polymer matrix. In addition to this, the decrease in the modulus can also be attributed to the decrease in the coefficient of friction between polymer chains at higher temperatures.

Loss Modulus (E″): The E″ curves, as shown in Fig. 6.10, also exhibit an increase in the E″ with increasing the fiber content. The E″ maximum peak value changes from 1.7 GPa for pure PC to 3.9 GPa for 15 vol% CF/PC composite. This is because as the fiber content in the polymer matrix increases, the interfacial shear strength losses due to the dominance of the frictional forces. With further increase in the temperature beyond the peak temperature, the E″ drops due to the increased mobility of polymer chains. In addition, the temperature corresponding to the peak value of E″ also varies slightly with fiber content. Since the E″ peak is a better indicator of Tg of the polymer composite [62], it may be concluded that the Tg decreases slightly with increasing fiber content showing a minimum at 5 vol% CF and then increases with further increase in the CF content. In polymer composites, several factors such as size, volume fraction, and the aspect ratio of the reinforcement/filler influence the Tg [63]. Rheological studies on the VGCF/PC and PAN-based CF/PC composites reveal that VGCF have a higher influence on the rheological behavior of the composites than the PAN-based CF. In addition, the matrix nature also affects the rheological properties of the composites as the poor fiber–matrix adhesion results to the fiber- matrix relative slip [64, 65]. Increasing the fiber content increases the viscosity, with a larger increment for higher aspect ratio fillers like VGCF and CNT [64, 66].

Damping (tan δ): Variation of damping (tan δ) for composites having different fiber contents follows trends similar to those of E″. With increasing the fiber content,

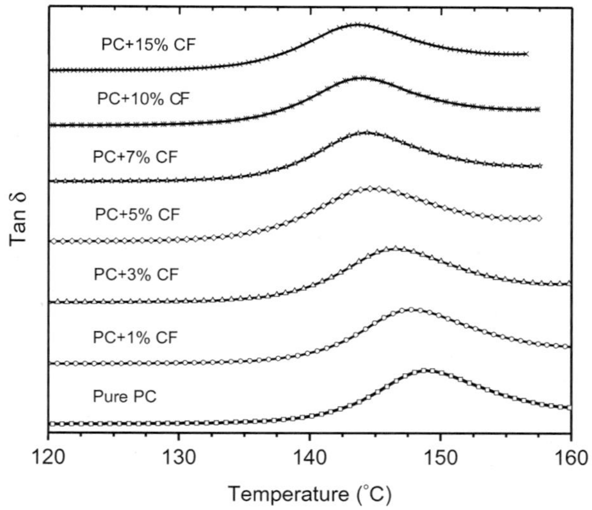

Fig. 6.11 Variation of tan δ with temperature

the peak becomes less prominent and broadened signifying a decrease in the damping (Fig. 6.11). In addition, the peak shifts toward lower temperature with increasing fiber content.

Thermal Analysis (TGA/DTA)

Figure 6.12a shows the TGA plots (variation in sample weight with temperature) for the composites with varying CF vol%. The pure PC as well as the composites exhibit at least two distinct plateau regions in the TGA plots. This is associated with the characteristic multistep degradation behavior of PC [67, 68]. Again, the residual weight above 600 °C increases with increasing CF vol%. This is because the thermal degradation of CF is negligible at this temperature. The derivative weight loss curve for pure PC (Fig. 6.12b) also shows two peaks at temperatures 495 °C and 605 °C corresponding to different degradation steps, which is supported by other studies as well. These peaks correspond to the maximum weight loss rate. In the first step of degradation, carbonate and isopropylene groups of the PC chain decompose, while the second step corresponds to the degradation of aromatic rings [69]. The derivative weight loss curves exhibit sharper peaks with increasing fiber content, which can be explained by the fast reaction rate caused by more active sites. Likewise, possible saturation of reaction rate at higher CF concentrations leads to the overlap of the TGA/derivative weight loss curves for 10 vol% and 15 vol% CF. The corresponding heat flow curves (DTA curves) for pure PC and composites are shown in Fig. 6.13. Incorporation of CF decreases the degradation temperature of CF/PC composites significantly. Inclusion of only 1 % CF in the polymer matrix decreases the first and second DTA peak temperatures by 4.4 % and 8.0 %, respectively. The possible reason behind the large decrease in the degradation temperature is the presence of active sites

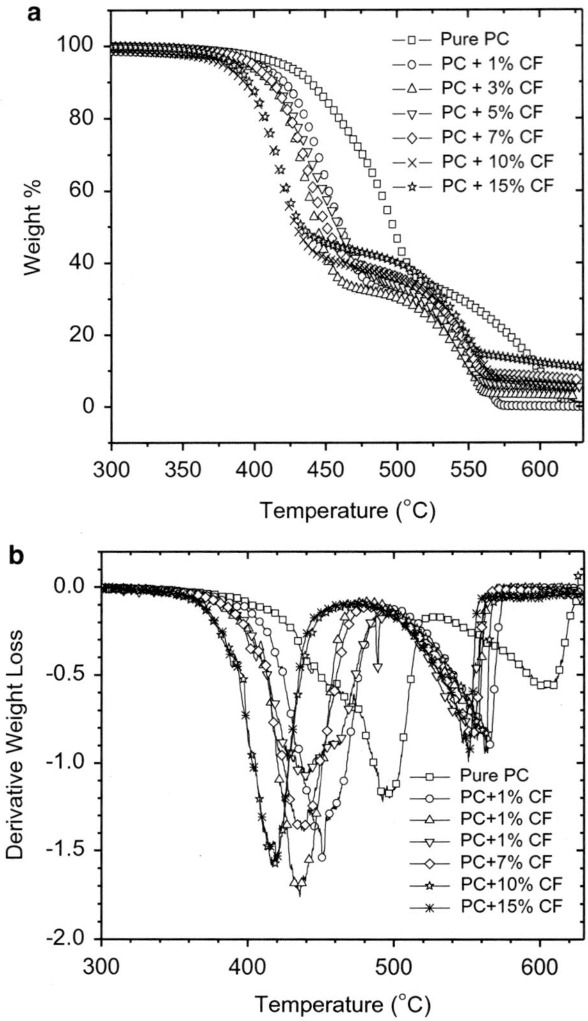

Fig. 6.12 (**a**) Weight% versus temperature and (**b**) derivative of weight loss versus temperature with varying fiber volume fraction

at the polymer chain- fiber interface facilitating low temperature degradation. With increasing fiber content, the increased number of active sites further decreases the degradation temperature slightly. Similar studies on short CF-reinforced polypropylene (SCFR/PP) composites suggest a significant increase in the thermal stability with increasing fiber length from 2 to 10 mm. In this case, increased number of fiber ends for smaller length fibers may act as the active sites for thermal degradation [22]. Other studies on short CF/PC composites also show that the thermal stability of the composite increases with increasing the CF length. Thus, for a constant fiber length, increasing fiber V_f increases the number of active sites at the fiber ends and the degradation temperature decreases. With increasing fiber content, the intensity of the

Fig. 6.13 Heat flow for (**a**) 0, 1, and 3 and (**b**) 5, 7, 10, and 15 vol% fiber

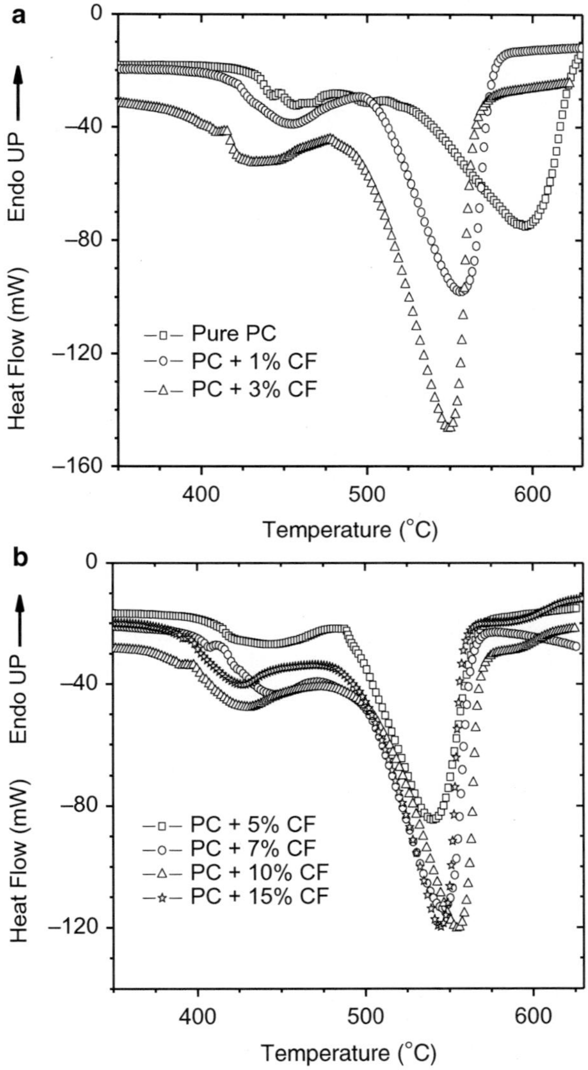

second peak of heat flow curve increases, whereas the peak width decreases. This is due to the increased rate of the highly exothermic degradation reaction.

Fragmentation/Fiber Pullout Tests

The performance of fiber-reinforced composites depends largely on the interfacial shear strength (τ) which determines the way loads can be transferred from matrix to the fiber or vice versa. The determination of τ is frequently done by means of single fiber pullout test and fragmentation test on single filament composites [38, 70]. Interfacial properties can be significantly enhanced by increasing the "surface activity" of the reinforcement [71]. The study concerning the effect of plasma treatment on the interfacial properties of carbon fiber–polycarbonate composites by Moran et al. reveal that the increased surface activity of the plasma-treated CF increases the interfacial shear strength, which, in turn, decreases the critical length significantly [72].

Concluding Remarks

Short carbon fiber-reinforced polycarbonate composites with different fiber content are made to study the effects of fiber concentration on the mechanical, viscoelastic, and fracture response of the composites. Melt mixing technique provides uniform dispersion of the fibers in the polycarbonate matrix as confirmed by SEM micrographs. With increasing the fiber content from pure PC to 15 vol% CF, the stiffness and strength of the composites improve significantly. As compared to the pure PC, considerable increments in the tensile modulus (~3 times), flexural modulus (3.5 times), tensile strength, and flexural strength (~1.5 times) are observed for 15 vol% CF composite. Observed reduction in percentage elongation of the composite material suggests ductile to brittle transition upon the incorporation of CF. The surface morphology of the fractured surfaces of the composites studied by the SEM micrographs confirms this transition. Charpy and Izod impact strengths decrease sharply with an increase in the CF fraction up to a threshold concentration of 5 % and then increase marginally. The shore hardness of the composite shows an increase with fiber loading. At lower temperatures (<Tg), the storage and loss moduli show large increments. As compared to the pure PC, for 15 vol % CF/PC composite, the storage and loss moduli increase by factors of 3 and 2, respectively. Above the Tg, the E' decreases with increasing the temperature. E'' first increases to reach a peak at the Tg and then decreases with increasing temperature. Broadening and shifting of the tan δ peak toward lower temperature is observed with increasing CF content. This suggests a poor fiber–matrix interaction, which must be improved by surface activation/functionalization. TGA/DTA studies show a decrease in the degradation temperature with increasing fiber vol%.

Acknowledgment The authors acknowledge the financial support provided by Indian Space Research Organization, India for carrying out this research work.

References

1. Kaw AK (2006) Mechanics of composite materials, 2nd edn. CRC Press/Taylor & Francis Group, Boca Raton
2. Aitasalo KMJ, Piitulainen JM, Rekola J, Vallittu PK (2014) Craniofacial bone reconstruction with bioactive fiber-reinforced composite implant. Head Neck 36:722
3. Li X, Yang Y, Fan Y, Feng Q, F-Z C, Watari F (2014) Biocomposites reinforced by fibers or tubes as scaffolds for tissue engineering or regenerative medicine. J Biomed Mater Res A 102:1580

4. Toldy A, Szolnoki B, Marosi G (2011) Flame retardancy of fibre-reinforced epoxy resin composites for aerospace applications. Polym Degrad Stab 96:371
5. Mallick PK (1993) Fiber-reinforced composites: materials, manufacturing, and design. Marcel Dekker, New York
6. Callister WD (2007) Materials science and engineering: an introduction, 7th edn. Wiley, New York
7. Gordeyev SA, Ferreira JA, Bernardo CA, Ward LM (2001) A promising conductive material: highly oriented polypropylene filled with short vapour-grown carbon fibres. Mater Lett 51:32
8. Chung DDL (1994) Carbon fiber composites. Butterworth-Heinemann, Boston
9. Donnet JB, Wang TK, Rebouillat S, Peng JCM (1990) Carbon fibers, 3rd edn. M. Dekker, New York
10. Morgan P (2005) Carbon fibers and their composites. CRC Press/Taylor & Francis Group, Boca Raton
11. Lee SM (1991) International encyclopedia of composites. VCH Publishers Ltd, New York, p 141
12. Mortazavian S, Fatemi A (2015) Effects of fiber orientation and anisotropy on tensile strength and elastic modulus of short fiber reinforced polymer composites. Compos Part B 72:116
13. Dyer SR, Lassila LVJ, Jokinen M, Vallittu PK (2004) Effect of fiber position and orientation on fracture load of fiber-reinforced composite. Dent Mater 20:947
14. Xie XL, Mai YW, Zhou XP (2005) Dispersion and alignment of carbon nanotubes in polymer matrix: a review. Mater Sci Eng R 49:89
15. Joshi M, Maiti SN, Misra A, Mittal RK (1994) Influence of fiber length, fiber orientation, and interfacial adhesion on poly(butylene terephthalate) polyethylene alloys reinforced with short glass-fibers. Polym Compos 15:349
16. Ho KC, Hwang JR, Doong JL (1996) Tensile properties of short glass fibre reinforced polycarbonate. Polym Polym Compos 4:563
17. Friedrich K (1985) Microstructural efficiency and fracture toughness of short fiber/thermoplastic matrix composites. Compos Sci Technol 22:43
18. Caprino G, Lopresto V (2001) On the penetration energy for fibre-reinforced plastics under low-velocity impact conditions. Compos Sci Technol 61:65
19. Zhou XF, Nairn JA, Wagner HD (1999) Fiber-matrix adhesion from the single-fiber composite test: nucleation of interfacial debonding. Compos Part A 30:1387
20. Lin T, Jia D, He P, Wang M, Liang D (2008) Effects of fiber length on mechanical properties and fracture behavior of short carbon fiber reinforced geopolymer matrix composites. Mater Sci Eng A Struct Mater Prop Microstruct Process 497:181
21. Fu S-Y, Lauke B (1996) Effects of fiber length and fiber orientation distributions on the tensile strength of short-fiber-reinforced polymers. Compos Sci Technol 56:1179
22. Rezaei F, Yunus R, Ibrahim NA (2009) Effect of fiber length on thermomechanical properties of short carbon fiber reinforced polypropylene composites. Mater Des 30:260
23. Bijsterbosch H, Gaymans RJ (1995) Polyamide-6 – long glass-fiber injection moldings. Polym Compos 16:363
24. Sarasua JR, Remiro PM, Pouyet J (1995) The mechanical-behavior of peek short-fiber composites. J Mater Sci 30:3501
25. Fu SY, Hu X, Yue CY (1999) Effects of fiber length and orientation distributions on the mechanical properties of short-fiber-reinforced polymers. Mater Sci Res Int 5:74
26. Weis EM, Wilke W (1992) Structure and mechanical-behavior of short glass fiber-reinforced ethylene tetrafluoroethylene copolymers. 1 Influence of the filler on mechanical-behavior and structure. J Mater Sci 27:1876
27. Denault J, Vukhanh T, Foster B (1989) Tensile properties of injection molded long fiber thermoplastic composites. Polym Compos 10:313
28. Biolzi L, Castellani L, Pitacco I (1994) On the mechanical response of short-fiber-reinforced polymer composites. J Mater Sci 29:2507
29. Tripathi D, Jones FR (1998) Single fibre fragmentation test for assessing adhesion in fibre reinforced composites. J Mater Sci 33:1
30. Manikandan V, Winowlin Jappes JT, Suresh Kumar SM, Amuthakkannan P (2012) Investigation of the effect of surface modifications on the mechanical properties of basalt fibre reinforced polymer composites. Compos Part B 43:812

31. Hughes JDH (1991) The carbon fibre/epoxy interface – a review. Compos Sci Technol 41:13
32. Fu H, Ma CQ, Kuang NH, Luan SL (2007) Interfacial properties modification of carbon fiber/ polyarylacetylene composites. Chin J Aeronaut 20:124
33. Ozkan C, Gamze Karsli N, Aytac A, Deniz V (2014) Short carbon fiber reinforced polycarbonate composites: effects of different sizing materials. Compos Part B 62:230
34. Bogoeva GG, Mader E, Hauler L, Sahre K (1995) Parameters affecting the interface properties in carbon fibre/epoxy systems. Composites 26:103
35. Thostenson ET, Li WZ, Wang DZ, Ren ZF, Chou TW (2002) Carbon nanotube/carbon fiber hybrid multiscale composites. J Appl Phys 91:6034
36. Li M, Gu Y, Liu Y, Li Y, Zhang Z (2013) Interfacial improvement of carbon fiber/epoxy composites using a simple process for depositing commercially functionalized carbon nanotubes on the fibers. Carbon 52:109
37. Lv P, Y-Y F, Zhang P, H-m C, Zhao N, Feng W (2011) Increasing the interfacial strength in carbon fiber/epoxy composites by controlling the orientation and length of carbon nanotubes grown on the fibers. Carbon 49:4665
38. Sager RJ, Klein PJ, Lagoudas DC, Zhang Q, Liu J, Dai L, Baur JW (2009) Effect of carbon nanotubes on the interfacial shear strength of T650 carbon fiber in an epoxy matrix. Compos Sci Technol 69:898
39. Sharma R, Yadav AK, Panwar V, Kar KK (2015) Viscoelastic properties of coil carbon nanotube-coated carbon fiber-reinforced polymer nanocomposites. J Reinf Plast Compos 34:941
40. Li F, Liu Y, Qu C-B, Xiao H-M, Hua Y, Sui G-X, Fu S-Y (2015) Enhanced mechanical properties of short carbon fiber reinforced polyethersulfone composites by graphene oxide coating. Polymer 59:155
41. Lee W, Lee JU, Cha H-J, Byun J-H (2013) Partially reduced graphene oxide as a multi-functional sizing agent for carbon fiber composites by electrophoretic deposition. RSC Adv 3:25609
42. Gibson AG, Manson JA (1992) Impregnation technology for thermoplastic matrix composites. Compos Manuf 3:223
43. Price R (1973) Inventor. Imperial Chemical Industries Limited (London), assignee. Production of impregnated rovings. US patent 3742106, 26 June 1973
44. Iyer S, Drazl L (1990) Manufacture of powder impregnated thermoplastic composites. J Thermoplast Compos Mater 3:325
45. Ye L, Kilinkuller V (1992) Impregnation and consolidation in composites made of GF/PP powder impregnated bundles. J Thermoplast Compos Mater 5:32
46. Lacroix FV, Werwer M, Schulte K (1998) Solution impregnation of polyethylene fibre/polyethylene matrix composites. Compos Part A 29:371
47. Rigdahl M, Westerlind B, Hollmark H, Ruvo AD (1983) Introduction of polymers into fibrous structures by solution impregnation. J Appl Polym Sci 28:1599
48. Wu GM, Schultz JM (2000) Processing and properties of solution impregnated carbon fiber reinforced polyethersulfone composites. Polym Compos 21:223
49. Smith FC, Moloney LD, Matthews FL, Hodges J (1996) Fabrication of woven carbon fibre/ polycarbonate repair patches. Compos Part A 27:1089
50. Ouml, Tschke P, Bhattacharyya AR, Janke A, Goering H (2003) Melt mixing of polycarbonate/ multi-wall carbon nanotube composites. Compos Interfaces 10:389
51. Pötschke P, Bhattacharyya AR, Janke A (2004) Melt mixing of polycarbonate with multiwalled carbon nanotubes: microscopic studies on the state of dispersion. Eur Polym J 40:137
52. Joseph PV, Joseph K, Thomas S (1999) Effect of processing variables on the mechanical properties of sisal-fiber-reinforced polypropylene composites. Compos Sci Technol 59:1625
53. Sennett M, Welsh E, Wright JB, Li WZ, Wen JG, Ren ZF (2003) Dispersion and alignment of carbon nanotubes in polycarbonate. Appl Phys B Lasers Opt 76:111
54. Dong S, Gauvin R (1993) Application of dynamic-mechanical analysis for the study of the interfacial region in carbon fiber/epoxy composite-materials. Polym Compos 14:414

55. Kim JK, Mai YW (1991) High strength, high fracture toughness fibre composites with interface control – a review. Compos Sci Technol 41:333
56. Hornbostel B, Potschke P, Kotz J, Roth S (2006) Single-walled carbon nanotubes/polycarbonate composites: basic electrical and mechanical properties. Phys Status Solidi B Basic Solid State Phys 243:3445
57. Zihlif AM, DiLiello V, Martuscelli E, Ragosta G (1995) Tensile properties and fracture behaviour of polycarbonate/PAN-based carbon fiber composite. Int J Polym Mater 29:211
58. Caldeira G, Maia JM, Carneiro OS, Covas JA, Bernardo CA (1998) Production and characterization of innovative carbon fiber polycarbonate composites. Polym Compos 19:147
59. Bijwe J, Rattan R, Fahim M (2007) Abrasive wear performance of carbon fabric reinforced polyetherimide composites: influence of content and orientation of fabric. Tribol Int 40:844
60. Tewari US, Bijwe J, Mathur JN, Sharma I (1992) Studies on abrasive wear of carbon fibre (short) reinforced polyamide composites. Tribol Int 25:53
61. Lhymn C, Tempelmeyer KE, Davis PK (1985) The abrasive wear of short fibre composites. Composites 16:127
62. Akay M (1993) Aspects of dynamic mechanical analysis in polymeric composites. Compos Sci Technol 47:419
63. Landry CJT, Coltrain BK, Landry MR, Fitzgerald JJ, Long VK (1993) Poly(vinyl acetate)/silica-filled materials: material properties of in situ versus fumed silica particles. Macromolecules 26:3702
64. Carneiro OS, Maia JM (2000) Rheological behavior of (short) carbon fiber/thermoplastic composites. Part 1: the influence of fiber type, processing conditions and level of incorporation. Polym Compos 21:960
65. Carneiro OS, Maia JM (2000) Rheological behavior of (short) carbon fiber/thermoplastic composites. Part II: the influence of matrix type. Polym Compos 21:970
66. Potschke P, Fornes TD, Paul DR (2002) Rheological behavior of multiwalled carbon nanotube/polycarbonate composites. Polymer 43:3247
67. Jang BN, Wilkie CA (2005) The thermal degradation of bisphenol A polycarbonate in air. Thermochim Acta 426:73
68. Li XG, Huang MR (1999) Thermal degradation of bisphenol A polycarbonate by high-resolution thermogravimetry. Polym Int 48:387
69. Ma CCM, Gu JT, Shauh LH, Yang JC, Fang WC (1997) Bisphenol-A polycarbonate/polydimethylsiloxane multiblock copolymers. I. Synthesis and characterization. J Appl Polym Sci 66:57
70. Pitkethly MJ, Doble JB (1990) Characterizing the fibre/matrix interface of carbon fibre-reinforced composites using a single fibre pull-out test. Composites 21:389
71. Peebles LH Jr (1995) Carbon fibers, formation, structure and properties. CRC Press, Boca Raton
72. Montes-Morán MA, van Hattum FWJ, Nunes JP, Martínez-Alonso A, Tascón JMD, Bernardo CA (2005) A study of the effect of plasma treatment on the interfacial properties of carbon fibre-thermoplastic composites. Carbon 43:1795

Ionic Polymer Metal Composites

7

Syed Nadeem Akhtar, Jayesh Cherusseri, J. Ramkumar, and
Kamal K. Kar

Contents

S.N. Akhtar • J. Ramkumar
Advanced Nanoengineering Materials Laboratory, Department of Mechanical Engineering, Indian Institute of Technology Kanpur, Kanpur, Uttar Pradesh, India

J. Cherusseri
Advanced Nanoengineering Materials Laboratory, Materials Science Programme, Indian Institute of Technology Kanpur, Kanpur, Uttar Pradesh, India

K.K. Kar (✉)
Advanced Nanoengineering Materials Laboratory, Materials Science Programme, Indian Institute of Technology Kanpur, Kanpur, Uttar Pradesh, India

Advanced Nanoengineering Materials Laboratory, Department of Mechanical Engineering, Indian Institute of Technology Kanpur, Kanpur, Uttar Pradesh, India
e-mail: kamalkk@iitk.ac.in

K.K. Kar (ed.), *Composite Materials*,
DOI 10.1007/978-3-662-49514-8_7

Abstract

Ionic polymer metal composites (IPMCs) are electro-active polymers with excellent electromechanical coupling properties. They are efficient candidates in many advanced technological applications such as actuators, artificial muscles, biomimetic sensors, etc. The manufacturing of electrodes for IPMCs is very critical in their electromechanical coupling. Force optimization, selection of cations and particle size distribution within the IPMC structure, etc. are the various factors, which determines their efficiency. In this chapter, we briefly discuss the structure, components and working mechanisms of IPMCs. The synthesis and characterizations of IPMCs are discussed in detail with the help of examples. A brief outlook on the modeling and potential applications of IPMCs is also included.

Keywords

Actuator • Artificial muscles • Damping • Elastic modulus • Electroding process • Electromechanical coupling • Ionic polymer metal composites • Microelectromechanical systems • Sensors • Polymer membrane

Introduction

Ionic polymer metal composites (IPMCs) are 'technologically matured' electroactive polymers (EAPs). IPMCs are frequently known as 'soft actuators-sensors' or 'artificial muscles' because of their electromechanical coupling behavior. The developments on the polymer/metal composites in the early 1930s have signaled the rise of IPMCs actively in research. Typically, an IPMC consists of a thin membrane of an ionic polymer structured between two metal electrodes on both the faces. They are capable in transducing the strain energy into electrical energy and vice-versa. Their synthesis and designing play a critical role in the efficiency of IPMCs based devices. They are found applications in a wide range of areas such as actuators, bio-mimetic sensors, artificial muscles, and many more. The light weight and flexibility are their added advantages, but still the huge price is making away from commercializing such technologies. This chapter briefly reviews the working principle of IPMC materials, manufacturing methods, designing, modeling, and characterization of IPMCs based structures, and their potential applications. Even though a full note on their applications is beyond the scope of this chapter, still a short description of applications in actuators, bio-mimetic sensors, artificial muscles, MEMS, fuel cells, energy harvesters, and some biomedical applications are also briefly discussed.

Significance of Ionic Polymer Metal Composites

The IPMCs represent a special group of EAPs with modified properties. They are synthetic composite materials, which can be cleverly manipulated for various

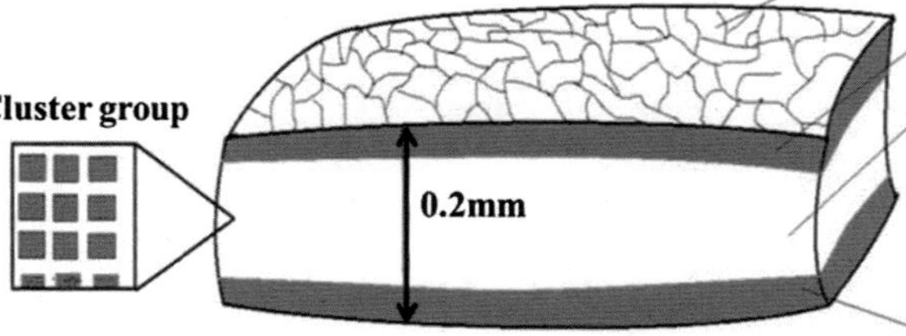

Fig. 7.1 Schematic showing the structure of an IPMC

applications. Their stunning properties such as actuation, sensing, etc. has become very useful in integrating with other advanced technologies for various applications. They are used in actuators for converting the electrical signals to mechanical bending of the material. Reversibly, as a smart material, they can produce voltage when a mechanical force is applied on their surface. The capability in working even in the fluid environments without sacrificing the respective properties has increased their importance. They become the functional components of MEMS due to their compatibility with such micro-mechanical systems. The journey of IPMCs is started from the basic application as actuators and has reached advanced applications as a laparoscopic surgical tool in the medical field. The imagination of their potential fields of implementation is beyond the limit of their today's applications, and is expected to perform in the future research.

Structure of Ionic Polymer Metal Composites

A schematic of a typical IPMC material is shown in Fig. 7.1. An IPMC consists of a polymer membrane of thickness 200 μm with electroless-deposited metal electrodes on both the faces. The metal electrode thickness can vary from 5 to 10 μm. The polymers employed in IPMCs are ionomers, which are the class of polymers, whose side branch typically has an ionic termination. The charge neutrality in the polyelectrolyte matrix is sustained by the amount of counter ions with anions, which are covalently fixed to the membrane.

Polymer Membranes The ionomeric membranes used in the construction of IPMCs are perfluorinated membranes. The commonly used ionomeric membranes for the manufacture of IPMCs are flemion from *Asahi Glass*, and Nafion® from *DuPont*, and aciplex from *Asahi Chemical*. Nafion® and aciplex have sulfonic acid groups as ionic termination ends, whereas flemion has carboxylic acid groups. Among the ionomeric membranes, Nafion® has achieved much popularity due to their exceptional properties. It is a copolymer of tetrafluoroethylene and sulfonyl fluoride vinyl ether discovered by Walther Grot in the late 1960s. It has been originally developed

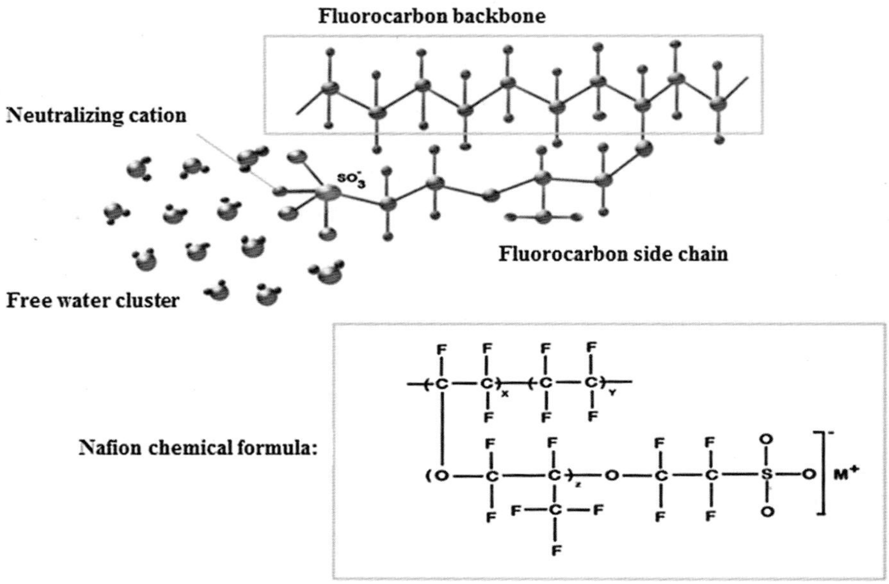

Fig. 7.2 Chemical structure of Nafion® membrane

for fuel cells as an ion-permeable membrane. Its importance has been hidden until 1992, where its transducing ability is discovered. Its chemical structure is shown in Fig. 7.2. It consists of a perfluorinated carbon backbone, a sulfonated side chain, mobile cations and hydrating water. The polytetrafluoroethylene backbone is helping to give it a three-dimensional structure. The long side chains of perfluorovinyl ether are regularly spaced and have ionic (sulfonate) terminations. Each of these acidic groups has an associated cation to neutralize it. Since the membrane combines hydrophobic (fluorocarbon) and hydrophilic (ionic phase) components, the orientation of the polymeric chains is strongly dependent on the water content of the polymer and cations, and its morphology comprises discrete hydrophobic and hydrophilic regions. The ionic groups, being hydrophilic, are aggregated in tightly packed interconnected clusters that are readily saturated by water. The membrane does not allow diffusion of anions through it whereas it is permeable to water and cations [1].

Metal Electrodes The ionomeric membranes are coated with metals such as Ag, Au, Pt, Pd, etc. Generally noble metals are selected due to the advantage of non-oxidizing nature, since the IPMCs are used in the fluid environments. Electroless deposition techniques are generally used for coating metals on top of ionic membranes. A thick overlayer of metal with thickness less than 10 μm is deposited on the membrane. These metal layers play an important role in the conductivity. As the thickness of the metal layer increases, the surface conductivity is also increases. This in turn helps in the membrane charging and subsequently leads to the

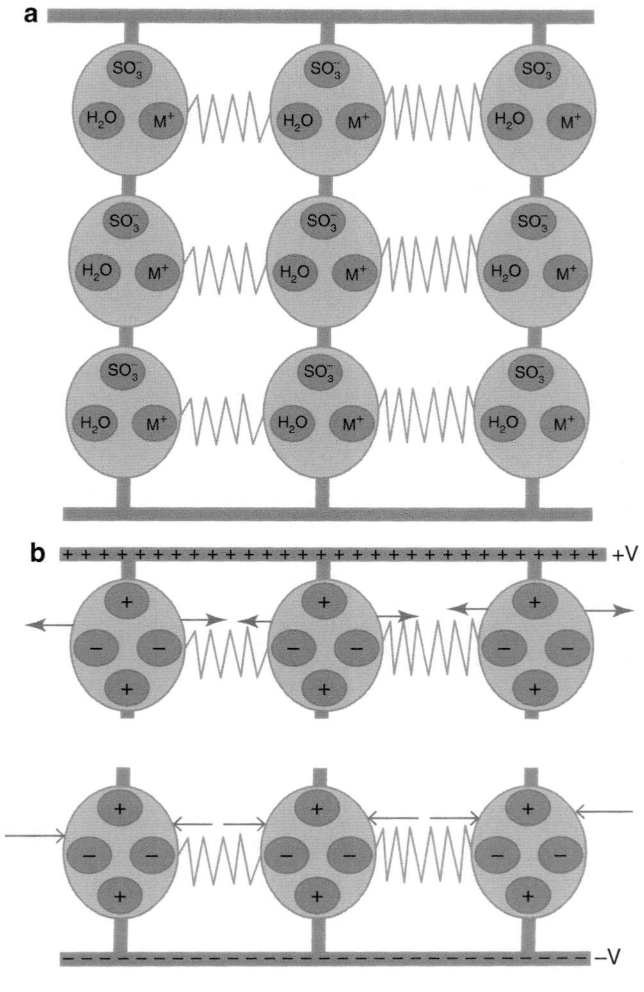

Fig. 7.3 Schematic of possible microstructure for (**a**) neutral state of hydrated Nafion®, and (**b**) hydrated Nafion® when a voltage is applied [2]

generation of bending due to actuation. On the other hand, the increased metal thickness increases the stiffness of the composite, which results in the need of increased force for the same displacement.

Working Principle of Ionic Polymer Metal Composites

A schematic of probable microstructure of hydrated Nafion® is depicted in Fig. 7.3a. It can be seen that the morphology comprises of discrete hydrophobic and hydrophilic regions. The hydrophobic region is composed of the polytetrafluoroethylene backbone and the hydrophilic region contains the ionic groups and cations. Note that while the cations are mobile within the hydrated clusters, the anionic groups are permanently attached to the fluorocarbon backbone, and hence, immobile. Bennett and Leo have reported that the actuation (electromechanical coupling) in IPMCs happens due to the electrically-induced motion of cations (note that the anions are immobile because they are covalently bound to the polymer backbone) [1]. Hence the mechanically–induced motion of cations should produce an electrical signal. Therefore, these ionic polymers are actually transducers, and can serve as both sensors and actuators. Since the extent of transduction is mainly dependent on the charge capacity (capacitance) of the transducer, increasing the capacitance increases the electromechanical coupling of the actuator. This is done by plating the membrane, on both sides, with interpenetrating metallic electrodes, typically Pt, using an 'impregnation/reduction' process. When a small step voltage (1–3 V, depending on the solvent) is applied to an IPMC in the hydrated state, the induced electric field exerts an electrostatic force on both the fixed anions and mobile cations. The cations diffuse towards the cathode under the electric field and get re-distributed, resulting in an imbalance of charges across the membrane. This gives rise to internal stresses in the polymer backbone and is schematically shown in Fig. 7.3b.

Nasser and Li have studied that formation of clusters leads to stretching of polymer chains throughout the polymer matrix [2]. When the cations migrate towards the cathode under the influence of electric field, polymer chains in the anode boundary layer relaxes and those at the cathode boundary layer continues further stretching. This leads to bending of the IPMC towards the anode. The cations are solvated by water molecules and the water molecules get re-distributed along with the cations. It is often suggested that this re-distribution of water produces differential swelling, as shown in Fig. 7.4. Consequently, the IPMC exhibits an initial fast bending towards the anode and a subsequent slow relaxation, generally in the opposite direction. These responses (magnitude, speed, direction) depend on the composition of the polymer backbone, the nature of the cations, the microstructure of the electrodes, and other conditions (e.g., the temporal profile of the applied voltage). It has been studied that Nafion® based IPMCs containing alkali metals or alkyl ammonium cations, typically bend first towards the anode, after application of a step voltage, and then relax towards the cathode even while the voltage is maintained. The relaxation tends to bend the IPMCs well beyond their starting positions. The initial bending of the IPMC can be retarded by gradually applying the voltage across its faces. As discussed earlier, sudden bending of an IPMC will generate a small voltage (typically of the order of millivolts) across its faces. Hence, IPMCs can serve as soft actuators and sensors.

Manufacturing of Ionic Polymer Metal Composites

An IPMC is a polymer membrane with metal electrodes plated on their both faces. The electrodes are either chemically plated or electro-plated. There are partially penetrated metal particles in the bulk of the polymer that serve to increase the

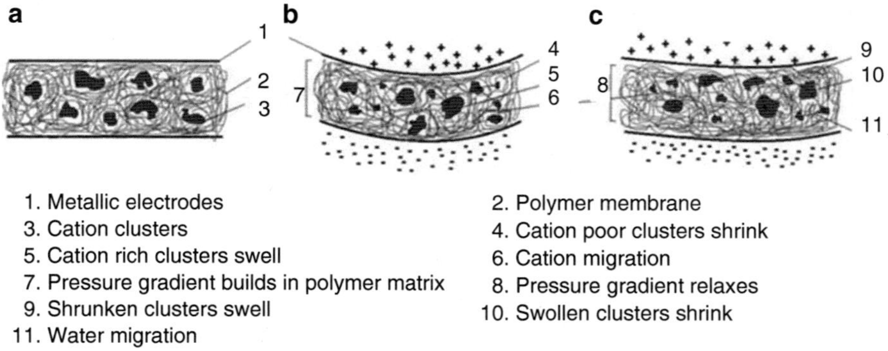

Fig. 7.4 Depiction of the process of IPMC actuation (**a**) neutral state: IPMC is flat, (**b**) applied potential causes cations to migrate to cathode: IPMC contracts, and (**c**) pressure gradient causes water to flow towards anode: IPMC relaxes

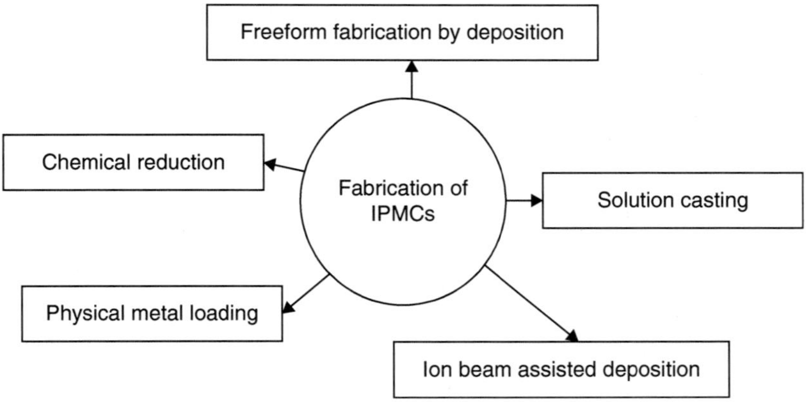

Fig. 7.5 Different fabrication methods for synthesizing IPMCs

conductivity of cations inside the hydrated ionomers. Application of an electric field applied across these electrodes causes the IPMC to bending in response. The electroding of the surfaces on both sides of the membrane is done by dipping it in a metal (typically Pt) salt solution for a considerable amount of time and reducing the metal salt to metal particles in the few micrometers thickness under the surface of the membrane. Pt is preferred because of high electrochemical potential and its high resistance to corrosion. This allows the application of higher voltages for actuation of IPMCs without damage. The dispersion of these conductive particles over a finite thickness under the surface, increases the thickness of conducting layer and hence the fraction of material, which experiences the electric field. Therefore this greatly affects the extent of transduction in IPMC. There are various methods adopted for the preparation of IPMCs. Fig. 7.5 shows the possible means of fabricating IPMCs.

Shahinpoor et al. have reported various techniques and experimental procedures for manufacturing of IPMCs [3]. The most followed procedure comprises four steps, namely, surface roughening of the membrane, ion exchanging, initial compositing, and surface electroding by reduction of the cations. The procedure starts with making available a strip of Nafion® 117 of the size 5 cm × 5 cm. The first step is to roughen the membrane so that better adsorption of metal ions takes place. Sandblasting and/or sandpapering is followed by ultrasonic cleaning and chemical cleaning (by boiling in dilute HCl) to remove impurities. The ion exchanging process involves dipping the membrane in a metal complex solution. The tetra amine Pt chloride hydrate is used for incorporating Pt cations on to the surface of polymer membrane. The strip is kept immersed for a few hours with occasional stirring of the solution. The third step is the initial compositing process that involves reduction of the metal (Pt) cations into metallic state by using a reducing agent such as a 5 % aqueous solution of sodium or lithium borohydride (LiBH4). Black or shiny metal layers are deposited on both sides of the strip indicating electroding. Shahinpoor et al. have described in detail the chemical reactions that take place during the compositing and electroding processes in the fabrication of IPMCs [4]. Pt is used as the electroding material and $LiBH_4$ as the reducing agent and found the optimal quantity of the reducing agent needed for complete reduction of Pt salt. $LiBH_4$ reacts with water as per the following reactions, generating hydroxyl ions in the process.

$$LiBH_4 + 4H_2O \rightarrow 4H_2 + LiOH\,(Li^+ + OH^-) + B(OH)_3$$

$$LiBH_4 + 8OH^- \rightarrow BO_2^- + Li^+ + 6H_2O + 8e^-$$

The solid form of $LiBO_2$ occasionally precipitates. The Pt complex cation further reacts to give Pt.

$$LiBH_4 + 4\left[Pt(NH_3)_4\right]^{2+} + 8OH^- \rightarrow 4Pt^0(s) + 16NH_3(g) + LiBO2 + 6H_2O(l)$$

The aim of surface electroding process is to reduce the surface resistivity by depositing noble metals on top of the initial Pt surface. The surface resistivity of the composite is found to decrease with increased Pt penetration and resulted in increased blocking force (Fig. 7.6).

The force density of the IPMCs has been optimized by optimizing multiple process parameters during the fabrication process, including solvent, temperature and concentrations of the metal containing salt and the reducing agent. Kim and Shahinpoor is reported an alternative economical method for fabrication of IPMCs that involves physical metal loading on to the substrate (membrane) followed by chemical and electro plating [3]. First, the polymer membrane was physically loaded with a conductive primary Ag powder forming a dispersed layer that functioned as a major conductive medium near boundaries. Next, these primary particles were secured within the surface with smaller particles (Pd or Pt) via a chemical plating process. Furthermore, electroplating was also carried out. Heterogeneous compositing technique has been viewed as a possible method of manufacturing IPMCs. Millet

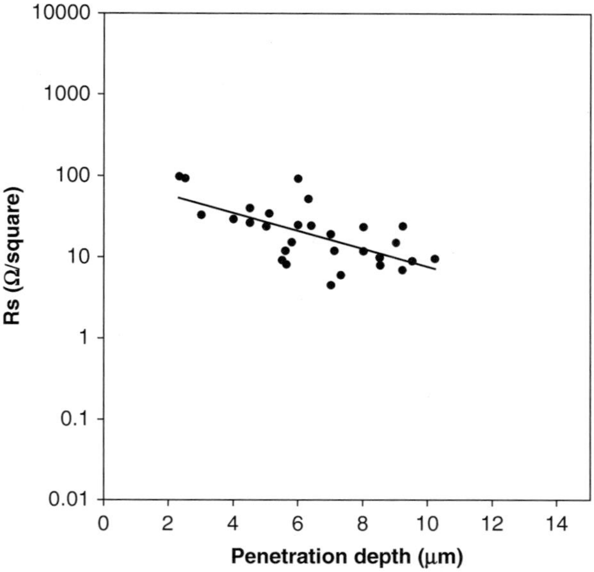

Fig. 7.6 Plot showing the variation of surface resistivity as a function of Pt penetration (Reprinted with permission from [3] Copyright (2003), IOP Publishing)

and Durand have studied the precipitation of Pt on to Nafion® membranes [5]. It has been observed that it is possible to limit the depth of precipitation to the membrane surfaces by optimizing the values for the parameters controlling the reaction. It is also observed that the metallic and polymeric constituents should interpenetrate strongly to improve the adherence of the electrodes and to lower electrochemical over potentials. The rate of incorporation and precipitation of metal particles is found to be proportional to the concentrations of the salt and the reducing agent in the solution respectively and depend on the hydrodynamic conditions at the interfaces. The concentration profiles of Pt across the membrane thickness depend significantly upon the reducer concentration in the solution and the number of loading-precipitation cycles. The membrane thickness of ionic polymers ranges from 100 to 300 μm. Although the IPMCs developed from these membranes show large displacements and force densities, still larger forces are required for their utilization in practical devices. Doubling the thickness of IPMCs increases the tip force by eight times. Kim and Shahinpoor have reported a solution recasting technique for manufacturing 3D IPMCs that can be scaled up or down from a membrane thickness of a micrometer to centimeter thickness (Fig. 7.7) [6]. The as-received membranes are dissolved in appropriate solvents and by carefully evaporating solvents out of the solution; re-casted ion-exchange membranes are obtained. Initially, during the solvent evaporation, the recast Nafion® developed surface cracks. Dimethylformamide (DMF) is used as an additive to make the mixture an azeotrope (constant boiling mixture). Following this, multiple layers of liquid Nafion® are successfully dried to produce 2 mm thick strips. The strips are annealed at a temperature of 70 °C to create crystallinity and to improve mechanical and chemical stability of the film. IPMCs are

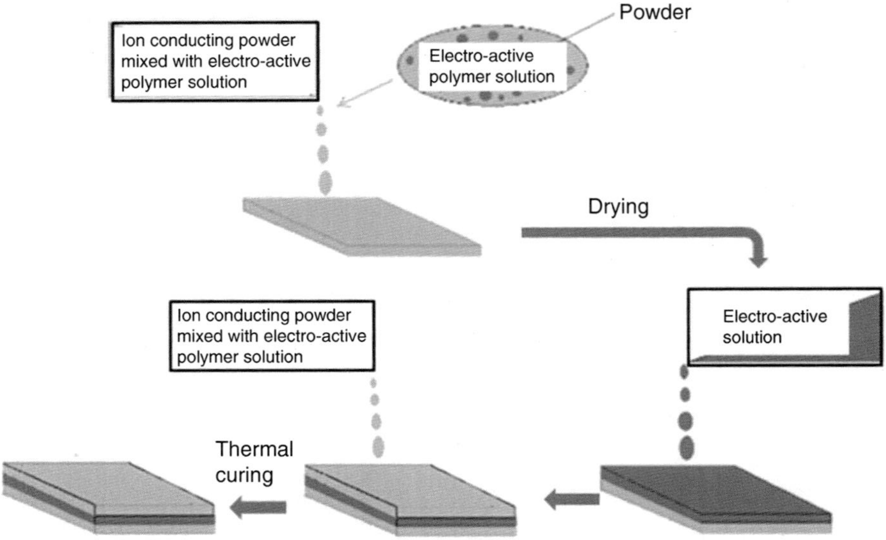

Fig. 7.7 An illustration of the process for making metallic powder coated electrode by solution recasting method [6]

fabricated by Pt electroding of these strips with particle penetration depths of up to 20 μm. Samples of size 15 mm × 5 mm × 2 mm are fabricated that exhibited generative tip forces in excess of 20 g-f and few millimeters of displacement.

Lee et al. have proposed an IPMC actuating system having a bio-mimetic function [7]. It has been fabricated that single and multi-layered IPMCs up to 800 μm thick with 178 μm thick individual layers to obtain higher actuation force. The IPMC is prepared by adsorption reduction method. Surface electrode fabrication is done by the ion beam assisted deposition (IBAD) and noble metals (Pt, Ir, and Au) are deposited on the surfaces of IPMC by an electric gun, achieving a deposition rate of 0.1 Å/s. The resulting Pt layer is more homogeneous and smooth than conventional chemical reduction method and its effect is evident in greater electromechanical transduction of the IPMC. The surface electric resistance of chemically plated IPMC is 2–5 times more than that of the electrode prepared by IBAD. It is also demonstrated that the IPMC could be efficiently actuated and controlled with electromyography signals generated from the human muscles. Malone and Lipson have attempted for making IPMCs through freeform fabrication method [8]. Freeform fabrication is a method of producing IPMC with varying geometry and allows the integration of IPMCs with mechanisms, which can yield linear or more complex motion. It has been suggested that a method of producing IPMCs by casting stacked layers, where the electrode layers are cast by ionomer dispersion with suspended metal particles. Initially, 5 wt% H+ Nafion® dispersion in mixed alcohol and water is used as the ionomer, and silver powder (99.9 %, 1–3 μm particles) is used as the electrode material. The cast film is annealed to eliminate

Table 7.1 Maximum bending curvatures of different samples of IPMCs (Reprinted from [9], Copyright (2001), with permission from Elsevier)

Specimen	I	II	III	IV
Imported cation	Li^+	Cu^{2+}	Li^+	Cu^{2+}
Electrode	Pt	Pt	Pt-Cu	Pt-Cu
Curvature (mm^{-1})	0.020	0.043	0.057	0.141

solubility of the ionomer without causing it to become brittle. A non-ionic surfactant is added to the electrode material to decrease the degree of hydrophobia followed by casting of a new layer on top of the cast electrode layers. To avoid swelling, buckling and cracking of the cast layer the subsequent layers are deposited atop prior layers that have not completely dried. It has been also experimented on developing a method of freeform fabricating a mold into which the IPMC materials can be cast, and on modifying the formulation of the electrode material to improve dispensability. Experiments to improve the service life and to reduce the actuating power requirements of freeform fabricated IPMCs are carried out by several researchers. Increasing the thickness of the central ionomer layer overcame the problem of internal shorting. Loss of solvent during exposure to the ambient, corrosion of the electrodes, and progressive electrical shorting are found to be the primary limitations on the durability of the IPMCs. In order to reduce the rate of solvent loss due to evaporation, impermeable electrode materials are surface coated on to the metal-dispersed ionomer electrode layers. Hydrin C thermoplastic (poly (epichlorohydrin-co-ethylene oxide)) and 'flexible silver ink' (EP403920-50, SRA Inc.) are mixed with silver powder for use as the coating material. Uchida and Taya have fabricated IPMCs using Li and Cu as mobile cations and Pt and Cu as electrodes [9]. IPMCs with Cu-Pt electrodes exhibit larger deflection compared to those with pure Pt electrodes (Table 7.1). The larger deflection is attributed to the lower electrochemical potential of Cu, as compared to that of Pt or Ag, that provide larger flux of mobile and hydrated Cu^{2+} ions.

The curvature is calculated using a cantilever configuration of the IPMC, as shown in Fig. 7.8. The curvature, ρ, (Eq. 7.1) is expressed as

$$\rho = \frac{1}{r}, \quad \text{where} \quad r = \frac{x^2 + y^2}{2x} \tag{7.1}$$

The symbols have their usual meaning. It is observed during the bending of strips that water on the anode side (bending side) is dried up and squeezed from the other surface (cathode). This indicated that the bending is primarily due to water transport and that the deposited electrode layer is porous.

Sheppard et al. have described in detail a procedure using a flow-through cell to deposit Pt on a Nafion® 117 membrane by diffusion through the membrane [10]. The effects of chloroplatinic acid concentration, solution agitation and surface roughening of the membrane on Pt deposition are studied. High acid concentration and membrane roughness are found to improve the Pt deposition. The flow-through cell

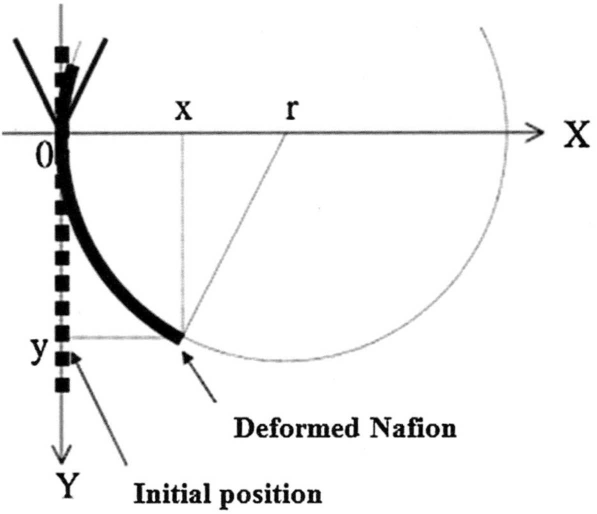

Fig. 7.8 Schematic of the IPMC in bent position (*solid line*). *Dashed line* represents the initial configuration (Reprinted from [9], Copyright (2001), with permission from Elsevier)

is preferred over stirring as it provides uniform deposition and high penetration of metal particles. For the IPMCs to work as actuators or sensors it is required that a solvent must be present inside the polymer network. When it becomes dried, the cations within the ionomer membrane are bound to the anionic sulfonate groups and are not mobile. This reduces the ionic conductivity of the membrane (a measure of the mobility of the ions within the polymer) to a level where electromechanical coupling (transduction) cannot occur. Therefore, a solvent must be present within the polymer network to allow the cations to move about within the polymer network by solvating both the cations and the anions (ionic sulfonate groups). Water, being protic (contains dissociable hydrogen ion) and highly polar, has been most popularly used as the solvent. It effectively solvates both the cations and anions, thus the ionic conductivity is increased and which leads to transduction can occur. However, water has several disadvantages as a solvent. First, it is volatile and any amount that is squeezed out of the membrane quickly evaporates from the surface of the membrane. This prohibits the use of these devices in air for extended duration without replenishing the lost water. Water-solvated IPMCs can't be used in space applications either. Second, with Pt electrode water undergoes electrolysis at a very low voltage of 1.23 V and hence application of higher voltages is not effective in such a system. In order to increase the spectrum of application of IPMCs, ionic liquids are explored as a new solvent that would not only maintain or improve the performance of water-based transducers but also allow long-term operation in dry environment. Ionic liquids are salts that are liquid state at low temperature and are attractive for use in IPMCs because they have a very low vapor pressure, a high inherent ionic conductivity and are stable over a large electrochemical potential. This could potentially allow these transducers to be operated in dry environment for long periods of time and could also allow application of higher voltages for actuation, thus increasing the transducer's energy density. Bennett and Leo have fabricated

IPMCs employing ionic liquids [1]. The IPMCs were found to work for more than 250,000 cycles in air. However, since ionic liquids are more viscous compared to water, the speed of response of the ionic liquid-swollen IPMCs is lesser compared to that of the water-swollen IPMCs. The response speed upon application of voltage is found to increase with increased content of ionic liquid. Also, the characteristic back-relaxation phenomenon is not observed in this case.

Characterizations of Ionic Polymer Metal Composites

Madden et al. have presented the fundamental properties of IPMCs as well as other actuator technologies [11]. Various characterization parameters like typical stress, strain, work density, power and efficiency of the IPMC have been listed. Some definitions are reproduced as follows:

(i) Stress: It is defined as the typical force per unit cross sectional area under which the actuator materials are tested. Peak stress is the maximum stress under which the material is able to hold its position (negligible deflection)
(ii) Strain: It is the displacement in the direction of actuation per unit length of the material
(iii) Strain rate: It is the average change in strain per unit time during an actuator stroke. Typically the maximum reported rate is quoted. These are observed at small strains and high frequencies
(iv) Bandwidth: It is the frequency at which the strain drops to half of its low frequency (maximum) amplitude
(v) Work density: It is the amount of work generated in one actuator cycle normalized by actuator volume, excluding electrolytes, counter electrodes, power supplies or packaging. “Specific power or power-to-mass ratio is the power output per unit mass of actuator material.”
(vi) Efficiency: It is the ratio of work output to the energy input.
(vii) Electromechanical coupling: It is the proportion of input energy that is transformed into work, including external work done by the actuator and stored internal mechanical energy generated in the actuator itself
(viii) Cycle life: It is the number of useful strokes that the material is known to be able to undergo. It is highly strain and stress dependent
(ix) Elastic modulus: It determines the actuator’s passive ability to reject load changes and disturbances.

Newbury and Leo have conducted various experiments to characterize IPMC behavior and identify the parameters of their model [12]. Force versus deflection experiments is performed to quantify the energy density of the IPMC. The maximum blocked force and energy density are observed to be 1.13 mN and 12.55 mJ/kg, respectively for a strip of length 22 mm. Shahinpoor et al. have tested the tensile properties of various samples with different cationic forms [3]. All IPMC samples (wet) have a Young’s modulus of approximately 50–100 MPa, at a normal strain of

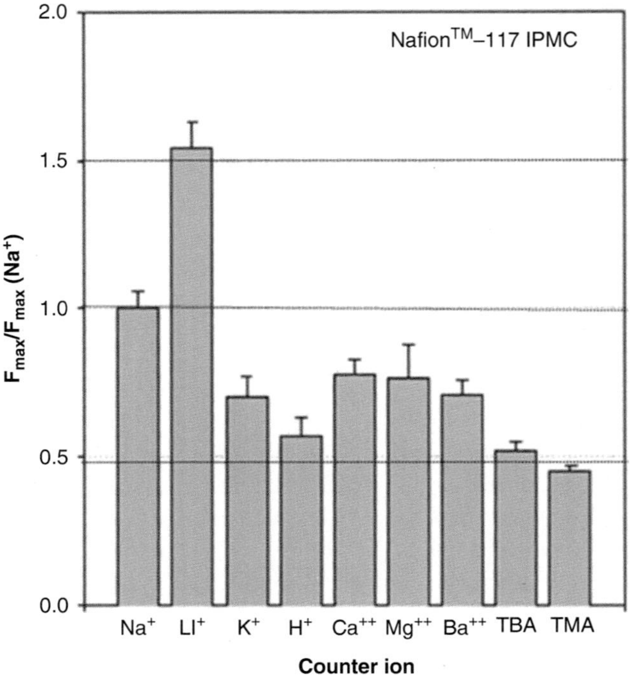

Fig. 7.9 Force response of IPMCs for different cations, with respect to Na^+ (Reprinted with permission from [3] Copyright (2003), IOP Publishing)

0.02, similar to that of wet Nafion®, and a tensile strength of 10–13 MPa. The dry Nafion® (H+ form) is stiffer than all other samples with the modulus lying between 220 and 260 MPa at a normal strain of 0.02. The greater stiffness in IPMC samples is attributed to the metallic electroded layer that interacts with the polymer matrix. It has been studied that the effect of nine different cations on the force generation capabilities of the IPMCs (Fig. 7.9). These IPMCs are compared for the maximum force generated at zero displacement. Strip length is 2.54 cm and the input voltage is 1.2 V sinusoidal at 0.5Hz. Li^+ containing IPMC generates the maximum blocking force implying that hydration of the cations, in addition to their mobility, play an important role in actuation of IPMCs. Electroding (concentration, penetration and size distribution) is found to directly affect the performance of IPMCs. The repetitive Pt reduction process is found to be effective to produce IPMCs exhibiting larger force generation. Porous electrodes for the IPMC allow effective transport of water within the thin electrode region and help to create larger strains. However, they can also allow water leakage from the surface that prevents the generated strains from causing larger forces. It is found that during reduction the just reduced Pt particles coagulate and grow large. Hence to reduce coagulation and enhance the dispersion of Pt particles within the ionomer, dispersing agents (additives) are added during the

Table 7.2 Comparison of the properties of IPMC, SMA and EAC (Reprinted with permission from [4] Copyright (1998), IOP Publishing)

Property	IPMCs	Shape memory alloys	Electro-active ceramics
Actuation displacement	>10 %	<8 % short fatigue life	0.1–0.3 %
Force (MPa)	10–30	About 700	30–40
Reaction speed	μs to s	s to min	μs to s
Density	1–2.5 g/cm^3	5–6 g/cm^3	6–8 g/cm^3
Drive voltage	4–7 V	NA	50–800 V
Power consumption	Watts	Watts	Watts
Fracture toughness	Resilient, elastic	Elastic	Fragile

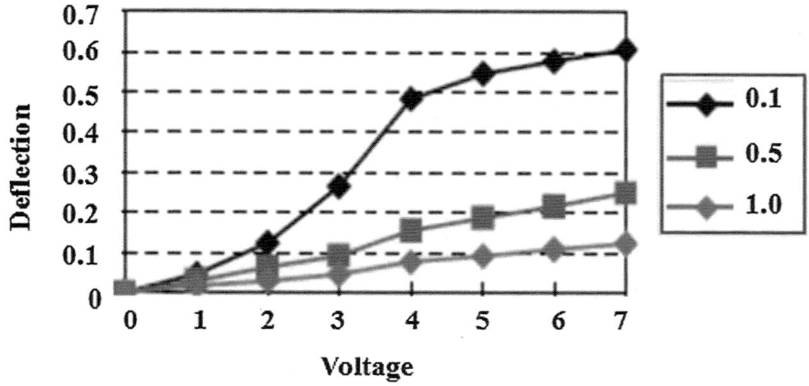

Fig. 7.10 Tip displacement of an IPMC as a function of applied voltage and frequency (Reprinted with permission from [4] Copyright (1998), IOP Publishing)

chemical reduction process. This increases the penetration of Pt particles in the surface with a well distributed smaller particle size thereby significantly reducing the water leakage. Pt penetration is expected to improve by stretching the base material prior to the Pt composition process that results in larger pores (higher permeability) relative to the starting materials.

The effect of such stretching significantly improves the generated force. Shahinpoor et al. have reported the sensing capability of IPMCs [4]. The comparison of the capability of IPMC with both electro-active ceramics and shape memory alloys is presented in Table 7.2.

The key parameters involving the vibrational characteristics of the IPMCs are identified. Load characterizations show that the IPMCs possess good force to weight characteristics under low applied voltages. Tip displacement of an IPMC as a function of applied voltage and frequency is plotted in Fig. 7.10. Under application of alternating voltage, the displacement level depends both on the voltage magnitude and the frequency of alternation. Lower frequencies lead to higher displacement.

Resonance is observed at about 20 Hz with a tip displacement of 7.5 mm for a strip of length 40 mm. The performance of the actuator is found to be strongly dependent on the water content, which gives mobility to the cation and the gradient of solvation across the film led to a pressure difference. The force generated by the IPMCs is measured and it is found that they could lift up to 40 times their own weight. The authors have studied the cryogenic properties and found that the IPMCs are able to operate at temperatures as low as 140 °C. Unlike other electric conductors, their resistance increased with decreasing temperature.

Comparative study of IPMCs in different cation forms have been done by Nasser and Wu [13]. Magnitudes of the initial fast bending displacements upon application of step voltage and time scales of fast bending and relaxation have been compared in Fig. 7.11, bending towards the anode is shown as positive. The dry masses of Nafion® in different cationic forms, the equivalent weights and the metal content have been reported. The ion-exchange capacity of an ionomer (i.e., bare Nafion®) is defined as the number of moles of sulfonate groups per unit volume of the ionomer. This is the same as the number of moles of monovalent cations in the ionomer. Upon application of a small voltage across the IPMC strip, it exhibits a fast bending response towards the anode, followed by a slow relaxation in the opposite direction (towards the cathode) and, for some cations, the back relaxation overshoots the initial position. Subsequent shorting of the two surfaces of the IPMC after relaxation causes a fast bending response towards the cathode and a slow relaxation towards the anode (as shown in Fig. 7.12, where a Nafion®-based IPMC in K+−form is actuated under 1 V DC voltage). Large alkyl-ammonium cations (e.g. TBA+), produce a gradual bending towards the anode. Such IPMCs have been reported by Shahinpoor and Kim has the tip displacement double that of the alkali-metal cation form, but at a lower speed [3]. The authors report that for TMA+− and Li+−forms, the relaxation is small compared to the initial fast displacement under a sustained DC voltage, whereas for Na+−, K+−, Rb+−, Cs+−, and particularly Tl+− form, the relaxation more than compensates the initial displacement and, in fact, bends the IPMC towards the cathode. From the study of the IPMC's initial fast motion, it is evident that the sample in Na+− form first bends quickly towards the anode, vibrates due to the initial jerk, and then a slowly relaxes towards the cathode, reversing the initial displacement. However, a similar IPMC in TBA+− form bends slowly towards the anode and does not show back relaxation. Fig. 7.12 shows superimposed images of an IPMC while in bending. 1 V DC voltage is applied across the surfaces of the strip. It bends quickly from the initial position A to B, and then relaxes back to C. Note that while relaxing the IPMC has overshot its initial position. When the surfaces are shorted at C, the IPMC bends moves quickly to D and then relaxes back to E. In this case, the IPMC has not attained its initial position and it has permanently deformed.

A comparative study of IPMCs coated with electrodes of different metals like Pt, Au and Ir are reported by Lee et al. [7]. Amongst Pt, Au and Ir, the displacement of IPMC plated with Pt by Ion Beam Assisted Deposition (IBAD) has

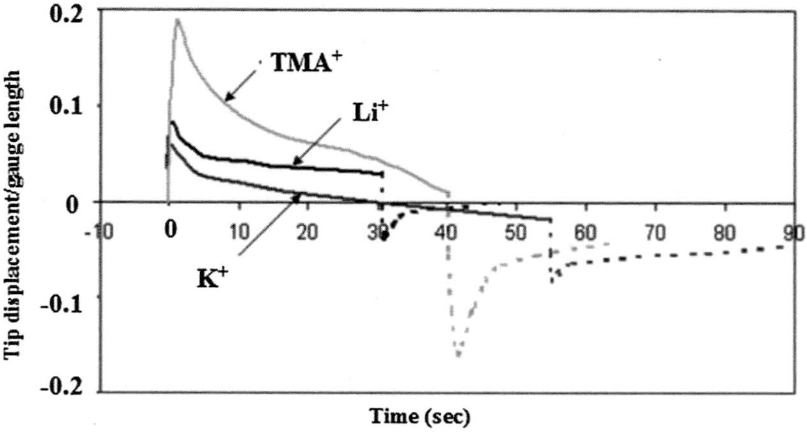

Fig. 7.11 Normalized tip displacement of Nafion®-based IPMCs as a function of time, in various cation forms (Reprinted with permission from [13] Copyright (2003), AIP Publishing LLC)

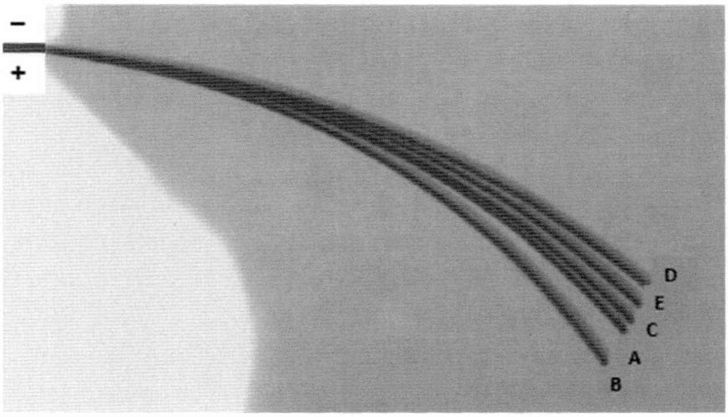

Fig. 7.12 Successive photos of bending of a Nafion®-based IPMC in K+ form when a 1 V DC voltage is applied (Reprinted with permission from [13] Copyright (2003), AIP Publishing LLC)

showed largest displacement. Long time displacements of IPMCs plated with Pt through chemical reduction and by IBAD under a voltage of 2 V DC have been observed. IPMCs with chemically plated electrodes show greater relaxation. On the other hand, the relaxation of the IPMC electroded by IBAD is retarded significantly apparently due to smoother and more homogeneous surface electrodes. The better performance of the IPMC by IBAD process, in terms of displacement and force, is linked to better surface electroding. The maximum force generated by the IPMC by IBAD process is over 1.0 g-f, while that of the chemically plated IPMC is 0.8 g-f. The high actuating force is attributed to the increased mechanical strength of the IPMC due to more homogeneous surface electrodes and the surface electrode density. Malone and Lipson have tested the blocking force in IPMCs manufactured through freeform fabrication [8]. The maximum blocked force is measured to be 0.678 mN and the RMS power delivered to the actuator is 9.4 mW.

Modeling of Ionic Polymer Metal Composites

An IPMC undergoes a large bending motion when a voltage is applied across its electroded faces. Nasser and Li have been successfully developed a micromechanical model to account for this electrostatically actuated bending of the IPMCs [2]. In the model, the Nafion® membrane is considered as isotropic, with permanently attached ionic sulfonate groups uniformly distributed throughout the bulk of the polymer. First, the fundamental field equations govern the response of the material are systematically developed and these are reproduced here [2]. The total stress field (Eq. 7.2) in the polymer, σ^*, is divided into an elastic part, σ (carried by the backbone polymer), and an eigen stress, σ^* (reflecting effect of the imbalanced charge density).

$$\sigma^T = \sigma + \sigma^* \tag{7.2}$$

The eigen stress is due to a combined effect of the electrostatic stress, σ^e, due to imbalanced charge density and the osmotic stress, σ^o. The final equation (Eq. 7.3) of the linearized model comes out to be

$$\nabla(\nabla.E) - \frac{C^- F^2}{\kappa_e RT}\left(1 - C^- \Delta V\right)E = 0 \tag{7.3}$$

Here, **E** is the electric field, $\mathbf{C}^-$ is the negative charge density (mol/m^3), **F** is the Faraday's constant (96,487 C/mol), $\boldsymbol{\kappa_e}$ is the effective dielectric constant of the polymer, **R** is the gas constant, **T** is the absolute temperature, and $\mathbf{\Delta V}$ is the net molar volume of the charges. The model couples the effects of ionic transportation across the bulk due to the electric field and elastic deformation of the polymer to predict the response. The interaction between an imbalanced charge density and the backbone polymer is represented by an eigen stress field. As a special case of bending of the IPMC strips, the linearized model (Eq. 7.4) is simplified to

$$\frac{\partial^2 E_x}{\partial x^2} - a^2 E_x = 0, \quad \text{where} \quad a^2 = \frac{C^- F^2}{\kappa_e RT}\left(1 - C^- \Delta V\right) = \frac{C^- F^2}{\kappa_e RT} \tag{7.4}$$

Assuming absence of any external load and ignoring the osmotic stress in comparison to the electrostatic stress, the resulting bending moment (Eq. 7.5) and the curvature (Eq. 7.5) come out to be

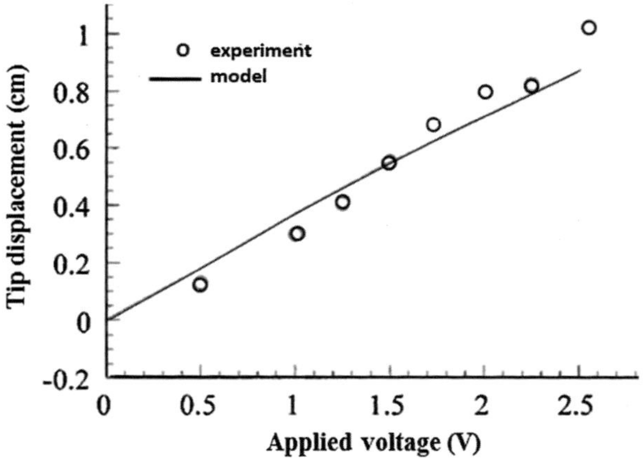

Fig. 7.13 Tip displacement of an IPMC strip with varying voltage. The Young's modulus is 0.5 GPa (Reprinted with permission from [2] Copyright (2000), AIP Publishing LLC)

$$M = \int_{-w/2}^{w/2}\int_{-h}^{h} x\sigma_{zz}dxdy = \frac{k_0\kappa_e\phi_0 a}{\sinh(ah)}\left[h\cosh(ah) - \frac{\sinh(ah)}{a}\right]w = k_0\kappa_e\phi_0 ahw$$

$$R_c = \frac{\overline{Y}I}{M} = \frac{2\overline{Y}h^2}{3k_0\kappa_e\phi_0 a} \tag{7.5}$$

Here, $\mathbf{k_o}$ is a parameter that depends on the cluster structure inside the membrane, **w** is the width of the membrane, **h** is the half-thickness, **Y** is the effective Young's modulus of the composite, $\boldsymbol{\phi_o}$ is the applied potential, **a** is the constant that depends on the mobile cation and **I** is the second moment of area.

The model can predict the relative effects of different cations on the force-displacement response of an IPMC strip. Several parameters in this model have been varied and the tip displacement of the IPMC has been studied [2]. The plots that follow show the effects of those parameters on the response of the IPMC. All plots are for a strip of dimensions 30 mm $\times$ 5 mm. Fig. 7.13 shows the tip displacement of an IPMC as a function of the applied voltage. The tip displacement increases linearly with the increase in voltage. The IPMC with Li^+ as mobile cations shows smaller deflection as compared to that with Na^+. This can be attributed to larger number of water molecules surrounding the Li^+ cation owing to higher charge density because of its small size. These water molecules make the effective size of the mobile species in case of Li^+ more than that in case of Na^+, thereby causing greater hindrance to the their movement inside the ionomer. It may be noted that this is not the only factor affecting the response of an IPMC and other effects may assume significance for cations with larger sizes. It has been studied that the tip displacement of an IPMC as a function of the thickness of the strip. The displacement varies inversely with the square of the thickness. It may also be noted that the blocked tip force of the IPMC varies directly with the cube of the thickness for a constant tip displacement, as shown by Kim and Shahinpoor [6]. Fig. 7.14 shows the

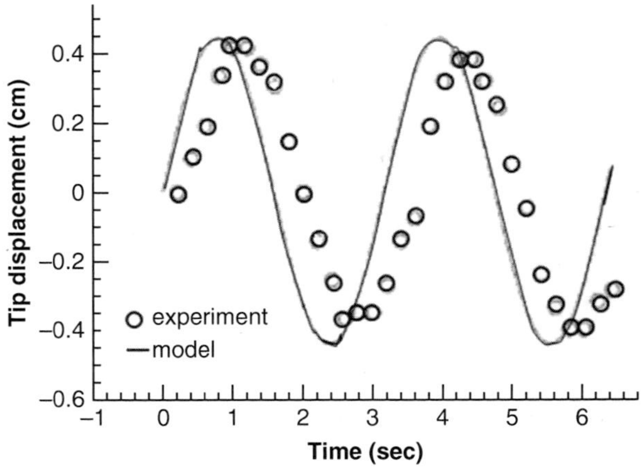

Fig. 7.14 Tip displacement of an IPMC strip as a function of time under a sinusoidal voltage. The length of the strip is 1.57 cm and the Young's modulus is 0.65 GPa (Reprinted with permission from [2] Copyright (2000), AIP Publishing LLC)

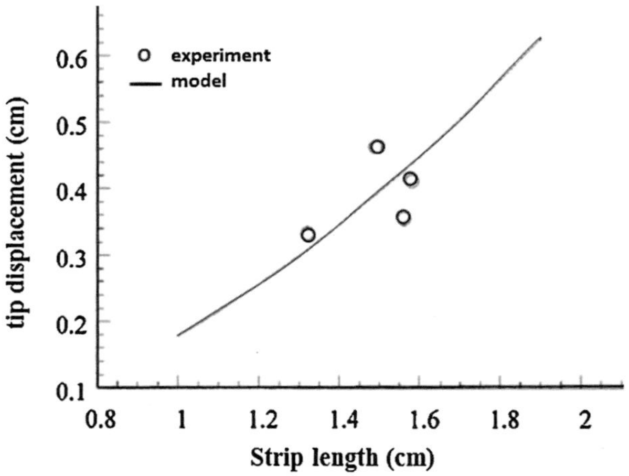

Fig. 7.15 The maximum tip displacement of an IPMC strip as a function of strip length under the application of a sinusoidal voltage. The Young's modulus of the IPMC strip is 0.65 GPa (Reprinted with permission from [2] Copyright (2000), AIP Publishing LLC)

tip displacement of the IPMC strip versus time, and Fig. 7.15 shows the tip displacement of the strip as a function of the strip length. The modeled results are in good agreement with the measurements, but a small difference in phase is observed. This phase difference is mainly due to the inertia, whereas the damping effects are ignored. This can be used for estimating the effective damping of the IPMC strip.

Applications

A very important application of IPMCs as distributed "soft" sensors and actuators are widely accepted. Recent developments on the preparation of IPMCs are made them capable of large deformations. The IPMCs is used as a transducer since they can act as integrated conversion and sensor devices, hence they can be efficiently integrate with MEMS for specific applications. A fast growing application of IPMCs is the energy harvesting using various IPMC structures from the fluid environments. The developments of artificial muscles show that the opportunities in the field of bio-related applications are yet to be explored. In reality, the description of all the applications of IPMCs is beyond the scope of this chapter. But still, the next section reviews IPMCs in actuators, sensors, artificial muscles, energy harvesters, fuel cells, MEMS, and also some of the biomedical applications.

Actuators

The IPMCs are behaved as active actuators since they undergo deformation to a large extent on the application of low applied voltage. The impedance possessed by them is also very low. The IPMCs are capable to work in humid as well as in dry environments, but the best performance can be observed in the humid environment. The mechanochemical engines developed by Kuhn et al. have found that water-swollen macromolecules are capable to convert chemical energy into mechanical energy [14]. The lower efficiency of IPMCs based actuators was the main problem in the past years. But the actuator efficiency has been improved in the recent years, which make them comparable with the other actuators. Kim et al. have reported the economically viable fabrication processes for synthesizing electrodes for spirally shaped IPMCs for functioning as linear actuators [3]. Various manufacturing techniques have been reported for the IPMCs in order to decrease the cost of actuators. Addition of conductive powder into the ionic polymer network in the first step of IPMC fabrication process is very important since this will act as the conductive medium in the IPMC. Chemical plating is done further in order to incorporate the smaller particles in to the network. The integration of these particles has been by selecting the electroplating process. Such a preparation procedure has reduced the cost of IPMCs up to 1/10th of the typical cost. It is a fact that the transduction process is highly dependent on the inherited layers of IPMC [15]. By controlling the properties of each layer, the IPMC transduction process can be enhanced. Varying parameters such as increasing the polymer content, increasing the thickness of electrode and intermediate layer, reducing the impedances, etc. will certainly enhances the transduction process. But the main problem in achieving these parameters is due to their inter-dependent nature. The work become extremely tedious, since varying one property will affect the other in a negative way. A true knowledge in the coupling of these parameters will decide the transduction process ultimately. Recent researches have shown that emphasizing on the by modification of electrode and polymer layers can give rise to the improvement in the transduction process of the IPMCs.

Sensors

The advancements in the synthesis polymer gels, which are stimulus responsive, have been used to detect the minor changes in the pH as well as temperature. The polymer gels are responding to the stimulus such as applied electric field and detect changes in the pH as well as in the temperature [16]. The developments in the manufacturing procedures of IPMCs have led to their application as biomimetic sensors [3]. The quasi-static or dynamic bending of the IPMC composite strip produces a voltage across the thickness of the IPMC strip; this can be treated as a kind of large motion sensing [4, 17]. A standard, calibrated IPMC sensor is capable of correlating the applied loads with the output potentials. It has been found that a linear relationship exist between the applied force and the output voltage. Shahinpoor has reported the dynamic sensing of ionic polymeric gels [18]. The ionic flexogelectric effect is manipulated for the sensing applications. Ion-exchange membranes have been used as a pressure sensor/damper, which is kept in a small chamber in order to make it as a prototype accelerometer. Bonomo et al. have reported the design and development of IPMC based sensor, which can measure the stiffness of a material it is in contact with [19]. The sensor is used in various applications in biomedical sector such as in laparoscopy, surgery, etc. the sensor has advantages such as low power consumption, light weight, low cost, etc.

Artificial Muscles

The IPMCs are demonstrated as actuators with both resistive and capacitive elements, capable to behave as artificial muscles. The developments on the IPMCs show that they are potential candidates for application in bio-related applications. These include bio-mechanical and bio-mimetic applications. The reports show that the disadvantage of using them as artificial muscles is that the reduced response time. The response time should be the same as that of live muscles, which can increase their practical applicability in the real-time applications. The slower response is due to the time taken for reconfiguring the ions inside the IPMC. The mechanical properties of IPMCs are solely determines their response time. Hence, the IPMCs with a quick responding ability only can make them successful in artificial muscles application. Researchers have been taken many efforts on the developments of IPMCs for use them as artificial muscles [20, 21]. The utilization of ionic polymeric gels in electrically controllable artificial muscles has been extensively studied in different dynamic environments by Shahinkapoor [22]. The IPMCs are capable to work under low pressure and low temperature environments as in the case of cryogenic chambers. Kim et al. have reported that the scaling possibility in the fabrication of IPMCs for artificial muscles [6]. The scaling can be done two ways-scale up and scale down. Depending on the scaling, the thickness of artificial muscle strips based on IPMCs varies from micrometers to centimeters. Various parameters will affect the actuation of an artificial muscle, such as type of cations, size of counter ions, etc. The effect of various cations is well studied by taking Na^{+}- containing IPMC artificial muscle [3]. The various samples containing different cations have been checked to find out the maximum force generated at a working voltage of 1.2 V_{rms}. The force generated is nothing but the blocking force generated by the IPMC at zero displacement. It has been observed that the Li^{+}- containing IPMC has superior performance than the other types of cations. The reason for its superior performance is attributed to the hydration processes by which the Li^+ cations played in the actuation behavior.

Energy Harvesters

The application of IPMCs as energy harvesters has achieved much popularity in the recent past. This is a fast growing field in which the energy can be harvested from the fluid environments with the help of IPMCs [23, 24]. The energy harvesting process can be from various kinds of flows including steady, unsteady, and oscillatory flows [25]. The capability of IPMCs to convert the strain energy to electrical energy is utilized in the flow environments. Different kinds of IPMC structures are used-heavy flapping flag hosting IPMCs, IPMC cantilevers, etc. Aureli et al. have reported the energy harvesting application of an IPMC strip cantilever, which is submerged in a fluid environment [26]. The base excitation is considered for the modeling of IPMC strip immersed in a fluid environment in which it is shunted. Experimental work also been done by checking the underwater vibration of the IPMC submerged and the corresponding electric responses are noted for various shunting resistors. The surface area of the IPMC is 1.5 cm^2 with a thickness of 200 μm. It has been found that the IPMC system is good for marine systems. The maximum power harvested by the IPMC is 1 nW when the base excitation is in the range of 1 mm. Panella et al. have developed a model, which is efficient in predicting the capability of the IPMCs as energy harvesting materials when the working environment is air [27]. It is required to be used very large IPMC samples for attaining a few micro-watts. The developments in the designing process of large samples will certainly help the energy harvesting devices. The IPMCs are known for their biocompatibility properties; hence they can be used even in the humid environments. It has been proposed that the potential use of IPMCs for energy recovery from micro-seismic or ocean waves. Software modeling is mandatory for the effective estimation of the IPMCs based systems for energy harvesting applications in the future.

Fuel Cells

Typically, IPMCs are known for their potential applications in fuel cells. IPMCs have been originally developed as fuel cell membranes in 1990s. They are capable to use when they are fully electroded with embedded electrodes in a fuel cell. The IPMC materials have been developed as membranes for solid polymer electrolyte fuel cell [28]. It is reported that the fabrication of IPMCs by molecular metallization process followed by surface plating for fuel cell applications by Kim et al. [29]. The molecular metallization process is found to be slow due to the increase in the mass transfer resistance and slow kinetics associated with it. Yoon et al. have filed a patent on synthesis of an IPMC for use as an electrolyte for a fuel cell where they used Pt nanoparticles dispersed Nafion® membrane [30]. As different from the conventional

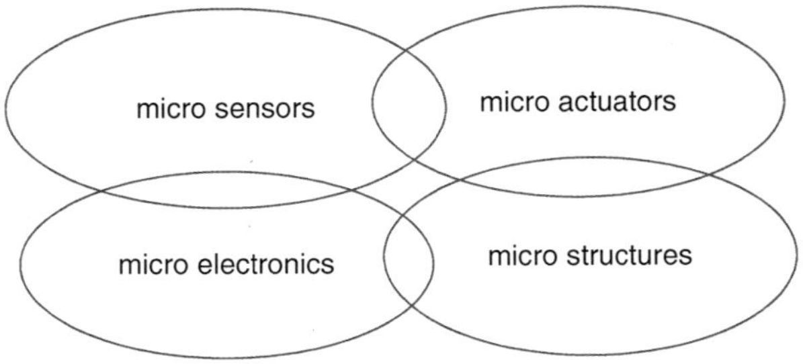

Fig. 7.16 Schematic of various components of MEMS

use of it alone as a polymer electrolyte, Pt/Nafion® IPMC has been used. This invention has helped to resolve the problem of cross-over of conventional membranes and hence successfully utilized it in a methanol direct fuel cell and the fuel cell has shown to have an enhanced open-circuit voltage.

Microelectromechanical Systems

Micro-Electro-Mechanical Systems (MEMS) technology constitutes various elements such as mechanical and electro-mechanical devices and structures. Microfabrication techniques have helped MEMS to achieve a miniature structure. The schematic of various components of MEMS is shown in Fig. 7.16.

A transducer is a device that converts energy from one form to another. As IPMCs can convert electrical signals to mechanical (act as an actuator) as well as convert this mechanical signal to an electrical output (act as a sensor), they can be successfully utilized in the MEMS. MEMS based applications include micro-sensing devices, which are capable to sense chemical species, radiation, magnetic fields, etc. The device should be made with accurate micromachining techniques, which are indeed to achieve the best results.

Biomedical

A critical application of IPMCs is in the biomedical applications. Their softness and flexibility has increased their demand on the integrated devices, which are used in the various biomedical applications. The limit of biomedical applications is beyond the scope of our imagination. The IPMCs are used as artificial ventricular or cardiac-assist muscles for the heart patients. IPMCs based heart compression device has been developed, which is implanted externally to the heart. Since the device is not interfering with the internal blood circulation, a common problem of 'thrombosis' can be avoided, which is impossible with other kinds of heart assist devices or the artificial heart. Another application of IPMCs is in the intra-cavity endoscopic surgery and diagnostics. The micro-catheters made by IPMCs can be navigated through the small internal cavities. The IPMCs are also used in peristaltic pumps. The pumps are made from the tubular sections of the IPMC membrane and locating the electrodes wherever needed. The volume trapped in the tube is controlled by changing the input voltage at a particular frequency. The other biomedical applications of IPMCs are in: artificial smooth muscle actuators, artificial sphincter and ocular muscles, incontinence assist devices, refractive errors of the human eyes, etc.

Concluding Remarks

The developments in IPMCs based devices and structures are very promising candidates for future generation integrated technologies, which can impart more outputs rather than the existing ones. The various factors determine the efficiency of an IPMC is: force optimization, dependency on cations for the electrochemical performance, synthesis of electrode and the particle size distribution, etc. The electromechanical coupling of IPMCs is enhanced by increasing the charge density, which is obtained by plating Pt like noble metal on either side of the polymer membranes. Several methods have been opted for the preparation of IPMCs, such as freeform fabrication by deposition, ion beam assisted deposition, solution casting etc. The electroding process during the preparation of IPMCs helps in reducing their surface resistivity. The various performance parameters of IPMCs are bandwidth, work density, efficiency, cycle life, etc. Modeling in IPMCs helps to understand the way they works and also can identify the various parameters that correlates with the experimental ones. The superior performance of IPMCs have made them important candidates for various applications such as coupled actuator/sensor devices, artificial muscles, energy harvesters, etc. The high manufacturing cost and high dependency on IPMC's force density generation capability determine their successful commercialization. Further developments are indeed in order to reduce the cost and hence acceptance of IPMC based devices.

Acknowledgement The authors acknowledge the financial support provided by Aeronautics Research & Development Board, India for carrying out this work.

References

1. Bennett M, Leo D (2006) Morphological and electromechanical characterization of ionic liquid/ Nafion polymer composites. In: Bar-Cohen Y (ed) Proceedings of SPIE 5759: smart structures and materials 2005: electroactive polymer actuators and devices (EAPAD), vol 5759. SPIE, Bellingham. doi:10.1117/12.599983
2. Nasser SN, Li JY (2000) Electromechanical response of ionic polymer-metal composites. J Appl Phys 87:3321
3. Kim KJ, Shahinpoor M (2003) Ionic polymer-metal composites: II. Manufacturing techniques. Smart Mater Struct 12(1):65
4. Shahinpoor M, Bar-Cohen Y, Simpson JO, Smith J (1998) Ionic polymer-metal composites (IPMCs) as biomimetic sensors, actuators and artificial muscles: a review. Smart Mater Struct 7:15

5. Millet P, Durand R, Dartyge E, Tourillon G, Fontaine A (1993) Precipitation of metallic platinum into Nafion ionomer membranes: I. Experimental results. J Electrochem Soc 140:1373
6. Kim KJ, Shahinpoor M (2002) Development of three dimensional ionic polymer-metal composites as artificial muscles. Polymer 43:797
7. Lee SJ, Han MJ, Kim SJ, Jho JY, Lee HY, Kim YH (2006) A new fabrication method for IPMC actuators and application to artificial fingers. Smart Mater Struct 15:1217
8. Malone E, Lipson H (2006) Freeform fabrication of ionomeric polymer-metal composite actuators. Rapid Prototyp J 12:244
9. Uchida M, Taya M (2001) Solid polymer electrolyte actuator using electrode reaction. Polymer 42:9281
10. Sheppard SA, Campbell SA, Smith JR, Lloyd GW, Walsh FC, Ralph TR (1998) Electrochemical and microscopic characterization of platinum-coated perfluorosulfonic acid (Nafion 117) materials. Analyst 123:1923
11. Madden PGA, Madden JDW, Anquetil PA, Vandesteeg NA, Hunter IW (2004) The relation of conducting polymer actuator material properties to performance. IEEE J Ocean Eng 29:706
12. Newbury KM, Leo DJ (2002) Electromechanical modeling and characterization of ionic polymer benders. J Intell Mater Syst Struct 13:51
13. Nasser SN, Wu Y (2003) Comparative experimental study of ionic polymer–metal composites with different backbone ionomers and in various cation forms. J Appl Phys 93:5255
14. Kuhn W, Hargitay B, Katchalsky A, Eisenberg H (1950) Reversible dilation and contraction by changing the state of ionization of high-polymer acid networks. Nature 165:514
15. Tiwari R, Garcia E (2011) The state of understanding of ionic polymer metal composite architecture: a review. Smart Mater Struct 20:83001
16. Osada Y, Gong J (1993) Stimuli-responsive polymer gels and their application to chemomechanical systems. Prog Polym Sci 18:187
17. Park K, Lee B, Kim H-M, Choi K-S, Hwang G, Byun G-S, Lee H-K (2013) IPMC based biosensor for the detection of biceps brachii muscle movements. Int J Electrochem Sci 8:4098
18. Shahinpoor M (1995) New effect in ionic polymeric gels: the ionic flexogelectric effect. In: Jardine AP (ed) Proceedings of SPIE, smart structures and materials 1995: smart Materials, vol 2441. SPIE, Washington, DC, p 42
19. Bonomo C, Brunetto P, Fortuna L, Giannone P, Graziani S, Strazzeri S (2008) A tactile sensor for biomedical applications based on IPMCs. IEEE Sens J 8:1445
20. Rossi DD, Parrini P, Chiarelli P, Buzzigoli G (1985) Electrically induced contractile phenomenon in charged polymer networks: preliminary study on the feasibility of muscle-like structures. Trans Am Soc Artif Intern Organs 31:60
21. Rossi DD, Domenici C, Chiarelli P (1988) Analogues of biological tissues for mechanoelectrical transduction: tactile sensors and muscle-like actuators. In: Dario P (ed) Sensors and sensory systems for advanced robots, NATO-AS1 series. Springer, Berlin, p 201
22. Shahinpoor M (1994) Micro-electro-mechanics of ionic polymeric gels as electrically controlled artificial muscles. In: Rogers CA, Wallace GG (eds) Proceedings of the second international conference on intelligent materials (ICIM94). Technomic Publishing, Switzerland, p 1095
23. Tiwari R, Kim KJ, Kim SM (2008) Ionic polymer metal composite as energy harvesters. Smart Mater Struct 4:549
24. Farinholt KM, Pedrazas NA, Schluneker DM, Burt DW, Farrar CR (2009) An energy harvesting comparison of piezoelectric and ionically conducting polymers. J Intell Mater Syst Struct 20:633
25. Porfiri M, Peterson SD (2013) Energy harvesting from fluids using ionic polymer metal composites. In: Elvin N, Erturk A (eds) Advances in energy harvesting methods. Springer, New York, pp 221–239
26. Aureli M, Prince C, Porfiri M, Peterson SD (2010) Energy harvesting from base excitation of ionic polymer metal composites in fluid environments. Smart Mater Struct 19:15003

27. Brufau-Penella J, Puig-Vidal M, Giannone P, Graziani S, Strazzeri S (2008) Characterization of the harvesting capabilities of an ionic polymer metal composite device. Smart Mater Struct 17:15009
28. Kordesch KV, Simader GR (1995) Environmental impact of fuel cell technology. Chem Rev 95:191
29. Kim KJ, Shahinpoor M, Razani A (2000) Preparation of IPMCs used in fuel cells, electrolysis, and hydrogen sensors. In: Bar-Cohen Y (ed) Proceedings of the SPIE smart structures and materials 2000: electroactive polymer actuators and devices (EAPAD), vol 3987. SPIE, Washington, DC, p 311
30. Yoon YS, Park HC, Yoon GJ, Jee SH (2008) Ionic polymer metal composite electrolyte for fuel cell. US Patent 20080003479 A1

Nanocomposites Based on Carbon Nanomaterials and Electrically Nonconducting Polymers

8

Soma Banerjee, Raghunandan Sharma, and Kamal K. Kar

Contents

S. Banerjee • R. Sharma
Advanced Nanoengineering Materials Laboratory, Materials Science Programme, Indian Institute of Technology Kanpur, Kanpur, Uttar Pradesh, India
e-mail: somab@iitk.ac.in; raghumsp@iitk.ac.in

K.K. Kar (✉)
Advanced Nanoengineering Materials Laboratory, Materials Science Programme, Indian Institute of Technology Kanpur, Kanpur, Uttar Pradesh, India

Advanced Nanoengineering Materials Laboratory, Department of Mechanical Engineering, Indian Institute of Technology Kanpur, Kanpur, Uttar Pradesh, India
e-mail: kamalkk@iitk.ac.in

K.K. Kar (ed.), *Composite Materials*,
DOI 10.1007/978-3-662-49514-8_8

Abstract

Polymer nanocomposites (PNCs) are the growing field of research having significant contribution in limitless application areas due to excellent combinations of tunable properties. Carbon-based PNCs are of recent research interest and cover numerous fields of applications including structural composites, drug delivery, shape memory polymers, etc. In this chapter, different allotropes of carbon, viz., carbon nanotubes, graphene, graphene oxide, fullerenes, metallofullerenes, carbon nanohorn, carbon nanodiamond, etc., are discussed in brief. Different types of polymeric materials used for the fabrication of PNCs are also introduced. Several types of carbon-based PNCs and their fabrication methodologies have been emphasized to represent a broad overview on carbon-based PNCs of nonconducting polymer matrices.

Keywords

Carbon • Polymer nanocomposite • Nanomaterials • Graphite • Carbon nanotubes • Nanorods • Graphene • Fullerene • Carbon nanofibers • Carbon nanohorn

Introduction

In recent era, polymer nanocomposites (PNCs) encompass almost all the growing fields of current research interest with significant applications in several emerging areas like structural, medical, energy, optical, etc. PNCs have tremendous commercial importance due to their low density, high specific strength, excellent barrier properties, and ease of tailoring different functional properties by just incorporating very small amount of nanofiller in the virgin polymers as per specific application requirements. Typically, these consist of two dissimilar entities, viz., polymer matrix and nanofiller. The nanofillers can be of several types such as reinforcing, non-reinforcing, or functional to meet the final requirements. The enhancement in properties is solely dependent on the extent of mutual interaction, interfacial bonding, and compatibility between the polymer matrix and nanofillers. Hence dispersion of nanofillers within the host polymer serves a crucial role in achieving the desired final properties of the PNCs.

Among various fillers, carbon-based fillers have been of significant importance owing to the diverse properties of their allotropes. Carbon-based PNCs are of significant research interest and cover a diverse field of applications including drug delivery, shape memory polymers, structural composites, etc. Till date, several carbon-based PNCs with conducting and nonconducting polymers were developed and analyzed by researchers of interdisciplinary fields to accomplish different functional requirements. These are fabricated with different allotropes of carbons, viz., carbon nanotubes (CNTs), graphene, graphene oxide, fullerene, metallofullerene, nanodiamond, carbon nanohorn, etc. In this chapter, different types of carbon materials of commercial importance will be discussed in view of their structural diversities. The nonconducting polymer-based nanocomposites with different carbon materials will be reviewed in brief with respect to their diverse applications to address a broad overview of the present stand of carbon-based PNCs.

Carbon Materials and Their Types

Carbon has been the element of interest since long back in history to the current age and serves as a valuable contributor in different fields. Elemental carbon plays an indispensable role in the human life since long back, even when the word "element" was coined hardly. In the age of 5000 BC, charcoal and soot derived from wood, allotropes of carbon having amorphous structure, were being utilized in metallurgy such as iron industries. Graphite, the other well-known form of carbon having a layered structure of honeycomb lattice, where the name has been derived from the word *graphein*, meaning "to write," is being used since long. Even in the ancient Egyptian age, graphite soot was used as the ink to make the pictographs. In the nineteenth century, the importance of elements fascinated the research interest to the scientists and different elements were discovered. These inventions later enable the scientists to distinguish between the inorganic and organic matter. More recently, CNTs, graphene, etc., the other allotropes of nanostructured carbon, have captivated significant scientific interest owing to their prospective applications in diverse areas.

Carbon, the sixth element of periodic table, is the main constituent element of the organic matters. It is the most essential element in the chemical world due to its capability of bonding with both electropositive and electronegative elements. In 1865 F. v. Kekule demonstrated benzene as a cyclic entity. Soon after in 1874, van t' Hoff and Bel threw light on fourfold coordination of carbon atoms (Fig. 8.1). This invention brought about revolution in science and opened the path to explain the chemical activity of the element.

Chemically, carbon bears six electrons with an electronic configuration of $1s^2\,2s^2, 2p^2$. Out of the six electrons, two remain closely bound to the nucleus, while the remaining four act as the valence electrons. It is generally tetravalent (sometimes bivalent, e.g., carbenes) due to the formation of four equivalent sp^3 hybrid orbitals as per the hybridization model. The four sp^3 hybridized orbitals of carbon directs toward the corners of the tetrahedron, leading to the formation of four sigma bonds with the other atoms. Depending on the degree of hybridization, carbon may form linear (sp hybridization), planar structures (sp^2 hybridization), or three-

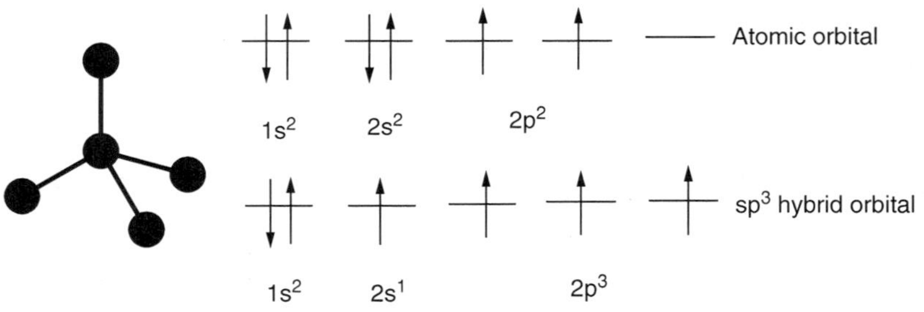

Fig. 8.1 Schematic of tetrahedral disposition of the carbon

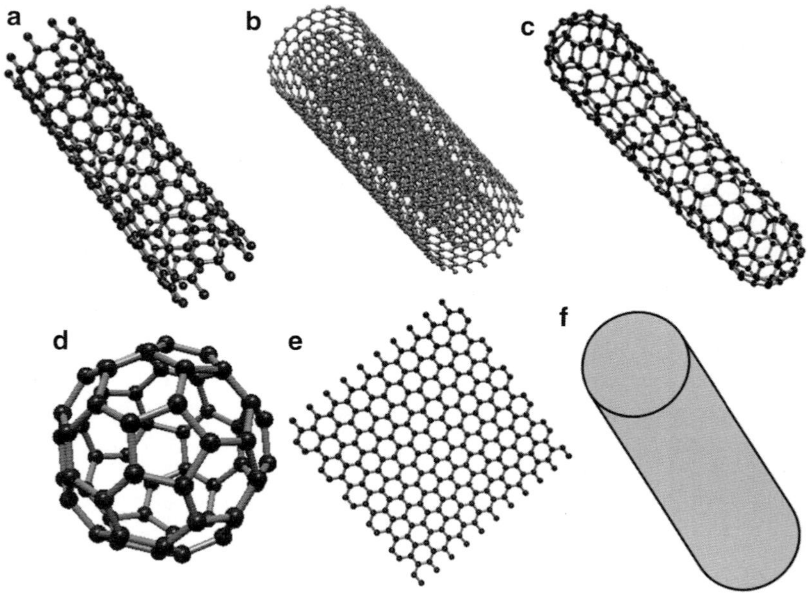

Fig. 8.2 Structures of different types of carbon materials, (**a**) single-walled CNTs (SWCNTs), (**b**) multiwalled CNTs (MWCNTs), (**c**) capped CNT, (**d**) fullerene, (**e**) graphene, and (**f**) carbon nanofibers

dimensional tetrahedron network (sp^3 hybridization). The atomic and hybridized orbitals of carbon are depicted in Fig. 8.1.

Carbon stands as an indispensable partner in formation of many compounds with the other elements and with same importance in the elemental form as well. The elemental carbon may exist in different forms (Fig. 8.2) such as graphite, diamond, soot, coke, pyrolytic carbon, glassy carbon, carbon black, CNTs, nanorods, graphene, fullerene, carbon nanofibers, nanodiamond, carbon nanohorn, activated carbon, etc. Each form has tremendous importance in the world of materials.

Graphite

Graphite is the most stable form of carbon under normal atmospheric conditions. Also known as writing stone, graphite is a semimetal mineral available in various forms, viz., flake graphite, amorphous graphite, and pyrolytic graphite. It has layered structure, where the carbon atoms are arranged in honeycomb lattice structure characterized by 0.335 nm distance between the layers [1]. The carbon atoms in graphite are sp^2 hybridized. They form strong sigma bonds with three adjacent atoms within a layer at a bond angle of 120° and held together by the weak van der Waals forces of interaction between the individual layers.

Exfoliated Graphite

Exfoliated graphite (EFG) is synthesized by exfoliating the graphite layers up to several hundred times along the c-axis resulting in highly flexible, heat-resistant filler material [2–4]. It is obtained by oxidation of the graphite layers followed by exfoliation via thermal expansion [2] or ultrasonication [3]. In the first stage, during the fabrication of the exfoliated graphite, intercalated compound is formed by inclusion of foreign species between the graphite layers. Later, these intercalated compounds are vaporized by thermal energy and thus the van der Waals forces of interaction between the graphite layers are overcome to attain the exfoliation of graphite layers along the c-axis.

Graphene Oxide

Graphene oxide (GO) is another important form of carbon obtained by treating graphite with strong oxidizing agents chemically composed of hydrogen, carbon, and oxygen in variable ratio. More recently, GO has fascinated as an intermediate compound for the synthesis of the graphene. The complete structural description of it is still not revealed due to the complex and irregular packing between the layers. GO is composed of oxygen bridges and other functional groups like carbonyl, phenol, hydroxyl, etc. in general, while the final properties of the material depend on the degree of oxidation. In GO, layers of graphite are expanded as the functional groups are added in between the graphite sheets.

Graphene/Reduced Graphene Oxide

Graphene is a one atomic graphite layer. It has grabbed tremendous attention in the year 2004 when the theoretical envision of the single-layer graphite was achieved practically [5]. It is transparent crystalline allotrope of carbon, where the carbon atoms are sp^2 hybridized forming a hexagonal honeycomb lattice structure. The structure of graphene enables it to function as excellent conductor of heat and electricity in combination with high strength and low density. Sometimes it is also known as reduced graphene oxide.

Graphene Nanoplatelets

Graphene nanoplatelets (GNPs) are promising carbon nanomaterial consisting of several graphene sheets of average thickness of 6–8 nm and diameter of 5–15 μm. They are characterized by platelet kind of morphology with a wide aspect ratio. When used as filler, the unique size and morphology of nanoplatelets provide excellent barrier properties and electrical and thermal conductivities to the composites. Additionally, they are also effective in improving the mechanical properties (e.g., stiffness, strength, and hardness) of the composites.

Carbon Nanotubes

Carbon nanotubes (CNTs) are the cylindrical nanostructures of carbon formed by rolling of a graphene sheet at a particular angle specifically termed as the *chiral angle*. Depending on the rolling angle and the radius of rolling, the nanotubes exhibit different electrical properties and act as a metal or semiconductor material. They are composed of sp^2 hybridized orbitals of carbon atoms similar to the graphite. The nanotubes can be classified broadly as single-walled or multiwalled CNTs of varying dimensions. They may be capped or non-capped, i.e., closed or open at the tube ends. CNTs can be also classified as zigzag, armchair, or chiral nanotubes depending on the rolling direction and angle employed during the formation of the cylindrical nanostructure. In addition, they also appear in other forms, viz., torus (CNTs bend into a doughnut form), nanobud (CNT combined with fullerene), peapod (combination of fullerene inside a nanotube), doped CNT, helically coiled CNT, branched CNT, etc.

Fullerene and Metallofullerene

Fullerenes are another important class of carbon allotrope structurally similar to graphite. They are formed by stacking of graphene sheets linked to the hexagonal rings and sometimes pentagonal and heptagonal rings, which restricts the sheet to become planar. They act like a radical sponge due to the strong electron affinity. Fullerenes create many structural organizations in the form of hollow sphere and ellipsoid. Buckminsterfullerene, C_{60}, is the most common form and the smallest of the class consisting of truncated icosahedron with a carbon atom at the hexagon's corner and bond at the edges. Metallofullerenes are the subclass of fullerenes also known as endohedral fullerenes, where metal atoms are trapped inside the fullerene structure.

Nanodiamond

Structurally nanodiamonds (NDs) are the same as the diamond except its small particle size (< ca. 50 nm). They have excellent combination of properties and high surface area. The crystal structure of NDs is composed of two closely packed interpenetrating cubic lattices of fcc types wherein one of the lattice is shifted with respect to the other by one-quarter of the cube diagonal [6]. They consist of both sp^2 and sp^3 hybridized carbon atoms that can be interchanged with each other. This salient feature of interchangeability makes them flexible in nature. They are synthesized commonly by detonation approach, an ecofriendly and commercially viable method of production [7]. The other synthesis methodologies employed are chemical vapor deposition, laser ablation, high-energy ball milling, autoclave synthesis, etc. [8]. The superior properties of NDs make them an attractive filler material for the fabrication of polymer nanocomposites. They are employed extensively in the biomedical applications, viz., drug delivery, tissue scaffold, protein mimics, surgical implants, etc., due to their non-toxicity, biocompatibility, and chemical stability [9].

Carbon Nanohorn

Carbon nanohorns (CNHs) are another interesting carbon nanomaterial, which have fascinated the field of nanomaterials recently. They are of horn-shaped closed graphene sheet aggregates obtained by laser ablation and arc discharge techniques in the absence of metal catalysts [10]. Dimension wise, a single-walled carbon nanohorn (SWCNH) is typically 40–50 nm in length and 2–3 nm in diameter with a conical cap composed of five pentagons [11–13]. The unique combination of properties such as high surface area, porosity, etc. makes them a promising material for numerous applications, viz., drug delivery, sensors, gas absorption, catalyst support for fuel cells, etc.

Carbon Nanoonion

Carbon nanoonions (CNOs) hold the attention in the recent years especially in the field of energy storage. These are the carbon nanoparticles consisting of closed carbon shells (viz., nested fullerenes). They represent a link between the multiwalled CNTs and fullerenes. The graphene shells with defects constitute closed shell structure in this material. They exhibit small diameter (~10 nm), easy dispersion ability, and excellent electrical conductivity. They are mainly employed as the conductive fillers for the battery or supercapacitor electrodes for energy storage devices. CNOs are the most stable form of carbon cluster due to the favorable surface to volume ratio and least number of dangling bonds.

Carbon Nanofibers

Carbon nanofibers (CNFs) are sp^2-hybridized, linear, noncontinuous carbon filaments synthesized either by chemical vapor deposition technique by passing carbon feedstock over metal particles or by electrospinning of organic polymer followed by thermal treatment. Properties such as reduced size (ca. 100 nm), high surface area, excellent strength with adequate flexibility, etc. make them a promising material for fabrication of the electrode material of energy storage devices. They are structurally different from CNT or carbon fibers and composed of truncated conical stacks in a regular manner along the length of the filament, which makes them suitable as catalyst support, photocurrent generator, and reinforcing filler in the polymer composites.

Polymer Systems Used for Fabrication of Carbon-Based Polymer Nanocomposites

Carbon-based polymer nanocomposites have enormous prospective in diverse applications. Inclusion of carbon nanofillers into polymer matrix enhances the functional properties of the PNCs to a great extent and ascertains their scientific, industrial, and academic importance. Different types of polymer matrices are used for the fabrication of carbon-based PNCs. Carbon nanomaterial-based PNCs are superior to the conventional PNCs due to improved mechanical strength, electrical and thermal conductivities, toughness, etc. Hence, they exhibit essential role in the field of electrochemical devices such as batteries, fuel cells, and supercapacitors. The polymer matrices used for fabrication of the PNCs include polyolefins [polystyrene (PS), polyethylene (PE), polypropylene (PP), polyvinyl chloride (PVC)], polyphenylene sulfide (PPS), poly vinyl alcohol (PVA), poly(propylene carbonate), poly(methyl methacrylate) (PMMA), poly(acrylonitrile butadiene styrene) (ABS), polyester, polycarbonate (PC), epoxy, thermoplastic polyurethane (TPU), polyamidoamine, polyacrylonitrile (PAN), polyacrylic acid (PAA), vinyl ester, etc. Figure 8.3 represents the structure of some common polymers used for the fabrication of the PNCs.

Fabrication Methodologies of Polymer Nanocomposites

Carbon-based polymer nanocomposites are fabricated by the general nanocomposite fabrication methodologies, viz., solution blending, melt blending, and in situ polymerization.

Solution blending is the common method used for the fabrication of PNCs involving three processing steps, i.e., dispersion of the carbon material in a solvent, incorporation of the polymeric material, and removal of the solvent. The compatibility between the polymer and the carbon nanomaterial in the solvent system plays decisive role in achieving uniform and homogeneous filler dispersion. Chemical

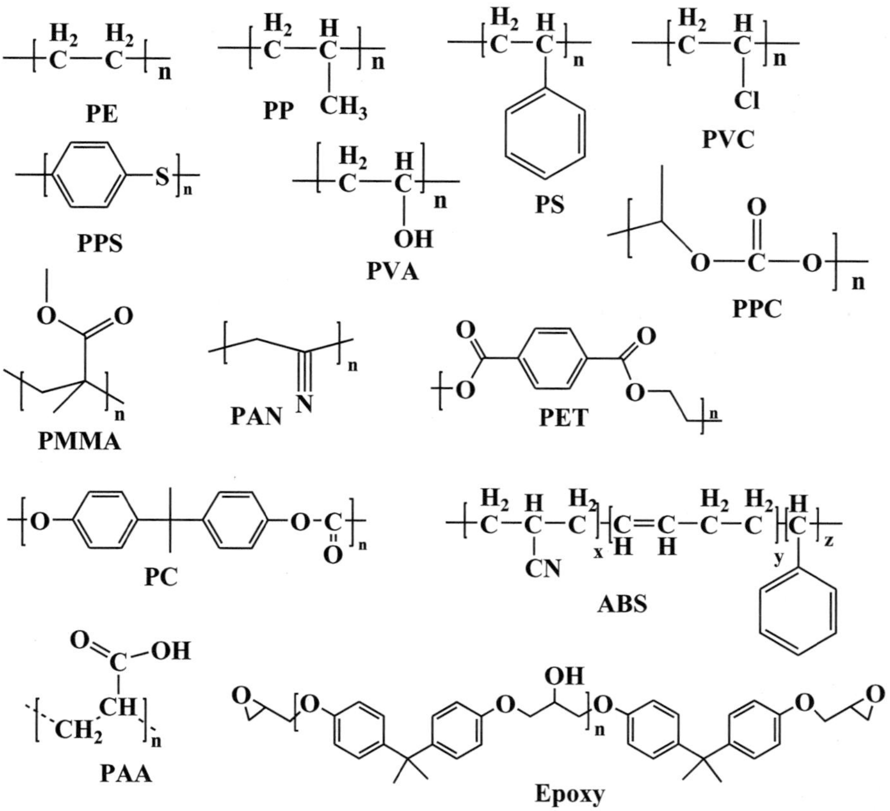

Fig. 8.3 Structures of different types of polymers

functionalization of the carbon materials is a general approach to improve their dispersion within the polymer matrix.

Melt blending is another approach employed for the dispersion of nanomaterials. From the industrial application point of view, melt mixing is preferred over the solution mixing due to the possibility of large-scale production and low manufacturing cost. In a typical method of fabrication, the polymer pellets are converted to a viscous liquid by application of heat in which the nanofillers are subsequently dispersed by applying high-shear forces. The main advantage of the melt mixing process over the solution blending is that it is devoid of use of solvent. However, the dispersion of the fillers at high loading becomes difficult due to the increased viscosity of the mix. Other major drawbacks of the system are buckling and rolling of some nanofillers (graphene, CNT, etc.) due to the application of the high-shear forces.

In situ polymerization is the third common approach applied for the fabrication of the carbon nanomaterial-filled PNCs. This is a convenient process for fabrication of the PNCs that otherwise cannot be prepared due to the lack of solubility and thermal stability of the polymer. In this process, filler is at first swollen in the monomer system in liquid form. Initiator is added in the reaction mixture containing filler in the liquid monomer. Further, polymerization of the monomer is conducted by application of heat or radiation. The in situ formation of polymer within the reaction mixture allows intercalation of the fillers yielding homogeneously dispersed filler-based PNCs. In situ polymerization approach also facilitates grafting or covalent bonding between the fillers and the polymer matrix. This approach further extents the possibility of grafting of the polymeric chains within the wall of the fillers. The main benefit of the in situ polymerization remains in the possibility of high filler loading within the polymer matrix with adequate filler dispersion.

Carbon-Based Polymer Nanocomposites

Carbon Nanofiber-Based Polymer Nanocomposites

Carbon nanomaterial-based PNCs are extensively used for the fabrication of the shape memory polymer (SMP) [14–16]. SMPs have the capability to attain preprogrammed shape when kept under the influence of external stimuli such as heat, electricity, magnetic field, light, etc. Electrically conductive SMPs triggered by Joule heating were studied by Lu et al. [17]. CNFs were assembled to form carbon nanopaper and enhance the bonding within the SMP. It assisted in transferring the effect of Joule heating from nanopaper to the polymer matrix. The addition of the carbon nanopaper proved to be beneficial for the improvement of glass transition temperature and electrical properties of the polymer nanocomposites. Figure 8.4 represents the SMP nanocomposites made up of four layers of CNF nanopaper. CNF/bio polyesters were fabricated by simple melt mixing approach [18]. CNFs were synthesized by vapor growth technique and were incorporated uniformly within the polymer matrix. The incorporation of CNF converted the nonconductive polymer matrix into electrically conductive one at 5 vol% percolation thresholds with significant enhancement in mechanical properties owing to the strong interfacial bonding between the CNF and polyester matrix. The nanocomposites exhibited excellent electrically induced shape memory effect and find potential application in the field of electroactive actuators.

Surface-treated CNF-reinforced PNCs based on vinyl esters were also investigated [19]. The CNFs were modified by purification and surface treatment to provide improved bonding between the CNF and the vinyl ester. Oxidation followed by functionalization of CNFs with vinyl ester provided improved interfacial bonding between the CNF and polymeric resin. The nanocomposite exhibited improved mechanical properties up to 1 wt% filler loading.

CNF-filled polypropylene (PP) nanocomposites were studied for electrical and electronic applications [20]. Dispersion of the CNF within the polymer matrix was achieved by boiling the CNF in PP/xylene solution for a sufficient time period. Thin nanocomposite films, fabricated by compression molding, showed high thermal stabilities due to the ability of CNF to serve as free radical absorber. The degree of

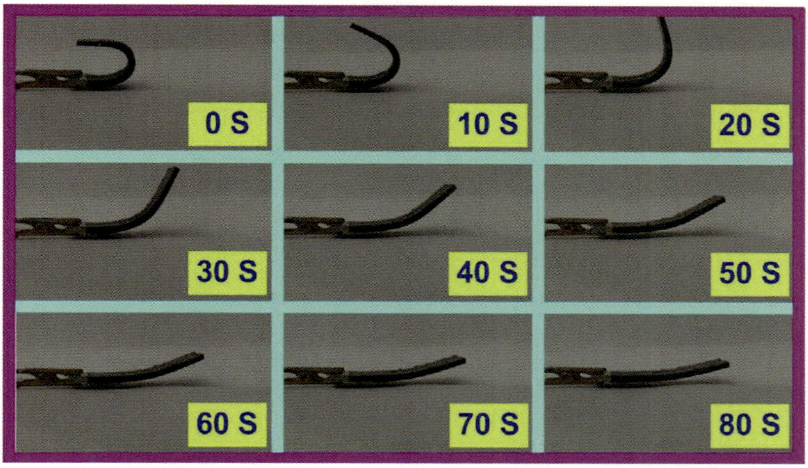

Fig. 8.4 Heat-induced SMP nanocomposites made of four-layer CNF nanopaper [17]

crystallinity of the nanocomposite was also improved significantly as CNF creates heterogeneous nucleating center for the formation of the nuclei. The nanocomposite also exhibited reduced electrical resistance at 5 wt% filler loading, which can be attributed to the cross-linked fiber structure.

CNFs were also incorporated into the polymer matrix to improve the thermal properties of the neat polymers [21]. Xu et al. studied high-density polyethylene (HDPE)/CNF nanocomposites prepared by melt blending and compressive processing techniques at different filler loadings of silane-treated CNF. Chemical treatment of the CNF with silane coupling agent improved the compatibility between the CNF and polymer matrix and resulted in enhanced thermal conductivity. The inclusion and uniform dispersion of CNF within the polymer matrix found to be advantageous in reducing the thermal contact resistivity as well as coefficient of thermal expansion of the PNCs.

Sun et al. studied electrical and mechanical properties (static and dynamic) of CNF/epoxy nanocomposites [22]. The nanocomposites exhibited high tensile strength at 1 wt% filler loading owing to the uniform distribution of the CNF within the polymer matrix. Above 1 wt% filler loading, the tensile strength of the nanocomposite was found to be decreased due to the aggregation of the CNFs. The nanocomposite at 1 wt% (0.578 vol%) filler loading displayed remarkable increase (~10 times) in conductivity due to the accomplishment of percolation threshold. The nanocomposites also exhibited 70 % higher storage modulus above glass transition temperature compared to that of the unfilled polymer. Zhu et al. demonstrated rheological, thermal, and mechanical properties of epoxy/CNF nanocomposites [23]. The epoxy composites with and without silanes (3-aminopropyltriethoxysilane (APTES)) modification were studied for the abovementioned functional properties. The nanocomposites were fabricated by dispersing the CNF within the polymer matrix. The amine groups formed bond with CNF via in situ silanization reaction in the presence of monomer. In situ silane functionalization provided improved dispersion of the CNF within the epoxy monomer, and hence a decrease in viscosity was observed with remarkable improvement in the mechanical properties. Silane-treated CNF-filled epoxy exhibited 12.6 % higher tensile strength at 0.1 wt% filler loading compared to that of untreated CNF at similar filler loading. The functionalization of the CNF also helped in enhancing the thermal stability of nanocomposites.

Carbon Nanotube-Based Polymer Nanocomposites

CNTs are extensively utilized for fabrication of nanocomposites due to their high mechanical strength [24, 25]. Principally, CNT-filled PNCs are prepared with the aim to improve the mechanical properties of the polymer matrices employed for structural applications. The success of the CNT/polymer nanocomposite is dependent on the compatibility and dispersion of the CNTs within the host polymeric material. The dispersion of the CNTs within the matrix can be achieved by the disaggregation of entangled CNTs by mechanical means or through functionalization approach. The chemical functionalization of CNTs breaks the van der Waals attraction between them and hence provides more available surface area for interaction with the host matrix. The dispersion and disaggregation of CNTs play a key role in attaining improved mechanical properties.

CNTs possess potential in the field of nanoelectronic applications. Patole et al. developed a self-assembled hybrid nanocomposite using combination of both functionalized graphene and CNT filler within the polystyrene matrix [26]. Graphene and CNT were functionalized together in the presence of the polymer matrix by utilizing water-based in situ microemulsion approach. The hybrid nanocomposite made of multiple carbon fillers formed an interconnected network wherein the graphene sheets provided a large surface area and the CNT created an interlinking with the polymer matrix by acting like a wire as represented in Fig. 8.5. These kinds of interconnected structures were responsible for reduction in sheet resistance and enhancement in the thermal and mechanical properties of the hybrid nanocomposite films compared to the virgin polymer.

The alignment of CNT within the polymer matrix plays an important role in determining the mechanical and electrical properties of the final PNCs. Haggenmueller and coworkers fabricated polymethylmethacrylate (PMMA)/SWNT polymer composite consisting aligned SWNT within the host matrix by solution casting and melt processing approach [27]. The CNTs were uniformly dispersed within the PMMA by using dimethyl formamide as the solvent. The nanocomposites exhibited high conductivity along the direction of flow due to the alignment of the CNTs. Elastic modulus and the yield strength of the nanocomposite were improved with nanotube content. In another study, Cooper et al. fabricated PNCs based on MWNTs, SWNTs, and carbon nanofibrils [28]. The investigation revealed that the tensile properties of the nanocomposites were unaffected by the addition of carbon nanofillers, but their impact strength was improved significantly

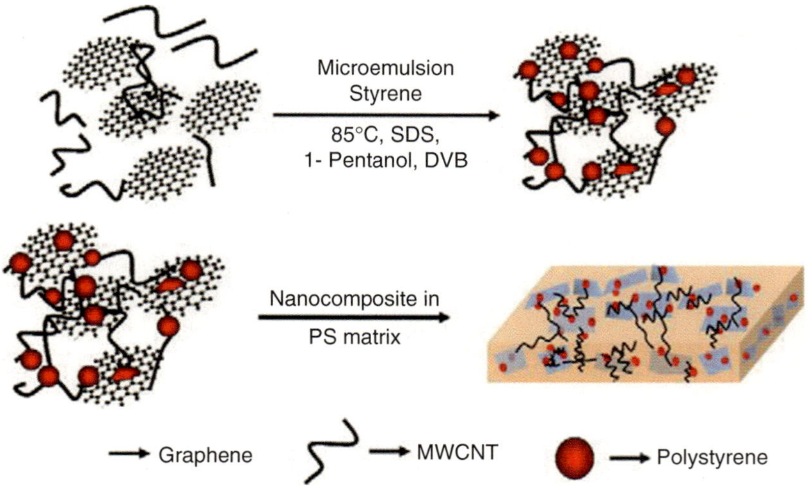

Fig. 8.5 Schematic of synthesis process for making conducting polystyrene-based graphene/MWCNT multicomposite films [26]

compared to the neat polymer matrix. The effect of MWNTs on thermal and electrical properties of PMMA was also evaluated [29]. The effect of irradiation on the properties of nanocomposites was examined by evaluating the hardness and dynamic mechanical properties of the nanocomposites after irradiating with Cesium-137 source. Study revealed that irradiation of the nanocomposites during fabrication improved microhardness, glass transition temperature, and mechanical properties due to the absorption of radiation in the conjugate bonds of carbon nanomaterial. Hence, damage of the PMMA matrix was suppressed by the use of nanofillers.

Allaoui et al. evaluated mechanical and electrical properties of MWNT/epoxy nanocomposites utilizing very small quantity of nanofiller [30]. The epoxy resin was made soft and ductile using overaged hardener. The soft and ductile nature of the epoxy matrix made it possible to evaluate the stress strain properties of the nanocomposites at high strain level. The addition of very small amount of nanotube (say from 1 % and 4 % with respect to the matrix polymer) exhibited twofold and fourfold enhancements in Young's modulus and yield strength of the nanocomposites, respectively. The nanocomposite also showed ninefold increment in conductivity as a result of attainment of percolation threshold of carbon nanotube in the polymer matrix when concentration was 0.5 and 1 wt%. In another study Schadler et al. studied tensile and compression behavior of MWNT/epoxy nanocomposites [31]. The nanocomposite was cured using triethylene tetramine cross-linking agent. The study revealed that the nanocomposite exhibits higher compression modulus than the tensile modulus as evidenced by Raman analysis. Load transfer behavior of the nanocomposites indicated that the strain developed in the carbon bonds only shifts to a considerable extent under the compression mode. This may be attributed to the fact that during compression all the nanotube walls contribute, whereas the outer walls only do so when subjected to tension. Yu et al. developed epoxy-based PNCs

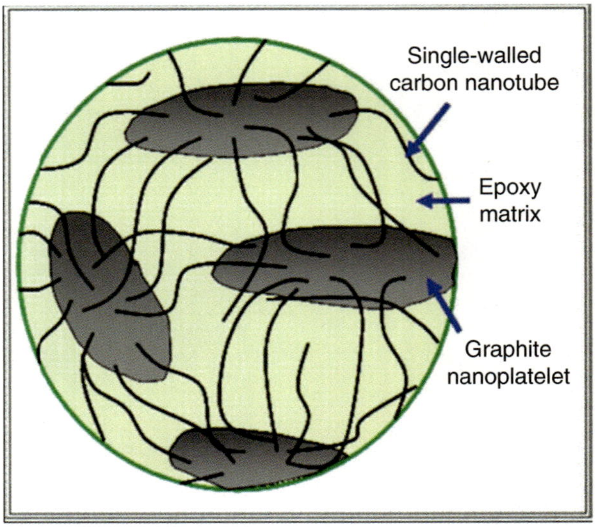

Fig. 8.6 Schematic representation of network structure formed by graphite nanoplatelet (GNP) and single-walled carbon nanotube (SWCNT)-based epoxy PNCs [32]

consisting of SWNT and graphitic nanoplatelets with the aim to improve thermal conductivity of epoxy [32]. The inclusion of hybrid nanofiller provided a network structure within the polymer matrix and hence thermal resistance at the interphase was reduced. Figure 8.6 represents the network structure formed by the incorporation of graphite nanoplatelet (GNP) and SWNT in the epoxy matrix. The incorporation of hybrid nanofillers was proved to be effective in enhancing thermal conductivity of PNCs, and hence the nanocomposites with improved processability and reduced viscosity could be prepared at low filler loading.

Santos and coworkers developed methyl-ethyl methacrylate copolymer-based CNT nanocomposite [33]. The dispersion of the nanotubes within the polymer matrix was improved by using nonionic surfactants. The nanocomposites exhibited 200 % improvement in mechanical properties along with 10 °C increment in the glass transition temperature. The addition of small amount (1 wt%) of the nanotubes was sufficient to improve mechanical and thermal properties of the nanocomposite. As the concentration of the CNTs increased beyond a certain value, the storage modulus of the composite lowered down due to the impurities added in the system. The lack of efficient contact between the filler and polymer matrix resulted in detrimental effect in the functional behavior of the nanocomposite.

Exfoliated Graphite-Based Polymer Nanocomposites

Poly(ethylene-2,6-naphthalate) (PEN)/EFG-based PNCs were fabricated by melt dispersion approach [34]. Functionalization of the graphene sheets was carried out by partially hydrolyzing the graphite oxide materials. Graphite is exfoliated along the crystallographic c-axis to yield EFG while maintaining its rigidity and flat structure even after the exfoliation. Hence, an improvement in the functional properties, viz., mechanical and thermal, is expected. The PNCs were prepared by melt blending the nanomaterial and polyester in a twin-screw extruder. Dispersion of the nanofiller within the matrix was examined by XRD studies and morphological analysis in electron microscope since disaggregation of filler within host matrix plays a crucial role in achieving enhanced properties in PNCs. Addition of very little amount of nanofiller, say 0.3 vol% of EFG, provided the interconnected networks for the effective conduction within the composite, whereas about 3 vol% non-functionalized nanofiller was required to form similar interconnected network in the PNCs. Electron microscopy revealed that the exfoliated morphology of the EFG was retained in the PNCs and the layers of graphite were unremoved even after melt processing.

Physical properties of poly(propylene carbonate) (PPC) were improved by inclusion of EFG having particle size of 7.4 μm and thickness of 30–50 nm [35]. EFG of nanometer size range was synthesized by heat treatment and ultrasonication method. The nanocomposite films were fabricated by solution blending approach at different EFG content. Study revealed that EFG was well dispersed in the PPC matrix at low filler content (2 %). However, the interaction between EFG and PPC was not strong enough to establish compatibility at high EFG loading. With an increase in EFG content, the oxygen and moisture permeability of PPC was found to be reduced considerably. The nanocomposite film exhibited excellent molecular ordering, which was evidenced from the improved mechanical properties of 2 % EFG-filled PPC composites. Figure 8.7 represents the variations in tensile strength and elongation at break for the nanocomposites with varying EFG content.

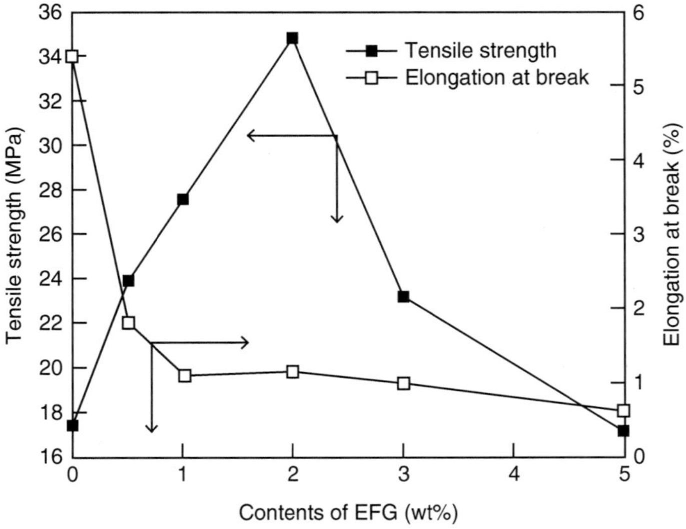

Fig. 8.7 Variation of tensile strength (in MPa) and elongation at break (%) with EFG content [35]

Graphene (Reduced Graphene Oxide)-Based Polymer Nanocomposites

The research on graphene has been triggered in recently due to its exceptional properties, which make it the material of interest in diverse areas. Graphene-based polymer composites are of industrial and academic interest as the functional properties of a polymer matrix can be enhanced by using graphene, even at low loading. Several polymers, viz., Nafion®, polyvinylidene fluoride (PVDF), polyaniline, poly (3,4-ethylenedioxythiophene) (PEDOT), epoxy, PS, polyethylene terephthalate (PET), polyurethane (PU), PC, PVA, etc., have been studied for the fabrication of PNC based on graphene [36].

Graphene as the carbon-based nanomaterial was incorporated in to the PP/bagasse composites with the aim to improve mechano-physical properties of composites [37]. The nanocomposites were fabricated by mixing via melt blending in an extruder followed by injection molding. Addition of graphene at low filler loading was found to be effective in maintaining the functional properties of composites. For example, at a very low graphene loading of 0.1 wt%, the nanocomposites exhibited maximum tensile and flexural properties with a slight reduction in impact strength. However, at high filler content (0.5–1 wt%), graphene got agglomerated easily.

Chaharmahali et al. reported highly conductive PVA-based reduced graphene oxide (rGO)-filled nanocomposites [38]. The high surface area allowed overlapping of rGO with each other to form continuous network of the nanofillers within the polymer matrix. Thermal reduction of large-area GO exhibited a huge improvement in the electrical conductivity of the nanocomposites due to the retention of its conjugated structure. The nanocomposites were characterized by very low percolation threshold of 0.189 wt%. Addition of rGO sheets of large surface area within the PVA matrix leads to electrical conductivity of 6.3×10^{-3} S m^{-1} at 0.7 wt% filler loading.

Wang et al. investigated dielectric properties of rGO/PP nanocomposites [39]. The nanocomposite exhibited percolation threshold at very low filler concentration of 0.033 vol%. The nanocomposites were prepared by in situ reduction of GO in the presence of PP. Electron microscopy and Raman spectroscopy revealed homogeneous dispersion of nanofiller within the PP due to strong interaction between the two. The nanocomposites exhibited significant change in dielectric properties near the percolation threshold. Figure 8.8 represents the effect of rGO content on the variation of conductivity of rGO/PP nanocomposites with frequency at room temperature. At a low rGO content ($\leq$0.03 vol%), conductivity of the composites was very much dependent on the frequency of measurement due to insulating nature of the polymer matrix. A transition from insulating to conducting nature was observed as the filler content exceeded 0.03 vol% and conductivity became nearly independent to the applied frequency.

Again, rGO-filled PNCs were also explored as actuator [40]. TPU-based PNCs were fabricated by dispersing thermally reduced GO via solution casting approach. Actuation properties of the nanocomposites were evaluated by using infrared

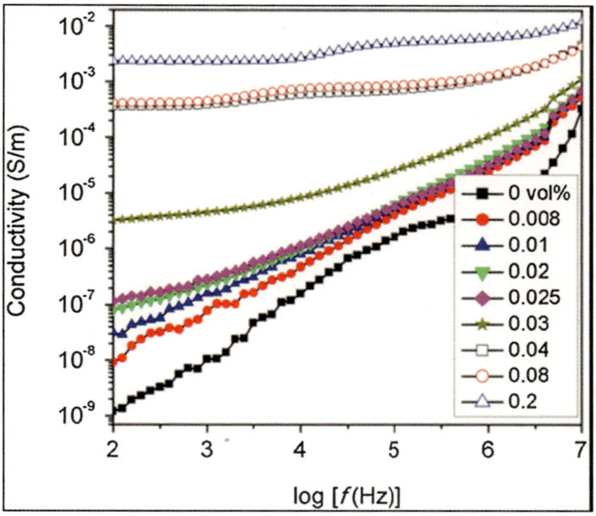

Fig. 8.8 Conductivity of rGO/PP nanocomposite with the rGO content with frequency at room temperature [39]

illumination technique. Photochemical actuation of PNCs can be controlled by controlling the filler content. At a filler content of 2 wt%, the nanocomposites exhibited high photochemical strain value. The PNCs emerge as promising materials for light activated actuation in robotics and biomedical devices, due to their excellent strain values.

Graphene Oxide-Based Polymer Nanocomposites

GO can be synthesized in large quantities by oxidation of graphite followed by exfoliation of pre-oxidized graphite sheet. These are chemically defined as the oxygenated graphene sheets containing hydroxyl, carboxyl, and epoxy functional groups. Goncalves et al. studied PMMA/GO nanocomposites by grafting GO onto the surface of polymer using atom transfer radical polymerization (ATRP) [40]. The nanocomposite films were fabricated by using both PMMA-grafted GO and neat GO. Evaluations of mechanical properties of the nanocomposites revealed that 1 wt% of PMMA-grafted GO (GPMMA) filler was sufficient to reinforce the polymer matrix. GPMMA was found to be more effective as reinforcing material compared to the neat PMMA or PMMA/GO nanocomposites. A filler loading of 1 wt% resulted in a tougher PMMA/GPMMA nanocomposite with considerable improvement in the elongation properties. Moreover, resulted nanocomposite films were readily dispersible in common organic solvents.

Etmimi et al. demonstrated fabrication of exfoliated PNC by incorporating modified graphite oxide via miniemulsion polymerization approach [41]. The GO nanofiller was functionalized with 2-acrylamido-2-methyl-1-propanesulfonic acid (AMPS) to expand the graphene layers and to ease the intercalation in between the nanogalleries. Surfactant-assisted emulsion polymerization of styrene and butyl

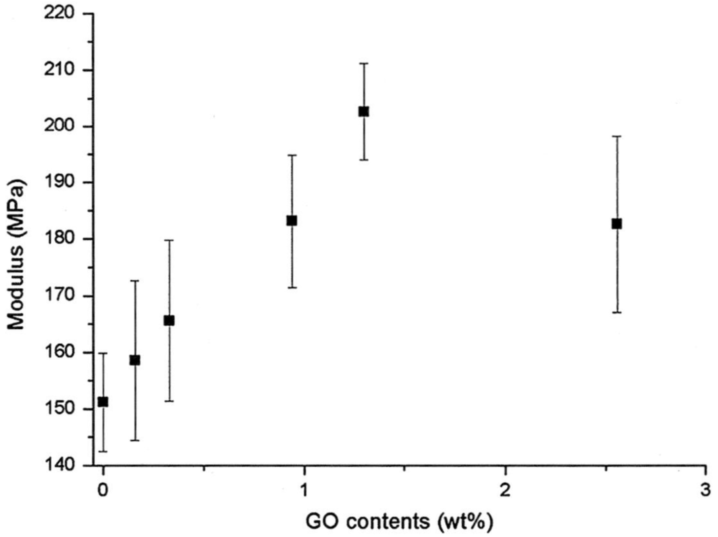

Fig. 8.9 Effect of GO content on Young's modulus of polysulfone/GO nanocomposite membrane [44]

acrylate monomers was conducted in the presence of AMPS-modified GO to synthesize latex particles composed of poly(styrene-butyl acrylate) and GO. Transmission electron microscopy (TEM) analysis of the nanocomposite showed huge exfoliation with final layer thickness of about 2–5 nm. The nanocomposites were characterized by the high thermal stability and glass transition temperature. At high GO loading (about 5 and 6 wt% with respect to monomer), nanocomposites exhibited enhanced storage and loss modulus, as shown by dynamic mechanical analysis (DMA).

The maximum effectiveness of GO in polymer matrix was obtained when the dispersion of the GO in the host matrix remained uniform. With the aim to improve dispersion of GO within the polymer matrix, Traina and Pegoretti developed a unique approach wherein reduction of GO was conducted in situ by thermal treatment [42]. The GO/polymer nanocomposites were prepared by dispersing the GO within PVA matrix followed by thermal treatment of the composite to perform reduction of GO [43]. Mild thermal treatment was carried out to prevent any further degradation of base polymer. These alternative methods of synthesis of GO-filled PNCs provided uniform dispersion of GO, which was otherwise difficult in the case of graphene-based PNCs due to extensive aggregation.

Lee et al. studied polysulfone (PSf)/GO nanocomposite membranes for wastewater treatment [44]. Figure 8.9 represents the effect of GO content on Young's modulus of PSf/GO nanocomposite membranes. With increasing GO content, the modulus exhibited an increasing trend for a filler concentration up to 1.30 wt%. With further increase in the GO content, the modulus of composite membranes was found to reduce due to stress concentration arising from aggregation of GO. Hence the nanocomposite membrane exhibited brittle failure at 2.6 % filler loading.

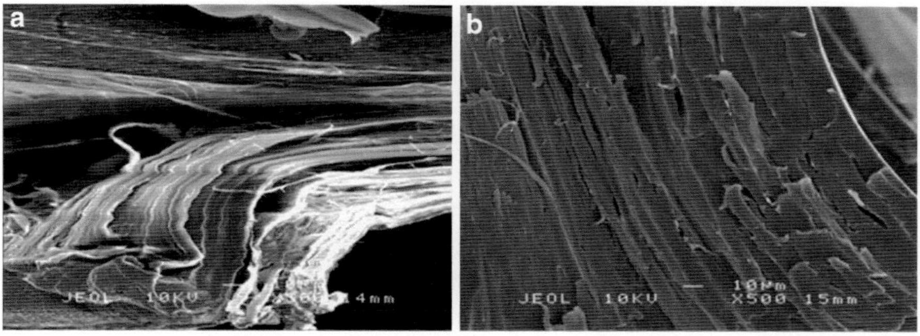

Fig. 8.10 Scanning electron microscopy of (**a**) poly(lactic acid) (PLA)/poly(ethylene glycol) (PEG) nanocomposites and (**b**) PLA/PEG/graphene nanoplatelet (0.3 wt%) nanocomposites [46]

Graphene Nanoplatelet-Based Polymer Nanocomposites

Similar to graphene and GO, graphene nanoplatelet (GNP), the other form of nanostructured graphitic carbon, has been studied as filler material for PNCs. Gaikwad et al. developed GNP-filled polyamide 11 nanocomposites for electrostatic charge dissipation and selective laser sintering applications [45]. GNPs were incorporated into the polyamide matrix by extrusion in a corotating twin-screw extruder at different filler content, viz., 1, 3, 5, and 7 wt%. The inclusion of GNPs was aimed to improve electrical conductivity, thermal stability, and mechanical properties (the tensile and flexural) of the matrix. Uniform dispersion of GNPs was achieved by high-shear extrusion in corotating twin-screw extruder, and hence improvement in functional properties of the nanocomposites was observed.

Chieng et al. demonstrated GNP-filled poly(lactic acid) (PLA)/poly(ethylene glycol) (PEG) polymer blends. The nanoplatelet-modified blends were fabricated by simple melt blending approach [46]. PLA was modified with PEG to improve the flexibility and reduce the brittleness of the same. In the subsequent step of fabrication, nanoplatelets were added into the plasticized PLA/PEG blends. The synthesized nanocomposites exhibited improved tensile properties at very low filler content of 0.3 wt%. Thermogravimetric analysis of the nanocomposites revealed that the inclusion of nanoplatelets was effective in improving the thermal stability of the host polymers. Figure 8.10 shows the scanning electron microscope (SEM) images of PLA/PEG blend and PLA/PEG/nanoplatelet nanocomposite containing 0.3 wt% filler. The SEM images exhibited homogeneous distribution of the filler within the PLA/PEG matrix (Fig. 8.10b), and hence an enhancement in thermal and tensile properties was obtained.

GNP was also added in PP with the aim to improve its impact properties by melt mixing and injection molding [47]. The nanocomposites were fabricated with 2, 4, 6, and 8 wt% GNP of different sizes in the presence of PP-graft-maleic anhydride coupling agent. The nanocomposites with and without the coupling agent were prepared and compared for different properties. Study revealed that the addition of nanofiller of small diameter led to the better improvement in mechanical properties like flexural and tensile strength. The presence of coupling agent favored the dispersion of GNPs within the matrix and improved their adhesion with polymer matrix and hence resulted in better mechanical properties compared to the composites fabricated without coupling agent.

Fullerene- and Metallofullerene-Based Polymer Nanocomposites

Incorporation of fullerene in polymer matrices has attracted recent research interest due to its potential application in numerous emerging areas. In combination with polymer, fullerenes can serve as promising materials in various fields, viz., optoelectronics [48], structural composites [49], organic photovoltaics, and biomedical imaging probes [50]. Fullerene improves the thermal stability of polymer matrix by preventing its degradation via the free radical chain scission process. For example, fullerene-based PNCs prepared via melt blending of high-density polyethylene (HDPE) exhibited improved thermal properties over pure HDPE [51]. At a very low loading level of 0.5 wt%, the nanocomposites showed a shift in degradation onset temperature from 389 °C to 459 °C. This effect became more pronounced with increasing C_{60} loading, due to efficient trapping of free radicals.

Similarly, mechanical properties of fullerene/epoxy nanocomposites at various weight ratios were also studied [52]. The nanocomposites were examined for their mechanical behavior such as Young's modulus, resistance to crack propagation, ultimate tensile strength, fracture energy, toughness, etc. The mechanical properties of the nanocomposites were enhanced by addition of small amount of filler (0.1–1 wt%). Hence fullerene/epoxy nanocomposites exhibited great potential in several application areas like structural, aeronautical and space applications, paints, electronics packaging, etc. PC/fullerene PNCs were also studied [49]. The nanocomposites consisting of fullerenes and polyhydroxylated fullerenes were examined for different functional properties such as mechanical, thermal, and optical properties. UV–vis spectroscopy analysis of the nanocomposites revealed that inclusion of hydroxylated fullerene did not affect the transmission of light in the range of 400–800 nm. Thermogravimetric analysis of the nanocomposites demonstrated that hydroxylated fullerene-based PNC exhibited higher thermal stability compared to the virgin PC due to the formation of rigid interphase as a result of hydrogen bonding. The unmodified fullerene-filled PC exhibited comparatively lower thermal stability and mechanical properties due to the aggregation of fullerenes within the polymer matrix, which acted as a point of failure during the mechanical testing. Good compatibility of the nanofiller with the polymer matrix is essential to obtain better thermal stability. As per the literature, PC/$C_{60}(OH)_{12}$ showed better thermal stability compared to PC/C_{60} and PC/$C_{60}(OH)_{36}$ composites. In this study, effect of addition of fullerene on the optical properties of PVA was also investigated. Both PVA- and PC-based nanocomposites exhibited good optical transparency due to the reduction in π-conjugation of the nanomaterials. Figure 8.11 represents the PC/fullerene interphase of different nanocomposite films. PC/C_{60} showed thin interphase region due to weak π-π stacking between the filler and polymer, whereas in PC/$C_{60}(OH)_{12}$,

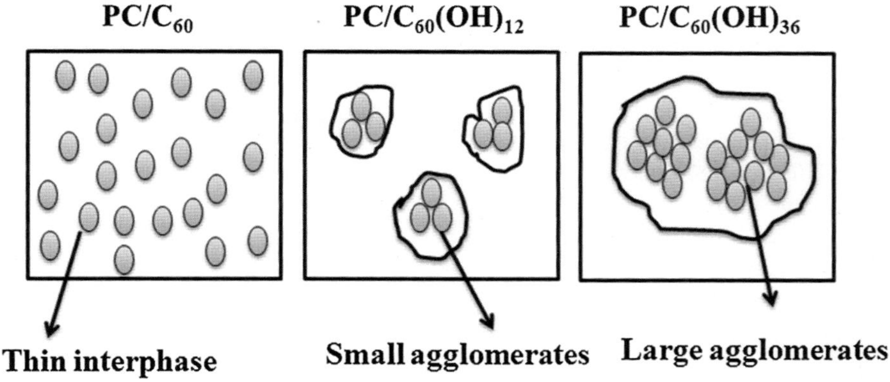

Fig. 8.11 Schematic representation of interphase in nanocomposite films [49]

this region was comparatively larger as a result of dipolar interaction between PC and $C_{60}(OH)_{12}$. Hence, an improved compatibility further contributed to thermal stability of the PC matrix. However, this trend was not observed for PC/$C_{60}(OH)_{36}$ and a decrement in stability was evidenced. As the hydroxyl content in nanofillers increases, two possible phenomena contributing to the thermal stability of the final composites may take place. Firstly, the additional hydroxyl groups might create large agglomerates which reduce effective surface area of the nanofiller and lead to poor interfacial bonding between filler and polymer. Again, heavy hydroxyl content leads to release of water molecules in huge amount (both surface and bound water). Such releasable water molecules from inside the composite lead to partial hydrolysis in PC backbone at the carbonate moiety and hence an overall decrement in thermal stability.

Bai et al. developed Gd-based fullerene (Gd@C_{82})/polystyrene (PS) nanocomposites via bulk radiation polymerization approach [53]. The core-shell morphology of the metallofullerene-based polymer originated due to the strong interaction between the metallofullerene units. Metallofullerene and metallic nitride fullerenes (MNF) exhibited exciting potential in the medical applications. Phillips et al. reported MNF-based PNCs, blends, and polymer networks by using Sc3N@C80 as MNF [54]. The MNF-based PNCs were explored for switchable adhesion application. Stimuli responsive fullerene-doped adhesives were also investigated. Chattopadhyaya et al. studied Ti-based endohedral metallofullerene (TEMF) polymer as a spin switch in an electrical field [55]. The density of state analysis of material revealed that the switching ability of TEMF arises due to the presence of opposite spin channels of spin states.

Nanodiamond-Based Polymer Nanocomposites

Nanodiamonds (NDs) have attracted much attention in the field of polymer composites due to the unique combination of properties such as outstanding thermal conductivity, low coefficient of friction, and excellent mechanical properties. NDs have become the materials of interest for the tribological application due to its superior mechanical properties. Neitzel et al. fabricated epoxy/ND-based nanocomposites and evaluated its potential for wear and frictional properties [56]. They studied the effect of as received and functionalized nanodiamonds on the tribological properties of the nanocomposites. The addition of either kind of NDs was found to be effective, while the aminated NDs outperformed the as received NDs due to their improved interfacial interaction with the host polymer matrix. Study revealed that the nanocomposites containing 25 vol% of NDs exhibited excellent Young modulus and hardness value and also showed the ability to wear the alumina body. This excellent combination of properties made them a promising candidate as an alternative to micron-sized diamond used in commercial abrasive applications. Figure 8.12 represents the AFM topography of neat epoxy and epoxy nanocomposites filled with 12.5 vol% NDs and aminated NDs. AFM topography of nanocomposites showed inconsistent behavior (as evidenced from the zones with lower and higher frictional forces) due to the presence of ND agglomerates within the matrix.

NDs hold great potential for the biomedical applications. Hikov and coworkers developed plasma-polymerized hexamethyldisiloxane/NDs composite for the medical implants [57]. Silanized and silver-modified ND particles were used for the evaluation of the properties of the nanocomposites. The cytotoxicity characterization of nanocomposites revealed that the cells were well grown and survived in the presence of ND particles. Study established that the addition of NDs was favorable for the development and growth of cells, and hence they can be efficiently used for biomedical implant applications.

Morimune et al. studied PVA-based nanocomposites using ND as the reinforcing fillers [58]. The nanocomposites were fabricated via casting method by dispersing the NDs into the aqueous solution of PVA. It exhibited 2.5-fold increment in Young's modulus at 1 wt% ND loading. The reinforcing properties of the nanocomposite were found to be promising and established them as nanomaterial against the conventional carbon nanofillers.

Carbon Nanohorn-Based Polymer Nanocomposites

CNHs are the promising material for biological application. Guerra et al. explored the suitability of polyamidoamine (PAMAM)/CNHs nanocomposites for drug delivery application [59]. CNHs effectively anchored the biologically active entities. PAMAM facilitated the electrostatic bonding with the ribonucleic acid and improved the solubility of CNHs in the system and made them biologically more compatible. It acted as the stage to hold the PAMAM dendrimers. The study revealed that the CNHs can serve as a useful nonviral agent for gene therapy.

Single-walled carbon nanohorn (SWCNH)-filled PAN nanocomposites are studied [60]. The SWCNHs were dispersed within the PAN matrix by mechanical stirring and ultrasonication and surfactant-assisted ultrasonication approaches. The

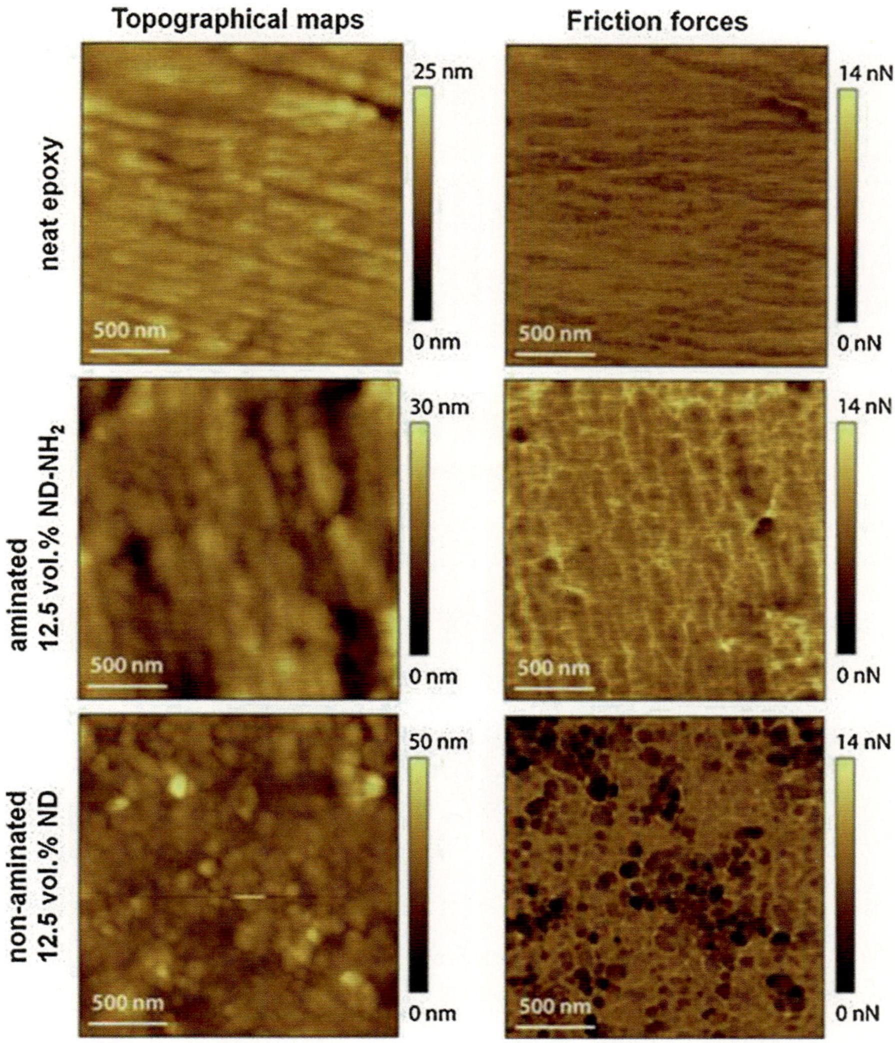

Fig. 8.12 AFM topography of neat epoxy and nanodiamond-filled epoxy nanocomposites filled with 12.5 vol% nanodiamond and aminated nanodiamond [56]

best possible dispersion of the SWCNHs was obtained with surfactant-assisted ultrasonication. The well-dispersed SWCNHs exhibited best mechanical properties. The dispersion led to the lowest size of carbon aggregates in the range of 50–250 nm. The inclusion of conducting carbon fillers provided a conductive pathway in the electrically insulating virgin PAN matrix due to the establishment of percolation threshold. 1 wt% SWCNHs were sufficient enough to bring about electrical conduction pathway in the PAN matrix.

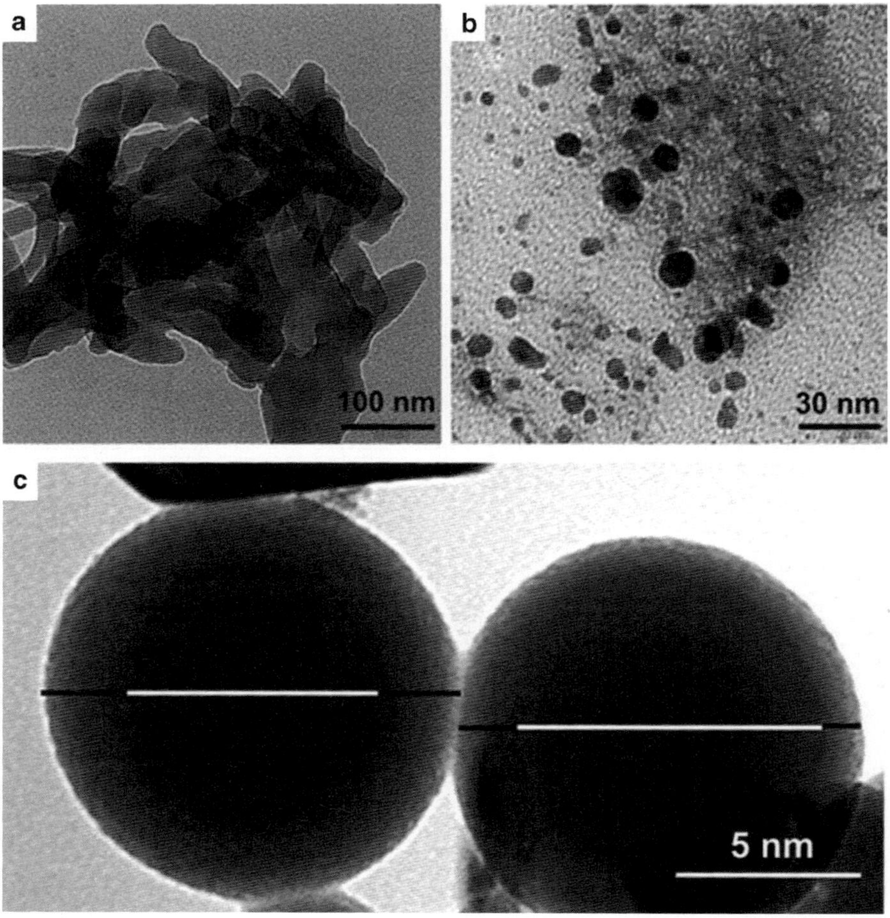

Fig. 8.13 TEM images of polyaniline/carbon nanoonion nanocomposites [61]

Carbon Nanoonion-Based Polymer Nanocomposites

Carbon nanoonion (CNO) was explored for the fabrication of nanocomposites in various fields and extensively studied for their contribution in the conducting polymer matrices [61, 62]. Figure 8.13 represents the TEM images of polyaniline/CNO nanocomposites. Nanocarbon materials, viz., nanoonions and NDs, are of significant potential for multifunctional composite applications [63]. Shenderova et al. developed polydimethylsiloxane (PDMS)- and polyurethane-based nanocomposites with CNO for electromagnetic absorption and electromagnetic interference shielding (EMI shielding) applications [63]. The inclusion of CNO was found to be beneficial for enhancement in the loss tangent of the composites. In general, addition of carbon nanomaterials such as nanodiamonds or nanoonions improved the multifunctional properties of the polymer matrix, viz., thermal conductivity and stability, EMI shielding, durability, and hence established them as a promising material for interdisciplinary research fields. In another study, Maksimenko et al. demonstrated the effect of CNO on the electromagnetic wave absorption properties of PMMA nanocomposite [64]. The electromagnetic wave attenuation of CNO and PMMA/CNO films was evaluated over the frequency range of 2–38 GHz. The potential of CNO in the abovementioned field and physical properties associated with EMI shielding application of the CNO and CNO-based polymeric films were discussed herein.

Applications

Carbon-based PNCs find potential applications in various fields, viz., conductive nanocomposites [39], high-strength composites [49], electromagnetic interference shielding [64], electrocatalysts [65], sensors [66], biomedicine [9], microwave absorption [67], power conversion and storage [68], etc. depicted in Fig. 8.14. CNT-based PNCs have potential applications in electrical circuit boards, heat sinks, lids and housing, etc. due to the enhanced thermal transport properties of the polymer nanocomposites originated from the excellent thermal conductivity of the CNTs. Polymer/CNT nanocomposites are also promising in the field of photovoltaic cells and devices. In addition to the electrochemical device application, CNTs also have attracted great deal of attention in the field of sensors. Carbon materials have been considered to be suitable element as transducer and mediator in the field of biosensors [69]. Inclusion of CNTs into the conducting polymer matrix results in improved electrical and mechanical properties, which makes them suitable for the sensor application. Recently, CNTs are also explored in photovoltaic devices. Curran et al. studied photovoltaic properties of the polymer nanocomposite made of MWNT and poly(m-phenylenevinylene-co-2,5-dioctoxy-p-phenylenevinylene) (PPV) [70]. The study revealed that dispersion of MWNTs within the polymer matrix was favored by the helical chains of polymer matrix and hence good dispersion of the nanofiller could be achieved. The light-emitting diodes (LEDs) fabricated with MWNTs also exhibited more stability in air as compared to the one without MWNT.

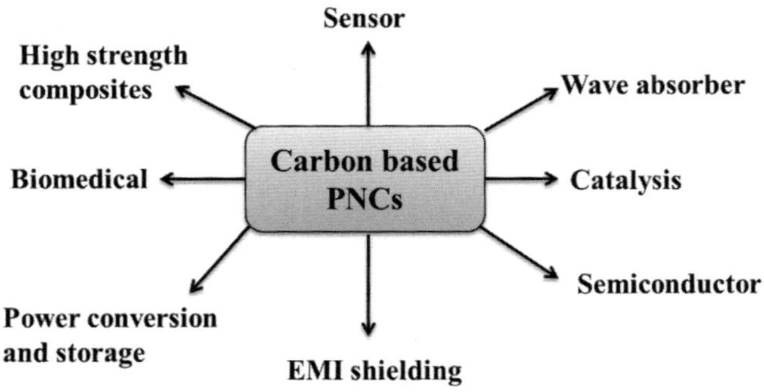

Fig. 8.14 Application areas of carbon-based PNCs

Carbon-based PNCs are widely used as catalyst by several researchers. Tavasoli and coworkers studied Co/CNT system to investigate the catalytic performance of the composite display at 15–40 wt% cobalt loading [71]. The hydrocarbon yield of the Fischer–Tropsch synthesis is much better than that of cobalt catalyst on other metal supports. Carbon nanomaterial-based polymer nanocomposites also have potential application in presence of magnetic fields. Li et al. reported nitrogen-containing nickel nanosphere with carbon encapsulation [72]. The nanospheres were composed of core-shell structure with a core formed by nickel and a shell of nitrogen-enriched carbon material. The material was reported to be suitable for the biomedical application. Nickel carbon nanostructures were also studied by Sunny et al. [73]. The nickel carbon interface was observed to create space charge polarization and an intrinsic magnetic loss, which resulted in improved microwave absorption properties.

Concluding Remarks

Carbon-based materials have immense potential in tuning the functional properties of polymer matrices for different emerging fields. Allotropes of carbon with different shapes and functional properties have been incorporated into numerous polymer matrices to obtain the synergistic effect in functional properties by mutual presence of individual components in a single material. Graphite, diamond, soot, coke, pyrolytic carbon, glassy carbon, carbon black, CNTs, nanorods, graphene, fullerene, carbon nanofibers, nanodiamond, carbon nanohorn, activated carbon, etc. are the different allotropes of carbon, which have been encountered in the literature till date. The polymer nanocomposites consisting one or more of these fillers have shown excellent mechanical, thermal, electrical, and optical properties. Key factors in achieving maximum benefit out of these fillers lie in the uniform dispersion of these fillers into the polymer matrix. Different approaches, like high-shear mixing, ultrasonication, in situ polymerization, and functionalization, are adopted to accomplish homogeneous dispersion in the polymer matrix. Polyolefins, acrylates, polyester, polyamide, PVA, PVC, PS, styrene-butadiene-styrene, epoxy, etc. are some the important nonconducting polymer matrices, which have been utilized to make the polymer nanocomposites. In this chapter, brief discussion about the classification of different allotropes and their nanocomposites with different nonconducting polymer matrices has been demonstrated in light of different functional properties along with possible applications.

Acknowledgments The authors acknowledge the Department of Science and Technology, Government of India, for granting the fellowship.

References

1. Delhaes P (2001) Graphite and precursors. CRC Press, France. ISBN 90-5699-228-7
2. Schniepp HC, Li J-L, McAllister MJ, Sai H, Herrera-Alonso M, Adamson DH, Prud'homme RK, Car R, Saville DA, Aksay IA (2006) Functionalized single graphene sheets derived from splitting graphite oxide. J Phys Chem B 110(17):8535

3. Stankovich S, Piner RD, Chen X, Wu N, Nguyen ST, Ruoff RS (2006) Stable aqueous dispersions of graphitic nanoplatelets via the reduction of exfoliated graphite oxide in the presence of poly(sodium 4-styrenesulfonate). J Mater Chem 16(2):155
4. Lueking AD, Pan L, Narayanan D, Burgess-Clifford CE (2005) Effect of expanded graphite lattice in exfoliated graphite nanofibers on hydrogen storage. J Phys Chem B 109(26):12710
5. (2009) Discovery of graphene. APS News Ser II 18(9):2
6. Bhosale RR, Osmani RA, Ghodake PP, Harkare BR, Shaikh SM, Chavan SR (2013) Nanodiamonds: a new-fangled drug delivery system. Indo Am J Pharm Res 3(12):1395
7. Mohamed El-Say K (2011) Nanodiamond as a drug delivery system: applications and prospective. J Appl Pharm Sci 01(06):29
8. Kazi S (2014) A review article on nanodiamonds discussing their properties and applications. Int J Pharm Sci 3:40
9. Mochalin VN, Shenderova O, Ho D, Gogotsi Y (2012) The properties and applications of nanodiamonds. Nat Nanotechnol 7:11
10. Iijima S, Yudasaka M, Yamada R, Bandow S, Suenaga K, Kokai F, Takahashi K (1999) Nano-aggregates of single-walled graphitic carbon nano-horns. Chem Phys Lett 309(3–4):165
11. Yodasaka M, Iijima S, Crespi VH (2008) Single-wall carbon nanohorns and nanocones. Top Appl Phys 111:605
12. Pagona G, Mountrichas G, Rotas G, Karousis N, Pispas S, Tagmatarchis N (2009) Properties, applications and functionalisation of carbon nanohorns. Int J Nanotechnol 6:176
13. Zhu SY, Xu GB (2010) Single-walled carbon nanohorns and their applications. Nanoscale 2(12):2538
14. Cho JW, Kim JW, Jung YC, Goo NS (2005) Electroactive shape-memory polyurethane composites incorporating carbon nanotubes. Macromol Rapid Commun 26:412
15. Lu HB, Huang WM (2013) Synergistic effect of self-assembled carboxylic acid-functionalized carbon nanotubes and carbon fiber for improved electro-activated polymeric shape-memory nanocomposite. Appl Phys Lett 102:231910
16. Le HH, Zulfiqar A, Mathias U, Ilisch S, Radusch HJ (2010) Effect of the cross-linking process on the electrical resistivity and shape-memory behavior of cross-linked carbon black filled ethylene-octene copolymer. J Appl Polym Sci 120:2138
17. Lu H, Liang F, Yao Y, Gou J, Hui D (2014) Self-assembled multi-layered carbon nanofiber nanopaper for significantly improving electrical actuation of shape memory polymer nanocomposite. Compos Part B 59:191
18. Tang Z, Sun D, Yang D, Guo B, Zhang L, Jia D (2013) Vapor grown carbon nanofiber reinforced bio-based polyester for electroactive shape memory performance. Comp Sci Technol 75:15
19. Plaseied A, Fatemi A, Coleman MR (2008) Influence of carbon nanofiber content and surface treatment on mechanical properties of vinyl ester. Polym Polym Comp 16:405
20. Chen X, Wei S, Yadav A, Patil R, Zhu J, Ximenes R, Sun L, Guo Z (2011) Poly(propylene)/carbon nanofiber nanocomposites: ex situ solvent-assisted preparation and analysis of electrical and electronic properties. Macromol Mater Eng 296:434
21. Xu S, Akchurin A, Liu T, Wood W, Tangpong XW, Akhatov IS, Zhong W-H (2014) Thermal properties of carbon nanofiber reinforced high-density polyethylene nanocomposites. J Comp Mater 49:795
22. Sun L-H, Ounaies Z, Gao X-L, Whalen CA, Yang Z-G (2011) Preparation, characterization, and modeling of carbon nanofiber/epoxy nanocomposites. J Nanomater 2011:1
23. Zhu J, Wei S, Ryu J, Budhathoki M, Liang G, Guo Z (2010) In situ stabilized carbon nanofiber (CNF) reinforced epoxy nanocomposites. J Mater Chem 20:4937
24. Bal S, Samal SS (2007) Carbon nanotube reinforced polymer composites–a state of the art. Bull Mater Sci 30:379
25. Velasco-Santos C, Martinez-Hernandez AL, Castano VM (2005) Carbon nanotube-polymer nanocomposites: the role of interfaces. Compos Interfaces 11:567

26. Patole AS, Patole SP, Jung S-Y, Yoo J-B, An J-H, Kim T-H (2012) Self assembled graphene/ carbon nanotube/polystyrene hybrid nanocomposite by in situ microemulsion polymerization. Eur Polym J 48:252
27. Haggenmueller R, Gommans HH, Rinzler AG, Fischer JE, Winey KI (2000) Aligned single-wall carbon nanotubes in composites by melt processing methods. Chem Phys Lett 330:219
28. Cooper CA, Ravich D, Lips D, Mayer J, Wagner HD (2002) Distribution and alignment of carbon nanotubes and nanofibrils in a polymer matrix. Compos Sci Technol 62:1105
29. Tatro SR, Clayton LM, O'Rourke Muisener PA, Rao AM, Harmon JP (2004) Probing multi-walled nanotube/poly(methyl methacrylate) composites with ionizing radiation. Polymer 45:1971
30. Allaoui A, Bai S, Cheng HM, Bai JB (2002) Mechanical and electrical properties of a MWNT/ epoxy composite. Compos Sci Technol 62:1993
31. Schadler LS, Giannaris SC, Ajayan PM (1998) Load transfer in carbon nanotube epoxy composites. Appl Phys Lett 73:3842
32. Yu A, Ramesh P, Sun X, Bekyarova E, Itkis ME, Haddon RC (2008) Enhanced thermal conductivity in a hybrid graphite nanoplatelet – carbon nanotube filler for epoxy composites. Adv Mater 20:4740
33. Velasco-Santos C, Martínez-Hernández AL, Fisher F, Rouff RS, Castaño VM (2003) Dynamical–mechanical and thermal analysis of carbon nanotube–methyl-ethyl methacrylate nanocomposites. J Phys D Appl Phys 36:1423
34. Kim H, Macosko CW (2008) Morphology and properties of polyester/exfoliated graphite nanocomposites. Macromolecules 41:3317
35. Lee Y, Kim D, Seo J, Han H, Khan SB (2013) Preparation and characterization of poly (propylene carbonate)/exfoliated graphite nanocomposite films with improved thermal stability, mechanical properties and barrier properties. Polym Int 62:1386
36. Das TK, Prusty S (2013) Graphene-based polymer composites and their applications. Polym Plast Technol Eng 52:319
37. Zhou TN, Qi XD, Fu Q (2013) The preparation of the poly(vinyl alcohol)/graphene nanocomposites with low percolation threshold and high electrical conductivity by using the large-area reduced graphene oxide sheets. Express Polym Lett 7:747
38. Chaharmahali M, Hamzeh Y, Ebrahimi G, Ashori A, Ghasemi I (2014) Effects of nano-graphene on the physico-mechanical properties of bagasse/polypropylene composites. Polym Bull 71:337
39. Wang D, Zhang X, Zha J-W, Zhao J, Dang Z-M, Hu G-H (2013) Dielectric properties of reduced graphene oxide/polypropylene composites with ultralow percolation threshold. Polymer 54:1916
40. Goncalves G, Marques PAAP, Barros-Timmons A, Bdkin I, Singh MK, Emamic N, Gracio J (2010) Graphene oxide modified with PMMA via ATRP as a reinforcement filler. J Mater Chem 20:9927
41. Etmimi HM, Sanderson RD (2011) New approach to the synthesis of exfoliated polymer/ graphite nanocomposites by miniemulsion polymerization using functionalized graphene. Macromolecules 44:8504
42. Traina M, Pegoretti A (2012) In situ reduction of graphene oxide dispersed in a polymer matrix. J Nanopart Res 14:1
43. Muralidharan MN, Ansari S (2013) Thermally reduced graphene oxide/thermoplastic polyurethane nanocomposites as photomechanical actuators. Adv Mater Lett 4(12):927
44. Lee J, Chae H-R, Won YJ, Lee K, Lee C-H, Lee HH, Kim I-C, Lee J-M (2013) Graphene oxide nanoplatelets composite membrane with hydrophilic and antifouling properties for waste water treatment. J Membr Sci 448:223
45. Gaikwad S, Tate JS, Theodoropoulou N, Koo JH (2012) Electrical and mechanical properties of PA11 blended with nanographene platelets using industrial twin-screw extruder for selective laser sintering. J Comp Mater 47(23):2973
46. Chieng BW, Ibrahim NA, Yunus WMZW, Hussein MZ (2014) Poly(lactic acid)/poly(ethylene glycol) polymer nanocomposites: effects of graphene nanoplatelets. Polymers 6:93

47. Duguay AJ, Nader JW, Kiziltas A, Gardner DJ, Dagher HJ (2014) Exfoliated graphite nanoplatelet-filled impact modified polypropylene nanocomposites: influence of particle diameter, filler loading, and coupling agent on the mechanical properties. Appl Nanosci 4:279
48. Kawauchi T, Kumaki J, Yashima E (2005) Synthesis, isolation via self-assembly, and single-molecule observation of a [60] fullerene-end-capped isotactic poly(methyl methacrylate). J Am Chem Soc 127:9950
49. Saotome T, Kokubo K, Shirakawa S, Oshima T, Hahn HT (2011) Polymer nanocomposites reinforced with C_{60} fullerene: effect of hydroxylation. J Comp Mater 45(25):2595
50. Kim J-W, Kim K-J, Park S, Jeong K-U, Lee M-H (2012) Preparation and characterizations of C_{60}/polystyrene composite particle containing pristine C_{60} clusters. Bull Kor Chem Soc 33(9):2966
51. Zhao L, Song P, Cao Z, Fang Z, Guo Z (2012) Thermal stability and rheological behaviors of high-density polyethylene/fullerene nanocomposites. J Nanomater 2012:1
52. Rafiee MA, Yavari F, Rafiee J, Koratkar N (2010) Fullerene–epoxy nanocomposites-enhanced mechanical properties at low nanofiller loading. J Nanopart Res 13:733
53. Bai X, Yue D, Zhao S, Dong J, Yang L, Ibrahim K, Wang J, Yang S, Hao J, Hu Z, Sun B (2011) Self-construction of core–shell structure by metallofullerene-containing polymer. J Nanosci Nanotechnol 11:2244
54. Phillips JP, Hoyle CE, Confait BS, McCluskey DM, Ahmed H, Stevenson S (2009) Synthesis, characterization, and applications of metallic nitride fullerene polymer nanocomposites. Polym Prepr 50(1):383
55. Chattopadhyaya M, Alam Md M, Sen S, Chakrabarti S (2012) Electrostatic spin crossover and concomitant electrically operated spin switch action in a Ti-based endohedral metallofullerene polymer. Phys Rev Lett 109:257204
56. Neitzel I, Mochalin V, Bares JA, Carpick RW, Erdemir A, Gogotsi Y (2012) Tribological properties of nanodiamond-epoxy composites. Tribol Lett 47:195
57. Hikov T, Pramatarova L, Kresteva N, Radeva E, Petrik P, Agocs E, Pecheva E, Presker R, Sabotinov O (2012) Study of nanocomposite layers based on polymers and nanodiamond particles: new materials for medical implants. Bulg J Phys 39:297
58. Morimune S, Kotera M, Nishino T, Goto K, Hata K (2011) Poly(vinyl alcohol) nanocomposites with nanodiamond. Macromolecules 44:4415
59. Guerra J, Herrero MA, Carrión B, Pérez-Martínez FC, Lucío M, Rubio N, Meneghetti M, Prato M, Ceña V, Vázquez E (2012) Carbon nanohorns functionalized with polyamidoamine dendrimers as efficient biocarrier materials for gene therapy. Carbon 50:2832
60. Fraczek-Szczypta A, Blazewicz S (2011) Manufacturing and physico-mechanical characterization of carbon nanohorns/polyacrylonitrile nanocomposites. J Mater Sci 46:5680
61. Plonska-Brzezinska ME, Mazurczyk J, Palys B, Breczko J, Lapinski A, Dubis AT, Echegoyen L (2012) Preparation and characterization of composites that contain small carbon nano-onions and conducting polyaniline. Chem Eur J 18:2600
62. Kovalenko I, Bucknall DG, Yushin G (2010) Detonation nanodiamond and onion-like-carbon-embedded polyaniline for supercapacitors. Adv Funct Mater 20:3979
63. Shenderova O, Tyler T, Cunningham G, Ray M, Walsh J, Casulli M, Hens S, McGuire G, Kuznetsov V, Lipa S (2007) Nanodiamond and onion-like carbon polymer nanocomposites. Diamond Relat Mater 16:1213
64. Maksimenko SA, Rodionova VN, Ya Slepyan G, Karpovich VA, Shenderova O, Walsh J, Kuznetsov VL, Mazov IN, Moseenkov SI, Okotrub AV, Lambin Ph (2007) Attenuation of electromagnetic waves in onion-like carbon composites. Diamond Relat Mater 16:1231
65. Okamoto M, Fujigaya T, Nakashima N (2009) Design of an assembly of poly(benzimidazole), carbon nanotubes, and Pt nanoparticles for a fuel-cell electrocatalyst with an ideal interfacial nanostructure. Small 5:735
66. Zhang W, Suhr J, Koratkar N (2006) Carbon nanotube/polycarbonate composites as multifunctional strain sensors. J Nanosci Nanotechnol 6:960
67. Hekmatara H, Seifi M, Forooraghi K (2013) Microwave absorption property of aligned MWCNT/Fe_3O_4. J Magn Magn Mater 346:186

68. Thomassin J-M, Kollar J, Caldarella G, Germain A, Jérôme R, Detrembleur C (2007) Beneficial effect of carbon nanotubes on the performances of Nafion membranes in fuel cell applications. J Membr Sci 303:252
69. Sotiropoulou S, Gavalas V, Vamvakaki VV, Chaniotakis NA (2003) Novel carbon materials in biosensor systems. Biosens Bioelectron 18:211
70. Curran SA, Ajayan PM, Blau WJ, Carroll DL, Coleman JN, Dalton AB, Davey AP, Drury A, McCarthy B, Maier S, Strevens A (1998) A composite from poly(m-phenylenevinylene-co-2,5-dioctoxy-p-phenylenevinylene) and carbon nanotubes: a novel material for molecular optoelectronics. Adv Mater 10:1091
71. Tavasoli A, Sadagiani K, Khorashe F, Seifkordi AA, Rohaniab AA, Nakhaeipour A (2008) Cobalt supported on carbon nanotubes – a promising novel Fischer-Tropsch synthesis catalyst. Fuel Process Technol 89(5):491
72. Li XL, Tian XL, Zhang DW, Chen XY, Liu DJ (2008) Solvothermal synthesis and characterization of nitrogen-enriched carbon-encapsulated nickel nanospheres. Materials science and engineering 151(3):220
73. Sunny V, Kumar DS, Mohanan P, Anantharaman MR (2010) Nickel/carbon hybrid nanostructures as microwave absorbers. Mater Lett 64(10):1130

Syntactic Foams for Multifunctional Applications

9

S. Sankaran, B. N. Ravishankar, K. Ravi Sekhar, Samudra Dasgupta, and M. N. Jagdish Kumar

Contents

S. Sankaran (✉)
Directorate of Extramural Research & Intellectual Property Rights, DRDO HQ, DRDO Bhawan, New Delhi, India
e-mail: ssankar_an@rediffmail.com

B.N. Ravishankar • K. Ravi Sekhar • S. Dasgupta • M.N. Jagdish Kumar
Aeronautical Development Establishment (DRDO), Bangalore, Karnataka, India
e-mail: bnrshankar@ade.drdo.in; kandalars@ade.drdo.in; samudradg@ade.drdo.in; MN.jagdishkumar@gmail.com

K.K. Kar (ed.), *Composite Materials*,
DOI 10.1007/978-3-662-49514-8_9

Abstract

Owing to their lightweight and high strength characteristics and tailorable end-use properties, rigid syntactic foams fabricated from polymeric binders, hollow micro-spheres, and other fillers, and their variants have assumed great importance in material design. The highly ordered syntactic foams have potential to be utilized in designs involving multifunctional requirements, viz., aerospace structures, radar transparency, microwave electronics, EMI shielding, etc. This overview article introduces syntactic foams and gives an insight into establishing multifunctional characteristics in them to meet the end applications and presents a collection of related literature on techniques and uses.

Keywords

Syntactic foam • Glass microballoons • Hollow glass microspheres • Rigid foams • Multifunctional foams • Foam sandwich composites

Foams

Solid foams form an important class of lightweight cellular engineering materials that are used in engineering design mostly as core materials in sandwich composite structures. These foams are formed by trapping of gas bubbles in a solid matrix which results in such lightweight materials.

Classification of Foams Based on Matrix

Solid foams can be synthesized by foaming metallic, ceramic, or polymeric materials, and accordingly, one has metallic foams, ceramic foams, and polymeric foams, respectively. Among all the varieties, polymeric foams are the most common, and they are also referred by various other nomenclatures like plastic foams, cellular plastics, foamed plastics, expanded polymers, etc. The dominant factors in determining properties of such foams are their density and the chemical nature of the solid polymer phases. The density of the polymeric foam is a function of the volume ratio of the gas filler to the solid polymer. In actual practice, the foam density may vary anywhere from 0.01 g/cm^3 to a value almost as high as that of the solid polymer itself [1, 2]. Epoxy, phenolic, cyanate ester, and polyurethane are some of the thermosetting, and polyvinyl chloride (PVC), polyethylene (PE), polypropylene (PP), and polyetherimide (PEI) are some of the thermoplastic foam materials.

Classification of Foams Based on Method of Fabrication

Based on the route of synthesis, foams can be classified as chemical foams or physical foams. In the former, chemical reactions between two and more ingredients evolve certain gaseous product, which gets trapped into a bubble phase forming the foam. On the other hand, in the case of physical foams, a gas is physically blown into the viscous polymeric resin, and the foam is formed by subjecting the system to a specified temperature and pressure regime, during which time it solidifies.

Classification of Foams Based on Cell Structure

The foams can be classified into two types based on their pore or cell structure. The first type is called open cell structured foams, which contain pores that are connected to each other forming an interconnected network. The second type of foams does not have interconnected pores and are called closed-cell foams. Normally, the closed-cell foams are denser and stronger compared to their open celled counterparts. They also possess higher-dimensional stabilities and lower moisture absorption coefficients.

The present chapter deliberates in detail upon still different a variety of foam materials called syntactic foams. These are classified under the category of "closed-cell physical foams," since they contain encapsulated voids that are introduced physically [3], in the form of microballoons or microspheres.

Syntactic Foams

Introduction

Syntactic foams are a class of composite materials formed by embedding hollow spherical particles, viz., microballoons or microspheres in a binder matrix denoted as the binder. Unlike the conventional foams, the syntactic foams can be a tertiary phase system and not necessarily be binary systems as the other foams [4–6]. Since the sizes of gas containing particles can be controlled within a close range resulting in foams of high order or regularity, the foam is called "syntactic" (in Greek, "syn" means same and "taktos" means ordered). These foams also are considered as reinforced plastics since glass and ceramic microballoons can be regarded as rigid reinforcing component. Compared to conventional foams, these foams have better strength to weight ratios and negligible water absorption properties and can withstand considerable hydrostatic pressure. However, these foams are heavier than conventional foams with apparent densities in the range 0.18–0.8 g/cm^3. An excellent review on the syntactic foams is authored by Shutov [7] wherein the preparation and properties of the syntactic foams are discussed in detail.

Even though metallic and ceramic syntactic foams are available, in the present chapter, it is intended to present a concise review of the published literature and current state-of-the-art technologies in the field of thermosetting polymer based syntactic foams in addition to a mention of structure and properties of the same.

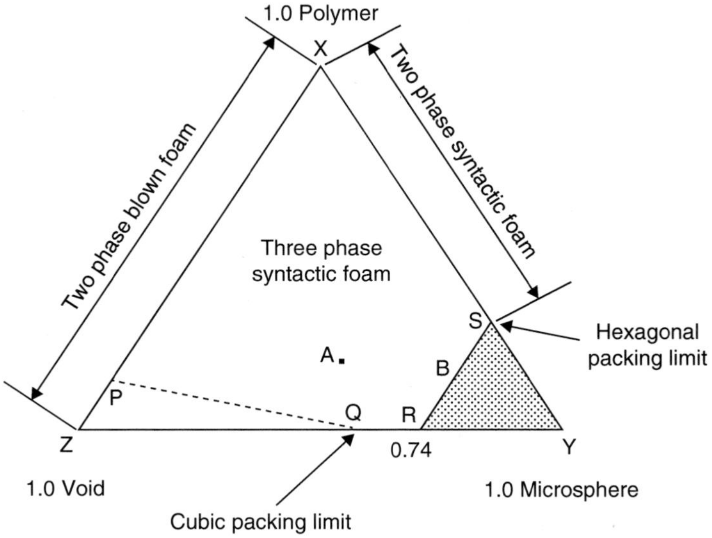

Fig. 9.1 Ternary phase diagram for three-phase syntactic foams [8]

Structure of Syntactic Foams

Structurally syntactic foams can be classified as (a) two-phase syntactic foam and (b) three-phase syntactic foam.

The three-phase syntactic foam system can be easily understood by a ternary-phase diagram given in Fig. 9.1 [8]. Polymer, hollow microballoons, and voids, the three components of the foam, represent the three vertices in the ternary diagram. Point A on the diagram denotes a composition having volume fraction values of 0.15, 0.60, and 0.25 of polymer resin, microballoons, and voids, respectively. The pure void-free syntactic foam as a two-phase composition will fall along the polymer-microballoon side, i.e., XY of the diagram. However, the syntactic foam composite is subjected to packing conditions of the spheres, which is mentioned in Table 9.1. Hexagonal close packing being the upper limit of packing for uniform-sized microballoons, packing factor cannot exceed 0.74 [9]. Similarly, the two-phase composite of microballoons and voids only (line YZ) (packed spheres without any binding resin) has a maximum limit of 74 % of the former. Cubic packing (volume fraction 0.52) restricts the lower limit for the microballoons for sphere-to-sphere contact. Below this limit, it requires some amount of resin to fill in between the spheres, since the structure is not self-supporting anymore. Another two-phase composite is chemically or physically blown foam composed of polymer and voids only. Its composition lies along the polymer-void side, i.e., XZ, and it too is limited by the maximum volume fraction of voids allowed while still maintaining the definition of foam. All the limits mentioned define the allowed compositions for the syntactic foams and determine the area within the diagram where they can fall. A limiting case is point B, which represents a composition of microspheres 0.74,

Table 9.1 Packing possibilities of microspheres [10]

	Coordination number	Packing	Volume fraction occupied
Uniform size spheres	6	Cubic	0.52
	8	Body-centered cubic	0.62
	12	Face-centered cubic	0.74
	12	Hexagonal	0.74
Nonuniform spheres	–	Random	0.90

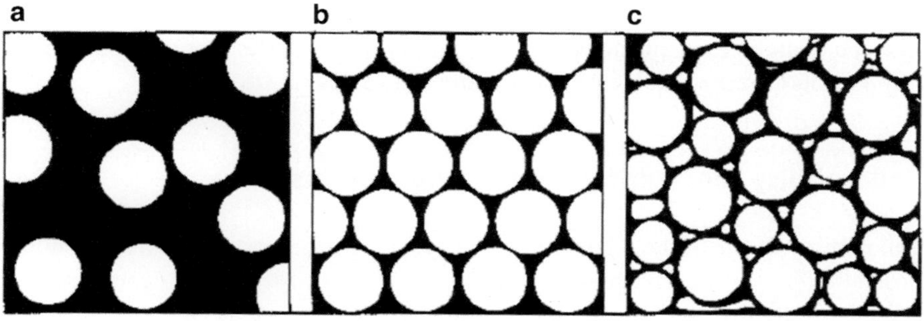

Fig. 9.2 Schematic representation of syntactic foam structures: (**a**) random dispersion of spheres (two-phase composite), (**b**) hexagonal close-packed structure of uniform-sized spheres (two-phase composites), (**c**) closely packed microspheres, dispersed voids, and binding resin (three-phase composite) [9]

polymer 0.11, and voids 0.15. In this case, the microspheres are arranged in the hexagonal close packing [10, 11].

The density of three-phase syntactic foam is determined by the composition represented by any point in the area marked in the diagram. The side XY of the phase diagram represents the two-phase syntactic foam. A schematic representation of the two-phase and three-phase syntactic foams is shown in Fig. 9.2.

Constituents of Syntactic Foams

The two essential constituent materials of syntactic foams are hollow microballoons (filler) and polymeric resin matrix (binder). Voids are incidental constituent of the two-phase foams, whereas they are essential or intentional in three-phase foams.

Reinforcing Microspheres

The reinforcing microballoons need to be spherical, non-cohesive, strong, and chemically resistant [12]. Glass microballoons, phenolic microballoons, carbonized microballoons, and metal (like silver, nickel, copper, etc.)-coated microballoons are used to be embedded in the polymer matrix. Microballoons synthesized from materials like polyurethane [13], polyimide [14, 15], unsaturated polyester [16, 17] poly vinylidene chloride (PVDC) [18], carbamide and melamine-formaldehyde oligomers, etc. [19] may also be used. Even microspheres containing an inert gas or some volatile liquid such as freons [19] can also be produced, so can be metal oxide microspheres [20, 21]. The present chapter concentrates mostly on hollow glass microspheres, which is the most widely used constituent for these classes of foams.

Polymeric Binder Materials

The polymeric materials used as binders for syntactic foam can be either thermoplastic or thermoset. The foams derived from thermoplastic resins are difficult to produce and are very limited in their use since they soften at comparatively lower temperatures. Hence, the main research effort is concentrated on thermoset resins like epoxy, polyimide, phenolic, silicone, and cyanate ester foams. A good number of publications are available on such thermosetting foams.

Tailorability for Multifunctional Applications

Tailorability is one of the biggest assets of syntactic foam materials. Apart from the possibility of selecting resin and microspheres from a wide range of commercially available options, these foams can also be tailored for specific applications by intelligent incorporation of different types of fillers, additives, flexibilizers, tougheners, etc. Often, a single syntactic foam itself will perform more than one function, thus behaving as a multifunctional material, which is essential for many structural, thermal, and electromagnetic applications. Hence, these foams have their own place carved out for various purposes wherein the specific functional requirements are an important design criteria in addition to weight saving.

Syntactic Foams for Structural Properties

The mechanical properties of syntactic foams required for any structural application is governed by the density and volume fraction of the constituents. Compressive property study is the major means of qualifying this class of materials for structural applications. Figure 9.3 shows the general shape of stress versus strain curve in compression for polymer syntactic foams [22]. The curve can be divided into three sections/regions. The first section, where the stress is proportional to strain, corresponds to the elastic behavior of the foam. In this region, if the load is released, the foam reverts back to its original shape. The compression yield point occurs followed by densification or consolidation of the sample, in the second section. In this region, the compressive stress remains nearly constant. The third region is characterized by an increase of load or stress, which is because of the resistance offered by the foams after the densification.

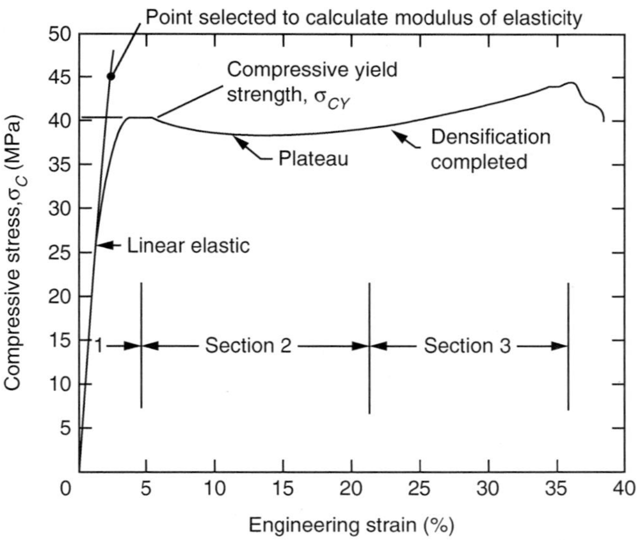

Fig. 9.3 Typical stress–strain curve for syntactic foam [22]

Figure 9.4 shows compressive stress versus compressive strain behavior of 0.39 g/cm^3 density foam samples (five specimens) manufactured using glass microballoons and epoxy-modified cyanate ester resin binder [23].

In general, compressive strength and modulus vary linearly with the foam density as is observed for epoxy syntactic foams [24], polyetherimide (PEI) syntactic foams [25], and vinyl ester syntactic foams [26]. Table 9.2 shows typical compressive strength and modulus values of epoxy [24] and epoxy-modified cyanate ester syntactic foams [23] of varying densities. The effect of different post-cure schedules on epoxy-modified syntactic foam structural properties has also been reported [23].

Comparison of dynamic compressive mechanical properties of these syntactic foams with quasi-static test data reveals their strain rate dependence of the mechanical properties [27]. Functionally, graded structure of syntactic foams is capable of withstanding compression loads up to 60–75 % strain [28], and the fracture features of these foams have been studied under different loading conditions by microscopic techniques [29].

The density of the syntactic foams can be tuned by varying either the volume fraction or wall thickness of the microballoons. The second approach is more useful, as it considerably increases the strength to the weight ratio [30].

Generally, compressive strength and modulus of syntactic foams increase with a decrease in internal radius of the microspheres. The peak compressive strength and modulus measured were higher for the specimens tested in flat wise orientation compared to that in edgewise orientation [31].

Syntactic foams can be further reinforced with different fibers [32, 33]. The process modifications for such reinforced foam systems decrease the amount of voids and increase the strength [34]. SEM study revealed the effect of specimen

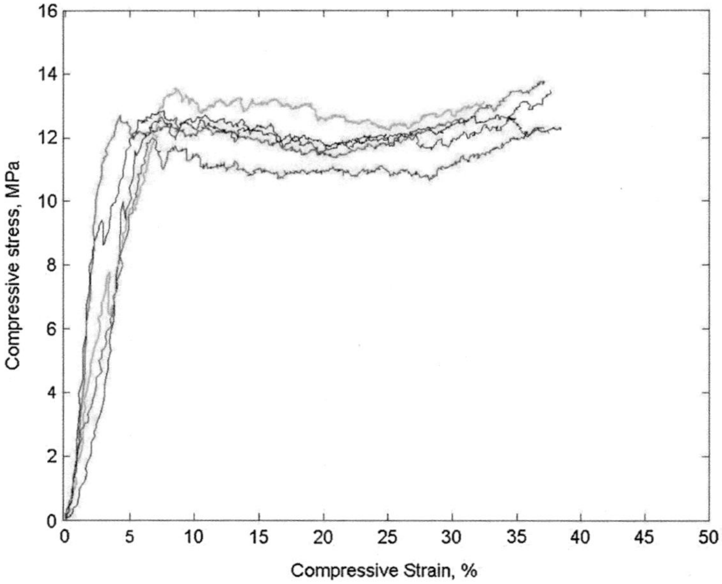

Fig. 9.4 Compressive stress versus compressive strain plots for CE-epoxy blend syntactic foam samples of density 0.39 g/cm^3 [23]

Table 9.2 Compressive properties of syntactic foams

Sample density (g/cm^3)	Epoxy syntactic foam [24]		Epoxy-modified CE syntactic foam [23]	
	Compressive strength (MPa)	Compressive modulus (GPa)	Compressive strength (MPa)	Compressive modulus (GPa)
0.20	–	–	2.99	0.13
0.23	–	–	4.00	0.19
0.27		–	6.48	0.22
0.34	–	–	9.35	0.37
0.39	–	–	12.26	0.53
0.85	67.90	1.97	–	–
0.88	87.90	2.28	–	–
0.89	102.2	2.33	–	–
0.93	105.2	2.51	–	–

aspect ratio on the compressive stress–strain curve and failure behavior of the syntactic foam materials [35]. In fact, in situ X-ray microtomography study was carried out during compressive testing of syntactic foams to understand the failure phenomenon [36].

For chopped glass fiber-filled syntactic foams of different compositions, although compressive modulus values mostly increase with the content of fibers in the system, there are contradicting literature indicating almost no effect of the same

[37–39]. Effects of water absorption on the compressive properties of the syntactic foams reinforced with chopped glass fibers have been studied by immersing them in various aqueous media [40]. Partial intercalation of surface-modified nanoclay particles in these foams leads to an increase in the toughness of the material by 80–200 % [41].

Apart from compressive properties, flexural properties of the reinforced syntactic foams increase with increasing percentage of short glass fibers [42–44]. Short-beam properties by the three-point bend mode also increase with increasing fiber fraction and decreasing microballoon fraction for the glass microballoon-filled epoxy foams. Such mechanical data could also be satisfactorily correlated with fractographic features by employing scanning electron microscope and studying the failure features of the tensile, compressive, and shear-dominated regions of the tested specimens. Short-beam test results on syntactic foams developed from paraffin oil-coated microballoons effectively bring out the effects of interface interactions and their role in influencing the mechanical properties of the foams [45–47].

In the case of two-phase syntactic foam, a systematic study of the mechanical properties and fractography of different regions of cast foam block reveals a clear flotation/settling of constituents leading to distinct vertical segregation [48].

Further, types of microspheres and volume fraction of microspheres used affect not only compressive properties but also tensile, flexural, and fracture behavior of syntactic foams [49–52]. The mechanical properties in flexural mode and fracture toughness are even affected by loading rate and moisture content in the foams [53].

When rubber latex-coated microballoons are reinforced with nanoclay and milled glass fiber in epoxy matrix, the resulting hybrid syntactic foams were found to have improved damage tolerance properties [54]. A detailed analytical and experimental study on rubber-modified epoxy syntactic foams and nanoclay-modified syntactic foams for their mechanical properties was carried out by Maharsia [55]. Figures 9.5 and 9.6, respectively, show the comparison of compressive strength and toughness of various foams (SF22, SF32, SF38, and SF46) reinforced with different percentages of nanoclay particles with respect to the unreinforced foams developed using epoxy binder and different commercially available grades of microballoons (M22, M32, M38, and M46). It can be noted that even though the addition of nanoclay decreases the strength, it has improved the toughness of the material significantly [55]. High strain rate (450–1000/s) characterization using split Hopkinson pressure bar (SHPB) apparatus has also been conducted for similar rubber-modified hybrid syntactic foams as well as their conventional counterparts [56, 57]. High strain rate mechanical properties of nanoclay syntactic foams with volume percentages of 0, 1, 2, and 5 of nanoclay have also been reported [58]. It is found that 1 % inclusion results in an optimum enhancement of the peak stress and modulus and the high strain rate testing exhibited higher peak stress and modulus. Enhancement in the flexural and dynamic mechanical properties on incorporating 2 % by weight of nanoclay in epoxy syntactic foams has also been observed [59]. Flexural strength improved by 42 % and modulus by 18 %, while in dynamic mechanical analysis (DMA), the storage modulus increased by 30 % and the loss modulus by 28 %.

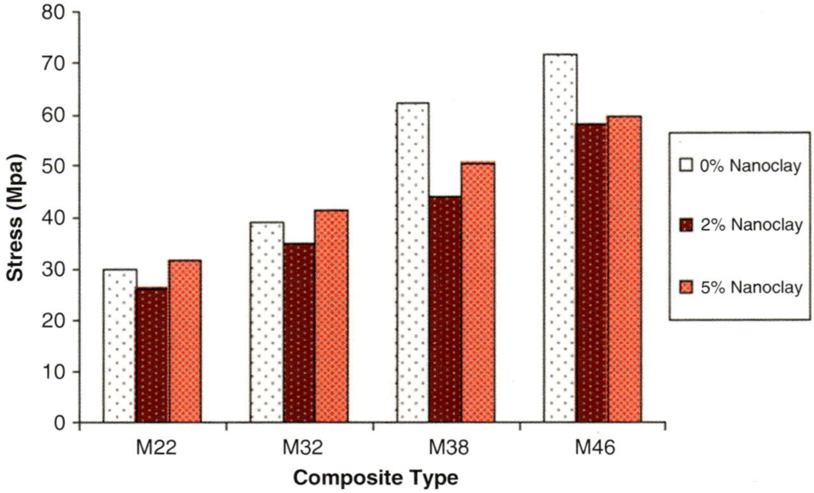

Fig. 9.5 Comparison of compressive strength of foams with different types of microballoons and nanoclay volume fractions [55]

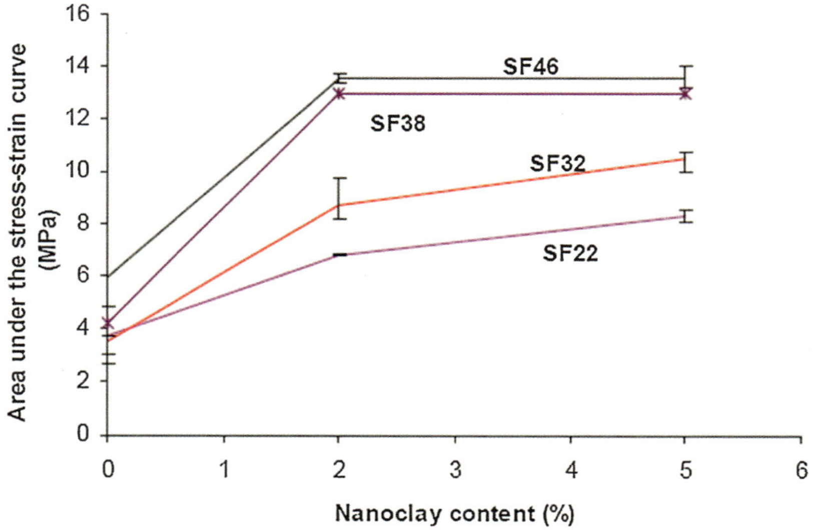

Fig. 9.6 Effect of nanoclay content on the toughness of syntactic foams [55]

Syntactic Foam Core Structural Sandwich Composites

Syntactic foams explained in the previous section are invariably used as lightweight core materials for sandwich constructions rather than as stand-alone structures [60]. It has been found that the microballoon radius ratio plays a very important role on the flexural and short-beam shear strength of the sandwich construction [61]. In fact, the aspect ratio of the sandwich test specimens also significantly affects the above properties [62]. Various mechanical properties of such syntactic foam core sandwich composites with different skin and binder materials have been exhaustively reported [63–65]. Even the core to skin thickness ratios can also be suitably optimized to design such a sandwich material with desired flexural modulus and bending rigidity [66].

Apart from varying the radius ratio of microballoons and densities of the foam core, the bending behavior of the sandwich composites has been improved by suitable material and structure modifications in the syntactic foam core. Sandwich composite panels with woven glass/epoxy facings and liquid rubber-modified epoxy syntactic foam core exhibited a reduction both in the failure energy and failure displacement but an enhanced residual energy and post failure displacement in the three-point bend tests, thereby indicating an improvement in the damage-tolerant properties in comparison to the unmodified epoxy syntactic foam core [67, 68]. Nanoclay modification of the foam core is found to result in improving the load-bearing capacity of the sandwich in the three-point and four-point bend tests, while a degradation is observed in the short-beam shear tests [69]. Maintaining a higher percentage of microspheres at the center and a lower percentage in the outer layers of a multilayered-syntactic-foam core sandwich composite resulted in higher stiffness under flexural loading in comparison to the homogenous percentage of microsphere in the core [70].

Syntactic Foams for Temperature Withstandability

Apart from achieving syntactic foams with desired structural properties, they also can be tailored for withstanding different desired temperature ranges as well. The temperature withstandability of thermosetting syntactic foams is predominantly controlled by the type of matrix resin used. Even though there are different thermal techniques available to characterize the polymeric materials, dynamic mechanical analysis (DMA) is one of the important techniques, which is extremely sensitive and reveals a wealth of information about the materials. Viscoelastic properties like storage modulus (measure of stored energy when the material is subjected to oscillatory loading), loss modulus (measure of energy lost in the material under oscillatory loading), tanδ (damping capacity of the material) as functions of temperature or frequency, and hence the glass transition temperature can be determined by this technique. DMA studies have been exhaustively carried out to characterize syntactic foams developed using different thermosetting resins. Studies on epoxies have gained major importance in recent past, since they are the workhorse of the aerospace industry.

Temperature Withstandability of Epoxy Syntactic Foams

Depending on the type of epoxy binder resin system employed for fabrication and cure schedule adopted, the glass transition temperature of the epoxy syntactic foam

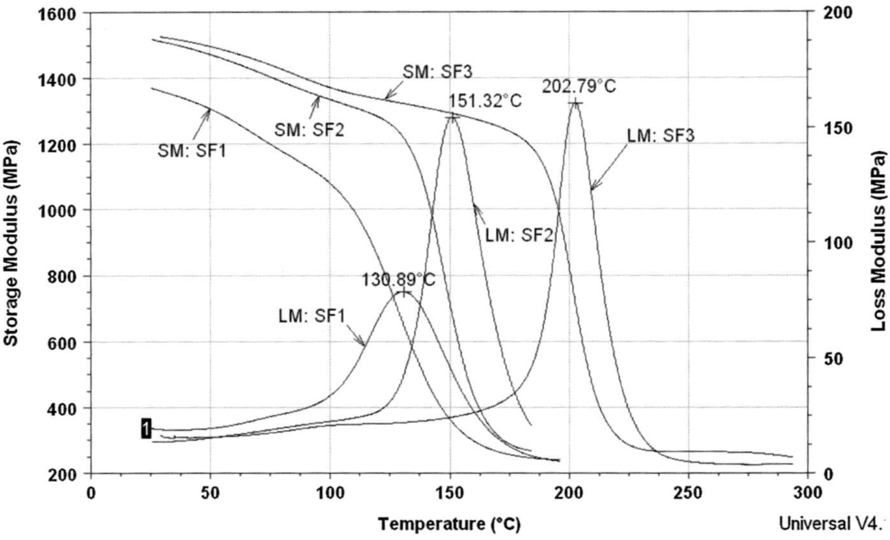

Fig. 9.7 DMA thermograms of different epoxy syntactic foams [71]

materials can vary from 80 °C to 200 °C. The higher T_g versions of epoxy syntactic foams can safely be used up to a maximum temperature of 175 °C.

Figure 9.7 shows a typical DMA thermogram showing the variation of SM and LM with temperature for various syntactic foams of similar density developed using different epoxy resin system formulations. It can be seen that DGEBA-based epoxy foams SF2 and SF3 cured with different aromatic diamine hardeners, respectively, had higher T_g compared to novolac (SF1 developed using cycloaliphatic amine hardener)-based epoxy foam. Again, among the DGEBA-based syntactic foams, SF3 had higher T_g (202 °C) compared to that of SF2 (151 °C) [71].

The tensile and compressive properties of these syntactic foam materials can also be tailored for dual functionality so that they can withstand both continuous operating temperature up to 150 °C and hydrostatic pressure up to 300 bar [72].

Temperature Withstandability of Cyanate Ester Syntactic Foams

Epoxy matrices however cannot be used for certain applications which demand very high temperature withstandability. High-performance thermosets like polyimides, cyanate ester (CE), etc. are employed for this purpose.

Table 9.3 compares the DMA results of DGEBA-based epoxy resin, bisphenol E-based CE resins, and their syntactic foams subjected to various post-cure schedules [73]. It is found that the glass transition temperature of epoxy resin cast and its syntactic foam is lower than that of their CE counterparts. Another interesting phenomenon to note is that the T_g values for epoxy resin cast and syntactic foam are only marginally different, whereas it is not so for CE system subjected to post-cure schedule P1. This indicates that the cure behavior of CE resin matrix in bulk and in composites is different, justifying the need for further careful studies of CE

Table 9.3 DMA details of epoxy and cyanate ester syntactic foams [73]

System	Sample		Cure schedule	T_g* (°C)
Epoxy	Resin cast		120 °C/2.5 h + 160 °C/3 h	169
	Syntactic foam		120 °C/2.5 h + 160 °C/3 h	175
Cyanate ester	Resin cast	P1	177 °C/6 h + 230 °C/2 h	278
	Syntactic foam	P1	177 °C/6 h + 230 °C/2 h	181
		P2	177 °C/6 h + 230 °C/4 h	200
		P3	177 °C/6 h + 260 °C/2 h	230
		P4	177 °C/6 h + 260 °C/4 h	331
		P5	177 °C/6 h + 280 °C/2 h	285

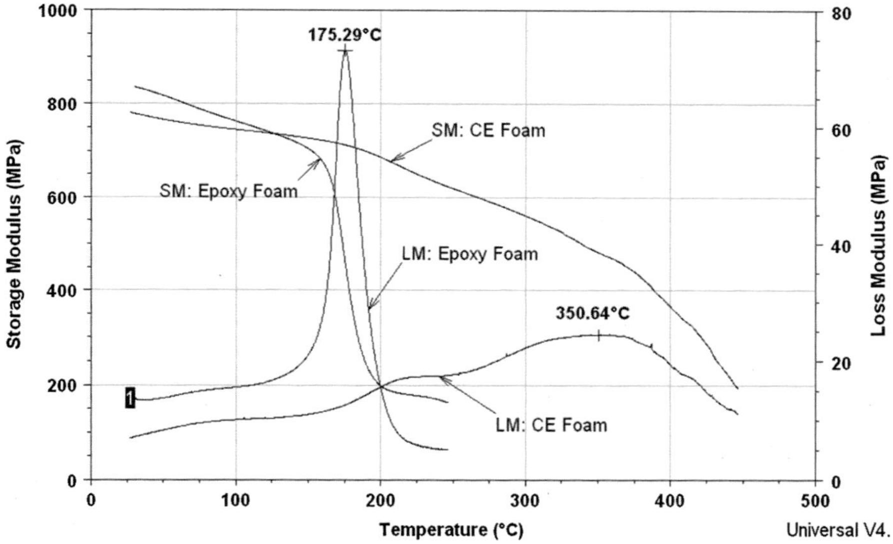

Fig. 9.8 DMA thermogram of epoxy and cyanate ester syntactic foams [74]

composite processing techniques. Based on observation of real-time curing of the foam during DMA run (above the glass transition temperature), the foam samples were subjected to other post-cure schedules (P2, P3, P4, and P5) and characterized further using DMA. This exercise yielded a higher T_g (>250 °C) foam material, which otherwise would not have been achieved using any of the epoxy resin systems. Thus, a careful study on cure and post-cure schedules of these CE resins and their syntactic foams allows them to be tailored for desired temperature withstandability.

For still higher temperature withstandability requirements, novolac-based CE system can be useful [74]. Figure 9.8 shows the results from DMA of syntactic foams developed using epoxy resin and CE systems. From the curve, it can be seen that with change in the resin system, the glass transition of the material increased from 175 °C for DGEBA-based epoxy resin system to 350 °C for novolac-based cyanate ester resin system, though there is a reduction in the modulus values to an extent of 35 %. This clearly illustrates the matrix dependence of the temperature withstandability of these foams.

CE-Epoxy Blend Syntactic Foams

It is to be noted that though CE system gives a higher T_g compared to epoxies, the inherent brittle nature of CE resins limits the material to be used for variety of applications. One way of toughening or reducing the brittleness of these systems is to blend suitably with epoxies leading to an optimal compromise on the glass transition temperature. Incidentally, blending with epoxy resin makes the CE resins easily processable. Cure behavior study of commercially available epoxy (diglycidyl ether of bisphenol A, DGEBA-based)-modified cyanate ester (1,1-bis(4-cyanato phenyl) ethane) and its syntactic foams using hollow glass microspheres at two different post-cure procedures reveals that the comparison of glass transition behavior and T_g values of the neat resin cast at two different post-cure procedures with those corresponding to the foams clearly indicated that there exists no correlation between the cure behavior of the neat resin to that of the resin binder in the foams [23, 75], possibly due to chemical reactions in the interphase. It has also been observed that the hollow glass spheres catalyze the polycyclotrimerization reaction of cyanate ester resin, which helps in lowering the processing temperatures of such syntactic foam systems [76].

Significant high-temperature-withstanding (>500 °C.) syntactic foams based on cyanate ester, phthalonitrile, and cross-linked acetylene systems form the theme of a patent disclosure [77].

Syntactic Foams for Microwave Transparency

Syntactic foams are also used in applications demanding transparency to microwaves. Interaction of electromagnetic waves with a material leads to three phenomena, viz., transmission, reflection, and absorption. For a material to be microwave transparent, obviously majority of the energy should transmit through it. On the other hand, the other two phenomena, viz., reflection and absorption, are important for electromagnetic shielding, which are discussed in subsequent sections. This section addresses the tailorability of syntactic foams for certain specific microwave transparent aerospace applications like broadband radomes (microwave transparent covers which protect the radar/antenna from external environment). It is extremely critical for a radome material to be transparent to the microwaves so that it does not hinder the communication function of the radar/antenna housed in it.

Microwave Transparency and Dielectric Properties

The microwave transparency of a product is predominantly governed by its dielectric properties, viz., dielectric constant and loss tangent of its material. Dielectric constant, which is numerically equal to the measure of stored part of electromagnetic

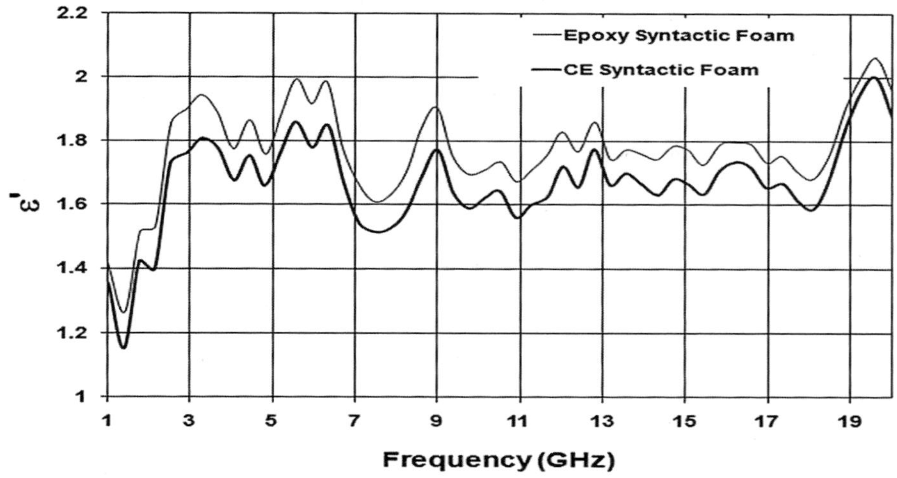

Fig. 9.9 Variation of dielectric constant with frequency for epoxy and CE syntactic foams by coaxial probe method [74]

energy incident on the material (ε′), can be defined as the ratio of the capacitance of a capacitor containing the dielectric medium to the capacitance when the medium is removed. Loss tangent (tanδ) is a measure of stored electromagnetic energy converted to leakage current and ultimately to heat. A microwave transparent material requires both ε′ and tanδ to be as low as possible, ideally matched with those of air/free space.

Effect of Constituents of Syntactic Foams on Dielectric Properties

As discussed, syntactic foam made from hollow glass microspheres and resin matrix is extensively used as core materials for microwave transparent applications. The hollow structure of glass microballoons in these syntactic foams results in low dielectric values that facilitate the function of microwave transparency. Further, these properties can be optimized by varying the density of the material, through volume fraction and/or the radius ratio of microballoons. The lesser the density of the foam, the more hollow the structure, and the more transparent it is to the microwaves [78]. To some extent, the dielectric properties of the foam can be controlled by the type of matrix resin as well.

Dielectric Properties of Syntactic Foams

Figure 9.9 shows the variation of dielectric constant with frequency measured by coaxial probe method for typical epoxy and cyanate ester (CE) syntactic foams of density around 0.33 g/cm^3, scanned over a frequency range of 1–20 GHz [74]. Table 9.4 shows the values of ε′ and tanδ for these foams at selected frequencies. Once again, it can be noted that ε′ value for both thermosetting foams was around 1.5, which is close to that of air (1.0). This could be attributed to the hollow structure of these foams, which makes them suitable for microwave transparent

Table 9.4 ε′ and tanδ values for epoxy and CE syntactic foams at selected frequencies by coaxial probe method [74]

	Epoxy syntactic foam		CE syntactic foam	
Frequency (GHz)	ε′	tanδ	ε′	tanδ
1	1.42	–	1.36	–
5.18	1.88	0.01494	1.76	0.00886
9.36	1.75	0.07611	1.64	0.07130
13.16	1.75	0.01094	1.66	0.01214
16.96	1.73	0.00652	1.66	0.01697

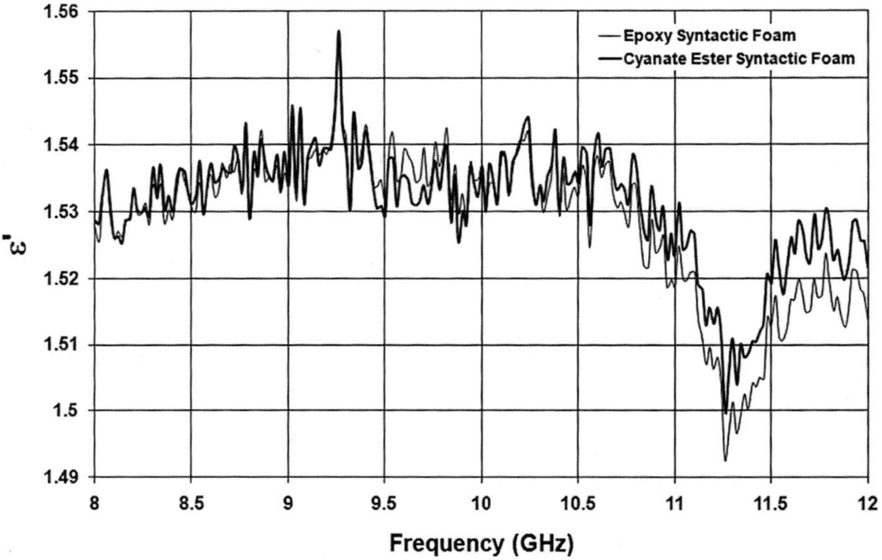

Fig. 9.10 Variation of dielectric constant with frequency for epoxy and CE syntactic foams by waveguide method: X band [74]

applications. The tanδ measurements of these foams suffer from experimental inaccuracies.

Figures 9.10 and 9.11, respectively, show the variation of dielectric constant and tanδ with frequency (measured by waveguide technique [69]) in X band (8–12 GHz) for typical epoxy and cyanate ester syntactic foams of density around 0.33 g/cm^3 [74]. Table 9.5 shows values of ε′ and tanδ for these foams at selected frequencies. Once again, it can be noted that ε′ value for both thermosetting foams was around 1.5. On comparing Figs. 9.9 and 9.10, in the frequency range 8–12 GHz, it can be understood that the waveguide technique had more consistent, repeatable, and hence reliable results. Tanδ values for both the foams also had comparable values within the characterization limits (Table 9.5).

Dielectric constant and loss for these foams generally decrease with increasing frequency and with decreasing temperature in the range of 40–140 °C [26]. Syntactic

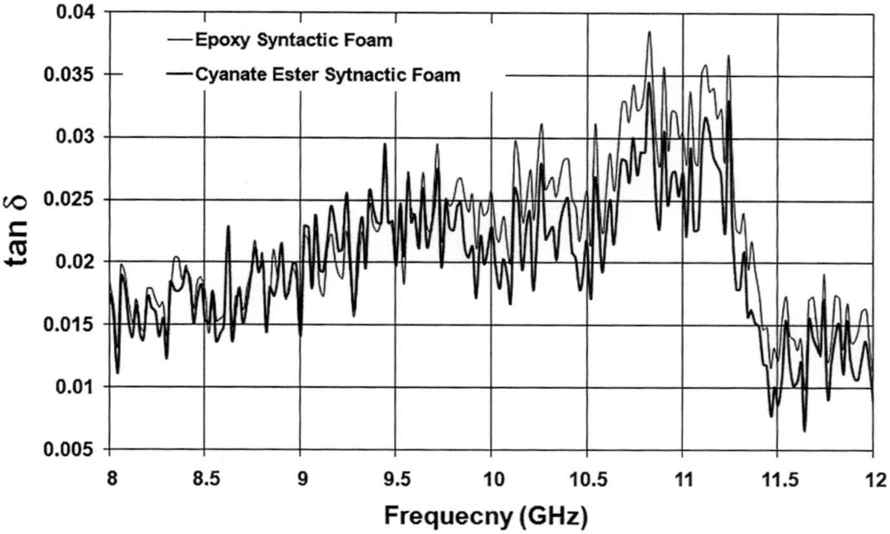

Fig. 9.11 Variation of loss tangent with frequency for epoxy and CE syntactic foams by waveguide method: X band [74]

Table 9.5 ε' and tanδ values for epoxy and CE syntactic foams at selected frequencies by waveguide method: X band [74]

	Epoxy syntactic foam		CE syntactic foam	
Frequency (GHz)	ε'	tanδ	ε'	tanδ
8	1.53	0.01841	1.53	0.01737
9	1.53	0.01449	1.53	0.01420
10	1.53	0.02572	1.54	0.02290
11	1.52	0.03047	1.52	0.02714
12	1.51	0.01111	1.52	0.00862

foams in the density range of 0.19–0.32 g/cm^3, with dielectric constant of around 1.3 and loss tangent of around 0.005 developed by M/s Cuming Microwave Corporation, USA, are available off the shelf in the market and can be readily used for microwave transparent applications [79]. Epoxy and cyanate ester syntactic foams with a wide range of applications and dielectric constant in the range 1.25–1.45 for foam densities 0.20–0.3 g/cm^3 are developed by M/s UDC, USA, as well [80]. M/s Tencate, USA, also supplies cyanate ester syntactic foams with trade names EX-1541-1 and BC −550 with a dielectric constant of around 2.0 [81]. M/s CRG, USA, has developed syntactic foam with the trade name Advantic, which has a density close to that of balsa wood, but has a significantly higher strength. Depending on the resin system used, foamed syntactic composite also has the potential for an improved transition temperature and a low dielectric constant, making it ideal for space use [82].

Table 9.6 Dielectric constant values (1 MHz) for syntactic foams of different densities [24]

Specimen type	Experimental density (g/cm^3)	Average experimental dielectric constant ε′
VE 220-30	0.84	3.36
VE 220-40	0.77	3.05
VE 220-50	0.68	3.04
VE 220-60	0.57	2.84

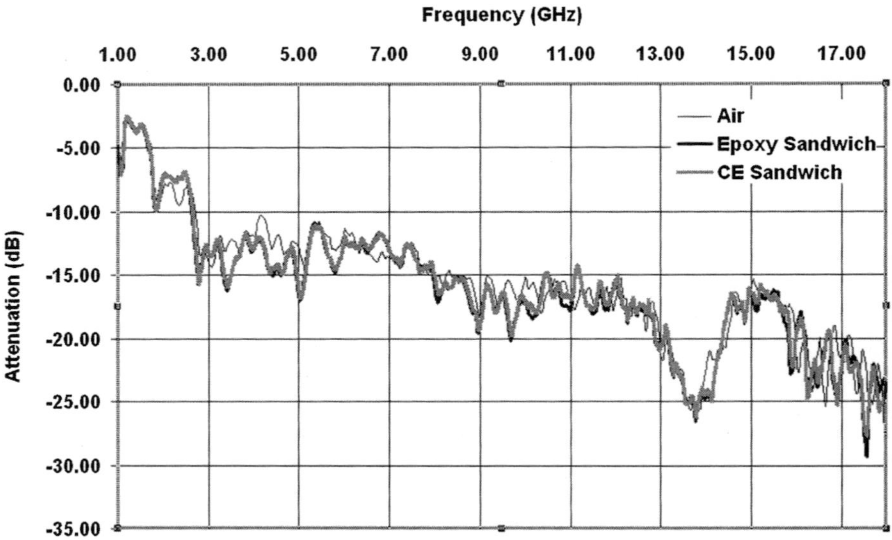

Fig. 9.12 Microwave attenuation or dB loss for epoxy and cyanate ester sandwich panels [74]

Apart from measuring the dielectric constant values in GHz region, similar measurements in MHz region have been reported with an anticipated utilization of the foams in integrated circuit design [83]. Table 9.6 gives the relationship between density of syntactic foam (0.57–0.84 g/cm^3) and dielectric constant values, measured at 1 MHz frequency. The inventors found a decrease in the dielectric constant value with increasing frequency in the measured range.

Syntactic Foam Sandwiches for Microwave Transparency

Generally, the syntactic foams mentioned above will be used as core materials in sandwich composites for the development of microwave transparent structures. The microwave transparency of such sandwiches can be evaluated by using a typical horn antenna setup in an anechoic chamber and measuring the insertion loss in the desired frequency range [74]. Figure 9.12 is a typical result showing the variation of microwave attenuation (in dB) for air (i.e., without any sample between transmitting and receiving horn antennae), epoxy, and cyanate ester syntactic foam sandwich panels over the frequency band 1–18 GHz. The results indicate a very good microwave transparency of the syntactic foam core sandwich panels.

A patented process for a multilayered radome, using cyanate ester resin matrix with quartz skin and syntactic foam core, is disclosed [84]. Syntactic foam core sandwich radome can also be developed using epoxy, bismaleimide and cyanate ester resin blend and glass/quartz/polyolefin reinforcements [85, 86]. Materials and process details regarding the development of foam sandwich radome are very well documented [87, 88].

Utilization of syntactic foams for a multiband, broadband, high-angle sandwich radome structure [89] and a B-sandwich radome structure that exhibits transparency over a wide variety of frequencies and incident angles with increased thermal insulation [90] is disclosed in patent literature.

Syntactic Foams for Electromagnetic Interference (EMI) Shielding/ Electromagnetic Compatibility (EMC)

As discussed in the preceding sections, even though the behavior of syntactic foams had been vastly studied by different researchers, the applications of these foams had been primarily restricted to the domain of electrical insulation and microwave transparency only. However, in today's age of miniaturization, when almost all electronic equipment designs are becoming extremely weight critical, lightweight polymer composites and foams are emerging as viable alternatives for conventional metals in various forms of electronic packaging. In contrast to metals, these composites are transparent to microwaves and hence do not offer any intrinsic electromagnetic interference (EMI) shielding/electromagnetic compatibility (EMC) to the system. This scenario has opened up an entirely novel field of research activities concentrating on enhancing the EMI/EMC capabilities of these materials by rendering them electrically conducting. The following sections deliberate upon the theories, principles, and practices pertaining to such unconventional composite materials in general and syntactic foams in particular.

EMI shielding effectiveness (SE) of a material may be physically defined as an index of its opacity to an incident electromagnetic wave at any given frequency band. Such opacity may be caused by reflection or absorption of the incident wave from the interface or within the bulk material. From fundamental electromagnetic wave equations (discussed in later sections), it can be easily understood that electrically conducting materials are intrinsically opaque to EM waves by virtue of both improved reflection and absorption mechanisms. It is precisely for this reason that the polymeric composites in general begin with a major handicap when it comes to EMI shielding applications and hence require their properties to be significantly tailored for this specific purpose.

One of the ways of tailoring the conventionally insulating syntactic foams for EMI shielding characteristics is to use carbon microballoons in various proportions along with the glass microballoons [91]. Another way of achieving the same is to incorporate judicious proportions of electrically conductive fibers or fillers within the material system. The possibilities of such fiber-reinforced syntactic foams have also been reported, but not all of them explore the electrical properties of such

Table 9.7 Composition of NSF and CSF [94]

Material description	Density (g/cm^3)	Volume percentages			
		Matrix binder	Microballoon	Chopped fiber	Void
NSF	0.65 ± 0.01	44.4	51.1	–	4.5
CSF	0.65 ± 0.01	42.7	48.9	2.8	5.6

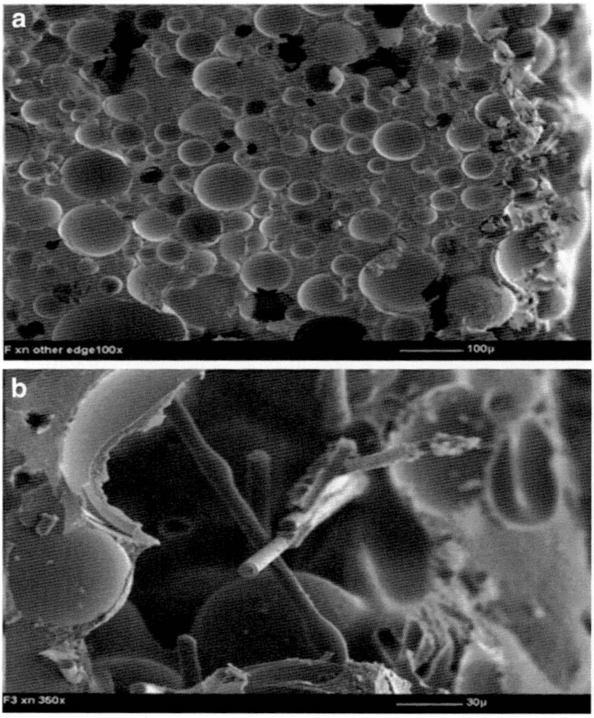

Fig. 9.13 SEM images of NSF and CSF [94]

materials for the applications being discussed [92–94]. The incorporation of carbon fibers not only imparts required EMI shielding properties to the material; they also improve the structural properties of the foam. A typical composition with volume fraction of the constituents of such carbon fiber-reinforced foam (CSF) vis-à-vis its unreinforced counterpart (NSF) is shown in Table 9.7.

It may be interesting to note from the typical SEM images (Fig. 9.13a, b) of such foams that not only the glass microballoons are homogeneously distributed in the neat foam but the chopped carbon fibers are also randomly oriented in all possible directions in the fiber-filled foam, leading to a near isotropic nature of these materials [94].

Table 9.8 Resistivity values of NSF and CSF [94]

Sample	Surface resistivity (ohm/square)	Volume resistivity (ohm-cm)
NSF	4.5×10^9	1.1×10^9
CSF	7.7×10^1	6.6×10^1

Measurement of Electrical Resistivity of Syntactic Foams

In terms of electrical resistivity values, the chopped carbon fiber-reinforced syntactic foams may be classified as semiconductive in nature. The electrical resistivity measurements of such semiconducting composite samples are extremely tricky in terms of establishing reliable ohmic contacts between the materials and the measuring probes and hence call for some special steps to be taken during the process [95]. Table 9.8 shows the surface and volume resistivities of neat and carbon fiber-filled syntactic foams (NSF and CSF) from which it is obvious that the resistivity values of fiber-filled foam were many orders lower than its neat counterpart [94].

Estimation and Evaluation of Shielding Effectiveness (SE)

The EMI shielding capability of a material is measured by a parameter known as "shielding effectiveness." Quantitatively, SE is basically a ratio of two field quantities and can be derived from the field theory approach using mathematical modeling techniques. Hence, given the electrical conductivity of the material, it is possible to predict SE of any structure theoretically [93].

Theoretical Estimation

Given below is an expression for estimating the shielding efficiency of such semiconductive materials [96, 97]. According to this, the reflection loss (R_{dB}) and absorption loss (A_{dB}) for a particular homogenous material can be estimated using the following equations:

$$R_{dB} = 168 + 10\ \log[(\sigma/\sigma_{Cu}).(1/\mu_r).(1/f)] \tag{9.1}$$

$$A_{dB} = 131.4d\ [(\sigma/\sigma_{Cu}).\mu_r.f]^{1/2} \tag{9.2}$$

where σ is volume conductivity of the material (S/m), σ_{Cu} is volume conductivity of copper (5.7×10^7 S/m), μ_r is relative magnetic permeability of the material, f is frequency of the incident EM wave (Hz), and d is thickness of the material coupon (m).

The overall electromagnetic shielding provided by any material configuration can be attributed to three different mechanisms, namely, reflection at the interface, absorption along the thickness, and attenuation and scattering caused by multiple internal reflections. In other words, an ideal EMI shielding material must either reflect or absorb any incident electromagnetic wave, thereby transmitting nothing through itself. Thus, the overall shielding effectiveness is calculated as

$$SE_{dB} = R_{dB} + A_{dB} + B_{dB} \tag{9.3}$$

Table 9.9 Theoretical estimates of SE for CSF [93]

Frequency	A_{dB}	R_{dB}	SE_{dB}
100 kHz	0.01	40.44	40.45
1 MHz	0.03	30.44	30.47
10 MHz	0.11	20.44	20.55
100 MHz	0.35	10.44	10.79
1 GHz	1.10	00.44	01.54

Table 9.10 Shielding effectiveness of CSF [94]

	Shielding effectiveness (dB)				
Material description	100 kHz	1 MHz	10 MHz	100 MHz	1 GHz
CSF	40.30	40.00	20.78	11.00	11.00

where B_{dB} (attenuation coefficient arising out of multiple internal reflections) is mostly assumed to be negligible. Hence, we simplify

$$SE_{dB} = R_{dB} + A_{dB} \tag{9.4}$$

The above equations offer an extremely useful insight toward understanding the behavior of any homogenous material in response to electromagnetic waves. It is evident that an increase in the electrical conductivity improves both reflection and absorption and hence the overall electromagnetic shielding effectiveness of the material. Table 9.9 gives the numerical data for the above quantities for the carbon fiber-filled syntactic foam of 2.5 mm thickness [93].

Experimental Evaluation of SE

Experimentally, the SE values of a test coupon can be measured using the coaxial jig method (as per ASTM 4935 – D). The measured SE values at 100 KHz, 1 MHz, 10 MHz, 100 MHz, and 1 GHz frequencies for the above carbon fiber-filled foam are presented in Table 9.10 [94].

In Fig. 9.14, it may be observed that the predicted values of SE for the same samples (Table 9.9) were more or less in agreement with the experimental values at most of the frequencies [94].

Effects of Conductive Reinforcements on Other Properties of EMI Shielding Syntactic Foams

Thermomechanical Properties

Figure 9.15 shows the overlay of DMA (single cantilever mode) thermograms for NSF and CSF, which shows that the storage modulus (SM) of the reinforced foam increased by about 23 % compared to its unreinforced counterpart. Glass transition temperature (T_g) of the CSF also recorded a very marginal increase which may be attributed to the slower response of the fiber-filled foam to stress displacement under the dynamic loading conditions [94].

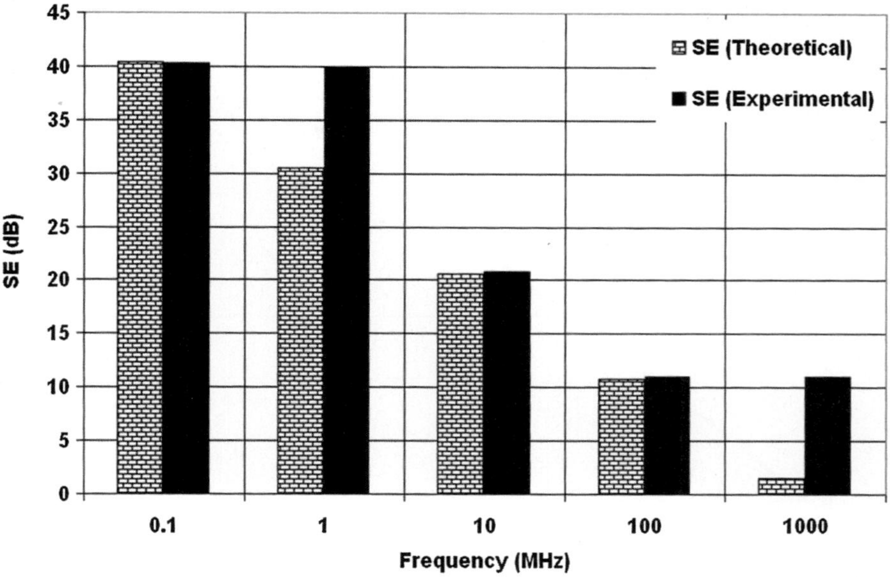

Fig. 9.14 Theoretical and experimental approach of SE measurements: a comparison [94]

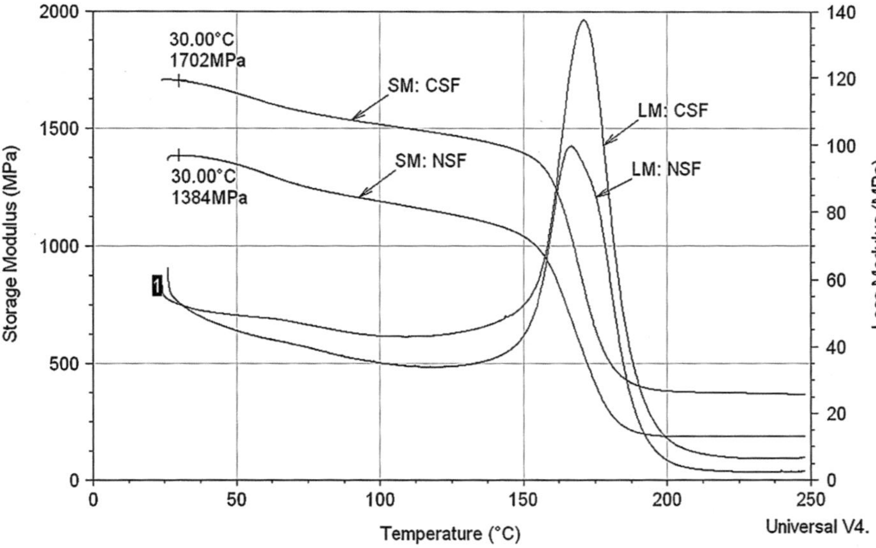

Fig. 9.15 DMA thermograms of NSF and CSF [94]

Flexural Properties

Figure 9.16 shows typical stress versus strain curves for NSF and CSF samples subjected to flexural loading under quasi-static conditions, whereas the flexural properties of the same are summarized in Table 9.11 [94].

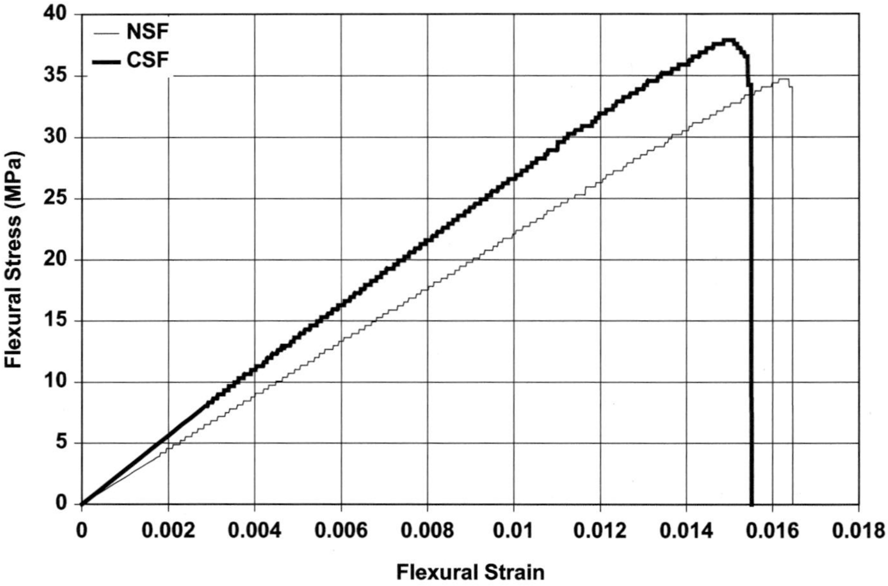

Fig. 9.16 Typical flexural stress versus flexural strain curves for NSF and CSF [94]

Table 9.11 Flexural properties of NSF and CSF [94]

Sample ID	Flexural properties	
	Strength (MPa)	Modulus (GPa)
NSF	31.05	2.22
CSF	38.66	2.74

From Table 9.11, it is evident that mean flexural strength and modulus of CSF samples were higher respectively by 24.5 % and 23.4 % than that of NSF samples which clearly indicate the mechanical reinforcing abilities of the chopped carbon fibers. Syntactic foams incorporating vapor-grown carbon fibers on carbon spheroids of different sizes have been disclosed to improve the flexural properties without compromising compressive strength [98]. A considerable enhancement in the mechanical properties, viz., compressive, flexural, and tensile strength for foams in the density range of 0.56–0.91 g/cm^3, was reported, wherein the foams were prepared from couplant-treated microspheres and carbon fibers [99]. The T_g showed a reduction due to the plasticizing effect of the coupling treatment which on the other hand enhanced the damping properties of the foam.

Syntactic Foam Core Sandwich Materials for EMI Shielding

The conductive syntactic foam described above can be used as lightweight core material in subsequently designed sandwich composites for EMI shielding applications [100–102]. While Table 9.12 details the configurations of two such sandwich

Table 9.12 Syntactic foam core sandwich configurations for EMI shielding applications [100]

ID	Description	t (mm)	ρ (g/cm^3)
SW-01	One layer Alumesh/epoxy + 1 layer glass/epoxy + insulating syntactic foam + 1 layer glass/epoxy + 1 layer Alumesh/epoxy	2.75	0.77[a]
SW-02	One layer Alumesh/epoxy + 1 layer glass/epoxy + conductive syntactic foam + 1 layer glass/epoxy + 1 layer Alumesh/epoxy	2.81	0.84[a]

t thickness, *ρ* density
[a]Effective densities obtained for unit volume of the sandwich

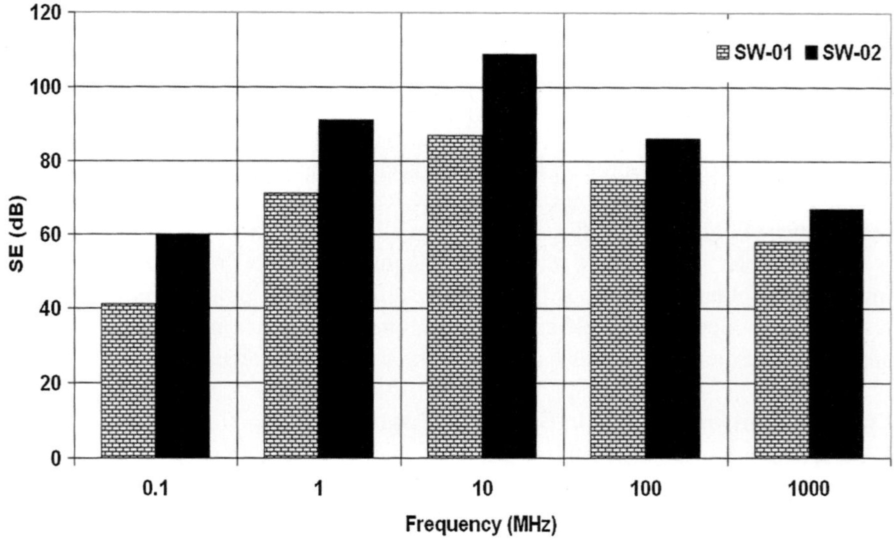

Fig. 9.17 Shielding efficiency of sandwich composites [100]

designs wherein the only variation was in the nature of the syntactic foam core, Fig. 9.17 offers the SE data of the same [100]. The ambient temperature SM value of the fiber-reinforced foam (3.19 GPa) also increases marginally compared to that of its unreinforced version (3.09 GPa).

All these results are indicative of the fact that the use of chopped carbon fiber-reinforced foam instead of its unreinforced version not only enhances the EMI shielding capabilities of the sandwich materials but also marginally improves the mechanical stiffness of the same.

Nanocomposite Syntactic Foams with Multifunctionalities

Speaking about multifunctional reinforcements in composite materials, one of the most promising exponents is believed to be the carbon nanotubes (CNTs). The CNTs are essentially long graphene sheets wrapped in different angles (chiral angles) and

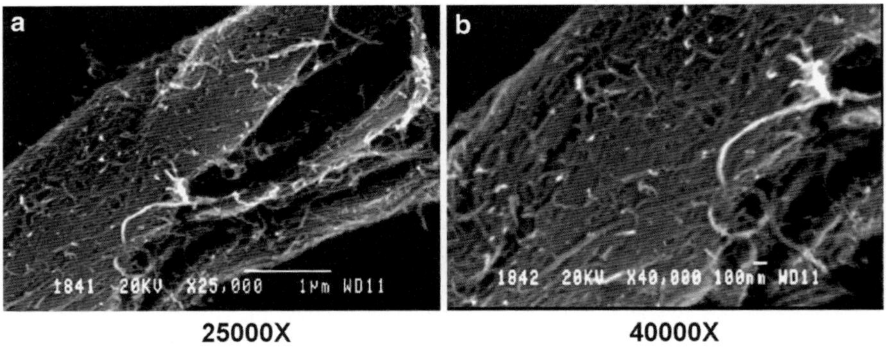

Fig. 9.18 SEM images of MWNT bundles at two magnifications [106]. (**a**) 25,000X (**b**) 40,000X

with different circumferential lengths (chiral vector), depending upon which they are able to offer a wide range of mechanical, electrical, and thermal properties. Based on the number of concentric shells, these carbon allotropes may be classified as single wall nanotubes (SWNT) or multiwall nanotubes (MWNT). Among its ever-increasing popularity in various fields of scientific research including particle physics, molecular electronics, field emitters, etc., one of the major areas of research and application of these nanotubes has been in the field of reinforced plastics.

CNTs as Reinforcements in Syntactic Foams

Very few mentions of CNT-filled syntactic foams can be found in the literature. Studying SWNT and MWNT as reinforcements in syntactic foams has shown a marginal improvement in ambient temperature stiffness, even at very high CNT loading (10–15 % by wt.). This may primarily be attributed to insufficient de-agglomeration of the CNT bundles in the composites, leading to less than optimal load transfer between the matrix and the reinforcements [103, 104]. Nanocomposite foams with various percentages of CNT loading in DGEBA-based epoxy matrix by using amine-functionalized MWNTs and employing high-speed shear homogenization in the binder resin have been found to result in significant improvements in the electrical and thermomechanical properties at very low CNT loadings [105, 106]. Figure 9.18a, b shows the SEM images at two magnifications 25,000X and 40,000X, respectively, of the MWNTs used in the experiments (source: M/s Nanocyl, Belgium).

Table 9.13 describes the detailed compositions of the nanocomposite syntactic foam sample. Figure 9.19a is a typical SEM image of such CNT-reinforced syntactic foams at a magnification 400X, which shows the distribution of microballoons in the resin matrix. Figure 9.19b is the SEM image of the same sample at a magnification of 20,000X revealing the random distribution of CNTs in the resin matrix.

Table 9.13 Compositions of syntactic foams [106]

Samples	Sample ID	Weight %			Density (g/cm^3)
		Resin system	Glass microballoon	CNT	
Neat foam	SF-I	66.68	33.32	–	0.48
0.6 % CNT foam	CNTF-I	65.31	34.11	0.58	0.48
1.1 % CNT foam	CNTF-II	65.10	33.77	1.13	0.49
2.2 % CNT foam	CNTF-III	64.00	33.76	2.24	0.50

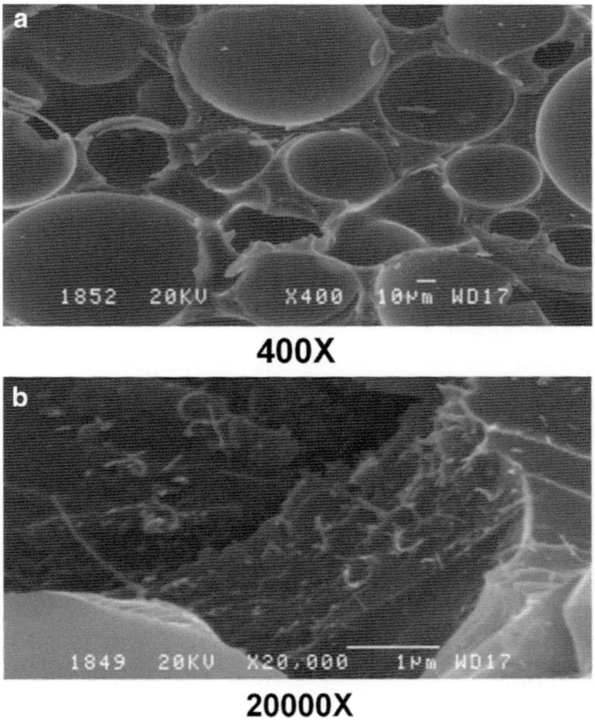

Fig. 9.19 SEM micrographs of CNT-reinforced syntactic foams at two magnifications [106]. (**a**) 400X (**b**) 20,000X

Properties of CNT-Reinforced Syntactic Foams

Electrical Resistivity

Figure 9.20 illustrates the sharp fall in the bulk resistivity of the nanocomposite foams with increasing weight percent of CNT addition. The percolation behavior is very evident in the foam system at about 0.5–1 wt% of CNTs.

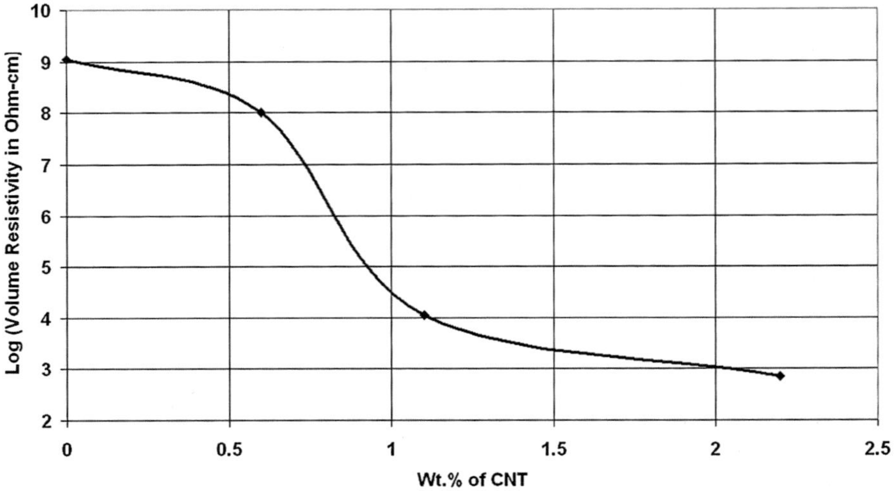

Fig. 9.20 Electrical percolation phenomenon in nanocomposite syntactic foams [106]

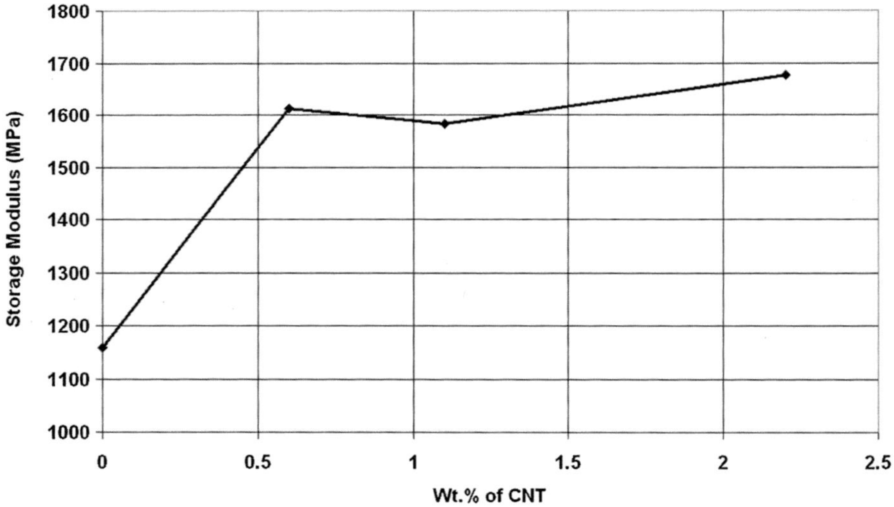

Fig. 9.21 Ambient temperature SM values of nanocomposite foams [106]

Dynamic Mechanical Analysis (DMA)

DMA study of the above foams in single cantilever mode reveals an almost 40–50 % increase in the ambient temperature SM of the nanocomposite foams, in comparison to the unreinforced version. Figure 9.21 depicts the trend in the same with respect to the weight percent of CNT additions in the nanocomposite foams.

It has been reported that preparing syntactic foam in high vacuum (0.2 KPa) from functionalized carbon nanotubes and microballoons resulted in a significant increase in compressive strength and apparent shear strength which otherwise was not observed in foams prepared under atmospheric pressure. An increase of 22 °C in the T_g was observed [107] in the vacuum processed foams. Microballoons with vertically aligned CNTs grown on their surface, when used as reinforcement in epoxy matrix syntactic foam, resulted in improving the compressive modulus by 35–41 % and damping coefficient values by 10–14 % [108]. An increase in tensile strength and modulus of syntactic foams on addition of graphene platelets has been reported [109] with the maximum strength enhancement of 15.9 % attained at a concentration of 0.1 % of graphene platelets.

Metal-Coated Hollow Microsphere-Embedded Syntactic Foams

An alternative approach toward achieving electrical conductivity and associated properties in syntactic foams has been to use metal-coated microballoons [110]. Silver-coated hollow glass microsphere-embedded novolac-epoxy-based syntactic foam (density = 490 kg/m^3) has been developed and characterized for electrical resistivity, wherein extremely low values in the order of $10^{-4}\Omega$-m have been recorded. The results were overall suggestive of the fact that such foams may find applications in design of EMI shielding sandwich composite structures.

Concluding Remarks

1. Syntactic foams are a special class of closed-cell, physical foams. The fundamentals and an overview of the current state of research and applications of thermosetting polymer-based syntactic foams have been presented.
2. The technologies involved in tailoring these foams to cater for diverse applications by optimal tuning of material chemistry and composition have also been addressed.
3. With reference to the literature available on syntactic foam materials, their composition and structural capabilities like compressive, flexural, short-beam, tensile, and impact properties of foams and their sandwiches are discussed.
4. Tuning the temperature withstandability of syntactic foams by suitable selection of binder matrix has been highlighted, with a special mention of different grades of epoxy and cyanate ester resin systems.
5. The complex dielectric properties of these foams and microwave transmission characteristics of their sandwich counterparts have been presented to explore their suitability for microwave transparent applications.
6. The techniques to transform these conventionally insulating and microwave transparent foams and their sandwich composites into electrically conducting materials for EMI shielding and related applications by judicious incorporation of conductive fibers/fillers have been detailed.

7. The different features of carbon nanotube-filled nanocomposite syntactic foams with multifunctional capabilities have also been discussed:

All the above discussions strongly consolidate the position of these syntactic foams as extremely important engineering materials for diverse multifunctional applications. With the advent of nanomaterials and technologies, greater scope of their tailorability has made these syntactic foams more promising for novel applications in the future.

References

1. Menges G, Knipschild F (1982) Stiffness and strength-rigid plastic foams. In: Hilyard NC (ed) Mechanics of cellular plastics. Applied Science Publishers, London, pp 27–72
2. Feldman D (1989) Polymer foams. In: Feldman D (ed) Polymeric building materials. Elsevier Science Publishers, New York, pp 283–355
3. Shutov FA (1981) Foamed polymers based on reactive oligomers. Adv Polym Sci 39:1–64
4. Shutov FA (1983) IUPAC 29th international symposium on macromolecules, Bucharest
5. Hylard NC, Young J (1982) Introduction. In: Hylard NC (ed) Mechanics of cellular plastics. Applied Science Publishers, London, pp 1–26
6. Luxmoore AR, Owen RJ (1982) Syntactic foam. In: Hylard NC (ed) Mechanics of cellular plastics. Applied Science Publishers, London, pp 359–391
7. Shutov FA (1986) Syntactic polymer foams. Adv Polym Sci 73(74):63–123
8. Price HJ, Nelson JB (1976) Phase relationship in three-phase composites which include a void phase. J Compos Mater 10:314
9. Shutov FA (1983) Foamed polymers. Cellular structure and properties. Adv Polym Sci 51:155–218
10. Puterman M, Narkis M, Kenig S (1980) Syntactic foams II. Preparation and characterization of three-phase systems. J Cell Plast 16:326–330
11. Narkis M, Gerchcovich M, Puterman M, Kenig S (1982) Syntactic foams III. Three-phase materials produced from resin coated microballoons. J Cel Plast 18(4):230–232. doi:10.1177/0021955X8201800402
12. Berlin AA, Shutov FA (1980) Chemistry and technology of gas-filled high polymers. Nauka, Moscow, p 503
13. Braun T, Farag AB (1978) Polyurethane foams and microspheres in analytical chemistry: improved liquid–solid, gas-solid and liquid-liquid contact via a new geometry of the solid phase. Anal Chim Acta 99(1):1–36
14. Rand PB (1973) An aromatic polyimide syntactic foam. J Cell Plast 9(3):130–133. doi:10.1177/0021955X7300900304
15. Weiser ES, Clair St. TL, Echigo Y, Kaneshiro H (2000) US Patent 6084000
16. Balyberdin GA, Orlov VA, Tarakanov OG (1974) Plast Massy 10:22–26
17. Sternfield A (1982) New types and sources of micro-spheres can help widen RP markets. Mod Plast Int 12(6):43–45
18. Cravens TE (1973) Syntactic foams utilising saran microspheres. J Cell Plast 9(6):260–267
19. Berlin AA, Shutov FA (1980) Strengthened gas-filled plastics. Khimia, Moscow
20. Mildner RC, Nacke KF, Veazey EW, Woodland PC (1970) Blocking multi-pair cable with plastic-microsphere syntactic foam. Mod Plast 47(5):98–99
21. Matthews RB, Swanson ML (1979) Fabrication of large (Th, U)O/sub 2/microspheres. Am Ceram Soc Bull 58(2):223
22. Bunn P, Mottram JT (1993) Manufacture and compression properties of syntactic foams. Composite 24(7):565–571

23. Ravi Sekhar K (2006) Studies on processing and characterisation of syntactic foams with epoxy modified cyanate ester binder and glass microspheres. Dissertation, Indian Institute of Science
24. Gupta N, Ricci W (2006) Comparison of compressive properties of layered syntactic foams having gradient in microballoon volume fraction and wall thickness. Mater Sci Eng A 427 (1–2):331–342
25. Douglas AM (1996) US Patent 5532295
26. Gupta N, Priya S, Islam R, Ricci W (2006) Characterization of mechanical and electrical properties of epoxy-glass microballoon syntactic composites. Ferroelectrics 345:1–12
27. Jadhav A (2003) High strain rate properties of polymer matrix composites. Dissertation, Louisiana State University
28. Gupta N (2007) A functionally graded syntactic foam material for high energy absorption under compression. Mater Lett 61:979–982
29. Gupta N, Woldesenbet E, Kishore (2002) Compressive fracture features of syntactic foams-microscopic examination. J Mater Sci 37(15):3199–3209
30. Gupta N, Woldesenbet E (2004) Microballoon wall thickness effects on properties of syntactic foams. J Cell Plast 40(6):461–480
31. Gupta N, Woldesenbet E, Mensah P (2004) Compression properties of syntactic foams: effect of cenosphere radius ratio and specimen aspect ratio. Compos A Appl Sci Manufact 35 (1):103–111
32. Nutt SR, Vaikhanski L (2005) US Patent 6864297
33. Nowak GP, Tegeler AF, Timmons TL (1997) US Patent 5665787
34. Gupta N, Karthikeyan CS, Sankaran S, Kishore (1999) Correlation of processing methodology to the physical and mechanical properties of syntactic foams with and without fibers. Mater Charact 43(4):271–277
35. Gupta N, Kishore WE, Sankaran S (2001) Studies on compressive failure features in syntactic foam material. J Mater Sci 36(18):4485–4491
36. Adrien J, Maire E, Gimenez N, Sauvant-Moynot V (2007) Experimental study of the compression behaviour of syntactic foams by in situ X-ray tomography. Acta Mater 55 (5):1667–1679
37. Karthikeyan CS, Sankaran S, Jagdish Kumar MN, Kishore (2001) Processing and compressive strengths of syntactic foams with and without fibrous reinforcements. J Appl Polym Sci 81 (2):405–411
38. Karthikeyan CS, Sankaran S, Kishore (2004) Elastic behaviour of plain and fibre-reinforced syntactic foams under compression. Mater Lett 58(6):995–999
39. Karthikeyan CS, Sankaran S, Kishore (2000) A comparison of compressive properties of fibre-free and fibre-bearing syntactic foams. J Reinf Plast Compos 19(9):732–742
40. Karthikeyan CS, Kishore, Sankaran S (2001) Effect of absorption in aqueous and hygrothermal media on the compressive properties of glass fiber reinforced syntactic foam. J Reinf Plast Compos 20(11):982–993
41. Gupta N, Maharsia R (2005) Enhancement of energy absorption in syntactic foams by nanoclay incorporation for sandwich core applications. Appl Compos Mater 12:247–261
42. Karthikeyan CS, Sankaran S, Kishore (2000) Influence of chopped strand fibers on the flexural behaviour of syntactic foam core system. Polym Int 49:158–162
43. Karthikeyan CS, Sankaran S, Kishore (2007) Investigation of bending modulus of fiber-reinforced syntactic foams for sandwich and structural applications. Polym Adv Technol 18 (3):254–256
44. Karthikeyan CS, Sankaran S, Kishore (2005) Flexural behaviour of fibre-reinforced syntactic foams. Macromol Mater Eng 290(1):60–65
45. Kishore, Shankar R, Sankaran S (2005) Short-beam three-point bend tests in syntactic foams. Part II: effect of microballoons content on shear strength. J Appl Polym Sci 98:680–686
46. Kishore, Shankar R, Sankaran S (2005) Short-beam three-point bend tests in syntactic foams. Part I: microscopic characterization of the failure zones. J Appl Polym Sci 98:673–679

47. Kishore, Shankar R, Sankaran S (2005) Short-beam three-point bend test study in syntactic foams. Part III: effects of interface modification on strength and fractographic features. J Appl Polym Sci 98:687–693
48. Kishore, Shankar R, Sankaran S (2005) Gradient syntactic foams: tensile strength, modulus and fractographic features. Mat Sci Eng A 412:153–158
49. Wouterson EM, Boey FYC, Hu X, Wong SC (2005) Specific properties and fracture toughness of syntactic foam: effect of foam microstructures. Comp Sci Technol 65:1840–1850
50. Wouterson EM, Boey FYC, Hu X, Wong SC (2007) Effect of fiber reinforcement on the tensile, fracture and thermal properties of syntactic foam. Polymer 48(11):3183–3191
51. John B, Nair C, Devi K, Ninan K (2007) Effect of low-density filler on mechanical properties of syntactic foams of cyanate ester. J Mater Sci 42(14):5398–5405
52. Devi KA, John B, Nair CPR, Ninan KN (2007) Syntactic foam composites of epoxy-allyl phenol-bismaleimide ternary blend – processing and properties. J Appl Polym Sci 105 (6):3715–3722
53. Capela C, Costa JD, Ferreira JAM (2007) Test conditions effect on the fracture toughness of hollow glass micro-sphere filled composites. Strain 44(2):141–146
54. Li G, Jones N (2007) Development of rubberized syntactic foam. Compos A Appl Sci Manufact 38(6):1483–1492
55. Maharsia RR (2005) Development of high performance hybrid syntactic foams: structure and material property characterization. Dissertation, Louisiana State University
56. Wehmer P (2008) High strain rate characteristics of rubber modified syntactic foams. Dissertation, Louisiana State University
57. Song B, Chen W, Frew DJJ (2004) Dynamic compressive response and failure behavior of an epoxy syntactic foam. Compos Mater 38(11):915–936
58. Peter S, Woldesenbet E (2008) Nanoclay syntactic foam composites – high strain rate properties. Mater Sci Eng A 494(1–2):179–197
59. Saha MC, Nilufar S (2009) Nanoclay-reinforced syntactic foams: flexure and thermal behavior. Polym Compos 31(8):1332–1342
60. Woldesenbet E, Mylavarapu P (2007) Dynamic modulus of syntactic foam core – a non destructive approach. In: Gdoutos EE (ed) Experimental analysis of nano and engineering materials and structures. Springer, Dordrecht, pp 716–762
61. Gupta N, Woldesenbet E (2005) Characterization of flexural properties of syntactic foam core sandwich composites and effect of density variation. J Compos Mater 39 (24):2197–2212
62. Gupta N, Woldesenbet E, Kishore, Sankaran S (2002) Response of syntactic foam core sandwich structured composites to three-point bending. J Sandw Struct Mater 4 (3):249–272
63. Gupta N, Kishore, Sankaran S (1999) On the characterization of syntactic foam core sandwich composites for compressive properties. J Rein Plast Compos 18(14):1347–1357
64. John B, Nair CPR, Mathew D, Ninan KN (2008) Foam sandwich composites with cyanate ester based syntactic foam as core and carbon-cyanate ester as skin: processing and properties. J Appl Polym Sci 110(3):1366–1374
65. Islam MM, Kim HS (2007) Properties and mechanical behaviour of novel sandwich composites made of syntactic foam and paper skin. In: Abstracts of the 5th Australian congress on applied mechanics, Brisbane, 10–12 Dec 2007
66. Sankaran S, Jagdish Kumar MN, Rajput C, Ravishankar BN, Dasgupta S (2007) Mechanical evaluation and weight optimization of CFRP skin-syntactic foam core sandwich composites. In: Abstracts of the 2nd international conference on recent advances on composite materials – ICRACM 2007, New Delhi, 20–23 Feb 2007
67. Sankaran S (1996) Studies on FRP composites with modified epoxies: structural and mechanical evaluation of syntactic foam sandwich composites. Dissertation, Indian Institute of Science

68. Sankaran S, Rajaiah K, Chanda M (1997) Failure modes and fractography of syntactic foam sandwich beams under static loading. In: Abstracts of the 6th NASAS, Bangalore, 07–08 Nov 1997
69. Woldensenbet E, Narendra S (2009) Flexural properties of nanoclay syntactic foam sandwich structures. J Sandw Struct Mater 11:425–444
70. Capela C, Ferreira JAM, Costa JD (2013) Effect of the foam core density on the bending response on sandwich composites. Fibers Polym 14(4):597–602
71. Sankaran S, Ravi Sekhar K, Raju G, Jagdish Kumar MN (2006) Characterization of epoxy syntactic foams by dynamic mechanical analysis. J Mater Sci 41:4041–4046
72. Fine T, Sautereau H, Sauvant-Moynot V (2003) Innovative processing and mechanical properties of high temperature syntactic foams based on a thermoplastic/thermoset matrix. J Mater Sci 38(12):2709–2716
73. Ravi Sekhar K, Ravishankar BN, Sankaran S (2006) Cure characterization of thermosetting resins in composite systems by dynamic mechanical analysis. In: Abstracts of the 5th ISAMPE national conference on composites (INCCOM 5), Hyderabad, 24–25 Nov 2006
74. Sankaran S, Ravishankar BN, Ravi Sekhar K, Jagdish Kumar MN (2005) Cyanate ester sandwich composites for radome applications. In: Abstracts of the 4th ISAMPE national conference on composites (INCCOM 4) Coimbatore, 9–10 Dec 2005 (Figures and tables pertaining to this reference in the article also include relevant additional data experimentally obtained by the authors subsequently)
75. Ravi Sekhar K, Kishore, Sankaran S (2008) Cure behavior of epoxy-cyanate ester blend in composite systems: evaluation studies in neat resin cast by thermal and FTIR techniques. J Appl Polym Sci 109:2023–2028
76. Wang J, Liang G, He S, Yang L (2010) Curing behavior and mechanical properties of hollow glass microsphere/bisphenol a dicyanate ester composites. J Appl Polym Sci 118 (3):1252–1256
77. Keller TM, Matthew L (2014) US Patent 8,288,454, 16 Jan 2014
78. Yung KC, Zhu BL, Yue TM, Xie CS (2009) Preparation and properties of hollow glass microsphere filled epoxy–matrix composites. Compos Sci Technol 69(2):260–264
79. Cyanate ester syntactic foams, Technical bulletin 370–2, Document control no N-10-000-501905-3 11/03/11, Cuming Microwave Corporation, 264 Bodwell Street, Avon, MA 02322, USA
80. Product literature on syntactic foams, Utility Development Corporation, 112 Naylon Avenue, Livingston, NJ 07039, USA
81. Technical data sheet, Tencate Advanced Composites USA, Inc., 18410 Butterfield Blvd. Morgan Hill, CA 95037, USA
82. Product data sheet, Cornerstone Research Group, Inc. 2750 Indian Ripple Rd., Dayton, OH 45440, USA
83. Gupta N, Chakravarthy V, Shumugasamy (2015) US Patent publication no US 2015/0031793 A1, 29 Jan 2015
84. Harrison ES, Melquist JL, Hemming LH (1997) WIPO Patent Application WO/1997/029907
85. BASF (1992) US Patent 5167870
86. Meades GF (1990) EP0359504 Patent
87. Sankaran S, Jagdish Kumar, MN, Govinda Raju (1996) Materials and processes for foam sandwich radomes. In: Abstracts of the national seminar on composites product development (NASCOMP), Bangalore, 22–23 Apr 1996
88. Speak SC, Sitt H, Fuse RH (1991) Novel cyanate ester based products for high performance radome applications. In: Abstracts of the 36th international SAMPE symposium, San Diego, California, April 15–18, 1991 vol 36, pp 336–347
89. Ziolkowski FP, Clark TJ (2014) US Patent 8,917,220 B2
90. Ziolkowski F, Clark T (2015) US Patent 7,420,523B1
91. Kenig S, Raiter I, Narkis M (2004) Three-phase carbon microballoon syntactic foam composites. Polym Compos 6(2):100–104

92. Mereer CL, Philipps TE (1996) US Patent 5587231
93. Dasgupta S, Ravi Sekhar K, Sankaran S (2006) Estimation and evaluation of EMI shielding capabilities of electrically conducting epoxy syntactic foam. In: Abstracts of the 5th ISAMPE national conference on composites (INCCOM 5), Hyderabad, 24–25 Nov 2006
94. Ravishankar BN, Ravi Sekhar K, Dasgupta S, Sankaran S (2007) Carbon fibre reinforced composite foam for multifunctional applications. In: Abstracts of the international and ISAMPE national conference on composites (INCCOM 6), Kanpur, 12–14 Dec 2007
95. Dasgupta S, Ravi Sekhar K, Sankaran S, Jagdish Kumar MN, Prasad V (2007) Challenges in electrical characterisation of lightweight polymer composites for electronic packaging applications. In: Abstracts of the international conference on polymer materials and power engineering (ICPMPE 2007), Bangalore, 4–6 Oct 2007
96. White DRJ (1971) EMI/EMC handbook series. Don White Consultants, Germantown, p 4
97. Krueger QJ (2002) Electromagnetic interference and radio frequency interference shielding of carbon-filled conductive resins. Dissertation, Michigan Technological University
98. Gladysz G, Carlisle KB, Mendoza D (2011) US Patent publication no 2011/0020630, 27 Jan 2011
99. Huang C, Huang Z, Qin Y, Ding J, Lv X (2015) Mechanical and dynamic mechanical properties of epoxy syntactic foams reinforced by short carbon fiber. Polym Compos. doi:10.1002/pc.23374
100. Sankaran S, Dasgupta S, Ravi Sekhar K, Jagdish Kumar MN (2006) Thermosetting polymer composites for EMI shielding applications. In: Abstracts of the 9th international conference on electromagnetic interference and compatibility (INCEMIC 2006), Bangalore, 23–24 Feb 2006
101. Dasgupta S, Ravi Sekhar K, Ravishankar BN, Jagdish Kumar MN, Sankaran S (2008) Sandwich composite approach for EMI shielding structures. In: Abstracts of the 10th international conference on electromagnetic interference and compatibility (INCEMIC 2008), Bangalore, 26–27 Nov 2008
102. Dasgupta S, Ravi Sekhar K, Jagdish Kumar MN, Sankaran S (2008) Syntactic foam core sandwich composites for electronic packaging applications. In: Abstracts of the ISAMPE national conference on composites (INCCOM 7), Bangalore, 4–5 Dec 2008
103. Sankaran S, Dasgupta S, Ravi Sekhar K, Ghosh C (2005) Carbon nanotubes reinforced syntactic composite foams. In: Abstracts of the 4th ISAMPE national conference on composites (INCCOM 4), Coimbatore, 9–10 Dec 2005
104. Dasgupta S, Ravi Sekhar K, Sankaran S (2009) Quality control issues: the other side of carbon nanotubes. J Aerosp Qual Reliab 3(2) 4(1–2) 5(1):1–10
105. Sankaran S, Dasgupta S, Ravi Sekhar K, Ravishankar BN (2014) US Patent 8,734,685 B2
106. Sankaran S, Dasgupta S, Ravi Sekhar K (2007) Electrically conducting nanocomposite foam. In: Abstracts of the international conference on syntactic and composite foams (SCF-II), Davos, 5–10 Aug 2007
107. Guzman ME, Rodriguez AJ, Minaie B, Violette M (2012) Processing and properties of syntactic foams reinforced with carbon nanotubes. J Appl Polym Sci 124(3):2383–2394
108. Zegeye EF, Woldesenbet E (2012) Processing and mechanical characterization of carbon nanotube reinforced syntactic foams. J Rein Plast Compas 31(15):1045–1052
109. Zegeye E, Ghamsari AK, Woldesenbet E (2014) Mechanical properties of graphene platelets reinforced syntactic foams. Compos B Eng 60:268–273
110. Dasgupta S, Ravishankar B N, Ravi Sekhar K, Ghosh C, Sankaran S (2011) Silver coated hollow glass microsphres embedded electrically conducting syntactic foam. In: Abstracts of the international conference on syntactic and composite foams (SCF-III), Cetraro, 29 May–3 June 2011

Advanced Carbon–Carbon Composites: Processing Properties and Applications

10

Raghunandan Sharma, N. L. Ravikumar, Kinshuk Dasgupta, J. K. Chakravartty, and Kamal K. Kar

Contents

R. Sharma
Advanced Nanoengineering Materials Laboratory, Materials Science Programme, Indian Institute of Technology Kanpur, Kanpur, Uttar Pradesh, India
e-mail: raghumsp@iitk.ac.in

N.L. Ravikumar
Advanced Nanoengineering Materials Laboratory, Department of Mechanical Engineering, Indian Institute of Technology Kanpur, Kanpur, Uttar Pradesh, India
e-mail: nlravi@iitk.ac.in

K. Dasgupta • J.K. Chakravartty
Materials Group, Bhabha Atomic Research Centre, Mumbai, Maharashtra, India
e-mail: kdg@barc.gov.in; jayanta@barc.gov.in

K.K. Kar (✉)
Advanced Nanoengineering Materials Laboratory, Materials Science Programme, Indian Institute of Technology Kanpur, Kanpur, Uttar Pradesh, India

Advanced Nanoengineering Materials Laboratory, Department of Mechanical Engineering, Indian Institute of Technology Kanpur, Kanpur, Uttar Pradesh, India
e-mail: kamalkk@iitk.ac.in

K.K. Kar (ed.), *Composite Materials*,
DOI 10.1007/978-3-662-49514-8_10

Abstract

Carbon–carbon composites (C/Cs) are of significant technological importance in various advanced applications, owing to their unique mechanical and thermal properties. C/Cs are composed of carbon fiber-reinforced carbon matrix. This chapter summarizes various aspects of C/Cs in terms of matrix and reinforcement precursors, fabrication of C/Cs and effects of various processing parameters on their mechanical and thermal properties, and their applications. Effects of processing parameters and architecture of carbon fiber reinforcement have been highlighted.

Keywords

Advanced composites • High-temperature materials • Carbon fiber • Carbon matrix • Thermal stability • Thermal conductivity • Densification • Crack propagation

Introduction

Carbon is one of the most important elements that remains present as a main constituent of all the living organisms and also is present in countless molecules essential in various aspects of human activity because of its extraordinary characteristics, i.e., it may combine with other elements or by itself to form different structural materials. Apart from its presence in form of various compounds, carbon, its elemental form, exists as various allotropes such as diamond, graphite, carbynes, and fullerenes. The allotropes of carbon exhibit a diverse range of properties and are important both scientifically and technologically. Among all, graphite is the most abundantly available allotrope, which consists of stacks of tightly bonded and hexagonally arranged basal planes (graphene layers) separated by a distance of 0.335 nm. The graphite structure is illustrated in Fig. 10.1.

The atoms within the graphene layer have a covalent bond strength of ~524 kJ/mol [1], while the bonding energy between the basal planes is ~7 kJ/mol because of weak van der Waals forces between them [2]. As a result, the crystalline graphite is anisotropic, being almost isotropic within the basal planes. Depending on the arrangement of the atomic bonds, carbon can form various structures with respect to microstructure, ranging from the randomly oriented two-dimensional amorphous or glassy carbon structure to the highly crystalline three-dimensional graphite structure, where the density ranges from 0.5 to 2.2 g/cm^3 with different shapes. The various types of graphites, synthesized under specific processing conditions, are also termed as synthetic or engineered carbons. Examples include porous carbons, activated carbons, glassy carbons, carbon blacks, cokes, carbon nanotubes, carbon fibers, carbon–carbon composites (C/Cs), etc.

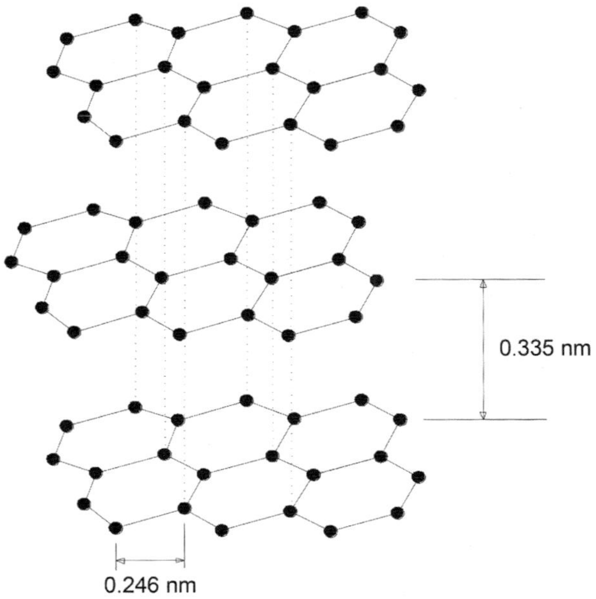

Fig. 10.1 Structure of graphite

Owing to the presence of strong covalent bonds, carbons in solid forms exhibit superior resistance to extreme environmental conditions of temperature or corrosion, which makes them suitable for structural applications under such conditions. Theoretically, the covalent bonding can provide them extremely high specific strength of up to 40–50 GPa over a broad temperature range. In nonoxidizing atmospheres, solid carbons can retain their high strength for temperatures as high as 1500 °C or more [3, 4]. However, owing to the presence of various defects, the normal synthetic carbons exhibit a strength of <2 % of the theoretical values. In order to achieve and utilize the covalent bond strength to the maximum in bulk carbons, carbon fibers having high strength and modulus values have been developed [5].

Carbon Fibers

Carbon fibers, consisting mainly of turbostratic carbon, are being employed as the main reinforcement in high specific strength composites for advanced applications including structural, sports, aerospace, etc. These fibers pose several advantages such as high strength and modulus, excellent thermal and electrical conductivity, creep resistance, thermal shock resistance, chemical inertness, and low thermal expansion, which enable their application with a broad range of matrix materials, i.e., polymers, ceramics, and metals, employing various processing techniques [6–8]. The demands for low density, high strength, and stiff composite structures from the aeronautical sector led to the interest for the development of carbon fiber-reinforced structural composites. These carbon fibers are the only choice for advanced applications such as high-temperature composites because of their high specific thermal and thermomechanical properties. The majority of products based on carbon fiber still belong to the high-technological applications such as space and aeronautics. Subsequently, the applications of carbon fibers are extended to civilian sectors such as sports goods, transport vehicles, biomedical sectors, etc. [8–10]. At present, all commercial carbon fibers are manufactured by thermal decomposition of various organic fiber precursors. The popular precursors are polyacrylonitrile (PAN) polymers, pitch, rayon, etc. In an alternative process using catalytic chemical vapor deposition (CCVD), a low-cost and discontinuous high-performance carbon fiber called as vapor-grown carbon fiber (VGCF) is also made. The associated processing steps and properties of carbon fibers are discussed in the subsequent sections.

Polyacrylonitrile-Based Carbon Fibers

Polyacrylonitrile (PAN) is an atactic linear polymer consisting of carbon backbone with polar carbon–nitrogen pendent groups. The PAN precursor for developing carbon fiber needs to have high molecular weight, minimum molecular defects, low comonomer content (2–5 %), high strength, high modulus, and high carbon yield. To produce carbon fibers from PAN precursor, generally four steps are considered [11, 12], i.e.:

1. Spinning and stretching of PAN precursor
2. Stabilization at 220 °C in air under tension
3. Carbonization at ~1400 °C in an inert atmosphere
4. Graphitization at 2500 °C in an inert atmosphere

Spinning and Stretching

In the first step, polyacrylonitrile plastic is prepared by using a conventional suspension or solution polymerization process where a mixture of acrylonitrile powder and another monomer such as methyl acrylate or methyl methacrylate is reacted in presence of a catalyst. The next step involves the formation of polyacrylonitrile fibers by spinning the plastic. This step is important as the internal atomic structure of the fiber is formed during the spinning process. The spinning of fibers can be performed by using any of the conventional methods based either on solution-based or thermal process. In solution-based processes, the plastic is mixed in an appropriate solvent, and the mixture is pumped through tiny jets into a chemical bath or quench chamber where the plastic coagulates and solidifies as fibers. In thermal processes, the plastic mixture is heated and pumped through tiny jets into a chamber, where the solvents evaporate to form solid fibers. In the final step, fibers of desired diameter are formed by stretching the spun fibers. This step also helps molecular alignment of the polymeric chains within the fiber to provide the basis for the formation of aligned structures during carbonization [13–15].

Stabilization

Before carbonization, the linear structure of fibers needs to be converted to ladder architecture, a thermally more stable structure, by heating them in air at 220 °C for 30–120 min. The process is termed as stabilization, during which the fibers react with oxygen and rearrange their molecular arrangement through a number of complex chemical reactions involving several steps. The process must be controlled carefully as the exothermic nature of the chemical reactions may lead to overheating of the fibers. Numerous techniques have been developed to perform the stabilization process. For example, a series of heated chambers can be employed to perform the stabilization of the fibers. Another technique involves stabilization by passing the fibers over hot rollers. Sometimes, a mixture of air and other gases heated to certain temperatures is also used to accelerate the stabilization process [16–19].

Carbonization

To form carbon fibers, the stabilized polyacrylonitrile fibers are carbonized by thermal treatment at elevated temperatures ranging between 900 °C and 1400 °C.

The fibers are heated for several minutes in a furnace having inert atmosphere and a pressure higher than the normal atmospheric pressure. During heat treatment, thermal decomposition of the polymer takes place, and lower hydrocarbons, water, and other volatile by-products such as NH_3, CO, CO_2, H_2, N_2, etc., are produced. The remaining carbon atoms are crystallized to graphitic/turbostratic layers aligned more or less parallel to the fiber axis. During this stage the high mechanical properties are introduced in the carbon fibers. The development of these properties is directly related with the formation and orientation of turbostratic graphite-like fibrils or ribbons within each individual fiber [20, 21].

The carbonized fibers have a comparatively inactive surface, which, when used in composites, cannot provide strong bonding with various resins/polymers used as the composite matrix. Hence, it necessitates surface modification of the carbon fiber to provide proper bonding properties. One of the routes to surface modification is surface oxidation, which leads to the addition of oxygen-containing functional groups and the enhancement of the surface roughness. Both of these help to provide better mechanical bonding. The surface oxidation can be attained by reacting the fibers with various gases such as air, CO_2, or ozone or liquids such as sodium hypochlorite or nitric acid. Apart from surface oxidation, coatings of various conductive materials can be applied on the fiber surface. The surface treatment may involve formation of tiny surface defects, which must be avoided by controlling the process parameters carefully as formation of defects may lead to reduced mechanical performance of the fibers. The surface-treated fibers are finally coated with specific chemicals, termed as sizing agents. The sizing agent protects the fiber surface from damage during winding or weaving. The sizing agents used in the process are such that they are compatible with the possible matrix materials to be used to form carbon fiber-reinforced composites [5, 14, 15].

Graphitization

In general, graphitization is performed by heating the carbonized fibers to relatively higher temperatures ranging in between 1900 °C and 2500 °C under inert atmosphere. Graphitization improves the crystalline order of the graphite-like crystallites within each individual fiber, which leads to improved tensile modulus of the fiber. The term "graphite fiber" is used for the fibers having a carbon content more than 99 %, whereas the term "carbon fiber" is used where the fiber has 80–95 % of carbon. The carbon content in the fiber is a function of the heat treatment temperatures. The density of carbonized and graphitized fibers generally ranges from 1.6 to 2.1 g/cm^3 as compared to that of 1.2 g/cm^3 for PAN fibers. The final diameters of fibers generally range from 5 to 10 μm [12, 15].

The tensile modulus of PAN-based carbon fibers depends mainly on the degree of heat treatment, because the modulus is determined by the orientation of the graphite-like crystallites along fiber axis, which is a function of the heat treatment temperature. The tensile modulus of a fiber, heat treated at 1100 °C, is found to be 210 GPa, whereas those with 345 GPa are processed at 2200 °C. Unlike tensile modulus, the tensile strength of carbon fiber is not only dependent on the heat treatment temperatures, and also it is not yet fully understood. The tensile strength of fibers may be controlled by the presence of discrete flaws both within the volume of the material as well as on the surface. The typical strain to failure of a low-modulus (210 GPa), high-strength (3275 MPa) fiber is approximately 1.5–1.6 %, whereas for a high modulus (450 GPa) with lower tensile strength (1620 MPa), the value is 0.36 % [11, 12, 15].

Pitch-Based Carbon Fibers

Pitch materials are generally by-products in the destructive distillation of coal and crude oils. This pitch is a very cheap and readily available source of carbon for use as a carbon fiber precursor. In addition to low cost, it also gives high carbon yield during carbonization. Pitch-based carbon fibers are produced by two processes. The first of these processes results low-modulus fibers unless the stress graphitization at extremely high temperatures is employed. The precursor for this process is a low softening point isotropic pitch. The processing scheme is as follows [11, 12]:

1. Melt spinning of isotropic pitch
2. Thermosetting at relatively low temperatures for long periods of time
3. Carbonization in an inert atmosphere
4. Stress graphitization at extremely high temperatures

Tensile strength as high as 2585 MPa and tensile modulus of elasticity in excess of 480 GPa can be attained in fibers produced by this technique. On the other hand, the non-stress-graphitized fibers produced by this process tend to have tensile modulus of elasticity values in the range of 35–70 GPa [11]. Since, the isotropic pitch process is not commercially viable, the most commonly used pitch is mesophase (anisotropic) pitch. For high-performance carbon fibers, mesophase pitch needs to have low ash content; 100 % anisotropic, softening point in the range of 230–280 °C; low viscosity at the spinning temperature; good spinning ability and thermal stability; high oxidation activity; and good carbon yield. The processing scheme for making mesophase pitch-based carbon fibers is as follows [12, 15]:

1. Preparation of mesophase pitch
2. Spinning
3. Thermosetting
4. Carbonization
5. Graphitization

Preparation of Mesophase Pitch

Heat treatment of commercial pitch above 350 °C for a prolonged time results in the succession of dehydrogenative condensation reaction among the molecules, thus forming large molecules, which further aggregate into liquid crystalline phase known as mesophase. It has higher surface tension than isotropic pitch [22–25].

Spinning and Thermosetting

Mesophase pitch can be spun into filamentary form by means of a variety of techniques. Melt spinning is the most common technique adopted for this purpose. In this process, the material to be spun (generally about 50–90 % mesophase) is loaded into either a monofilament or a multifilament die, heated in the temperature range of 250–300 °C, and pushed through the die with pressurized inert gas. The final diameter, which usually ranges from 10 to 15 μm, can be varied by varying take-up rates of the fibers [11, 25, 26]. Since the mesophase pitches are somewhat thermoplastic, it is necessary to change into thermoset state prior to carbonization in order to obviate relaxation tendencies at high temperatures. The thermoset fibers can be made either by heat treating them at approximately 300 °C for a short period of time (2.5 h.) in an oxygen-containing atmosphere or by immersing them in strong oxidizing liquids. During these treatments, the large plate-like molecules that formed in the mesophase are linked together via oxidative polymerization reactions; as a result, the fibers are stabilized against melting, which, in turn, allows them to be carbonized [5, 11, 15].

Carbonization and Graphitization

The carbonization and graphitization steps are similar to that of PAN-based carbon fibers. These heat treatments result in conversions of the mesophase pitch to coke then carbon and ultimately graphite. The tensile modulus increases rapidly as the processing temperature is increased. This is a result of higher degree of preferred orientation imparted to the fibers by the higher heat treatment temperatures. The fibers, which are heat treated at 1700 °C, exhibit tensile modulus values of approximately 210 GPa and in the case of heat-treated fibers at 3000 °C can have modulus values of 880 GPa. The ultimate tensile strength also apparently increases linearly with processing temperature, ranging from an average of 1380 MPa for fibers heat treated at 1700 °C to 2205 MPa for those processed at 3000 °C. The ultimate tensile strength values are relatively low, and, as a result, the strain to failure for the fibers is also low. This could be a major limitation with respect to their use in advanced composites [12, 14, 15]. In general, the fibers derived from PAN have higher strength than the pitch-derived fibers and later exhibit higher elastic moduli. The commercially available fibers can be categorized as high-strength (HT), high-modulus (HM), and intermediate-modulus (IM) fibers, based on their specific properties. Table 10.1 lists the mechanical properties of some of the commercially available both PAN- and pitch-based carbon fibers [15].

Rayon-Based Carbon Fibers

Rayon is a cellulosic material produced by wet spinning after extracting solid products from the wood pulp. Unfortunately, the carbon fibers produced from this precursor are rather weak in properties due to the high internal void contents and modifiers present in the precursor. Also, the yield of carbon is only 10–36 % upon carbonization. These are the main reasons for rayon-based carbon fibers not to become popular and being less commercially available [5, 12, 14].

Table 10.1 Properties of few commercially available carbon fibers

Manufacturer	Product name	Precursor	Density (g/cm^3)	Tensile strength (MPa)	Tensile modulus (GPa)	Strain to failure (%)
Amoco (USA)	T300	PAN	1.75	3310	228	1.4
	P55	Pitch	2.0	1730	379	0.5
	P75	Pitch	2.0	2070	517	0.4
	P100	Pitch	2.15	2240	724	0.3
Hercules (USA)	AS-4	PAN	1.78	4000	235	1.6
	IM-6	PAN	1.74	4880	296	1.7
	IM-7	PAN	1.77	5300	276	1.8
	UHMS	PAN	1.87	3447	441	0.8
Mitsubishi Kasei (Japan)	K135	Pitch	2.1	2550	540	0.5
	K139	Pitch	2.12	2750	740	0.4
Tonen (Japan)	FT500	Pitch	2.14	3000	490	0.6
	FT700	Pitch	2.17	3220	690	0.5
Toray (Japan)	T300	PAN	1.76	3530	230	1.5
	T800H	PAN	1.81	5490	294	1.9
	T1000G	PAN	1.80	6370	294	2.1
	T1000	PAN	1.82	7060	294	2.4
	M46J	PAN	1.84	4210	436	1.0
	M40	PAN	1.81	2740	392	0.6
	M60J	PAN	1.94	3920	588	0.7

Vapor-Grown Carbon Fibers

Vapor-grown carbon fibers (VGCFs) are discontinuous or short fibers produced by decomposing a hydrocarbon gas such as methane, benzene, etc. on a heated substrate in the presence of transition metal such as iron, nickel, cobalt, etc. [27, 28]. The filaments first nucleate on the metal particles and then grow away from the hot substrate. As transition metal is employed, VGCFs are sometimes referred as CCVD carbon filaments. Depending on the type and distribution of catalyst, temperature, and substrate position during the vapor deposition process, different types of VGCFs can be grown [5]. Low-temperature deposition (>900 °C) normally results in vermicular filaments with low density and nonuniform ring-like structures, while high temperatures (1500–2500 °C) favor the growth of long and straight filaments. These are also graphitized when heat treated to 3000 °C. The diameters of the VGCFS vary from 1 to 100 μm and lengths range up to 100 mm. As a result of the above, the mechanical properties vary widely. Since the VGCF does not exhibit catastrophic failure, they may prove to be attractive low-cost reinforcements for applications such as C/C brake pads and other related applications [5, 14].

Carbon Fiber Architecture

Once the reinforcing phase has been decided, one must find a way to place the fibers in the desired directions to obtain best properties. Textile preforming is a method of placing reinforcing fibers in a desired arrangement prior to form a composite structure. Starting with linear assemblies of fibers in continuous and/or discrete form, these micro-fibers can be organized into two-directional (2-D), three-directional (3-D), or multidirectional (4-D to 11-D) structures by means of textile processing technology such as interlacing, intertwining, or interlooping [5, 29]. These preform structures are also referred as fiber architecture. In addition, the fiber placement method also creates textile preforms that possess a wide spectrum of pore geometries and their distribution, a broad range of structural integrity, fiber volume fraction, and its distribution, as well as a wide selection of formed-shape and net-shape capabilities. Thus, proper selection of geometric arrangement and/or fiber placement methods can tailor the structural performance of composites.

In C/Cs, the fiber architecture not only imparts rigidity to the composites, but also in combination with fiber properties, it determines the other properties of composites. Therefore, the choice of preform or fiber architecture is made on the basis of intended applications. On the basis of structural integrity and fiber linearity and continuity, Ko et al. have classified fiber architecture into four categories: discrete, continuous, planar interlaced (2-D), and fully integrated (3-D) structures [30]. Figure 10.2 shows some of these types of fiber architectures.

Discrete Fibers

The composites using discrete (random) fibers are produced either by spray layup or pulp molding. It is difficult to control precisely the orientation of fibers. Fiber volume fraction and strength translation efficiency in these composites made of discrete fiber systems are also quite low. The discrete fiber products are available in the form of chopped strand mat and tissues [14].

Unidirectional Fibers

The second category of fiber architecture consisting linear continuous fibers (i.e., unidirectional system) shows a highest level of property, i.e., load translation efficiency. Theoretically, these systems may have the highest fiber packing efficiency, in the order of 90 % (although only 50–60 % fiber volume fraction is the optimum), and possess low and controlled porosity. These fibers are either processed by filament winding or hand layup or by angle ply tape layup. The disadvantage is that the composites made out of these fibers show poor out-of-plane strength [5, 14].

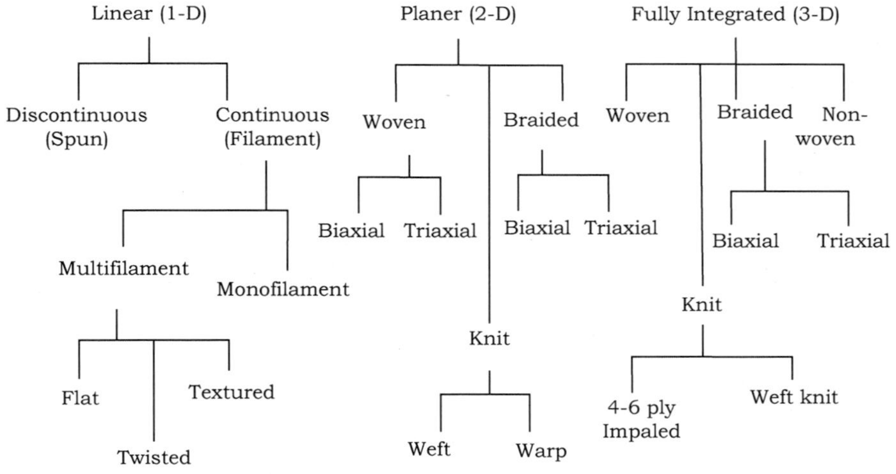

Fig. 10.2 Classification of fiber architecture

Planar (2-D) Preforms

For applications where more than one fiber orientation is required, a fabric combining 0° and 90° fiber orientation is useful. The majorities of these products are woven. The fabrics consist of a regular pattern or weave style produced by interlacing of warp (0°) and weft (90°) fibers. The integrity of fabric is maintained through the mechanical interlocking of fibers. There are mainly three basic weave geometries: plain, satin, and twill weave [5, 8]. Plain weave, which is the most highly interlaced, is the tightest basic fabric design and shows the most resistant to in-plane shear movement. Satin weave represents a family of constructions, woven as standard, four-, five-, or eight-harness satin with the least amount of yarn interlacing. In satin weave, a warp yarn skips over a few weft yarns and goes under one weft yarn. Like in four-harness (4H) satin, the warp yarn skips over three weft yarns and goes under one. The satin weave constructions offer few advantages like high volume fractions, higher strength and modulus, and higher stress translation efficiency over plain fabrics. In addition to these advantages, the satin fabrics are preferred over plain fabrics due to its ability to conform complex contours [31, 32]. Twill weave has patterned in between plain weave and satin weave with warps and wefts interlacing regularly. Figure 10.3 shows the structural geometry of more commonly found biaxially woven fabrics. Although the interlaminar failure problems associated with the unidirectional fiber architecture are taken care in planar structures, the interlaminar strength is still governed by the matrix properties, which can be improved by placing the fiber in the thickness or third direction of the structure.

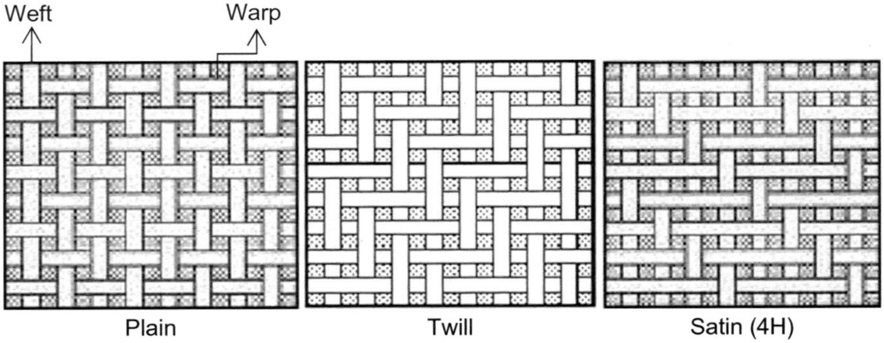

Fig. 10.3 Structural geometry of commonly found biaxially woven fabrics. (**a**) Plain, (**b**) twill, (**c**) satin (4H)

Three-Directional (3-D) Preforms

The 3-D preforms are the most important and critical component of advanced C/Cs, especially for aerospace applications, since the fiber architecture controls the processing as well as the properties of final C/Cs. These are also termed as through-the-thickness reinforcements; the fibers are laid in three directions using braiding, knitting, weaving, and nonwoven techniques. These reinforcements overcome some of the limitations of 2-D composites, especially the out-of-plane strength and interlaminar shear strength [5, 12, 30].

Multidirectional Structures

The composites with higher fiber volume fractions and isotropic properties can be fabricated by multidirectional reinforced preforms. It has been seen in practice that the 4-D to 11-D composite structures can be obtained by placing fibers in angular positions or along the diagonals depending on the final applications. The main drawback of multidirectional reinforcements is that it enhances the off-axis properties at the expense of on-axis properties [5].

Carbon–Carbon Composites

Availability of carbon fibers in the late 1950s led to the development of an improved class of materials, which are now known as carbon–carbon composites (C/Cs). C/Cs form a new class of engineering materials of ceramic nature but exhibit brittle to pseudoplastic behavior. Carbon fiber reinforcements when embedded in carbonaceous matrix material result in C/Cs (Fig. 10.4). These composites are a family of materials consisting of carbon (or graphite) fiber-reinforced carbon (or graphite) matrix. The composites exhibit the properties of both the carbon and the composites.

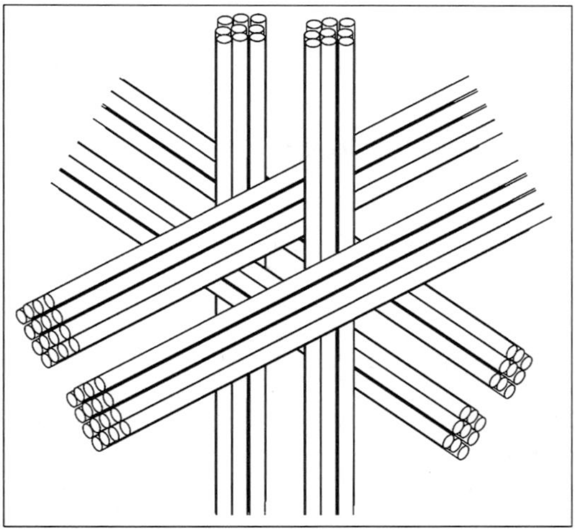

Fig. 10.4 Schematic of carbon–carbon composites

The C/C family is unique in that it is the only elemental composite. The carbon fiber reinforcement makes them stronger, tougher, and highly resistant to thermal shock as compared to the bulk carbons. Again, owing to low density of carbon, their specific properties (property/density), i.e., specific strength, specific modulus, and specific thermal conductivity, remain the highest among traditional composites. Moreover, low thermal expansion coefficient (~0) and retention of properties at elevated temperature (>2000 °C) in nonoxidizing atmosphere are the other unique features of the C/Cs, making them more attractive than any other traditional engineering material in high thermal management areas [33–35]. Figure 10.5 depicts the variation of specific strength of C/Cs including other engineering materials with temperature. While the strength decreases with increasing temperature for other engineering materials, C/Cs exhibit an opposite behavior. The increase in strength of C/Cs with increasing temperature is attributed to the fact that when exposed to high temperatures, the internal structure of the composites changes [36]. C/Cs can range from simple unidirectional fiber-reinforced structures to the complex woven three-directional structures. A variety of carbon fibers and multidirectional weaving techniques are now available, which allow tailoring of C/Cs to meet complex design requirements. By selection of fiber type, layup (or fiber weave), matrix, and composite heat treatment, the properties can be suited to different applications.

Importance of Carbon–Carbon Composites

These composites are one of the best among all high-temperature materials because of their exceptional thermal properties such as high thermal stability (melting point >3000 °C), high thermal conductivity, and low thermal expansion coefficient, which makes them highly resistant to thermal shock. They can retain their mechanical strength even at very high temperatures. Also, these composites maintain good

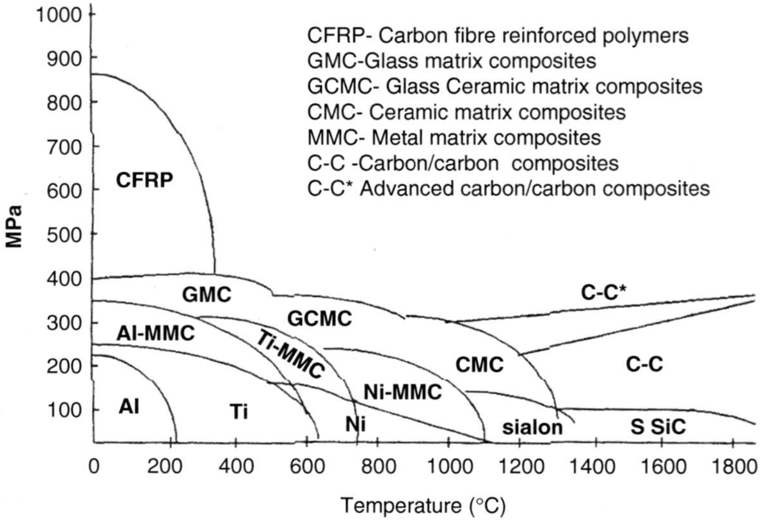

Fig. 10.5 Variation of strength of engineering materials with temperature [36]

frictional properties over the entire temperature range with low wear. They have high fracture toughness and do not fracture in a brittle manner like conventional ceramics. A multimode mechanism of fracture occurs where the fibers break and bonding between fibers and matrix ceases to exist. The brake disks for high-speed aircrafts like Mirage 2000, Concorde, and Airbus-320 are some of the examples where the favorable frictional properties were put into the use. The first-generation C/Cs held the limitation of proneness to oxidation over long exposures to atmosphere, heat, etc. However, with the advent of second-generation oxidation-resistant composites, this limitation was overcome. These unique features of C/Cs have made them the favorite materials for reentry nose tips, leading edge material for space shuttle wings, rockets, nozzles, thrust vectoring nozzles using C/C ball, socket joints, high-performance turbojet engines, nose cones of intercontinental ballistic missiles, brake systems in racing cars, etc.

Current International Status of C/Cs

C/Cs have their origin in the jet vanes used in the German V 2 rockets. The jet vanes are made up of graphite, which erodes rapidly and has limited lifetime. Subsequently pyrolytic graphite A TJ from UCAR Carbon UCAR® is used, which when reinforced with carbon fibers gives birth to C/Cs. Currently, major work is going on in the USA, France, and Russia. There is a great demand for C/C brake disks for aircraft, trains, trucks, and even racing cars. The pioneers in this field were Bendix in the USA and Dunlop in the UK; Germany and Japan tried to use these composites in industrial applications. Israel, Taiwan, and Egypt are also reported to have initiated some R&D activities. C/Cs are an important material for glass industry, furnace, and

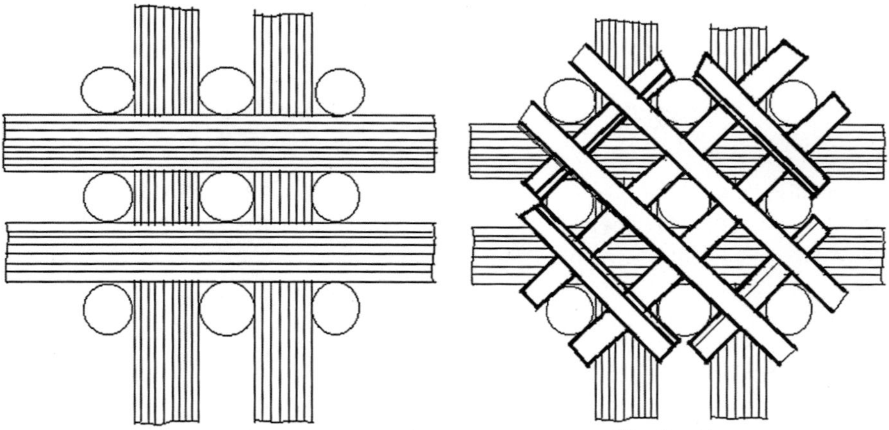

Fig. 10.6 Schematic showing three-directional (3-D) and 5-directional (5-D) arrays of carbon fibers

semiconductor industry and also for preventing corrosion in chemical plants. One interesting and innovative use is as tooling material for metallurgical superplastic stage forging process. Other high-tech applications are as heat exchanger tubes for helium-cooled high-temperature nuclear reactors, high-temperature crucibles, fasteners, load-bearing plates, rods, and heating elements. Elemental carbon is known to have the best biocompatibility with blood and soft tissues. Thus it finds use in hip bone end prosthesis, bone plates, osteosynthesis, and artificial heart valves.

Structural Designs of C/Cs

One of the main advantages of multidirectional C/Cs is the freedom to orient the selected fibers. The multidirectional composites may be formed by using the carbon fiber arrays of desired directionality (Fig. 10.6). Multidirectional preform fabrication technology provides the means to produce tailored and net-shape composites, in order to meet the directional properties required.

Mechanical, thermal, and physical properties can also be controlled by the appropriate design of substrate parameters such as fiber orientation, volume fraction of fibers in the required direction, etc. Preform weaving technology provides the ideal approach to tailor the structural composites. The simplest type of multidirectional structure is based on a three-directional (3-D) orthogonal construction, which consists of multiple yarn bundles located within the structure described in Cartesian coordinates. In any direction, fiber bundles are straight in order to obtain the maximum structural capability of fiber. The type of fiber, the number of fiber per site, the fiber bundle spacing, the volume fraction distributions, and the woven bulk densities characterize the preform. These characteristics can be calculated for a

Table 10.2 Micro-/macrostructural features of C/Cs and their effects [36]

Nature	Scale	Position	Effect
Micromechanical cracking	Fiber diameter scale	Cracking in matrix Fiber/matrix interface	Load transfer among fibers Transverse properties of fibers bundles Porosity
Minimechanical cracking	Cloth layer thickness scale	Fiber bundle Interface between fiber bundles Inter yarn pocket of unreinforced matrix Interface between fiber bundles and matrix pocket Matrix/matrix interface	Load transfer among fiber bundles and laminas Major influence on mechanical and thermal properties

typical unit cell of the preform. Several weave modifications to the basic orthogonal designs can be made to form more isotropic structure. To enhance the composite properties between the planes and diagonal, yarns can be introduced. The conventional weaving with dry yarns and pierced fabrics and assembly of pre-cured rods on manual, semiautomated, or automatic loom setups can be made. The weaving technology and defect characterization techniques are developed to realize defect-free preforms.

Structural Aspects of C/Cs

The important factors to achieve tough and high thermal conductive C/Cs include the choice of reinforcing carbon fibers, the carbon matrix microstructure, the density and microstructure of CCs, the processing routes, and the choice of carbon precursor. Properties of the composites are governed significantly by these factors. For example, the choice of reinforcing carbon fibers affects the strength of the final product, while the microstructures of the matrix and the C/Cs affect the failure mechanism. Table 10.2 lists some of the micro-/macrostructural features of C/Cs that are shown in the SEM micrograph of Fig. 10.7. The presence of voids and cracks on carbon fibers is found to depend on their fabrication route, with less number of voids in pitch-derived fibers processed through hot isostatic press (HIP) as compared to that in the CVD or phenolic resin-derived carbon fibers. These structural features of carbon fibers also influence the thermal transport mechanism of the C/Cs, and hence, it is desired to reduce the micro-/macrocracking. Among other approaches, coating of pyrolytic carbon on the surface of the fibers is used to obtain desired fiber/matrix bonding and to reduce the cracking. Similar to fiber microstructural features, the matrix microstructure also affects the performance of C/Cs. For high strength, a carbon matrix having semicrystalline or amorphous nature is desired, which can arrest the crack propagation effectively. On the other hand, for high thermal and electric conductivity, highly graphitic nature of the matrix is desired. The development of the amorphous carbon matrix may be obtained through thermosetting polymer route, while the crystalline carbon matrix is formed through CVD route or pitch route.

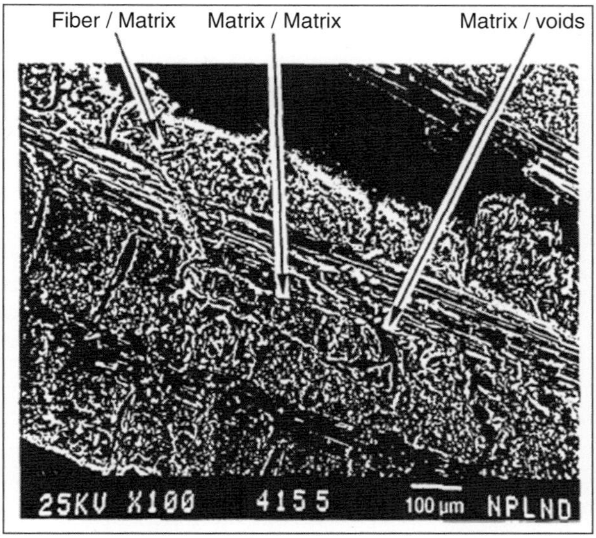

Fig. 10.7 SEM micrograph of 2-D C/C composite showing fiber/matrix and matrix/matrix cracks and voids [36]

Fabrication of Carbon–Carbon Composites

The classical method for fabricating carbon materials involves combining solid particles of pure carbon such as filler coke with a carbon precursor that can be carbonized to serve as a binder. In C/Cs, the carbon preforms are used as primary carbon instead of filler coke. During carbonization, the mass loss and shrinkage of the matrix precursor takes place, which results porosity in the final material as porous skeleton. In order to densify, repeated infiltration–pyrolysis is carried out, which is called as a densification process. This is achieved by impregnating liquid or gaseous carbon precursor and subsequent carbonization/pyrolysis. A key factor in the selection of a matrix carbon precursor involves the ability to achieve a high char yield and to fully densify the preform.

There are three basic methods, and the classifications are based on the type of matrix precursor used for fabricating C/Cs. Figure 10.8 shows a schematic for processing of C/Cs [5, 36]. The first two methods are based on thermal degradation of a thermosetting resin or a thermoplastic pitch. The third method involves deposition of carbon into a fibrous preform using gas-phase or chemical vapor infiltration (CVI), which uses hydrocarbon compounds such as methane, ethane, propane, and benzene as precursors. As discussed in following sections, the choice of fabrication method depends to a major extent on the geometry of the part being processed. Thin sections are prime candidates for CVI; however, since this method tends to preferentially deposit in and on the surface layers, it is not suitable for the fabrication of thick sections. Thick sections are produced using resin or pitch infiltration.

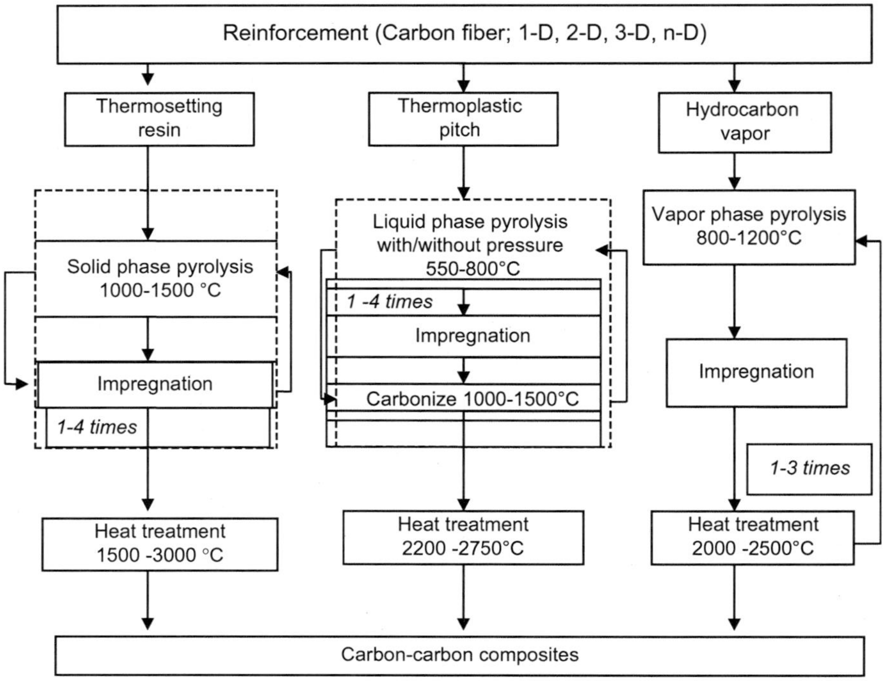

Fig. 10.8 Processing of carbon–carbon composites [36]

Chemical Vapor Infiltration

The CVI process is also known as chemical vapor deposition (CVD). For carbon deposition it uses volatile hydrocarbon precursor gases such as methane, propane, benzene, and others with low molecular weight. Thermal decomposition of any gas is achieved on the hot surfaces of the substrate (Fig. 10.9), resulting in deposition of pyrolytic carbon and emission of volatile by-products, which consist mainly of hydrogen. An important feature of the deposit is its anisotropy and its adherence to the contour of substrate, so that in preform all of the filaments are coated with carbon. However, the overall deposition rate is highly sensitive to the substrate surface area and the concentration of cracking gas. In addition, factors that influence the structure, uniformity, and rate of deposition of a CVI matrix include the nature of the substrate, the carrier gas temperature and pressure, and the geometry, particularly the thickness of the final structure [37]. There are three methods of forming carbon by CVI process, namely, isothermal, thermal, and pressure gradient.

Isothermal Chemical Vapor Deposition

In this process, a carbon preform is placed within a susceptor of induction furnace and is heated uniformly. The pressure and the temperature of the furnace are kept constant at typical values of 6 kPa and ~1100 °C, respectively. The flow rate of hydrocarbon gas is

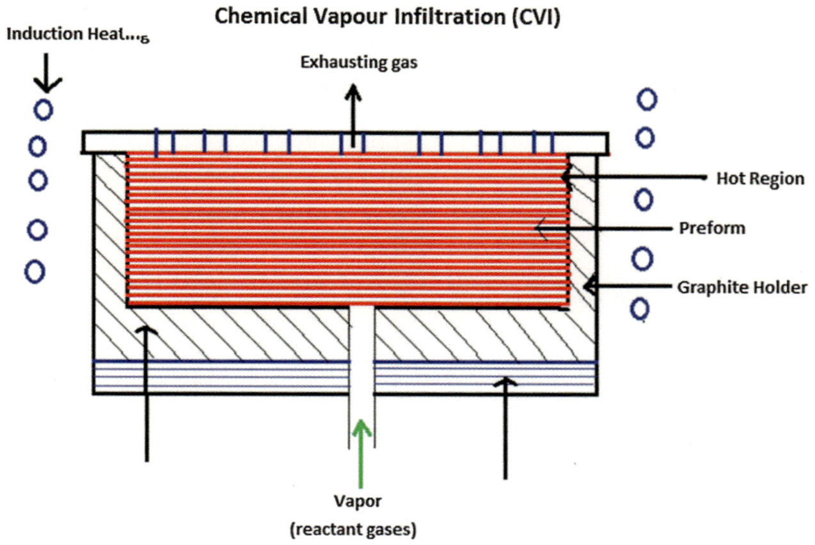

Fig. 10.9 Schematic of chemical vapor infiltration setup

predetermined depending on the surface area of the substrate. Machining of the surface is required in between because the CVD leaves a crust on the outer surface of the substrate. The machining cycle is repeated until the desired density is achieved [31]. Although this process is quite costly and time consuming, it is still used in production of a large number of parts.

Thermal Gradient Vapor Deposition

In this technique, the carbon preform is supported on a mandrel. Inductive coils heat the surface of mandrel to a temperature of about 1200 °C. The hottest portion of substrate is in contact with the mandrel, while the other side is in contact with the reacting gas, which is cooler. Thus, a thermal gradient through the substrate thickness is created. As the hydrocarbon gas passes through the furnace at atmospheric pressure, carbon is deposited on the hottest region of structure. This hot section migrates through the thickness of structure as the densified region grows toward the colder surface. This technique prevents the formation of crust on the outer surface of preform. Thus, the machining step is eliminated. Unfortunately, the process tends to be restricted to large individual parts [5].

Differential Pressure Chemical Vapor Deposition

Differential pressure CVD is a variation of the isothermal technique in which the inner portion of fiber preform is sealed off from the furnace chamber at the base. A pressure higher than that of the furnace chamber is used to feed the hydrocarbon gases into the inner cavity. The pressure difference thus created forces the hydrocarbon flow through the pores, where they decompose to deposit carbon. The by-products such as hydrogen are produced across the wall of structure [31]. This technique also prevents the formation of crust on the surface of preform and facilitates the densification uniformity.

Moreover, the densification process and final density of the composite using CVD technique not only depend on the temperature, pressure, gas concentration, and location of the substrate but also on the geometrical configuration of the fiber preform, initial porosity, and fiber packing density in the initial composite structure [5, 15]. The lower the fiber volume content with higher initial porosity, the higher the ultimate density. Generally, 30–50 % of fiber volumes appear to be the most suitable for achieving high-density C/Cs. In woven structures, depending on the geometric arrangement of the fibers, the matrix will form within the geometric interstices and fiber bundles. However, as discussed above, the formation of surface cluster and final density of the composite can be controlled by pressure and temperature..

Liquid-Phase Infiltration

This process involves impregnation with liquid impregnates. The criterion for the selection of impregnates is based on the characteristics like viscosity, carbon yield, matrix microstructure, and matrix crystalline structure, which are considerably influenced by the time–temperature–pressure relationships during the process. Two types of liquid impregnants are used to fabricate C/Cs. The first includes thermosetting resins, and the other is derived from thermoplastic pitch, which may be coal tar- or petroleum-based material. Both types of liquid impregnants are used because they have suitable viscosities, and the carbon yields are high enough to achieve high-density (up to 1.9 g/cm^3) C/Cs. Thus, the two general categories are aromatic, ring-structured, conventional thermosetting resins such as phenolic, furans, and advanced resins like pyrenes, ethynyl, or pitches based on coal tar, petroleum, and their blends using the multiple impregnation, carbonization (1000 °C), high-pressure (1000 bars) carbonization (HIP), and graphitization (2750 °C). During carbonization the thermosetting resins remain solid; however, pitches soften and tend to flow from carbon preform at high temperatures. Therefore, they require additional reinforcing agent during carbonization. In atmospheric pressure carbonization, the carbon yields obtained from pitch are only around 50 %, comparable to those from high-yield thermosetting resins. The carbon yields from pitch can be enhanced to 90 % by carbonizing it under high pressure (1000 bars). Hence, the use of high pressure makes the carbonization process more efficient. Pressure applied during pyrolysis also affects the matrix microstructure. High pressure aids formation of the more coarse and isotropic microstructure due to the suppression of gas formation and escape. High pressure also helps in lowering the temperature of mesophase formation in pitch, which results into highly oriented crystalline structure. Complex shapes are difficult to fabricate using either pitch matrix or CVI materials owing to the difficulty of maintaining the dry fiber preform shape during initial carbonization. To avoid this difficulty in the process, hybrid densifications are sometimes practiced. In

Table 10.3 Carbon yield of various thermoset precursors

Precursor	Carbon yield (%)
Polyacetals	85
Polyphenylenes	85
Polyimides	60
Phenolic resins	55–65
Polybenzimidazole	74
Resorcinol furfuraldehyde	60
Oxidized polystyrene	55
Polyacrylonitrile	45
Polysulfone	48
Furan resins	50–60
Epoxy resin	5

this first, a rigid structure is made using the resin prepreg autoclave molding processes. After first carbonization, subsequent infiltrations of C/C skeletons are made with low viscosity resin or pitch or CVI materials in the next densification [5, 14, 15].

Thermosetting Resins

Thermosetting resins have been widely used as impregnants because of their easy availability in the liquid state. These have moderate viscosity at room temperature and the ability to polymerize to form a highly cross-linked solid when heated at around 200 °C. Also, they will not flow from the fibrous preforms during the first carbonization. Phenolic and furan resins have been used as binders and impregnates for a long time. In addition to these, wide ranges of new types of resin have also been tried as precursors because of their importance in C/Cs. The char yield or carbon content of some of the resins is listed in Table 10.3 [5, 15].

Although some new resin precursors have been developed that have higher carbon content, because of their high price and problems in the manufacture of initial polymer composites for processing of C/Cs, in general, phenolic resins are used for the initial impregnation and carbonization cycles, while furan resins are used as an impregnate in the subsequent densification cycles [5, 15].

Phenolic–Carbon Fiber Composites

Phenolic resins are a group of a thermosetting resin formed in a condensation polymerization of phenol and formaldehyde. The phenolic resins exhibit large variation in their structural and mechanical properties, which depend on the processing conditions during synthesis of the prepolymer. The parameters such as temperature, condensation time, pH, catalyst, formaldehyde to phenol molar ratio, etc. are the governing parameters. These are mainly two types, novalac and resol. The latter resin has a higher potential as a matrix material than the former because of its low cost, higher carbon yield, high nonvolatile content, higher alkali resistance, high dimensional stability, wear resistance, good chemical resistance and good interfacial adhesion with carbon fibers, and better moldability [38, 39].

When the resin is incorporated with carbon fiber, the formed composites not only serve as green composites (starting material) for fabrication of C/Cs but also as potential candidates for other structural applications. As mentioned earlier, carbon fibers present several advantages such as high strength and modulus, excellent thermal and electrical conductivity, creep resistance, thermal shock resistance, chemical inertness, and low thermal expansion. Utilizing the advantages of the resin and carbon fibers, polymer composites (PCs) have shown the properties comparable with those of other traditional composite materials and metals. These are increasingly finding their applications in thermal protection systems of reentry vehicles and rocket engine components due to their excellent ablation resistance and mechanical properties [40–45]. Also, these are used in some of the automotive parts like rotors, pistons, and bearings since they have non-seizure characteristics and self-lubricating properties [45–49]. Furthermore, in other applications, the resin matrix is also modified with ~5–8 wt% of nanosized fillers like zinc oxide (ZnO), silicon carbide (SiC), and carbon black to improve the wear and frictional resistance of PCs [45, 50].

Processing of Carbon–Carbon Composites with Pitch Matrices

The basic fabrication method for production of C/Cs using pitch matrix is to use pressure to force pitch into an evacuated cavity that contains the dry carbon fiber preform. Pitches are thermoplastic in nature, which soften and tend to flow from carbon preform at high temperatures, but must be carbonized very slowly under pressure. In a typical process, the carbonization of the pitch matrix is carried out by heating to 1000 °C at atmospheric pressure and then densification is done with additional pitch under a very slight partial pressure of nitrogen (i.e., 2 MPa). It is also known as liquid-phase impregnation (LPI). The carbon yield in this process is around 50 wt% for ordinary pitch and 80–85 % for mesophase pitch [51]. In this process carbonization and impregnations are carried out as distinct steps. Sometimes, after carbonization, the composites are heat treated to 2800 °C (intermediate graphitization) to open more and more pores for subsequent impregnations [52]. Using this technique, unidirectional and 2-D composites of density values of 1.7–1.85 g/cm^3 can be obtained. However, in multidirectional and big-sized composites, the achievable densities are limited to 1.6–1.8 g/cm^3.

Significant increases in the carbon yield can be obtained by the use of high pressure, i.e., 100 MPa, during carbonization. Yields as high as 90 % have been observed [15, 32, 53]. Pressure is applied by using HIP where both carbonization and impregnation take place in a single step. Hence, the process is also known as hot isostatic pressure impregnation–carbonization (HIPIC) process. At the higher pressure, more coarse and isotropic microstructure is formed probably due to suppression of the gas formation and escape during carbonization. Above 400 °C, the isotropic liquid pitch consists of spheres, known as mesophase, exhibiting a highly ordered structure similar to liquid crystals. These spheres coalesce on prolonged heating to form larger regions of lamellar order. Upon subsequent heating to ~2500 °C, the lamellar regions convert preferably to graphitic carbon [54]. The formation of mesophase takes place at comparatively lower temperatures when high pressure is applied. The optimum applied pressure has been found to be ~100 MPa, as the coalescence of mesophase does not take place at very high pressures (~200 MPa) [15, 55]. The density of HIPIC products remains higher compared to that of the LPI products, requiring a smaller number of cycles to achieve a density of 1.9 g/cm^3.

Low-Cost Fabrication of Carbon–Carbon Composites

The fabrication of C/Cs using the above methods is not only very expensive but also time consuming because every method has difficulties in its own way. As discussed earlier, in CVI process, it is difficult to infiltrate effectively thick parts or complex fiber performs; therefore, it is limited to thin structure. The rate of deposition of pyrolytic carbon is also very slow, which leads to large consumption of energy and high costs. In addition, the initial setup cost is high and requires highly skilled operators. On the other hand, in polymer pyrolysis routes, both resin and pitch are carbonized at high temperatures to convert the polymers into carbon matrix. Weight loss due to evolution of volatile materials and the shrinkage of precursors during carbonization result in matrix cracking and subsequently a highly porous structure accompanied by low strength. In order to improve the density of composites for better final properties, in conventional densification process, multiple cycles of re-infiltration and re-carbonization of the liquid carbon precursors are carried out, which are again time consuming and costly processes. Thus, C/Cs have been used only in a limited number of applications despite their earlier said unique material properties.

In order to expand the area of the applications of C/Cs, researches have been focused through cost reduction within the manufacturing steps. One of the effective routes to reduce the manufacturing cost is to increase the otherwise low rate of carbonization of resins. However, the rate of carbonization cannot be set too high as during carbonization, a number of processes such as cleavage of C–H and C–C bonds to form reactive free radicals, molecular rearrangement, polymerization, condensation, and elimination of side products (e.g., H_2) take place, which lead to shrinkage, cracking, and development of thermal stresses [56, 57]. To minimize these adverse effects, low carbonization/heating rates (typically <10 °C/min) are usually required. However, change in the heating rate affects the final properties of the composites [58–63].

Another way to reduce the manufacturing costs of C/Cs is to increase the carbon yield by modifying the matrix precursors, especially through polymer pyrolysis routes. It is observed that the modification of the thermosetting resin or thermoplastic pitch matrix by adding fine carbon fillers such as coke powder, carbon black, and graphite fillers allows the manufacturing of C/Cs to a satisfactory density in one step or less number of steps with improvement in some of the mechanical and physical properties depending on the type and amount of fillers and heat treatment temperatures used during processing of C/Cs [5, 15].

Furthermore, references may be made to the published articles or patents in the literature for bringing down the processing cost of C/Cs. Howdayer et al. have described a process of densification by immersing the porous structure in a boiling liquid of hydrocarbon and heating the same by induction to temperatures in between 1000 °C and 1300 °C [64]. It is found that the carbon or pyrolytic graphite is deposited in the pores or cavities of the initial porous structures. Density of 1.75 g/cm^3 is achieved in 2 h by this method compared to in 80 h by normal CVD method.

Takabatake has invented a process for producing high-density and high-strength C/Cs, which does not require high-cost secondary treatment [65]. This process includes impregnating bundle of carbon fibers with liquid dispersions prepared by secondary carbonaceous particles of graphite, carbon, and carbon black in carbonaceous liquid such as phenolic resin or pitch and then impregnated with phenolic resin, furan resin, or pitch. The advantage of this process is that carbonization can be done easily even with pitch under atmospheric pressure because formation of pores during carbonization is greatly reduced. Further, using fine graphite powder of size 5–0.1 μm, the high-density, high-strength, and less brittle composites are obtained due to the formation of fine mosaic matrix structure. But the reduction in strength is observed when larger- and too-small-size particles are used. This is due to the nonuniform distribution of particles at larger size and at very small size, which causes agglomeration of particles. This further causes a tendency to prevent impregnation of matrix.

Upadhya and Hoffaman have described a process in which a rapid densification of porous article is achieved by a single processing cycle instead of multiple cycles [66]. In this process, the article surface is kept in proximity to a plasma containing densifying species for carbon deposition. An electric bias, with respect to the plasma, is applied to the article such that the energized densifying species are attracted toward the article. The process, which requires about 3–4 days to complete, is highly efficient as compared to the usual processes of densifying and graphitizing a carbon preform, requiring typically ~6 months to complete. They have achieved the density of 1.8–2.2 g/cm^3 by this method of CVI.

Thurston et al. have described a process in which a preform is densified by heating while immersed in a precursor liquid [67]. Heating is achieved by passing a current through the preform or by an induction coil immersed in the liquid. In this way, the total cost of densification is reduced. Final bulk density achieved in case of 1.5″ tubular and 6″ long preform is 1.83 g/cm^3, and the porosity is around 6 %.

A process for low-temperature densification of carbon fiber preforms has been reported by Sheehan, where colloidal graphite is impregnated in the preform followed by mechanical consolidation [68]. The process may be used to reduce the cost of C/C fabrication significantly. In this, a highly porous carbon fiber preform is impregnated by a liquid suspension of colloidal graphite and dried in air at low temperatures. Multiple impregnations and drying provide soft and flexible C/Cs. Subsequent mechanical consolidation methods such as pressing or rolling are used in combination with rigidation by pitch or resin impregnation and carbonization. From this method, the bulk density of around ~2 g/cm^3 is obtained.

A process for making C/Cs with improved properties in a single step has been invented by Withers et al. [69]. In this, at first, they have prepared the matrix comprised essentially of a mixture of a noncrystalline carbon particulate such as green carbon coke (gcc) that is soluble in an organic solvent (isopropanol alcohol, furfuryl alcohol, and methyl ethyl ketone) and a binder comprised of an organic carbon precursor that has a liquid phase such as phenolic resin and pitch. Then, the prepared matrix mixture is combined with the reinforcing materials to form an uncured structure. The uncured structure is carbonized in an inert atmosphere to produce the desired carbon composite structures, for example, piston for an internal combustion engine or a sheet or other shape for thermal management. It is noticed that the carbon yield of the matrix mixture is around 90 %. Also, the liquid matrix precursor is able to penetrate within the carbon architectures and produces a low porosity matrix within the reinforcement as well as in the spaces between reinforcement. With this method, they have obtained the final density of composite ranging from 1.68 to 2.0 g/cm^3 depending on the final heat treatment in one single step. In addition, it is observed that the smaller the particle size, the greater the strength of the composites. Also, the strain to failure of the cured carbon composite structure, wherein the coke particles are dissolved in organic solvent, is found to be twice the strain to failure of cured carbon composite structures that are made by mixing the gcc and phenolic resin without an organic solvent. Finally, from the study, they have concluded that the strength of the composite is affected by the size of particles, solvent used, molding and curing conditions, pyrolysis rate, and pyrolysis temperature.

Wielage et al. have described a process of densification using catalytic effects of metal carbonyl on the cross-linking efficiency of polymers used for a rapid and cost-effective densification [7]. In this process, the open pores in the matrix of C/Cs obtained via polymer pyrolysis are effectively closed in one-shot densification process by applying a dicobalt octacarbonyl [$Co_2(CO_8)$]-modified polysilane. They have achieved flexural strength of 125 MPa with 63 % fiber volume content in 2-D C/Cs. Besides the successful densification of C/Cs, using the modified polysilane, improvement in the oxidation behavior of C/Cs at elevated temperatures is also recorded.

A method for developing C/Cs of high density in a single step using a special technique has been invented by Bahl et al. [70]. In this, a special type of the sample holder is designed for heat treatment of polymer composite material up to 1000 °C. Then, the green composites are placed in the sample holder and covered from both the sides by two other plates. Finally, these three plates are sandwiched containing composite samples in the center. After that the sample holder is kept for carbonization up to 1000 °C in an inert atmosphere at the desired heating rates. Now, these C/Cs are further heat treated to 2600 °C in an inert atmosphere at the desired heating rates to obtain the C/Cs of 1.8–1.85 g/cm^3 density, which can be used for various structural applications.

Liu et al. have introduced a densification mechanism, known as chemical liquid deposition (CLD) process [71]. In their work, low-cost C/Cs are prepared by taking industrial fuel as the precursor for carbon matrix and carbon fiber felt as the reinforcement. The results show that the properties of CLD C/Cs are the same as those of CVD C/Cs, i.e., the bulk density is up to 1.6 g/cm^3, and the axial compressive strength is up to 92 MPa. But the densification rate of CLD process is five times faster when compared with the conventional CVD process. Consequently, the deposition time is also less. CLD carbon matrix includes both rough laminar pyrocarbon and smooth laminar pyrocarbon. Also, improvement in mechanical properties is observed due to the carbon matrix having an onion structure at fiber/matrix interface.

Another method for manufacturing of C/Cs by hot pressing has been invented by Huang et al. [72]. In this method, a mixture of carbon fibers, pitch, and optional additives for frictional applications is compressed in a cavity of vessel using a single- or dual-action ram. During compression, in order to form a preform, current is applied to the mixture to generate a temperature of at least 500 °C by means of resistance offered by the mixture. Once the preform is formed, carbonizable material such as phenolic resin or pitch is impregnated into voids in the compressed preform to form an impregnated preform. Then, the product is carbonized in an inert atmosphere. The impregnation and baking steps are optionally repeated. The impregnated preform is then graphitized to a final temperature of at least about 2000 °C to form the composite material. The advantage of this method is that the heating rate used is preferably at least 100 °C/min and as high as about 1000 °C/min or even higher. Depending on the heating rate, the initial heating and pressing steps just take nearly 3 h to less than 10 min, which is a much shorter time than the days required in conventional heating/pressing systems. Additionally, the density of the preform formed in this step is preferably at least 1.3 g/cm^3 and as high as 1.7 g/cm^3 without densification procedures. This is much higher than the density generally achieved in conventional methods, where the density of composite preform is about 0.6–1.3 g/cm^3. As a consequence, one or two infiltration cycles are required to achieve a final density of 1.5–1.9 g/cm^3 depending on the impregnant such as phenolic resin or pitch with this hot pressing method compared with the conventional methods.

Alternatively, the resin transfer molding (RTM) process can also be used for fabricating C/Cs of desired density in a just two cycles of reimpregnation and carbonization. In this process, the manufacturing of resin matrix composite part as well as impregnation of the carbonized parts is carried out simultaneously. RTM densification is about twice as fast as the conventional resin impregnation method, and it is environmentally friendly. The reported density and mechanical properties of just two time-densified C/C panels are superior or nearly same as the reported data in the literature by other processes [73, 74].

Properties of Carbon–Carbon Composites

C/Cs are not a single material but rather a family or class of materials whose properties can be varied depending on how and from where they are made. Therefore, it is important to know the behavior of its various systems under different tests and operating conditions before using in particular applications. The properties, which are essential and often addressed in the literature, are strength, stiffness, fracture toughness, thermal conductivity, thermal expansion, ablation resistance, etc. The operating mechanisms for all these properties are quite different. The mechanical properties are determined by the fiber/matrix bond strength and crack propagation mechanism, whereas the thermal properties are governed by thermodynamic and thermal transport phenomena [36]. In addition, properties of C/Cs depend on many other factors such as type of reinforcing carbon fibers, fiber architecture, carbon matrix microstructure, density, and macrostructure of the composites such as type, size, and quantities of defects, i.e., defects, pores, cracks, etc. [5, 36].

For most composites, the primary consideration for a given application is the properties of the reinforcing fibers. The strength and modulus of carbon fibers are also related to the physical properties such as thermal and electrical conductivity and coefficient of thermal expansion. The choice of a fiber based on one property usually determines the value of the other properties. For example, the microstructure of very-high-modulus fibers usually consists of perfectly aligned basal planes parallel to the fiber axis. As a result, the transverse modulus will be relatively low, the thermal and electrical conductivity will be high in the longitudinal direction, and the thermal expansion coefficient will be small or negative [15].

On the other hand, a composite matrix usually serves to protect the reinforcing fibers from damage or reaction with the environment, to provide some amount of support in compression, to provide adequate matrix-dominated properties, and to provide a continuity of material. This last property is important in electrical and thermal applications and is particularly important in mechanical applications since load must be transferred to the fibers through the matrix. In this respect, a load can be transferred to the fibers across a chemically or physically bonded interface or across a mechanically interlocked one formed by the matrix shrinking onto and thereby gripping the fiber surface during processing of C/Cs.

The tensile and flexural properties of C/Cs are fiber dominated, whereas the matrix-dominated properties such as transverse tensile strength and modulus, transverse flexural strength and modulus, compression strength and modulus, interlaminar shear strength (ILSS), thermal coefficient of expansion, and thermal conductivity are mainly affected by the density and matrix morphologies. The amorphous or semicrystalline carbon matrix structures, being partially or fully randomly oriented, can arrest the crack propagation to enhance the strength as well as modulus of the composites [75]. On the other hand, the graphite matrix structure can provide high thermal as well as electrical conductivities and high oxidation resistance [36]. However, the graphitic matrix exhibits brittle nature.

Elastic Modulus

In C/Cs, the value of Young's modulus could be much greater than calculated from the rule of mixture [76]. One of the regions for this is that improvement in crystallinity of the fiber due to the process induced thermal stresses and repeated thermal cycling. As mentioned before, the commercial carbon fibers are manufactured from PAN and pitch precursors, and their properties vary considerably again depending on raw materials and processing conditions used. Low-modulus carbon fibers are expected to exhibit more improvement in the crystallinity than highly heat-treated high-modulus carbon fibers [77].

Similarly, depending on the heat treatment temperature and precursor material, the microstructure of the matrix also varies from amorphous to graphitic with random or parallel alignment. For example, the amorphous matrix obtained by thermosetting resin route may be converted to graphitic form by heating at temperatures >3000 °C. On the other hand, since the matrices derived through pitch/CVD routes provide comparatively higher graphitic order than that of the thermosetting resin-derived matrices, they can be graphitized by heating at comparatively lower temperatures of ~2100 °C. On the other hand, in CVD process, depending on the concentration or partial pressure of the reactant gas and the processing temperature, the formed matrix will be isotropic and laminar (smooth and rough) structures. The former is nongraphitic and latter is graphitic [78–80].

However, the modulus of the matrix is much lower than the fibers. For a given volume fraction of fibers with similar bonding, the matrix contributes significantly to the stiffness of the composites, if it is highly crystalline, and planes of graphite crystal are arranged parallel to the fibers. This type of matrix structure can be obtained easily by using pitch precursor and choosing proper processing conditions [76]. Matrix in the interfilament region, with possibly large variance in microstructures, possesses a range of mechanical properties. A tensile modulus of 3–40 GPa has been obtained for the randomly arranged carbon or graphite matrix, whereas for a highly aligned graphitic structure, it is 400–700 GPa [5]. The effects of the same matrix precursor can be further enhanced by using carbon fibers with a highly aligned graphitic structure (high-modulus pitch carbon fibers). Whereas in thermosetting resin-derived composite, the modulus improves due to the formation of lamellar-type structure around the fibers because of stress-induced graphitization of the matrix at the fiber/matrix interface, when the composite is heat treated at around 2200 °C [5, 15].

Strength

The strength of C/Cs is governed by the Cook–Gorden theory for strengthening of brittle solids as the simple law of mixtures cannot explain their behavior. Thus, it depends on fiber/matrix bond strength, fiber microstructure, fiber volume fraction and its architecture, amount of porosity, matrix microstructure (degree of graphitization), and processing parameters [81–84]. In C/Cs the fiber/matrix interface bond strength must be optimal as too high bond strength may lead to extremely brittle nature, catastrophic failure, and poor strength of the composites, while too low bond strength leads to composites with poor fiber strength translation, exhibiting a pure shear-dominated failure [85, 86]. Thus, among (1) non-surface-treated unsized carbon fibers (too low in bond strength), (2) non-surface-treated sized carbon fibers

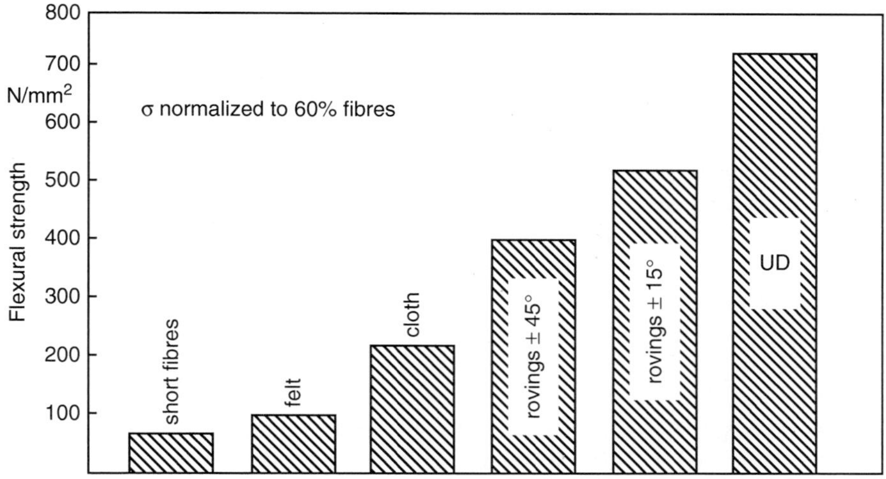

Fig. 10.10 Strength of carbon–carbon composites with fiber orientation [36]

(optimum), (3) surface-treated unsized carbon fibers (too high in bond strength), and (4) surface-treated sized carbon fibers (too high in bond strength), the non-surface-treated sized carbon fibers give C/Cs of higher strength. These optimum composites fail in a mixed mode fracture. Similarly, carbon fibers that have been oxygen treated give higher strength than plasma treated ones and those which have not been treated at all. Moreover, carbon fibers, which have been oxidized by nitric acid and then plasma treated, give composite of higher strength than those oxidized but not treated [85, 87].

It is very difficult to reach theoretical longitudinal strength of composites due to the twisting and distortion of the fibers, variations in fiber orientation, and stress concentration associated with the test method. Generally, 50–60 % volume fraction of fibers is optimum for attainment of good strength in C/Cs. The key factor in optimum fiber utilization is good alignment with respect to the axis of the load. In the case of woven fibers, there is misalignment due to the crimping (fiber crossover region), which occurs as a result of weaving process. The fibers are not aligned with the principal stress axis, and thus effective strengths are reduced [15, 85, 88]. Variation of strengths of C/Cs with fiber orientation has been shown in Fig. 10.10.

The other major factor, which determines the strength of C/Cs, is the density of the material, which depends on the porosity or void content. The pores and voids present in the composites act as stress concentration sites and reduce the strength of the composites, but these can be filled effectively during densification cycles. It is observed that even after four or more cycles, the composites still contain a large degree of porosity. Despite this, large increase in strength is obtained due to the reduction in the amount of porosity, and the reimpregnation tends to smooth the geometry, which reduces the stress concentration effects of pores and voids. Moreover, the intermediate graphitization treatments during densification cycles also influence the final strength of composites than the single graphitization after repeated densifications because the former treatment helps in filling the pores effectively, which improves the bulk density of the composites [5].

The weave pattern of carbon fabric also affects their densification [89]. For example, the 8H satin weave is favored over the plain weave; microcracks develop beneath the bundle crossover points in plain weave because of the inhomogeneous matrix distribution. Hence, the composites consisting of plain weave exhibit catastrophic failure due to bundle pullout. On the other hand, shear-type failure with fiber pullout is observed in the composites made with the satin weave fabric. Again, with densification, the flexural strength increases appreciably for the satin weave composites compared to marginal increments observed for the composites made with plain weave fabric.

Furthermore, the types of matrix precursor and processing parameters also affect the strength of composites. As mentioned earlier, depending on the nature of fiber surface, the choice of precursor for the matrix is also important because it controls the binding strength and adhesion strength. The processing parameters such as heating rate and final heat treatment also affect the properties of composites. In all cases, the bulk density increases with slow heating rates, which further increases the strength of composites [76].

The heat treatment temperatures also govern the mechanical properties of composites by changing the failure modes of composites. In general for carbonized composites that are heat treated around 1100 °C, the matrix-dominated failure prevails and depends on the fiber surface, and the fracture energy and strength of the composites change accordingly [36, 90]. Whereas in the intermediate temperature range 1200–2200 °C, mixed mode failure prevails, because the interface weakens in this temperature range; the strength of the composite increases till interface bond strength becomes an optimum to deflect the formed cracks without harming the fibers and strong enough to effectively utilize the stress transfer capability of the matrix with low fiber pullout [36]. On further heating to 2500 °C and above, the matrix will be in the form of graphite. The graphitized matrix has low shear strength and fail in a jagged fashion. The fiber/matrix interface strength also becomes very low (Fig. 10.11). As a result, the matrix is decoupled from the fibers, and therefore, the matrix contributes least to the composite strength. The failure is dominated by the fiber failure [36, 87, 90].

Fracture Toughness

In addition to strength and stiffness, the toughness of the composite is also important from a structural point of view. The fracture toughness (energy release per unit crack area) for carbon in both bulk and fiber alone is in the range of 0–200 J/m^2. However, more value of higher toughness and work of fracture can be obtained in case of C/Cs with optimum fiber/matrix bonding. The key parameters governing the toughness of C/Cs include the choice of reinforcement, carbon matrix microstructure, etc. The micromechanical process, which imparts toughness to C/Cs, includes matrix cracking, interface debonding, fiber fracture, friction at fiber/matrix interface, bridging effects, and fiber pullout. The 3-D orthogonal carbon fiber woven composites exhibit

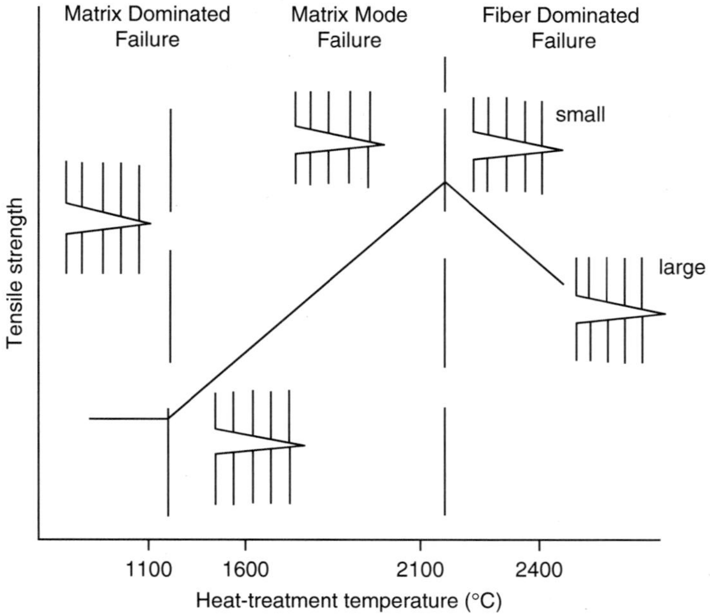

Fig. 10.11 Effect of heat treatment temperature on strength and fracture mode of carbon/carbon composites [36]

the highest work of fracture. Because the cracks diffuse in tortuous manner, probably by tracking pre-existing voids or microcracks, and the failure of the composites takes place by propagation of a series of stable cracks across the matrix and yarn bundles, until the cracks becomes unstable, propagation of which leads to failure [91]. The failure takes place through two dominating mechanisms, namely, the fiber bundle breakage and the matrix cracking, whereas C/Cs made using carbon felt exhibit the lowest value for work of fracture, while those made with random fibers or 2-D cloth exhibit intermediate values [5, 15].

Matrix-Dominated Properties

The matrix-dominated properties are also dependent on fiber/matrix interface bond strength, fiber architecture and volume fraction, density and microstructure of matrix, etc. The better properties are obtained in case of composite with good fiber/matrix interface bond strength [92]. In addition, the volume fractions of the fiber also affect the properties. In the case of well-bonded composite, the transverse strength will be independent of volume fraction of fibers and approximately equal to the strength of the matrix. Conversely, in poorly bonded composite, a dependence on volume fraction will be observed, and strengths are appreciably lower than the strength of the matrix.

The increase in density of the composite also improves the properties. In composites, the compressive behavior is controlled by ILSS. As the density increases,

ILSS of the samples also increases, but up to certain value, which is again clear from this that the shear strength of the matrix governs the ILSS of composites [53]. ILSS for carbonized pitch-densified materials appears to be superior to resin–char matrices [77]. Thus, it is difficult to interpret matrix-dominated properties by varying one parameter while keeping other factors constant. These properties can be maximized by optimizing all the parameters. However, the greatest improvement in these properties is obtained by placing fibers in the appropriate directions.

Thermal Properties

Being composed of complex microstructures of reinforcing fibers and matrix, along with embedded micro-/macropores and voids, estimation of thermal properties of the C/Cs becomes difficult. However, the thermal properties of C/Cs may be tailored by proper choice of constituents, their configuration, and processing conditions [36]. In C/Cs, the thermal conductivity in the fiber direction is high and low in perpendicular direction (Fig. 10.12). Also, it depends on whether fibers are derived from PAN- or pitch-based composites because they possess different heterogeneous microstructures along their cross section. Pitch-based fibers are more conductive than the PAN-based fibers along the fiber axis. Moreover, the vapor-grown carbon fibers also exhibit high thermal conductivity. Using these, high thermal conductivity composites can be fabricated conveniently [5, 15].

The type of matrix microstructure also contributes to the thermal conductivity. If graphite matrix orients around the fibers, the planes oriented in the fiber direction contribute to the thermal conductivity. In the case of thermosetting resin-derived

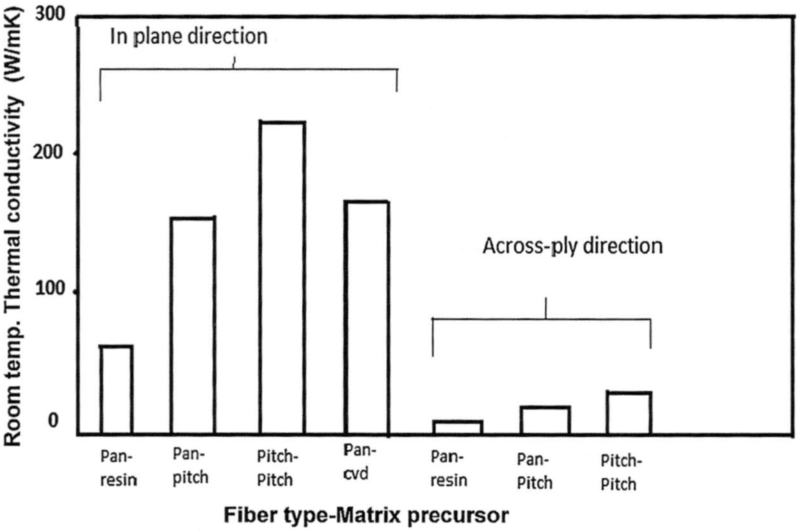

Fig. 10.12 Thermal conductivity of carbon/carbon composites with different fiber/matrix combinations [36]

C/Cs, the increase in thermal conductivity is observed with heat treatment temperature due to stress graphitization of the matrix. In contrast to thermoset-derived composites, the pitch precursor-derived composites exhibit high thermal conductivity and are attributed to the graphitic microstructure of the matrix among the filaments where all of the graphitic layers lie parallel to the fibers. This type of matrix structure can easily obtained by the use of mesophase pitch and HIPIC process. The improvement in thermal conductivity is more than 70 % when PAN fibers are replaced by pitch-based fibers in pitch-derived matrix, whereas in CVD-derived carbon matrix, it exhibits a wide range of microstructures, from isotropic to highly anisotropic rough laminar structures. Accordingly, the thermal conductivity of C/Cs with a CVD matrix also varies. The C/Cs with rough laminar microstructure exhibit highest thermal conductivity [5, 15].

The voids, pores, and cracks also disrupt the heat flow, thus decreasing the thermal conductivity. If the formation of matrix defects could be controlled then the thermal conductivity of matrix could be increased by sixfold. The fiber/matrix interface bond strength also contributes to thermal conductivity. In the case of good bond strength, the defects will be in the region of matrix due to its shrinkage on the fibers, whereas in the case of poor bond strength, the matrix defects tend to reduce. Hence, the latter composites will exhibit higher thermal conductivity both perpendicular and parallel to the fiber axis than the former. In addition, the thermal conductivity of composite material also increases with fiber volume fraction and density [5, 93, 94].

Similar to thermal conductivity, the amount of heat absorbed (specific heat) in C/Cs depends on voids, cracks, and defects as well as microstructure, i.e., the type of fiber and matrix present in the composites and structural arrangements of the two [93, 94]. The high-modulus pitch-type fibers exhibit higher specific heat than the PAN-based fibers, because high-ordered graphitic structure of the former scatters less thermal energy and absorbs more energy as compared to the latter. Similarly, matrix with highly ordered graphitic structure exhibits high specific heat than the less ordered graphitic structure. The voids, pores, and cracks also disrupt the heat flow, thus decreasing the specific heat. If the formation of matrix defects could be controlled, then specific heat of the composites could be increased.

Similarly, the thermal expansion of C/Cs in the direction of reinforcement is mainly controlled by the fibers, whereas in the direction perpendicular to it, the type and content of matrix and porosity have major influence [5, 93, 94].

Effects of Heating Rate on the Properties of C/Cs

Initial carbonization is an important step in the manufacturing of C/Cs derived from thermosetting resins. During carbonization, thermal degradation of polymers takes place, generating various volatile by-products, which lead to generation of temperature and pressure gradients within the composites [95–98]. The degradation of polymer chains leads to the irreversible physicochemical transformations in the composite. The composite experiences several macroscopic changes such as formation of cracks and voids by delamination, warping, and damages in the matrix. Again, the high-temperature carbonization leads to reduced density and changed thermal properties such as heat capacity and thermal conductivity [98]. In order to diminish these adverse effects, generation of both pressure as well as temperature gradients during carbonization should be minimized by controlling the rate of heating, which eventually controls the rate of production of volatile by-products. However, as low heating rates lead to high manufacturing costs, an optimization is generally adopted. Hence, the initial carbonization emerges as the rate determining step in the manufacturing process of C/Cs.

Studies of the carbonization behavior of resin-based composites have been performed either through experimental investigations or by developing numerical model. Experimental investigations mainly focus the effects of heating rates on the mechanical properties of C/Cs. Ko and Chen have studied the effect of heating rate on pyrolysis of a plain-woven PAN-based carbon fabric/phenolic resin composite by varying the heating rates between 0.1 and 5 °C/min [58]. The study suggested no significant effect of heating rate on the yield of the C/Cs carbonized at 1000 °C for the heating rates below 3 °C/min. Similarly, Roy has found that the interlaminar strength of 8H satin-woven PAN-based carbon fabric/phenolic resin composite remains unaffected by changing the heating rates between 0.03 and 0.8 °C/min [59]. On the other hand, Chang et al. have revealed that the density and flexural strength of plain-woven PAN-based carbon fabric/phenolic resin composites decreases by 0.7 % and 6.6 %, respectively, by changing the heating rate from 0.5 °C/min to 3 °C/min [60]. Again, no effect of heating rate on the weight loss and flexural modulus of the composites is observed.

Apart from experimental studies, numerical studies also predict the effect of heating rate on the properties of C/Cs. For example, using 1-D numerical model, Nam and Seferis have studied the carbonization behavior of 8H satin-woven PAN-based carbon fabric/phenolic resin composite when carbonized to 1000 °C at a heating rate of 2–7 °C/min [95]. The model predicts the variation of different parameters such as the pressure buildup, the temperature distribution, and the degree of degradation with the processing time and the through-thickness position in the composites. It is observed that a higher heating rate and a larger thickness of the composite give a larger pressure buildup and temperature gradient within the composites. Hence, higher heating rate could lead to a nonuniform carbonization and the development of internal stresses through the laminates due to thermal as well as pressure gradients, leading to localized delamination of the matrix.

Effects of Carbon Additives on the Properties of Carbon–Carbon Composites

As mentioned earlier, the pyrolysis produces cracks, pores, and voids in the composites due to the anisotropic shrinkage of matrix during carbonization. Researchers have observed that the modifications of the matrix by adding fillers such as fine powder of coke, graphite powder, and carbon black fillers to precursor matrix minimize the shrinkage, which not only reduce the manufacturing cost of C/Cs by improving their density in one step or less number of steps but also improve some of the properties of composites depending on the type and amount of fillers and heat treatment temperature used. Recently, carbon nanofibers (CNFs), carbon nanotubes (CNTs), and nanospheres have also been used as additives for thermosetting resins in very small quantities (1–5 %). These nanocarbon additives are preferred over coarse carbon materials due to their larger surface area as well as their inherent properties, which are translated to ultimate properties, but uniform distribution of these is a challenge. Also, costs are generally higher than the coarse carbon materials [5, 15, 34].

Furthermore, references may be made to published articles or patents in the literature to study the effects of carbon additives on the properties of C/Cs. Manocha et al. have reported that the addition of carbon black or graphite fillers to the resin matrix precursor increases the properties like density, strength, stiffness, and quasitensile fracture behavior of C/Cs, because the filler loading inhibits shrinkage cracks during carbonization of polymer and changes the microstructure of the bulk carbon matrix of the C/Cs from nongraphitizable to fine-grained graphitic carbon when it is treated at high temperatures [34].

Fitzer et al. have mentioned that the addition of graphite powder of 50 % to phenolic resin reduces the shrinkage of the composite samples from 20 % to 5 % after carbonization and exhibits accessible pores and cracks in case of composites made with surface-treated carbon fibers, which can be filled more effectively during subsequent densification cycles [5, 76]. Hence, less number of densification cycles is required compared to composites without fillers. In addition, the graphite filler loading also increases the failure strain as well as the fracture toughness of the composites.

Savage has mentioned that the interlaminar fracture toughness (ILFT) of 2-D C/Cs can be increased by more than a factor of 2 by the introduction of particulate graphite [15]. The graphite powder reduces the shrinkage of the matrix during carbonization that increases the fracture toughness by crack deflection and blunting mechanism. Optimization of filler content and particle size has resulted in the toughness value as high as 250 J/m^2.

In other studies, Kowbel et al. have observed the C/Cs of improved density, flexural strength, and shear strength by adding 70 % of carbon black fillers to thermosetting resin carbon fabric composites [99]. The improvement in density is observed above the heat treatment temperature of 1700 °C because of progressive graphitization of the matrix, which further improves the flexural strength and shear strength of the composites. Yasuda et al. have added the graphite powder up to 45 % to thermoset resin matrix and have made composites using unidirectional carbon fiber with or without surface treatment [100]. Then, the composites are carbonized at 1000 °C and graphitized at 3000 °C. The C/Cs of improved flexural strength are obtained at lower filler content (5–10 wt%). With further loading of fillers, the strength of C/Cs decreases depending on the fiber treatment and heat treatment temperatures. Also, the fracture pattern of the samples changes. The effect of carbon black filler loading on the phenolic-PAN fiber felt-based C/Cs has been studied by Ko et al. [101]. At 2500 °C, the addition of 10 wt% of carbon black filler improves the graphitization of the matrix and exhibits nearly 300 % improvement in flexural strength.

In another study, Hu et al. have studied the effect of mesocarbon microbead loading (0–30 %) on various properties of C/Cs prepared from oxidative PAN fiber, a resol-type phenolic resin, and mesophase pitch derived from coal tar [102]. C/Cs containing 10–30 wt% mesophase exhibit higher density, greater stacking size, and higher preferred orientation than those without any mesophase. Incorporation of mesophase enhances the flexural strength and flexural modulus by ~20 % to 30 % and ~15 % to 31 %, respectively. These composites also show improvement in thermal conductivity from ~12 % to 31 %.

Fitzer and Hutner have also reported a steady increase of flexural strength in the case of phenolics as precursor with increasing the graphite filler addition, whereas in the case of pitch, the flexural strength of the composites moves through a maximum [76]. Decrease in the strength with higher graphite contents is due to the reduction in the number of carbon binder bridges between the fibers. Hence, to achieve better strength, the graphite filler addition should not affect the binder bridges. From the study, it is also observed that the graphite grain size of 2 μm is too high to guarantee a homogeneous distribution in the composite, especially at higher filler content.

Moreover, the addition of nanostructured carbons in thermoplastic pitches also changes their microstructures. The added carbon introduces the oriented structures in isotropic pitches. This shows an increase in graphitic ordering and hence mechanical properties, while nanoscale mosaic structures in mesophase pitches reduce the planar cleavages and shear failure [34, 103]. Further, it is also observed that the addition of carbon black to coal tar pitch increases the carbon yield and strength, improves the thermal properties, and reduces the porosity of pitch-based cokes [104, 105].

The effect of carbon additives on the ILSS of 2-D C/Cs has been reported by Seok et al. [106]. These composites are prepared by compression molding method and then carbonized at 1400 °C. Additives such as graphite powder, carbon black, milled carbon fiber, and carbon fiber mat are added during preparation. The composites with 9 vol% of graphite powder show the greatest values of ILSS and flexural strength because of improvement in the density of composites. Whereas, in case of carbon black, the composites show a slight improvement of ILSS at 3 vol%, but the flexural strength decreases. When milled carbon fiber and carbon fiber mat are added, the lack of resin and the heat shrinkage during the carbonization cause the delamination in the composites, which further results in decrease of density, ILSS, and flexural strength of composites.

Manocha et al. have studied the interfaces in nanostructured C/Cs and their effect on matrix microstructure and thermophysical properties of the composites [107]. Composites are made with vapor-grown carbon fibers (VGCFs) and petroleum-derived nanospheres filled with thermosetting (polyfurfuryl alcohol) matrices. The amount is varied in between 1–5 wt% in the case of VGCF and 5–20 wt% in the case of nanospheres. The composites are heat treated at 1000 °C and 3000 °C. The structural transformations are studied as a function of heat treatment through X-ray diffraction (XRD), scanning electron microscopy (SEM), and Raman spectroscopy. It is observed that the addition of these nanostructured materials brings nanocrystalline anisotropic orientations at the interfaces, which increases the electrical conductivity of the composites.

Jain et al. have investigated the influence of carbon nanofibers (CNFs) on the microstructure and interlaminar shear strength (ILSS) of C/Cs by using different loadings of 0 %, 2 %, and 5 % (w/w) of CNFs [108]. It is observed that the CNFs provide bridging mechanism for matrix microcracking and reduce the matrix shrinkage that occurred during the first carbonization.

Dae et al. have studied the effect of the addition of CNTs on the tribological behavior of C/Cs. C/Cs are fabricated by using CNT/carbon composite layers having CNT loadings ranging between 0 wt% and 20 wt% [109]. With increasing CNT loading, the wear loss is found to decrease significantly, while the friction coefficient increases lightly. Chen et al. have studied the effects of multi-walled CNTs (MWCNTs) on the microstructure of resin carbon and thermal conductivity of C/Cs [110]. In this, they have added MWCNTs into furan resin. Then, unidirectional preforms are densified with the nanotube-doped furan resin by impregnation–carbonization cycles. The results show that MWCNTs induce the ordered arrangement during heat treatment and enhance the graphitization degree of resin carbon. Also, incorporation of small amounts of MWCNTs enhances the thermal conductivity of C/Cs significantly. The enhancement is found more significant for the direction perpendicular to the fiber axis. Higher concentrations of MWCNTs are found to decrease the thermal conductivity due to their improper dispersion. Sheikhaleslami et al. have developed a method for preparing a nanohybrid resin comprised of carbon nanomaterials such as carbon black and carbon nanotubes (CNTs) and phenolic resin [111]. In this, at first, carbon nanomaterials are dispersed in an organic solvent such as ethylene glycol and/or 1, 2 propandiol. Then, the dispersed organic solvent is mixed with phenolic resin by 10–15 % by weight. The advantage of introducing carbon nanomaterials is to increase the carbon yield of the resin and to form graphitic phase in residual carbon after carbonization in order to overcome the shortcoming of glassy structure of residual carbon from the resin without nanomaterials, because glassy carbon exhibits low mechanical strength and poor oxidation resistance. The evolution of graphite phase is characterized by X-ray diffraction pattern after heating the hybrid resin system at 1100 °C. It is observed that, among both the nanomaterials, CNTs show better results as a graphitizing agent.

Li et al. have studied the effect of carbon nanofiber (CNF) additive on the mechanical properties of C/Cs by using carbon composites containing 0.5 %, 10 %, 15 %, and 20 % of CNFs [112]. Both flexural strength and elastic modulus of the composites are found to increase with CNF incorporation. For 5 % loading, flexural strength and elastic modulus are found to increase by 76.3 % and 55.5 % compared to those of the composite without CNFs. However, the properties (flexural properties and ILSS) are found to decrease for higher (20 %) CNFs containing composite, which may be attributed to the presence of voids in the structure. Jinsong and Ruiying have studied the effects of CNFs on the flexural properties and interlaminar shear strength of CNF-reinforced C/Cs [113]. The flexural strength, modulus, and ILSS of the composite containing 5 wt% of CNFs exhibit enhancements of 21.5 %, 33.5 %, and 40.7 %, respectively, as compared to those of the composite without CNFs. Again, the mechanical properties decrease for higher loading (20 %) of CNFs.

Machining of Carbon–Carbon Composites

C/Cs are increasingly finding their use in the aeronautical, aerospace, nuclear, biomedical, and automotive industries because of their superior properties. Though the composites are made to near net shape, due to the high fabrication costs, any deficiency on conformity with the design specification of a product results in serious performance/financial losses. In order to meet the stringent manufacturing demands and for better performances, the need to machine these materials adequately increases.

Machining is the final operation on the fabrication of C/Cs in which the dimension precision and the surface finishing are determined. A very little information on machining of C/Cs has been found in the literatures as the machining of C/Cs is difficult by using conventional methods due to their non-homogeneity, anisotropy, variable hardness gradient within structure, and intrinsic brittleness. Ferriera et al. have studied some aspects of the C/Cs machinability using turning tests [114]. The performances of different tool materials like ceramics, cemented carbide, cubic boron nitride, and diamond are studied. Results show that only diamond tools are suitable for use in finishing of materials. In rough turning, the carbide tools can be used with some restrictions in parameters.

Mueller et al. have used laser machining for creating the grid features in the carbon–carbon grid blanks for ion engines [115, 116]. In addition, Gureev et al. have studied the possibility of pattern cutting of sheet made of C/Cs using laser radiation [117]. It is experimentally demonstrated that the continuous-wave laser radiation could be efficiently used only for pattern cutting of C/Cs with a thickness up to 1.5 mm. For larger thickness of the composites, the use of pulse-recurrent radiation with high pulse-repetition rate in the multipass-cutting mode is recommended.

Hocheng et al. have studied the feasibility of electrical discharge machining (EDM) in C/Cs [118]. In this, the material is machined by electrical discharge sinker. In this process, the important parameters include the pulse current and pulse duration time, which control the material removal rate, the surface topography, and the recast layer that remained on the workpiece surface. Experimental investigations by George et al. have revealed empirical relations as a function of machining variables to analyze the machinability of C/Cs [119, 120]. The machinability parameters (response functions) include relative circularity of hole, overcut, electrode wear rate, and material removal rate (MRR), while pulse current, pulse on time, and gap voltage are the machining variables.

Applications of Carbon–Carbon Composites

As mentioned before, C/Cs are considered to be a class of materials having a wide spectrum of properties and applications. C/Cs with desired shape and properties required for particular application can be produced by meticulously choosing the type, architecture, and amount of carbon fiber and matrix precursor and processing conditions. Moreover, the ongoing development of high-performance carbon fibers and newer matrices based on pitch or advanced high-char-yielding resins continuously adds to the spectrum of C/C properties and products, which offer a high-performance engineering material. Therefore, in addition to the defense, aircraft, and spacecraft applications, a steady interest is also growing in civilian applications [5, 15]. In the following sections, general applications of C/Cs are mentioned according to the various specific properties of the composites.

Carbon–Carbon Composites as Brake Disks

The development of high-speed and large-capacity aircrafts requires improved braking systems. The requirements are lightweight materials with a smooth frictional behavior and capable of withstanding high temperature. This is particularly important during takeoffs of aircraft. During this, the temperature of the frictional surface rises to more than 2000 °C in about 20 s [5, 121]. The conventional steel brake disk, if successful in stopping the aircraft in an emergency, is likely to get destroyed by warping or melting due to the intense heat generation during stoppage of vehicles. C/Cs, which are stable at high temperatures and have excellent friction and wear characteristics, are the obvious solution to this problem. Higher heat capacity (2.5 times that of steel) and high strength at elevated temperatures of C/C-based brakes make it possible to reduce the brake weight by 40 % and to increase their service life by a factor of two [15].

The potential for significant growth in C/C brakes is further underlined by the interest shown by manufacturers of racing cars, heavy-duty surface transport systems, and passenger cars in their advanced automotive braking systems [5, 15]. Figure 10.13 shows images of brake disks made of C/Cs.

Carbon–Carbon Composites for Aerospace Industries

C/Cs are continuously finding their applications in the aerospace field. Examples include thermal protection system of space shuttles, exit cones, nose tips of the reentry vehicles, etc. Thermal, thermophysical, mechanical, thermomechanical, and structural properties of C/Cs are utilized in these applications.

These are used in wing leading edges and nose caps of space shuttle because it encounters temperatures subzero in outer space and close to 2000 °C during reentry operations, which is well beyond the operating temperature of the metals. These

Fig. 10.13 Brake disks made of carbon–carbon composites (Image Courtesy to carid.com)

components provide thermal protection to the instrumentation from the searing heat of reentry and maintain structural integrity during operation [122, 123]. In addition to the space shuttles, nuclear missiles also employ nose cones and heat shields. For such applications, C/Cs are ideal candidates due to their high thermal conductivity and high specific heat so that the components operate as a heat sink and absorb the heat flux without any problem. The porosity should be low and uniform so as to have a uniform low ablation/erosion. These stringent requirements are achieved through three-directional (3-D) fibrous reinforcement and the HIPIC densification route [5].

However, the leading edges are formed from 2-D carbon fabric with carbon matrix derived from phenolic resin/pitches. Furthermore, these are generally protected against oxidation resistance coatings for the high-temperature applications. Other components used for aircraft industries are gas rudders and thrust deflectors for military aircrafts and expansion nozzle of hypersonic propulsion unit [5, 121].

In rocket motors with a solid propellant system, the throat and exit cone are made of C/Cs. The exhaust gases from the propellant chamber pass out through the throat and then finally out through the exit cone. These also necessitate the choice of high-density 3-D C/Cs [5, 14, 15].

Carbon–Carbon Composites for Aeroengine and Turbine Components

The efficiency of heat engines is greatly improved by operating them at high temperatures and permitting high combustions. The choice of C/Cs in jet engine rotors and stators offers the possibility of operating at higher temperatures than those used in conventional high-temperature alloys. Furthermore, the use of C/Cs results in a significant reduction in a weight, engine size, and fuel consumption. The various jet engine components include turbine wheels, combustions chambers, and exhaust nozzles [5, 14, 15, 121]. Images of C/C-based turbine rotors have been shown in Fig. 10.14.

Fig. 10.14 Turbine rotor made of carbon–carbon composites (Image Courtesy to grabcad.com)

Carbon–Carbon Composites in Nuclear Reactor

Carbon-based materials are of significant interest in nuclear reactor design owing to their desirable neutronic properties and are being used in the nuclear reactors since the development of very first nuclear reactor CP-1. Low atomic weight coupled with high neutron scattering cross section and low neutron absorption cross section has made carbon a suitable moderator and reflector material in high-temperature nuclear reactors. For Generation IV nuclear reactors, C/Cs are being considered for better tailor-made properties, mainly with respect to thermal conductivity and strength. The high-temperature reactors, designed by India, propose to use C/Cs in fuel tube, downcomer, and reflector blocks (Fig. 10.15) [124].

However, the irradiation properties of C/Cs are not well established. Burchell et al. irradiated 1-D, 2-D, and 3-D C/Cs at 600 °C and to damage doses up to 1.5 dpa. 3-D C/Cs have more isotropic dimensional changes than that of 1-D or 2-D composites [125]. Pitch fiber composites are more dimensionally stable than PAN fiber composites, and high graphitization temperatures are found to be beneficial.

Though graphite is used in high-temperature reactors, where its irradiation defects get annealed out, it cannot be used in low-temperature thermal nuclear reactors due to accumulation of Wigner energy, which is the stored energy in graphite lattice due to displacement of atoms from lattice position by neutron or particle irradiation [126]. In such cases amorphous carbon–carbon composites have been proposed [127]. Dasgupta et al. have shown that carbon black–phenolic resin-derived carbon composite is a candidate material for such application [128, 129]. Figure 10.16 shows the flow sheet for making such composite.

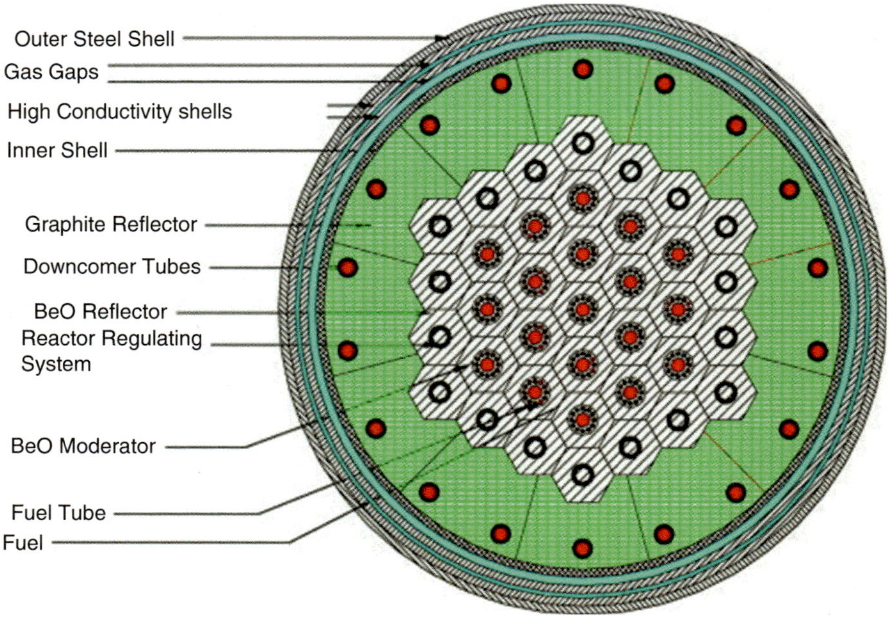

Fig. 10.15 Cross-sectional view of compact high-temperature reactor [124]

Pyrolytic graphite, artificial fine-grained graphite, and C/Cs have been adopted as plasma-facing components in fusion devices. Owing to their high thermal conductivity and high thermal shock resistance, C/Cs are one of the ideal choices compared to the conventional graphite materials for the fusion reactors. The use of C/Cs as plasma-facing materials has been reported in Tokamak Fusion Test Reactor (TFTR), Joint European Torus (JET), Tore Supra, and JT-60 in Japan. Tokamak fusion devices utilize carbon materials for their first-wall linings, limiter, and armor on their plasma-facing components (PFC) as shown in Fig. 10.17. C/Cs possess a number of attributes such as low atomic number, high thermal shock resistance, high sublimation temperature, and high thermal conductivity, which makes it a good choice in the fusion reactors. These C/Cs may be the choice for International Thermonuclear Experimental Reactor (ITER) which must endure severe environment including high-heat fluxes, high armor, surface temperature, and eddy-current-induced stresses during plasma disruption. The plasma-facing C/Cs will suffer structure and property degradation as a result of carbon atom displacements and crystal lattice damage, caused by impinging high-energy fusion neutrons and energetic helium ions for carbon transmutations. As C/Cs are infinitely variable family of materials, the processing and design variables such as (1) architecture, i.e., 1-D, 2-D, 3-D, or random fiber distribution; (2) fiber precursor, i.e., pitch, polyacrylonitrile (PAN), or

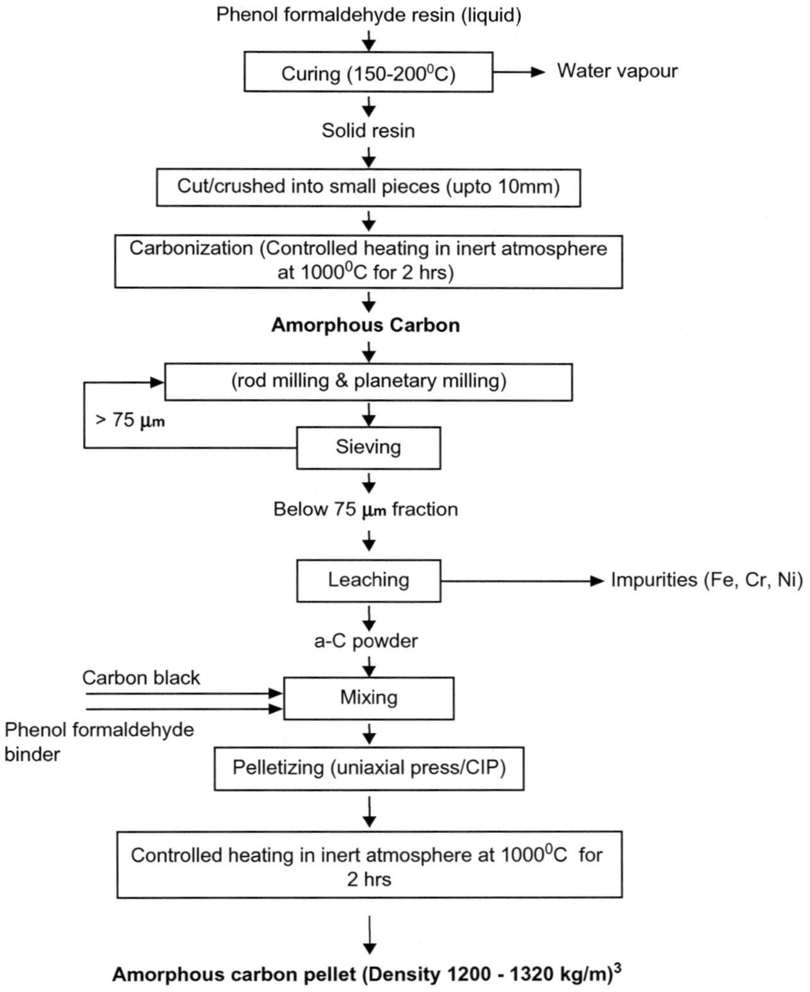

Fig. 10.16 Flow-sheet for making amorphous C/Cs [129]

vapor grown; (3) matrix, i.e., liquid impregnation (pitch or resin) or CVI; and (4) final graphitization temperature will influence the properties and behavior of C/Cs.

Carbon–Carbon Composites for Diesel Engine Components

Piston-driven engines such as diesel engines could also operate with increased efficiency and reduced weight if cooling requirements are minimized through the use of materials, which function efficiently at high temperatures. C/Cs are currently

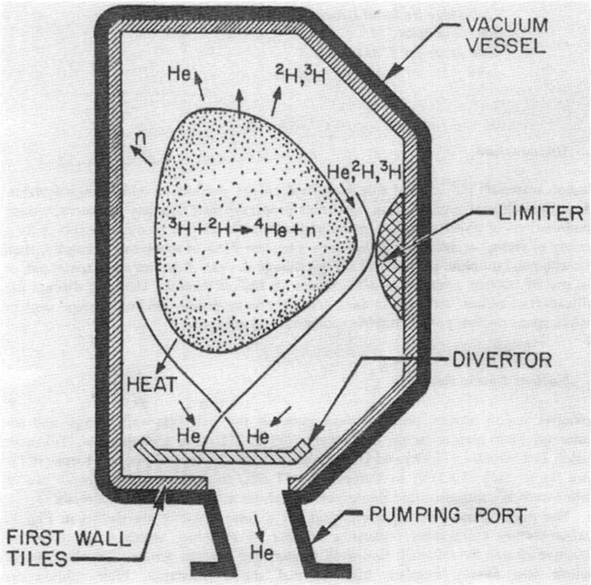

Fig. 10.17 Schematic of the plasma-facing components (PFC) of a Tokomak fusion reactor [125]

being evaluated for many such applications. The various components include piston crown valves, cylinder liners, and connecting rods [5, 14, 130].

Carbon–Carbon Composites as Refractory Materials

It is known that polygranular carbon and graphite materials are widely used as refractory materials in an oxidizing environment. With carbon fiber reinforcement, the mechanical properties of these carbon refractories can be further improved. Thus, C/Cs are increasingly finding their applications in industries also.

Hot Press Dies

The traditional material used in die manufacturing for high-temperature, high-pressure sintering of ceramics and metals is thick-walled fine-grained polycrystalline graphite. As an alternative, thin-walled fine-grained C/C dies with a high hoop strength have been tested for molding in powder metallurgy and have been found to withstand repeated applications. For example, the die for hot molding of cobalt-based metal powders and tools for superplastic forging of titanium are made of entirely with C/Cs [5, 121]. Furthermore, cylindrical heating elements made of C/Cs for hot isostatic presses provide a constant resistance; hence, fewer power adjustments are required during operations, and costs are reduced. This not only allows a high-temperature application but also increases the life span of the mold by a factor of 10 compared with graphite mold [5, 15].

High-Temperature Mechanical Fasteners

Bolts, screws, nuts, and washers made of C/Cs are used where high temperature and severe chemical resistance are present. The strength and stiffness at high temperature guarantees high fastening stability. Also, due to the anisotropic coefficient of thermal expansion, the systems made of C/Cs are self-fastening at high temperatures. The parts are used in the semiconductor industry, furnace construction, and other high-temperature apparatus and equipments. Examples include liners, plates, crucible sleeves, and other auxiliary aids [5, 14].

Carbon–Carbon Composites for Glassmaking

Carbon and graphite are not easily wetted by molten glass. Additionally, C/Cs possess a higher thermal shock resistance and impact toughness, and their porosity and thermal conductivity may be controlled; hence, there is no impairment of glass surfaces. Therefore, C/Cs are used in various parts of glass container forming machines, i.e., in channeling systems to carry the gobbets of molten glass, as molds for crystal glass products and as an asbestos replacement for hot-end glass contact elements for moving hot glassware articles [5, 15].

Carbon–Carbon Composites for Use at Intermediate Temperatures

C/Cs having high thermal and electrical conductivities and resistance to most chemicals are used as electrodes and other structural components in fuel cells. Also, these are the materials of choice even for vanes for rotary vane compressors and vacuum pumps. Moreover, these may be preferred for nuclear waste storage containers and also used in laser shields to protect space-based satellite systems from the heat of high-powered laser beams [5, 14, 121].

Carbon–Carbon Composites for Chemical Industries

C/C packings are highly effective for separation columns used to separate liquid mixtures in distillation–rectification plants. They are also used for stirrers, feed pipes, support grids, filter plates, and thin-walled heat exchangers [5, 14, 121].

Carbon–Carbon Composites as Biocompatible Materials

Elemental carbon is considered to be the most biocompatible among all known materials. It is compatible with blood, soft tissues, and bones. Therefore, C/Cs are used as bone plates in osteosynthesis and endoprosthesis. Also, these are used in artificial heart components [5, 121, 131].

Emerging Applications

C/Cs can be made in any complex shape. With their highly flexible properties, these are materials of choice for newly evolving applications, i.e., advanced aircrafts and thermal management systems. Examples include parabolic radio frequency (RF) antennae, reflectors for satellite communications, RF limiters, plasma-facing surfaces, cladding elements, and diverter plates for nuclear fusion reactors [5, 14, 15].

Oxidation Protection Mechanism

Carbon–carbon composites have excellent mechanical and thermal properties at elevated temperatures, but some of the potential applications like turbine structural components, which require long-term exposure to high temperature, are restricted due to the inherent reactivity of carbon with oxygen beyond 450 °C. For sufficient lifetime of C/Cs at elevated temperatures in normal atmospheres, their protection from oxidation is essential. A number of different oxidation protection mechanisms have been explored to improve the oxidation resistance of C/Cs. The developed techniques can be categorized as:

(a) Surface coatings – single layer/multilayers, using chemical vapor deposition, pack cementation, physical vapor deposition (PVD), and plasma spray.
(b) In-depth protection includes sol–gel process impregnation with inorganic salts but for limited temperature range and melt impregnation or in-depth deposition of SiC matrix. With the external protection methods, the thermal expansion mismatch between carbon material and possible refractory coatings is the main problem to be overcome. Microcracks developed in refractory layers have to be sealed with glassy coatings. The best oxidation resistance is achieved when CVD surface coatings are formed in addition to in-depth protection.

Internal protection methods include:

(i) Direct removal and/or deactivation of catalytic impurities
(ii) Incorporation of oxidation inhibitors and total or partial substitution of matrix material

A successful protection system comprises a coating, internal inhibitor, and a compatible substrate since C/Cs constitute a diverse class of materials with a wide range of mechanical, thermal, and morphological properties. Selection of appropriate fiber, preform fabrication technique, matrix precursor, and densification processing method is essential if good oxidation resistance as well as physicochemical compatibility between substrate and coating is to be achieved. For higher oxidation resistance, we introduce a ceramic matrix like SiC instead of carbon matrix in the carbon fiber preform, which gives higher oxidation resistance than that of oxidation-resistant C/C. These composites are known as C/SiC composites. It provides a good trade-off between the high-temperature capability of carbon fibers and the high oxidation resistance of ceramic matrices. Extensive work has been carried out on C/SiC composites for liquid propellant rocket and air breathing engines, thrust vectoring nozzles, hot gas valves and tubes, and space plane thermal structures. The third family of thermo-structural composites, viz., SiC/SiC, employs ceramic fibers (SiC) and ceramic matrix (SiC). These composites provide an excellent oxidation resistance for long durations and are capable of withstanding thermal cycling for reusable structures. SiC/SiC composites are used for liquid propellant rocket engine chambers, jet engines, gas turbine components, and space thermal structures. However, SiC/SiC composites start losing the mechanical strength beyond 1200 °C unlike carbon–carbon composites.

Prospective and Challenges

C/Cs have many advantages and can be tailored to meet the specific functional requirements, yet these are used only in special applications. This is due to cost associated with the fabrication of composite materials using conventional methods. In order to expand the area of applications, researchers have focused their works through cost reduction within the manufacturing steps. Logically, one simplest way is to increase the carbonization rate/heating rate in the resin pyrolysis technique. This is because generally complex shapes are difficult to fabricate using either pitch matrix or CVI materials due to the difficulty of maintaining the dry fiber preform shape during the initial carbonization step. To avoid this difficulty in the process, hybrid densifications are sometimes practiced. In this process, first a rigid structure is made using the resin pyrolysis technique. After first cycle of carbonization, depending on the final applications, subsequent infiltrations of C/C skeletons are made with low viscosity resin or pitch or CVI materials in the densification cycles. In the resin pyrolysis technique, PCs are widely used as green composites (starting materials) for fabrication of C/Cs because of low cost and high carbon yield of phenolic resin. During carbonization, the phenolic matrix material absorbs energy (heat) from the surrounding area, resulting in degradation reactions and further causing volatile generation and weight loss of solid material, which in turn form the pressure and temperature gradients within the composites. As a result of these, shrinkage, cracking, and thermal stresses may build up during carbonization. Also, as described earlier, from the studies, it has generally been agreed that the high carbonization rates can deteriorate the mechanical performance of the composites. To minimize the adverse effects, low carbonization/heating rates are generally preferred. But the low heating rates lead to high processing cost because of long cycle times. Thus, the initial carbonization step becomes rate determining step in the manufacturing of C/Cs and more detailed investigation of this is essential.

Furthermore, PCs not only serve as green composites for fabrication of C/Cs but also as potential candidates for other structural applications. When these composites are exposed to high temperature, changes in the properties of composites such as decrease of density and non-monotonic behavior of heat capacity and thermal conductivity are expected, which further affect the performance of the composites. Thus, detailed investigations of these properties of PCs under high temperatures at different heating rates are essential.

Another way to reduce the manufacturing costs of C/Cs is to increase the carbon yield by modifying the matrix precursors, especially through polymer pyrolysis routes. As described earlier, modification of the matrix by adding carbon additives such as fine powder of cokes, graphite powders, carbon black fillers, CNFs, CNTs, and nanospheres to the resin matrix minimizes the shrinkage, which not only reduces the manufacturing cost of C/Cs by improving their density in one step or less number of steps but also improves some of the properties of composites depending on the type and amount of fillers and heat treatment temperature used. But uniform distribution of these fillers is a challenge. Also, costs are generally higher than the coarse carbon materials. Furthermore, the modification of the resin matrix with fillers shows an improvement of wear and frictional characteristics of precursor PCs. But no study has clearly discussed the effects of graphite fillers loading on the mechanical properties of precursor PCs and their C/Cs when these are heat treated at different temperatures, because the laminate thickness and density of the precursor composites change when these are made under similar processing conditions due to viscosity of the resin, which increases as the filler loading increases. Here, the advantage of using graphite filler is to get its low cost benefits.

As mentioned before, C/Cs are widely used in high-technological applications because of their superior properties compared to other traditional materials. Though these composites are made to near net shape, these materials have to be machined to meet the manufacturing demands and for better performances. A very little work on machining of C/Cs has been found in the literature. This is because of difficulties to machine the composites by conventional methods due to their non-homogeneity, anisotropy, variable hardness gradient within structure, and intrinsic brittleness. However, researchers have established that these can be machined by some of the nontraditional methods such as laser machining and electrodischarge machining. Since the composites are difficult to machine, a detailed study of their machinability features is essential using alternative methods.

Acknowledgment The authors acknowledge the financial support provided by the Board of Research in Nuclear Science, India, for carrying out this research work.

References

1. Kanter MA (1957) Diffusion of carbon atoms in natural graphite crystals. Phys Rev 107 (3):655
2. Dienes GJ (1952) Mechanism for self-diffusion in graphite. J Appl Phys 23(11):1194
3. Pierson HO (1993) Handbook of carbon, graphite, diamond and fullerenes- properties, processing and applications. Noyes Publications, Park Ridge
4. Burchell TD (1999) Carbon materials for advanced technologies, 1st edn. Elsevier, Oxford
5. Fitzer E, Manocha LM (1998) Carbon reinforcements and carbon/carbon composites. Springer, Germany

6. Park SJ, Jang YS (2001) Interfacial characteristics and fracture toughness of electrolytically Ni plated carbon fiber-reinforced phenolic resin matrix composites. J Colloid Interface Sci 237 (1):91
7. Wielage B, Odeshi AG, Mucha H, Lang H, Buschbeck R (2003) A cost effective route for the densification of carbon-carbon composites. J Mater Process Technol 132(1):313
8. Morgan P (2005) Carbon fibers and their composites. CRC Press, Boca Raton
9. Edie DD (1998) The effect of processing on the structure and properties of carbon fibres. Carbon 36(4):345
10. Manocha LM (2001) Carbon fibres. In: Buschow KJ (ed) Encyclopedia of materials science and technology. Elsevier, Amsterdam/New York, p 906
11. Lubin G (1990) Hand book of composites. Van Nostrand Reinhold Company, New York
12. Donnet JB, Wang TK, Reboulillat S, Peng JCM (1998) Carbon fibers, 3rd edn. Marcel Dekker, New York
13. Ram MJ, Riggs JP (1972) Process for the production of acrylic filaments. US Patent 3657409
14. Buckley JD, Edie DD (1992) Carbon-carbon materials and composites. Noyes Publications, Park Ridge
15. Savage G (1993) Carbon-carbon composites. Chapman & Hall, London
16. Clarke AJ, Bailey JE (1973) Oxidation of acrylic fibres for carbon fiber formation. Nature 243 (5403):146
17. Bahl OP, Manocha LM (1974) Characterization of oxidised PAN fibres. Carbon 12(4):417
18. Fitzer E, Muller DJ (1975) The influence of oxygen on the chemical reactions during stabilization of PAN as carbon fiber precursor. Carbon 13(1):63
19. Thorne DJ (1985) Manufacture of carbon fibre from PAN. In: Watt W, Perov BV (eds) Strong fibres. Elsevier, Amsterdam, p 475
20. Morita K, Miyachi H, Hiramatsu T (1981) Stabilization of acrylic fibers by sulfur atoms mechanism of stabilization. Carbon 19(1):11
21. Riggs DM, Shuford RJ, Lewis RW (1982) Graphite fibers and composites. In: Lubin G (ed) Handbook of composites. Van Nostrand Reinhold Company, New York, p 196
22. Diefendorf RJ, Tokarsky E (1975) High performance carbon fibers. Polym Eng Sci 15(3):150
23. Singer LS (1977) High modulus high strength carbon fibers produced from mesophase pitch. US Patent 4005183
24. Diefendorf RJ, Riggs DM (1980) Forming optically anisotropic pitches. US Patent 4208267
25. Nazem FF (1982) Flow of molten mesophase pitch. Carbon 20(4):345
26. Edie DD, Dunham MG (1989) Melt spinning of pitch-based carbon fibers. Carbon 27(5):647
27. Tibbetts GG (1990) Vapor-grown carbon fibers. In: Figueiredo J, Bernardo CA, Baker RTK, Hiittenger KJ (eds) Carbon fibers filaments and composites. Kluwer, Dordrecht, p 79
28. Baker RTK (1990) Electron microscope studies of catalytic growth of carbon fibers. In: Figueiredo J, Bernardo CA, Baker RTK, Hiittenger KJ (eds) Carbon fibers filaments and composites. Kluwer, Dordrecht, p 405
29. Scardino FL (1989) Introduction to textile structures. In: Chou TW, Ko FK (eds) Textile structural composites. Elsevier, New York, p 1
30. Ko FK (1989) Three-dimensional fabrics for structural composites. In: Chou TW, Ko FK (eds) Textile structural composites. Elsevier, Tokyo, p 129
31. McAllister LE, Lachman WL (1983) Multidirectional carbon-carbon composites. In: Kelly A, Mileiko ST (eds) Fabrication of composites. Elsevier, Amsterdam, p 109
32. Klein AJ (1986) Which weave to weave. Adv Mater Process 3:40
33. Tzeng SS, Chr Ya-ga (2002) Evolution of microstructure and properties of phenolic resin-based carbon/carbon composites during pyrolysis. Mater Chem Phys 73(2–3):162
34. Manocha LM (2005) Introduction of nanostructures in carbon–carbon composites. Mater Sci Eng A 412(1–2):27
35. Diwedi H, Mathur RB, Dhami TL, Bhal OP, Monthioux M, Sharma SP (2006) Evidence for the benefit of adding a carbon interphase in an all-carbon composite. Carbon 44(4):699
36. Manocha LM (2003) High performance carbon-carbon composites. Sadhana 28(1–2):349

37. Kaae JL (1985) The mechanism of the deposition of pyrolytic carbons. Carbon 23(6):665
38. Knop A, Pilato LA (1985) Phenolic resins chemistry applications and performance future directions. Springer, Berlin, p 91
39. Aierbea GA, Echeverrıab JM, Martina MD, Etxeberriac AM, Mondragona I (2000) Influence of the initial formaldehyde to phenol molar ratio (F/P) on the formation of a phenolic resol resin catalyzed with amine. Polymer 41(18):6797
40. Mottram JT, Taylor R (1987) Thermal conductivity of fibre-phenolic resin composites. Part I: thermal diffusivity measurements. Compos Sci Technol 29(3):189
41. Mottram JT, Taylor R (1987) Thermal conductivity of fibre-phenolic resin composites. Part II: numerical evaluation. Compos Sci Technol 29(3):211
42. Wang S, Adanur S, Jang BJ (1997) Mechanical and thermo-mechanical failure mechanism analysis of fiber/filler reinforced phenolic matrix composites. Compos Part B 28(3):215
43. Sreejith PS, Krisnamurthy R, Narayanasamy K, Malhotra SK (1999) Studies on the machining of carbon/phenolic ablative composites. J Mater Process Technol 88(1):43
44. Sreejith PS, Krisnamurthy R, Malhotra SK, Narayanasamy K (2000) Evaluation of PCD tool performance during machining of carbon/phenolic ablative composites. J Mater Process Technol 104(1–2):53
45. Zhang ZZ, Su F-H, Wang K, Jiang W, Men X, Liu WM (2005) Study on frictional and wear properties carbon fabric composites reinforced with micro and nano particles. Mater Sci Eng A 404(1–2):251
46. Kim SS, Park DC, Lee DG (2004) Characteristics of carbon fiber phenolic composites for journal bearing materials. Compos Struct 66(1–4):359
47. Kim JW, Kim HG, Lee DG (2004) Compaction of thick carbon/phenolic fabric composites with autoclave method. Compos Struct 66(1–4):467
48. Park DC, Lee DG (2005) Through-thickness compressive strength of carbon–phenolic woven composites. Compos Struct 70(4):403
49. Park DC, Lee SM, Kim BC, Kim HS, Lee DG (2006) Development of heavy duty hybrid carbon–phenolic hemispherical bearings. Compos Struct 73(1):88
50. Park DC, Kim SS, Kim BC, Lee SM, Lee DG (2006) Wear characteristics of carbon-phenolic woven composites mixed with nano-particles. Compos Struct 74(1):89
51. White JL, Sheaffer PM (1989) Pitch-based processing of carbon-carbon composites. Carbon 27(5):697
52. Fujiura R, Kojima K, Kanno K, Mochida I, Korai Y (1993) Evaluation of naphthalene-derived mesophase pitches as a binder for carbon-carbon composites. Carbon 31(1):97
53. Fitzer E, Terwiesch B (1973) The pyrolysis of pitch and the baking of pitch bonded carbon/ carbon composites under nitrogen pressure up to 100 bar. Carbon 11(5):570
54. Marsh H, Menendez R (1989) Mechanisms of formation of isotropic and anisotropic carbons. In: Marsh H (ed) Introduction to carbon science. Butterworths, London, p 37
55. Forrest MA, Marsh H (1983) The effects of pressure on the carbonization of pitch and pitch/ carbon fibre composites. J Mater Sci 18(4):978
56. Lewis IC (1982) Chemistry of carbonization. Carbon 20(6):519
57. Lausevic Z, Marinkovic S (1986) Mechanical properties and chemistry of carbonization of phenol formaldehyde resin. Carbon 24(5):575
58. Ko TH, Chen PC (1991) Study of the pyrolysis of phenolic resin reinforced with two-dimensional plain-woven carbon fabric. J Mater Sci Lett 10(5):301–303
59. Roy AK (1993) Effect of carbonization rates on the interlaminar tensile stiffness and strength of two-dimensional carbon–carbon composites. Thermomechanical Behavior of Advanced Structural Materials ASME AD-vol34/AMD-vol173
60. Chang WC, Ma CCM, Tai NH, Chen CB (1994) Effect of processing methods and parameters on the mechanical properties and microstructure of carbon/carbon composites. J Mater Sci 29 (22):5859
61. Gupta A, Harrison IR (1994) Small-angle X-ray scattering (SAXS) in carbonized phenolic resins. Carbon 32(5):953

62. Olsen RE, Reese HF, Backlund SJ (1997) Process for forming carbon–carbon composite. US Patent 5686027
63. Sastri SB, Armistead JP, Keller TM (1999) Carbon-based composites derived from phthalonitrile resins. US Patent 5965268
64. Howdayer M, Gieres SJ, Van TD (1984) Process for the densification of a porous structure. US Patent 4472454
65. Takabatake M (1990) Process for producing high strength carbon-carbon composite. US Patent 4975261
66. Upadhya K, Hoffaman PW (1995) Densification of porous articles by plasma enhanced chemical vapor infiltration. US Patent 5468357
67. Thurston SG, Suplinskas JR, Carroll JT, Connors FD, Scaringella DT, Krutent CR (1998) Method for densification of porous billets. US Patent 5733611
68. Sheehan JE (1999) Low temperature densification of carbon fiber performs by colloidal graphite impregnation and mechanical consolidation. US Patent 5993905
69. Withers CJ, Loutfy OR, Kowbel W, Bruce C, Vaidyanathan R (2000) Process of making carbon-carbon composites. US Patent 6051167
70. Bahl OP, Mathur RB, Dhami TL, Chauhan SK (2006) A single step process for the preparation of high density carbon-carbon composite material. Indian Patent 2005DE00567
71. Liu HL, Jin ZH, Hao ZB, Zeng XM (2007) Research of low cost preparation of carbon/carbon composites: chemical liquid deposition process. J Solid Rocket Technol 30(6):529
72. Huang D, Snyder D, Lewis TR, Lewis CI (2007) Manufacture of carbon-carbon composites by hot pressing. US Patent 7207424
73. Abali F, Shivkumar K, Hamidi N, Sadler R (2003) An RTM densification method of manufacturing carbon/carbon composites using primaset-P30 resin. Carbon 41(5):893
74. Shivakumar NK, Avva SV, Sundarresan JM, Abali F, Cunningham A, Sadler LR (2008) Apparatus and method for forming densified carbon-carbon composites. US Patent 7332112
75. Weisshaus H, Kenig S, Sivegmann A (1991) Effect of materials and processing on the mechanical properties of C/C composites. Carbon 29(8):1203
76. Fitzer E, Huttner W (1981) Structure and strength of carbon/carbon composites. J Phys D Appl Phys 14:347
77. Fitzer E, Huttner W, Manocha LM (1980) Influence of process parameters on the mechanical properties of carbon/carbon-composites with pitch as matrix precursor. Carbon 18(4):291
78. Bokoros JC, Walker PL (1969) Chemistry and physics of carbon. CRC Press, Boca Raton
79. Pierson HO, Liebermann ML (1975) The chemical vapor deposition of carbon on carbon fibers. Carbon 13(3):159–166
80. Oh SM, Lee JY (1988) Structure of pyrolytic carbon matrices in carbon-carbon composites. Carbon 26(6):763
81. Hatta H, Aoi T, Kawahara I, Kogo Y, Shiota I (2004) Tensile strength of carbon-carbon composites: I- effect of density. J Compos Mater 38(19):1667
82. Hatta H, Aoi T, Kawahara I, Kogo Y (2004) Tensile strength of carbon-carbon composites: II- effect of heat treatment temperature. J Compos Mater 38(19):1685
83. Hatta H, Goto K, Ikegaki S, Kawahara I, Aly-Hassan MS, Hamada H (2005) Tensile strength and fiber/matrix interfacial properties of 2-D and 3-D carbon/carbon composites. J Eur Ceram Soc 25(4):535
84. Kogo Y, Kikkawa A, Saito W, Hatta H (2006) Comparative study on tensile fracture behavior of monofilament and bundle C/C composites. Compos Part A 37(12):2241
85. Manocha LM, Yasuda E, Tanabe Y, Kimura S (1988) Effect of carbon fiber surface-treatment on mechanical properties of C/C composites. Carbon 26(3):333
86. Manocha LM (1994) The effect of heat treatment temperature on the properties of polyfurfuryl alcohol based carbon-carbon composites. Carbon 32(2):213
87. Kowbel W, Shan CH (1990) The mechanism of fiber/matrix interactions in carbon-carbon composites. Carbon 28(2–3):287

88. Aglan MA (1993) The effect of intermediate graphitization on the mechanical and fracture behavior of 2-D C/C composites. Carbon 31(7):1121
89. Manocha LM, Bahl OP (1988) Influence of carbon fiber type and weave pattern on the development of 2-D carbon-carbon composites. Carbon 26(1):13
90. Zaldivar RJ, Rellick GS, Yang JM (1993) Fiber strength utilization in carbon-carbon composites. J Mater Res 8(3):501
91. Algan H (1992) The effect of microstructural heterogeneities on the fracture behavior of 3-D carbon-carbon composites. J Mater Sci Lett 11(4):241
92. Perry JL, Adams DF (1974) An experimental study of carbon-carbon composite materials. J Mater Sci 9(11):1764
93. Manocha LM, Warrier A, Manocha S, Sathiyamoorthy D, Banerjee S (2006) Thermophysical properties of densified pitch based carbon/carbon materials-I. Unidirectional composites. Carbon 44(3):480
94. Manocha LM, Warrier A, Manocha S, Sathiyamoorthy D, Banerjee S (2006) Thermophysical properties of densified pitch based carbon/carbon materials-II. Bidirectional composites. Carbon 44(3):488
95. Nam JD, Seferis JC (1992) Initial polymer degradation as a process in the manufacture of carbon–carbon composites. Carbon 30(5):751
96. Trick KA, Saliba TE (1995) Mechanisms of the pyrolysis of phenolic resin in a carbon/phenolic composite. Carbon 33(11):1509–1515
97. Trick KA, Saliba TE, Sandhu SS (1997) A kinetic model of the pyrolysis of phenolic resin in a carbon/phenolic composite. Carbon 35(3):393
98. Dimitrienko YI (1997) Thermomechanical behavior of composite materials and structures under high temperatures: 1. Materials. Compos Part A 28(5):453
99. Kowbel W, Chellapa V, Withers JC (1996) Properties of C/C composites produced in one low cost manufacturing step. Carbon 34(6):819
100. Yasuda E, Tanabe Y, Manocha LM, Kimura S (1988) Matrix modification by graphite powder additives in carbon fiber/carbon composite with thermosetting resin precursor as a matrix. Carbon 26(2):225
101. Ko TH, Kuo WS, Han WT, Day TC (2006) Modification of carbon/carbon composites with a thermosetting resin precursor as a matrix by addition of carbon black. J Appl Polym Sci 102 (1):333
102. Hu HL, Ko TH, Kuo WS (2005) Changes in the microstructure and characteristics of carbon/carbon composites with mesophase mesocarbon microbeads added during graphitization. J Appl Polym Sci 98(5):2178
103. Chollona G, Sirona O, Takahashia J, Yamauchib K, Maedab K, Kosakab K (2001) Microstructure and mechanical properties of coal tar pitch-based 2D-C/C composites with a filler addition. Carbon 39(13):2065
104. Aggarwal RK, Bhatia GJ (1978) Physical characteristics of baked carbon mixes employing coal tar and petroleum pitches. J Mater Sci 13(8):1632
105. Menendez R, Fernandez JJ, Bermejo J, Cebolla V, Mochida I, Korai Y (1996) The role of carbon black/coal-tar pitch interactions in the early stage of carbonization. Carbon 34(7):895
106. Seok JS, Hun KC, Jong HJ (2000) The improvement of interlaminar shear strength for low density 2-D carbon/carbon composites by additives. Polymer (Korea) 24(6):845
107. Manocha LM, Manocha S, Raj R, Yasuda E, Tanabe Y (2005) Effect of nanosized additives in carbonaceous precursors on microstructure of thermosetting resin based carbon-carbon composites. In: Singh M (ed) Proceedings of the international conference on high temperature ceramic matrix composites (HTCMC 5). American Ceramic Society Publication, Westerville, p 137
108. Jain R, Vaidya KU, Haque A (2006) Processing and characterization of carbon-carbon nanofiber composites. Adv Compos Mater 15(2):211
109. Dae SL, Jeong WA, Hwack JL (2002) Effect of carbon nanotube addition on the tribological behavior of carbon/carbon composites. Wear 252(5–6):512–517

110. Chen J, Xiong X, Xiao P (2009) The effect of MWNTs on the microstructure of resin carbon and thermal conductivity of C/C composites. Solid State Sci 11(11):1890–1893
111. Sheikhaleslami MS, Golestanifard F, Sarpoolaky H (2009) Method of preparing phenolic resin/carbon nano materials (hybrid resin). US Patent 0318606
112. Li JS, Luo RY, Ma TT, Bai SL (2006) Effect of carbon nanofibers additive on densification and mechanical properties of carbon/carbon composites. Tansu Jishu Bianjibu Publ 25(4):1–5
113. Jinsong L, Ruiying L (2008) Study of the mechanical properties of carbon nanofiber reinforced carbon-carbon composites. Compos Part A 39(11):1700
114. Ferriera JR, Coppini NL, Levy NF (2001) Characteristics of carbon-carbon composite turning. J Mater Process Technol 109(1–2):65
115. Mueller J, Brophy JR, Brown DK (1995) Endurance testing and fabrication of advanced 15 cm and 30 cm carbon-carbon composite grids. In: AIAA 95-2660 31st AIAA/ASME/SAE/ASEE Joint propulsion conference and exhibit, San Diego
116. Mueller J, Brophy JR, Brown DK (1996) Design fabrication and testing of 30 cm diameter dished carbon-carbon ion engine grids. In: AIAA 96-3204 32nd AIAA/ASME/SAE/ASEE Joint propulsion conference and exhibit, Lake Buena Vista
117. Gureev MD, Kuznetsov SI, Petrov AL (1999) Laser beam pattern cutting carbon based composites. J Russ Laser Res 20(4):349
118. Hocheng H, Guu YH, Tai NH (1998) The feasibility analysis of electrical discharge machining of carbon-carbon composites. Mater Manuf Process 13(1):117
119. George PM, Raghunath BK, Manocha LM, Warrier AM (2004) Modeling and machinability parameters of carbon-carbon composites a response surface approach. J Mater Process Technol 153–154:920
120. George PM, Raghunath BK, Manocha LM, Warrier AM (2004) EDM machining of carbon-carbon composite- a Taguchi approach. J Mater Process Techol 145(1):66
121. Fitzer E (1987) The future of carbon-carbon composites. Carbon 25(2):163
122. Curry DM (1988) Carbon-carbon materials development and flight certification experience from space shuttle. In: Howard GM (ed) Oxidation-resistant carbon-carbon composites for hypersonic vehicle applications. NASA CP-2051, Washington, DC, p 29
123. Martin J (1988) Creating the platform of the future NASA-The national aerospace plane. Def Sci 7(9):55
124. Dulera IV, Sinha RK (2008) High temperature reactors. J Nucl Mater 383:183
125. Burchell TD (2001) Carbon materials for energy production and storage. In: Rand B et al (eds) Design and control of structure of advanced carbon materials for enhanced performance. NATO series E, vol. 374, Kluwer Academic Publishers, Antalya, Turkey, pp 277–294
126. Marsden BJ (1998) Irradiation damage in graphite (The works of Professor B.T. Kelly). In: Graphite moderator lifecycle behavior. Proceedings of a specialists meeting, IAEA-TECDOC-901, Bath, United Kingdom, p-32.
127. Dasgupta K, Roy M, Tyagi AK, Kulshrestha SK, Venugopalan R, Sathiyamoorthy D (2007) Novel isotropic high density amorphous carbon composite for moderator application in low temperature thermal reactors. Compos Sci Technol 67:1794
128. Dasgupta K, Barat P, Sarkar A, Mukherjee P, Sathiyamoorthy D (2007) Stored energy release behaviour of disordered carbon. Appl Phys A 87:721
129. Dasgupta K, Prakash J, Tripathi BM (2014) Novel low Wigner energy amorphous carbon-carbon composite. J Nucl Mater 445:72
130. Taylor A (1989) Fabrication and performance of advanced carbon-carbon piston structures. In: Buckley JD (ed) Fiber-Tex 1988. NASA CP-3038, Washington, DC, p 375
131. Claes L, Fitzer E, Hiittner W, Kinzl L (1980) Torsional strength of carbon fiber reinforced composites for the application as internal bone plates. Carbon 18(6):383

Metal Matrix Composites: Theory, Techniques, and Applications

11

Sumit Pramanik, Jayesh Cherusseri, Navajit Singh Baban, L. Sowntharya, and Kamal K. Kar

Contents

S. Pramanik
Advanced Nanoengineering Materials Laboratory, Materials Science Programme, Indian Institute of Technology Kanpur, Kanpur, Uttar Pradesh, India

Department of Biomedical Engineering, Faculty of Engineering, University of Malaya, Kuala Lumpur, Malaysia

J. Cherusseri • N.S. Baban • L. Sowntharya
Advanced Nanoengineering Materials Laboratory, Materials Science Programme, Indian Institute of Technology Kanpur, Kanpur, Uttar Pradesh, India

K.K. Kar (✉)
Advanced Nanoengineering Materials Laboratory, Materials Science Programme, Indian Institute of Technology Kanpur, Kanpur, Uttar Pradesh, India

Advanced Nanoengineering Materials Laboratory, Department of Mechanical Engineering, Indian Institute of Technology Kanpur, Kanpur, India
e-mail: kamalkk@iitk.ac.in

K.K. Kar (ed.), *Composite Materials*,
DOI 10.1007/978-3-662-49514-8_11

Abstract

Numerous researches have been carried out on the developments of metal matrix composites (MMCs) in various applications. However, the main problem related to their processing is that controlling the balance between different parameters, such as ductility, strength, toughness, and so on, has created impediments on the way of MMC development due to the lack of knowledge in theory and proper processing technique for a particular application. As a result, its areas of applications become limited. Therefore, proper selection of matrix and reinforced materials and suitable fabrication technique could be fruitful for desired applications. Thus, crucial related theories, processing techniques, most important properties, and various advanced applications of MMCs have been focused in this chapter.

Keywords

Coefficient of thermal expansion • Thermal conductivity • Specific strength • Modulus • Wear • Metal matrix composite • Powder metallurgy • Interfacial bonding • Automobile • Orthopedic

Introduction

In 1957, during the reentry of the Soviet Sputnik satellite into the earth's atmosphere, the temperature attained by its outer periphery exceeded 1500 °C, crossing the limit of sustainability of any composite or monolithic material known then. Thus, a need for a suitable material arose, and scientists started experimentation on the metal matrix composites (MMCs) for a solution. To incorporate refractoriness in the lightweight metals and to minimize their thermal expansion coefficient, the use of inorganic fibers, ceramic fibers, or particulate phase is highly appreciated. Strength and stiffness can also be augmented with the help of reinforcement, but toughness tends to decline as compared to its monolithic metal. The momentum is being gained in the field of development of carbon and boron fibers instead of copper fibers due to the inception of space race between the different developed and developing countries. In the mid-1970s, new fibers like silicon carbide (SiC) have come into existence, and coatings for carbon and boron assisted them to work in tandem with the metallic matrices. In the 1980s, SiC whiskers have become a hot topic due to its extreme functionality and low cost. By the 1990s, heterogeneity has prevailed in the field of MMCs as far as spacecraft application is concerned. In rocket's combustion chamber, carbon-reinforced copper has been used; for rocket nozzles, SiC-reinforced copper has been used; fuselage-employed (alumina Al_2O_3)-reinforced aluminum (Al) has been used for wings and blades; SiC-reinforced Al and graphite–Al composite have been used for the antenna boom on the Hubble Space Telescope (HST) [1]. Particulate-reinforced MMCs (PRMMCs) recently have become a topic of interest because of their cost-effectiveness, when compared with the fiber-reinforced MMCs (FRMMCs). A higher volume fraction of particulates can be incorporated into metal matrices as compared to the whisker reinforcement. The MMCs are comprised of minimum two constituents – one is a metal and the other can be metal, ceramic, intermetallic, carbide, nitride, oxide, or an organic compound. Various types of composite systems have been developed in the recent years with different microstructures.

Components

Composites are the materials, which are fabricated from two or more constituent materials with significantly different physical or chemical properties, in order to achieve a better characteristic from their individual components. The individual components of MMCs are metal matrices and varying reinforcements. A brief discussion on components of MMCs is given in this section.

Matrix

The continuous framework into which the reinforcements are inserted is called "matrix," implying that any point in the material can be connected by a path through

Table 11.1 Matrix alloys and reinforcing materials for MMCs [2]

Matrix alloys		Reinforcements			
Aluminum alloy series		Short fiber		Particulates	
2xx	6xxx	SiC	Boron	SiC	AlN
3xx	7xxx	Al_2O_3	TiB_2	Al_2O_3	TiB_2
2xxx	8xxx	Al_2O_3–SiO_2	W	B_4C	Si_3N_4
Cu alloys		Carbon (graphite)		Carbon (graphite)	
		Fiber		Whisker	
Ti alloys		SiC	Boron	SiC	B_4C
		Al_2O_3	TiB_2	Al_2O_3	Si_3N_4
Mg alloys		Al_2O_3–SiO_2	W	Carbon (graphite)	
		Carbon (graphite)			

the matrix. This is not seen in the sandwiched type of materials. Matrix material usually consists of lighter counterparts such as Al, magnesium (Mg), titanium (Ti), copper (Cu), and their alloys, intermetallic compounds, and so on in order to provide a support to the secondary phase or reinforcement.

Reinforcement

Reinforcement is a material, which is inserted into the matrix of a composite, to provide the structural integrity to the composites by modifying the properties, including thermal conductivity, wear resistance, and so on. Examples of widely used reinforcement are Al_2O_3, boron nitride (BN), SiC, tungsten carbide (WC), carbon fiber, glass fiber, metals, organic materials, and so on. The reinforcement can be employed by continuous or discontinuous fibers or by particulates. The discontinuous MMCs follow isotropic trend, which has led them to undergo standard metalworking techniques. In the preparation of continuous reinforcement, fibers of SiC, carbon, or monofilament wires can be incorporated in the matrix. The fibers are embedded in such a way that sometimes its concentration or direction varies with distance, which ultimately results in anisotropy. Hence, the alignment of fibers plays a crucial role in deciding the performance of the composite. The use of whiskers, short fibers, or particles can act as discontinuous reinforcement, which leads to isotropic composites. One of the oldest materials used as reinforcing filler for making MMCs is boron filament. Both the matrix and reinforcing material have distinguished roles in determining the properties of the resulting composite structure. A brief description of the various types of metal matrix alloys and reinforcements for MMCs is illustrated in Table 11.1.

The typical values of coefficient of thermal expansion (CTE, within a temperature range of 25–200 °C), thermal conductivity, density, stiffness of metal matrices, and reinforcements of MMCs and their typical applications are given in Table 11.2.

Table 11.2 Typical values of CTE, thermal conductivity, density, stiffness of metal matrices, reinforcements, and MMCs and their applications [3–9]

Materials	CTE (ppm/°C)	Thermal conductivity (W/mK)	Density (g/cc)	Stiffness (GPa)	Application(s)
Metal matrices					
Cu	17.6–17.9	391–398	8.91–8.96	117–140	Packaging
Al (1100)	23.6	238	2.7	50–60	Packaging
Mg	24.8	156	1.74	45	–
Kovar (Ni–Fe)	5.2	11–17	8.1	131	Packaging
Solder	25	51	–	16	Electronic substrate
Gallium arsenide (GaAs)	6.5	54	5.23	–	IC
Silicon (Si)	4.2	151	2.3	112	IC
CuW (10–20 % Cu)	6.5–8.3	180–200	15.7–17	367	Packaging
CuMo (15–20 % Mo)	7–8	160–170	10	313	Packaging
Invar (Cr–Fe)	1.88	–	–	–	–
Reinforcements					
Carbon fiber	−1.5–28.6	7.5–450	1.80	3.6–1060	–
SiC	2.7	200–270	3.2	415	Substrate
AlN	4.0	170–200	3.3	310	Substrate
DBC AlN	7.0	180	–	>600	–
Al_2O_3	6.5	20–30	3.98	350	Substrate
Beryllia (BeO)	7.6	250	3.9	345	Substrate
Diamond	2.3	2000	3.5	–	–
BN	3.5	600	3.7	–	–
TiC	7.4	–	–	–	–
Stainless steel (SS316L)	17	–	7.5	–	–
W	4.4	166	19.3	406	
Metal matrix composites					
Al–SiC	6–12	130	3.0	150–240	Packaging, electronic substrate
Al2618 + 8 % TiB_2	19.5	138–295	–	–	–
Cu–C – SFRM	5.87–20	–	–	–	Heat sinks
Cu–C CFRM	4.4–18	39.7–219.7	–	–	–
Cu–Invar–Cu	5.7	132	8.3	–	–
Cu–carbon fiber	7–8.5	110–145	4.5–5.5	–	–
15Cu–85 W	6.5	167	16.6	248	Electronic component
Al + Al_2O_3	7.3–18.7	–	–	–	Structural
Alloy A359 + 20 % SiC	17.5	181.2	–	–	Brake components

(*continued*)

Table 11.2 (continued)

Materials	CTE (ppm/°C)	Thermal conductivity (W/mK)	Density (g/cc)	Stiffness (GPa)	Application(s)
3 M–MMC	6.6–9	–	3.2–3.4	–	Automotive pushrods, guidance fin
Invar–TiC	2.5–6	–	–	–	Laser-assisted SDM
Al alloy (A356.2)/rice husk ash (RHA)/SiC hybrid composite	17.44–15.06	182–258	296–356	6.25–4.9	Packaging, electronic substrate

Types

There are mainly three divisions in the MMCs with respect to the type of reinforcement used. These are (i) continuous or long-fiber MMCs, (ii) discontinuous or short-fiber MMCs, and (iii) particulate MMCs.

Continuous or Long-Fiber Metal Matrix Composite

In this type of MMC, a metal or an alloy is used as a matrix material, and the reinforced materials are continuous or long fibers. Research has been extensively carried out on the processing of long fibers for MMC in various applications. However, the main limitation related to their processing is that maintaining the balance between various parameters, such as ductility, strength, toughness, etc., has created impediments on the way of MMC development and hence its applications remain limited. The fibers having comparatively smaller diameter (~5–30 μm), which are sufficiently limber to be managed as tows or bundles, are termed as *multifilaments*. The materials incorporated are carbon, SiC, and various oxides, those can be woven, braided, filament wound, etc. Fibers having large diameter (~100–150 μm) are termed as "monofilaments" and are mostly processed by chemical vapor deposition (CVD) of either SiC or B having a core of carbon fiber or tungsten (W) wire. These are much less flexible than the multifilaments and hence managed as single fibers in place of use as bundles. The damage is caused through the incorporation of sharp curvature during handling; hence, meticulous precautions are needed [10]. Like stiffness, the strength of continuous fiber-reinforced MMCs is also anisotropic and is a function of loading direction like longitudinal direction, as shown in Fig. 11.1. Typical stress–strain curves of matrix metal or alloy, fiber, and MMC are shown in Fig. 11.2. In this figure, at stage I, both matrix and fiber remain

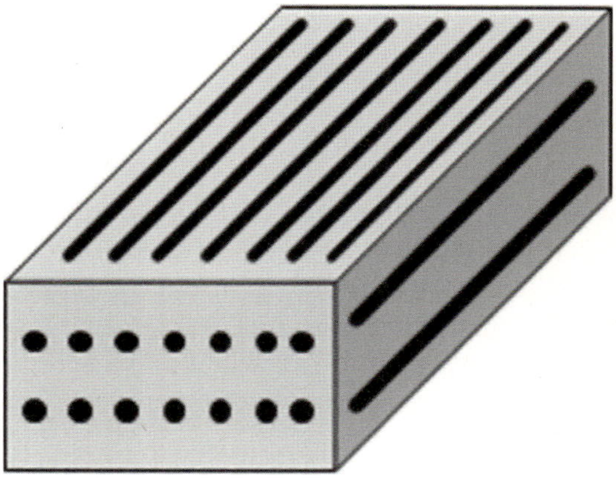

Fig. 11.1 Continuous or long-fiber-reinforced MMC

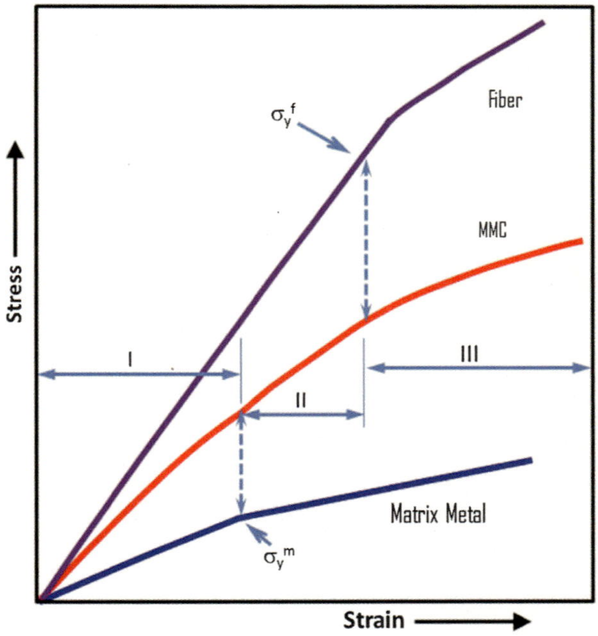

Fig. 11.2 Typical stress–strain curves of matrix metal or alloy, fiber, and MMC

elastic. At stage II, the matrix is deformed plastically and the fiber is deformed elastically, and at stage III, both matrix and fiber remain plastically deformed.

When the fiber-reinforced composite is loaded along the fiber direction and at isostrain condition, the strains in the matrix, fiber, and composite will be identical. Then the longitudinal stress (σ_{11}, Eq. 11.1), Young's modulus in longitudinal direction (E_{11}, Eq. 11.2), and Young's modulus in transverse direction (E_{22}, Eq. 11.3) of MMCs can be written [11] as

$$\sigma_{11} = (E_f \cdot A_f + E_m \cdot A_m)\,\varepsilon_{11} \tag{11.1}$$

$$E_{11} = E_f \cdot V_f + E_m \cdot V_m \tag{11.2}$$

$$\frac{1}{E_{22}} = \frac{V_f}{E_{ft}} + \frac{V_m}{E_m} \tag{11.3}$$

where, A_f, A_m, E_f, E_m, ε_{11}, V_f, V_m, and E_{ft} are the area of fiber, area of matrix metal, stiffness of the fiber in longitudinal direction, stiffness of the matrix metal in longitudinal direction, strain in the composite, volume fraction of fiber, matrix metal volume fraction ($=1 - V^f$), and transverse modulus of the fiber, respectively.

The common strengthening mechanisms of MMCs can be represented as follows [12]:

(i) The yield strength and flow stress can be increased with fiber volume fraction and fiber aspect ratio.
(ii) The yield strength and flow stress are strongly dependent upon the fiber orientation with respect to the loading direction.
(iii) The yield stress during compression is larger than that in tension due to the residual stress in MMC, which is caused due to imparity in coefficient of thermal expansion (CTE) between the matrix metal and fiber, likely during the fabrication process.

Discontinuous or Short-Fiber Metal Matrix Composite

In this type, the matrix material is metal or alloy, and the reinforced materials are discontinuous or short fibers. Fracture strength of this composite is very complicated due to the intricate morphology of the short fibers. Short-fiber MMCs can be divided into two types – (i) randomly oriented and (ii) aligned. These are schematically shown in Fig. 11.3.

The stress distribution is represented in Fig. 11.4. Here, the dash–dot line shows a linearization of the buildup portion of the length $L_c/2$, where L_c is the critical fiber length. If the fiber length $L < L_c$, the fiber cannot carry load to its full extent, whereas for $L \geq L_c$, the middle section of the short fiber endures the maximum stress, σ^f_L. According to the simplified stress distribution and rule of mixtures, the Eqs. 11.4 and 11.5 are used to predict the strength (σ_L) of aligned short-fiber MMC [12]:

$$\sigma_L = V^f.\sigma^f_L.\left(1 - \frac{L_c}{2L}\right) + V^m.\sigma'^m \qquad \text{for}\, L \geq L_c \tag{11.4}$$

$$\sigma_L = V^f.\sigma^f_L.\frac{L_c}{2L} + V^m.\sigma^m \qquad \text{for}\, L < L_c \tag{11.5}$$

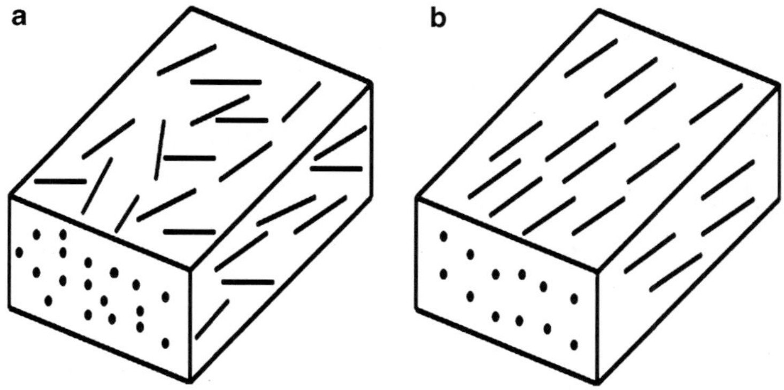

Fig. 11.3 Schematic of discontinuous or short-fiber MMCs with (**a**) randomly oriented and (**b**) aligned short fibers

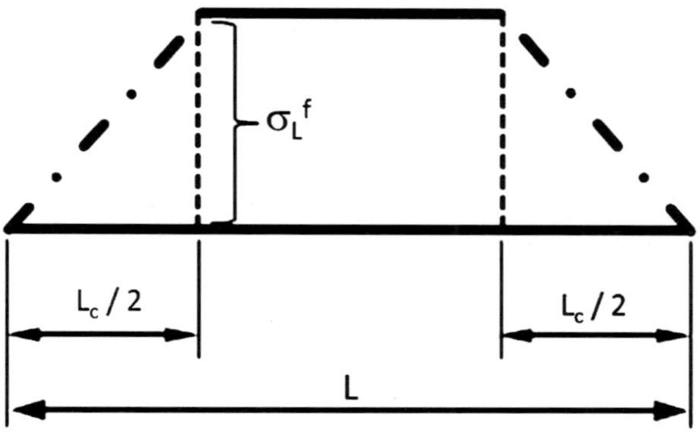

Fig. 11.4 Axial stress distribution in a short fiber (*solid line*) with buildup region of length $L_c/2$

where σ^m and σ'^m are the strength of the matrix metal and flow stress of the matrix metal at the failure strain of the fiber, respectively.

Prediction of Stiffness of the Short-Fiber MMCs: According to the modified shear-lag (MSL) model, the stiffness of short-fiber MMC can be predicted [13]. The predicted model of the short-fiber MMC is shown in Fig. 11.5.

The MSL predicted model equations for the short-fiber MMC (Eqs. 11.6, 11.7, 11.8, 11.9, 11.10, 11.11, and 11.12) are given below [13]:

$$\frac{E^c}{E^m} = \frac{V^w.E^w}{E^m} + \frac{V^m.B}{E^m.\varepsilon^c}\sinh(ns) + V^m + \frac{V^g}{E^m.\varepsilon^c.n'.s'}[A'\{\cosh(n's')\text{-}1\} + B'\sinh(n's')] \quad (11.6)$$

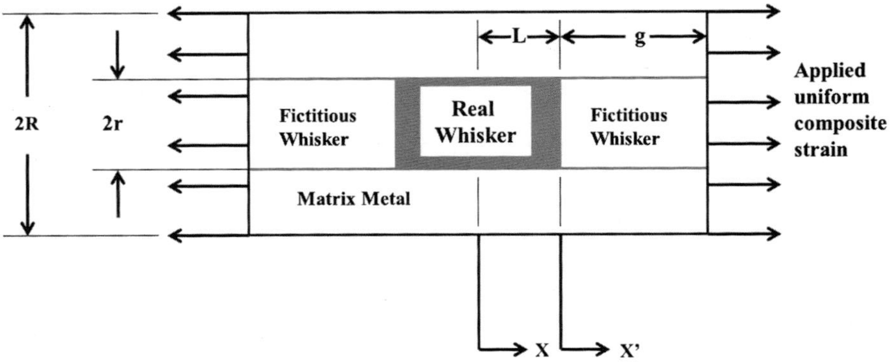

Fig. 11.5 The modified shear-lag predicted model for short-fiber MMC [13]

where

$$n^2 = \frac{E^m}{E^m(1+\upsilon^m)\ln\left(\frac{R}{r}\right)} \tag{11.7}$$

$$n'^2 = \frac{1}{(1+\upsilon^m)\ln\left(\frac{R}{r}\right)} \tag{11.8}$$

$$A = 0 \tag{11.9}$$

$$B = \frac{(E^w\text{-}E^m).\varepsilon^c}{\cosh(ns) + (n/n')\sinh(ns)\coth(n's')} \tag{11.10}$$

$$A' = \text{-}B'.\tanh(n's') \tag{11.11}$$

$$B' = \text{-}B(n/n')\sinh(ns)\coth(n's') \tag{11.12}$$

where R, r, s, s′, E^w, E^m, E^c, ε^c, ν^m, V^w, and V^g are the unit cell radius, radius of the whisker, fiber aspect ratio (= L/r, where "2 L" is the length of the real whisker), aspect ratio of the fictitious fiber (= g/r, where "g" is the half spacing between the whisker and composite ends), Young's modulus for real whisker, Young's modulus for matrix metal, Young's modulus for MMC, composite strain, Poisson's ratio of the matrix metal, volume fraction of the real whisker, and volume fraction of the fictitious whisker, respectively.

Now, the stress concentration factor (K_t) in a nearest-neighbor fiber can be written [14] as in Eq. 11.13

$$K_t = \frac{\sigma_i}{E_f.\varepsilon} \tag{11.13}$$

where σ_i, E_f, and ε are the whisker end stress, fiber modulus, and strain, respectively.

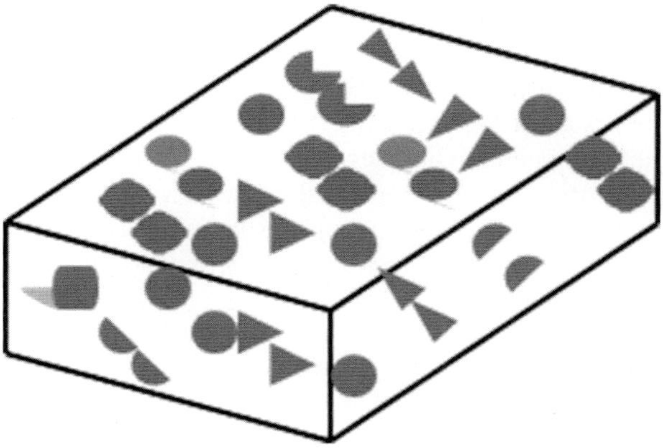

Fig. 11.6 Schematic of a particulate MMC

Short fibers have diameters in the order of few microns. These are first produced in lengths comprising of several hundreds of micrometers with a fine-grained polycrystalline microstructure. For the production of short fibers, melt infiltration process is used. The degree of reaction during processing partly decides the interfacial properties, which are influenced by the fiber's capabilities related with its surface chemistry. Better mechanical properties can be achieved if single crystal is used rather than a fine-grained polycrystal. The term "whiskers" is used in this case, usually having diameters ≤ 1 μm with aspect ratio around several hundreds, augmenting their tensile strengths to a higher degree. The tendency of these extremely fine fibers to form tenacity-induced ball-like particles creates handling problems and is cumbersome to handle its orientation in a controlled way, but most importantly it is concerned with perceived health hazard; for instance, its fragments of submicron size can be readily carried away by air destined to reach and harm the lungs [10].

Particulate MMC

In this type, the reinforcements are having rough, equiaxial geometry with aspect ratio <5. Particulates are adaptable to various shapes (spherical, angular, or plate-like), possessing both mono- and polycrystalline microstructures, and are normally more than 1 μm in diameter. Schematic of a particulate MMC is shown in Fig. 11.6. Particulate MMCs are suitable for entire gamut of industrial usage, the one having Al alloy matrices has achieved much popularity. Also, Ti-, Fe-, and Mg-based materials are being experimented based on their specificity and functionality. Most commonly, the particulates comprising of SiC or Al_2O_3 are the materials of much interest. The materials such as W, WC, B_4C, BN, TiB_2, TiC, SiO_2, ZrO_2, etc. are under investigation. During processing, chemical reaction can take place in some cases, and SiC can create impediments in Al- and Ti-based MMCs. The common manufacturing processes for particulate MMCs are melt incorporation, casting, powder blending, and consolidation. Reactive processes are some alternative processes to manufacture the same [10].

Interface

A bidimensional region where there is an alteration in one or more material parameters is termed as an interface. This is a boundary, which recognizes discontinuity in various material properties, and there is always some volume associated over which a change occurs in the material parameters gradually. The interface is important in a composite as (a) the interface constitutes a very large proportion of composites, and (b) the system formed by reinforcement and matrix is not in equilibrium thermodynamically.

Importance of Interface

For a fraction volume, the entire area of fiber–matrix–metal interface is increased with a decrease in fiber diameter according to the Eqs. 11.14, 11.15, and 11.16 [15, 16]:

$$V^{f} = \frac{\text{Volume of Fibers}}{\text{Volume of Composite}} = \frac{N\pi\, d^{2} l}{lwh} \tag{11.14}$$

$$A_{int} = N\pi\, ld \tag{11.15}$$

$$A_{int} = \frac{4lwh}{d} V^{f} \tag{11.16}$$

where N, h, l, w, d, and A_{int} are the number of continuous and unidirectional aligned fibers, height of the composite, length of the composite, width of the composite, diameter of the fiber, and area of the fiber–matrix–metal interface, respectively. From Eq. 11.16, it is clear that the area of fiber–matrix–metal interface is inversely proportional to the fiber diameter.

Crystallographic Nature of Interface

The crystallographic nature of ceramic–metal–matrix interfaces in composite is coherent having high energy, which can be highly credible vacancy sinks and act as a source of rapid diffusion paths [17]. These are also a destination of various segregation sites, heterogeneous precipitation sites, and precipitation-free sites. On fracture surfaces of the Al–diamond MMCs, the best properties were observed due to the preferential adhesion between Al matrix and diamond which occurs on the (1 0 0) faces of diamond [17].

Interfacial Bonding

In order to transfer load from the matrix to reinforcements, there must be a good degree of bonding between them. This bonding is different from the wettability in a sense that wettability only provides an idea about the degree of intimation of communion between liquid and solid. It can be treated as a necessary condition and not an indispensable condition for strong bonding criteria. Nice wetting is highly determined by the extent of pure mechanical bonding or weak van der Waals bonding, which rules out the possibility of formation of voids at the interface, and also the strength parameter is affected by chemical and mechanical bonding of the constituents [16, 18, 19]. It has been shown that depending on the interfacial energies between the particles and matrix, the growing crystal can either engulf or reject particles during solidification of an MMC [19]. The interface between reinforcement and metal matrix in MMCs can rely on interfacial mechanical bonding as found in W wire/Al MMC. Kennedy and Wyatt, 2001, had found the change in elastic modulus with increasing plastic strain, particle–matrix bonding in Al–TiC MMCs [20]. The rates of damage accumulation were lowest; damage initiation stresses were highest due to the strong interfacial bonding in cast and hot isostatically pressed Al–TiC MMCs. The strong interfacial bonding occurred due to intimate contact between Al matrix and TiC reinforcement during casting and cast and hot isostatically pressing processes via nucleation of solid Al on the particle surfaces. The interfacial bonding not only influences the mechanical properties but also modulates thermal and other properties of the MMCs. For example, the interface between the AlSi7 (i.e., Al alloyed with 7 wt% Si) matrix and diamond particles enhances thermal properties of AlSi7–diamond MMCs [17].

Chemical Bonding: Thermodynamically, most of the MMCs are nonequilibrium systems, and there persists a chemical potential gradient across the fiber–matrix–metal interface suggesting that for favorable kinetic conditions (generally a very high or very low temperature), diffusion occurs, which results in a chemical reaction between the entities. Eq. 11.17 [11] suggests a diffusion-driven growth in a controlled manner for the infinite diffusion coupled with a planar interface:

$$x^2 = Dt \tag{11.17}$$

where, x, D, and t are the thickness of the reaction zone, diffusivity, and time, respectively.

In MMCs, chemical bonding is much more common. It has been found that a chemical reaction between Al and carbon occurred to produce Al_4C_3, i.e., $4Al(l) + 3C(s) \rightarrow Al_4C_3(s)$, by forming a strong crystallographic preference on (1 0 0) planes of diamond in Al–diamond MMC. This chemical bond has significantly improved the thermal conductivity of Al–diamond MMC prepared by gas pressure infiltration (GPI) method (670 W/mK) compared to Al metal (240 W/mK) [17].

Mechanical Bonding: The interfacial bonding between the surfaces is highly affected by the mechanical bonding between the constituents. In the case of fiber-reinforced MMCs, longitudinal direction or fiber direction is the major direction to show the effect of mechanical bonding; transverse direction of the fiber displays only minor effect. Stronger adhesion between the particles and matrix interfaces improves the load transfer ability, yield strength, and stiffness [20]. The mechanical strength and ductility (or toughness) can be improved by using a third-phase material. The ultimate tensile strength of up to 708 MPa was observed for cryomilled Al 5083–boron carbide (Al–B_4C)/unmilled Al MMCs due to a clear interfacial bonding. On the other hand, a significant improvement in ductility (i.e., elongation of 2.3–3.7 %) with a reduction in strength (708–619 MPa) was observed as the volume percent of unmilled Al phase increased from. Both the elongation values of Al–B_4C/unmilled Al MMC were significantly higher compared to the Al–B_4C (1.8 %) [21]. This is because in the Al–B_4C composite, the high density of sharp particle edges coupled with strong stress concentrators leads to localization of strain at or near the Al–B_4C interface where cracks are nucleated and grown. Thus, the overall ductility of this composite was found to be limited.

Energy of Solid–Solid Interface

In a theoretical analysis during solidification process, the nucleation phenomenon can be expressed by Eqs. 11.18 and 11.19 [22]:

$$\Delta G = nV_a\Delta G_v + Mn^{\frac{2}{3}}\Delta\sigma \tag{11.18}$$

$$\Delta\sigma = \sigma_{sl} + \sigma_{sp} - \sigma_{pl} \tag{11.19}$$

where ΔG, ΔG_v, V_a, M, σ_{sl}, σ_{sp}, and σ_{pl} are the free energy required to form the system containing “n” numbers of atoms, free energy change per unit volume due to solidification of primary phase, atomic volume, geometrical factor, interfacial energy between the solidifying primary phases and liquid alloy, interfacial energy between the solidifying primary phases and particle reinforcement, and interfacial energy between the particle reinforcement and liquid alloy, respectively. Here, ΔG and shape of the nuclei depend on σ_{sl}, σ_{sp}, and σ_{pl}.

Fracture energy can be written [22] as Eq. 11.20:

$$\sigma_{sp} = \sigma_{sv} + \sigma_{pv} - W_{ad} \tag{11.20}$$

where it is assumed that the fracture energy between the materials is equal to the work of adhesion (W_{ad}). σ_{sv} = interfacial energy between the solidifying primary phases and vapor phase, where σ_{pv} is the interfacial energy between the particle reinforcement and vapor phase.

Equation of state approach [22] is given by Eqs. 11.21 and 11.22:

$$\sigma_{ps} = \frac{\left(\sqrt{\sigma_{pv}} - \sqrt{\sigma_{sv}}\right)^2}{1 - K\sqrt{\sigma_{pv}\sigma_{sv}}} \tag{11.21}$$

$$K = \frac{1}{\sqrt{\sigma_{pv}\sigma_{sv}}}\left[1 - \frac{\left(\sigma_{pv} - \sigma_{sv}\right)^2}{\sigma_{ps}}\right] \tag{11.22}$$

where K is a constant.

When the matrix wets the reinforcement particle, the matrix makes strong contact with the reinforcement and causes the good bonding with the reinforced materials [23]. For metal matrix systems, the equation of state approach is valid, and hence, work of adhesion (W_{ad}) can be calculated using London–van der Waals dispersion equation [22] as Eq. 11.23:

$$W_{ad} = 2\Phi\left(\sigma_{sv}.\sigma_{pv}\right)^{\frac{1}{2}} \tag{11.23}$$

where Φ is the contact angle. The contact angle, Φ, is 0.25 for metal-reinforced systems. The thermodynamic criteria, $\sigma_{psl} < \sigma_{pl}$, may not be satisfied apriority in a given system. A primary phase is nucleated around the reinforced dispersoids, and debonding of the particles from the matrix may not immediately result in the unstable cracks [24].

In a silver (Ag) matrix composite with carbon nanotubes (CNTs), the nanotubes contributed less to the bend strength due to the less wetting of the matrix to the reinforced CNTs. The main mechanism involves to it is that the weak van der Walls bonding between the nanotube and the Ag matrix [25].

Syntheses

Various processing routes exist for fabricating MMCs. Among these, liquid-state processing and solid-state processing have achieved much importance. Apart from these two, vapor state processing and deposition techniques are also taken care of. A flowchart showing various processing routes available for MMCs is given in Fig. 11.7. A comparison of different liquid state techniques used for fabrication MMC and their important feature are illustrated in Table 11.3.

Liquid-State Processing

Stir Casting: Stir casting, one of the most common liquid metallurgical techniques, is used to produce MMCs in recent days [26]. This involves filtration of particles or fiber bundles by liquid metals or alloys. Initially, the solid-reinforced materials are stirred in the metallic matrix melt, and the mixture is then allowed to solidify in a specific mold or ingot [19]. This is attained by using conventional processing machineries on a continuous or semicontinuous background having good particle wettability. The homogeneity of the MMCs depends on the stirring speed and time. For Al–SiC-based MMCs, it has been found that the increasing of stirring speed leads to get more homogeneous MMC along the deposition direction. Good wetting in the molten metal also depends on the degree of mixing that can be controlled by different shapes of mixer blade [23]. The schematics of stir casting and different mixer

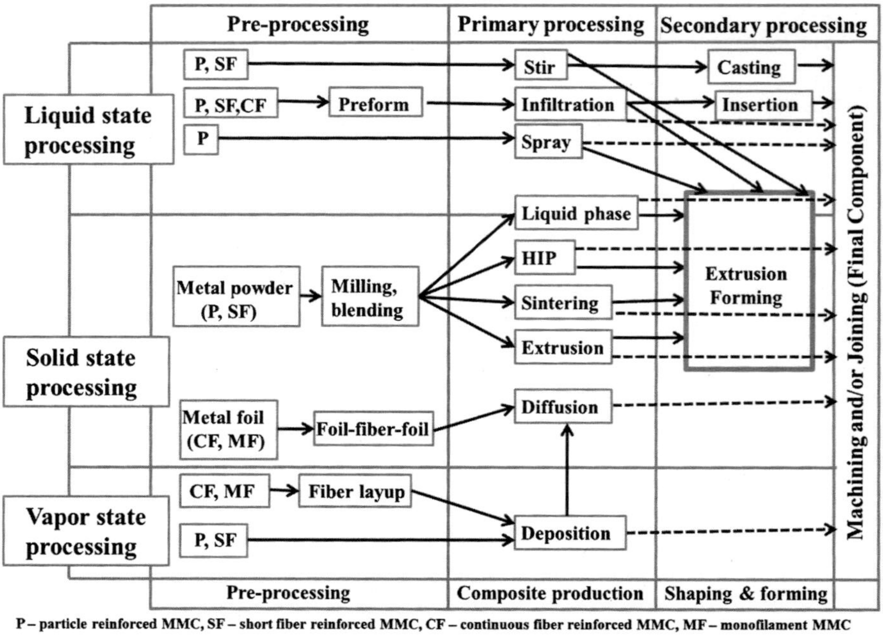

Fig. 11.7 Flowchart of various routes for processing MMCs

Table 11.3 Different liquid-state techniques used for fabrication of MMCs

Process of MMC fabrication	Range of shape and size	Metal yield	Range of volume fraction	Damage to reinforcement	Cost
Stir casting (liquid metallurgy)	Wide range of shapes, larger size, more than 0.5 t	Very high, >90 %	Max. 0.3	No damage	Least expensive
Squeeze casting	Limited by preform shape, up to 2–10 cm height	Low	Max. 0.45	Severe damage	Moderately expensive
Spray deposition	Limited shape, large size	Medium	0.3–0.7	–	Expensive
Reactive technique	Limited by preform shape, restricted size	High	0.2–0.5	No damage	Expensive
Powder metallurgy (P/M)	Wide range, restricted size	High	0–1	Reinforcement fracture	Expensive
Vapor-state processing	Complex geometry, wide range, restricted size	High	0–1	–	Very expensive

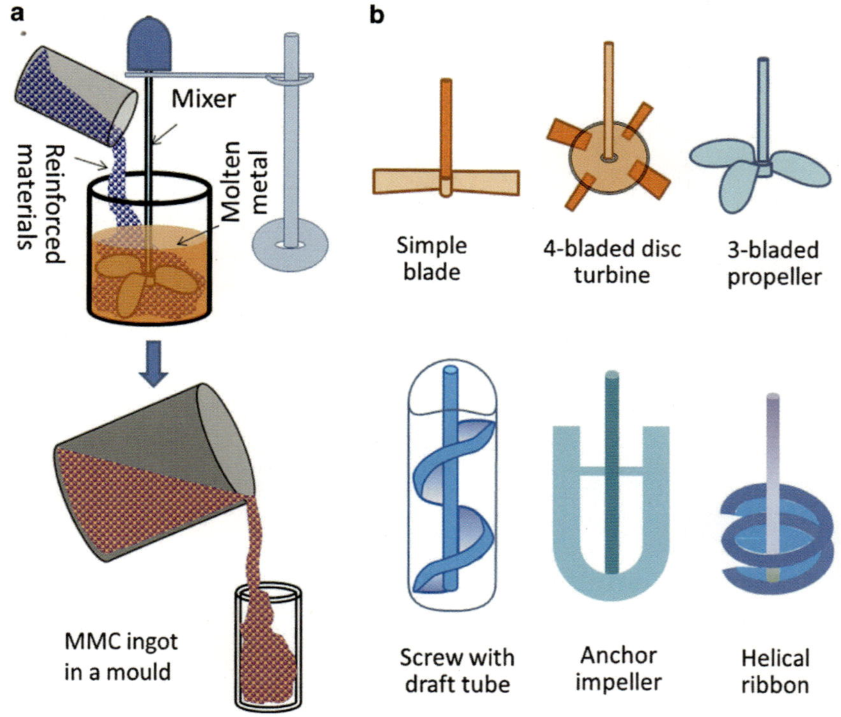

Fig. 11.8 (**a**) Stir casting of MMC and (**b**) different types of mixer blades

blades are depicted in Fig. 11.8a and b, respectively. Substantial interfacial reaction can take place because of prolonged liquid–ceramic contact. Various microstructural inhomogeneities like particle agglomeration and sedimentation become immanent in the melt. An advancing solidification front can be redistributed by the action of particle pushing, creating impediments in the process. In order to pacify this problem, the solidification is carried out rapidly so that the critical growth velocity is met, over which solid particles are enveloped instead of getting pushed, leading to refinement in the scale of structure.

Squeeze Infiltration: The squeeze infiltration processes is schematically depicted in Fig. 11.9. It can be done mainly by two ways – (a) pressure-driven infiltration, where external pressure is required, and (b) spontaneous infiltration, where no external pressure is required. In this process, a preform is tailored to a relevant shape. A "preform" is nothing but the cluster of short fibers (usually contains interstices), which could act as infiltrating sites for the liquid metal in order to form an articulate part of the final product [27]. The short-fiber sedimentation from liquid suspension is the source of fabrication of the preforms. The particulate MMCs are also produced by this process. Binders are used to retain the shape and integrity of the preforms. "Meniscus curvature" is the parameter, which decides how much pressure is required for infiltration to take place (in the range of MPa); corrections

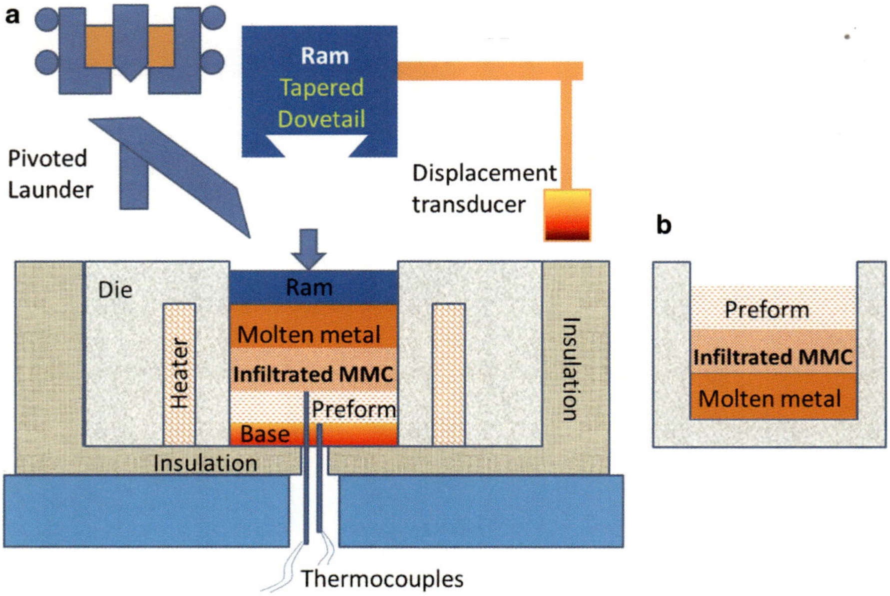

Fig. 11.9 A schematic of squeeze infiltration process – (**a**) pressure-driven infiltration and (**b**) spontaneous infiltration

regarding melt–fiber wetting can also be made by adopting various means. A strong interfacial bonding can be obtained by making contact with the fiber–melt with solute enrichment, for a prolonged time, under high hydrostatic pressure condition. Porosity, local variations in fiber content, average alloy composition, etc. are some of the common defects in squeezed infiltrated composites. In this context, an MMC, comprising of short Al_2O_3 fibers embedded (20 vol.%) in an $Mg_{10}Al_{0.4}Zn$ alloy, produced by squeeze infiltration process showed markedly increased stiffness, tensile strength, and hardness but slightly lower ductility and toughness [28].

Spray Deposition: A spray deposition technique is schematically depicted in Fig. 11.10. The graphite-reinforced MMCs prepared by spray deposition technique were demonstrated in a patent work [29]. This process comprises of building bulk metallic material onto a substrate by the action of atomized stream of droplets. The technique depends on two definite assisting processes: (i) production of the droplets from a molten bath and (ii) a regular supply of metal having low temperature into an ambience of rapid heat inoculation. The average velocity of droplet is typically about 20–40 m/s. There may be presence of superficial layer of liquid, or semisolid, on the top of ingot during its formation, leading the MMC material to have inhomogeneous distribution of ceramic particles. Porosity is typically about 5–10 % at this state. The several differences between thermal spraying and melt atomization processes could be cited comparatively in terms of deposition rates (~1 g/s) and particle speed (~50–400 m/s), which are slower and rapid, respectively. It can be very high in terms of quenching

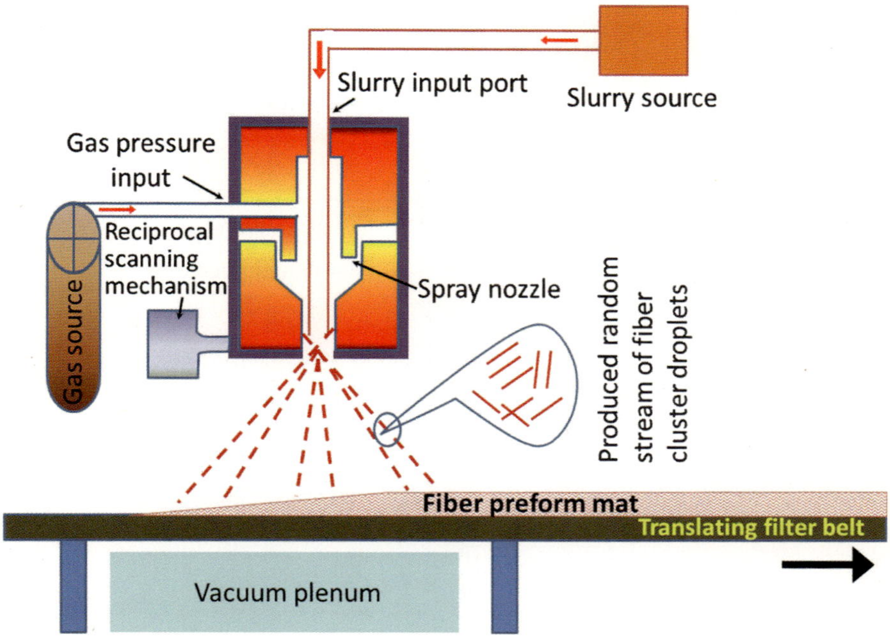

Fig. 11.10 A schematic of spray deposition technique

rates for an individual splat (~106 K/s). However, problems regarding maintenance of uniform fiber distribution and low void content by spraying onto fiber arrays persist in the making of MMCs by this technique.

Reactive Processing: This is an in situ process. It is a directional oxidation of Al, also known as DIMOX process. It involves blending of several elemental powders, and the reaction takes place by heating, leading to transient formation of liquid phase. Another route is introducing the liquid metal and oxidizing it progressively. Making a suitable powder compact, which serves as a destination for liquid metal infiltration, can make proximate shape forming possible, as a result of reaction with the preform. A dried blend metal matrix powder or alloy containing short whiskers, chopped filaments or particles, and fiber tows has undergone hot isostatic press (HIP) in order to generate randomly oriented fiber MMCs. An effective MMC often contains a residual unreacted metal. An in situ reactive process in a vertical tube electric furnace is schematically depicted in Fig. 11.11. The reactive process is also used to prepare refractory reinforcements in the Al alloy matrix. Generally, the metallic phase, e.g., Mg alloy, is placed on the top of ceramic preform in a crucible. In a gas–liquid reactive process, the TiC is reinforced Al matrix composites. For example, carbonaceous gas such as methane, acetylene, etc. is used by bubbling into Al–Ti melt at elevated temperature to produce Al–TiCp composites. This method also has been used to fabricate Al–Si/SiCp composite material by bubbling methane in Al alloy melt [30].

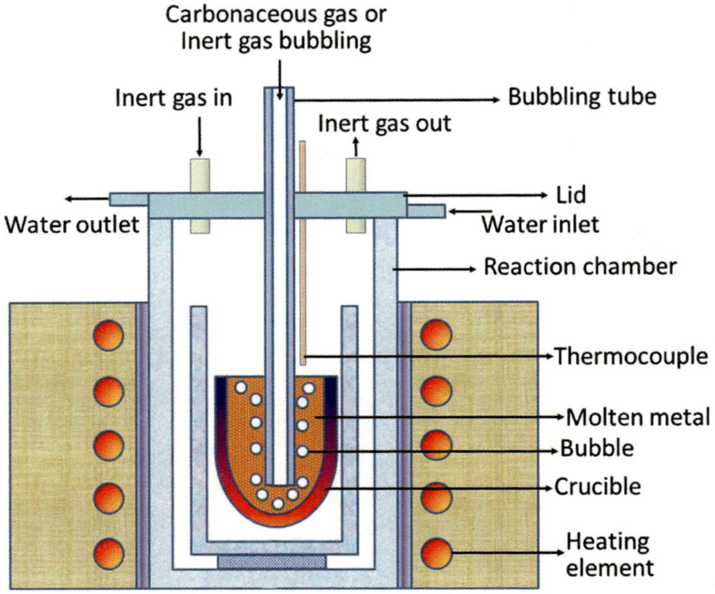

Fig. 11.11 A schematic of an in situ reactive process in a vertical tube electric furnace

Solid-State Processing

Powder Processing: This includes cold pressing and hot pressing for the fabrication of MMCs. During blending, the green body of cold pressing is dense about 80 %, which can be easily handled. The blend powders are canned and pressed uniaxially or isostatically in order to produce more dense MMCs. In hot forming processes, HIP produces no deviatoric stresses, which minimize the microstructural or macroscopic defects. A schematic of a powder metallurgical process, including mixing of raw powders, compaction to a desired shape, and sintering, is depicted in Fig. 11.12. Recently, carbonaceous materials such as graphite, CNTs, etc. have been used to improve the mechanical properties, thermal conductivity, and electrical conductivity and to reduce the coefficient of thermal expansion (CTE) and coefficient of friction in many MMCs by powder metallurgical technique [31–33].

Diffusion Bonding of Foils: This is a frequent solid-state bonding stratagem to join alike or like metals. The foil–fiber–foil or foil–particulate–foil (diffusion bonding) is one of the routes for the production of monofilament-reinforced MMCs. A schematic of diffusion bonding of foils is depicted in Fig. 11.13. Due to high pressure and/or heat, the matrix foil layer is disrupted thoroughly and a thin diffusion layer is formed in the composites at the interface. A good metallurgical bonding has formed at the interface between the filler and matrix materials. It occurred through solid-state grain diffusion mechanism. In solid-state diffusion bonding, the fibers or particulates are discontinuously reinforced to MMCs, and thus, a barrier created by matrix metal would make diffusion bonding process more difficult. To overcome this

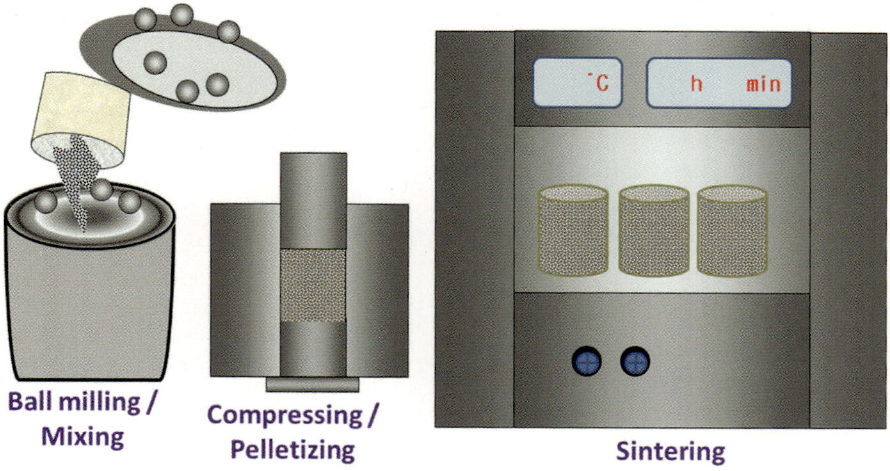

Fig. 11.12 A schematic of powder metallurgical process

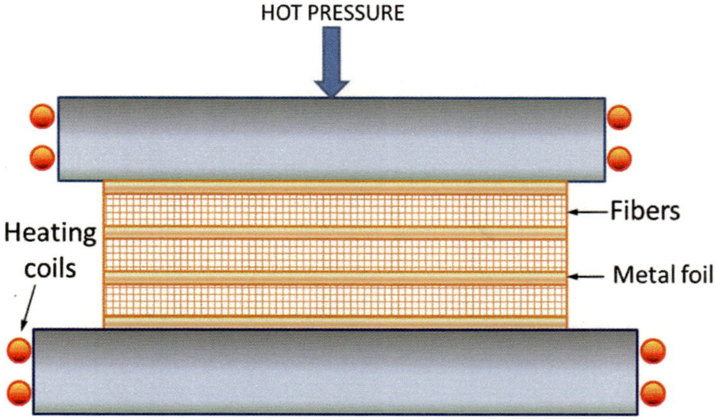

Fig. 11.13 A schematic of diffusion bonding of foils

problem, transition liquid phase diffusion bonding can be used by using a metal foil inserted between the surfaces to be joined [34, 35]. The diffusion bonding method deals with surface of the fiber onto which evaporation of relatively thick layers of metal matrix material takes place. Placing of rows and columns of fibers between thin metallic foils followed by hot pressing, often involving a filament-winding operation, is the commercially accepted production technique for long-fiber-reinforced Ti. Failure of surface oxide layer at a high temperature in controlled ambience leads Ti to form diffusion bonds readily. Fabrication of laminates having alternate layers of metals and ceramics also makes use of diffusion bonding. The advantages of this technique are (i) a wide variety of matrix metals can be processed and (ii) the fiber direction and volume fraction can also be controlled in an efficient manner.

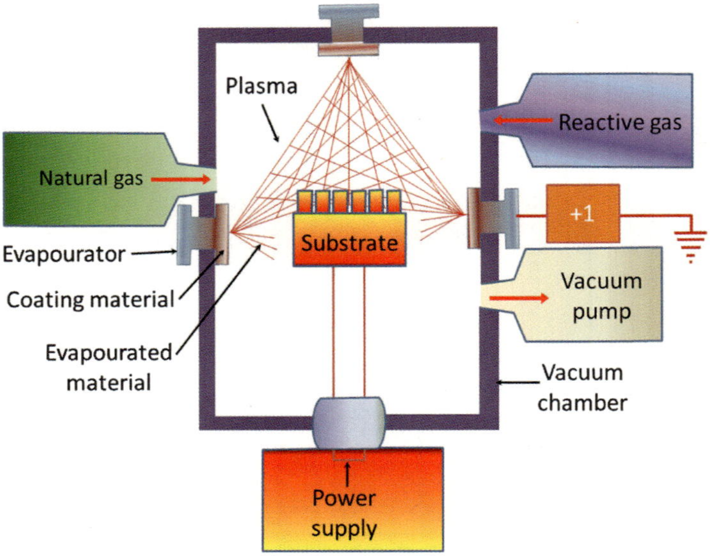

Fig. 11.14 A schematic of PVD setup

Controlling excess chemical reaction at the fiber–metal interface is one of the main problems that it faces. The production of parts with complex shapes creates difficulties, as the processing time gets increased, elevating the cost related to maintenance of high temperature and pressure. Diffusion bonding of metallic foils is generally used for fabrication of simple shape components such as plates, tubes, etc.

Vapor-State Processing

This technique involves coatings of individual fibers in a tow with the matrix metal to form MMCs. The microstructure will be more homogeneous than the cast composite. This is because the composite is prepared of identical units and the extent of interfacial bonding is easily controllable. Thin filament wound monolayer tapes can be produced, which are easier to mold into various shapes than its other precursor forms. Though the handling is easy, the disadvantage of this technique is the time consumption.

Physical Vapor Deposition: Physical vapor deposition (PVD) is a relatively slow process and involves thermal vaporization of the target species in vacuum, making use of the evaporation process. The rate of evaporation depends on the target species on which the thermal vaporization is deposited in a vacuum. A schematic of a PVD setup is depicted in Fig. 11.14. The fiber is allowed to pass through an ambience having a high vapor pressure of the metal to be deposited, and a subsequent condensation process produces a surface coating. This coating is thick enough with the deposition rate of about 5–10 μm/min. A highly intense (~10 kW) electron beam is directed on to

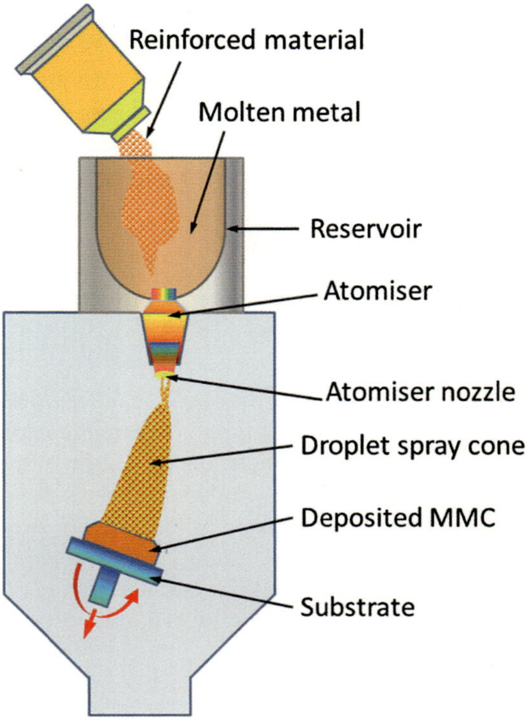

Fig. 11.15 A schematic showing spray-forming particulate MMC

the extremes of a solid bar feedstock to produce vapors. In this process, the produced fibers are very uniformly and homogeneously distributed and the fiber contents of up to about 80 %. The fiber volume fraction can also be accurately maintained via thickness of the deposited coating in this technique. Sintered steel matrix composites reinforced with hard carbide phases and deposited with gradient (Ti,Al)N coatings by PVD have been developed as different tools for industrial uses [36].

Spray Forming of Particulate MMCs: A schematic of spray-forming particulate MMC is depicted in Fig. 11.15. In this process, for the fabrication of particle-reinforced MMC products, a spray gun is used to atomizing the molten metal or alloy matrix. An example is Al alloy into which heated particles (e.g., SiC) are injected. The benefit of this process is its adjustability in making different types of composites. Some other processes are also developed recently. These include Lanxide's process or pressureless molten metal infiltration process (PRIMEX), Martin Marietta's process or XD process, etc. A pressureless infiltration of metal into a ceramic preform is achieved in PRIMEX technique. The ceramic preform can be produced by tape casting, molding, or machining of a green body. The heat-liberating reaction between the entities helps to produce a third component and this constitutes the XD process, also known as self-propagating high-temperature synthesis (SHS) process. The detailed descriptions on processing MMCs are available in literatures [10, 11, 13].

Properties

In this section, various properties of MMCs are discussed. These include mechanical properties (fracture and fracture toughness, fatigue, creep, etc.), tribological properties (wear, erosion), electrical properties, and thermal properties. The mechanical properties of different MMCs produced by different techniques are illustrated in Table 11.4.

Effects of Reinforcement Geometry on Mechanical, Elastic, and Plastic Behaviors

In 1995, Mammoli et al. have used unit cell theories to understand and model the behavior of Al–SiC and irregularly reinforced MMCs [40]. Stress–strain curve obtained through a simulation of tensile test on this composite is shown in Fig. 11.16a, and the stress differential versus strain of the composite, in which matrix is not reinforced, is depicted in Fig. 11.16b. A typical behavior has been observed that the matrix materials show strain-hardening phenomenon [40]. The rule of mixtures kind of relationship does not hold good to predict the strength of MMCs due to the structure sensitivity. The microstructural changes in the matrix metal during processing because of different processing routes, which include thermal stress and/or chemical reaction at the interface, can make it difficult to predict the strength of MMCs. In another study, it has been found that the mechanical strengths and hardness of powder metallurgical (P/M) extruder Ti/CNT MMCs significantly increased with the increase of CNT content from 0 to 0.35 wt% besides slight reduction in tensile strain. This is mainly obtained due to extremely fine grains (nearly 6 μm) along with large oxygen content [33]. Later, this result is also strongly supported by another study, where the mechanical properties were strongly changed from 654 $\pm$ 6.6 to 1182 $\pm$ 15.9 MPa (ultimate tensile strength) by reducing the grain size of Ti–TiC/CNT MMCs prepared by P/M method from 11.6 to 5 μm due to addition of vapor-grown CNTs from 0 to 1 wt% (Fig. 11.16c) [31].

Fracture

Babout et al. have proposed the nuances between the interface decohesion and particle cracking as a basis of basic deterioration nucleation phenomena in heterogeneous materials [41]. This model has helped to analyze the major failure criteria in model composites with matrix containing Al alloys and reinforcement of spherical-shaped zirconia/silica. Fig. 11.17 shows the pure Al matrix decohesion which dominates minimum yield stress and intermediate work-hardening rate, whereas for a matrix containing Al2124 alloy with an elevated yield stress and strain-hardening rate, particle fracture is the main failure criteria. In another study, in a fracture surface of pure Ti prepared by P/M extruder, the fine dimple pattern was observed without any primary grain boundaries. This indicated that this Ti powder

Table 11.4 Mechanical properties of different MMCs prepared by different techniques

Material	Method	YS (MPa)	UTS–UCS (MPa)	Strain (%)	Hardness	Ref.
Pure Ti	P/M, SPS, hot extrusion	484–591	472–654	29–36.2	261 Hv (0.05 N)	[31, 33]
Ti + 0.18–0.35 wt% CNT	-do-	671–764	578–710	31.9–39.9	275–285 Hv (0.05 N)	[33]
Ti 1.0 CNT	-do-	585 ± 10.5	697 ± 11.6	22 ± 5.8	–	[31]
Ti 0.4 CNT	Planetary ball milling (PBM)	795 ± 8.5	887 ± 5.2	25 ± 1.5	–	[31, 37]
Ti 0.8 CNT	-do-	1017 ± 16.7	1026 ± 11.5	19 ± 1.1	–	
Ti 1.0 CNT	-do-	1179 ± 15.7	1182 ± 15.9	15 ± 0.5	–	
Al alloy (A356.2)	Stir casting	168	263	7.35	68 (BHN)	[9]
Al alloy (A356.2)/2–8 % rice husk ash (RHA)/ 2–8 % SiC hybrid composite		182–258	296–356	6.25–4.9	74–104 (BHN)	
Al6061 + 1 vol.% SiC + 0.5 vol.% B_4C	Ultrasonic cavitation based solid casting process	200	270	5	100 (BHN)	[38]
Invar 18 vol.% TiC	Laser-deposited fuse mixing	500	–	5	20 (RC)	[3]
Ti-55Ni/calcium silicate (CS) (0–100 wt%)	P/M, cold isotactic press	–	84–795	0.9–4.8	0.73–7.9GPa	[39]
Ti6Al4V/CS (0–100 wt%)	-do-	–	33–604	1.3–5.0	0.3–4.3 GPa	

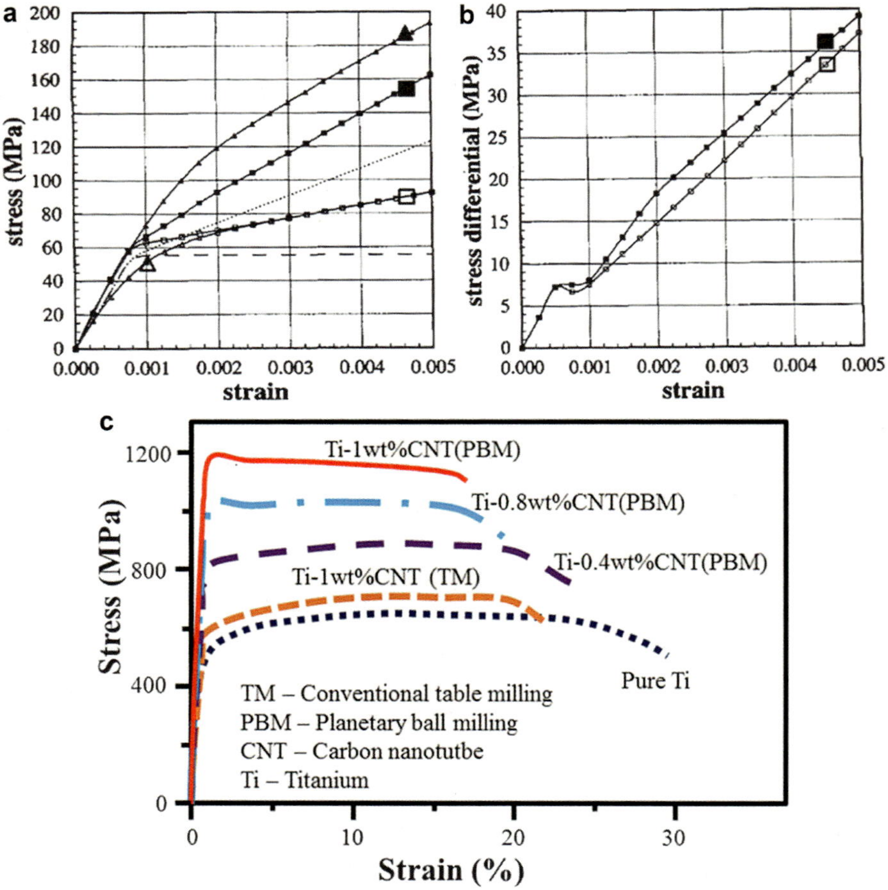

Fig. 11.16 (**a**) Stress–strain behavior for a 10 % volume fraction cubic reinforcement composite with (□) perfectly plastic matrix, (Δ) perfectly plastic matrix and thermal residual stresses, (■) strain-hardening matrix, (▲) strain-hardening matrix and thermal residual stresses; (**b**) stress differential between composite and unreinforced matrix, for (□) perfectly plastic and (■) strain-hardening matrices. The dotted line and the dashed lines show strain-hardening matrix and perfectly plastic behavior, respectively [40]; and (**c**) schematic stress–strain behavior of the Ti–CNTs MMCs [31]

bonding is metallurgically good, and the fracture occurs inside the grains when applying SPS and hot extrusion to consolidate Ti powders, whereas in the Ti/CNT MMCs, the unbundled CNTs which remained in matrix had retained the original size and shape at the surface of the composite along with some TiC particles, which were formed by the reaction of CNT with Ti. It has been found due to using of solid-state sintering condition at 800 °C by spark plasma sintering (SPS) and preheating at 1000 °C before extrusion. These two sintering conditions had taken a key role to prepare the Ti composite including both of CNT and TiC particles, which have influenced the improvement of mechanical properties [33].

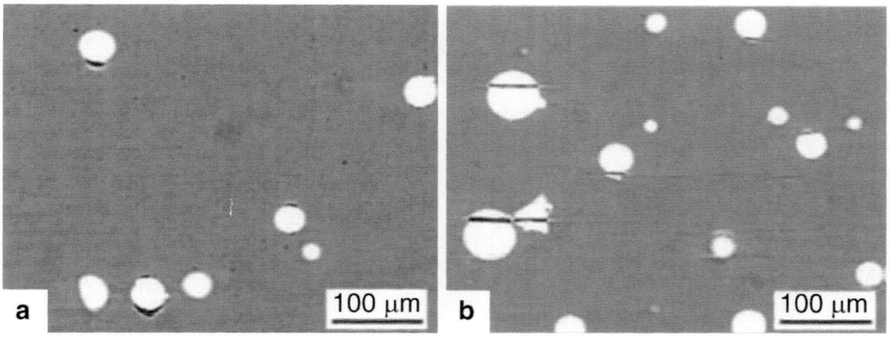

Fig. 11.17 X-ray tomographic image during tensile test for different MMCs reinforced by 4 % volume fraction of ZrO_2–SiO_2 spherical particles. Deformed states of composites: (**a**) pure Al matrix + 4 % ZrO_2–SiO_2 (E_p = 0.275 – σ = 120 MPa) and (**b**) Al2124 (T6) alloy matrix + 4 % ZrO_2–SiO_2 (E_p = 0.088 – σ = 430 MPa) (E_p: plastic deformation in the matrix) [41]

Toughness and Fracture Toughness

Toughness can be defined as the material's ability to absorb energy and plastically deform without fracturing it. This property is a measure of material resistance to fracture when it is stressed. For a given MMC, the toughness depends mainly upon (i) the matrix alloy compositions and microstructure, (ii) the type of reinforcement, (iii) size, and orientation. The processing can affect the microstructural parameters such as porosity, segregation, distribution of reinforcement, etc. McLelland et al. have reported that the laminated structure, which has more toughness than reinforcements, has homogeneous character [42]. It has been synthesized by taking layers of Al matrix alloy 2014 (Al–4Cu–0 · 8Si–0 · 8Mn–0 · 5 Mg) and matrix alloy reinforced with ~17 vol.% SiC particles and has used spray-forming technique. The impact test for toughness (by Charpy and Izod) has displayed marginal enhancement with the laminated materials as compared to the reinforced MMCs, which are homogeneous. It has been observed that the toughness is nearly doubled in slow crack growth during an estimation of the mean energy of fracture from the load–crosshead displacement curve. The reason is that at low strain rates, there is a reduction in the energy regarding crack propagation. For the structures, this energy reduction phenomenon has increased the potential in terms of slowing or stopping the crack propagation, particularly for the unreinforced layer [42]. In SiC/Al MMCs with different size of SiC-reinforced particles, it has been found that fracture toughness (42.1 kJ/m^2) is significantly better for finer (0.6 μm) size-reinforced particles compared to coarse-grained reinforced particle (32.4 kJ/m^2 for 9.5 μm) [43]. The crack initiation and crack propagation energy for the fine-grained reinforced particles was also compared to coarser reinforced particles. Qian et al. have observed that the crack initiation and fracture in SiC/Al MMC with the coarse reinforced particle were mainly related to void formation in the matrix around the individual particles ahead of main crack tip. On the other hand, the boundaries between the particle clusters and surrounding matrix may be associated to the failure SiC/Al MMC with the fine reinforced particle. Fracture toughness can be defined as the material's ability to resist fracture, when it contains a crack. It is a very important design parameter in many applications. In the case of particulate-reinforced MMCs (PRMMCs), measurement of acceptable plane-strain fracture toughness (K_{IC}) is an important step for developing the product up to the mark. Accepted standards for metals are ASTM-E399 and BS-7448, as there is no worldwide accepted fracture toughness testing technique available particularly for PRMMCs till date. The calculation of an acceptable toughness value can be obtained only when the test is finished for the plane-strain K_{IC} fracture toughness test. It is available from the load–displacement curve and can also measure the fatigue pre-crack length. The following equation [44] depicts the provisional fracture toughness K_Q, which is given in Eq. 11.24:

$$K_Q = f\left(\frac{a}{W}\right)\left[\frac{P_Q}{BW^{\frac{1}{2}}}\right] \tag{11.24}$$

where P_Q, B, W, and a correspond to the load for delineated augmentation of crack length, thickness of the specimen, width of the specimen, and the crack length, respectively. The function $f\left(\frac{a}{W}\right)$ is a geometry-related factor that signifies the compliance of the specimen to the ratio of crack length and width.

Fatigue

Fatigue can be defined as the progressive and localized structural damage, which occurs due to cyclic loading. It is a crucial phenomenon for MMCs as many high-frequency applications of MMCs involve repeated loading conditions, as in automobile components. This phenomenon is meticulously analyzed to identify the large incidences of failures in metals, which are not shown by high-strength brittle fiber, e.g., carbon or boron. For the composites having fibers as reinforcement and these are continuous and uniaxially distributed, the testing is done in the direction of the fibers, which gives a maximum output in terms of strength and toughness gain. The composites containing carbon or boron fibers with large volume fraction and aligned along the stress axis show high fatigue strength. Particles on the short fibers can be easily served as crack initiation sites. Generally, MMCs show a gradual crack growth rate than that of matrix with no reinforcement. The higher stiffness of the composite leads to the improvement of S-N curve behavior in terms of disciplined cyclic loading or high cyclic fatigue. A schematic of crack growth (da/dN) due to the cyclic intensity factor (ΔK) for a monolithic alloy and MMC is shown in Fig. 11.18. A choice of ideal size of particle and relevant volume fraction contained in the tidy matrix alloy would result in improved fatigue properties [15].

Creep

Creep can be defined as the tendency of a solid material to deform permanently under the influence of mechanical stresses. It arises when the material is under

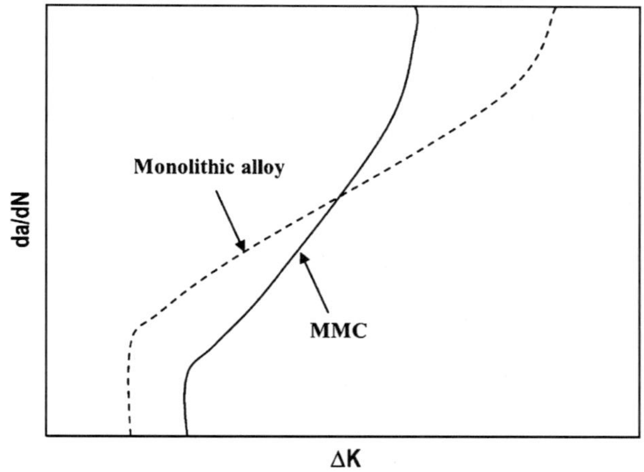

Fig. 11.18 Schematic of crack growth with cyclic intensity factor for monolithic alloy and MMC [15]

long-term exposure to high levels of stress. It is more severe in materials that are subjected to heat for a long period of time. It always increases with temperature. The combination of power law and exponential relationship can describe the creep behavior of metals and alloys. Characteristics regarding low to intermediate stress application and at elevated temperatures are efficiently explained by power laws, while the exponential relationship specifically deals with high stresses and comparatively at low temperatures. The basic governing equation of creep [45], which is found to applicable over a wide range of stresses, can be expressed by Eq. 11.25:

$$\dot{\varepsilon}_s = A(\sigma_a)^n \exp\left(\frac{-Q_{app}}{RT}\right) \tag{11.25}$$

where $\dot{\varepsilon}_s$, A, σ_a, n, Q_{app}, R, and T are the steady-state creep rate, constant, the applied stress, the stress exponent, the apparent activation energy, universal gas constant, and temperature in Kelvin, respectively.

The steady-state creep rate can be changed by the incorporation of reinforcement into the metal matrices. A typical characteristic of steady-state creep rate with stress for W wire in Ag matrix is shown in Fig. 11.19. The reduction in creep rate for steady state is attributed to the action of W in Ag.

Tribological Properties

Wear: Wear is defined as the erosion of material from a solid surface by the action of another solid. It depends on various parameters like type of contact (e.g., rolling or sliding), environment, operating conditions, the properties of test materials and the mating materials, etc. The hard intermetallic (e.g., Ni_3Al, $MoSi_2$) and ceramic (e.g., SiC) reinforcements act as wear-resisting sites in MMCs. These are more wear tolerant than the matrix comprising of metals, which is having a role of keeping reinforcement intact and transferring the applied load to the reinforcement [2].

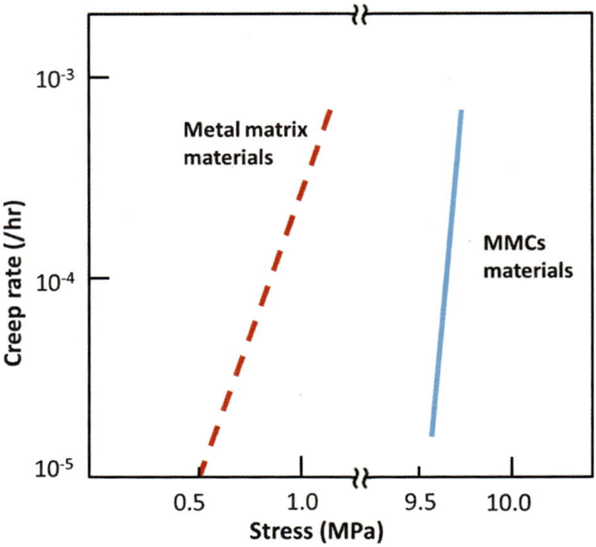

Fig. 11.19 Plot of creep rate versus stress for the incorporation of W-wires in Ag matrix [15]

The main variables in the wear standard test are temperature, speed, time, and sliding distance. The wear rate (mm^3/min) and sliding speed (m/s) can be calculated from Eqs. 11.26 and 11.27 [46, 47]. The wear rate and related parameters of different MMCs are illustrated in Table 11.5:

$$\text{Wear rate} = V_{\text{loss}} \times F_{\text{appl}} \tag{11.26}$$

$$\text{Sliding speed} = \frac{\pi DNT}{60,000} \tag{11.27}$$

where V_{loss} is volume loss during a given time (mm^3/min) and F_{appl} = applied load per unit distance (N/m); D is diameter of the wear tract in pin on disk test (mm), N is disk speed (rpm), and T is test duration (s).

With increasing applied normal load, initially the wear rate increases gradually; after a critical load, the wear becomes abruptly high. The wear rate increases with sliding speed, and after reaching a maximum, it becomes reduced at higher speed. The wear rate and critical load can also be different for different volume fractions of reinforcements used in MMCs, as it was shown in different studies [46–48].

Erosion: Erosion is defined as the damage of material surfaces by the impact of very high speed particles. This happens either by tiny particle impingement or by fluid impingement. Examples are slurry (liquid carrying solid particles) impact on the surfaces of many important machining components (e.g., impeller, valves, turbine blades, pipes, oil field equipments, etc.) in industrial applications. It reduces the life of machine components. The Al-based MMCs with different orientations of Al_2O_3 fiber reinforcement are having better resistance toward erosion than that of unreinforced Al alloy matrices.

Table 11.5 Wear rate and related parameter of MMCs

Material	Method	Wear rate (mm^3/min)	Conditions	Ref.
Al alloy (LM25)	Stir casting	0.020	At load: 30 N, speed: 1.04 m/s, for sliding distance: 628 m	[46]
		0.021	At load: 40 N, speed: 1.04 m/s, for sliding distance: 628 m	
LM25 + 7.5 % SiC	-do-	0.019	At load: 30 N, speed: 1.04 m/s, for sliding distance: 628 m	
		0.020	At load: 40 N, speed: 1.04 m/s, for sliding distance: 628 m	
LM25 + 7.5 % SiC+ 2.5–5 % TiO_2	-do-	0.018–0.017	At load: 30 N, speed: 1.04 m/s, for sliding distance: 628 m	
		0.019–0.018	At load: 40 N, speed: 1.04 m/s, for sliding distance: 628 m	
Al6061/10 wt% (SiC/Al_2O_3)/5 wt% graphite	Stir casting	0.0056569	At load: 30 N, speed: 1.5 m/s, for sliding distance: 2100 m	[47]
	-do-	0.0363433	At load: 30 N, speed: 2.0 m/s, for sliding distance: 2100 m	
	-do-	0.0100904	At load: 30 N, speed: 2.5 m/s, for sliding distance: 2100 m	
	-do-	0.0404794	At load: 25 N, speed: 2.5 m/s, for sliding distance: 2100 m	
	-do-	0.07123	At load: 35 N, speed: 2.5 m/s, for sliding distance: 2100 m	
Al/10 wt% TiB_2	Stir casting	4.5×10^{-3}	At load 30 N, speed: 300 rpm, for sliding distance: 1000 m	[48]
Al matrix	-do-	13.8×10^{-3}	At load 30 N, speed: 300 rpm, for sliding distance: 1000 m	
Al/10 wt% TiB_2	-do-	6×10^{-3}	At load 30 N, speed: 300 rpm, for sliding distance: 1000 m	
Al matrix	-do-	61×10^{-3}	At load 30 N, speed: 300 rpm, for sliding distance: 1000 m	

Weldability

Al alloy MMCs have shown a remarkable enhancement in the material properties over monolithic alloys in terms of higher strength and stiffness, better refractory properties, hardness, and minimal thermal expansion. Keeping off the precedence record, Al alloys reinforced with Al_2O_3 particulate are found to be stable in the molten Al. The technique like MIG welding can be successfully applied to it, finding its application in sporting goods. Use of these advanced materials may impart cost factor but functionality of these materials are not compromised. Packaging materials prepared with high volume fraction of SiC particulate match its CTE with the Si devices, and the bonding or brazing must be applied appropriately, when these packages are hermetically sealed. The investigation on the mechanical properties

Fig. 11.20 SEM images of Al–Al_2O_3 composite containing (**a**) angular particles (size ~ 30 μm) with 45 vol.% nonconducting phase and (**b**) equiaxed particles (size ~ 18 μm) with 56 vol.% nonconducting phase [51]

of Al matrix composites reinforced with continuous Al_2O_3 has revealed five possible failure mechanisms such as (i) onset of generalized interlayer yield, (ii) dismantling of the fiber tip/matrix metal interface, (iii) onset of cavitation in the metal due to unstableness caused because of high triaxiality, (iv) generalized fiber fracture of composite, and (v) load limited by plasticity in the mean interlayer [49].

Electrical Conductivity

The prediction of electrical conductivity of MMCs comprises the stratagem, which involves the replacement of equiaxed particles by spheres. Weber et al. have shown that the electrical conductivity of MMCs is dependent on several factors such as effect of particle size, spherical inclusions, spheroidal inclusions, etc. [50]. The conductivity of the composites with equiaxed particles is found to be about 30 % (at 55 vol.%) to 40 % (at 63 vol.% nonconducting phase), and these values are more than the composites containing angular particles for a given volume fraction of nonconducting phase. The SEM images of composites prepared of angular and equiaxed particles are shown in Fig. 11.20a and b, respectively.

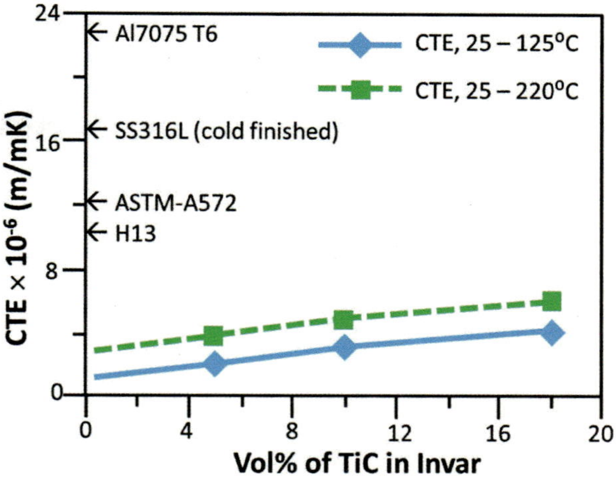

Fig. 11.21 Thermal expansion coefficient (CTE) of Invar with different TiC content in comparison with metals at different temperature ranges of 25–125 °C and 25–220 °C [3]

Thermal Properties

One of the most hectic problems with any composite is its mismatch in thermal expansion. This arises mainly due to the differences in the CTEs of matrix and fibers. The thermal stresses in fiber-reinforced composites are analyzed by various groups of researchers [52–56]. It should be noted that the thermal stresses will not disappear even if a uniform temperature is maintained throughout the composite volume. However, the thermal expansion properties of a composite can be controlled by balancing the amount of fibers and the matrix with desired dispersal of fibers. In 1960, Paul has studied the elastic and thermal expansion characteristics of two-phase composites [51]. Li has studied the thermal behavior of Invar–TiC MMCs [3]. The MMCs having low CTE (like Invar) can effectively minimize the thermal stress, as Invar (36 % Ni, 64 % Fe) is an alloy with very low CTE and it becomes nearly zero below 300 °C. The measurement of average CTE is being done between two temperature domains 25–125 °C and 25–220 °C. The average CTE of the composite remains satisfactory in the range regarding the CTE of Invar and TiC which is shown in Fig. 11.21.

Advantages and Disadvantages

Advantages over Monolithic Metals

- Refractory resistance
- Reduced CTE
- Higher strength-to-density ratios
- Higher stiffness-to-density ratios
- Higher fatigue resistance

- Better strength and reduced creep rate, at high temperature
- Higher wear resistance

Advantages over Polymer Matrix Composites

- Better temperature enduring capability
- Better transverse stiffness and strength
- Not easily inflammable
- Moisture repelling
- Higher electrical and thermal conductivities
- Enhanced radiation resistance
- No outgassing
- Efficient fabrication of whisker and PRMMCs with conventional metalworking equipment

Disadvantages

The MMCs also have several disadvantages, as follows:

- High density
- High processing temperature
- Corrosion at the interface between reinforcement and matrix, although it can be minimized by using appropriate coating or chemicals onto the reinforced surfaces

Factors Influencing Metal Matrix Composite Characteristics

- Volume fraction of the reinforcing filler
- The properties of reinforcing filler
- Form and geometric arrangement of the filler
- The properties of fiber–matrix interface
- The properties of matrices
- Porosity effect
- Residual stresses
- High-temperature chemical reactions resulting in possible degradation of the reinforcement
- Mechanical damage during processing, impact, etc.

Applications

In this section, various applications of MMCs are discussed. These include automotive components (automotive drive shaft, brake disk and calipers, etc.), spacecraft (antenna waveguides), aerospace (aircraft sturt, fins, etc.), military, electronic packaging, etc. [57].

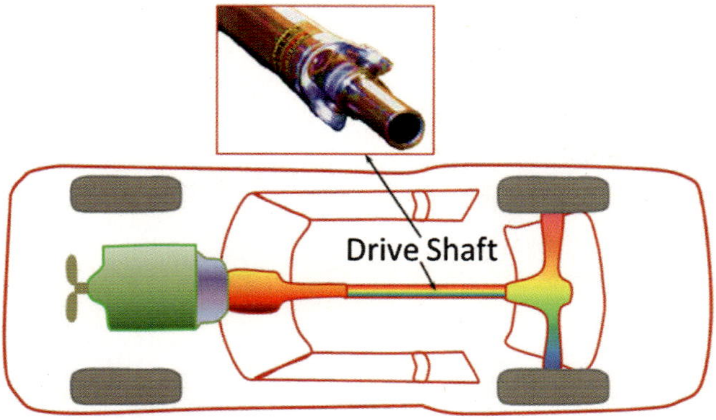

Fig. 11.22 Automotive drive shaft made of Al matrix composite (Courtesy to www.dennysdriveshaft.com)

Automotive Drive Shaft

It consists of a short-fiber preform of Al_2O_3 or Al silicate. Thermal fatigue in the inner portion of the piston crown in which the combustion chamber of a diesel engine is present, which is the main culprit to cause cracks leading to deterioration. Carbon fiber wound Al shafts are used by Renault to have an increment in stiffness, whereas Ford and Dodge vans use MMC with 20–30 % SiC extruded tube to achieve the same in an efficient way. An automotive drive shaft prepared of Al matrix composite for vehicles is shown in Fig. 11.22. The main advantages of using the MMCs in this area are high wear resistance, lower density, elevated temperature strength, fatigue resistance, CTE, and lower cost.

Ground Vehicle Brake Disks and Calipers

MMC brake calipers have found its use in racing high-duty sports cars. But due to their increased cost, its use in family cars is not feasible. Al PMMCs give much efficient cooling than cast iron because of its high thermal conductivity as compared to the former, making it to run at 500 °C. The development of brakes made of PMMC is originally meant for racing motorcycles. For the rotor having homogeneous reinforcement, SiC particulate has been used. Experimentation is ongoing to enhance the wear characteristics, making use of fiber and whisker preform technology. The brake disk and rotors made of Al MMC are found to be very efficient and are shown in Fig. 11.23.

Spacecraft

MMCs have been used in spacecraft due to its lightweight, low density, and very high strength-to-weight ratio [59]. Blisks and blings, or integrally bladed compressor

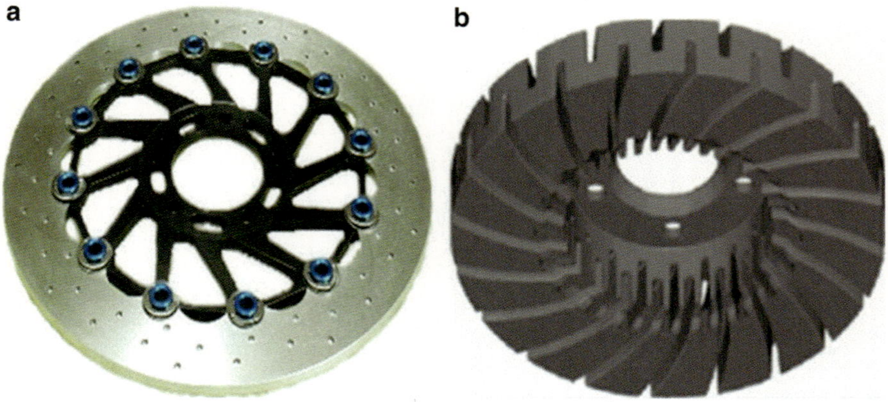

Fig. 11.23 Solid model designed by Visteon specifically for Al MMC (**a**) brake disk and (**b**) brake rotors [58]

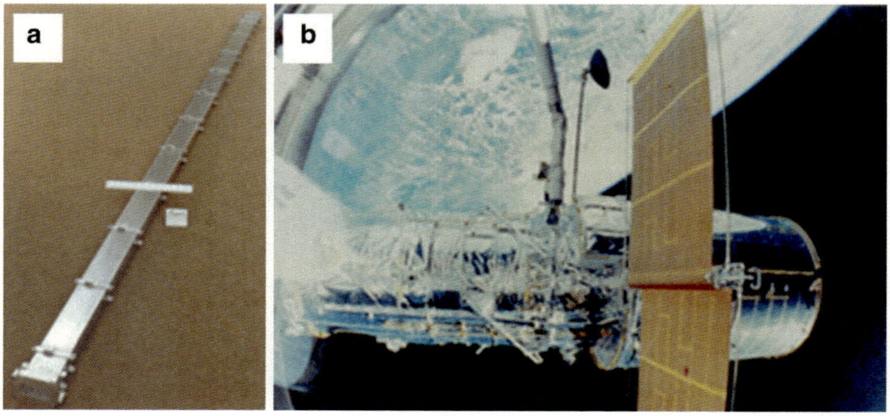

Fig. 11.24 P100/6061 Al high-gain antenna waveguides/boom for the HST (**a**) before integration in the HST and (**b**) before integration in the HST; it deployed in low earth orbit from the space shuttle orbiter [1]

rings of the spacecraft, are basically manufactured by CVD using SiC. Ti-reinforced SiC has become the topic of interest in aerospace applications. Fig. 11.24 shows the P100/6061 Al high-gain antenna waveguides/boom for the HST.

Aerospace Structures

Proximate net-shaped Al MMCs are developed for critical support structures for missiles as these are cheaper than the heavily machined Ti and are also lightweight. Other qualities, which make Al MMC SiCs more acceptable than Ti, are its low CTE and high modulus. An aircraft strut made of Al MMC (developed by Materion

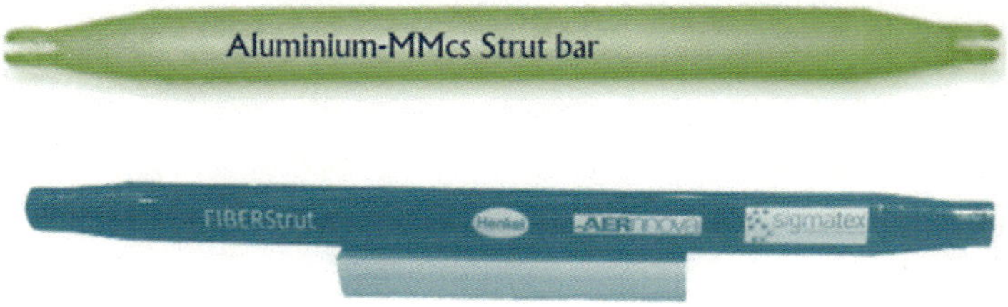

Fig. 11.25 Aircraft strut made of Al MMC (Courtesy to www.amc-mmc.co.uk)

Corporation) is a good replacement for carbon fiber composites (produced by Aernnova and Sigmatex) and is shown in Fig. 11.25. The major advantages of the MMCs in aerospace industries are high thrust-to-weight ratio for engines, high stiffness, low density, high specific strength, controlled thermal expansion, and high wear resistance [26]. Several other MMCs such as graphitic fiber-reinforced Al, Mg, and Cu MMCs have been used owing to their high specific strength (strength-to-density ratio).

Military Tank Track Shoes

Al MMC tank track shoes are developed for the vehicles of military use. The weight can be reduced up to 25 % over the present forged steel track shoes weighing nearly three tons per set, by using SiC whiskers as reinforcement. Test reports suggest that Al MMC wear resistance in the SiC-reinforced domains is consistent to steel. Improved heat dissipation characteristic of the Al MMC has paved the way to increase shoe bushing life by up to 200 %.

Electronic Packaging Applications

Discontinuously reinforced Al (DRA) composites are beneficial due to combination of high thermal conductivity, its intimacy in terms of CTE with electronic materials like GaAs, Al_2O_3, SiC, less density materials, etc. [59]. These find applications in thermally oriented machineries and in electronic industries for packaging. Several cast graphite–Al composite components are shown in Fig. 11.26. These are also used in communication satellites as well as in global positioning system (GPS) satellites. The main advantages of the MMCs to use as electronic housing are high stiffness, high heat dissipation capacity, controlled CTE, and low density.

Electronic Substrates

The MMCs act as sources of heat sink for a Si device, which is the reason behind why SiC particulate Al MMCs are used in substrate application for electronic

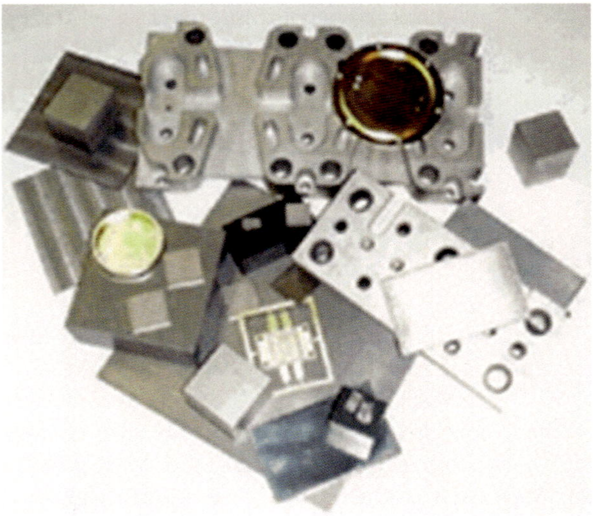

Fig. 11.26 Discontinuously reinforced Al MMCs for electronic packaging applications: cast graphite–Al composite components [1]

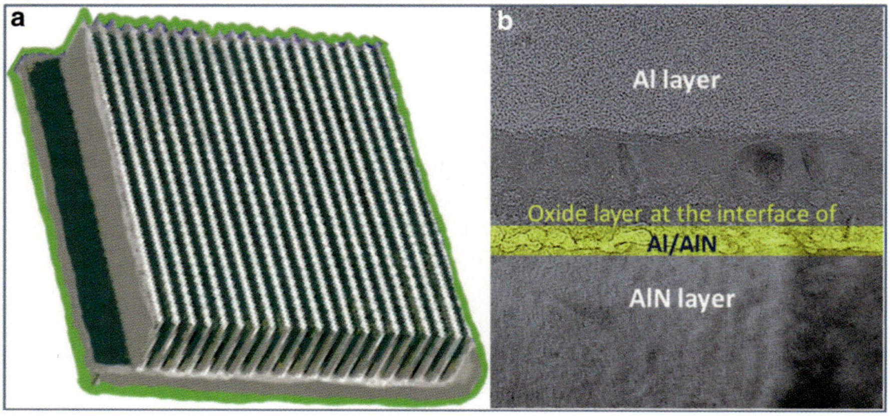

Fig. 11.27 (**a**) Heat sink and (**b**) electronic substrates made of Al MMCs [61]

industries. The MMCs of diamond/Al, SiC/Al, W/Cu, carbon fiber/Cu, and so on have been produced by different techniques such as squeeze casting, P/M, SPS method, etc. [8, 59]. One of the important parameters for electronic materials is thermal conductivity, which can be controlled by tuning the volume fraction of nonconducting reinforced particles [60]. The electronic heat sink finds its use in domestic electrical requirements using the principle of electrochemical conversion of gases. The intimacy between Si and the MMC in terms of CTE reduces the stresses, leading to device cracking or debonding, and also enhanced heat dissipation takes place because of high thermal conductivity. Some electronic substrates of Al MMCs are shown in Fig. 11.27.

Explosion Engine Components

Ceramic-reinforced Ti-based alloy MMCs can be used as explosion engine components due to their high refractoriness.

Space Satellite

Due to high strength-to-weight ratio, Ti-based and Al-based MMCs are generally used in space satellites.

Jet Fighter Aircraft Fins

The MMCs are found application in jet fighter aircraft fins. Particularly, Al- or Mg-based MMCs are mostly used. These provide a unique medley of lightness, high specific stiffness, better fatigue properties, better electrical and thermal conductivities, and low CTEs.

Sports Industry

The MMCs are also being used in sports industries to develop some sports component such as bicycle frame, tennis racket, golf shaft, and so on. The main advantages of the MMCs to use as sports goods are high stiffness, lightweight, low density, and high fatigue resistance [62].

Biomedical Industry

Recently, the different advanced MMCs are being used in various biomedical applications owing to their high load-bearing properties, good elastic modulus, lightweight, low density, high corrosion resistance, and so on [39, 63, 64]. Biomaterials of MMCs are desired specifically in load-bearing orthopedic implants owing to their higher mechanical properties. Several metallic alloys such as stainless steel (SS316L), Ti, cobalt–chromium–molybdenum (Co–Cr–Mo), Ni–Ti, Mg, and so on have been tried as matrix materials with various reinforcement materials such as hydroxyapatite (HA), CS, Al_2O_3, etc. for orthopedic biomedical applications [65–68]. Recently, carbonaceous nanomaterial-reinforced MMCs are being attempted to be used as advanced biomaterials for many potential applications [25, 31–33]. Therefore, industries of surgical instruments and artificial orthopedic implants and limbs are growing mainly due to the technological advancements in the MMCs.

Concluding Remarks

Different fabrication techniques of MMCs have been discussed in ordered to develop some select proper method for particular materials for specific applications. Basic theories have also been demonstrated to select proper geometry, size, and volume fraction of reinforcement in the MMCs in order to achieve desired combination of properties for target applications. The characteristics of MMCs can also be molded for specific applications. The MMCs have satisfied the designer conceptions, in such a way to act as functional materials for advanced applications. These composites can solve the major problems associated with many applications only if these are designed to achieve the standards. The MMCs have to be competing with the advanced powder metallurgy technologies in terms of cost performance index. The high cost of MMCs is attributed to the cost of reinforced materials and matrix metals and the processing expenses. At present, several vapor deposition processes are being employed in the fabrication of MMCs, but still the cost factor is keeping them away from day-to-day applications. Novel, cost-effective methods have to be accomplished in order to achieve the market for MMCs.

Acknowledgment The authors acknowledge the financial support provided by the Department of Science and Technology, India, for carrying out this research work.

References

1. Rawal SP (2001) Metal-matrix composites for space applications. J Mater 53:14–17
2. Legzdins C, Samarasekera I, Meech J (1997) MMCX – an expert system for metal matrix composite selection and design. Can Metallurg Q 36:177–202
3. Li XC, Stampfl J, Prinz FB (2000) Mechanical and thermal expansion behavior of laser deposited metal matrix composites of Invar and TiC. Mater Sci Eng A Struct 282:86–90
4. Occhionero M, Hay RA, Adams RW, Fennessy KP (1999) Aluminum silicon carbide (AlSiC) thermal management packaging for high density packaging applications. In: Proceedings-SPIE the international society for optical engineering, 1999
5. Saums DL (2004) Developments in selective high thermal conductivity orientation in CTE-compatible substrate and package component materials. In: Semiconductor thermal measurement and management symposium, twentieth annual IEEE, 2004
6. Korb G, Neubauer E (2001) Thermophysical properties of metal matrix composites, vol 7, MMC-Assess Thematic Network, Seibersdorf.
7. Korab J, Korb G, Sebo P (1998) Thermal expansion and thermal conductivity of continuous carbon fibre reinforced copper matrix composites. In: Electronics manufacturing technology symposium (IEMT), Europe, 1998
8. Zweben C (1992) Metal-matrix composites for electronic packaging. J Mater 44:15–23
9. Prasad DS, Shoba C, Ramanaiah N (2014) Investigations on mechanical properties of aluminum hybrid composites. J Mater Res Technol 3:79–85
10. Clyne T (2001) Composites: MMC, CMC, PMC. In: Mortensen A (ed) Encyclopaedia of materials: science and technology. Elsevier, USA
11. Chawla KK (2006) Metal matrix composites. In: Materials science and technology. Wiley online library. doi: 10.1002/9783527603978.mst0150
12. Taya M, Arsenault RJ (1989) Metal matrix composites: thermomechanical behavior. Elsevier, Burlington
13. Nair S, Kim H (1992) Modification of the shear lag analysis for determination of elastic modulus of short-fiber (or whisker) reinforced metal matrix composites. J Appl Mech 59:S176–S182
14. Landis CM, McMeeking RM (1999) Stress concentrations in composites with interface sliding, matrix stiffness and uneven fiber spacing using shear lag theory. Int J Solids Struct 36:4333–4361

15. Chawla KK (1993) Ceramic matrix composites. Chapman and hall, London
16. Ochiai S, Osamura K (1990) Influences of matrix ductility, interfacial bonding strength, and fiber volume fraction on tensile strength of unidirectional metal matrix composite. Metallurg Trans A 21:971–977
17. Ruch PW, Beffort O, Kleiner S, Weber L, Uggowitzer PJ (2006) Selective interfacial bonding in Al (Si) – diamond composites and its effect on thermal conductivity. Compos Sci Technol 66:2677–2685
18. Ibrahim I, Mohamed F, Lavernia E (1991) Particulate reinforced metal matrix composites – a review. J Mater Sci 26:1137–1156
19. Hashim J, Looney L, Hashmi M (2002) Particle distribution in cast metal matrix composites – Part I. J Mater Process Technol 123:251–257
20. Kennedy A, Wyatt S (2001) Characterising particle-matrix interfacial bonding in particulate Al–TiC MMCs produced by different methods. Compos Part A: Appl Sci 32:555–559
21. Yang H, Topping TD, Wehage K, Jiang L, Lavernia EJ, Schoenung JM (2014) Tensile behavior and strengthening mechanisms in a submicron B4C-reinforced Al trimodal composite. Mater Sci Eng A Struct 616:35–43
22. Rohatgi PK (1994) Nucleation phenomenon during solidification of metal matrix composites. In: Final report to ONR. University of Wisconsin. Available via http://www.dtic.mil/dtic/tr/fulltext/u2/a317340.pdf. Accessed 10 July 2015
23. Hashim J, Looney L, Hashmi MSJ (1999) Metal matrix composites: production by the stir casting method. J Mater Process Technol 92–93:1–7
24. Rohatgi PK (1993) Microstructure formation during solidification of metal matrix composites. Minerals, Metals and Materials Society, Warrendale
25. Feng Y, Yuan HL, Zhang M (2005) Fabrication and properties of silver-matrix composites reinforced by carbon nanotubes. Mater Charact 55:211–218
26. Umanath K, Palanikumar K, Selvamani ST (2013) Analysis of dry sliding wear behaviour of Al6061/SiC/Al_2O_3 hybrid metal matrix composites. Compos Part B: Eng 53:159–168
27. Clyne TW, Mason JF (1987) The squeeze infiltration process for fabrication of metal-matrix composites. Metallurg Trans A 18:1519–1530
28. Purazrang K, Kainer KU, Mordike BL (1991) Fracture toughness behaviour of a magnesium alloy metal-matrix composite produced by the infiltration technique. Composites 22:456–462
29. Cornie J, Cornie S, Zhang S (2005) Spray deposition apparatus and methods for metal matrix composites. US Patent US20060086434A1, 27 Apr 2006
30. Mitrică D, Moldovan P (2012) In-situ synthesis of Al-Si/SiCp composites by reactive gas injection method. UPB Sci Bull, Ser B 74:185–194
31. Li S, Sun B, Imai H, Kondoh K (2013) Powder metallurgy Ti–TiC metal matrix composites prepared by in-situ reactive processing of Ti-VGCFs system. Carbon 61:216–228
32. Neubauer E, Kitzmantel M, Hulman M, Angerer P (2010) Potential and challenges of metal-matrix-composites reinforced with carbon nanofibers and carbon nanotubes. Compos Sci Technol 70:2228–2236
33. Kondoh K, Threrujirapapong T, Imai H, Umeda J, Fugetsu B (2009) Characteristics of powder metallurgy pure titanium matrix composite reinforced with multi-wall carbon nanotubes. Compos Sci Technol 69:1077–1081
34. Muratoğlu M, Yilmaz O, Aksoy M (2006) Investigation on diffusion bonding characteristics of aluminum metal matrix composites (Al/SiCp) with pure aluminum for different heat treatments. J Mater Process Technol 178:211–217
35. Zhang X-P, Ye L, Mai Y-W, Quan G-F, Wei W (1999) Investigation on diffusion bonding characteristics of SiC particulate reinforced aluminium metal matrix composites (Al/SiC p-MMC). Compos Part A: Appl Sci 30:1415–1421
36. Matula G (2009) Study on steel matrix composites with (Ti, Al) N gradient PVD coatings. J Achiev Mater Manuf Eng 34:79–86
37. Li S, Sun B, Imai H, Mimoto T, Kondoh K (2013) Powder metallurgy titanium metal matrix composites reinforced with carbon nanotubes and graphite. Compos Part A: Appl Sci 48:57–66

38. Poovazhagan L, Kalaichelvan K, Rajadurai A, Senthilvelan V (2013) Characterization of hybrid silicon carbide and boron carbide nanoparticles-reinforced aluminum alloy composites. Procedia Eng 64:681–689
39. Ataollahi Oshkour A, Pramanik S, Shirazi SFS, Mehrali M, Yau Y-H, Abu Osman NA (2014) A comparison in mechanical properties of cermets of calcium silicate with Ti-55Ni and Ti-6Al-4 V alloys for hard tissues replacement. Scientific World Journal. doi:10.1155/2014/616804
40. Mammoli A, Bush M (1995) Effects of reinforcement geometry on the elastic and plastic behaviour of metal matrix composites. Acta Metallurg Mater 43:3743–3754
41. Babout L, Brechet Y, Maire E, Fougeres R (2004) On the competition between particle fracture and particle decohesion in metal matrix composites. Acta Mater 52:4517–4525
42. McLelland A, Atkinson H, Anderson P (1999) Thixoforming of a novel layered metal matrix composite. Mater Sci Technol Ser 15:939–945
43. Qian L, Kobayashi T, Toda H, Goda T, Wang Z-g (2002) Fracture toughness of a 6061Al matrix composite reinforced with fine SiC particles. Mater Trans 43:2838–2842
44. ASTM E399-90 (1997) Standard test method for plane-strain fracture toughness of metallic materials. Annual Book of Standards, ASTM International, West Conshohocken
45. Ochiai S (1993) Mechanical properties of metallic composites. CRC Press, USA
46. Elango G, Raghunath BK (2013) Tribological behavior of hybrid (LM25Al + SiC+ TiO_2) metal matrix composites. Procedia Eng 64:671–680
47. Kundu S, Roy B, Mishra AK (2013) Study of dry sliding wear behavior of aluminium/SiC/ Al_2O_3/graphite hybrid metal matrix composite using taguchi technique. Int J Sci Res Publ 3:1–8
48. Sreenivasan A, Paul Vizhian S, Shivakumar N, Muniraju M, Raguraman M (2011) A study of microstructure and wear behaviour of TiB2/Al metal matrix composites. Lat Am J Solids Struct 8:1–8
49. Burr A, Yang J, Levi C, Leckie F (1995) The strength of metal-matrix composite joints. Acta Metallurg Mater 43:3361–3373
50. Weber L, Dorn J, Mortensen A (2003) On the electrical conductivity of metal matrix composites containing high volume fractions of non-conducting inclusions. Acta Mater 51:3199–3211
51. Paul B (1959) Prediction of elastic constants of multi-phase materials. In: Technical report no. 3. Brown University. Available via. http://www.osti.gov/scitech/biblio/4273941
52. Hull AW, Burger E (1934) Glass-to-metal seals. J Appl Phys 5:384–405
53. Chawla KK, Metzger M (1972) Initial dislocation distributions in tungsten fibre-copper composites. J Mater Sci 7:34–39
54. Scherer GW (1992) Relaxation in glass and composites. Krieger Publishing Company, Florida
55. Dvorak GJ (1991) Inelastic deformation of composite materials. Springer, New York
56. Hsueh CH, Becher PF, Angelini P (1988) Effects of interfacial films on thermal stresses in whisker-reinforced ceramics. J Am Ceram Soc 71:929–933
57. Surappa MK (1997) Microstructure evolution during solidification of DRMMCs (discontinuously reinforced metal matrix composites): state of art. J Mater Process Technol 63:325–333
58. Garfinkel GA, Myers DC, Gianaris NJ, Hashmi SAA (2003) Vented disc brake rotor. US Patent 6,536,564
59. Gui M, Kang S, Euh K (2004) Al-SiC powder preparation for electronic packaging aluminum composites by plasma spray processing. J Therm Spray Technol 13:214–222
60. Lee HS, Jeon KY, Kim HY, Hong SH (2000) Fabrication process and thermal properties of SiCp/Al metal matrix composites for electronic packaging applications. J Mater Sci 35:6231–6236
61. Kida M, Weber L, Monachon C, Mortensen A (2011) Thermal conductivity and interfacial conductance of AlN particle reinforced metal matrix composites. J Appl Phys 109:064907
62. Froes F (1997) Is the use of advanced materials in sports equipment unethical? J Mater 49:15–19
63. Kutz M, Adrezin RS, Barr RE, Batich C, Bellamkonda RV, Brammer AJ, Buchanan TS, Cook AM, Currie JM, Dolan AM (2003) Standard handbook of biomedical engineering and design. McGraw-Hill, New York

64. Tanigawa H, Asoh H, Ohno T, Kubota M, Ono S (2013) Electrochemical corrosion and bioactivity of titanium–hydroxyapatite composites prepared by spark plasma sintering. Corros Sci 70:212–220
65. Geetha M, Singh A, Asokamani R, Gogia A (2009) Ti based biomaterials, the ultimate choice for orthopaedic implants-a review. Prog Mater Sci 54:397–425
66. Ning C, Zhou Y (2008) Correlations between the in vitro and in vivo bioactivity of the Ti/HA composites fabricated by a powder metallurgy method. Acta Biomater 4:1944–1952
67. Fei S, Jie L, Bin F (2011) Corrosion behavior of extrusion-drawn pure Mg wire immersed in simulative body fluid. Trans Nonferr Metal Soc 21:258–261
68. Pramanik S, Agarwal AK, Rai K (2005) Chronology of total hip joint replacement and materials development. Trends Biomater Artif Organs 19:15–26

Effect of Concentration Gradient on the Magnetic Properties of Functionally Graded Styrene Butadiene Rubber Composites

12

Raghunandan Sharma, Sandeep S. Ahankari, and Kamal K. Kar

Contents

Abstract

Functionally graded materials (FGMs), due to their characteristic spatial gradient of structural/compositional features, exhibit unique properties not possessed by their constituent components. Specific applications of FGMs cover diverse areas

R. Sharma
Advanced Nanoengineering Materials Laboratory, Materials Science Programme, Indian Institute of Technology Kanpur, Kanpur, Uttar Pradesh, India
e-mail: raghumsp@iitk.ac.in

S.S. Ahankari
Department of Mechanical Engineering, Annasaheb Dange College of Engineering and Technology, Sangli, Maharashtra, India
e-mail: asandeeps@gmail.com

K.K. Kar (✉)
Advanced Nanoengineering Materials Laboratory, Materials Science Programme, Indian Institute of Technology Kanpur, Kanpur, Uttar Pradesh, India

Advanced Nanoengineering Materials Laboratory, Department of Mechanical Engineering, Indian Institute of Technology Kanpur, Kanpur, Uttar Pradesh, India
e-mail: kamalkk@iitk.ac.in

K.K. Kar (ed.), *Composite Materials*,
DOI 10.1007/978-3-662-49514-8_12

including biomedicine, tissue engineering, structural composites for advanced applications, etc. This chapter deals with a case study of synthesis of and characterization of FGMs consisting of soft magnetic iron/nickel particles as filler in styrene-butadiene rubber matrix. Effect of concentration gradient on the physical, mechanical, and magnetic properties of the composites has been reported.

Keywords
Functionally graded composites • Electromagnetic interference • Electromagnetic shielding • Nickel • Iron • Styrene-butadiene rubber • Saturation magnetization • Coercivity • Modulus • Hardness • Tensile strength

Introduction

Functionally graded materials (FGMs) are a class of materials characterized by a structural and/or compositional gradient in the bulk, leading to a similar variation of properties [1–3]. A composite, having a spatial concentration gradient of the filler, is an example of FGMs (Fig. 12.1). Owing to their unique physical, thermal, mechanical, and electronic properties, FGMs are used in advanced technological fields such as structural applications (components for combustion chambers; high-pressure cylinders; parts of transport vehicles such as propulsion shafts, flywheels, shock absorbers, and brakes; etc.), biomedical applications (artificial ligaments, tissue engineering), microwave absorbers, fuel cells, etc. [4–6]. FGMs based on various matrix/filler combinations such as ceramic/ceramic, ceramic/metal, polymer/ceramic, polymer/metal, and many other polymers have been synthesized for different applications. More recently, the use of nanomaterial fillers in various matrices is being investigated [7–9]. Among others, polymer matrix-based FGMs are of significant importance for diverse areas such as bioengineering [10], mechanical engineering [11], electromagnetic interference (EMI) shielding [6], etc.; based on the

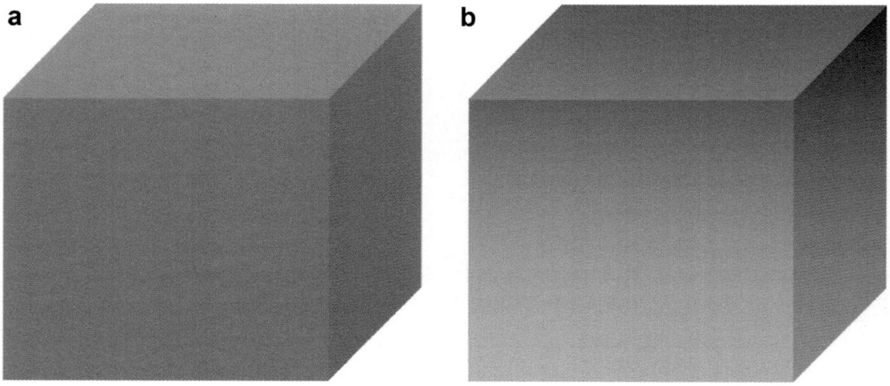

Fig. 12.1 Schematic of (**a**) uniformly graded and (**b**) functionally graded (linearly) composites

Table 12.1 Formulation of mixes for preparing UDPCs/FGPCs

Ingredient	Phr[a]
SBR 1502	100
TQ[b]	1.5
Stearic acid Zinc oxide	2.0 5.0
MBT[c]	0.8
Sulfur	2.5
Iron/nickel	0, 10, 50, 100, 200, 300 up to 1000 in steps of 100 onward

[a]Parts per 100 g of rubber
[b]1, 2-dihydro-2, 2, 4-trimethyl quinoline
[c]2-mercaptobenzothiozole

end-user applications, various types of filler may be used. For example, for gradation of mechanical properties, concentration of reinforcing phase is varied spatially, while for gradation of magnetic properties, composites having concentration gradient of magnetic fillers are employed.

This chapter deals with a case study of synthesis of and characterization of soft magnetic iron/nickel particle/styrene-butadiene rubber (SBR) functionally graded composites. Effect of filler concentration gradient on the physical, mechanical, and magnetic properties of the composites has been reported. Again, a comparison has been made between the properties of the uniformly dispersed elastomeric composites (UDECs) prepared by uniform dispersion of iron/nickel powder in SBR and the functionally graded elastomeric composites (FGECs), having gradient filler concentration along thickness of the sample.

Processing of Functionally Graded Composites

Iron/nickel powder was mixed with KOSYN 1502 SBR (styrene-butadiene rubber). Various loadings of filler content were added to SBR apart from the regular recipe as shown in Table 12.1. All these ingredients were homogeneously mixed in a two roll mixer and then vulcanized at 150 °C in an electrically heated hydraulic press according to ASTM D-15 to make sheets of ~2.5 mm thickness. The processing method adopted for the preparation of functionally graded iron/nickel-SBR composites was somewhat different. Herein, the thin layers of uncured homogeneous mixes containing increasing or decreasing volume fractions of iron/nickel were stacked on each other, and the stacked layers were molded and vulcanized in the press. With an application of pressure, the interfaces between the subsequent thin layers disappear so as to form a smoothly graded composite. Toroidal-shaped samples were made by shear cutting the sheet in a specially designed punch and die.

Various UDECs having different volume fractions of the filler volume fraction ranging from 0.18 to 0.56 were prepared. Different coordinated sequences of filler gradations were also prepared wherein the loading was varied from 0.18 to

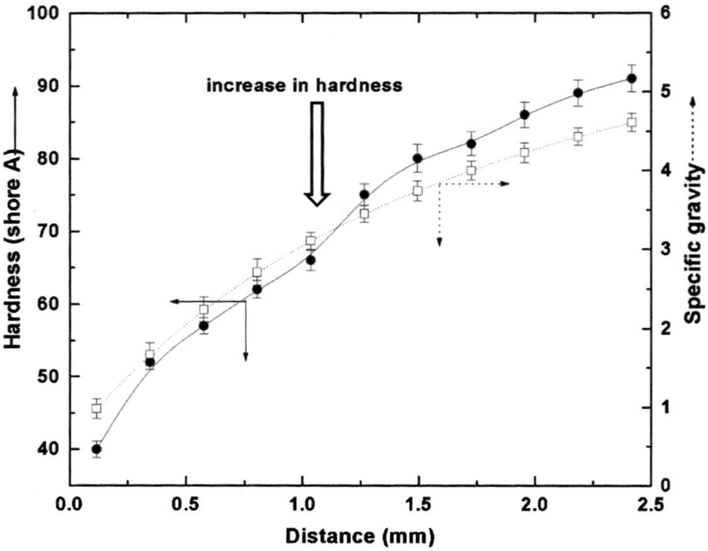

Fig. 12.2 Variation of hardness and specific gravity of iron-filled FGPC as a function of thickness

0.56 along the sheet thickness. The average volume fraction of the filler in FGECs can be changed by varying the loading in each layer of the stack and even by changing the number of layers per stack but maintained equal while comparing it with its UDEC counterpart. The volume fraction of the filler in the composite was maintained within ± 2 % deviation for higher volume fractions of the filler. The filler volume fraction was calculated by thermogravimetric analysis (TGA). Finally, all these composites are experimented with physical, mechanical, and magnetic characterizations.

Physical Properties of Magnetic Particle-Filled FGPC

The hardness pattern is studied for various loadings of iron in SBR-FGPC along the thickness direction. With increasing distance along thickness, the hardness increases as shown in Fig. 12.2. This increase in hardness is attributed to the increase in the volume fraction of iron. There are three distinct regions wherein one can see the difference in the rate of change in the hardness value. The first rate of increment in hardness is steeper (up to 0.4 mm) that corresponds to initial filler loading (100–150 phr iron), moderates up to the depth of 1 mm (400 phr iron), and again increases significantly at a distance of 1.26 mm (500 phr iron) along thickness. The rate of increment becomes moderate again beyond 1.26 mm with surplus addition of filler to the matrix (up to 1000 phr iron). The hardness values vary from 40 to 94 Shore A in the span of 2.5 mm sheet. For the top two layers (at the lower loading zone), the

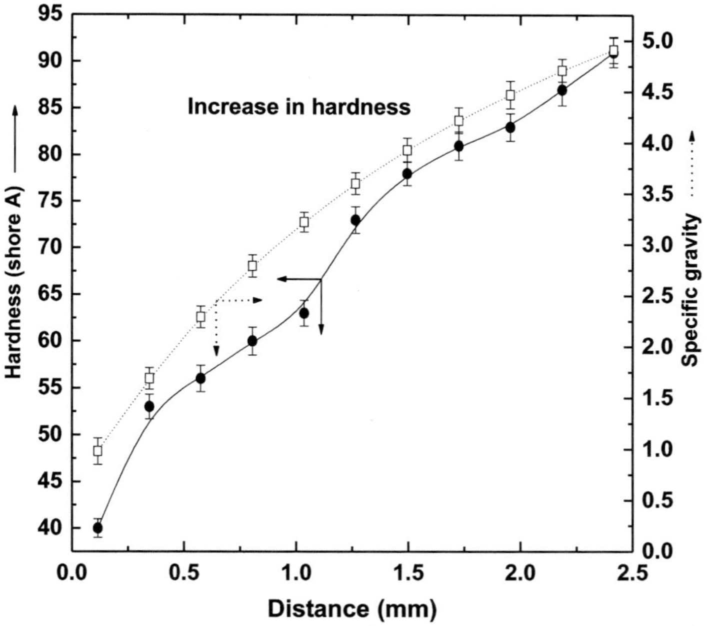

Fig. 12.3 Variation of hardness and specific gravity of nickel-filled FGPC as a function of thickness

hardness value jumps from 40 to 52 Shore A. With additional loading of the hard filler along distance, it increases at a moderate rate up to 1 mm (400 phr iron). This is the maximum limit of filler addition having uniform dispersion with hardly any clusters formed. For the layers present between 0 and 1 mm along thickness, when the hardness indenter is pressed on the surface, hard filler particles will not deform but get displaced in the softer matrix, and hence the rate of hardness increment is moderate. With further filler addition, at and above 1.26 mm (500 phr iron) along distance, the probability for formation of particle clusters increases due to the increasing volume fraction of the filler. These filler clusters will not get displaced easily in the matrix, and it shows further increase in the hardness of the composite. Specific gravity is also measured for the layers employing different weight fractions of iron and is plotted against distance as shown in Fig. 12.2. Increment in the specific gravity is gradual and varies from 0.98 to 4.6 along 2.5 mm thick sheet as it mostly depends on the weight fraction of the dense filler.

Nickel (Ni) particles are also dispersed in the SBR matrix in different amounts along the thickness, and the hardness of these stacking layers of FGPC is measured (Fig. 12.3). The corresponding plots of hardness and specific gravity against distance along which loading of nickel is varied from 0 to 1000 phr show similar kind of trend. Hardness varies from 40 to 91 Shore A while the specific gravity increases from 0.98 to 4.91 along distance. It is expected that the clusters of filler are mainly

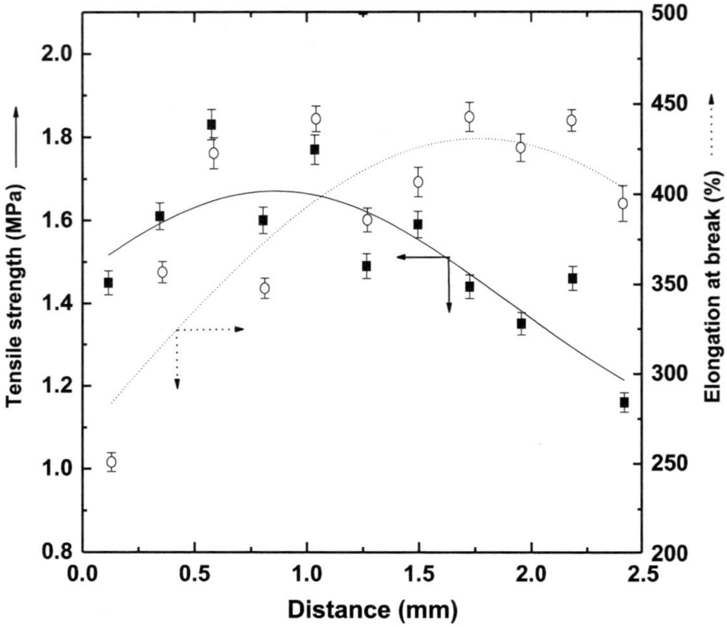

Fig. 12.4 Variation of tensile strength and elongation at break as a function of thickness in iron-filled FGPC

responsible for the changes in many other properties that affect the performance of composite.

Mechanical Properties of Magnetic Particle-Filled FGPC

Mechanical properties of stacking layers of FGPCs are prerequisite, as many applications demand for certain extents of mechanical loading. The effect with layers employing superfluous addition of fillers on FGPCs in modifying these properties is to be known properly. Tensile test is carried out to measure the ability of the material to sustain the tensile force and elongation before it breaks. Tensile strength, modulus, and elongation at break are the important properties to study the serviceability of the composite.

The mechanical properties of polymer composites strongly depend on the interfacial interaction between filler and the matrix. It is also influenced by the filler size, shape, dispersion, and its amount in the matrix. The change in tensile strength and elongation at break along the thickness of iron-filled FGPC is shown in Fig. 12.4. The tensile strength increases with distance for the first few layers and decreases again with increasing filler loading along the thickness. In addition to this, the tensile strength of the top layer is less, as it lacks self-reinforcing (due to stress-induced crystallization) qualities of NR. Further, the tensile strength of stacking layers

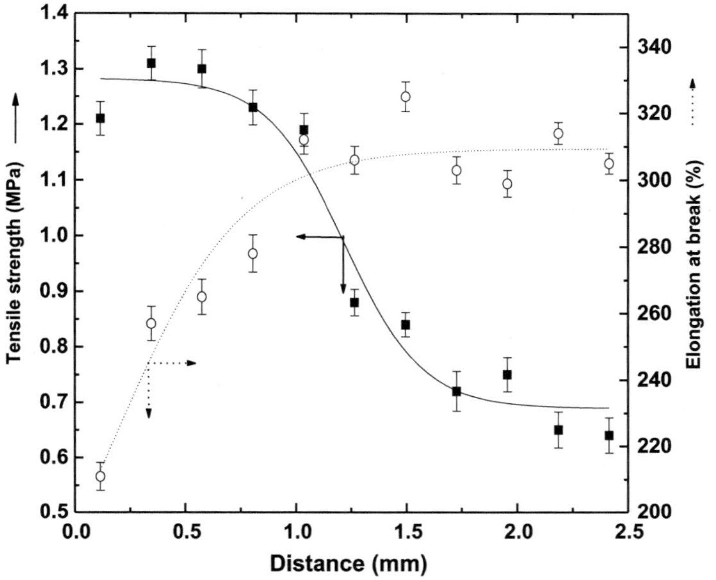

Fig. 12.5 Variation of tensile strength and elongation at break as a function of thickness in nickel FGPC

present between 0 and 1 mm along thickness (employing the loading variation from 0 to 400 phr) increases from 1.45 to 1.85 MPa and falls down slowly to 1.15 MPa at the end. With lesser amount of filler in the top few layers, the dispersion is quite uniform throughout the matrix, each particle acts as an anchor point in the matrix, and hence tensile strength increases with increasing filler content along the distance. With further increasing filler loading beyond 1 mm, the probability of particle cluster formation and hence the formation of bigger-sized flaw becomes more prominent [12]. It should apparently decrease elongation at breaking point with increasing distance. But the elongation at break is found to increase continuously along the span of 2.5 mm thickness. It is 250 % for the top layer (~0.12 mm), reaches its maximum to 450 % in the middle (~1.73 mm), and again decreases with superfluous filler loading at the end. It is assumed that the process of cross-linking might have interfered because of higher concentration of filler. With variation of Ni particles from 0 to 1000 phr within the span of 2.5 mm, the tensile strength decreases in the same fashion from 1.2 to 0.64 MPa, while the elongation % at break increases from 210 to 310 throughout the loading (Fig. 12.5).

Figure 12.6 shows the variation of modulus at 100 and 200 % elongations along the thickness of an iron-filled FGPC sheet. Both moduli at 100 and 200 % increase with distance for initial filler loading and decrease again with further loading along the distance. It follows the same trend as the tensile strength. The modulus at 100 % increases from 0.92 to 1 MPa, while the modulus at 200 % increases from 1.11 to 1.27 MPa with thickness variation from 0 to 1 mm (i.e., up to 400 phr iron) and decreases correspondingly from 1 to 0.74 MPa and from 1.27 to 0.84 MPa when the

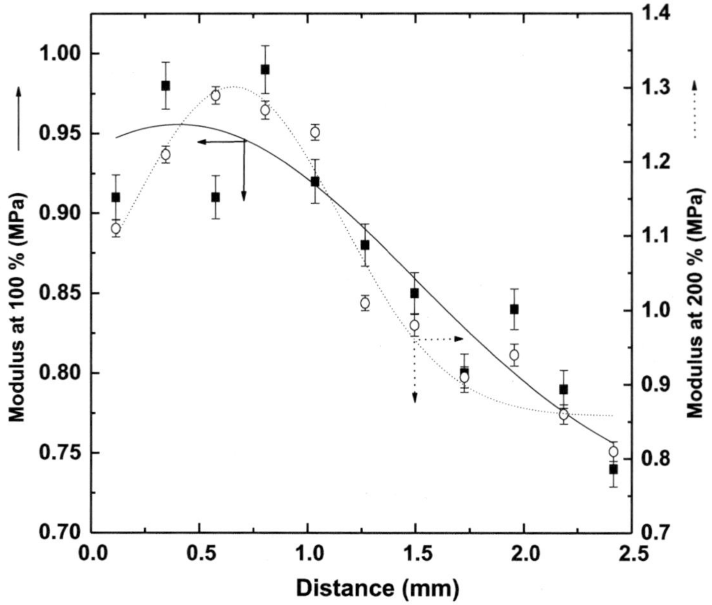

Fig. 12.6 Variation of modulus at 100 % and 200 % elongation as a function of thickness in iron-filled FGPC

distance increases from 1 to 2.5 mm (i.e., up to 1000 phr iron), respectively. At higher filler loading, the elongation at breaking point increases. This leads to the drop in modulus for the end layers containing higher loading. The mechanical characterizations reveal that the magnetic properties can be imparted without compromising on the flexibility even at much higher loading.

The variation of change in modulus at 100 % and 200 % elongations for the nickel-filled FGPC also follows the same trend (Fig. 12.7). The anchoring effect of particles at the initial filler loading increases the modulus at 100 % and 200 % elongations from 0.94 to 1.18 MPa and from 1.18 to 1.37 MPa, respectively. Beyond an average 1 mm along distance, the modulus at 100 % and 200 % elongations decreases from 1.18 to 0.81 and from 1.37 to 1.02 MPa, respectively.

Microanalysis of Magnetic Particle-Filled FGPC

Figure 12.8 shows the SEM fractographs of stacking layers of FGPC employing increasing iron content in the SBR matrix. It is evident that at the top few layers, particles are dispersed uniformly, oriented haphazardly, and separated from each other throughout the matrix as shown in Fig. 12.8a–c. With increasing iron content in the matrix along distance (Fig. 12.8d–f), the interparticle distance decreases, and the probability of formation of clusters becomes more. Figure 12.8d shows the formation of few clusters of particles for the layer present at an average distance of 1 mm,

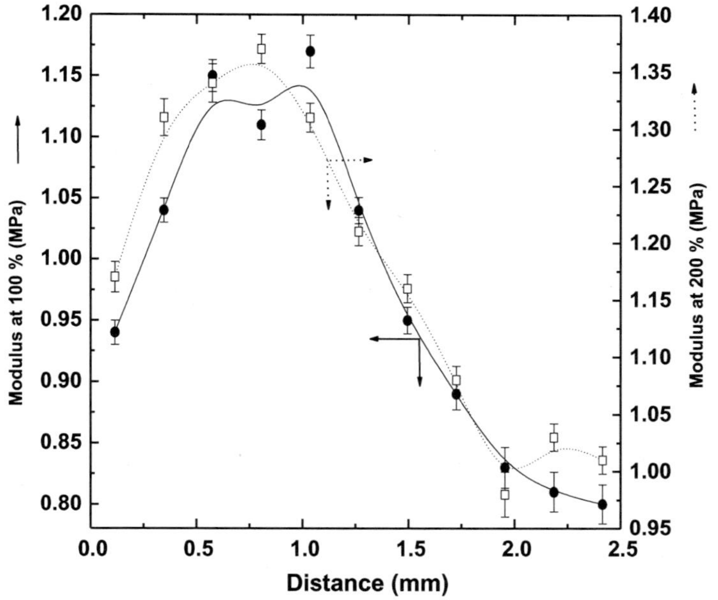

Fig. 12.7 Variation of modulus at 100 % and 200 % elongations as a function of thickness in nickel-filled FGPC

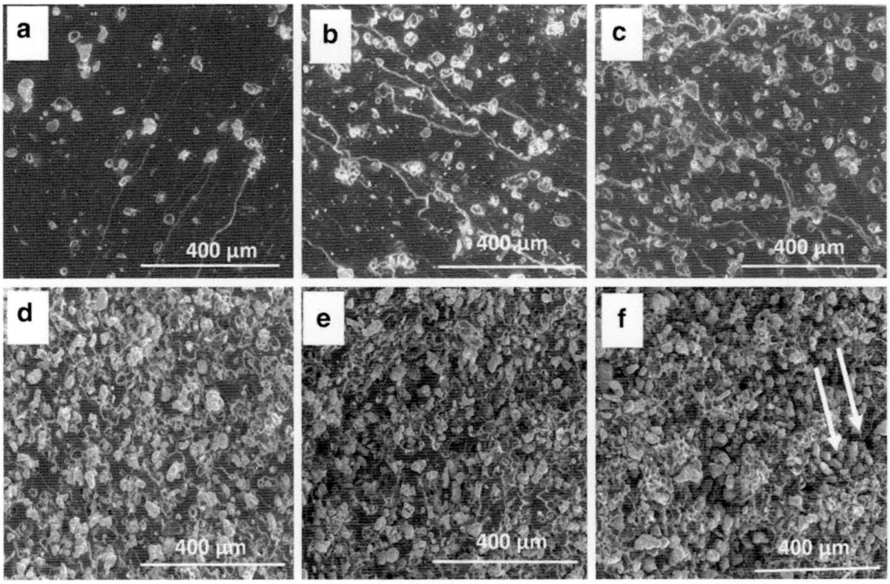

Fig. 12.8 SEM fractographs of layers at different distances in iron-filled SBR: (**a**) 10 phr, (**b**) 0.35 mm (100 phr), (**c**) 0.58 mm (200 phr), (**d**) 1 mm (400 phr), (**e**) 1.26 mm (500 phr), (**f**) 2.42 mm (1000 phr). *Arrows* in (**f**) show the clusters of particles chipped off

and the amount of clusters formed increases exorbitantly in the next layer at 1.26 mm (Fig. 12.8e). The irregular fracture surfaces of layers employing higher loading (Fig. 12.8d–f) depict the distinct phase separation that occurred due to the formation of clusters, and that phase has been removed from these locations during failure of FGPC. The removal of small particles as well as particle clusters from their respective positions in the layers leaves behind the pores in the matrix. This reveals a poor interaction between the matrix and filler. Figure 12.8f shows a highly irregular fractured surface of the last layer present at 2.42 mm with deep depressions that corresponds to the chipping off of the big clusters during fracture.

Magnetic Properties of Magnetic Particle-Filled FGPC

The information regarding matrix-filler interaction, dispersion of the filler, and percolation threshold can possibly be mustered combinedly from the mechanical as well as magnetic measurements. Magnetic measurements are carried out for various stacking layers of FGPCs employing different loadings of iron. Different parameters like saturation magnetization (Ms), coercivity (Hc), and remanent magnetization (Mr) are determined as a function of distance along the thickness of FGPC.

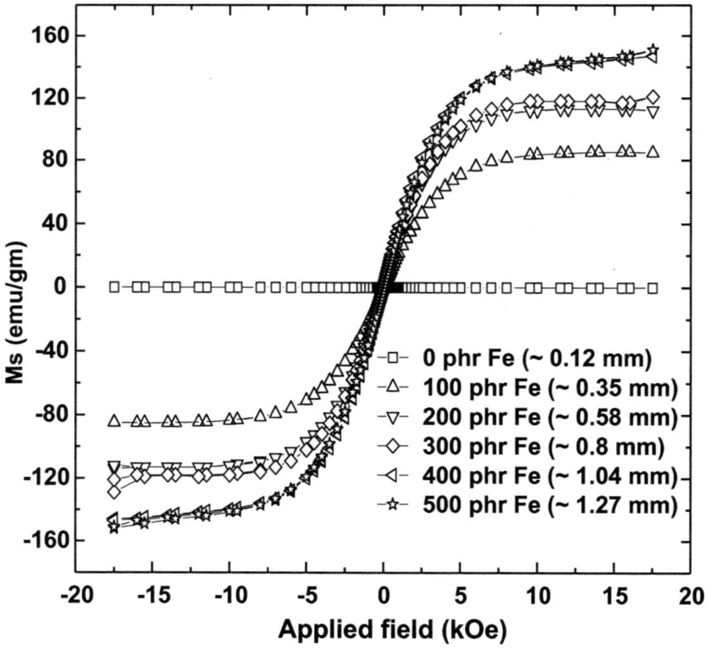

Fig. 12.9 Variation of saturation magnetization of stacking layers in iron-filled SBR (up to 500 phr) as a function of applied field

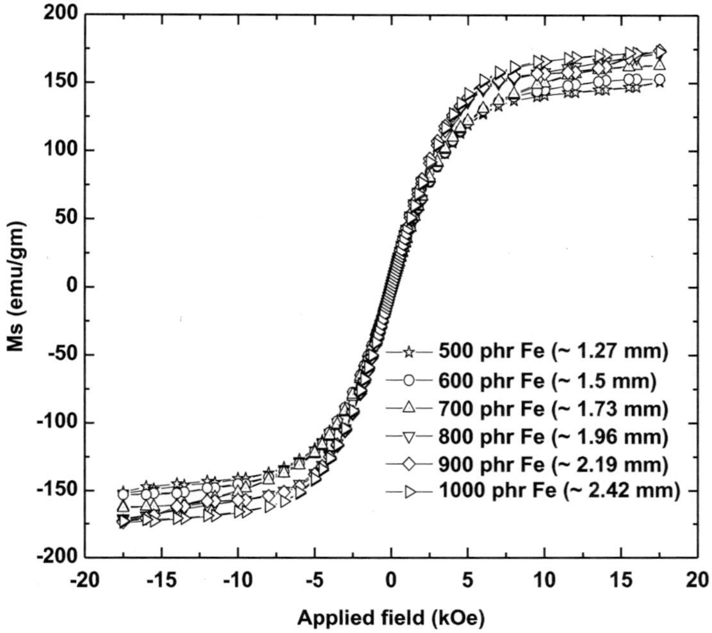

Fig. 12.10 Variation of saturation magnetization of stacking layers in iron-filled SBR (from 500 to 1000 phr) as a function of applied field

From the hysteresis curves for various layers of FGPC, it is observed that the saturation magnetization increases with distance. This is attributed to the increase in the filler loading along distance. Figure 12.9 shows the hysteresis curves for the first half layers of FGPC containing variation of iron loadings from 0 to 500 phr. It is clear that the rate of increment in the saturation magnetization is rapid for the top layer and is moderate up to the depth of 1 mm. From Fig. 12.10, one can see that with increasing distance from 1.26 mm onward, the rate of increment of saturation magnetization becomes lenient. The formation of clusters of magnetic fillers might be the cause of this lenient incremental rate. With increasing distance, the filler loading becomes superfluous, and the probability of cluster formation increases that result in the cancelation of magnetic lines of forces within the system, and hence the resultant saturation magnetization hardly differs. The saturation magnetization values corresponding to the layers present at an average distance of 0, 0.34, 0.57, 1, and 2.4 mm are 0, 85, 114, 146, and 177 emu/g, respectively, while that for pure iron is 210 emu/g. All the rest of the parameters are noted and plotted against distance.

Similarly, Figs. 12.11 and 12.12 show the similar features of saturation magnetization curves for nickel. Saturation magnetization values corresponding to the layers present at an average distance of 0, 0.34, 0.57, 1, and 2.4 mm are 0, 23.5, 29.3, 38.2, and 45 emu/g, respectively, while that for pure nickel, it is 55 emu/g.

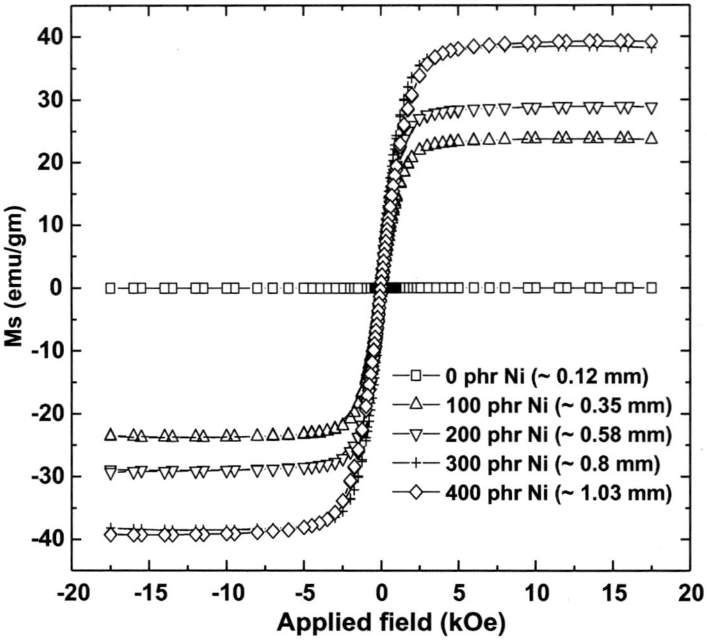

Fig. 12.11 Variation of saturation magnetization of stacking layers in nickel-filled SBR (up to 400 phr) as function of applied field

Figure 12.13 shows the variation of saturation magnetization as a function of thickness of FGPC. It shows that the increment in the magnetization with distance is not linear. There is a rapid increment in the saturation magnetization for the first few layers employing lesser filler loading, and the degree of steepness slowly decreases with increasing loading along distance. Beyond 1.26 mm along thickness, the saturation magnetization increases mildly. Attempts have been made to calculate the saturation magnetization of layers of FGPC from the only information of Ms value and weight fraction of the filler present in a given layer. Rule of mixture is introduced to calculate the magnetization values of stacking layers employing different loadings of filler.

$$M_{PC} = W_f M_F + W_p M_P \tag{12.1}$$

where W_f, W_p and M_F, M_P are the weight fractions and saturation magnetizations of the filler and polymer matrix, respectively, while M_{PC} is the magnetization of a given layer of FGPC.

As the matrix is nonmagnetic, Eq. 12.1 reduces to

$$M_{PC} = W_f M_F \tag{12.2}$$

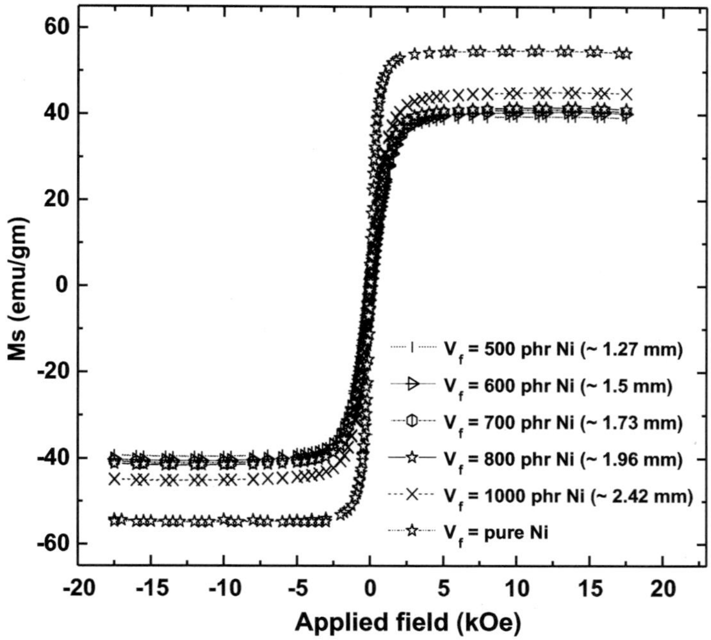

Fig. 12.12 Variation of saturation magnetization of stacking layers in nickel-filled SBR (from 500 to 1000 phr) as a function of applied field

The calculated and measured Ms values of the layers containing varied concentration of filler are plotted against distance as shown in the Figs. 12.13 and 12.14. The calculated Ms values from the rule of mixture are much closer to the experimental Ms values for the top few layers (Fe and Ni both) till the curve follows linearity. With the deviation of curve from linearity, the difference between the calculated and experimental Ms values magnifies. For superfluous filler loading along distance, the rule of mixture is not applicable to predict the Ms value of layers of FGPC. From the experimental data of Ms for various stacking layers, it is observed that the Boltzmann fit gives the least error ($R^2 \sim 0.98$) and is represented by Eq. 12.1:

$$(Ms)_{W_F} = \frac{A_1 - A_2}{1 + e^{\frac{\left(W_F - W_{F_0}\right)}{dW_F}}} + A_2 \tag{12.3}$$

where $(M_S)_{W_F}$ is the saturation magnetization of any given layer of FGPC corresponding to any given weight fraction of the filler, W_F is the weight of filler in phr and remaining A_1, and A_2, W_{F0}, and dW_F are constants. This equation is checked using Ni as different filler (with same particle size) in SBR matrix. It is observed that the experimental plot of Ms with various loadings of Ni along the thickness of sheet follows the Boltzmann fit only [13].

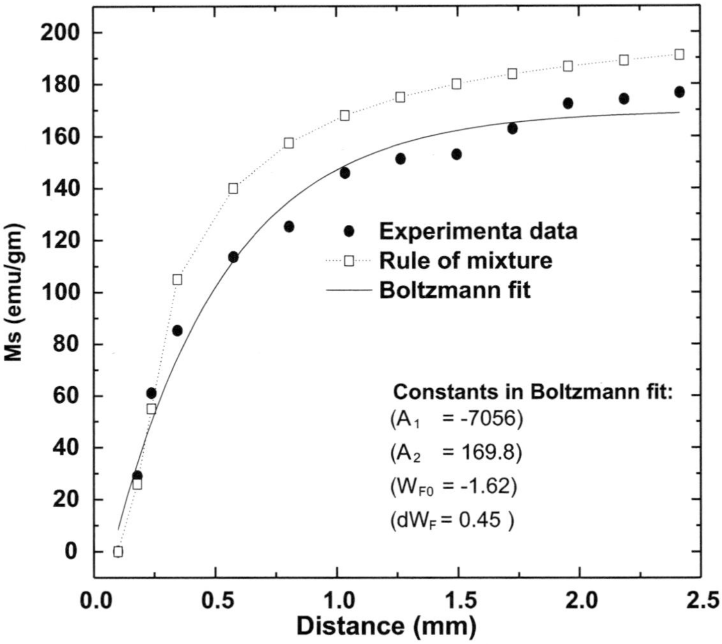

Fig. 12.13 Variation of saturation magnetization along the thickness of SBR-iron FGPC

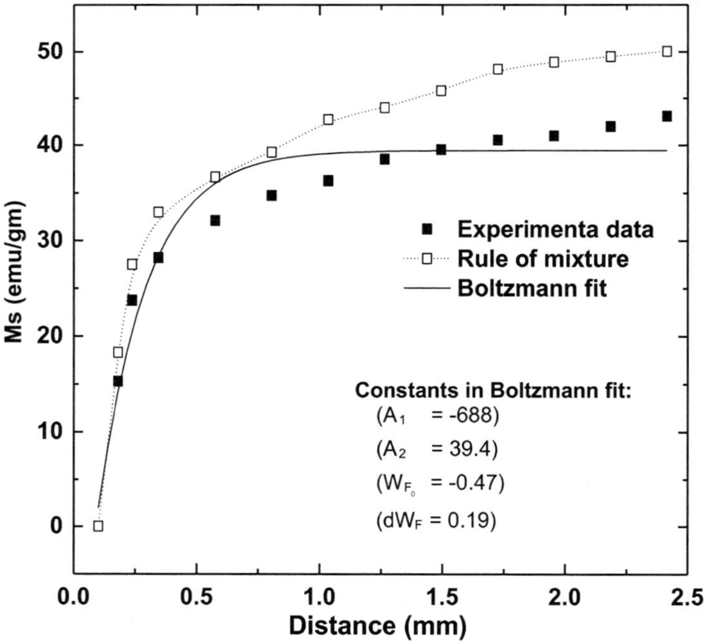

Fig. 12.14 Variation of saturation magnetization along the thickness of SBR-nickel FGPC

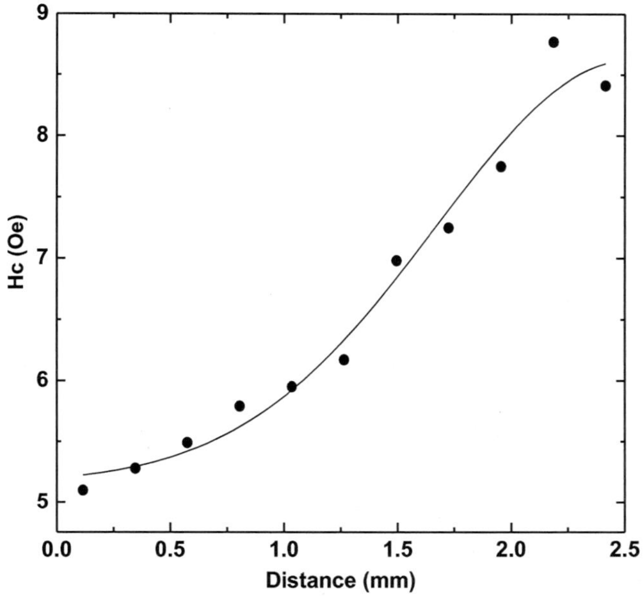

Fig. 12.15 Variation of coercivity along the thickness of SBR-iron FGPC

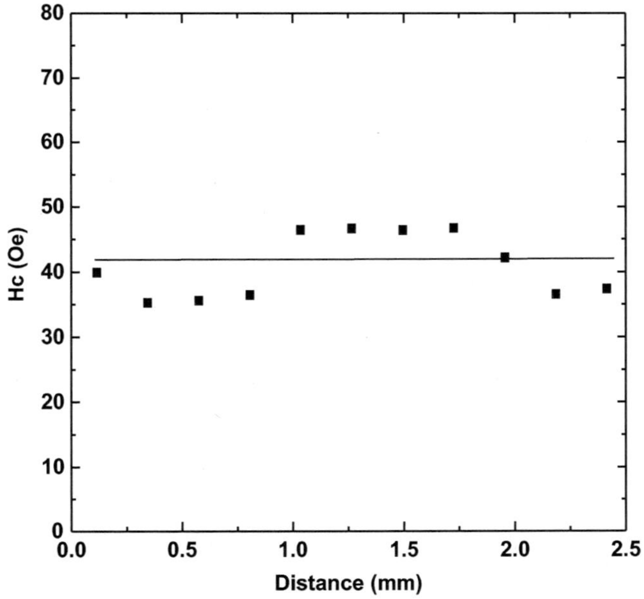

Fig. 12.16 Variation of coercivity along the thickness of SBR-nickel FGPC

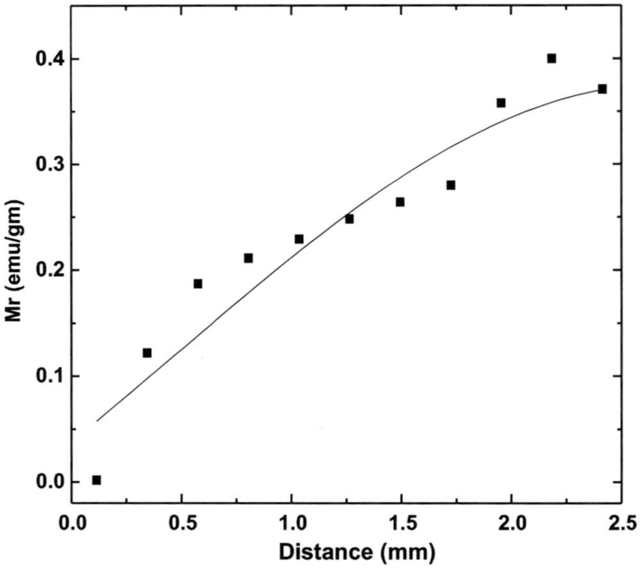

Fig. 12.17 Variation of remanent magnetization along the thickness of SBR-iron FGPC

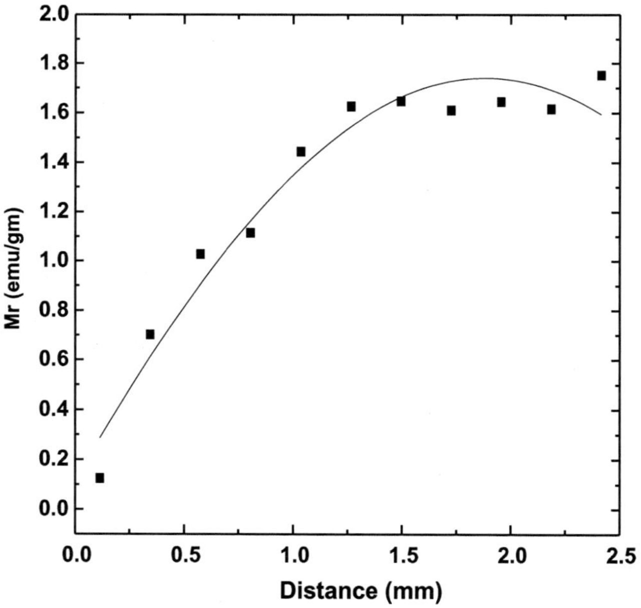

Fig. 12.18 Variation of remanent magnetization along the thickness of SBR-nickel FGPC

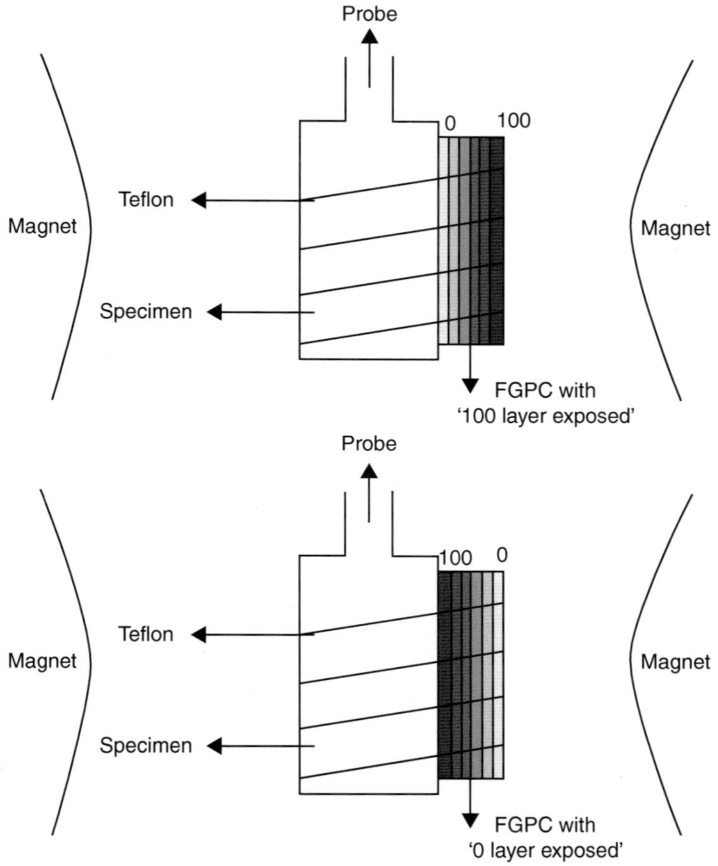

Fig. 12.19 Placing of a 0-20-40-60-80-100 FGPC with "100 layer exposed" and "0 layer exposed" in a given magnetic field

A closer agreement between the measured and experimental values of saturation magnetization assists in tailoring the different appropriate materials (with known saturation magnetization values) and its loading required in a given polymer to fulfill the material properties needed for a given application.

Coercivity is defined as the intensity of magnetic field required to reduce the magnetization of that material to zero after the magnetization of the sample has reached saturation. Figure 12.15 shows the variation of coercivity along distance. With increasing filler loading along distance, the coercive field is increasing. Figure 12.8 reveals that the iron particles are uniformly dispersed and oriented haphazardly. The particles are trapped within the polymer matrix, and an easy direction of iron particles will not get aligned in the direction of magnetic field. This might be the cause of slight increment in the coercivity of the stacking layers. Figure 12.16 shows that the coercivity is hardly varying along distance and hence hardly depends upon the loading of nickel in the polymer matrix. Conclusively, the ineffective difference

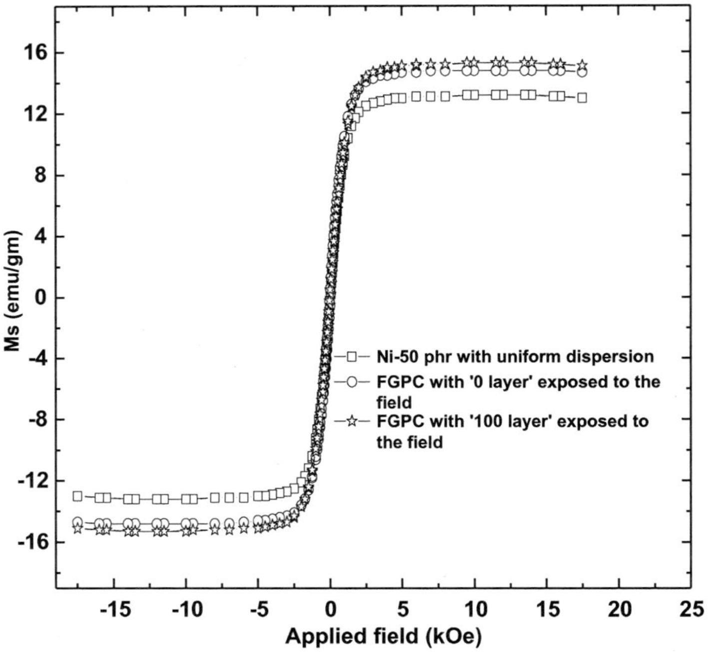

Fig. 12.20 Comparison of hysteresis curves of UDPCs and FGPCs employing an average 50 phr nickel

between the coercivity values of pure magnetic fillers (either iron or nickel) and the FGPC layers implies weak interaction between fillers and the matrix.

Remanent magnetization is the permanent magnetization that remains after the external field is removed. Figures 12.17 and 12.18 show an increasing trend of remanent magnetization with distance along thickness. With increasing distance, the saturation magnetization increases, which increases the remanent magnetization. Once the saturation magnetization saturates (~1 mm as shown in Figs. 12.13 and 12.14), a corresponding saturation of remanent magnetization is also observed. Figures 12.14 and 12.18 follow the same trend.

Comparison of Magnetic Properties of UDPCs and FGPCs

The restriction of carrying out a comparative study of magnetic properties of UDPCs and FGPCs is the mass of the composites. The "vibrating sample magnetometer" does not allow the mass of the specimen to be more than 4 g. With this restriction, samples of UDPCs and FGPCs with average loading of 50 phr nickel are taken.

FGPC sample is placed in such a way that the variation of filler occurs in the perpendicular direction of applied field. Positions with "0 layer exposed" and "100 layer exposed" are shown in Fig. 12.19. Figure 12.20 displays a plot of hysteresis curves for 50 phr UDPC and 0-20-40-60-80-100 FGPC. It shows that FGPCs with both the layers exposed show higher saturation magnetization than UDPCs. The saturation magnetization values for FGPCs with "0 layer exposed" and "100 layer exposed" are nearly similar and are 14.9 and 15.2 emu/g, respectively. The corresponding Ms value for UDPC is 13.4 emu/g. The saturation magnetization of FGPCs is nearly 12 % higher than its UDPC counterpart. It is worth to mention that the volume fraction of both UDPCs and FGPCs is constant.

Concluding Remarks

Iron powder-filled SBR composites having gradient concentration along thickness are synthesized, and their physical, structural, mechanical, and magnetic properties are studied. While the hardness increases for increasing filler concentration along the thickness, the tensile strength and modulus exhibit a decrease for higher filler concentration. This could be attributed to increased agglomeration and formation of cracks at higher filler concentration. Again, the saturation magnetization increases with increasing filler concentration and reaches to saturation at higher loading. With similar volume fraction, FGPCs exhibit higher values of saturation magnetization compared to their UDPC counterparts.

Acknowledgment The authors acknowledge the financial support provided by the Department of Science and Technology, India, for carrying out this work.

References

1. Ahankari S, Kar KK (2012) Mechanical properties of functionally graded carbon black–styrene butadiene rubber composites: effect of modifying gradation and average filler loading. J Appl Polym Sci 125:3469
2. Ahankari SS, Kar KK (2008) Processing and characterization of functionally graded materials through mechanical properties and glass transition temperature. Mater Lett 62:3398
3. Ahankari SS, Kar KK (2010) Processing and mechanical behavior of carbon black graded rubber compounds. J Appl Polym Sci 115:3146
4. Udupa G, Rao SS, Gangadharan KV (2014) Functionally graded composite materials: an overview. Procedia Mater Sci 5:1291
5. Müller E, Drašar Č, Schilz J, Kaysser WA (2003) Functionally graded materials for sensor and energy applications. Mater Sci Eng A 362:17
6. Varadan VK, Varadan VV, Ma Y, Hall W (1986) Design of ferrite-impregnated plastics (PVC) as microwave absorbers. IEEE Trans Microw Theory Technol 34:251
7. Liew KM, Lei ZX, Zhang LW (2015) Mechanical analysis of functionally graded carbon nanotube reinforced composites: a review. Compos Struct 120:90
8. Chung H, Das S (2008) Functionally graded Nylon-11/silica nanocomposites produced by selective laser sintering. Mater Sci Eng A 487:251
9. Kwon H, Bradbury CR, Leparoux M (2011) Fabrication of functionally graded carbon nanotube-reinforced aluminum matrix composite. Adv Eng Mater 13:325

10. Erisken C, Kalyon DM, Wang H (2008) Functionally graded electrospun polycaprolactone and β-tricalcium phosphate nanocomposites for tissue engineering applications. Biomaterials 29:4065
11. Shen H-S (2009) Nonlinear bending of functionally graded carbon nanotube-reinforced composite plates in thermal environments. Compos Struct 91:9
12. Boonstra B, Medalia A (1963) Effect of carbon black dispersion on the mechanical properties of rubber vulcanizates. Rubber Chem Technol 36:115
13. Kar KK, Sharma SD, Kumar P (2007) Effect of rubber hardness on the properties of fiber reinforced plastic composites made by the newly proposed rubber pressure molding technique. Polym Compos 28:618

Mechanical, Electrical, and Tribological Properties of Copper–Graphite Composites 13

K. Rajkumar and S. Aravindan

Contents

Abstract

Recently copper-based metal matrix composites (MMC) are finding usage in high current density and low-voltage electrical sliding contact applications. Copper-based composites are generally processed through powder metallurgy (P/M) route. Mechanical and wear resistance properties were enhanced with microwave heat treatment as a result of reduction in porosity of powder metallurgy product.

K. Rajkumar
Department of Mechanical Engineering, SSN College of Engineering, Kalavakkam, Tamil Nadu, India
e-mail: rajkumark@ssn.edu.in

S. Aravindan (✉)
Department of Mechanical Engineering, Indian Institute of Technology Delhi, New Delhi, India
e-mail: aravindan@mech.iitd.ac.in

K.K. Kar (ed.), *Composite Materials*,
DOI 10.1007/978-3-662-49514-8_13

Keywords
Metal matrix composite • Copper matrix • Graphite • Wear • Sliding contact

Introduction

Tribology is defined as the science and technology of interacting surfaces. It comprises the study of friction, wear, and lubrication. Friction and wear problems cost the society a huge amount of money each year. Friction and wear have been observed in electromechanical systems when there is a relative movement between two solid bodies. Friction mainly results in loss of energy and wear results in material loss. The proper understanding of friction and wear mechanisms provides practical solutions to tribological-related problems. Electromechanical systems consist of brushes and commutator/slip ring. Brushes are components that transfer electrical current between the stationary and rotating elements in electrical motors and generators as shown in Fig. 13.1. In such application, conventional lubrication like thick lubricating film cannot be used. In such cases self-lubricating plays an important role in reduction of friction and wear [1]. The brushes are composites of graphite and a conductive metal such as copper. Copper–graphite composite is a tribological composite that can be used in sliding electrical contact applications requiring low friction and wear in addition to high electrical conductivity [2]. Graphite lubrication is one of the important factors affecting the friction and wear behavior of copper–graphite electrical sliding contacts. Graphite also has reasonably good conductivity and good arc-resisting characteristics. Also graphite retains its properties at high temperature. Addition of graphite particles to copper matrix reduces density, increases stiffness, raises the service temperature, and provides a mechanism for tailoring the coefficient of thermal expansion.

Improvement of properties of existing composite materials constantly takes place using the newer processing methodology. In the case of conventionally sintered composites, the presence of pores affects the physical and mechanical properties [4]. Inherent porosity decreases the strength, ductility, and modulus of composites. There is an urgent need to minimize the porosity to realize an overall improvement in properties. By adopting the modern heating technique like microwave heating, the properties of composites can be also improved. Recent research studies on microwave processing of metal powders elucidated the improved properties of materials through a finer grain size and smaller number of rounded and uniformly distributed pores [5, 6]. Generally larger and angular pores were seen in conventionally sintered ones. Microwave heating of materials is fundamentally different from conventional heating in that the heat is generated internally within the material instead of originating from an external heating and subsequent radiative transfer. In actual practice, the sample in fact becomes the source of heat during microwave processing.

Fig. 13.1 Electrical brush-rotor system [3]

Microwave heating has gained a lot of significance in recent times for synthesis of materials and sintering mainly because of its intrinsic advantages such as rapid heating rates, reduced processing time, and improved properties, finer microstructures, and being environmentally more clean. In conventional heat-treating process, energy is transferred to the material from the surface of the material, whereas in microwave heating, the energy is directly delivered to material through the molecular interaction with the electromagnetic field [7]. This difference makes the microwave heat treating as a potential method. Different types of copper–graphite/carbon composites and their processing route and applications are presented in the following section. Tribological characteristics of untreated and treated composites were also presented in the subsequent sections. This gives a comprehensive knowledge on copper–graphite composite processing and tribological characteristics.

Types of Copper–Graphite Composites

According to the type of reinforcement, copper–graphite composites can be classified into three types: (i) particulate-reinforced, (ii) whisker/short fiber-reinforced, and (iii) continuous fiber-reinforced composites.

Particulate-Reinforced Copper–Graphite Composites

Particulate or particles with aspect ratio (length/diameter) less than 5 are reinforced in the metal matrix to have a dispersed phase of roughly equiaxed particles. Graphite particles are reinforced in a copper matrix from 5 % to 40 % in the volume fraction, which can be used for structural and wear resistance applications [8]. In general, these composites are manufactured either by powder metallurgy route or through liquid metallurgy route. Moreover, these manufacturing routes are cost-effective when compared to continuous fiber-reinforced composite. Mechanical properties of these kinds of composites are inferior to whisker/continuous fiber-reinforced copper matrix but superior to unreinforced copper alloys. These composites were isotropic in general. These kinds of composites generally necessitate secondary manufacturing operations such as rolling, forging, extrusion, and heat treatment to impart better mechanical properties.

Whisker-Reinforced Copper–Graphite Composites

Whiskers are described as either elongated particles or short fiber with aspect ratio greater than 5. Simple shaped components of short fiber composites are manufactured by direct squeeze casting method. Indirect squeeze casting is similar to pressure die casting. It is used for thinner casting sections. The melt is poured into the shot sleeve of a vertical casting machine and followed by tilting to the original position. The melt is then injected into the die through relatively larger gates. This is accomplished through relatively slow velocity (less than 0.5 m/s). The melt in the die cavity is then solidified under pressures, ranging from 55 to 300 MPa. The application of indirect squeeze casting makes it possible to manufacture more complex parts, but it results in more expensive die and tools for casting. Whisker-reinforced composites can also be manufactured by infiltration method. Whisker-reinforced composites have attractive combinations of strength and thermal stability relative to those of particulate-reinforced materials. But extensive commercialization of whisker-reinforced MMC is at a slower pace owing to the high costs associated with currently available whiskers. Also the health hazards associated with high aspect ratio particulates are higher [9].

Continuous Fiber-Reinforced Copper Matrix Composites

Copper matrix contains long fiber reinforcement that is available as tows of several hundreds or thousands of fiber. The diameter of the fiber ranges from 5 to 20 μm. Continuous fibers are exceptionally strong and they have higher stiffness along the direction of the fiber, which results in anisotropy. They conduct more heat energy along fiber direction than other directions because of its anisotropy. These composites are used in thermal management in aerostructure for directional heat transfer [10]. Copper matrix having fiber volume fraction up to 40 % is produced by squeeze infiltration and/or pressure infiltration route.

Processing of Copper–Graphite Composites

A wide variety of fabrication methods are employed for the production of copper–graphite composites. The selection of processing route depends on various factors including type and volume fraction. Basically fabrication processes of composites are classified into two categories such as solid state processing and liquid state processing.

Liquid State Processing

The simplest liquid state technique referred to as melt stirring is to mix graphite particle reinforcements with copper molten metal and cast in the conventional manner [11]. Lack of good wetting between graphite particles and liquid copper melt is a problem here, and the addition of wetting agent during melting such as Ti or chromium will improve the wettability. The volume fraction reinforcement has a limit of 20 vol.%. The effective dispersion of the reinforcement becomes difficult above this level. A schematic diagram of setup for stir casting for manufacturing copper–graphite composite is shown in Fig. 13.5. Stir casting involves producing a copper melt in graphite cubicle which is located inside the resistance furnace which continuously supplies heat to melt and followed by the introduction of graphite particle reinforcement into the melt. The uniform dispersion of the reinforcing material in melt was carried out by thorough stirring. Stirrer is normally made of steel with specific ceramic coating so that contamination of stirrer material in the melt can be avoided. The thermocouple was used to give a feedback of the temperature inside the furnace. The inert gas is continuously supplied to protect the melt to avoid oxide formation.

Since graphite and copper are exhibiting poor wetting behavior, the interfaces are weak, i.e., it induces lack of wettability. Interface between copper and graphite composite is very important in determining the ultimate properties of the composite. Wettability can be improved by utilizing copper-coated graphite. It will increase interfacial strength and reduce the pullout of graphite particle from the copper matrix during service. The wettability of solid graphite by a copper molten metal can be also improved by the addition of alloying elements like carbide-forming elements, i.e., Ti or Cr to the matrix composition. This improvement of wettability is attributed mainly to these added reactive elements. They induce reaction at interface, i.e., between matrix and reinforcement which leads to reduce the solid–liquid interfacial energy.

Centrifugal casting process: Centrifugal casting is one of the methods used for production of copper–graphite composite. This casting process imparts better tribological properties to inner periphery of copper alloy cylindrical casting, where graphite particles are segregated [12]. Centrifugally cast copper–graphite composites are widely used as bearing components due to their good tribological properties. The

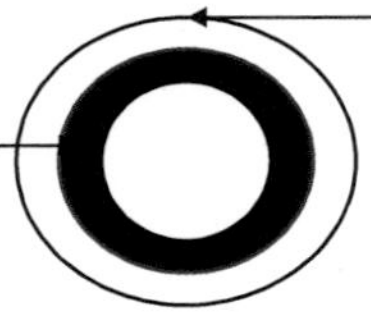

Fig. 13.2 Typical centrifugally cast copper–graphite composite

tribological properties of a material depend on its microstructure. Centrifugally cast copper–graphite composites have a unique microstructure of matrix and distribution of graphite particles. A rich graphite particle zone is observed at the inner periphery. Due to the presence of graphite at the inner peripheries, normally this zone is termed as a graphite-rich zone, with larger porous sites. Under certain lubrication condition, these porous sites can act as a reservoir for lubricants which enhances the tribological performance. Typical centrifugally cast copper–graphite composite is shown in Fig. 13.2.

Infiltration Process: Infiltration of fiber preform requires good wetting of fibers by liquid metal. Liquid copper neither wets graphite nor induces chemical reaction at the interface. The lack of adhesion between copper and graphite surfaces can be overcome either by coating of graphite fiber with copper or by adding a wetting agent to copper melt [13]. Moreover, infiltration depends on the nature of reinforcement and its volume fraction, with or without the application of pressure. These composites, having reinforcement volume fraction ranging from 10 % to 60 %, can be produced using an infiltration technique. The pressure should be optimal so that it will be sufficient to aid the infiltration, and when the pressure is increased beyond the optimal range, it results in damaged fibers.

Solid State Processing

Powder Metallurgy Route (P/M): P/M technique is useful for particles and whisker reinforcement to copper matrix. Powder metallurgical processes offer the possibility of obtaining uniform products and of reducing the tedious and costly machining processes for advanced materials. P/M methods also offer the unique advantages of lower manufacturing temperature that avoids strong interfacial reaction between matrix and reinforcement. P/M processed copper–graphite composite contains graphite particles in volume fraction ranges from 5 % to 50 % and consolidation pressure ranges from 490 to 875 MPa, sintering temperature is 1000 °C, and sintering time is three hours including cooling time of samples [14].

Preparations of Copper–Graphite Particulate Composites for Microwave Heat Treatment: Copper–graphite composites were manufactured through powder metallurgy (P/M) route. The P/M processes offer the possibility of obtaining uniform products. Block diagram of powder metallurgy process for production of copper–graphite composite is shown in Fig. 13.3. Starting raw materials for copper–graphite composites were of electrolytic copper powder and graphite

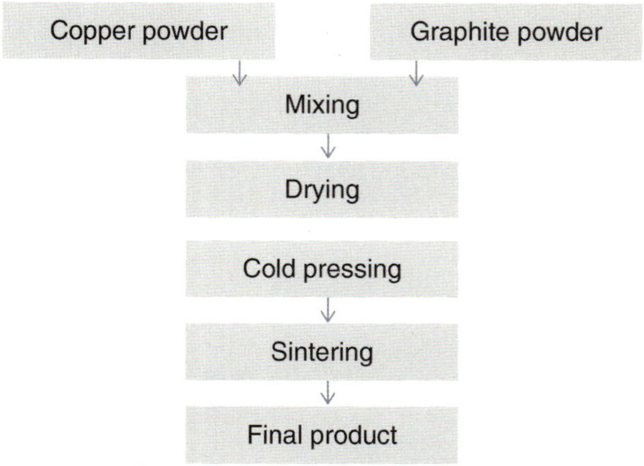

Fig. 13.3 P/M process for manufacturing of copper–graphite composite

Fig. 13.4 Typical copper–graphite composite processed through P/M route

powder. Copper powder with the grain size and purity level of copper used here are 12 μm (at least 90 %) and 99.5 %, respectively. Graphite powder (99.8 % purity level) has a grain size of 50 μm.

Centrifugal ball miller was used for mixing powders, initially correct weight percentage of powders were blended for two hours with speed of rotation of 20 rpm to ensure good mixing. For uniform mixing, ball to powder ratio was taken 15:1. Prior to compacting, the mixed powders are heated to a temperature 150 °C, to evaporate any volatile components. Cylindrical components of height 10 mm and diameter 30 mm were prepared using a simple die with the help of single action hydraulic press at a pressure of 600 MPa. The specimens were sintered in induction furnace at the temperature of 900 °C under inert atmosphere. After the definite holding time, specimens were allowed to cool in the furnace itself. Copper–graphite composite samples are shown in Fig. 13.4, processed through P/M route.

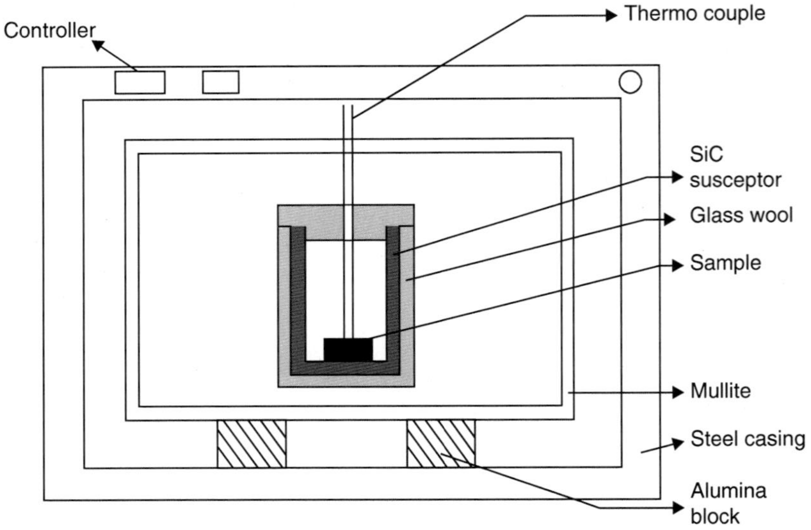

Fig. 13.5 Microwave heat treatment setup

Microwave Heat Treatment

The copper-5 wt% graphite specimens were subjected to microwave heat treatment with suitable arrangement. The schematic diagram of microwave heating is shown in Fig. 13.5. Hybrid heating setup was built inside the microwave furnace (3.2 kW, 2.45 GHz) with the outer transparent layer to microwave (Alumina wool) and the inner absorbing layer (Silicon carbide fencing). Silicon carbide fencing not only provides hybrid heating facility to green pellets but also reduces the thermal gradient to promote the formation of crack-free components. The duration of heat treatment was 10 and 20 min at an isothermal holding temperature of 800 °C.

Results and Discussion

Microstructure Analysis

The tribological properties depend on the structure of matrix as well as the distribution of graphite in matrix. To obtain the understanding of distribution graphite particles in the copper matrix and effect of the microwave heating over the interface between graphite reinforcement and copper matrix, microstructural studies were carried out. An energy-dispersive X-Ray (EDX) was used to analyze the surface

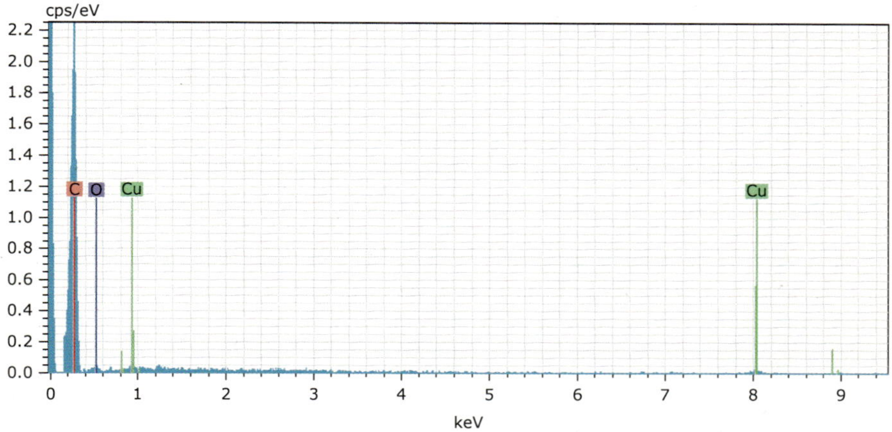

Fig. 13.6 Typical energy-dispersive X-Ray analysis of sintered copper-5 wt% graphite composites

chemistry. The EDX result of copper-5 wt% graphite composite is shown in Fig. 13.6. Peaks of carbon and copper elements could be observed.

Microwave heat-treated specimens and untreated specimens were mirror polished using a standard metallographic technique and then the specimens were subjected to SEM study. SEM images of untreated copper-5 wt% and microwave heat treated for different durations are presented in Fig. 13.7. From Fig. 13.7, the distribution of reinforcement of graphite particles in the copper matrix can be observed. Graphite particles are seen as black particles. These graphite particles are uniformly distributed throughout copper matrix phase. The mapping element data of copper–graphite composite is presented in Fig. 13.8. This confirms that the uniform distribution of graphite particles in copper matrix and presence of impurities are of very negligible amount. From Fig. 13.7a, b, it can be seen that a good amount of porous sites are available in the untreated composites. It can be also seen that the pores are distributed uniformly throughout the matrix. SEM pictures of the 10-min microwave heat-treated specimens are presented in Fig. 13.7c, d. It can be observed that the distribution of graphite reinforcement is fairly uniform. Graphite particles extended from the surface due to pushing of copper matrix, which has been attributed to microwave heating. Due to microwave exposure, the pore closure happens so that the number of pores is lower than untreated specimens. SEM pictures of the 20-min microwave heat-treated specimens are presented in Fig. 13.7e, f. From these figures, the graphitic film produced over tribosurface can be observed, which is essential for reduction of wear rate and coefficient of friction. Longer microwave exposure of specimens makes the copper matrix stronger and pushes the graphitic particles enough to be extruded from the subsurface to tribosurface. Moreover, it has improved the interface bonding between graphite and copper matrix through closure of pores.

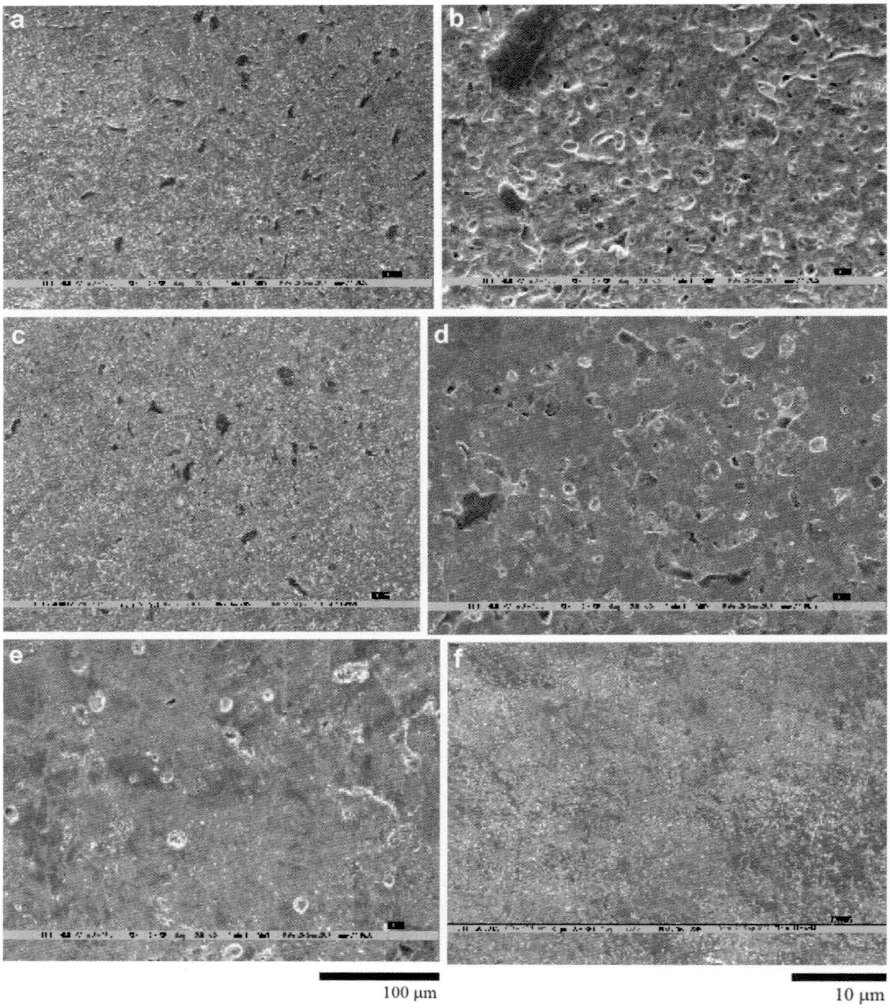

Fig. 13.7 SEM of sintered copper-5 wt% graphite composite. (**a**) Untreated (250 X). (**b**) Untreated (2000 X). (**c**) 10 min heat treated (250 X). (**d**) 10 min heat treated (2000 X). (**e**) 20 min heat treated (250 X). (**f**) 20 min heat treated (2000 X)

Evaluation of Mechanical and Electrical Properties

Copper–graphite composite specimens fabricated and heat treated were tested for their physical and mechanical properties. The sintered specimens were evaluated for their porosity, density, and hardness. Porosity and density were calculated by Archimedes principle, in machined and polished condition. The samples were weighed using an electronic balance of having a least count of 0.001 g. The hardness of sintered composites was measured using a Rockwell hardness testing machine (B scale). The load for the B scale was 100 kg and the indenter was 1/16″ diameter

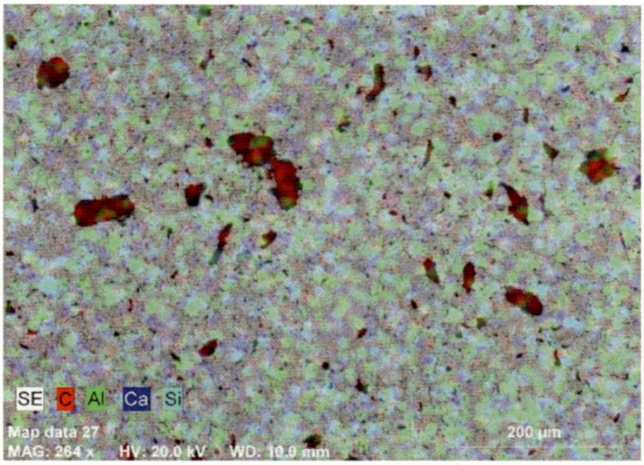

Fig. 13.8 Copper–graphite composites elements mapping data

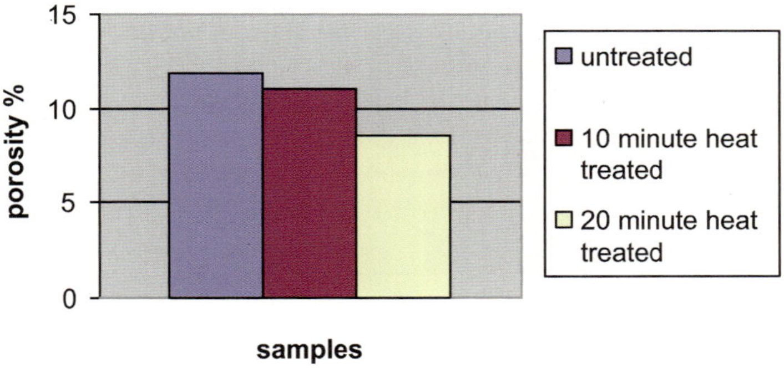

Fig. 13.9 Porosity of copper-5 wt % of graphite

steel ball. Porosity is a condition of trapped pockets of air, gas, or vacuum within a solid material, usually expressed in percentage. Porosity is the measuring factor which influences the mechanical and electrical properties of the copper–graphite composite processed through powder metallurgy. Variation of porosity of composites with the microwave exposure time is presented in Fig. 13.9.

From Fig. 13.9, it is observed that the increase in microwave exposure time leads to decrease in porosity of composite due to closure of pores. Also graphite particles are pushed into pore location when the neck is growing between copper particles. Copper–graphite composites are multiphase materials where one of the phases is a high-loss material while the other is a low-loss material. This mixed microwave-absorbing condition gives one of the significant characteristics of microwave heating, that of selective heating. Porosity reduction is resulted due to internal and selective heating phenomenon. The inside of the sample is allowed to achieve higher density than surface. Internal porosity is minimized which leads to fewer trapped pores. Due to the changes in dielectric constant and losses of pores and grains, pores selectively couple with microwaves. This causes the closure of pores.

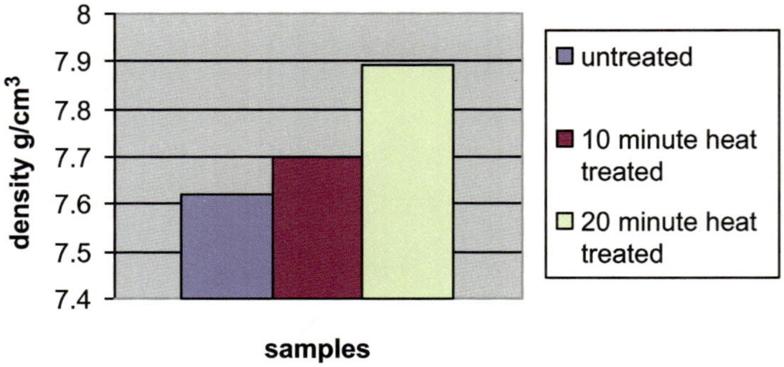

Fig. 13.10 Density of copper-5 wt % of graphite

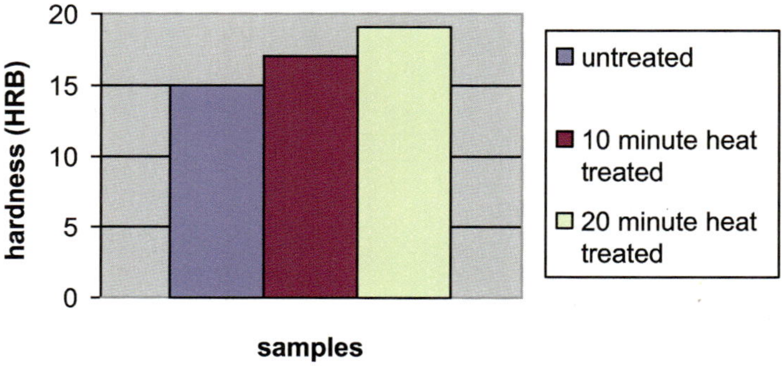

Fig. 13.11 Hardness of copper-5 wt % of graphite

Sintered density is the major factor influencing the mechanical properties of the material processed through powder metallurgy. Density of copper–graphite composite is influenced by the graphite content, pressing load, sintering temperature, and secondary operation like heat treatment. Density of untreated and treated samples of copper-5 wt% graphite is presented in Fig. 13.10. Compared to untreated composites, microwave-treated composites exhibited increased density. 20 min-treated composites exhibited higher densities comparable to theoretical density. Selective heating of microwaves resulted in pore closure which in turn increased the density.

The wear performance of copper–graphite composites is strongly affected by their hardness. Hardness of copper-5 wt% graphite with different microwave heat-treated time is presented in Fig. 13.11.

It is observed from Fig. 13.11 that the hardness of the composites increases with the increasing amount of microwave heat-treating time. This can be attributed to volumetric heating and selective coupling which induce the interface strengthening. This leads to an improvement in hardness of composites. The improved mechanical properties of composites are useful in sliding electrical contacts, which in turn have a significant effect on frictional force and wear rate.

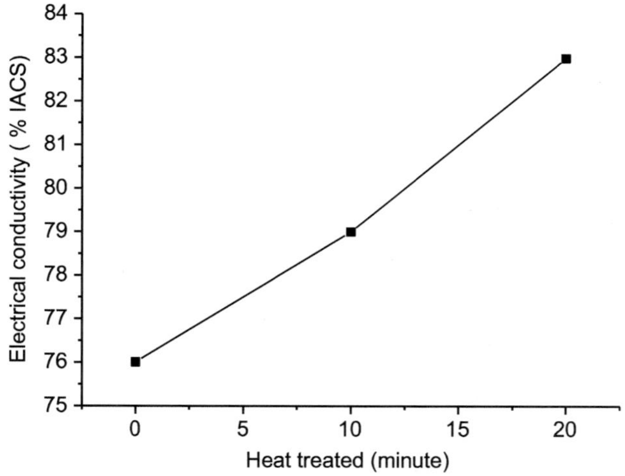

Fig. 13.12 Electrical conductivity of heat-treated composites

Electrical conductivity of developed composites was measured by four-point probe DC method and it was reported in percent IACS (International Annealed Copper Standard). Two silver wires were attached to the end of cylindrical specimen using a silver paste. Two more silver wires were also connected to the known distance in the cylindrical specimen. The area of cross section (A) of the end faces through which constant current (I) enters and leaves the specimen using PGSTAT30 instrument. The voltage (V) across the known distance is measured using a Sanwa 277D multimeter. The electrical conductivity heat-treated copper–graphite composites are shown in Fig. 13.12. The electrical conductivity of heat-treated copper–graphite composite is higher when compared to untreated composites due to the presence of continuous conductive phase across the sample as a result of reduction in porosity.

Tribological Investigation on Copper–Graphite Composites

Recently copper–graphite composites are finding application in sliding contacts. Studies on tribological performance of these composites are necessary for such application. Dyachhenko et al. [15] studied the tribological properties of copper–graphite composite. They reported that the coefficient of friction decreased with the increase in percentage of graphite in copper matrix. Electron diffraction studies were carried out on the tribosurface. Copper with lower percentage graphite composites exhibited catastrophic wear at sliding speed of 11 m/s and relatively higher percentage graphite-reinforced composite exhibited reduction in wear and friction coefficient. This improvement in wear performance was attributed to the surface films formed during sliding. The surface film characterization studies confirmed the uniform distribution of graphite particles with preferential orientation with the basal planes parallel to the friction surface. Okada et al. [16] investigated the copper–graphite lined with porous copper material. It reduces the vibration during sliding. Compared with conventional copper–graphite material, this specially made material exhibited reduced friction coefficient, specific wear rate, and contact voltage drop. However, though the reduction observed in friction coefficient is marginal, the contact voltage drop for the newer lined copper–graphite material is appreciable, when compared with the existing ones. Moustfa et al. [17] investigated the tribological properties of copper-coated and uncoated graphite (8, 15, & 20 wt%) through the powder metallurgy technique. Copper was coated to graphite powder by electroless coating method. Pin-on-disc wear tester was used for evaluation of tribological properties. Coated composites exhibited lower coefficient of friction and lower wear rate when compared with uncoated composite. However, severe wear rate was reported for both the composites at higher load. Kestursatya et al. [18] made an attempt to study the tribological properties of lead free copper–graphite (up to volume 10 %) with low melting point element 8 % tin and 4 % zinc composite which was manufactured by the centrifugal casting. This result was compared with leaded copper alloy. In copper lead alloy, higher intensity copper peaks in EDAX was observed, i.e., copper metal transfer from pin to disc surface was more at higher load due to softer nature of copper lead alloy sliding against a harder counter surface. Coefficient of friction of both the composite and copper lead alloy was comparable at lower load. At higher load the coefficient of friction was slightly higher for composite than alloy. The wear rate of copper lead alloy was higher due to more amount of material transfer than composite. Keun Hyung Cho et al. [19] studied the wear performance of copper–graphite with polyphenolic resin binder composites manufactured by hot pressing technique. Reinforcement in a matrix varies from 0 % to 75 % volume fraction. Coefficient of friction of developed composite was observed to be decreased with the increase of graphite volume fraction. Voltage drop increases with increasing of graphite content due to hindrance of graphite to regular copper network. Zhao et al. [20] investigated the wear performance of copper–graphite composite manufactured by electroforming technique. Adhesive and delamination wear were responsible for quick wear-out of copper and copper–graphite composite. Copper–graphite composites can be also used in marine applications. For such applications, understanding its corrosion behavior is important. The corrosion behavior of electroformed copper–graphite composite was reported. Corrosion behavior was examined in aerated 3.5 % NaCl solution. Based on the impedance spectroscopy analysis and the products of corrosion, corrosion mechanism also was proposed. Compared to pure copper, only a fewer corrosion sites were observed in composite surface. This revealed a different corrosion mechanism. The established mechanism on corrosion of copper composite is that the corrosion process proceeds along the interface between copper matrix and graphite particles. According to this study, corrosion proceeds along the grain boundary. When corrosion path contacts with graphite particles, the corrosion process is hindered. The inert graphite particles in the composites diminish the exposed metallic area, improving the corrosion resistance. Dewidar and Lim [21] fabricated the copper/(2.5–15 wt%) graphite composites through high-frequency induction heated sintering (HFIHS) technique. It was reported that high dense product (relative density 97 %) could be achieved within a short duration. The relative density decreased with the increase in amount of graphite addition, and the reason for the same was attributed to the fewer copper–copper interparticle contacts. This leads to higher porosity of developed composites. The tribological properties of developed composites were studied using a pin-on-disc tribometer. The wear tests were carried out at four different applied loads (50, 100, 150, and 200 N) and up to a sliding distance of 2000 m. The wear resistance increased with the increase in graphite. Mechanism of formation of tribo-layer was not studied. Author also had not reported any results on friction coefficient. Hardness of the composite with different weight percentage of composites was not reported. The high cost of a high-frequency generator and the need for a proper alignment between the heating coil and mold are the limiting factors to use this system for complicated geometry of products.

For electrical sliding contacts, good electrically conducting material with self-lubricating properties is needed. This application requires good wear resistance, good lubrication properties, and good electrical contact conductive performance while sliding. There are no pure metal or their alloys possessing a combination of properties to satisfy the complete requirements of electrical sliding contact. Metal matrix composite materials will be of promising material for the abovementioned application; particularly, copper-based matrix material with graphite reinforcement will be fulfilling the requirements. Copper–graphite composites have shortage of its mechanical and tribological properties. In order to improve their properties, secondary processing such as rolling and extrusion are needed. This will increase the tooling cost and also increase cost of the product. Attention is needed to focus on alternative techniques for improved properties of sintered copper–graphite composites. Microwave heat treatment (novel heat treatment method) is promising for improving the performances of copper–graphite composites. The advantages observed with microwave heat treatment are reduced processing cost, better production quality, new materials and products, improved human health, reduced hazards to humans and the environment, and enhanced quality of life [22].

Copper–graphite composites are finding applications in electrical sliding applications. Sliding wear tests are needed to evaluate the performance of the developed composites. The specimens were machined to have the form of cylindrical pins of 14 mm diameter and 10 mm height. These pins were tested against medium carbon steel disc of surface hardness of 62 HRC. Pin-on-disc type machine was used to study the tribological behavior. Prior to the test, all contacting surfaces were polished, cleaned with acetone, and dried. The tribological tests were carried out under dry sliding condition at normal loads of 5 N, 10 N, and 15 N in all experiments with a constant sliding speed of 0.92 m/s for duration of 30 min. Pin-on-disc equipment consists of rotating spindle on which the disc is fixed. A pivoted lever arm has the facility to fix the pin. This lever arm has a provision of adding desired load for corresponding wear test. The pivoted lever arm can be moved to the desired track diameter at the surface of disc. The LVDT probe can be positioned to measure the displacement and the load cell was used for measuring the tangential force. Pin-on-disc wear tester is shown in Fig. 13.13.

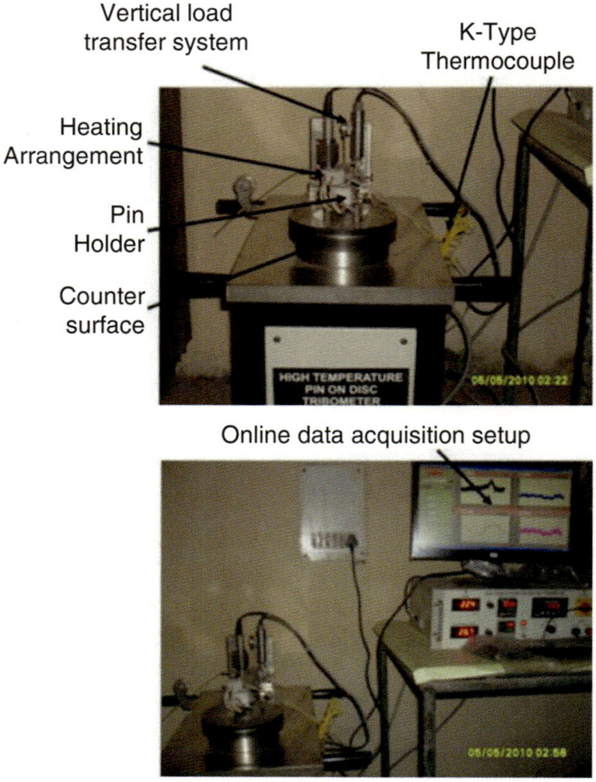

Fig. 13.13 Pin-on-disc wear tester

The tribological aspects of copper-5 wt% graphite particulate composites (microwave heat treated and untreated) were studied through the pin-on-disc test setup. Figure 13.14 shows the relationship between volumetric wear rate and normal load for heat treated and untreated composites. The wear rate is generally expressed in terms of volume loss for specific sliding distance. Wear rates were calculated from the measured weight loss of pin after conducting pin-on-disc wear test. The wear rate measurements were taken in the steady-state region. From Fig. 13.14, it is observed that the untreated copper-5 wt% graphite pin exhibits higher wear rate for all ranges of applied normal loads. It can be seen that the wear rate of the microwave heat-treated composites is minimized. Moreover, the wear rate of 20-min heat-treated composite is much lower than the 10 min heat treated. Microwave heat treatment has influenced the tribological properties by way of rapid and selective heating, finer microstructure, and improved mechanical properties. Microwave exposure, because of its selective heating nature, leads to improved bonding between graphite-reinforced particles with copper matrix. The stronger bonded matrix and the formation of graphitic film over the tribosurface are responsible for the reduction in the wear rate. This graphitic film was highly adherent toward the contact region during sliding. Hence, the graphite film protects the contact region during the wear test [23].

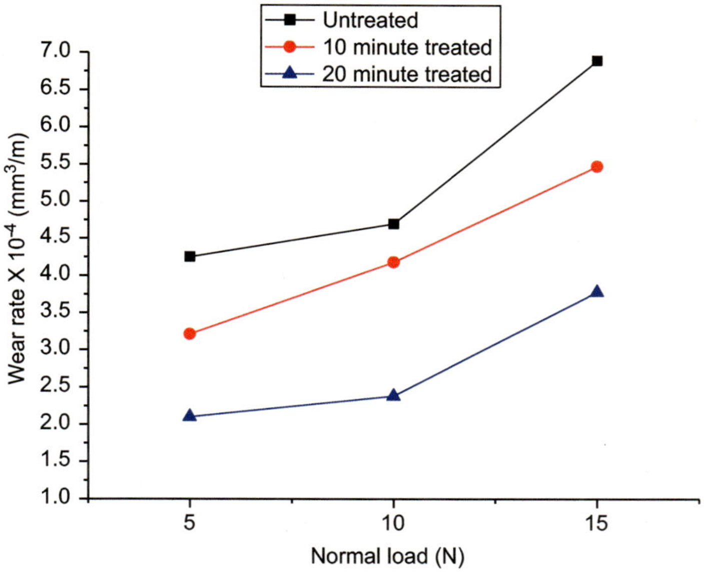

Fig. 13.14 Typical variation of wear rate with the applied load at 0.92 m/s

In the case of untreated specimens, graphite particles are less adherent, and bonding between graphite particles and copper matrix is considerably weaker; hence, fast removal of graphitic particles from the pin surface has occurred when the pins are subjected to high normal loads during the wear test. From Fig. 13.14, it is also observed that the wear rate of untreated composite at the transition region (mild to severe) is 5.1×10^{-4} mm^3 m^{-1} at the normal load of 10 N. The wear rate of 20 min microwave heat-treated specimen (mild to severe) wear regime is of lower order, around 2.5×10^{-4} mm^3 m^{-1}. When the duration of microwave exposure is increased from 10 min, the copper to iron disc adhesion is gradually decreased by carbonaceous (graphitic film) layer between pin and counter surface (disc). It is evident that heat-treated specimens are having the capacity of enduring higher loads than the untreated specimens.

Coefficient of friction can be calculated by dividing the tangential load with the applied load. During the wear test, the data logger continuously records the tangential load with the help of a load cell. The variations of coefficient of friction with time under various normal loads (5 N, 10 N, and 15 N) of untreated and microwave heat-treated copper-5 wt% graphite specimens are shown in Figs. 13.15, 13.16, and 13.17. The average value of friction coefficient was calculated from the wear test. Variation of coefficient of friction of treated and untreated composites with the normal loads of 5 N, 10 N, and 15 N is presented in Fig. 13.18. From Figs. 13.15, 13.16, and 13.17, it is observed that 20 min microwave heat-treated specimens exhibited the lowest coefficient of friction with sliding distance. The amount of graphite film at the contact surface is larger when the heat treatment time exceeds 10 min. The presence of smeared graphite layer at sliding surface of the wear

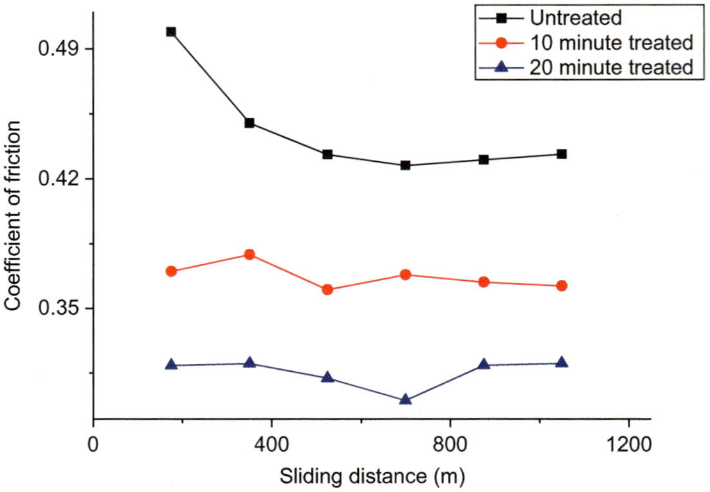

Fig. 13.15 Variation of coefficient of friction with sliding distance (load of 5 N at 0.92 m/s)

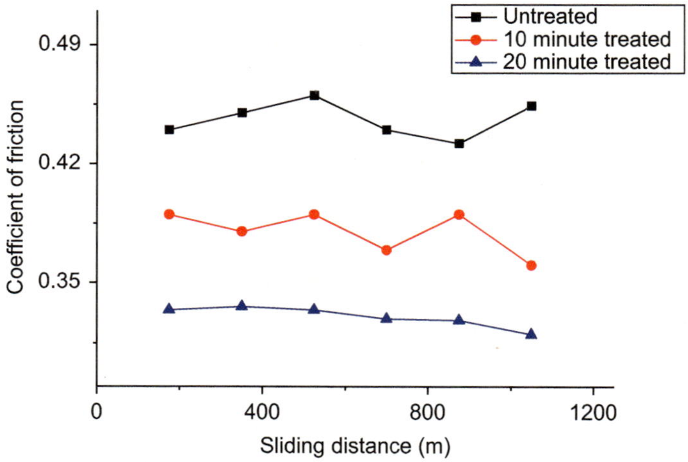

Fig. 13.16 Variation of coefficient of friction with sliding distance (load of 10 N at 0.92 m/s)

specimen acts as solid lubricant and reduces the metal-to-metal contact area between pin and disc. This smeared graphitic film becomes thicker with the increase in microwave exposure time as well as highly adherent and compacted between tribo-components. In microwave heat treatment, energy is transferred at molecular level and also this heating has some additional advantages such as selective coupling of materials that has been attributed to formation of graphitic film. It was evident from Fig. 13.7f; the composite specimens have graphitic film spread over the surface. Also the pushed-out graphite particles tend to occupy the pores that

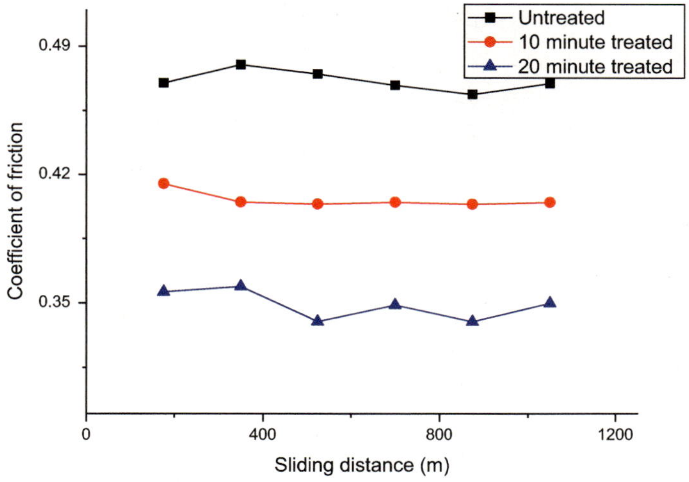

Fig. 13.17 Variation of coefficient of friction with sliding distance (load of 15 N at 0.92 m/s)

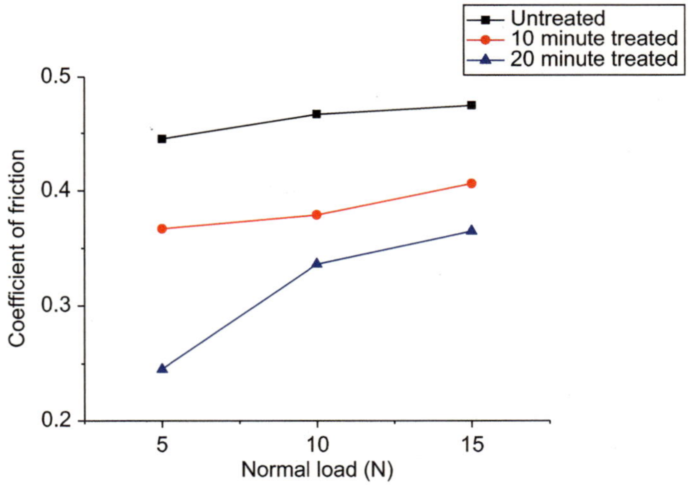

Fig. 13.18 Variation of coefficient of friction with normal load

surround it. In the case of untreated copper–graphite composite, the graphite layer is not that much adherent to matrix material and the bond between the copper matrix and graphite reinforcement is weaker. This caused the faster removal of graphite particles. These particles tend to agglomerate after certain definite period of sliding distance and they are expelled away from the contact region [24]. There is low bonding strength between graphite layers with pin (untreated specimen) metallic surface, which leads to metal-to-metal contact. This increases the coefficient of friction.

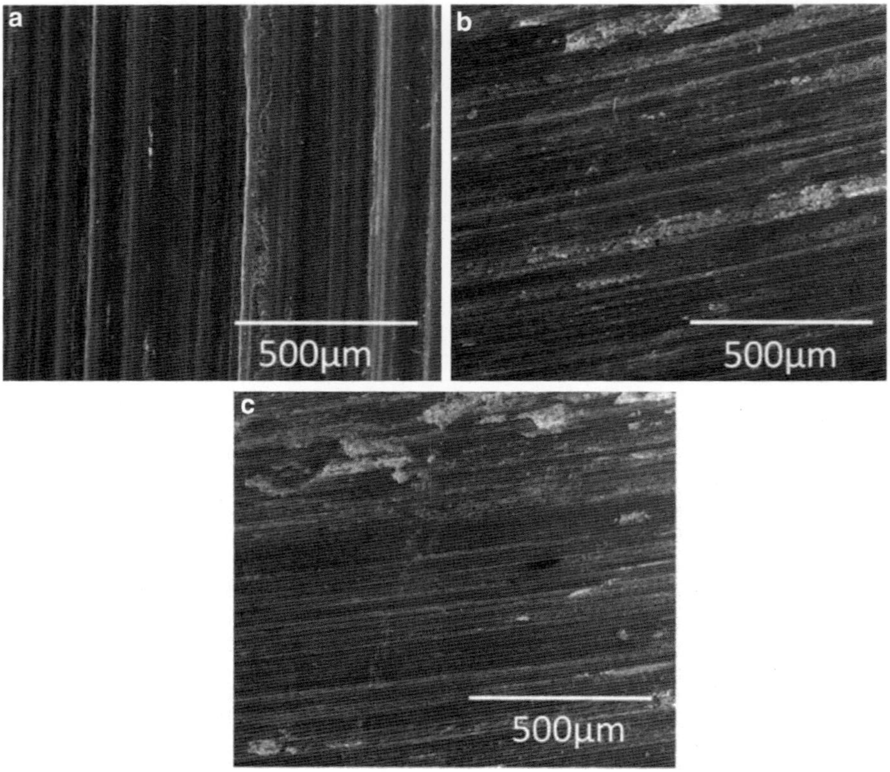

Fig. 13.19 (**a**–**c**) SEM image of worn surfaces of (**a**) untreated, (**b**) 10 min heat treated, (**c**) 20 min heat treated at 15 N

Variation of coefficient of friction of treated and untreated composites with the normal loads of 5 N, 10 N, and 15 N is presented in Fig. 13.18. With increase in normal load, irrespective of the status of the specimens (treated and untreated), the coefficient of friction is increasing. It can be seen from Fig. 13.18. In all the loading condition, the coefficient of friction exhibited by 20 min microwave-treated specimens is of lower order. Wear rate of these treated composites is also of lower order. Hence microwave heat treated is suitable for enhancing the tribological properties of copper–graphite composites.

SEM image of worn surfaces of untreated, 10 min heat treated, and 20 min heat treated are shown in Fig. 13.19a–c, respectively. The untreated sample shows higher wear-generated grooves when compared to heat-treated samples. This shows that large-scale plastic deformation on the worn surface was occurred, as shown in Fig. 13.19a. As a result, considerable amount of plastic deformation along the sliding direction leads to large wear fragmentation. This is also due to that cumulative effect of weaker interface between copper matrix and graphite. Due to effect of microwave heating, the tendency to form large-sized wear grooves is limited because of the strong integrity between the matrix and graphite particles and increased hardness.

This increased hardness can reduce the plastic deformation of the matrix during the sliding process. It clearly shows a little plastic deformation along the sliding direction on the worn surfaces, as shown in Fig. 13.19b–c. It can be understood from these SEM images of worn surfaces that plastic deformation mechanical wear mechanism is operating in copper–graphite composites.

Concluding Remarks

- Importance of copper–graphite composites is discussed.
- Copper matrix composites reinforced with 5 wt% graphite are manufactured by powder metallurgy route for wider tribological application.
- Microwave heat-treated composites exhibited improved physical and mechanical properties such as density, porosity, and hardness.
- Microwave heat treating is helpful for the formation of graphite layer at the contact region of tribo-component and also it enhances the bonding between copper matrix and graphite reinforcement.
- The untreated copper–graphite composites exhibited the highest wear rates and they can withstand only lower load conditions.
- The wear rate was of lower order for treated composites than untreated composites. It is observed that the wear rate of untreated composite at the transition region (mild to severe) is of 5.1×10^{-4} mm^3 m^{-1} at the normal load of 10 N whereas wear rate of 20 min microwave heat-treated specimen (mild to severe) wear regime is of lower order, around 2.5×10^{-4} mm^3 m^{-1} due to the highly adherent graphitic film toward the contact region.
- Microwave heat treating is very effective in reducing the wear rate and coefficient of friction of the composites.

Future Scope

Copper–graphite composite is a typical electrical sliding contact material and consists of two different components, namely, copper and graphite. Graphite has unique antifriction properties but also possesses some shortage including low strength and current-carrying capacity. The electrical conductivity and hardness of electrical sliding materials decrease as the increase of graphite content. It is reasonable to expect some new reinforcements to increase both the electrical conducting and strength of sliding contact materials. Improvement of properties of existing composite materials is constantly taking place using the newer reinforcements. Among them, carbon nanotube is finding an important place. Carbon nanotubes (CNTs) are increasingly attracting scientific and technological interest by virtue of their properties and potential applications. Compared to conventional composites (micron level particle reinforcement), CNT-reinforced nanocomposites exhibit better mechanical properties [25]. Their high length and diameter ratio, strength, elastic modulus, flexibility, and electrical conductivity along with other properties have led to the use of carbon nanotube as a novel reinforcement for a variety of composite materials. Electrical sliding applications require the good wear resistance materials, good lubrication properties, and good electrical conductive performance while sliding. Carbon nanotube (CNT)-reinforced copper metal matrix satisfies the demanding properties of electrical sliding applications [26]. Carbon nanotube (CNT)-reinforced copper metal matrix exhibited better tribological and mechanical properties for its electrical sliding use [27]. Though carbon nanotube is an important reinforcement in composite, stringent restrictions are there to completely exploit it. The following are the major impediments in using CNTs as reinforcement for copper matrix:

- Higher cost of carbon nanotubes
- Safety related issues in handling CNTs
- Dispersion of CNT in the matrix phase
- The damages to CNT, during the compaction preparation
- Poor wettability and bonding between CNT (reinforcement) and matrix

Studies are initiated all over the world to evolve methods to surpass these problems.

References

1. Bravunovic M, Konchits VV, Myshkin NK (2007) Electrical sliding contact, fundamentals, applications and technology. CRC Press, Boca Raton
2. Myshkin NK, Konchits VV (1992) Fiction and wear of metal-graphite composite electrical contacts. Wear 158:119–140
3. Shin W-G, Lee S-H (2010) An analysis of the main factors on the wear of brushes for automotive small brush-type DC motor. J Mech Sci Technol 24:37–41
4. Ott RD, Blue CA, Santella ML, Blau PJ (2001) Wear of iron–aluminide intermetallic-based alloys and composites by hard particles. Wear 251:868–874
5. Roy R, Agrawal D, Cheng J, Gedevanishvili S (1999) Full sintering of powdered-metal bodies in a microwave field. Nature 399:668–670
6. Ankelekar RM, Agrwal D, Roy R (2001) Microwave sintering and mechanical properties of PM copper steel. Powder Metall 44:355–362
7. Thostenson ET, Chou TW (1999) Microwave processing: fundamentals and applications in composites. Appl Sci Manuf 30:1055–1071
8. He DH, Rafel Manory A (2001) Novel electrical contact material with improved self lubrication for railway current collectors. Wear 249:626–636
9. Ibrahim IA, Mohamed FA, Lavernia EJ (1991) Particulate reinforced metal matrix composites – a review. J Mater Sci 26:1137–1156
10. Korb G, Buchgraber W and Schubert T, Proc.Conf. "International Electronics Manufacturing Technology symposium (IMET-Europe)", Berlin, pp.98-103, 1998.
11. Rohatgi PK (1993) Metal-matrix composites. Def Sci J 43:323–349
12. Jim JK, Kestursatya M, Rohatgi PK (2000) Tribological properties of centrifugally cast copper alloy-graphite particle composite. Metallurg Mater Trans A 31:1283–1290
13. Mortimer DA, Nicholas M (1970) The wetting of carbon by copper and copper alloys. J Mater Sci 5:149–155

14. Yeoh W, Persad C, Elizer Z (1997) Dimensional responses of copper–graphite powder composites to sintering. Scr Mater 37:271–277
15. Dyachhenko LI, Paderno VN, Baranov NG, Britun VF, Pilyankevich AN (1987) Substructure of the friction surface of copper–graphite materials. Poroshkovaya Metallurg 300:66–71
16. Okada K, and Yoshida M, Copper–graphite lined with porous metals, Electrical Contacts, Proceedings of the Thirty-Seventh IEEE Holm Conference, pp.149-152, 1991
17. Moustfa SF, Badry EI, Sannad AM (2002) Friction and wear of copper–graphite composite made with Cu-coated and uncoated graphite powders. Wear 253:699–710
18. Kestursatya M, Kim JK, Rohatgi PK (2003) Wear performance of copper–graphite composite and a leaded copper alloy. Mater Sci Eng A 339:150–158
19. Keun Hyung Cho, Ui Seok Hong, Kang Sun Lee, Ho Jang (2007) Tribological properties and electrical signal transmission of copper–graphite composites. Tribol Lett 27:301–306
20. Zhao H, Liu L, Wu Y, Hu W (2007) Investigation on wear and corrosion behavior of Cu-graphite composites prepared by electroforming. Compos Sci Technol 67:1210–1217
21. Dewidar MM, Lim JK (2007) Manufacturing processes and properties of copper–graphite composites produced by high frequency induction heating sintering. J Compos Mater 41:2183–2194
22. Clark DE, Folz DC, West JK (2000) Processing materials with microwave energy. Mater Eng A 287:153–158
23. Hirotaka K, Masahiro T, Yoshiro I, Kazuo W, Yoshinori S (2003) Wear and mechanical properties of sintered copper–tin composites containing graphite or molybdenum disulfide. Wear 255:573–578
24. Kovacik J, Emmer S, Bielek J, Kelesi L (2008) Effect of composition on friction coefficient of Cu–graphite composites. Wear 256:417–421
25. Goh CS, Wei J, Lee LC, Gupta M (2006) Development of novel carbon nanotube reinforced magnesium nanocomposites using the powder metallurgy technique. Nanotechnology 17:7–12
26. Lim B, Kim C-j, Kim B, Shim U, Seyoung O, Sung B-h, Jee-oon C, Seunghyun B (2006) The effects of interfacial bonding on mechanical properties of single-walled carbon nanotube reinforced copper matrix nanocomposites. Nanotechnology 17:5759–5764
27. Tu JP, Wang YZ, Wang LY, Ma XC, Zhang XB (2001) Tribological properties of carbon-nanotube-reinforced copper composites. Tribol Lett 10:225–228

Current Advancements in Ceramic Matrix Composites

14

Sumit Pramanik, Ayan Manna, Ashis Tripathy, and Kamal K. Kar

Contents

S. Pramanik (✉)
Advanced Nanoengineering Materials Laboratory, Materials Science Programme, Indian Institute of Technology Kanpur, Kanpur, Uttar Pradesh, India

Department of Biomedical Engineering, Faculty of Engineering, University of Malaya, Kuala Lumpur, Malaysia
e-mail: prsumit@gmail.com

A. Manna • A. Tripathy
Department of Biomedical Engineering, Faculty of Engineering, University of Malaya, Kuala Lumpur, Malaysia

K.K. Kar
Advanced Nanoengineering Materials Laboratory, Materials Science Programme, Indian Institute of Technology Kanpur, Kanpur, Uttar Pradesh, India

Advanced Nanoengineering Materials Laboratory, Department of Mechanical Engineering, Indian Institute of Technology Kanpur, Kanpur, Uttar Pradesh, India

K.K. Kar (ed.), *Composite Materials*,
DOI 10.1007/978-3-662-49514-8_14

Abstract

Ceramic materials have excellent strength and thermal stability. However, the brittleness is a major barrier to these monolithic ceramics. Therefore, various particulates, i.e., whisker, fibers, or fabric materials, with different forms and orientations that are used to enhance the toughness of ceramic matrix composites (CMCs) are being used at elevated temperature in several advanced applications. To achieve the desired properties in any target application, selection of fabrication technique, ceramic matrix, and reinforcing agent along with its morphology or orientation are extremely crucial parameters. The trend of advanced CMCs is moving toward the nanocomposites. Since at nanoscale, the high specific surface area of substance can tune different interesting physical, chemical, biological, and optical properties. This chapter focuses on processing, properties, and applications of CMCs.

Keywords

Nanocomposite • Advanced ceramic • Silicon carbide (SiC) • Coefficient of thermal expansion (CTE) • Fiber • Reinforcement • Implant • Component • Aerospace • Automotive • Chemical • Cutting tool • Machining • Chemical vapor infiltration

Introduction

Advancement in ceramic matrix composite (CMC) is emerging as an essential study to the researchers, engineers, and industries in the fields of aerospace, automotive, and biomedical engineering. The ceramic industry has a very large international market with sales amounting to over $100 billion per year [1]. One of the most common applications of the advanced ceramic matrix composites (CMCs) is cutting tools. Another advanced application of CMCs is high-temperature spacecraft components. The global market of ceramics has now reached to 8.5 % for cutting tools. It was expected to grow from an estimated US$ 905 million in 2010 to $1.7 billion by 2015 with a compound annual growth rate of 14.1 % over the last 5 years specifically for cutting tools with diamond and diamond-like coatings [2]. Recently, CMCs have been widely used in automotive engineering, brake systems [3–5], and thermal protection systems or gas turbines [6].

The advanced CMCs consist of a single-phase ceramic (oxides or non-oxides) or multiphase ceramics as a matrix material combined with reinforcing agent of ceramic, glass, metallic, or polymeric materials. Plenty of ceramic materials, oxide (such as alumina, silica, mullite, calcium aluminosilicate, lithium aluminosilicate, barium aluminosilicate, beryllium oxide, and so on) or non-oxide (such as carbides, nitrides, phosphates, silicates, halides, and so on), have been used as matrix materials, and different shapes of particulates and fibrous materials of ceramics, metals, or polymers have also been used as reinforcing materials in various CMCs. Different types of CMCs are depicted in Fig. 14.1. Ceramics are nonmetallic and mostly crystalline compounds. It is distinguished from the glass material via crystallinity – ceramics are highly crystalline, whereas glasses are mostly amorphous. Ceramics have very high Young's modulus, compressive strength, excellent hardness, light weight, excellent temperature resistance, high wear resistance, high corrosion resistance, very good chemical resistance, good biocompatibility, etc. It has a wide range of optical transparency from ultraviolet (UV) to infrared (IR), high thermal resistance, and very high electrical resistivity. Thus, CMCs are attractive in wide range of applications, i.e., automotive to biomedical. However, ceramics have some prominent demerits, such as very low fracture toughness, low damage tolerance, low ductility, highly brittle, and low fatigue endurance limit, which hinder it to be used in many advanced applications. For comparison, the typical

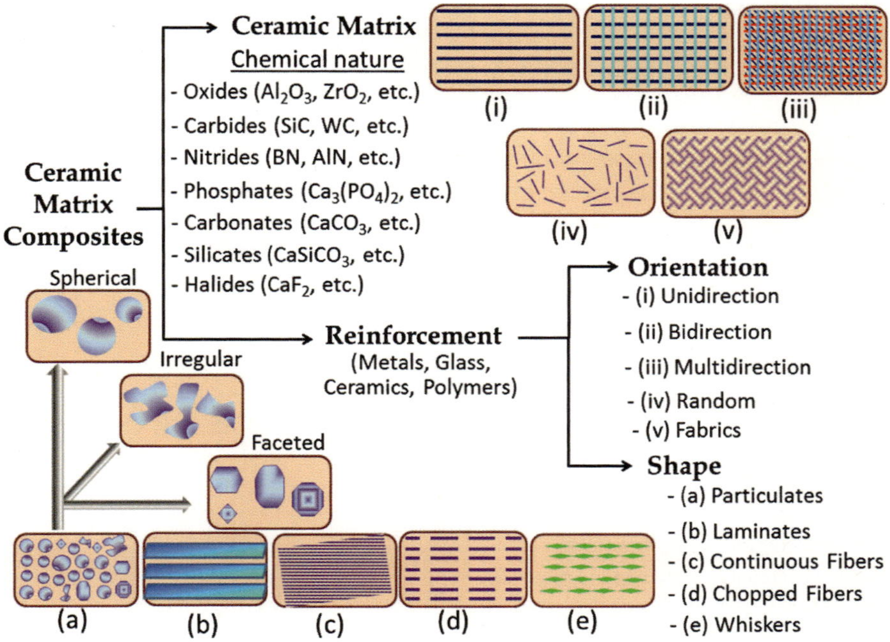

Fig. 14.1 Classifications of CMCs. The uniform background color (orange accent 6 (60 %)) in the images represents ceramic matrix and different front colors indicate reinforcing materials. Different morphologies (i.e., spherical, irregular, and faceted) of the reinforcing particles in a particulate composite are indicated by wider gray arrows

Table 14.1 Typical properties of ceramic matrices, reinforcement, and ceramic matrix composites (CMCs) [7–12]

Materials	Young's modulus (GPa)	Tensile strength (MPa)	Coefficient of thermal expansion, 20–1000 °C (10^{-6}/°C)	Thermal conductivity (W/mK)	Density (g/cc)
Matrix ceramics					
Silicon nitride (Si_3N_4)	307	410	2.3	–	3.2
Silicon carbide (SiC)	400	310	4.8	–	3.2
SiC fiber (chemical vapor deposited)	364	450	2.9–4.6	175	2.95
Alumina (Al_2O_3)	380	250	8.5	–	3.9
$3Al_2O_3{\cdot}2SiO_2$	143	83	5.3	–	–
Zirconia (ZrO_2)	17–25	130	7.6	–	5.6
Borosilicate glass	60	100 (MOR)	3.5	–	2.3
Reinforcing agents					
Carbon fiber (CF)	350	1380–270	0 (axial) 8 (radial)	–	1.9
Boron fiber (B)	440	2800–3400	8.2		2.3
SiC fiber	480	2100	4.8		3.35
Nicalon (65%SiC, 20% SiO_2, 15%C) fiber	185	2700	–	–	2.55
CG-Nicalon™ fiber (Si:C:O = 57:32:12)	–	3000	3.2–3.9 (radial)	–	2.55
Hi-Nicalon™ fiber (Si: C:O = 62:37:0.5)	–	2800	3.5 (radial)	–	2.74
Hydridopolysilazane (HPZ) (59 % silicon, 28% nitrogen, 10 % carbon, 3 % oxygen) fiber	150	2200	–	–	2.35
99 % Al_2O_3 (axial) fiber	380	1400	–	–	3.9
84 % Al_2O_3 + 15% SiO_2 fiber	224	2300	–	–	3.05
Silica (SiO_2)	70	32	0.65	1	2.2
Aluminum (Al)	70	270	27	237	2.7
Beryllium (Be)	300	324	11.4	150	1.85
Silicon (Si)	113	50–80	2.6	150	2.33
Ceramic matrix composites (CMCs)					
A359-T0 aluminum – 54 vol% SiC	340	386	12.24	180	2.99
A2000-T0 aluminum – 68 vol% SiC	186	214	7.56	170	3.04
Reaction-bonded carbon nanotube (CNT)–reinforced SiC	374	–	2.72	185	3.06

(*continued*)

Table 14.1 (continued)

Materials	Young's modulus (GPa)	Tensile strength (MPa)	Coefficient of thermal expansion, 20–1000 °C (10^{-6}/°C)	Thermal conductivity (W/mK)	Density (g/cc)
Reaction-bonded SiC (80 %)–Si	380	290	2.9	185	3.03
Reaction-bonded SiC–carbon fiber (CF)	100–250	–	±0.5	100–200	2.4
Si–B_4C (70 %)	383	270	4	100	2.57
C/C–SiC (short fiber) (25%SiC)	25	65 (flexural)	–	14 (⊥) 25 (‖)	2
C/C–SiC (2D reinforcement) (25%SiC)	55	130 (flexural)	–	20 (⊥) 30 (‖)	1.95

properties of different ceramic matrices, ceramic reinforcement, and CMCs materials are illustrated in Table 14.1.

Classifications of CMCs and Toughening Mechanism

CMCs can be classified mainly according to the degree of reinforcement. A fundamental classification of CMCs depending on the reinforcement is given in Fig. 14.1.

Particulates

One of the efficient methods to improve the fracture toughness of ceramics is to use strongly bonded and uniformly distributed reinforced particles in the matrix. At normal atmospheric conditions, the fracture mode of ceramic matrix is cleavage. Thus, the toughening mechanisms of the reinforced particles for improving the fracture toughness are related to both the crack flank bridging and crack front trapping. When a crack meets an array of strongly bonded particles, the crack front initially penetrates between the particles. In this mechanism, an additional driving force for crack growth is required to get nonhomogeneous distribution of stress intensity along the verge of propagation. If the particle–matrix bonding is perfect, after reaching a critical depth of penetration, the crack overcomes the trapping effect. Then, the reinforced particle is left behind and forms bridge across the crack flanks. Eventually, the reinforced particles are broken since the applied stress intensity rises above the critical value [13]. The overall energy release rate (G_{total}) is a function of critical energy release rate for crack trapping (G_{ct}) and energy release rate for crack bridging (G_{cb}), which can be calculated by Eq. 14.1 [14]:

$$G_{total} = G_{ct} + G_{cb} \tag{14.1}$$

The particulate CMCs can have large self-crack-healing ability, high strength, and high limit of heat resistance temperature [15]. Depending on the morphological structure of the second phase, i.e., particles in the CMC, the particulates can be further classified in the three categories such as spherical, irregular, and faceted, and they are depicted in Fig. 14.1a. Different geometry of the particles has significant influence on the interfacial bonding with the matrix.

Mechanisms of crack front propagation with a part of the crack being trapped by strongly bonded particles in a CMC are shown in Fig. 14.2. The nanocomposites of particulate CMCs follow mainly three strengthening mechanisms such as (i) flaw size reduction [16], (ii) grinding effect [17], and (iii) grain boundary strengthening effect [13, 18]. The flaw size reduction mechanism mainly enhances the mechanical strength of composites. This mechanism is also associated with the strengthening phenomena due to the dislocation network formations. Therefore, the strength could be further enhanced by the formation of subgrain boundary through dislocations owing to post-annealing treatments. The grinding effect is associated with the strengthening of nanoceramic matrix composites (nanoCMCs) owing to the compressive stress at the surface during grinding with abrasive particles. This compressive stress at the surface of the nanocomposite cannot be released even after post-annealing. However, the surface cracks developed due to the grinding can be mended easily [19]. Thus, the strength of nanoCMCs can be further improved by annealing. It has been found that the grain boundary strengthening mechanism has significant effect of increasing strength in the nanoCMCs [13, 18]. A typical comparison in stress–strain behavior of particulate CMC (Fig. 14.3a) with pure ceramics (Fig. 14.3b) and other types of reinforced CMCs is depicted in Fig. 14.3.

Laminates

Laminated structures are also used to fabricate laminated CMCs for improving the brittleness of ceramics. The laminated CMC is depicted in Fig. 14.1b. It can be

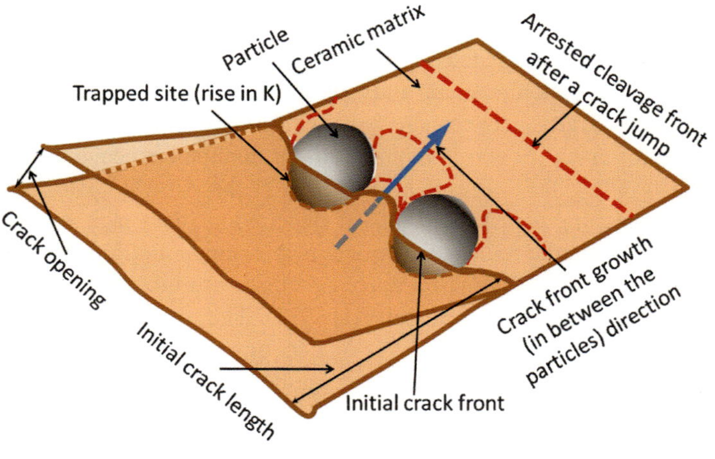

Fig. 14.2 Mechanisms of crack front propagation with a part of the crack front being trapped by strongly bonded particles in CMC

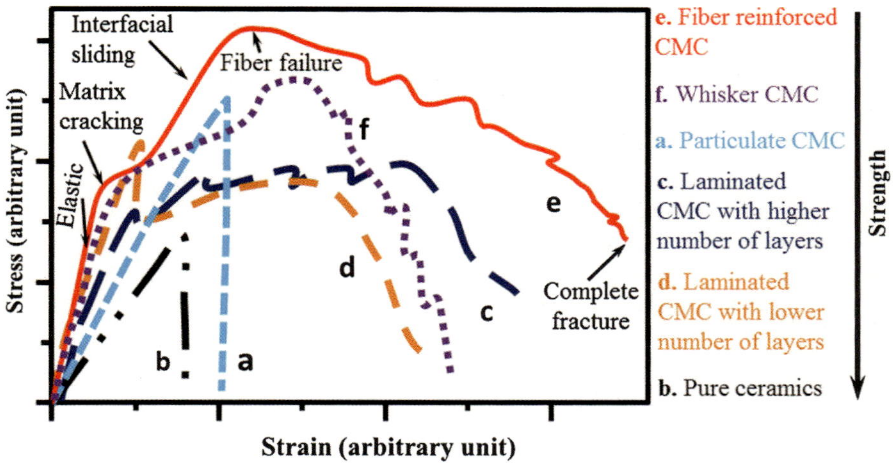

Fig. 14.3 Typical stress–strain behaviors of pure ceramics and different types of reinforced CMCs

oriented in different directions to minimize the anisotropy. The laminated CMCs show stepped stress–strain behavior with higher apparent toughness and work of fracture than monolithic ceramic. The ceramic–metal laminated CMCs have many advantages over ceramics and metals [20, 21]. When cracks are encountered in the metallic interlayer, owing to the ductility of the metal, they can be deflected or bridged, which generally improves the toughness of composites [22]. Incorporation of metallic laminates as a second phase into the ceramic matrix is an alternate method to increase the toughness and strength of ceramic [23, 24]. However, owing to the mismatch of thermal expansion between the ceramic matrix and second reinforced phases, a residual compressive stress is generated and it enhances the strength and toughness [25]. The most effective toughening mechanism in this kind of CMCs is crack bridging, which is obtained by addition of metallic laminates in ceramic matrix [21]. It has been found that in a higher number of reinforced layers (in Fig. 14.3c), the ductility or fracture toughness of laminated CMCs increases in comparison to the lower number reinforced layers (see orange accent 6 (25 %) squire dot line in Fig. 14.3d).

Fibrous

Reinforcement with the help of fibrous materials is used to fabricate most effective high-strength CMCs. However, it has been found that the strong anisotropy and inhomogeneity of fiber-reinforced CMC (FRCMC) may cause many specific problems, such as fiber pullout, matrix craze, interfacial debonding, and fiber fracture during machining [26]. Therefore to control the machining of FRCMC surface in the manufacturing industry, several forms of fibers with different orientations are used to reinforce the ceramic matrices. Fiber-reinforced CMCs can be classified mainly according to the size and orientation of the reinforcement, and they are depicted in Figs. 14.1c–d and Fig. 14.1i–v, respectively.

Shape of Fiber

Mainly two types of fibers are used in the CMCs. These are continuous fibers (see Fig. 14.1c) and chopped fibers (see Fig. 14.1d).

Continuous fiber: Continuous fiber-reinforced CMCs are the most promising materials as high-temperature structural materials in many advanced applications. In addition to the intrinsic properties (such as high specific compressive strength, excellent heat and corrosion resistances, and so on) of monolithic ceramics, incorporation of these fibers enhances other mechanical properties of the CMCs such as its fracture toughness, wear resistance, fatigue strength, and creep resistance. The unique and attractive combinations of material properties can make the CMCs excellent prospective materials as mechanical components in aerospace, automobile propulsion, and power system [27]. Owing to the high stiffness of elastic fibers, both the mean normal stress and deviatoric stress within the matrix relative to the applied stress increase locally [28]. The maximum mean normal compressive stress developed in the matrix at steady-state condition can be high enough to debond the interface. At the higher stress level and at steady state, the deformation mechanism is related to dislocation creep of polycrystalline anorthite [27]. At lower applied stress, the CMC (see firm red line in Fig. 14.3e) behaves like elastic materials or pure ceramics (Fig. 14.3b). In Fig. 14.3e, after elastic limit at higher applied stress, the cracks are initiated in ceramic matrix and stress decreases. Then, the cracks propagate and evolve into fiber-bridged matrix cracks. Thus, the increment of stress gradually decreases and follows nonlinear behavior. With further increasing of stress, the matrix crack opening reaches to saturation and remains nearly constant with additional applied stress. Then a primary deformation occurred owing to elongation of fibers, and fiber–matrix interfacial sliding becomes dominant as a fracture mechanism. The stress–strain plot follows linear behavior with lower tangential modulus compared to initial modulus until the failure of the reinforced fibers [29].

Chopped fiber: Chopped fiber-reinforced CMC is depicted in Fig. 14.1d. The main advantage of the chopped fibers compared to continuous fibers is that the chopped fiber has lower fabrication cost and high adaptability for conventional manufacturing techniques and thus economically favorable [30]. Since the physical properties of continuous fibers are different in different axes, a distinct anisotropy is revealed in mechanical, thermomechanical, and electrical properties. To the contrary, the anisotropic properties can be minimized or eliminated by fabricating a composite having homogeneous and random distribution of chopped fibers in ceramic matrix [31]. Crack deflection and bridging are very common mechanisms for this kind of composite [32].

Orientation of fiber: As depicted in Figs. 14.1i–v, the ceramic composites can be divided into mainly five categories such as unidirectional [33], bidirectional [6, 34], multidirectional [35], randomly oriented [35], and fabrics [36] according to the fiber orientations such as unidirection, bidirection, multidirection, random, and fabrics, respectively.

Unidirection: Both short and long fibers are used in CMCs. The main purpose of using these fibers is to increase the toughness for allowing a graceful rather than catastrophic failure of the composites. However, the addition of continuous small-diameter fibers shows most efficient composites owing to their orientation in the direction of the principal load, which significantly enhances the matrix cracking strain as well as the ultimate load-carrying capability of the composites. Continuous fiber composites are often made into laminates by stacking single sheet of continuous fibers in different directions. It provides the desired strength and stiffness properties to the CMCs. Dense CMCs (mostly of glass or glass ceramic matrix) reinforced with continuous unidirectional fibers or unidirectional plies of longitudinal fibers display multiple cracks in the matrix perpendicular to the fibers that leave them unaffected [37].

Bidirection: The mechanical properties of the bidirectional composites are inferior to the unidirectional composites. The different grain structures influence the properties of the fiber-dominated CMC, because the microstructure at the junction point depends on the orientation of fiber. This kind of CMCs shows creep behavior at high-temperature applications [6].

Multidirection: The main advantages of this composite are high specific strength and stiffness. When the fibers are laid in near-net-shape structure having fiber in multiple directions as reinforcement, the composite provides an ideal structure to withstand high temperature [35]. Thermal shock can be reduced using multidirectional fiber-reinforced CMCs in spacecraft applications.

Random: Randomly oriented fibers are used in CMCs to make composites with homogeneous and highly isotropic structure. This kind of CMCs shows very high fracture toughness. Randomly oriented short-fiber composites in complex shapes can readily be fabricated after processing of conventional fibers with the aspect ratio of maximum 100 using carbon nanotubes (CNTs) or carbon nanofibers (CNFs) [38]. Chopped fibers are normally used to fabricate this kind of CMCs. Thus, the same crack deflection and bridging mechanisms as chopped fiber-reinforced CMCs are exhibited in this kind of CMCs [32].

Fabrics

In this composite, first preforms are made from fabric by several infiltration techniques and then three-dimensional (3D) form is given. The composites have high strength, high stiffness, low density, and high thermal and chemical stability to withstand long exposure time at high temperatures [36]. But the fiber orientations in a fabric play vital role in the machining or grinding of surfaces of CMCs made of woven fabric. It has been found that the highest surface quality is observed when the fibers are orientated at 90° due to the machining or grinding, but the poorest at a fiber orientation of 0° [26]. The surface quality also depends on the weaving rules of reinforced fibers in the CMC materials. The machining or grinding parameters such as machining speed, grinding wheel speed, depth of cut, and so on have also significant effects on the surface microtopography of CMCs.

Whiskers

As a class of CMC, whisker-reinforced CMCs are produced to be used in various structural applications because of their significant improvements in mechanical properties (see Fig. 14.3b) as compared to the monolithic ceramic matrix materials. The incorporation of whiskers into ceramic matrices increases its strength, fracture toughness, thermal conductivity, thermal shock resistance, and high-temperature creep resistance [39, 40]. Although SiC whiskers were initially developed in the 1960s by vapor–liquid–solid (VLS) method, the first application of whisker reinforcement in CMCs was started around the 1980s. The mass production of industrial-scale SiC whisker includes carbothermic reduction reactions between carbon precursors and low-cost silica, such as rice hulls – silica. The whiskers are 0.1–5 μm in diameter with aspect ratio of 1–10, typically needle-like or acicular-shaped, discontinuous, and single crystal-like structure. These unique features that differ them from other chopped or short fibrous materials are polycrystalline and amorphous and may consist of multiple coatings. The SiC whiskers are the most widely used as commercial-graded reinforcing agent in the ceramic matrices. In addition, whiskers of titanium carbide (TiC), titanium nitride (TiN), Al_2O_3, mullite, silicon nitride (Si_3N_4), boron carbide (B_4C), and hollandite ($Ba_{1.23}Al_{2.46}Ti_{5.54}O_{16}$) are also produced and being incorporated in CMCs [41, 42].

Toughening behavior of whisker-reinforced composites is analyzed with many ceramic matrix materials such as Al_2O_3, zirconia (ZrO_2), celsian ($BaAl_2Si_2O_8$), and so on [41, 42]. The major toughening behavior, which is responsible in the whisker-reinforced CMCs, shows that the crack–whisker interaction results in crack bridging (see Fig. 14.4a) and whisker pullout (see Fig. 14.4b). The debonding along the matrix–whisker interface is often associated with crack deflection, and it is shown in Fig. 14.4c [43]. It must occur during crack propagation and allows the whiskers to bridge the crack in order to increase the fracture toughness of CMCs. The toughness of CMCs ($K_{IC}{}^{C}$) is a function of matrix toughness ($K_{IC}{}^{M}$) and whisker reinforcement ($K_{IC}{}^{WR}$), which can be calculated by Eq. 14.2:

$$K_{IC^C} = K_{IC^M} + K_{IC^{WR}} \tag{14.2}$$

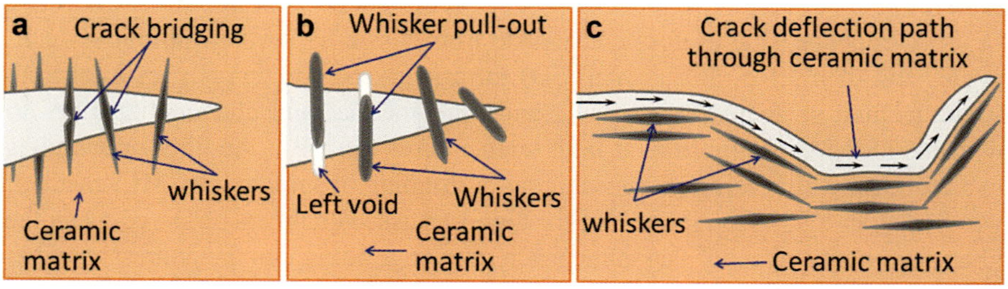

Fig. 14.4 Fracture mechanisms in whisker-reinforced CMCs: (A) crack bridging, (B) whisker pullout, and (C) crack deflection path through ceramic matrix

The increase in fracture toughness (ΔK_{IC}^{C}) owing to incorporate whisker reinforcement in ceramic matrix can be expressed by Eq. 14.3 [44, 45]:

$$\Delta K_{IC}^{WR} = \sigma_f \sqrt{\frac{V_f . r_{WR}}{B(1-\upsilon^2)} \cdot \frac{E_C}{E_{WR}} \cdot \frac{\gamma_m}{\gamma_i}} \quad (14.3)$$

where:

σ_f = fracture strength of whisker.

V_f = volume fraction of whisker.

r_{WR} = whisker radius.

υ = Poisson's ratio of whisker.

γ_m = fracture energy of matrix (m).

γ_i = fracture energy of matrix–whisker interface (i).

B = constant that depends on the bridging stress profile, for example, nearly 6 for SiC whisker–Al_2O_3 matrix CMC.

From Eq. 14.3, it is obvious that the surface chemistry of whisker has a significant role on the matrix–whisker interface bonding, which is attributed to γ_m/γ_i in Eq. 14.3. It should be mentioned that the whiskers developed by various manufacturers have different surface characteristics and can affect their performance as reinforcement.

Fabrication

The properties of CMCs, especially fiber reinforced, depend strongly on their microstructural composition and porosity. Their properties also depend on the respective manufacturing processes. The non-oxide CMCs are widely used in advanced applications. Although oxide ceramic matrices are often considered more mature and environmentally stable, non-oxide ceramic markets are rapidly changing, owing to their superior structural properties, hardness, and chemical and corrosion resistances. Several routes have been developed to fabricate the CMCs. Only important advanced and widely used techniques for fabricating of advanced CMCs are illustrated here.

Solid State

Solid-state route is one of the easiest techniques and economically favorable batch process to fabricate particulates, whisker, or short-fiber-reinforced CMCs. Plenty of CMCs have been developed using this technique. Recently, advanced metal-reinforced CMCs (called as cermets), i.e., particulate-reinforced titanium and nickel alloys with calcium silicate matrix, have been developed for biomedical implant applications using this route [46]. Solid-state method for fabrication of CMCs can be

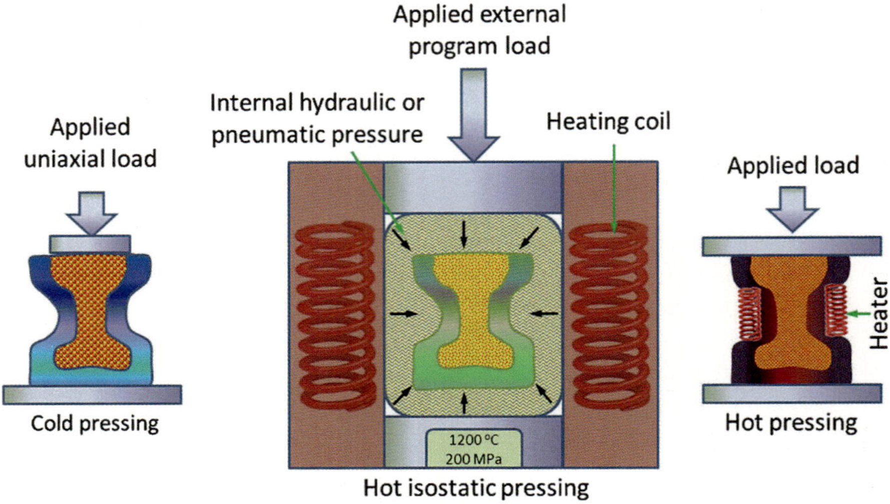

Fig. 14.5 Solid-state routes including cold pressing, hot isostatic pressing, and hot pressing for the particulate-reinforced CMCs

done by several processes such as hot pressing, hot isostatic pressing, solid-state hot pressing, solid-state sintering, pressureless sinter, and so on. The solid-state route including cold pressing, hot isostatic pressing, and hot pressing for the particulate-reinforced CMCs is depicted in Fig. 14.5. Although it has been used for solid-state and powder metallurgical applications, the laminated composites can also be produced using this technique (Fig. 14.5).

Sol–Gel

Sol–gel process is one of the cheapest and oldest processes to produce CMCs. Here, sol–gel infiltration technique is used to prepare various composites. This technique is used to prepare matrix from a colloidal suspension (in water or organic solvent) of fine ceramic particles (sol), which first soaks and then converts into solid (gel). The colloidal sol of fine ceramic particles reacts more effectively when the particle size becomes less than 100 nm. Colloidal sol should be prepared as low viscosity to infiltrate easily into the preform. At elevated temperature, organometallic compounds (e.g., alkoxides) of sol undergo cross-linked polymerization by either polycondensation or hydrolysis mechanism. The polymerization process turns the sol into gel. Gels are converted into ceramic matrix at relatively low temperature, which reduces the probability of fiber damage. Basic steps of a sol–gel process are schematically depicted in Fig. 14.6.

For example, Al_2O_3 matrix CMCs are generally prepared from alumina gel, which forms aluminum alkoxides during hydrolysis, i.e., decomposition in presence of water [47]. Since the presence of ceramic amount in gel is relatively low, it

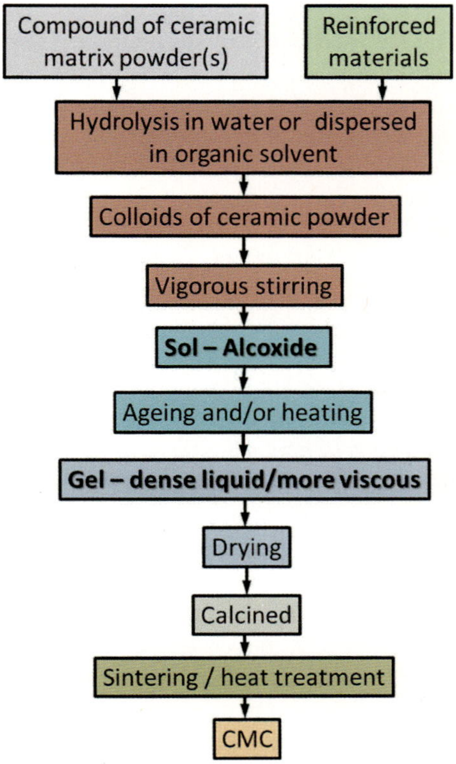

Fig. 14.6 Primary essential steps in sol–gel process for fabrication of CMCs

undergoes significant shrinkage after drying. The densification can be improved by repeating the infiltration-drying cycles by several times until a desired density is reached. The sol–gel can be done by an addition of ceramic particles for further improvement in yield of composite. The addition of ceramic particles also controls the formation of cracks in the drying stage. Recently, Rezaie et al. have developed Al_2O_3–SiC nanoCMC using $AlCl_3$ as precursor by sol–gel method. The mechanical proprieties are found to increase with the distribution of SiC in the Al_2O_3 matrix [48]. 2-D and 3-D structures can also be fabricated by sol–gel infiltration method.

To prepare 2-D structure of CMCs by sol–gel infiltration, fabrics are used as preform. Once the sol is prepared, the binders are added to it to enhance the coating ability on the fabrics and to increase yield; further fillers can be added. Then, a prepreg is made by dipping the fabrics into the sol or fully coated with sol. The freshly prepared prepreg is then (i) cut, (ii) stacked, and (iii) cured in autoclave. These three processing steps are similar to the polymer composite fabrication process, which has very high yield up to 100 %. Unlike polymer composites, the preform of ceramic composite has very poor yield due to the poor binding properties of its matrix phase. Therefore, ceramic matrix preforms often need some additional infiltration processes to achieve high density. The porosity level present in these materials is 20–25 %.

To prepare 3-D structure of CMCs by sol–gel infiltration, woven (or braided) fabric is used to prepare the preform. The prepared sol is poured into a mold attached with a vacuum system. The 3-D preform is then fully immersed into a sol at a low applied vacuum pressure to remove the excess air bubbles and to help sol to reach into the interior places of the 3D weaves. Subsequently the sol is converted to gel by applied heat and/or suitable catalyst. Then infiltration step is repeatedly employed until the expected density is achieved. After infiltration, several heat treatments are employed from 100 to 500 °C (depending on the used sol) on the gelled product to remove the adsorbed volatile matters such as water, alcohols, and any residual organic and to densify the products. The densification is essentially used to improve the physical and mechanical properties of the products. Further densification is carried out based on the need by low-pressure (50–100 psi) autoclave. Autoclave provides close dimensional tolerances. Then, a freestanding firing cycle is used to produce a crystalline CMC with a desired density. The key advantage of 3-D weaves compared to 2-D lay-ups for CMCs is a high structural strength due to the 3-D reinforcement. The 3-D structure of fabric in the through-thickness direction of the 3-D weave provides the extra strength to the materials. The densification of the 3-D CMCs can be improved by some conventional methods using vacuum bagging, autoclave molding, and so on and repeating the infiltration process. The low-viscous sol can easily spread into the interior part of the 3-D preforms through the reinforced filament axis. A thin coating is formed on the filaments by each infiltration. Then a thicker filament coating film is produced by the following subsequent infiltrated films until the complete infiltration of the two bundles. As a result, density reaches to a desired value after reducing the surface porosity.

Melt Infiltration

Melt infiltration or reactive melt infiltration (RMI) is basically a combination of fiber preforms and matrix fabrication process. A molten form or precursor related to ceramic matrix is used to infiltrate through residual porosity in preformed fiber. The fiber reinforcement is covered with the interfacial and protective barrier coatings prior to infiltrate the matrix. The molten or precursor material of this method is a combination of infiltrated material(s) and some solid secondary compounds. A flowchart of this process is shown in Fig. 14.7. It is mainly used to fabricate the fiber-reinforced CMCs. The preparation of fiber reinforcement includes two main steps such as shaping of fibers into desired shape(s) and interface treatment (e.g., boron nitride, BN, carbon, etc.) and barrier (SiC, Si_3N_4, etc.) coatings. The interfacial coating helps to transmit the load from the matrix to fiber in order to improve the mechanical properties. The barrier coating protects the interface and fibers from damage due to the high-temperature molten materials.

In contrast to other matrix densification process, the melt infiltration technique has several advantages such as dense matrix, less open porosity, superior fiber/interface protection, greater thermal conductivity, higher proportional limit, lower oxidation, less natural degradation, and significantly lower processing period. The

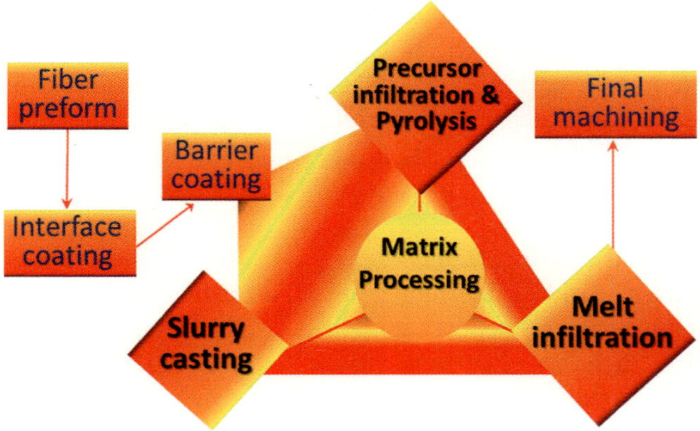

Fig. 14.7 Processing steps for melt infiltration for fiber-reinforced CMCs

matrix fabrication method is used to incorporate the compounds in the pores of a coated preform. The matrix fabrication step may be involved through several methods as explained in this section below.

Resin Infiltration and Pyrolysis Reaction

In this process, the preform of coated fibers is infiltrated using a liquid precursor (e.g., polymeric resin, polydimethylsiloxane, etc.), which is pyrolyzed to provide the reactants (such as carbon) in the pores of preform. During infiltration, the reactants react with molten or precursor materials to form the secondary components of matrix.

Slurry Casting

In slurry casting, the secondary reactant materials related to matrix are directly placed in the porous preform of fibers. The reactant materials help to provide the infiltration to make the final matrix into an integral part. At the end step of metal infiltration, the prepared preform is heated above the melting temperature of the selected ceramic matrix under vacuum. This molten ceramic material is supplied on to the preform in such a manner that it goes into pores. Eventually, the preform is cooled and the molten matrix solidifies leaving a rigid composite structure.

Direct Metal Oxidation (DiMOx)

The DiMOx process involves a formation of the matrix through the reaction of a molten metal in an oxidizing gas (e.g., oxygen gas). The preform of dispersed phase of reinforced fibers or particles is kept on the surface of parent molten metal in an atmosphere of oxidizing agent. The essential two conditions for conducting this DiMOx process are that the dispersed phase is to be wetted by the melt and must not be oxidized in the oxidizing atmosphere. Therefore, a thin layer (0.1–1 μm) of a debonding phase (pyrolytic carbon or hexagonal boron nitride) is always pre-deposited on the surface of second phase (particles or fibers) by chemical vapor infiltration (CVI) method. When the liquid metal is in contact with oxygen, it is oxidized and formed a thin layer of ceramic along with the incorporation of dispersed phase in it. At this time, a capillary effect forces the melt to penetrate via porous ceramic layer to the reaction site. At the reaction front, the molten metal reacts with the oxidizing gas. As a result growth of ceramic matrix layer takes place.

Liquid Silicon Infiltration (LSI)

LSI process is a type of RMI technique. In LSI, the ceramic matrix is formed due to the chemical interaction between the infiltrated liquid metal into a porous reinforcing preform and the solid or gaseous substance surrounding the melt. LSI has been widely used in fabrication of SiC matrix CMCs. To prepare SiC matrix CMCs, the carbon (C) microporous preform is infiltrated with molten silicon (Si) at a temperature more than 1414 °C. The surface of carbon preform is wetted by Si liquid. The Si liquid soaks into the microporous preform through capillary forces. The melt reacts with carbon to form silicon carbide according to the reaction depicted in Fig. 14.8. The produced SiC fills the preform pores and forms the ceramic matrix. Polymer-impregnated fibers (tow, tape, weave, etc.) are impregnated with a resin and then dried or cured to B-stage (partial curing). The resin reacts with molten silicon. The prepreg took desired shape using proper tooling or mold. Pyrolytic decomposition is performed at a temperature of 800-1200 °C in an inert (argon or helium) atmosphere. A porous pyrolytic carbon structure is produced after removing of volatile products. Then a primary machining is carried out on the product. Infiltration of the porous prepreg is placed into a furnace with liquid Si. Here, the porous carbon structure is infiltrated with the melt. The porous preform may be fabricated by some other methods such as pyrolysis of a polymer or resin, CVI, and so on. The optimum microstructure of preform is an important factor to obtain a homogeneous complete infiltration. Large pores help to get good complete infiltration; however, it may form a structure with high residues, which are free of silicon and unreacted carbon as a result of incomplete chemical interaction. On the other hand, small preform pores result in more complete chemical reaction, but an incomplete infiltration due to the blockage (chock-off) of the channels, which leads to inferior properties.

The ceramic matrices developed by LSI are fully dense (have zero or low residual porosity) as compared to the composites fabricated by other fabrication method such as polymer infiltration and pyrolysis (PIP), CVI, and so on. Additionally, the interphases may be strengthened by CVI, and a protective barrier from pyrolytic carbon may be formed by PIP unlike LSI process. The aerospace materials of carbon/carbon–SiC (CC/SiC) are developed by LSI technique [12].

Generally, the properties of fiber-reinforced ceramics depend strongly on their microstructural composition and therefore also on the respective manufacturing method.

For non-oxide CMC, three processing methods are currently used for space applications: chemical vapor infiltration (CVI), liquid polymer infiltration (LPI or PIP), and liquid silicon infiltration (LSI).

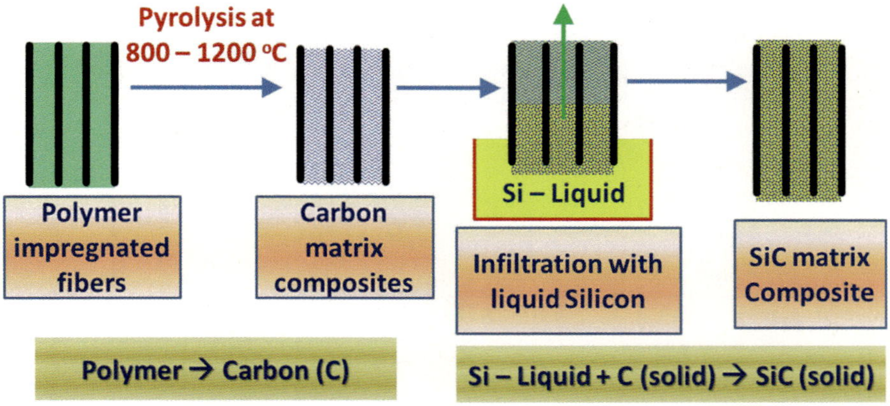

Fig. 14.8 Preparation route of liquid silicon infiltration (LSI)

Polymer Infiltration Pyrolysis (PIP)

PIP is also known as liquid polymer infiltration (LPI), one of the simplest and cost-effective fabrication route of CMCs [49]. Polycrystalline SiC fibers are used to produce the fiber preforms. The fibers are first desized (removal of organic coatings) at suitable temperature, generally above 700 °C. Then, the desized fiber having roving structure is wounded onto differently sized mandrels at a certain winding angle according to a customized winding program. During the winding process, the roving is infiltrated with a polymeric precursor that comprises a desired ceramic matrix material, a binder, and a sintering aid. A desired volume fraction of powder is measured before loading in the slurry. The slurry is then dried and made preform using a suitable precursor (polysilazane Si–C–N). The polymer is cured (cross-linked) at 250 °C. Then, subsequently pyrolysis is performed at suitable temperature in inert atmosphere. Conversion takes place from the liquid polymeric precursor to a desired ceramic matrix. The produced ceramic matrix is amorphous in nature, and significant shrinkage takes place in this process. The shrinkage leads to formation of pores and cracks in the matrix. A defined amount of fiber spacing is obtained due to the presence of powders between the individual fibers introduced during slurry infiltration in the winding process. Several pyrolysis and reinfiltration steps are employed in order to minimize the porosity of matrix. A proper sintering condition(s) is(are) performed in an atmospheric furnace at suitable temperatures depending on the product. In this process, CMCs can also be joined or repaired. A schematic preparation route of PIP is depicted in Fig. 14.9. Recently, C/SiC [4], SiC_f–SiCN [49], and many advanced CMCs are developed using this process. The design of combustion chamber/nozzle (thruster) component made of this C/SiC using LPI process is complex. A Vulcain subscale C/SiC nozzle produced by LPI before and during 40-bar hot-firing test is depicted in Fig. 14.10. It shows that the high-temperature resistance of the thruster by 40-bar hot-fire testing using hot gas is generally used with hydrogen (H_2)–oxygen (O_2) propellant.

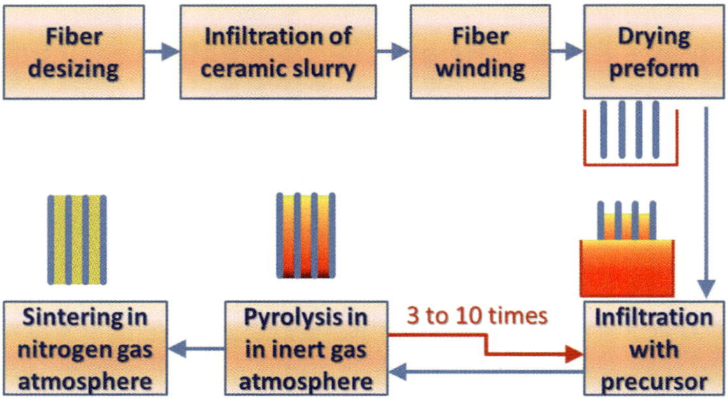

Fig. 14.9 A schematic preparation route of polymer infiltration pyrolysis

Fig. 14.10 Vulcain subscale C/SiC nozzle produced by LPI (a) before and (b) during 40-bar hot-firing test [4]

In the transition range from free- to restricted-shock separation, it survives up to 2027 °C temperature, which is detected by thermography [4].

Chemical Vapor Infiltration (CVI)

The CVI method is one of the advanced industrially used techniques to manufacture fiber-reinforced CMCs. In CVI process, a reactive gas mixture is used to expose an antecedent porous preform material. The interconnected open porosity present in the preforms can be filled with matrix by the MI, sol–gel, or polymer infiltration pyrolysis methods [50]. Then, it is thermally decomposed as deposits to occupy the close pores of the preforms. Since it has many advantages over other techniques, many aerospace and automotive industries use the CVI for advanced complicated CMC products [51]. The main advantages of this technique are the manufacturing of near-net or complex net-shape components at comparatively low temperature and to modify and control the entire morphology of the matrix. This method also allows the product to get tough and strong composites. It is suitable for the production of CMCs with high endurance to erosion, corrosion, and wear properties. Near-net-shape structures made by this CVI minimize the loss of mechanical properties of the fibers owing the use of much lower temperature and pressure compared to other processes. The matrix produced by CVI is significantly pure compared to the other methods such as hot pressing (where impurities come due to the sintering additives). However, this process is quite slow, takes from a day to more than a week, and has poor reproducibility [52]. The most common CMC fabricated by the CVI technology is SiC matrix reinforced with SiC fibers (SiC_f/SiC). The CVI can produce controlled porous preform. The microstructure of preform is important for complete infiltration. A schematic of CVI process is depicted in Fig. 14.11. The CVI equipment consists of mainly three systems such as reagent delivery supply, reactor chamber for preform infiltration, and the discharged gas treatment. The reagent precursors and auxiliary gas are supplied through the reagent delivery system to the reactor chamber. A heat

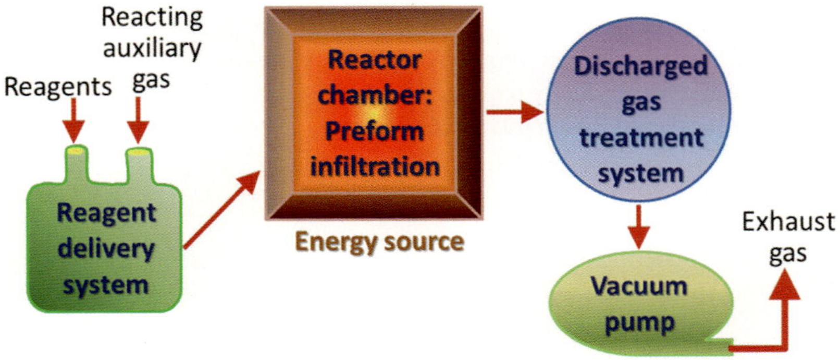

Fig. 14.11 Schematic of chemical vapor infiltration process

energy source is attached with the reactor chamber to heat and allow the precursors to react or decompose. The deposition of composite takes place at the bottom of the reactor chamber. The effluent gases and volatile by-products are going out through the discharged gas treatment system using vacuum pump. The composites are collected through mandrels after finishing a whole cycle.

The energy source can be used by different processes such as microwave heating (volumetric heating), radio frequency heating (induction heating), radiant heating (halogen lamps), resistive heating (tube furnaces), ultraviolet–visible (UV–vis) light, lasers, and so on. The energy sources are selected owing to the used precursor or matrix materials for CVI. For example, using a standard furnace, the temperature can be varied from 1600 to 2200 °C to develop the paralytic graphite (PyC). The hot zones used in the CVI vacuum furnaces are made of graphite. The low-density fibrous graphite heating elements are used for lighter and higher heat resistance insulation.

Generally, a single ingredient is provided to the infiltrated preform with the precursor reagent, and the volatile reaction by-products are evaporated during the process. In special instances, more than one compound may be provided with precursor to simplify the delivery system with minimum quantity of reactants needed by the infiltration. In CVI, the methyltrichlorosilane (MTS), CH_4, $TiCl_4$, CH_3SiCl_3, $SiCl_4$, BCl_3, $AlCl_3$, NH_3, different organometallic compounds, complexes, and ligands are widely used as precursor materials [53].

Several kinds of ceramic matrix compounds are used in CVI such as carbonaceous material (C), carbides (SiC, B_4C, TiC, etc.), nitrides (BN, Si_3N_4, etc.), borides (TiB_2, etc.), and oxides (Al_2O_3, ZrO_2, etc.). Generally, the deposits of ceramics develop radially on the surfaces of fiber filaments. A homogeneous fine crystalline structure of β-SiC is deposited by the CVI process as SiC matrix. The coating of the fibers such as carbon or BN has a similar graphitic crystal structure having planes parallel to the surface of filament.

The common volatile by-product gases are flammable, toxic, corrosive, or explosive; hence, these are discharged to the atmosphere after appropriate treatment.

The near-net-shape components with high curing temperature (>1400 °C) and excellent mechanical properties can also be produced by CVI. Although near-net-shape components are produced by CVI, they usually require a subsequent proper machining to provide desired geometry up to assembling tolerance and surface integrity. Grinding is especially required to achieve high-dimensional accuracy and surface finish for these CVI products [26, 54].

Reaction Processing

In this reaction process, the matrix of CMCs is formed with more than one phase. The preform of this method comprises reinforcing materials that are densely compacted by matrix precursor. The precursor can be fabricated by various techniques, such as the powder curtain process (PCP) and other slurry operations. In the PCP, the resin powder is mechanically deposited onto the fiber tow bundles during pulling through a series of process. The resin powder used on the fiber tow bundles in prepreg process overcomes the number of obstacles in the tape manufacturing for the fabrication of a placement grade product [55]. Then, the densely packed compact is allowed to react with the molten material, supplied gas, and/or itself at high temperature. This reaction process can be used for powders or slurries, which have low to moderate melting temperatures. This is a great advantage over the conventional ceramic processes, which are generally employed at elevated sintering conditions, and thus, it has low shrinkage. The other advantages of reaction process of CMC fabrication are the following: it has the ability to produce near-net and complex shapes, can densely solidify in one step, and may produce zero porosity. Since a new matrix phase is produced upon reaction, besides the dense structure, the volume also increases, and thus the product CMC exhibits lower density than that of the starting materials. Near-net-shape process is a crucial part of this technique. An example of the fundamental techniques for manufacturing a larger product with low-dimensional change at a high yield by a near-net-shape technique is depicted in Fig. 14.12. In the near-net-shape process, nonwoven fabric can be used for developing a product with a complex shape. It has high accuracy and dimensional stability. The near-net-shape process can eliminate additional machining or grinding process by producing exact precise geometry of the CMC products.

On the other hand, reinforcing fibers can be turned into disrepute at higher atmospheric temperatures or by the exothermic reactions. The determination of working temperatures is very important to minimize the fiber's damage, where the

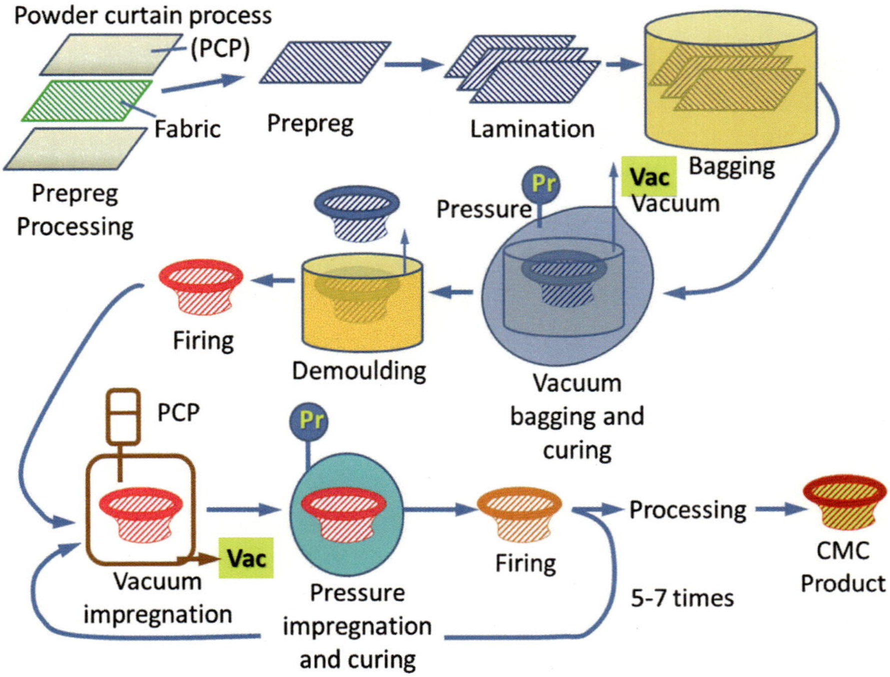

Fig. 14.12 Reaction process: basic steps of a near-net-shape technique

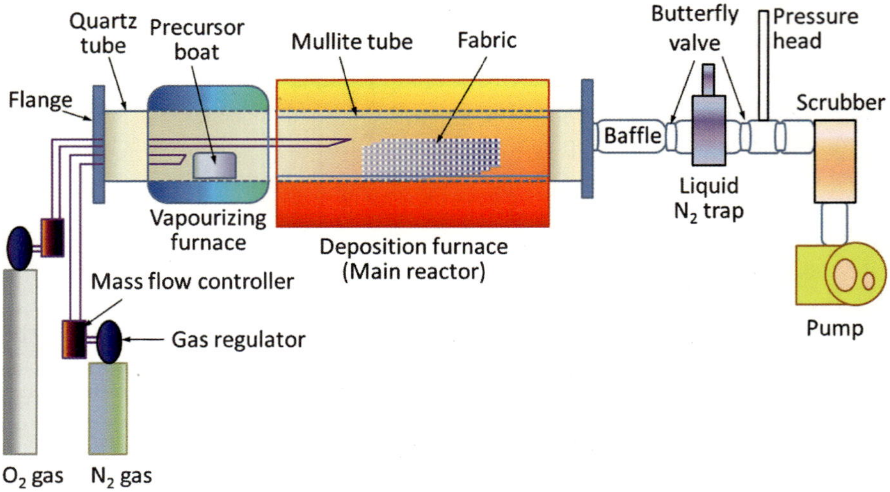

Fig. 14.13 A schematic of CVD setup for deposition of metal oxide on the fabric

chemical affinity for the fiber is very low. In some instances, the protective interfacial layer is coated to the fibers, where the fibers are chemically bonded by the reactants. For example, reaction-bonded silicon nitride (RBSN) is one of the widely used protective interface coated fibers and has been used in many high-temperature applications. RBSN is compacted by Si particles with a desired shape. Then the porous compact is placed in a nitrogen atmosphere at 1200–1400 °C. The silicon and nitrogen are exothermically reacted over a period of time, i.e., few days to a week to form silicon nitride having increased weight up to 60 % owing to the growth of Si_3N_4 into the porous structures. Therefore, its relative density (i.e., actual/theoretical) increases to 80–85 % with a small change (<1 %) in external dimensions. Since this reaction is exothermic, the whole process must be carefully controlled to avoid the melting phase of silicon. Furthermore, there is possibility to damage the reinforcing fibers during fabrication of CMC owing to excessive exposure time at higher temperature.

Chemical Vapor Deposition (CVD)

In CVD, mainly metallic oxides are deposited on the fabric-based materials. It has been used widely to develop nitrogen-bonded silicon carbides, zinc oxide (ZnO)-coated Nicalon fabrics, silicon-doped boron nitride, and so on [11, 56, 57]. At first, a substrate is placed at an appropriate distance into the mullite tube. Then, a metal oxide coating is deposited on the fabrics using a hot-walled isothermal CVD reactor (main reactor) and a hot-walled vaporizing furnace as shown in Fig. 14.13. The main reactor consists of a fused silica (quartz) tube (with an inner diameter of 7.62 cm) and a mullite with lower diameter (6.35 cm). The mullite is used to protect the quartz tube from the deposited metal oxide. Stainless steel endcap flanges of the reactor are completely sealed with O-rings (fluoroelastomer). The mass flow controllers (MFCs) are used to control the flow of ultrahigh purity (UHP) nitrogen and oxygen gases into the system. Two absolute pressure transducers are used to monitor the pressure inside the reactor. Then, a liquid nitrogen trap and a particulate trap are used to collect the by-products, if any using a vacuum pump. The deposition parameters are to be set for a typical experiment. A carrier metallic precursor is generally used with the target material precursor. For example, an aluminum foil precursor boat can be filled with the appropriate amount of zinc acetate dihydrate precursor in order to deposit ZnO on the Nicalon TM fabrics [11]. The fabric is placed in the mullite tube, and the quartz tube contains the zinc precursor at measured distance. Then, the tube is evacuated down to less than 1 Torr. The main reactor and vaporizer furnaces are then brought up to the desired temperatures. The temperature of main reactor is higher than that of vaporizing furnace. The starting point of the reaction is considered when the UHP nitrogen is allowed to flow over the precursor boat.

Continuous fibers have shown very high directional properties at high temperatures particularly above 800 °C [11]. It is extremely advantageous for some specific applications. In order to improve the fabrication technique for future applications of advanced CMCs, some advantages and disadvantages of the existing techniques should be known properly. Therefore, some typical advantages and disadvantages of some important fabrication process of continuous fiber-reinforced CMCs are mentioned in Table 14.2.

Besides the above important frequently used techniques, several other fabrication methods, which have been used to develop CMCs, are viscous phase hot pressing, slip casting, tape casting, slurry, polymer precursor route, gas–metal reaction, self-propagated high-temperature synthesis or combustion synthesis, reaction sintering, pressureless sintering, low-pressure sintering, and so on. Some important ceramic matrix materials and their important processing routes are listed in Table 14.3.

Specific Properties, Materials, and Techniques

It has been widely accepted that nanomaterials have a very significant role on many aspects of material properties owing to their extremely high surface areas. The ceramic matrix nanocomposites (nanoCMCs) have shown significant improvements of their various properties not only with respect to the respective monolithic materials but also compared to the respective micro or macro CMCs [59]. It can be seen that the fracture strength is noticeably higher because of the higher interfacial interaction between the particles in nanoCMCs compared with its micro counterpart. Using up to 15 vol% SiC particles (SiC_p) in Al_2O_3, the nanoCMC exhibits superficial grooves of plastic deformation compared to the intergranular fracture observed in monolithic materials [60, 61]. In another example, the SiC_p/Si_3N_4 system has strength of 1080 MPa up to 1400 °C, whereas the strength of the monolithic sample

Table 14.2 Advantages and disadvantages of various process used for continuous fiber-reinforced CMCs [58]

Processing method	Advantages	Disadvantages	Fiber	Matrix	Processing temperature range (°C)
I. Sol–gel					
Sol–gel and polymer processing	Good matrix composition control Easy to infiltrate fibers Lower densification temperature	Low yields. Very large shrinkage Would require multiple infiltration/densification steps No promising results reported	Nicalon	Nonoxide, alumina, and silicates	800–1200, 800–1400
II. Slurry infiltration					
a) Glass ceramic matrix	Commercially developed Good mechanical properties	Limited maximum temperature due to matrix Needs to be hot pressed, expensive Formation of complex shapes is difficult	Graphite Nicalon	Glass ceramic	800–1000
b) Ceramic matrix: 1. Sintered matrix	Potentially inexpensive Could produce complex shapes	Shrinkage during sintering cracks matrix Temperature limit due to glassy phase	–	Alumina SiC, Si_3N_4	800–1400, 800–1600, 800–1500
2. Cement-bonded matrix	Inexpensive. Ability to produce large complex shapes Low-temperature processing	Relatively poor properties to date	Graphite Nicalon "new" fibers	Cements	400–1400
3. Reaction-bonded matrix	Good mechanical properties Pressureless densification	Required hot pressing of Si powder in silicon nitride system prior to reaction bonding Simple shapes only	Nicalon "New" fibers	Si_3N_4	800–1500, 800–1600
III. Melt infiltration					
a) Ceramic melt	Potentially inexpensive Should be easy to infiltrate fibers Lower shrinkage on solidification	High melting temperatures would damage fibers	Graphite Nicalon "new" fibers	Alumina, oxides	800–1100

(continued)

Table 14.2 (continued)

Processing method	Advantages	Disadvantages	Fiber	Matrix	Processing temperature range (°C)
b) Metal melt, followed by oxidation	Potentially inexpensive Cermet-type material	Difficult to control chemistry and produce all ceramic system Difficult to envision in use for large, complex parts for aerospace applications	Graphite Nicalon "new" fibers	Alumina, B_4C, SiC	800–1200
IV. Chemical vapor infiltration					
a) General approach	Has been commercially developed Best mechanical properties Considerable flexibility in fibers and matrices High-quality matrix, very pure Little fiber damage In situ fiber surface treatment Ability to fill small pores	Slow and expensive Requires iterative process Never achieved full density Capital intensive	Nicalon Nextel	SiC, HfC, nitrides, oxides, borides	800–1600, 800–1800
b) Lanxide	Ability to produce complex shapes Properties dominated by ceramic Porous grain boundaries Systems include AlN/Al, TiN/Ti, ZrN/Zr	Slow reaction and growth kinetics Long processing time & high temperature limits chemistry Wetting and reaction are limitations	Graphite Nicalon	Alumina, AlN, TiN, ZrN	800–1200

Table 14.3 Some matrix materials and their corresponding processing techniques

CMC processing technique	Matrix material
Solid-state hot pressing	SiC, Si_3N_4
Cold pressing and cold isostatic pressing, sintering	Calcium silicate
Hot pressing	Oxides
Sol–gel	Oxides
CVI	Carbide, nitride, oxide boride
LMI	Si→SiC
Reaction (gas–metal)	Oxides (Al), nitrides (Al, Zn, Ti)
Polymer precursor	SiC, Si_xN_y, $Si_xC_yN_z$
Prepreg curing and pyrolysis	SiC, Si_3N_4

Table 14.4 Comparisons in fracture strength and toughness between CMCs and nanoCMCs

Properties of materials	Material system (processing route)	Nanophase (particle size)	Composites (CMC)	Nanocomposite (nanoCMC)	Reference
Fracture strength (MPa)	SiC_p/Al_2O_3 (hot pressing)	15 vol%SiC particle (SiC_p) (~250 nm)	106–283	549–646	[60, 61]
Fracture toughness ($MPa.m^{1/2}$)	SiC_p/Al_2O_3 (hot pressing)	15 vol%SiC_p (~250 nm)	2.4–6.0	4.6–5.5	[60, 61]
Fracture strength (MPa)	SiC_p/Si_3N_4 (pyrolysis and hot pressing)	10 vol%SiC_p (<300 nm)	700	1300	[62, 63]
Fracture toughness ($MPa.m^{1/2}$)	SiC_p/Si_3N_4 (pyrolysis and hot pressing)	10 vol%SiC_p (<300 nm)	5.3	7.0	[62, 63]

decreases considerably at high temperatures [62, 63]. Matrix failure at 1400 °C in tension is found to occur at 0.3 % strain after 0.4 h at 200 MPa, whereas no failure is observed in the 10 % SiC/Si_3N_4 nanoCMC even after 1000 h at 1.5 % strain. A typical comparison in fracture strength and toughness between CMCs and nanoCMCs is illustrated in Table 14.4. Advanced CMCs are also growing rapidly in the markets along with other competitive materials. Therefore, an additional typical comparison in specific strength as a function of temperature between CMCs and other class of materials is necessarily and depicted in Fig. 14.14 [4]. It can be seen from Fig. 14.14 that the specific strength of the metal matrix composites and fiber-reinforced polymer matrix composites is very high compared to CMCs at lower temperatures typically below 1000 °C. However, the CMCs have very high demand in many advanced applications owing to their retention of properties at very high temperatures, i.e., above 2000 °C. Since the ceramics are cost-effective compared to the other materials, CMCs have been widely exploited in various industries.

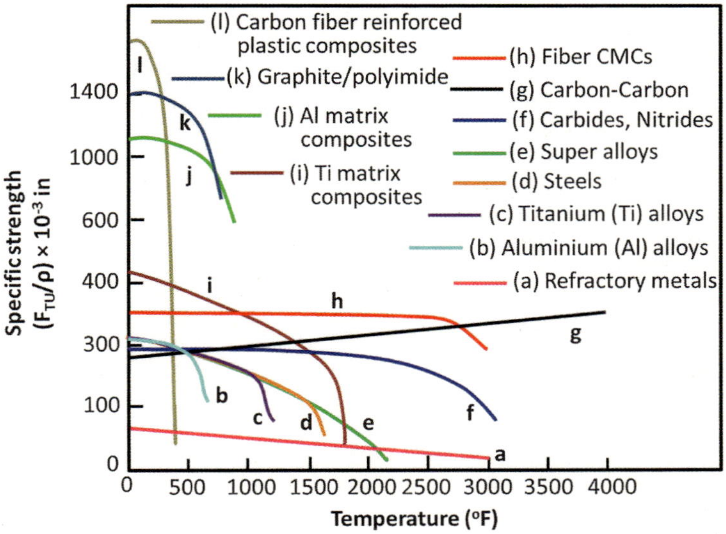

Fig. 14.14 A comparison of specific strength as a function of temperatures with different class of materials [4]

Products Made of Advanced Ceramic Matrix Composites

Various components developed by different CMCs and different processing techniques for several advanced applications are given in Table 14.5. The ceramic materials that are being replaced by advanced CMCs in various advanced fields of applications are, namely, mechanical, aerospace, automotive, defense industry, biomedical, chemical processing, electrical, magnetic, nuclear industry, oil industry, electric power generation, optical engineering, and thermal engineering [58]. The typical applications of CMCs are explored in this section below.

Mechanical

The CMCs used in mechanical engineering applications must have high strength, sufficient toughness, high wear resistance, low coefficient of thermal expansion (CTE), and good lubrication or self-lubrication properties. Cutting tool is one of the crucial materials used in machining process. CMCs exhibit significant improvement in the performance of turbine engines. However, improvements in design methodology are still needed in order to develop high-performance engine for prolonged use. Another importation part of advanced CMCs application is machining tools. A ZrO_2-based CMC drill for high-speed dry machining of gray cast iron is depicted in Fig. 14.15. It should have high hot hardness properties because a high temperature is always developed at cutting tip during operation owing to dynamic and static friction. Therefore, metallic materials are not suitable as a cutting tool or

Table 14.5 Developed CMC components, materials, processing routes, and their test temperature

Component	Reinforcement/matrix	Processing route	Test temperature (°C)	Ref.
Advanced friction systems: Porsche ceramic composite brake (PCCB) and Porsche ceramic clutch PCCC	C/C–SiC (short fiber) (25%SiC)	LSI	1450–1600	[12]
Advanced friction systems: Porsche ceramic composite brake (PCCB) and Porsche ceramic clutch PCCC	C/C–SiC (2D reinforcement) (25%SiC)	LSI	1450–1600	[12]
Thermal shock resister	Nicalon fibers/ calcium aluminosilicate	Lamination	800	[37]
High-strength material as ceramic gas turbine (CGT) rotor	LRC: CF/SiC SRC: SiC_w/Si-Al-O-N, $Si_3N_{4\ is}/Si_3N_4$ W: whisker LRC: long-fiber-reinforced CMC SRC: short-fiber-reinforced CMC is: in situ	Continuous CVD coating method (Reactants: $SiCl_4$, BCl_3, CH_4, NH_3, N_2, and H_2, reaction conditions: temp $<$ 1500 °C, press $<$ 1 atm)	1200	[64]
CGT backplate	SRC: SiC_w/Si-Al-O-N, Si_3N_{4is}/Si_3N_4	Continuous CVD	1200	[64]
High-temperature oxidation-resistant material as CGT orifice liner extension liner	SRC: TiB_{2p} · CF/SiC/SiC LRC: Si-CF/SiC/SiC LRC: Si-CF/Si-N-C/SiC p: powder	Continuous CVD	1350–1450	[64]
High-toughness material as CGT inner scroll support	LRC: Si-N_f, Si-CF/Si-NC/SiC, CF/Si-C/SiC f: fiber	Continuous CVD	1250	[64]
Thermal conductivity as cathodoluminescence (CL) spectra semiconductor	AlN/SiC matrix	Solid-state diffusion reaction	1100	[65]
High mechanical strength and thermal shock absorber	SiC_p/Si_3N_4	Powder metallurgy	1800	[66]

(*continued*)

Table 14.5 (continued)

Component	Reinforcement/matrix	Processing route	Test temperature (°C)	Ref.
High strength, stiffness, and thermal stability for future generations of aircraft engines	BN/SiC coated SiC/Sylramic SiC_f	CVI and MI	Room temperature; this material can also work at 800–1600	[50]
High strength, stiffness, and thermal stability for future generations of aircraft engines	BN/SiC coated Hi-Nicalon SiC_f	CVI and MI	Room temperature; this material can also work at 800–1600	[50]
Notch-insensitive advanced aerospace requires structural component such as turbine engine components, spacecraft reentry thermal protection systems, and hypersonic flight vehicles	Hi-Nicalon™/ SiC–B_4C laminated composites	CVI	1200	[67]
Nuclear applications	SiC_f/SiC (continuous fiber)	–	800–1000	[68]

die. In contrast, the non-oxide ceramic materials, such as WC, SiC, Si_3N_4, and so on, have shown better cutting properties. The CMCs with non-oxide ceramic reinforcement are being used in recent advanced studies. The advanced CMCs produced by different studies have shown higher mechanical properties, good creep resistance, reliable thermal shock, and particle impact damage. These are suitable for potential engine components and thermal shock absorber [26, 64].

Different components made of CMCs for mechanical engineering applications are:

- Cutting tools and dies
- Abrasives
- Precise instrument parts
- Molten metal filter
- Turbine engine components
- Low-weight components for rotary equipment
- Wearing parts

Fig. 14.15 A ZrO_2-based CMC drill for high-speed dry machining of gray cast iron (Courtesy to https://www.mtm.kuleuven.be/Onderzoek/Ceramics/Materials)

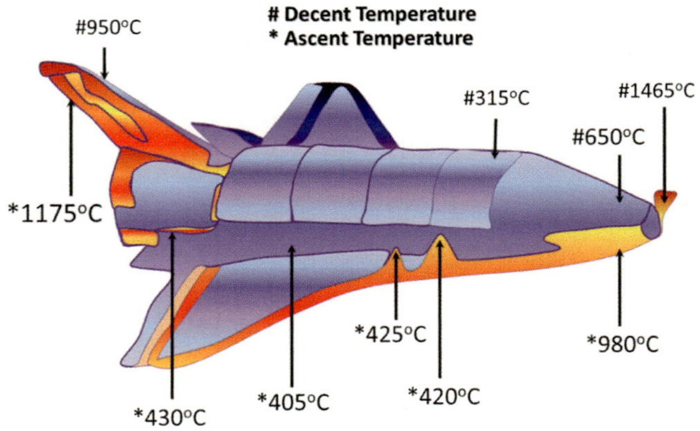

Fig. 14.16 Indication of ascent and descent temperatures in a spacecraft (Courtesy to http://matse1.matse.illinois.edu/ceramics/prin.html)

- Bearings
- Seals
- Solid lubricants

Aerospace

Advanced aerospace applications of the CMCs are turbine engine components, spacecraft reentry thermal protection systems, hypersonic flight vehicles, etc. [67]. These are structural CMCs that exhibit superior long-term mechanical properties and retained properties under high pressure, high temperature, and different environmental conditions, including moisture. The essential properties of CMCs required for the aerospace applications are low density, high-temperature strength, hot hardness, high performance-to-weight ratio (e.g., specific strength, specific stiffness, etc.), refractoriness, high insulation, heat collection, and low thermal conductivity. In this context, C/C–SiC composites show superior tribological properties for advanced friction systems in spacecraft applications in addition to lower densities and higher damage tolerance [12]. Particularly in spacecrafts, different temperatures are exposed in different parts of its surface. The descent and ascent temperatures exposed at different surface parts of a spacecraft are depicted in Fig. 14.16. Currently, several engine manufacturers and research organizations are stepping up their activities toward the use of CMCs in rocket engine thrust chamber components [4]. Several kinds of reinforced materials such as fibers, whiskers, and particles with ceramics produced by CMCs have been widely used in aerospace industries in Europe and the United States (US) as high fracture-resistant materials for fatal defect, especially "brittleness" of monolithic ceramic materials [64]. Fiber-reinforced CMCs (FRCMC) of C/SiC, SiC/SiC, etc. composites produced by CVI have been widely used in aeronautic and space applications owing to their high-temperature resistance, higher curing point (>1400 °C), very high mechanical strength and toughness, and excellent corrosion resistance [26, 54].

Different components made of CMCs for advanced aerospace applications are:

- Fuel systems and valves
- Power units
- Low-weight components
- Fuel cells
- Thermal protection systems
- Turbine engine components
- Combustors
- Bearings
- Seals
- Structures

Automotive

Automotive industries use CMC products. The important properties such as high strength, good toughness, high wear resistance, low CTE, good lubrication or self-lubrication properties, refractoriness, high insulation, heat collection, and good thermal conductivity related to automotive industries are essential for CMC products. The carbon fiber (CF), silicon nitride fiber (Si_3N_{4f}), silicon carbide fiber (SiC_f), silicon carbide whisker (SiC_w), in situ Si_3N_4, and titanium boride (TiB_2)/milled CF are used as reinforcement and SiC, Si_8N_8C, SiAlON, and Si_3N_4 as matrix in CMC composites [64].

Different components made of CMCs for automotive industrial parts are:

- Gas turbine
- Heat engines
- Catalytic converters
- Drivetrain components
- Fixed boundary recuperators
- Turbocharger rotors
- Low heat rejection diesels
- Water pump seals

Defense Industry

The defense industries are growing rapidly with the advanced CMCs. The important properties needed for defense components are light weight, high specific strength, excellent fracture toughness, high wear resistance, good thermal conductivity, and so on [58]. The defense industries also include the aerospace, automotive, and armor components. The armor for structures, machines and personnel, and, more particularly, to an integrated, layered armor incorporating FRCMC material layers are few examples [69, 70].

Different components made of CMCs for defense industries are:

- Tank power trains
- Submarine shaft seals
- Improved armors
- Propulsion systems
- Ground support vehicles
- Fuel injection components
- Military weapon systems
- Military aircraft (airframe and engine)
- Wear-resistant precision bearings

Biomedical and Chemical Industries

Biomedical studies are rapidly growing with the ceramic-based advanced composite materials. Recently, Pramanik et al. have developed newly bovine apatite-based ceramic composites using novel impregnation technique for tissue engineering applications [71]. Several implant materials have also been fabricated with the CMCs [46, 72]. The desired properties of CMCs for biomedical, biological, and chemical applications are high biocompatibility, good strength, excellent fracture toughness, good catalysis adsorption, high photocatalytic properties, high chemical resistance, and corrosion resistance, whereas in chemical industries, photocatalytic properties are very important in order to degrade the organic compound. The present trend in these materials is growing toward new CMCs [73].

Different components made of CMCs for biomedical and chemical industries are:

- Artificial teeth, bones, and joints
- Catalysts and igniters
- Heart valves
- Heat exchanger
- Reformers
- Recuperators
- Refractories
- Nozzles

Electrical and Magnetic Engineering

CMCs have also been used in electrical application as electrical insulators. The magnetic nanoCMCs are being used in different advanced applications [74]. The essential characteristics of the CMCs for these applications are electrical insulation, electrical conductivity, semiconductivity, piezoelectric, magnetic, and dielectric properties.

Different components made of CMCs for electrical and magnetic engineering applications are:

- Memory element
- Resistance heating element
- Varistor sensor
- Integrated circuit substrate
- Multilayer capacitors
- Advanced multilayer integrated packages

Nuclear Industry

SiC_f/SiC CMCs have been widely used as refractory material in nuclear engineering. SiC_f/SiC CMCs can be used in extremely harsh environments. On the other hand, pure SiC provides exceptional stability under neutron radiation. Thus, pure SiC can be used in advanced fission and fusion reactors. To accomplish the CMCs in nuclear industries, the high-temperature mechanical, thermal chemical, and radiation properties are extremely crucial. Ceramic composites made from SiC_f/SiC are used as potential materials for nuclear industries since they have excellent high thermal, mechanical, and chemical stability and high-temperature strength, fracture, creep, corrosion, and thermal shock resistance. Particularly, for nuclear applications, the radiation resistance of the β-phase of SiC imparts shows excellent radiation resistance. The β-SiC has a saturation swelling value of about 0.1–0.2 % at 800–1000 °C, suggesting that SiC_f/SiC CMCs have the excellent radiation stability [68]. The other properties needed for other parts of this industry are excellent radiation resistance, high-temperature strength, refractoriness, high insulation, heat collection, and thermal conductivity.

Different components made of CMCs for nuclear industries are [58]:

- Nuclear fuel
- Nuclear fuel cladding
- Control materials
- Moderating materials
- Reactor mining

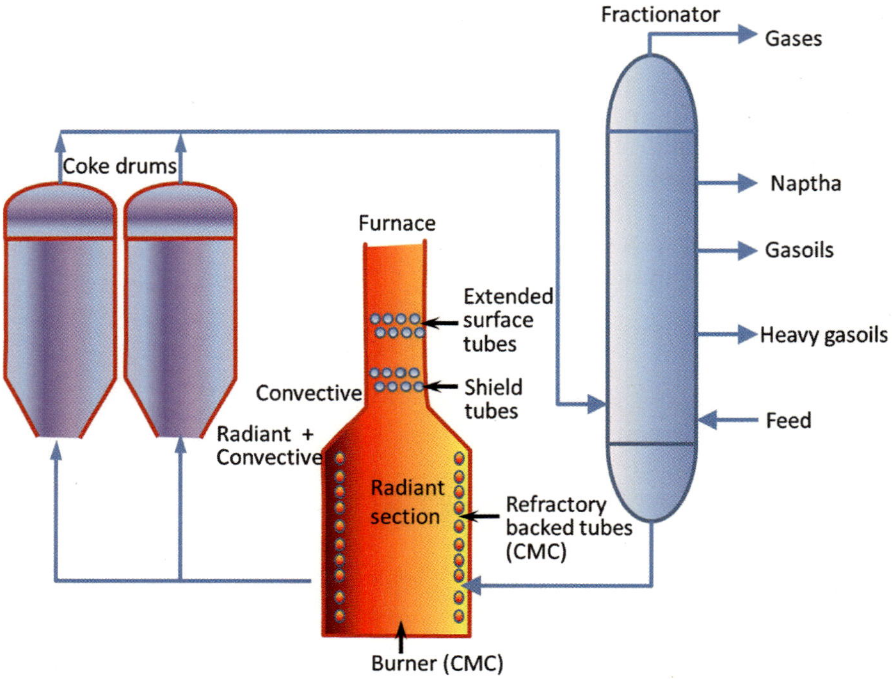

Fig. 14.17 Use of CMCs at oil refinery plant in oil industry

Oil Industry

The main properties of CMCs that are needed for oil industries are refractoriness, high insulation, heat collection, thermal conductivity, high strength, and low thermal expansion coefficient. Several application areas of CMCs at refinery plant in the oil industries are depicted in Fig. 14.17 [75, 76]. It can also be used in liner of furnace, coke drums, and fractionator units.

Different components made of CMCs for oil industries are [58]:

- Bearings
- Flow control valves
- Pumps
- Refinery heater
- Blast sleeves

Electric Power Generation

The excellent electrical insulation, refractoriness, heat collection, thermal conductivity, good strength, high wear resistance, low CTE, and good lubrication or self-lubrication properties are also needed for electrical power industries.

Different components made of CMCs for electrical power generation applications are [58]:

- Bearings
- Ceramic gas turbines
- High-temperature components
- Fuel cells (solid oxide)
- Filters

Thermal Engineering

The excellent thermal stability, low CTE, refractoriness, and good thermal conductivity are necessary properties of CMCs that are used in the thermal engineering applications. Commercially available CMCs have several high-temperature applications. Conventional CMCs are limitedly exposed to temperatures ~ 1200 °C for a prolonged time. However, the CMCs have lower thermal conductivity than metallic materials, and therefore, their uses in cooling arrangements owing to manufacturing constraints are limited. As a result, the conventional CMCs cannot be effectively cooled under high temperatures above 1400 °C or high heat flux conditions. In this context, several researches are continuing to develop a suitable CMC material that can be used to insulate moderate temperature along with erosion resistant, thermal shock resistant, and can have coefficients of thermal expansion relatively similar to that of CMCs [77].

Different components made of CMCs for thermal engineering applications are [58]:

- Electrode materials
- Heat sink for electronic parts
- High-temperature industrial furnace lining

Optical Engineering

The most essential properties of the CMCs for optical engineering are optical condensing, fluorescence, translucence, and optical conductivity. Wide bandgap semiconductors are doped with rare earth elements to increase the optical properties of the CMC materials in photonic applications [78]. A large bandgap (i.e., >3 eV) material allows rare earth emissions in the whole visible range, covering the colors blue (Tm^{3+}), green (Tb^{3+}), and red (Eu^{3+}). The conventional semiconductors such as silicon emit any rare earth emission at room temperature and thus are not suitable for specific technical applications [79]. Recently, AlN/SiC CMCs which are doped with some rare earth materials have been investigated and show a potential candidate to overcome the abovementioned problems [65, 80]. A new class of praseodymium-doped CMCs displays electrostriction (i.e., the conversion of electrical energy into mechanical energy), mechanoluminescence, and electroluminescence at a time.

These are environmentally friendly and may find application in multifunctional sensors or devices [81].

Different components made of CMCs for optical engineering applications are [58]:

- Laser diode
- Optical communication cable
- Heat-resistant translucent
- Light-emitting diode porcelain

Concluding Remarks

Ceramic matrix composites (CMCs) exhibit significant possibility in improving the high performance-to-weight ratio in many advanced applications enabling higher life cycles at elevated temperatures. The high specific strength, refractoriness, and high thermal and chemical resistances of CMCs are the most attractive properties in the field of structural, chemical, and biomedical applications. The incorporation of various reinforcement into ceramic matrix using selected fabrication techniques, many properties of the CMCs can be tailored very easily in order to achieve the desired properties. In the recent past, most components using CMCs have been designed by "trial and error" due to the lack of mathematical models. To understand the key parameters, controlling response, and the exact mechanics of failure, the precise development of structural failure models is more required in future. The addition of second phases in ceramic matrices with optimized interfacial bonding improves the fracture toughness and decreases the sensitivity of the preexisting flaws in the brittle matrix. The reinforced second phase has a variety of shapes, ranging from nearly spherical particles, through whiskers, chopped fibers, and continuous fibers with various aspect ratios and to fabrics.

Acknowledgment The authors acknowledge the financial support provided by the Indian Space Research Organization, India, for carrying out this research work.

References

1. Campbell J (1997) Forty-first mellor memorial lecture: opportunities for ceramics industry. Br Ceram Trans 96(6):237–246
2. Szutkowska M, Jaworska L, Boniecki M, Stobierski L, Rozmus M (2015) Mechanical behavior of diamond matrix composites with ceramic Ti_3(Si, Ge)C2 bonding phase. Int J Refract Met Hard Mater 49:302–306
3. Ohnabe H, Masaki S, Onozuka M, Miyahara K, Sasa T (1999) Potential application of ceramic matrix composites to aero-engine components. Compos Part A Appl Sci Manuf 30(4):489–496
4. Schmidt S, Beyer S, Knabe H, Immich H, Meistring R, Gessler A (2004) Advanced ceramic matrix composite materials for current and future propulsion technology applications. Acta Astron 55(3–9):409–420
5. Gumula T, Blazewicz S (2013) Thermal conversion of carbon Fibers/polysiloxane composites to carbon fibers/ceramic composites. Ceram Int 39(4):3795–3802

6. Volkmann E, Tushtev K, Koch D, Wilhelmi C, Grathwohl G, Rezwan K (2014) Influence of fiber orientation and matrix processing on the tensile and creep performance of Nextel 610 reinforced polymer derived ceramic matrix composites. Mater Sci Eng A 614:171–179
7. Mukerji J (2013) Ceramic Matrix Composites. Defence Sci J 43(4):385–395
8. Chamlee TC (1999) Particulate field distributions in centrifugally cast composites. US Patent US5980792 A
9. Karandikar P, Evans G (2007) Aghajanian M Carbon nanotube (CNT) and carbon fiber reinforced high toughness reaction bonded composites. Ceram Eng Sci Proc 6:53–63
10. Yang W, Araki H, Noda T, Park JY, Katoh Y, Hinoki T, Yu J, Kohyama A (2002) Hi-Nicalon fiber-reinforced CVI-SiC matrix composites: I Effects of PyC and PyC-SiC multilayers on the fracture behaviors and flexural properties. Mater Trans 43(10):2568–2573
11. Coons TP, Reutenauer JW, Mercado A, Kmetz MA, Suib SL (2013) The characterization of an oxide interfacial coating for ceramic matrix composites. Mater Sci Eng A 573:190–196
12. Krenkel W, Berndt F (2005) C/C–SiC composites for space applications and advanced friction systems. Mater Sci Eng A 412(1–2):177–181
13. Chen RZ, Tuan WH (1999) Pressureless Sintering of Al_2O_3/Ni Nanocomposites. J Eur Ceram Soc 19(4):463–468
14. Kong X, Qiao Y (2005) Fracture in ceramic-matrix composites reinforced with strongly bonded metal particles. Mech Compos Mater 41(3):205–216
15. Ando K, Kim BS, Chu MC, Saito S, Takahashi K (2004) Crack-healing and mechanical behaviour of Al_2O_3/SiC composites at elevated temperature. Fatigue Fract. Eng Mater Struct 27(7):533–541
16. Niihara K (1991) New design concept of structural ceramics – ceramic nanocomposites. Nippon Seram. Kyo Gak Ron J Ceram Soc Jpn 99(10):974–982
17. Zhao J, Stearns LC, Harmer MP, Chan HM, Miller GA, Cook RF (1993) Mechanical Behavior of Alumina–Silicon Carbide Nanocomposites. J Am Ceram Soc 76(2):503–510
18. Jiao S, Jenkins M, Davidge R (1997) Interfacial fracture energy-mechanical behaviour relationship in Al_2O_3/SiC and Al_2O_3/TiN nanocomposites. Acta Mater 45(1):149–156
19. Thompson AM, Chan HM, Harmer MP, Cook RE (1995) Crack healing and stress relaxation in Al_2O_3 SiC nanocomposites. J Am Ceram Soc 78(3):567–571
20. He JL, Li WZ, Li HD, Liu CH (1998) Plastic properties of nano-scale ceramic–metal multilayers. Surf Coat Technol 103–104:276–280
21. Zuo KH, Jiang DL, Lin QL, Y-p Z (2007) Improving the mechanical properties of Al_2O_3/Ni laminated composites by adding Ni particles in Al_2O_3 layers. Mater Sci Eng A 443 (1–2):296–300
22. Zuo KH, Jiang DL, Lin QL (2006) Mechanical properties of Al_2O_3/Ni laminated composites. Mater Lett 60(9–10):1265–1268
23. Wei GC, Becher PF (1984) Improvements in mechanical properties in SiC by the addition of TiC particles. J Am Ceram Soc 67(8):571–574
24. Oh S-T, Sekino T, Niihara K (1998) Fabrication and mechanical properties of 5 vol% copper dispersed alumina nanocomposite. J Eur Ceram Soc 18(1):31–37
25. Zuo KH, Jiang DL, Lin QL (2006) Mechanical properties of Al_2O_3/Ni laminated composites. Mater Lett 60(9-10):1265–1268
26. Cao X, Lin B, Zhang X (2015) Investigations on grinding process of woven ceramic matrix composite based on reinforced fiber orientations. Compos Part B Eng 71:184–192
27. Nair BG, Cooper RF, Almquist JN, Plesha ME (2001) Off-axis creep of a ceramic-matrix/continuous-ceramic-fiber composite: Experimental evaluation. J Mater Sci 36(9):2133–2144
28. Meyer DW, Cooper RF, Plesha ME (1992) Rheological modeling of ceramic composites: An indirect method of interfacial mechanical property measurements. Int J Solids Struct 29 (20):2563–2582
29. Sørensen B, Holmes J (1996) Fatigue of continuous fiber-reinforced ceramic matrix composites: review of mechanisms and models. In: Fatigue under thermal and mechanical loading: mechanisms, mechanics and modelling. Springer, Petten, The Netherlands, pp 487–499

30. He X, Guo Y, Yu Z, Zhou Y, Jia D (2009) Study on microstructures and mechanical properties of short-carbon-fiber-reinforced SiC composites prepared by hot-pressing. Mater Sci Eng A 527 (1):334–338
31. Yang F, Zhang X, Han J, Du S (2008) Processing and mechanical properties of short carbon fibers toughened zirconium diboride-based ceramics. Mater Design 29(9):1817–1820
32. Zhang Y, Li S, Han J, Zhou Y (2012) Fabrication and characterization of random chopped fiber reinforced reaction bonded silicon carbide composite. Ceram Int 38(2):1261–1266
33. Xia Z, Riester L, Curtin WA, Li H, Sheldon BW, Liang J, Chang B, Xu JM (2004) Direct observation of toughening mechanisms in carbon nanotube ceramic matrix composites. Acta Mater 52(4):931–944
34. Gonon MF, Hampshire S (1999) Comparison of two processes for manufacturing ceramic matrix composites from organometallic precursors. J Eur Ceram Soc 19(3):285–291
35. Magnant J, Pailler R, Le Petitcorps Y, Maillé L, Guette A, Marthe J, Philippe E (2013) Fiber-reinforced ceramic matrix composites processed by a hybrid technique based on chemical vapor infiltration, slurry impregnation and spark plasma sintering. J Eur Ceram Soc 33(1):181–190
36. Bansal NP (2006) Handbook of ceramic composites, vol 200. Springer Science & Business Media, NY, USA
37. Kastritseas C, Smith P, Yeomans J (2008) Thermal shock behaviour of angle-ply and woven dense ceramic-matrix composites. J Mater Sci 43(12):4112–4118
38. Cho J, Boccaccini AR, Shaffer MS (2009) Ceramic matrix composites containing carbon nanotubes. J Mater Sci 44(8):1934–1951
39. Tiegs T (2005) SiC whisker reinforced alumina. In: Handbook of ceramic composites. Springer, NY, USA, pp 307–323
40. Becher PF, Wei GC (1984) Toughening Behavior in Sic-Whisker-Reinforced Alumina. J Eur Ceram Soc 67(12):C-267–C-269
41. Lansmann V, Jansen M (2001) Application of the glass-ceramic process for the fabrication of whisker reinforced celsian-composites. J Mater Sci 36(6):1531–1538
42. Lin X, Darrell Ownby P (2000) Pressureless sintering of B_4C whisker reinforced Al_2O_3 matrix composites. J Mater Sci 35(2):411–418
43. Boccaccini AR, Winkler V (2002) Fracture surface roughness and toughness of Al_2O_3-platelet reinforced glass matrix composites. Compos Part A Appl Sci Manuf 33(1):125–131
44. Becher PF, HSUEH CH, Angelini P, Tiegs TN (1988) Toughening Behavior in Whisker-Reinforced Ceramic Matrix Composites. J Eur Ceram Soc 71(12):1050–1061
45. Evans AG (1990) Perspective on the development of high-toughness ceramics. J Eur Ceram Soc 73(2):187–206
46. Ataollahi Oshkour A, Pramanik S, Shirazi SFS, Mehrali M, Yau Y-H, Abu Osman NA (2014) A comparison in mechanical properties of cermets of calcium silicate with Ti-55Ni and Ti-6Al-4 V alloys for hard tissues replacement. Sci World J 2014:616804
47. Tan HB, Guo CS (2011) Preparation of long alumina fibers by sol-gel method using malic acid. Trans Nonferrous Met Soc China 21(7):1563–1567
48. Rezaie HR, Mohammad-Rahimi R, Nemati A, Samadani M (2011) Synthesis and characterization of Al_2O_3-SiC nano composite by sol-gel method and the effect of TiO_2 on sintering. J Nano Res 13:7–19
49. Herrmann M, Schönfeld K, Klemm H, Lippmann W, Hurtado A, Michaelis A (2014) Laser-supported joining of SiC-fiber/SiCN ceramic matrix composites fabricated by precursor infiltration. J Eur Ceram Soc 34(12):2913–2924
50. Bhatt RT, Chen Y, Morscher GN (2002) Microstructure and tensile properties of BN/SiC coated Hi-Nicalon, and sylramic SiC fiber preforms. J Mater Sci 37(18):3991–3998
51. Leuchs M (2008) Chemical vapour infiltration processes for ceramic matrix composites: manufacturing, properties, applications. Ceram Matrix Compos: Fiber Reinforced Ceram Appl :141–164
52. LAzzeri A (2012) CVI processing of ceramic matrix composites. In: Bansal NP, Boccaccini AR (eds) Ceramics and composites processing methods. Wiley, Hoboken, p 313

53. Besmann T, Sheldon B, Lowden R, Stinton D (1991) Vapor-phase fabrication and properties of continuous-filament ceramic composites. Science 253(5024):1104–1109
54. Papakonstantinou C, Balaguru P, Lyon R (2001) Comparative study of high temperature composites. Compos Part B Eng 32(8):637–649
55. Belvin HL, Cano RJ (2003) Dry process for manufacturing hybridized boron fiber/carbon fiber thermoplastic composite materials from a solution coated precursor. US Patent 6514370 B1
56. Kmetz M (2009) Multilayered boron nitride/silicon nitride fiber coatings. US Patent US7510742 B2
57. Yun GNMHM (2009) High temperature si-doped bn interphases for woven SiC/SiC composites. Adv Ceram Mater Struct A Ceram Eng Sci Proc 26:295
58. Rosso M (2006) Ceramic and metal matrix composites: Routes and properties. J Mater Process Technol 175(1):364–375
59. Camargo PHC, Satyanarayana KG, Wypych F (2009) Nanocomposites: synthesis, structure, properties and new application opportunities. Mater Res 12(1):1–39
60. Ferroni L, Pezzotti G, Isshiki T, Kleebe H-J (2001) Determination of amorphous interfacial phases in Al_2O_3/SiC nanocomposites by computer-aided high-resolution electron microscopy. Acta Mater 49(11):2109–2113
61. Anya C (1999) Microstructural nature of strengthening and toughening in Al_2O_3-SiC (p) nanocomposites. J Mater Sci 34(22):5557–5567
62. Mabuchi H, Tsuda H, Ohtsuka T, Matsui T, Morii K (1999) In-situ synthesis of Si_3N_4-SiC composites by reactive hot-pressing. HTHP Abstr 31(5):499–506
63. Sternitzke M (1997) Structural ceramic nanocomposites. J Eur Ceram Soc 17(9):1061–1082
64. Kaya H (1999) The application of ceramic-matrix composites to the automotive ceramic gas turbine. Compos Sci Technol 59(6):861–872
65. Weingärtner R, Guerra Torres JA, Erlenbach O, Gálvez de la Puente G, De Zela F, Winnacker A (2010) Bandgap engineering of the amorphous wide bandgap semiconductor $(SiC)_{1-x}(AlN)_x$ doped with terbium and its optical emission properties. Mater Sci Eng B 174(1–3):114–118
66. Tian L, Zhou Y, Zhou W-L (1998) SiC-nanoparticle-reinforced Si_3N_4 matrix composites. J Mater Sci 33(3):797–802
67. Ruggles-Wrenn MB, Kurtz G (2013) Notch sensitivity of fatigue behavior of a Hi-Nicalon™/SiC-B_4C composite at 1,200 °C in air and in steam. Appl Compos Mater 20(5):891–905
68. Jones RH, Henager CH (2005) Subcritical crack growth processes in SiC/SiC ceramic matrix composites. J Eur Ceram Soc 25(10):1717–1722
69. Miguel AS (1983) Composite floor armor for military tanks and the like. US Patent US4404889 A
70. Strasser TE, Atmur SD (2000) Fiber reinforced ceramic matrix composite armor. US Patent US6135006 A
71. Pramanik S, Ataollahi F, Pingguan-Murphy B, Oshkour AA, Osman NAA (2015) In vitro study of surface modified poly(ethylene glycol)-impregnated sintered bovine bone scaffolds on human fibroblast cells. Sci Rep 5:9806
72. Pramanik S, Agarwal AK, Rai K (2005) Chronology of total hip joint replacement and materials development. Trend Biomater Artif Organ 19(1):15–26
73. Feng X, Guo H, Patel K, Zhou H, Lou X (2014) High performance, recoverable Fe_3O_4ZnO nanoparticles for enhanced photocatalytic degradation of phenol. Chem Eng J 244:327–334
74. Amir M, Kurtan U, Baykal A (2015) Rapid Color degradation of organic dyes by Fe_3O_4@His@Ag recyclable magnetic nanocatalyst. J Ind Eng Chem 27:347–353
75. Díaz-Mateus F-A, Castro-Gualdrón J-A (2010) Mathematical model for refinery furnaces simulation. CT&F-Ciencia, Tecnol Futuro 4(1):89–99
76. Filho RM, Sugaya MF (2001) A computer aided tool for heavy oil thermal cracking process simulation. Comput Chem Eng 25(4–6):683–692
77. Merrill GB, Morrison JA (2000) High temperature insulation for ceramic matrix composites. US Patent US6013592 A
78. Steckl AJ, Zavada JM (1999) Photonic applications of rare-earth-doped materials. MRS Bull 24 (09):16–20

79. Favennec P, L'haridon H, Salvi M, Moutonnet D, Le Guillou Y (1989) Luminescence of erbium implanted in various semiconductors: IV, III-V and II-VI materials. Electron Lett 25 (11):718–719
80. Weingärtner R, Erlenbach O, Winnacker A, Welte A, Brauer I, Mendel H, Strunk H, Ribeiro C, Zanatta A (2006) Thermal activation, cathodo-and photoluminescence measurements of rare earth doped (Tm, Tb, Dy, Eu, Sm, Yb) amorphous/nanocrystalline AlN thin films prepared by reactive rf-sputtering. Opt Mater 28(6):790–793
81. Wang X, Xu CN, Yamada H, Nishikubo K, Zheng XG (2005) Electro-Mechano-Optical Conversions in Pr^{3+}-Doped $BaTiO_3$–$CaTiO_3$ Ceramics. Adv Mater 17(10):1254–1258

Advanced ZrO_2-Based Ceramic Nanocomposites for Optical and Other Engineering Applications

15

S. Ram and G. P. Singh

Contents

S. Ram (✉)
Materials Science Centre, Indian Institute of Technology Kharagpur, Kharagpur, West Bengal, India
e-mail: sram@matsc.iitkgp.ernet.in

G.P. Singh
Centre for Nanotechnology, Central University of Jharkhand, Ranchi, Jharkhand, India
e-mail: gpsinghcuj@gmail.com

K.K. Kar (ed.), *Composite Materials*,
DOI 10.1007/978-3-662-49514-8_15

Abstract

Although the traditional ceramics cover the major production of ceramic industries (in terms of volume as well as the selling price), a variety of new ceramics has been immersing progressively with better properties and performances. The chapter addresses zirconia and its nanocomposites, and their performances in thermal and chemical barrier coating, buffer layer for high temperature superconducting films, ball heads for hip replacements, and automobiles. Specific applications include optical systems and devices, optoelectronics, optical waveguides, optical data storage systems, optical communications, optical displays, and series of optical and/or biological sensors.

Keywords

Zirconia • Composites • Thermal stability • Absorption • Emission • Polymeric liquids • Nanogels • Thermoluminescence

Introduction

Ceramics and derivatives in forms of composites include a wide range of advanced materials for structures, tools, devices, optical systems, biomaterials, sensors, and catalysts for domestic usages, defense, science and technology, medicals, and other disciplines. There is strong correlation in structure (distribution of atoms and also electrons), properties, and performance for selective applications of a specific material, i.e., a natural product, a derived product from the nature, or a synthetic one from preexisting elements and compounds using known strategic methods. Since as early as civilization started, people are using materials along with energy with progressively growing standard of our living. Materials provide energy as well as convert energy desirably from one form to others useable of pertinent applications. Of those which are synthesized or developed in selective forms for specific applications at economic value are often referred as the "engineering materials." Constant research and development are adding newer materials day-to-day. For the best utilization, the engineers should be aware with the up-do-date materials (also by-products if any) and their thermal stabilities, structures, biocompatibility, and other functional properties so that they can explore the most suitable ones for specific purpose and be able to develop the best processing method(s) to design or manufacture in specific shape and size of the final form of a product, a tool, or a device. In general, in terms of the electronic structure and applications, the engineering materials are classified into three primary categories,

(a) Metals and alloys
(b) Polymers (and plastics)
(c) Ceramics
(d) Classical composite materials

Metals and Alloys

A classical definition of conventional metal is that it consists of continuous distribution of electrons in the various shells and subshells with no energy gap between filled and valence subshells. Delocalized electrons span over an infinitely large space. Examples are gold, iron, nickel, copper, and aluminum of infinitely large size (bulk solid). Metals in general are good conductors of both heat and electricity. A nonmetal (such as carbon, nitrogen, oxygen, etc.) has localized distribution of electrons at atomic scale, and it is characterized by an optical bandgap E_g, which separates the conduction band from the valence band of the electrons. A metal alloy is a combination of two or more metals or a metal and a nonmetal (s) at atomic scale, i.e., one kind of atoms distributes through other kinds of atoms in specific sites. It may be amorphous or crystalline in nature. In general, an alloy is short of a composite material, which behaves as metal or nonmetal according to the final electronic structure. A metal or an alloy can retain the metallic character only when it is sufficiently large of three-dimensional geometrical size well above the mean electron free path, e.g., ~50 nm in noble metals [1–3].

Polymers (and Plastics)

A polymer consists of an interconnected network from a certain molecular unit, called "monomer," with regular chemical bonding of the monomer. The number of monomers "n" in such polymer structure varies over a wide range, usually 10^3–10^6, determining its molecular size. Most popular engineering polymers consist of organic compounds. Structurally, organic polymers are often noncrystalline of nature, but they can be recrystallized partially (using thermal annealing, cold rolling, laser annealing, and other specific methods) [4–6], forming a hybrid composite consisting of a crystalline phase with part of the amorphous phase. Very few are known of inorganic polymers, e.g., some inorganic gels, vitreous glasses, and vitreous ceramics such as zirconia, alumina, and derivatives [7–9]. The polymeric structure disappears upon crystallizing such inorganic materials. In general, polymeric materials (both organic and inorganic) have low specific densities and relatively low softening or decomposition temperatures.

Poor thermal stability and high biodegradability in ambient air atmosphere are a major barrier for many applications of polymer and plastics. Because of the prominent covalent chemical bonding, they are mostly poor conductors or insulators of electricity (also heat). In view of industrial applications in electronic and optical devices, sensors, biomaterials, and other disciplines [10–12], attempts are being made to design novel conducting polymers with electron doping (using nonbonding electrons, metals, or ceramics) in form of so-called a kind of "hybrid nanocomposite materials."

Ceramics

Ceramics are defined as inorganic nonmetallic materials. Phenomenologically, they consist of two kinds of metallic and nonmetallic elements in form of a simple compound (such as Al_2O_3, ZrO_2, SiO_2, etc.), a mixed compound (such as zircon, mullite, magnetite Fe_3O_4, other spinels, hexagonal ferrites, perovskites, etc.), a complex compound (clays, minerals, cements, porcelain, feldspar), or simply a solid solution of one ceramic phase dispersed through another ceramic phase (s), e.g., glasses and glass-ceramics (also called vitroceramics), foams, and some mixed oxides, or ceramic composites in a more general term. This definition of ceramics (an extension of the Greek word "*Keramos*") is wider than simply the art and science of making and using solid articles formed merely by heating earthy materials or derived compounds and is far broader from a common dictionary definition such as "pottery" or "earthenware."

Advances in newer synthetic methods of synthesizing and/or manufacturing materials of tailored properties close to specifications for specific applications make traditional definition too restrictive for the practical purposes. The ceramics can be crystalline, amorphous (including conventional glass, viz., a material obtained by rapidly cooling a homogeneous liquid so that it does not crystallize during the cooling), or a combination of crystalline and amorphous phases in a single material. In general, in terms of the band structure and properties, they belong to the group of the semiconductor or insulator. Most ceramics usually stand very high temperature with high mechanical hardness, good heat and wear resistance, low friction, and insulative properties.

In an integral part of our daily life, ceramics cover a wide range of engineering products in almost all disciplines of materials science and engineering (e.g., automobiles, architectures, robotics, turbine engines, electronics, ferroelectricity, electric motors and generators, capacitors and devices, transducers, solid electrolytes, optical systems, or photonics) including biomaterials, medicines, drug delivery agent, and various kinds of sensors. Such wide diversity of ceramic products and application does not permit simple classification of them neither on the basis of their properties nor applications wise. However, for simplicity, we consider ceramics in two primary classes as follows;

(i) Traditional ceramics
(ii) Advanced ceramics
(iii) Hybrid ceramics and composites

Traditional Ceramics

The traditional ceramics are manufactured from natural raw materials, i.e., clay products, minerals (sands, stones, zeolites, soda ash, fly ash), and cements. Forming wares and trolleys by firing clays in specific shapes have been practiced for quite long from the earliest civilization. Casting silicate glasses from liquid of natural sands are another ancient art. Portland cement is being manufactured for about the nineteenth century. The Romans prepared natural hydraulic cement by combining burned lime with volcanic ash. In these days, the largest part of ceramic industries is the production of glasses, with lime and cement products as the next largest products (especially for architecture). White wares, structural clay products, and enamels (which are silicate glasslike coatings on selective metals) constitute diverse group of traditional ceramics. The materials include pottery and porcelainlike compositions, bricks, tiles, and similar products such as sewer pipes. About 40 % of ceramics in refractory consists of fired clays, and another 40 % consists of heavy refractories, e.g., magnetite, chromite, zirconia, and similar compositions. Special refractories such as silicon carbide, boron carbide, and aluminum oxide for abrasives impart sizable production of ceramic industries.

Advanced Ceramics

Although we cannot deny the fact that the traditional ceramics still cover major production of ceramic industries (in terms of volume as well as the selling price), a variety of new ceramics has been immersing progressively with better properties and performances as follows. Monolithic oxides such as Al_2O_3, ZrO_2, ThO_2 (thoria), BeO (beryllia), MgO (magnesia), $MgAl_2O_3$, and Mg_2SiO_4 (forsterite) and some nitrides/carbides are well-known electrical and thermal insulators and refractories. UO_2 (uranium dioxide) is an excellent nuclear fuel in nuclear reactor. Lithium niobate ($LiNbO_3$) and rare-earth-modified titanates provide a medium by which electrical information can be transferred to optical information or by which optical functions can be operated by an electrical signal. Ferroelectric titanates, having extremely high dielectric constant, are particularly useful for capacitors and electronic devices. Spinel ferrites, hexagonal ferrites, and garnets are some of important magnetoceramics for magnet technology, magnetic memory devices, bubble memory devices, high-frequency communications with low power loss, microwave absorption in radar, and other applications [13–15].

A series of ceramic materials which comprise both magnetic and ferroelectric properties in a single system (called multiferroics) are getting renewed interest nowadays for multifunctional devices. Ceramics for superconductors, colossal magnetoresistance (CMR), electrodes, and gas sensors are of particular interest in different disciplines of electronic industries. Many of such ceramics behave unusually as a ferromagnetic metal below a certain temperature, often called "half-metallic ceramics" in view of a dual nature (metallic and semiconductor) of band structure [13, 14, 16, 17].

Many ceramics (semiconductors and insulators) serve as a good optical host, i.e., facilitate optical emission of selective wavelengths (λ_{em}) from optical species (ions and color centers), which are doped thereon purposely in a dispersed microstructure, upon exposing to a suitable wavelength of light of excitation (λ_{ex}). The best optical host for such purpose would be an amorphous or nanocrystalline insulator of low energy phonons, e.g., Al_2O_3, Y_2O_3, ZrO_2, TiO_2, zeolites, some nitrides, or SiO_2 and their derivative compounds. Trivalent rare-earth cations R^{3+} (also R^{2+}) and Cr^{3+} and architect defect centers (interstitial ions or vacancies) are identified optical species in well-known optical materials or lasers. Other common hosts for efficient emission include oxide glasses (also vitroceramics), ZnO, perovskites ABO_3, perovskite-type fluorides ABF_3, spinels AB_2O_4, garnets $A_3B_5O_{12}$, silicates such as $Sr_2MgSi_2O_7$ or $Sr_4Al_{14}O_{25}$, and sulfides such as ZnS, SrS, CaS, and CdS, where A and B refer to metal cations of requisite valances.

A co-doping with two or more kinds of optical species tunes selective light emission with desired band intensity and other properties in the above ceramics. In particular, the silicate Sr_2SiO_4, $Sr_2MgSi_2O_7$, or $Sr_4Al_{14}O_{25}$ allows the optical signal to glow desirably longer for several hours after removing the light source [18–21]. Sometimes, coating optical species in thin films onto insulator particles (a support or host) reveals surface-enhanced light emission [22–25] in discrete structure of the optical species of our interest in this article.

The rare-earth ions with empty, filled, or a half-filled f-shell, viz., $4f^0$, $4f^7$, or $4f^{14}$ valence electronic configurations, which no longer involve so low energy 4f–4f electronic levels in the visible or near UV regions unlike other rare-earth optical species, can be used as a transparent light host in these specific regions. The $4f^0$, $4f^7$, or $4f^{14}$ electrons would share cooperative magnetic interactions with selective optical doping of the other kinds of the rare-earth ions useful for spintronics. Moreover, they are promising wideband UV light or far UV light sources.

As will be discussed further with specific examples of light emitters, a nanoporous host or support of either amorphous or nanocrystalline ceramics offers similar improvements in light emission signals in a specific design of the optical species, i.e., the rare-earth ions or the valence $4f^n$ electrons. Spins in paramagnetic interstitial ionic sites or vacancies assume a similar localized distribution. Especially nanoporous ceramics such as ZnO, ZrO_2, SiO_2, or Al_2O_3 and their derivatives have efficient light emissions in the impurity levels (color centers) [26–35]. In doped ceramics, the electrons $4f^n$ and $3d^{n'}$ in the dopants recombine the spins from the color centers in the host, modifying average spin configuration in "exchange-coupled ion pairs." Ultimately, it is the self-localized structure of the optical species that appears in model electronic transitions of discrete energy levels of isolated optical species embedded in a dielectric host medium. A microstrain led by the electromagnetic coupling between the magnetic and dielectric fields mediates the electron–phonon coupling, which enforces the electronic transitions as per the induced mechanical oscillation of the optical species by interaction with the light signal imposed to induce the light emission [35, 36].

Hybrid Ceramics and Composites

These are very special ceramics with multifunctional engineering properties in a single material applicable for multifunctional activities under selective experimental conditions. Two or more kinds of selective properties are united one another through cooperative interactions among them. Examples include a variety of ceramics with structural, optical, magnetic, magnetocaloric, dielectric, and other properties which support mutually one another in a cooperative manner. Such ceramics can be designed in several ways which include (i) dispersing one kind of ceramic phase into another kind of ceramic phase(s) in a specific microstructure, (ii) surface coating ceramic particles of a specific phase by using a different material (which may be a ceramic, polymer, or a metal/alloy) in thin layers, (iii) incorporating nanopores in a given ceramic phase in a specific microstructure, (iv) incorporating interstitial ions and/or vacancies (or dislocations) of selective valence states, and (v) designing ceramic nanoparticles of a specific core–shell structure. Basic shape and size of small ceramic particles play a crucial role in tailoring desired properties and applications.

In all the above possible examples of hybrid ceramics, electronic structure and derived physical properties in the basic species (particles, ions, or vacancies) undergo a kind of "exchange coupling," leading to modify the final structure and properties out of a single hybrid composite system of material. In optical materials, the "exchange coupling" leads to alter the primary electronic levels in the optical species and their excitation processes in part of the influence of the dielectric and/or magnetic fields surrounding the optical species. In terms of thermodynamics, it either suppresses or promotes the electron–phonon and the spin–phonon couplings and thus in turn the vibrational and spin entropies in part of the internal energy (or lattice heat energy) of the combined system.

Classical Composite Materials

According to the structure and properties, we can define a composite material as follows. Classically, a composite material forms by dispersing one kind of phase of a material into another distinct phase (or a system of two or more distinct phases) of distinct materials, with strong interfacing and macroscopic interactions between the distinct structural ingredients over an extended scale of the common interfaces or boundaries. A physical mixture of the same origin and chemical composition differs very distinctly from this purposely modulated structure in the sense that it involves exclusively no such extensive interactions among the ingredients at so large scale. In a nanocomposite, at least one of the ingredients has a finite dimension of a few nanometers, i.e., far above the bulk equilibrium energy state. The distinct phases (which may be from pure elements and inorganic as well as organic compounds) may be as solid, liquid, or a combination of both the solid and liquid phases. In other words, in terms of the structure, they may be amorphous, semicrystalline, crystalline, or a complex system consisting of all these possible phases. As mentioned in limited reference books and review articles, high-tech composites are being fabricated in part of a global R&D in several interdisciplinary fields to operate simultaneously multifunctions in many devices [37–39].

Moreover, using suitable thermal/chemical treatments, a nanocomposite (especially a nanoporous) can be explored for producing homogeneously large-grained materials which can offer superior properties compared to the conventional bulk materials or a conventional bulk composite. Useful functional properties of such nanocomposites are affected sensitively by the individual properties of the ingredients according to the final microstructure and the synthetic method used in obtaining the final product. In terms of microstructure, when average particle size decreases below a certain critical value, the concomitant growth of the effective volume fraction of the surface interfaces (i.e., the increase in the effective number of atoms in the interfaces) leads to drive the final properties as per the promoted exchange interactions among the ingredients.

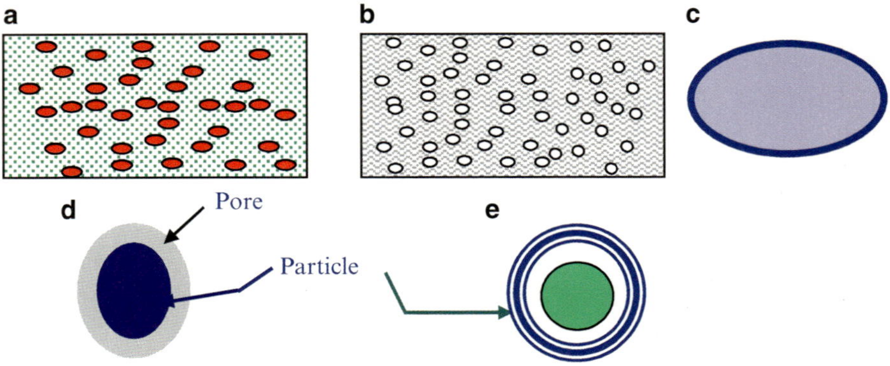

Fig. 15.1 A model diagram of a composite with two main phases of the (**a**) fillers (the *red* balls) and (**b**) pores (*white* balls) and the matrix phase (*greenish* background) of a continuous host, (**c**) a nanoparticle with a difference surface layer of its own from the core, (**d**) a nanoparticle embedded in a pore, and (**e**) a nanoparticle with thin surface layers of a core–shell structure

As demonstrated schematically with a model diagram in Fig. 15.1a, a classical composite in general consists of two main phases, namely, (i) a matrix phase and (ii) a filler. Fibers of ceramics, carbon, and polymers serve to be the best fillers for this purpose. As the filler, limited amounts (typically 2–10 wt%) of such fibers, small molecules, or particulates are admixed to and dispersed in a selective material of choice in part of the primary phase, often called a matrix phase. In case of a porous system, the pores themselves can be regarded as the fillers (Fig. 15.1b). The shape and size of such pores can be excellent material parameters to control the final hybrid composite structure and derived properties. In either class of these examples, depending upon their chemistries, one structural component mediates and supports dispersion of the other component cooperatively through each other in a thermodynamic equilibrium.

As can be seen from a model diagram in Fig. 15.1c, in view of comprising a significant fraction of the surface atoms, a tiny particle itself consists of two structural components, i.e., the central part (core) and the surface region, which does not retain the same regular distribution of atoms with the same coordination numbers as the core. Phenomenologically, such particle can be considered as a composite particle. Average structure and properties in such composite particles vary depending on the shape, size, and exchange coupling in the two components. Models (d) and (e) in this figure demonstrate formation of other two kinds of composite particles on recombination with a pore and multiple surface layers in part of a material with a kind of core–shell structure. In both the cases, the sample stores effectively large microstrain and macrostrain (such kinds of strains are described in the handbook of nanostructure materials and nanotechnology edited by Nalwa [40]) which serve to be good variables for tailoring local structure and hybrid composite properties for variable light emission, gas sensors, catalysts, and other applications.

Classification of Composites

In general, according to the chemistry, structure, properties, and applications, an engineering composite material of interest of modern science and technology can be classified into three major categories of materials as follows:

- Polymer matrix composite (PMC)
- Metal matrix composite (MMC)
- Ceramic matrix composite (CMC)

Let us describe them briefly with selective examples and characteristic features for specific applications of the interest in this investigation as follows.

Polymer matrix composites: This kind of composites consists of a polymer matrix. The fillers may be metals, alloys, or compounds which include ceramics and organic molecules. When the filler phase is of metallic nanoparticles (dispersed phase) embedded in a polymer matrix (dispersion medium) in form of small polymer molecules or thin nanolayers, the obtained composite is called a "hybrid metal–polymer composite (MPMC)." Similarly, ceramic–polymer and polymer–polymer composites have similar demands in developing newer and newer materials and technologies of the future. Evidently, it is a fast-growing area of research nowadays because of the ability of the synthetic methods to obtain a precise control over the microstructure at nanometer or subnanometer scale. Despite tremendous technology demands [38, 41], only very limited number of methods have been developed so far for preparing such specific class of nanocomposites at a large scale. In metal–polymer nanocomposites, these techniques usually consist of highly specified approaches, which can be classified as in situ and ex situ synthesis methods as follows.

In the in situ methods, the metal particles of desired specifications are generated inside a polymer by chemical decomposition (using thermolysis, photolysis, radiolysis, or other processes [39, 42]) or chemical reduction of a metallic precursor or a metal–polymer complex dissolved into a desired polymer in a liquid medium. In the ex situ approach, separately prepared nanoparticles are dispersed in a polymer in predetermined ratio, with proper stirring such as mechanical structuring, sonication, etc. Usually, the preparative scheme allows us to obtain metal particles whose reactive surfaces get passivated by a monolayer of selective organic molecules. The surface passivation inhibits particle aggregation, unwanted surface oxidation, and contamination reaction on small particles in ambient atmosphere. Passivated metal particles behave to be hydrophobic easier to mix with selective polymers. The in situ methods have advantages over ex situ methods in exploring high-quality products. Nevertheless, a rather high level of loading of fillers is possible only with an ex situ method to meet wider specifications.

Metal matrix composites: A metal matrix composite (MMC) is formed by reinforcing a soft ductile metal/alloy as a matrix by using particulates or fibers of selective ceramics or other hard materials. The reinforcement proceeds with exothermic reaction between the ingredients, i.e., a metal, alloy, or compound. In this approach, several MMCs with a wide range of hosts (such as aluminum, titanium, copper, nickel, and iron) and a second phase of particulates (such as borides, carbides, nitrides, oxides, and their combinations) are being manufactured in tailoring functionalized properties [43–47]. Stable reinforcements interbridging on stable interfaces yield in situ MMCs with excellent mechanical, electrical, and other useful properties for applications. Rigid ceramic reinforcements fixed within characteristically ductile host recombine the metallic (high ductility and bending toughness) and hard ceramic properties of high strength and modulus, leading to modulate superior mechanical strength and service temperature.

In particular, lightweight and tailorable physical properties make carbon or boron fiber-reinforced aluminum and magnesium and silicon carbide-reinforced aluminum for use in aerospace and automobiles [9, 48–50]. It includes interest of producing MMCs that incorporate nanoparticles and nanotubes for structural applications. Such inclusions trigger their physical and tribological properties further. Carbon nanotubes in a metal matrix in particular, that have much higher strength, stiffness, and electrical conductivity than many metals, can tailor such properties out of hybrid MMCs. Such materials are being explored worldwide for novel tools and structural applications in the defense, aerospace, MEMS sensors, energy devices, and automobiles.

Another open scope of exploring MMCs includes magnet technology such as choke coils, magnetic valves, transformers, step motors, magnetic refrigeration, magnetic sensors, magnetoresistance, robotics, medical diagnosis, biocatalysts for biological reactions, cancer therapy, etc. In this case, ferromagnetic metals (also CrO_2 like half-metallic ferromagnetic ceramics [13, 14, 16, 17]) and alloys should be taken in part of a matrix phase with magnetic ceramic particles (or fibrils) as the fillers. Here, special polymers with magnetic or ferroelectric properties are advised to incorporate in part of the filler (which would also serve as a binder in fabricating a specific shape from the product). As a shape former, it helps interfacing distinct magnetic and/or ferroelectric phases in a specific pattern of them in the final product in a given shape (such as disk, sheet, ring, rod, wire, and other engineering shapes) for applications. In such examples, a "spring exchange-coupled magnet" forms out of a kind of magnetic exchange interactions between a hard magnetic phase and a rather soft magnetic phase, evolving multifunctional properties in a single system of device of ferroics for specific applications [13, 14, 17].

Ceramic matrix composites: A structural modification of a ceramic phase by incorporating particulates or fibers from ceramics (of the same or different group in terms of polymorphic structure, chemistry, and other properties), metals, alloys, or polymer molecules can tailor useful functional properties out of a modified product called "ceramic matrix composite (CMC)." Many monolithic ceramics are identified as candidate materials for using as structural materials under the conditions of high temperature, wear, and chemical attack that are otherwise too severe for virgin metals and alloys. Inherent brittleness of ceramics however limits these widespread applications. This is overcome to a major extend by the CMCs because they are become rather practical with much extended flow tolerant. Self-reinforced silicon nitrides present a good example of such toughened ceramics, which were developed as early as 1970s [9]. Unfortunately, not much study is available on optical properties, which can be an indirect probe to develop the "toughness mechanics" in terms of the local structural changes in such examples.

Metal nanoparticles dispersed in a ceramic host result in a metal–ceramic nanocomposite (MCMC) or cermet. In such examples, it is found that part of the additives either segregate at the grain boundaries with a gradient concentration or precipitate as molecular or cluster size particles within the grains or at the grain boundaries [7–9]. There are reports that dispersing selective metal nanoparticles (copper, silver, gold, Mn, Er, Eu) in Al_2O_3, ZrO_2, silicon carbides, and silicon nitrides improves the ductility, tensile stress, and fracture toughness [8, 38]. In MCMCs, a wide variety of properties (e.g., magnetic, electrical, and optical properties) can be tailored due to the size effects of the nanosized dispersions of the metal and ceramic ingredients and the surface interfaces. Usually, powder metallurgical methods and chemical processes like sol–gel chemistry and coprecipitation methods have been used frequently to prepare MCMCs [7–9]. The obtained product is then sintered in a reductive atmosphere to render homogeneous dispersions of the metal nanoparticles within the ceramics. As described in a reference book edited by Kelsall et al. [51] in 2005, the CMCs can be grouped into three different types, i.e., (i) intragranular nanocomposites, (ii) intergranular nanocomposites, and (iii) nano/nano composites. In the first two kinds of the examples, the nanosized particles are dispersed mainly within the matrix grains or at the grain boundaries of the matrix, respectively.

Light-Emitting Materials and Applications

As light-emitting materials, the semiconductors and insulative ceramics reserve their place with large optical density important for data storage systems and other devices. As we pointed out above in the section of advanced ceramics, an insulative host from well-known ceramics (preferably amorphous or nanocrystalline, with and without a specific porous structure) Al_2O_3, Y_2O_3, SiO_2, ZrO_2, TiO_2, zeolites, $CaMg_2O_4$, $SrMg_2O_4$, $CaIn_2O_4$, $Y_3Al_5O_{12}$ (YAG), $Sr_2MgSi_2O_7$, or $Sr_4Al_{14}O_{25}$ offers the best light emission properties after suitable optical doping. Some non-oxide semiconductors provide similar light emission properties in visible and near-infrared regions. For example, in optical memory systems, the storage density of compact disks (CDs) based on GalA lasers (recording wavelength (RW) at 780 nm) is 0.25 Gbit in^{-2}, while the storage density of digital versatile disks (DVDs) based on GaAlInP is 2. 0 Gbit in^{-2} (RW $=$ 630–650 nm), and the storage density of high-density (HD) DVDs based on III–V nitrides is 10–20 Gbit in^{-2} (RW $=$ 430–500 nm). Short-wavelength blue lasers and ultraviolet lasers play important roles in laser printing information storage [52–56]. The shorter the wavelength of the operating laser, the higher is the information storage density of CD.

A great shortcoming of lasers based on III–V nitrides is that nitrides get oxidized in air. Oxides with room temperature light emission are thus preferred because they have higher chemical stability and heat capacity. Occurring intense light emission in ultraviolet region has renewed interest in developing oxide materials with improved optical storage density of CDs in the near future. As ideal building blocks for such devices, especially nanowires and nanotubes render efficient transport of electrons and optical exitons. Furthermore, luminescent materials can be used in cathode rays tubes, fluorescence lamp, vacuum fluorescent display devices, color plasma display

Table 15.1 Optical bandgap and dielectric constant in ZrO_2 in selected forms

Samples	E_g (eV)	ε	References
t-ZrO_2	5.78, 5.87	–	[70, 71]
m-ZrO_2	5.83, 5.42	–	[70, 72]
c-ZrO_2	6.1		[70]
ZrO_2 bulk powder	5.85	–	[73]
ZrO_2 thin films	5.26	25	[70, 74–76]
Rare-earth-doped ZrO_2 thin film	4.88–5.44	–	[77]
ZrO_2-16%Y nanocrystalline thin films	5.62	–	[78]
La doped ZrO_2 nanocrystalline thin film	5.6	–	[79]
$ZrTiO_4$ thin films	3.72	–	[76]
YSZ thin films	–	27	[70]
ZrO_2–Si thin films	4.75	32	[80, 81]
ZrO_2 nanoparticles	5.6	–	[82]
35 % ZrO_2– 65 % polypyrrole	4.2	7827	[82]

panel, and electroluminescent flat panel displays [57–60]. There is tremendous commercial demand to produce efficient and long-lasting blue light-emitting diodes and short-wavelength laser diodes. It has been demonstrated that wide bandgap oxides with short-wavelength photoluminescence emission may be used as compact disk (CD) and read heads [54–56].

In metal–oxide–semiconductor (MOS) device scaling, moderate dielectric constant (ε) oxides $SrTiO_3$, Ta_2O_5, TiO_2, Al_2O_3, ZrO_2, HfO_2, etc. have special attention as potential candidates for replacing conventional SiO_2 gate oxide [61–65]. As can be seen from Table 15.1, in this purview, ZrO_2-based dielectrics (ε ~25) [61] are considered to be one of futuristic oxides for optoelectronics and other devices with good thermal stability [66], large band offset in direct contact with a substrate such as silicon [67], and large bandgap (~5.8 eV) [68]. In view of these specifications, heavy metal oxides ZrO_2 and HfO_2 with low phonons are being studied nowadays widely in characterizing dielectric and optical properties in light of other useful properties in developing optical materials for multifunctional devices. Recent report from Cao et al. [69] reveals that ionized vacancies in ZrO_2 nanowires exhibit a strong light emission at 388 nm at room temperature. It might be useful in inducing light emission from impurity levels after optical doping with rare-earth and other species. This is described with crystal structure in small particles as follows.

Monolithic and Doped ZrO_2 of Hybrid Composites

ZrO_2 Polymorphs

Let us first examine polymorphic structures of bulk ZrO_2 in monolithic form in ambient pressure. As early as 1926, Yardley reported X-ray diffraction of ZrO_2 of baddeleyite structure [83], i.e., a monoclinic (m) crystal structure of $P2_1/c$ space

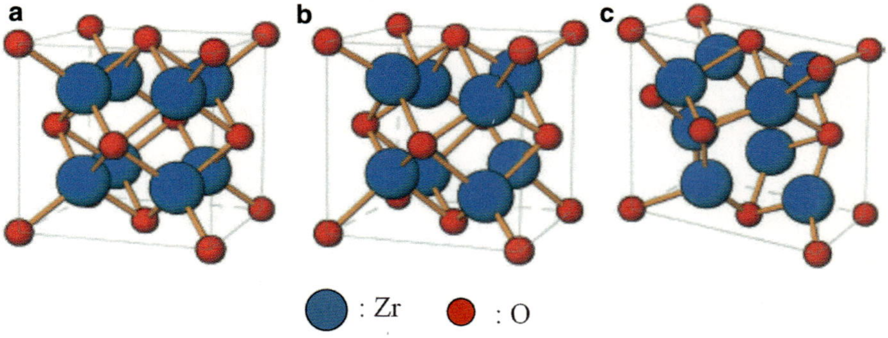

Fig. 15.2 Schematic representations of atomic distributions in (**a**) Fm3m c-ZrO_2, (**b**) $P4_2/nmc$ t-ZrO_2, and (**c**) $P2_1/c$ m-ZrO_2 [93]

group, with z = 4 ZrO_2 units. It is described as a distorted lattice of the cubic fluorite (CaF_2) structure [83, 84]. In 1929, Ruff and Ebert [85] observed that another polymorph appears of $P4_2/nmc$ tetragonal (t) crystal structure (slightly distorted from the fluorite structure) over 1200–1950 °C during heating while 1950–900 °C during the cooling [86]. In 1961, Intrater and Hurwitt [87] observed another high temperature ZrO_2 phase of an Fm3m cubic (c) structure using high temperature X-ray diffraction. Occurrence of a doublet peak in the (200) and (002) lattice reflections is the best indicator of this polymorphic transformation. A single monolithic ZrO_2 cubic phase persists only above 2285 ± 15 °C.

Thus, at ambient atmospheric pressure, pure ZrO_2 exists in three well-identified polymorphs, i.e., $P2_1/c$ m-ZrO_2, $P4_2/nmc$ t-ZrO_2, and Fm3m c-ZrO_2. Being the lowest Gibbs free energy equilibrium bulk structure at low temperatures, m-ZrO_2 is taken as the thermodynamic standard state of polymorphic zirconia and derivatives. When heating, it transforms successively to high temperature polymorphs, i.e., t-ZrO_2 at 1170 °C followed by c-ZrO_2 at 2370 °C [86–88]. These transformations are irreversible. The m → t-ZrO_2 transformation involves a large volume decrease by 3–5 %, inducing a large stress which invites cracks while cooling [91–94]. This prevents refractory properties of ZrO_2 being used in structural ceramics. Following the work of Hannink et al. [93], the atomic distribution in these phases is given in Fig. 15.2. In c-ZrO_2, the O^{2-} ions arrange in simple cubic packing with half interstices occupied by the Zr^{4+} ions. A fully ionic packing requires that the ratio of the ionic radius of the Zr^{4+} to that of O^{2-} is 0.73, which stands larger above the real value 0.59, at room temperature. Such large misfit in their ionic radii leads to destabilize this c-ZrO_2 phase in monolithic form. At low temperatures, the t- or m-ZrO_2 structure is favored on lowered coordination number on the Zr^{4+} sites.

Under pressure, m-ZrO_2 transforms to a series of orthorhombic (o) phases [94–98]. The first one (Pbca space group) starts at an applied pressure of ~3 GPa depending upon the crystallite size and is observed to exist up to ~22 GPa when another o-phase (cotunnite structure with Pnam space group) sets in as shown in

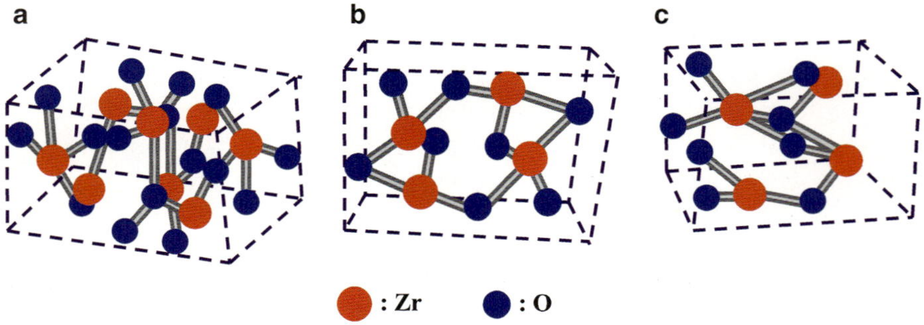

Fig. 15.3 Crystal structures of o-ZrO_2 polymorphs of (**a**) Pbca (brookite), (**b**) Pnma (cotunnite), and (**c**) $Pbc2_1$/space groups [97]

Fig. 15.3. From theoretical studies, it has been suggested that the cotunnite phase may be the hardest phase of monolithic zirconias, at least as learned from high value of the bulk modulus [96]. This phase also has the highest oxygen coordination [97]. A third high-pressure $Pbc2_1$ o-phase has been observed in some cases of partially stabilized ZrO_2 [97]. A metastable o-ZrO_2 phase ($P2_12_12_1$ space group) is reported to exist also under ambient pressure [94, 99]. It is highly unstable and converts to m-ZrO_2 on heating above 300 °C or when grinding in a mortar with a pestle. Crystal structures, lattice parameters, and density values in important ZrO_2 polymorphs are given in Table 15.2. The o-ZrO_2 crystal structures are very similar to the tetragonal one but in a higher symmetry.

Limitations with Bulk ZrO_2 Polymorphs

The different ZrO_2 polymorphs have their own intrinsic physical and chemical properties, which determine merits and utilities of them of selective applications. High hardness, high wear resistance, low coefficient of friction, high elastic modulus, chemical inertness, good ionic conductivity, low thermal conductivity, and high melting point are some of their common properties. A displacive t $\rightarrow$ m-ZrO_2 phase transformation, which encounters around 950 °C with a shear strain of ~16 and a volume expansion of ~4 % on cooling from a high temperature [93, 103], limits applications of m-ZrO_2 as refractory or other engineering ceramics. It invokes a catastrophic fracture. Some of these problems can be resolved to a major extent if replacing m-ZrO_2 by the other phases in so-called stabilized zirconia in small composite particles after adequate ceramic additives. Total phase stabilization in c-ZrO_2 suppresses this disruptive transformation, but it is generally recognized that the most useful mechanical properties arise in multiphase system, i.e., "partially stabilized zirconia (PSZ),"

It is well established that a fine-scale m-ZrO_2 precipitate in a stabilized c-ZrO_2 matrix enhances the strength of PSZ [104]. In a famous article on "ceramic steel,"

Table 15.2 A comparison of crystal structure, lattice number z, lattice parameters, lattice volume V_o, and density ρ in monolithic ZrO_2 polymorphs

Polymorphs	z	Lattice parameters (nm)		V_o (nm^3)	ρ (g cm^{-3})	Ref.
Ambient pressure:						
c-ZrO_2	4	a	0.5090	0.1319	6.21	[99]
		a	0.5042			
o-ZrO_2 ($P2_12_12_1$)	4	b	0.5092	0.1350	6.06	[94, 100]
		c	0.5257			
		a	0.3340			
o-ZrO_2 (pmnb)	4	b	0.5535	0.1177	6.96	[101]
		c	0.6364			
t-ZrO_2 ($P4_2$/nmc)	2	a	0.3596	0.0671	6.10	[102]
		c	0.5189	(0.0662)	(6.13)	
		a	0.5143			
m- ZrO_2 ($P2_1$/c)	4	b	0.5194	0.1396	5.86	[101]
		c	0.5298	(0.1405)	(5.82)	
High pressure:						
		a	0.5067			
Orth-I Pbca	8	b	0.5156	0.2590	6.32	[102]
		c	0.9915			
		a	0.5068			
Orth-III $Pbc2_1$	4	b	0.5260	0.1353	6.05	[97, 101]
		c	0.5077			
		a	0.3328			
Orth-V Pmnb	4	b	0.5565	0.1204	6.80	[99]
		c	0.6503			

V_o and ρ values in the parentheses in the ambient pressure phases are reported from Ref. [102]

Garvie et al. [105] pointed out that a dispersion of metastable t-ZrO_2 in c-ZrO_2 is also possible, and in fact it involves more powerful transformation toughness of mechanical strength. In view of a vast range of properties available in alloys, PSZ can be considered a ceramic analogue of steel. Monolithic t-ZrO_2 is known to exist at room temperature only in small particles below ~30 nm diameter. As shown in Fig. 15.4, thermal expansion exhibits no discontinuity in heating up to 1500 °C. Length changes sensitively in m-ZrO_2 present.

ZrO_2 Polymorphs in Small Particles

It is quite phenomenological that high temperature polymorphs c/t-ZrO_2 can be made to persist at room temperature in small particles and that in even monolithic state in the absence of any foreign additive in order to promoting the stabilization. This happens simply by increasing the total surface energy $\Omega = \sigma A$ in individual particles (where σ is the surface energy per unit area in the particle of surface A and

Fig. 15.4 Dilatometer traces obtained from PSZ with the secondary phase from (a) t-ZrO_2 and (b) m-ZrO_2 [105]

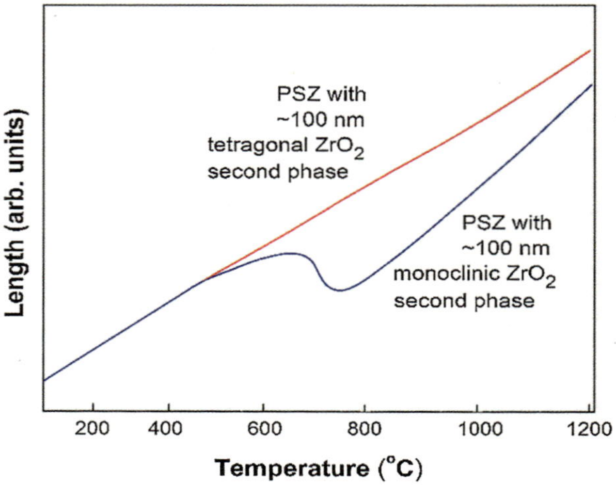

volume V) when decreasing the size (r) below a certain critical value R_c, i.e., the bulk demission. In thermodynamics, the surface-to-volume ratio minimizes in as compact particle structure as possible. A model spherical particle of radius (r) thus assumes a specific surface area

$$\mathrm{Sp} = \left(\frac{3\sigma}{\mathrm{r}}\right)\frac{1}{\rho}. \tag{15.1}$$

In given r, an equilibrium S_p-value (minimal) involves a maximal ρ-value of specific density (Eq. 15.1). As a result, a high ρ-value polymorph develops on a kind of a lattice compaction or surface hardening, i.e., a high-pressure effect in small particles, $r < R_c$. Note that $S_p \rightarrow 0$ when $r \rightarrow \infty$, i.e., the bulk value R_c. In other words, a bulk particle has negligible Ω-value in comparison to the volume Gibbs free energy G_v. Obviously, a decreasing r-value in a particle arises in a monotonically increasing G-value as follows (Eq. 15.2):

$$G = (\Delta G_v)V + \Omega \tag{15.2}$$

which meets selective values in the high temperature (also high-energy kT, with k the Boltzmann constant) ZrO_2 polymorphs. Consistent to the ρ-values in Table 15.2 in the various ZrO_2 polymorphs, it occurs with modification (decrease) in average coordination number in zirconium sites in the lattice.

In addition to decrease of r-value, another way of increasing average G-value in a sample is to involve (i) structural architect defects (which include ionic inclusions, oxygen vacancies, lattice imperfections, and lattice strain as per the shape, size, and topology of the crystal lattice) which raise average ΔG_v value, (ii) small pores which increase effective V-value, and (iii) surface coating in form of core–shell composite particles that increases average Ω-value on the additional surface interfaces. This is exactly what is obtained in a doped sample of stabilized zirconias with selective

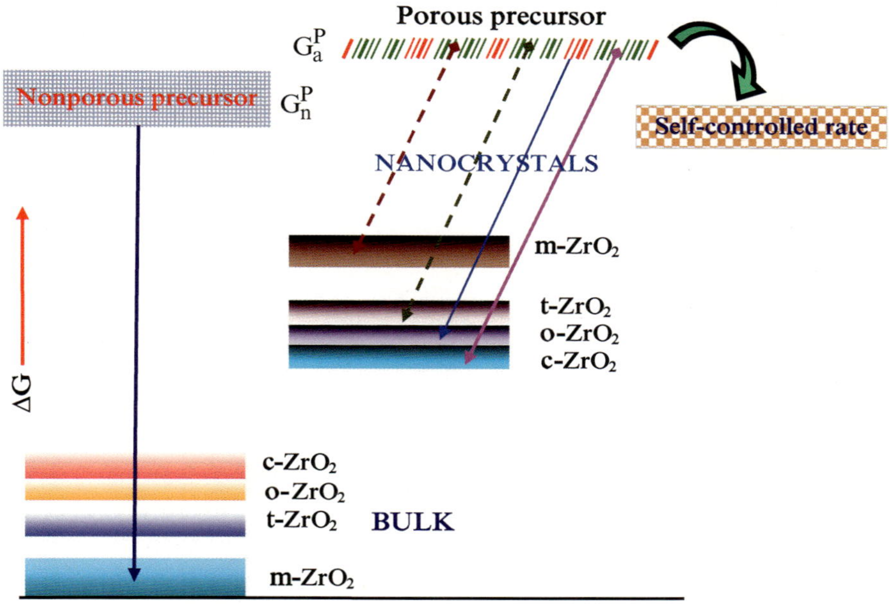

Fig. 15.5 Schematic Gibbs free energy level diagram in selective polymorphs in bulk and ZrO_2 nanoparticles. High-energy composite particles appear in the high-energy polymorphs as a better stable structure in processing from an amorphous precursor of a further higher energy value

dopants such as MgO, CaO, Y_2O_3, Y_2O_3, CeO_2, Mg_3N_2, AlN, etc. [91, 106–110]. A model energy diagram in Fig. 15.5 describes the sequence in which the various ZrO_2 polymorphs can occur in small particles and bulk particles in stabilized zirconias when increasing their average size by thermal annealing. In thermodynamics, an amorphous polymer precursor G_a^p as well as a nonpolymer precursor G_n^p has a further larger G-value as included in this figure. Note that at an effectively small processing temperature, a polymer structure of precursor (especially stabilized through small pores in an interconnected network) undergoes a self-controlled phase transformation to final ceramic particles. As we proposed earlier, it is a necessary condition in order to operate a phase transformation from a precursor to one of the metastable ZrO_2 polymorph states by releasing the excess structural energy when heating at elevated temperature in air [111]. These results help in understanding and designing the ceramic processes of formation and existence of selective polymorphs out of stabilized ZrO_2 of hybrid composite particles with adequate doping or surface coating.

Stabilized ZrO_2 with Small Inclusions

In general, a suitable ceramic additive to a sample of ZrO_2 particles through an in situ chemical method controls the size and morphology of surface-modified ZrO_2 in a high-energy hybrid composite microstructure. Different kinds of inclusions are

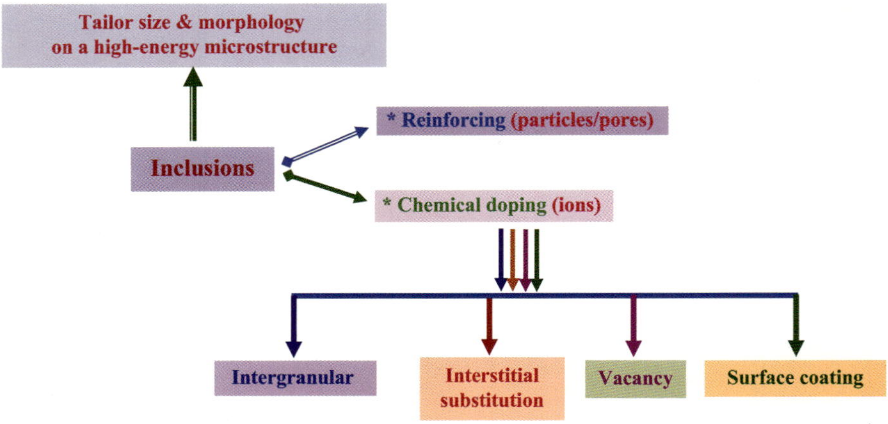

Fig. 15.6 Classification of four major groups of inclusions explored in stabilizing selective ZrO_2 polymorphs to exist at room temperature

classified in Fig. 15.6. According to ionic size and chemistry, such inclusions can be broadly classified in three different groups as follows: an oversized or undersized dopant (or inclusion) in general occupies the positions in grain surfaces or grain boundaries of primary phase. In this case, it is called an intergranular dopant or inclusion. In another case, it favorably occupies the interstitial sites in the final lattice, and hence it is called an interstitial inclusion or dopant. In general, a primarily intergranular doping is more effective in stabilizing a high-energy phase of c- or t-ZrO_2. The third possibility is that a major part of the additive is utilized in coating individual ZrO_2 particles in the form of a thin adhesive layer. A coating material of interest in this case may be a metal, polymer, or a ceramic gel. It is important especially to design camouflage in part of surface-enhanced optical properties of selective colors in the surface species. As mentioned above, vacancies and tiny pores serve another kind of inclusions useful to stabilizing specific ZrO_2 polymorphs with tailored color centers and other useful properties.

For example, Lee et al. [112] studied the influence of the size, distribution, and interparticle distances of Al_2O_3 inclusions (under sized) on the grain boundary conduction in 15 mol% CaO-stabilized ZrO_2. The Al_2O_3 inclusions (1 mol%) had gradually dissolved into grain boundaries on heating at 1600 °C for 2 h. They play two different roles of (i) improving grain boundary conduction, at temperature below 1500 °C by scavenging siliceous phase, while (ii) deteriorating it by forming a more ion blocking intergranular phase at temperatures above 1550 °C. As shown in Fig. 15.7, a modified ZrO_2 in this example undergoes a phase transformation process. The c $\rightarrow$ t-ZrO_2 transformation is displacive and of non-martensitic type [113, 114], while the t $\rightarrow$ m-ZrO_2 is displacive and martensitic in nature [93, 95, 106]. The transformation type retains in other kinds of the inclusions. Three forms of stabilized ZrO_2 can be achieved. Generally, the addition of more than 16 mol% MgO, or 8 mol% Y_2O_3, into ZrO_2 is needed to form a fully stabilized c-ZrO_2

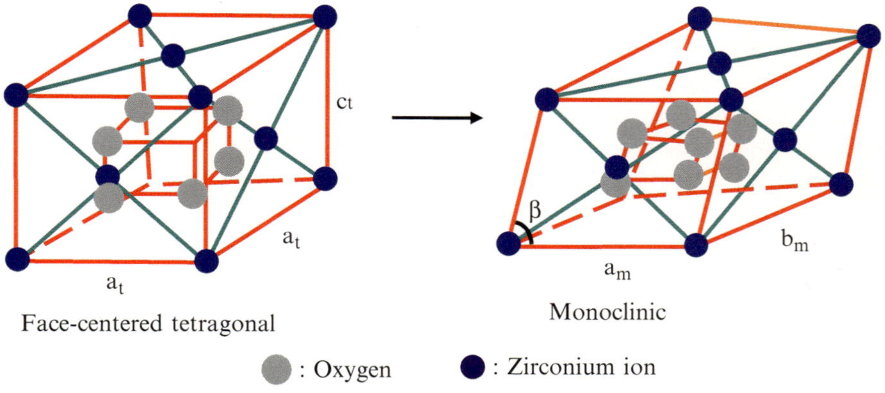

Fig. 15.7 A model t → m-ZrO_2 phase transformation

[115]. A partially stabilized ZrO_2, sometimes also called "t-ZrO_2 polycrystals (TZP)," occurs with fine t-ZrO_2 crystallites embedded in c-ZrO_2 matrix [116, 117]. Usually, in this example, such PSZ consists of larger than 8 mol% of MgO, 8 mol% of CaO, or 3–4 mol% of Y_2O_3 in form of a transformation-toughened material of "ceramic steel." Understanding and modeling the stabilization mechanisms of metastable ZrO_2 polymorphs with selective inclusions at moderate temperatures is still a subject of debate. Accepted processes proposed by different groups of research are described as follows.

Influence of Particle Size and Surface Energy

In Eq. 15.1, the validity of a smaller S_p-value in a metastable ZrO_2 phase relative to the equilibrium m-ZrO_2 phase implies that the metastable phase must be of a relatively smaller σ-value. Garvie [118] proposed that t-ZrO_2 in order to occur over m-ZrO_2 at low temperature requires satisfying the relation (Eq. 15.3)

$$(G_t\text{-}G_m) + A^t\sigma_t\text{-}A^m\sigma_m \leq 0. \tag{15.3}$$

Here, G_t and G_m are the molar Gibbs free energies, and A^t and A^m are the surface areas in crystallites in the two t- and m-ZrO_2 phases, respectively. An analysis of the variation of the total surface and Gibbs free energy as a function of size in model spherical particles predicts that the t-ZrO_2 phase would exist only below 5 nm of crystallite size (D). Experimental $\sigma_t = 0.77\ \mathrm{J\ m^{-2}}$ and $\sigma_m = 1.13\ \mathrm{J\ m^{-2}}$ values are used in the respective phases [119, 120]. Bailey et al. [121] tested the validity of this conjuncture by milling m-ZrO_2 with the observation that it undergoes a progressive polymorphic transformation to t-ZrO_2. Murase and Kato [122] have studied the influence of water vapor atmosphere in the t → m- ZrO_2 transformation, and that favors the *D*-value and reduces the critical size D_c of the transformation. The effects are explained in terms of the reduction in the difference in σ-values in the two phases. Al^{3+}-stabilized c-ZrO_2 (5–30 mol%), prepared by hydrazine method, encounters due transformation c → t-ZrO_2 at 1000 °C [107]. This can be analyzed using the thermodynamic parameters as follows.

The Gibbs free energy change in forming an ideal crystallite (spherical shape and free from any constrain) in a specific polymorph i following an amorphous precursor can be written in a simple form as (Eq. 15.4)

$$\Delta G^i = \frac{4}{3}\pi r^3(\Delta G_v) + 4\pi r^2\sigma, \tag{15.4}$$

in which ΔG_v is the Gibbs free energy of formation of the crystallite of infinite size. The difference in the values $\Delta G^{t\text{-}m}$ in two polymorphs t- and m-ZrO_2 is expressed as (Eq. 15.5)

$$\Delta G^{t\text{-}m} = \frac{4}{3}\pi r^3(\Delta G_v) + 4\pi r^2(\sigma_t - \sigma_m). \tag{15.5}$$

Assuming the fact that $\sigma_t < \sigma_m$ in relation (15.3) allows to reduce the $2r$-value to the critical value D_c at which $\Delta G^{t\text{-}m} \to 0$ at some temperature below the normal transformation temperature T_b, i.e., the t-ZrO_2 phase regime ($\Delta G^{t\text{-}m} \leq 0$). At $\Delta G^{t\text{-}m} = 0$, a D_c-value can be rewritten in this approximation as follows (Eq. 15.6):

$$D_c = \frac{6(\sigma_t - \sigma_m)}{\Delta H^{t\text{-}m}(1 - T/T_b)}, \tag{15.6}$$

where $\Delta H^{t\text{-}m}$ is the heat of t → m-ZrO_2 transformation of an infinite crystal. Using $\sigma_t = 0.77\ \text{J m}^{-2}$, $\sigma_m = 1.13\ \text{J m}^{-2}$, $\Delta H^{t\text{-}m} = -2.82 \times 10^8\ \text{J m}^{-3}$ (determined from calorimetry), T = 27 °C and T_b = 1175 °C, Garvie had estimated a D_c = 10 nm value. First principle study of ZrO_2 surfaces provides that the (111) and (111) facets are the most stable surfaces in these two t-/m-ZrO_2 phases, respectively, and also that the σ-value is anisotropic [123]. This is well supported with HRTEM images in these two phases [124].

Influence of Structural Similarities

Several authors attribute t-ZrO_2 metastability to structural similarities between the precursor amorphous phase used and the t-ZrO_2 phase which forms when heating the precursor at an elevated temperature [125–127]. X-ray or neutron diffraction analysis of amorphous ZrO_2 gives similar Zr–Zr and Zr–O interatomic distances as those in the t-ZrO_2 structure [126]. According to it, as also proposed by Tani et al. [126], at moderate temperature, a kind of a topotactic t-ZrO_2 crystallization occurs in support over a template of amorphous ZrO_2 precursor, forming self-stabilized t-ZrO_2 of small crystallites. Mondal and Ram [101] prepared a metastable o-ZrO_2 (D = 15 nm) from a mesoporous precursor $ZrO(OH)_2{\cdot}xH_2O$. No other additives except mesopores have been used in part of foreign inclusions. It is a modified skeleton which appears on slowly relieving H_2O molecules by heating the precursor at effectively low temperature such as 500 °C for 2 h in air.

Stabilization by Lattice Strain

It is suggested that high angle domain boundaries (which impose a lattice strain on the lattice enclosed by the boundaries) inhibit t → m-ZrO_2 transition over extended temperatures [111, 128]. Existence of an active nucleation site and mobility of grain boundaries are strongly controlled in the presence of inclusions from pores, dopants, and particles of a secondary phase. The concept of the stabilization by internal pressure has been used extensively for the stabilization of high-pressure ZrO_2 phases of nanocrystallites [129–131]. Highly curled surfaces of ZrO_2 particles that occur in support with tiny pores in a specific mesoporous structure involve a similar kind of internal pressure. It is a novel structural parameter to design monolithic ZrO_2 in high-energy polymorphs for selective applications such as oxygen conductors, gas sensors, catalysts, biomaterials, and optical materials.

Stabilization by Oxygen Vacancies

The most straightforward mode of a dopant inclusion into a ZrO_2 lattice is in the form of a substitutional ion at the Zr^{4+} sites with compensating oxygen vacancies. Such process of dopant substitution can be represented by the following defect Eqs. 15.7 and 15.8 [132]

$$MO + Zr_{Zr}^{x} + O_{O}^{x} \rightarrow M_{Zr}^{''} + V_{O}^{++} + ZrO_2 \text{ and} \tag{15.7}$$

$$M_2O_3 + 2Zr_{Zr}^{x} + O_{O}^{x} \rightarrow 2M_{Zr}^{'} + V_{O}^{++} + 2ZrO_2, \tag{15.8}$$

where $M_{Zr}^{''}$ and $M_{Zr}^{'}$ represent di- and trivalent dopant substitutions. Osendi et al. [129] postulated that the initial nucleation of m-ZrO_2 is favored by anionic vacancies with trapped electrons. Srinivasan et al. [133] developed m-ZrO_2 particles of much smaller D-value (10–20 nm). This is against the concept of the t-ZrO_2 phase stabilization in smaller particles on lowered σ_t value. According to them, anionic O^{2-} vacancies present on the surface control the t → m-ZrO_2 transformation on cooling and that an oxygen adsorption triggers t-ZrO_2 formation. The t-ZrO_2 stability can be described by the interaction between neighboring O^{2-} species rather than the covalency of Zr–O bonds [93, 134].

In general, two conditions are thought to be necessary for stabilizing c-ZrO_2. Firstly, inclusion of a cation that has a larger ionic radius r_i, such as Ce^{4+} ($r_i = 0.097$ nm) or U^{4+} ($r_i = 0.100$ nm) than the Zr^{4+} [126, 135, 136]. Such doping leads the c-ZrO_2 lattice to expand with an increased ratio in their r_i values. Secondly, a more notable cubic-stabilizing effect occurs from doping with cations of lower valency such as rare earths than that of Zr^{4+} cations [137–139]. This involves O^{2-} vacancies to maintain the charge neutrality. Ho [140] pointed out that c-ZrO_2 stabilized with CaO or Y_2O_3 involves decreased coordination number on Zr^{4+} sites. Li et al. [141] analyzed crystal structure of Y^{3+}-stabilized ZrO_2 with extended X-ray absorption fine structure spectroscopy and confirmed that the O^{2-} vacancies are created at the nearest site to Zr^{4+} rather than Y^{3+}. This supports that the stabilization occurs on decrease of the coordination number.

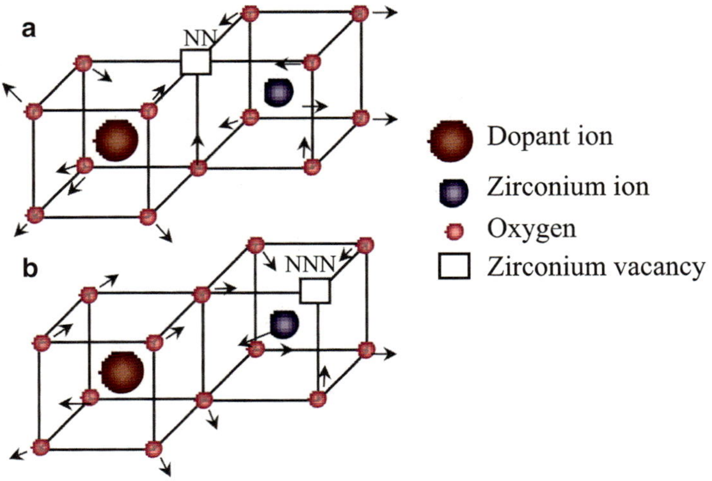

Fig. 15.8 Displacement of ions: (**a**) an O^{2-} vacancy at NN site to the dopant and (**b**) an O^{2-} vacancy at a NNN site to the dopant [132]

As shown in Fig. 15.8, a set of clusters involves an O^{2-} vacancy located at either a nearest-neighbor (NN) or a next-nearest-neighbor (NNN) site to the dopant ion. In general, oversized dopants are more strongly bound to an O^{2-} vacancy at the NNN site, whereas the smaller trivalent metal ions generally favor the cluster with a NN vacancy. The bigger dopants prefer to be eightfold coordinated and thus impose a cubic symmetry on the surrounding anion sublattice. This is one of the reasons why the bigger dopants act as good stabilizer for the c-ZrO_2 phase. Chemical factors such as adsorbed atoms and purity of raw materials play important roles in the stability of small ZrO_2 crystallites, and they may completely alter the original surface energy consideration. As a matter of fact, apparently, none of the above proposed mechanisms of stabilization seem to obey a general validity especially in monolithic ZrO_2 of small crystallites.

Source of ZrO_2 and Composites

Natural Source of ZrO_2 and Derivatives

Commercial ZrO_2 is extracted from two major ores (i) baddeleyite (ZrO_2) and (ii) zircon ($ZrO_2 \cdot SiO_2$). Both are found abundant in India, Australia, the USA, Brazil, and South Africa. Beneficiation process involves separating away undesirable components, e.g., silica in the case from zircon and low-level impurities such as titanium and iron from baddeleyite. Breaking down refined ores through a reaction with sodium hydroxide is the most versatile method of extracting pure ZrO_2 [142, 143]. Well-known methods of producing pure ZrO_2 can be briefed as follows.

Chlorination of Zircon

Refined zircon ($ZrO_2 \cdot SiO_2$) is heated with carbon at 800–1200 °C in the presence of Cl_2 gas in the simple chemical reactions (Eq. 15.9),

$$ZrO_2 \cdot SiO_2 + 4C + 4Cl_2 \xrightarrow{800-1200^\circ C} ZrCl_4 + SiCl_4 + 4CO \uparrow . \tag{15.9}$$

After chlorination, the product $ZrCl_4$ is separated by liquid–liquid extraction process and then allowed to react with H_2O (Eqs. 15.10 and 15.11) [143, 144],

$$ZrCl_4 + 9H_2O \rightarrow ZrOCl_2 \cdot 8H_2O + 2HCl \tag{15.10}$$

$$ZrOCl_2 \cdot 8H_2O \rightarrow ZrO_2 + 2HCl + 7H_2O . \tag{15.11}$$

By-product hydrochloric acid goes away into the atmosphere along with water vapor.

Alkali Reaction of Zircon

In this process, $ZrO_2 \cdot SiO_2$ (ore) is transformed into sodium zirconate and sodium silicate (frit) by a chemical reaction with NaOH or Na_2CO_3 as follows (Eqs. 15.12 and 15.13) [143, 145]:

$$ZrO_2 \cdot SiO_2 + 4NaOH \rightarrow NaZrO_3 + Na_2SiO_3 + 2H_2O \tag{15.12}$$

$$ZrO_2 \cdot SiO_2 + 2Na_2CO_3 \rightarrow Na_2ZrO_3 + Na_2SiO_3 + 2CO_2 \uparrow . \tag{15.13}$$

In hot water used in these reactions, the frit is leached out leaving behind the zirconate, which is then hydrolyzed to hydrated ZrO_2 on reaction in the water (Eq. 15.14)

$$\begin{array}{c} Na_2ZrO_3 + nH_2O \rightarrow ZrO_2 \cdot (n-1)H_2O + 2NaOH \\ \downarrow Calcination \\ ZrO_2 + (n-1)H_2O. \end{array} \tag{15.14}$$

Thermal Dissociation of Zircon

Thermal decomposition of zircon is the simplest way of producing bulk zirconia. In open air, it starts at temperature as early as 1800 °C. The process completes at 2126 °C, resulting in a pure ZrO_2 (Eq. 15.15) [143, 146]

$$ZrO_2 \cdot SiO_2 \rightarrow ZrO_2 + SiO_2 \uparrow . \tag{15.15}$$

Usually, the by-product SiO_2 is separated as a slag. It is also possible to recover in a value-added by-product such as oxide glasses. Using boron oxide during the process may help to separate ZrO_2 out of a glass by-product at moderate temperature.

Method of Synthesis of Zirconia and Composites

A number of synthetic methods have been developed within the last three decades in an attempt to obtain ZrO_2 ceramics of controlled particle size and/or morphology.

Selective ZrO_2 polymorphs of functional properties for pertinent applications are derived from calcinations of inorganic or organometallic precursors, usually zirconium hydroxide $Zr(OH)_4 \cdot xH_2O$, oxyhydroxide $ZrO(OH)_2 \cdot xH_2O$, zirconium alkoxides, or derived polymer precursors at an elevated temperature 300–200 °C in air. In general, the following methods are used for the synthesis of monolithic ZrO_2 or derived ceramic composites;

- Mechanochemical process
- Vapor phase reactions
- Hydrothermal method
- Sol–gel method
- Pechini or Pechini method
- Combustion or autoignition method
- Spray pyrolysis method
- Precipitation method
- Emulsion precipitation method
- Chemical reduction method

Each of these methods has its own merit in designing the product in specific forms and properties. A chemical route which operates with a solution leads to form an amorphous precursor. Formation of a gel-like sample inhibits any phase separation during the processing. When heating a recovered precursor in air, at crystallization temperature with reaction species from the precursor, diffusional constraints promote resulting product to pass through series of metastability and intermediate microstructures. The initial precursor structure and thermodynamics of the reaction species render the kinetics of the process in a specific structure following the final temperature and other experimental conditions. Specific methods can be briefed one by one as follows.

Mechanochemical Process

In this process, a chemical precursor is subjected to mechanical milling in a closed container. Milling induces chemical reaction on refined species from the precursor. If necessary, subsequent heat treatment is given to complete the reaction at low temperature, forming refined nanocomposite ceramic particles embedded within a salt matrix [147, 148]. Sometimes, the ceramic part is recovered by a simple washing procedure, and then heating at an elevated temperature leads to transform desired ZrO_2 polymorph. Using such processing, Y_2O_3-, CaO-, and MgO-stabilized c-ZrO_2 (D = 10–20 nm) had been prepared successfully [149]. In another experiment, 5 mol% Cr^{3+}/Cr^{4+} easily stabilized c-ZrO_2 ($D \sim 18$ nm) [150] on milling (20 h) dispersed CrO_3 in amorphous $ZrO(OH)_2 \cdot xH_2O$ followed by heating at 500 °C for 2 h in air. Inclusion of impurities, sample agglomeration, and long milling period are major demerits in this process.

Vapor Phase Reactions

This process deals with dissociation vaporization of precursor followed by vigorous quenching of the vapors onto a cold metal surface (template) in a refined deposit of the sample, which is removed periodically from the deposition zone to avoid further aggregation and coalescence in the particles. Generally, the rate and the temperature of decomposition determine the reaction kinetics and rates at which the resulting product can crystallize on the templates [101, 110, 151]. $D \sim 5$ nm ZrO_2 nanoparticles appeared by oxidation of zirconium tertbutoxide in O_2 gas at 1000 °C. In gas condensation method, zirconium monoxide is evaporated in a helium atmosphere (250–1500 Pa pressure) and then oxidized of deposited ZrO_2 particles ($D = 4$–8 nm). Xia et al. [109] prepared t-ZrO_2 (near-spherical particles, $D = 5.8$ nm) by vapor phase hydrolysis of $ZrCl_4$ and H_2O at 320 °C and 530 °C in a carrier gas N_2 and dry air.

A limitation with this method is that it applies only to selective precursors that easily disperse and polycondense at desired temperature. It is not easy to obtain a high purity and high consolidation level of the product. Monolithic ZrO_2 can be prepared in this method easily, but selective phase stabilization is not so easy to achieve [152]. It is too expensive process for commercial applications. Inert gas condensation [153], sputtering [154], laser ablation [155], and plasma spraying [156] include important vapor phase reaction methods. Chemical vapor reactions are receiving special attention in producing high-grade zirconias of selective phases in electronic industries [157].

Hydrothermal

In hydrothermal process, a precursor of reactants – often metal salts, oxides, hydroxides, or metal powders in a solution or a suspension in a solvent such as water – is heated at specific temperature and pressure in a sealed vessel. This process successfully yields nanocrystalline ZrO_2 from $ZrOCl_2 \cdot 8H_2O$ [158], zirconium alkoxides [91], zirconium hydroxide [159], or hydrous ZrO_2 [160] in acidic [91, 161] or alkaline solutions [158, 159]. Alkaline medium is preferred in preparing stabilized ZrO_2. Y^{3+}-stabilized ZrO_2 was prepared with $D \sim 10$ nm and 170–220 $m^2\ g^{-1}$ surface area in such medium (pH ~ 14) [161]. Monolithic m-ZrO_2 (D = 12 nm) grows in (111) direction on refluxing an acidic 1 M $ZrCl_4$ solution at 1–2.5 pH [100]. D = 6 nm m-ZrO_2 forms on heating a $Zr(NO_3)_4$ solution in HNO_3 at 150 °C for two days [162]. Komarneni et al. [163, 164] proposed that using microwaves during hydrothermal synthesis promotes the kinetics of crystallization by 1–2 orders of magnitude. A t-ZrO_2 nanopowder has been obtained by adding NaOH to an aqueous $ZrOCl_2$ in this process [123]. Well-crystallized 3 mol% Y_2O_3-stabilized c-/t-ZrO_2, D = 11.6 nm and surface area 100 $m^2\ g^{-1}$, has been obtained by hydrothermal precipitation at 220 °C under 7 MPa for 5 h [160]. A pure t-ZrO_2 turns up on heating at 800 °C in air.

Other similar processes used in preparing ZrO_2 powders include hydrothermal decomposition, hydrothermal oxidation, hydrothermal electrical method, hydrothermal electrochemical method, hydrothermal reactive electrode submerged arc method, etc. [165–168]. They offer many advantages: (i) crystallization temperature is lower than that in the heat treatment process, (ii) particle agglomeration is suppressed because the sample forms under high pressure, (iii) as-received product prior to calcinations or milling may guarantee a high-quality sample, and (iv) it allows preparing nanosized particles of selective size and shape. However, this is a slow process limited to only small-scale production. In case the sample forms at or below the boiling points of water, open reaction vessels can be used at economic cost and safety purpose.

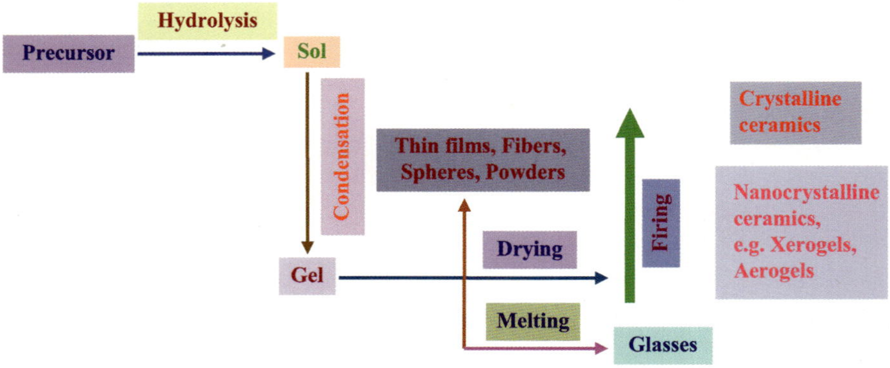

Fig. 15.9 General proceeding in the sol–gel synthesis of ceramic materials

Sol–Gel Method

In conventional sol–gel method, metal ions are dispersed and bonded in a specific amorphous structure with the help of a polymeric medium of interconnected network. A metal compound(s) is taken in form of a liquid solution in a proper solvent or a suspension of very tiny particles in a liquid (referred to "sol"), in part of a starting precursor. After proper mixing, chemical treatment, aging, and polycondensation, a viscous gel forms. As briefed in Fig. 15.9, general procedure proceeds in steps as follows:

(a) Synthesis of a homogeneous precursor solution by dissolving either (i) a requisite metallorganic compound(s) in an organic solvent that is miscible with water or (ii) an inorganic salt(s) in water as per the final ceramic composition
(b) Forming a sol from the precursor solution with a chemical treatment (if necessary) by using a suitable reagent of acidic or basic nature
(c) Aging of the sample during which the sol changes into a gel by polycondensation or self-polymerization process as per the sample chemistry
(d) Shaping the gel in a desired form such as thin films, fibers, spheres, etc.
(e) Firing the ceramic gel in specific conditions in air (or any other desired atmosphere) in producing the ceramic product of amorphous or crystalline state

Controlled hydrolysis of zirconium alkoxide, e.g., isopropoxide, propoxide, butoxide, etc., is often used in preparing ultrafine ZrO_2 ceramics. A gel $ZrO(OH)_2 \cdot xH_2O$ hydrolyzed from $ZrOCl_2$ with ethylene oxide C_2H_4O yields m-ZrO_2 ($D \sim 12$ nm) after firing at temperature as low as 600 °C, i.e., sufficient to burn out the by-products [169]. ZrO_2-Al_2O_3 solid solutions (m-, t-, and c-ZrO_2 phases) have been prepared by co-hydrolysis of aluminum tri-sec-butoxide and zirconium propoxide in 2-propanol (at ~7 pH) [170]. 3 mol% Y^{3+}-doped ZrO_2-Al_2O_3 (0–40 mol%) nanopowders have been prepared successfully [171]. α-Al_2O_3 appears along with t-ZrO_2 at 1000 °C. Stabilized t-ZrO_2 (up to 73 wt%) is obtained by co-doping 2 mol% Sm_2O_3 or Tb_2O_3 by annealing at 1000 °C [172].

The sol–gel method is highly capable of fabricating coatings. Major advantages include a simple process, an excellent control of stoichiometry, a maintenance of homogeneous film over a large area, a relatively low cost method, and flexible deposition parameters [173–177]. High-quality films are being prepared of nano-crystalline rare-earth-stabilized ZrO_2 [178–180]. Sol–gel method makes possible to design shape of such ceramics without a powder processing. Nevertheless, it involves an organic precursor, which is highly moisture sensitive and hydrolyzes so quickly that it is difficult to check the same at the "sol" formation, and a phase separation encounters immediately in some metal oxides. In this situation, a chelating ligand used to facile the replacement of the alkoxide groups, bonded to the cations, slows down the hydrolysis process undesirably to the stage of "sol" formation. Chelating ligand invites high cost, a long processing time, and a large shrinkage with possibility of forming hard agglomerates.

Pechini or Pechini Method

This method was developed long ago in 1967 by Pechini in preparing titanates and niobates for capacitors [181]. After slight modification it is also referred to a citrate gel process or an amorphous citrate process. Metal cations from starting materials, e.g., carbonates, nitrates, chlorides, or alkoxides, are complexed using α-hydroxycarboxylic acids such as citric acids in aqueous medium [182]. A transparent gel turns up, on heating with a polyhydroxy alcohol such as ethylene glycol, which is then subjected to autocombustion (in solid form) to burn out organic species. Resin as a template renders microstructure and properties of resulting ceramic particles.

This method is capable of engineering good chemical homogeneity of so obtained product. It is applied in preparing Al_2O_3-, CaO-, MgO-, CeO_2-, Y_2O_3-, Gd_2O_3-, or Sc_2O_3-stabilized c-/t-ZrO_2 [182–187]. δ-Al_2O_3 precipitates were shown in Al_2O_3-modified t-ZrO_2 at temperatures below 1000 °C. Citric acid was added to aqueous solution in $ZrOCl_2 \cdot 8H_2O$ and $AlCl_3 \cdot 6H_2O$. The usage of water modifies this method over the Pechini-type process in which ethylene glycol is used as the solvent. A partial t → m-ZrO_2 transformation encounters with α-Al_2O_3 at temperature as early as 1200 °C in air. Substantially stabile Sc^{3+}:c-ZrO_2 over 1000–1400 °C has been synthesized in three independent Pechini-type gel routes, viz., polyvinyl alcohol (PVA)-containing process, polyethylene glycol- and formic acid-containing process, and in situ polymerizable complex method. m-ZrO_2 impurities persist in the first route [187]. Stabilized t-/c-ZrO_2 with 10 mol% Gd_2O_3 forms from a polymeric precursor.

Combustion or Autoignition Method

The basic principle of this process is of using a redox mixture to induce a self-ignition of it at a temperature during which the reaction species from the ignition recombine instantaneously in the form of a finely divided loose ceramic powder. A gel combustion route follows gelling and subsequent combustion of an aqueous solution consisting of desired metal salts (usually nitrates which serve to be oxidizer) and some organic fuel (e.g., urea, carbohydrazide, oxalic dihydrazide, citric acid, glycine, alanine, camphor, etc.). Self-ignition propagates as soon as the sample is being evaporated to dryness following preheating near 150 °C in open air. An exothermic redox reaction between nitrate ions and the fuel drives the combustion process [188, 189]. Large volume of gases which evolve during this reaction causes the precursor to disintegrate into very tiny species and those ultimately reform only small ceramic particles. Sometimes, post-thermal annealing leads to tailor-desired microstructure and properties.

Spray Pyrolysis Method

This method is applied in synthesizing selective ceramics and composites in selective forms. It consists of (i) preparing a salt solution, (ii) removal of part of solvent from a precursor solution, and ultimately (iii) formation of aerosol droplets. A ceramic powder appears when aerosol droplets are decomposed in a stream in a reactor under specific conditions. Monolithic and doped ZrO_2 using Y_2O_3 or Al_2O_3 has been obtained from $ZrOCl_2{\cdot}8H_2O$, $Zr(NO_3)_4{\cdot}5H_2O$, or zirconium oxalate in this method [190–193]. Spraying a solution or sol onto a substrate followed by heating the specimen at elevated temperature performs the experiments. The final product consists of t-ZrO_2 and γ- or α-Al_2O_3 at temperatures up to 1400 °C [88]. t-/m-ZrO_2 forms on heating hydrolyzed Zr^{4+} sols with aluminum sulfate (0–90 mol%) at 500–1300 °C [138]. Organometallic precursors such as $Zr(C_5H_5O_2)_4$ and $Zr(n\text{-}OC_3H_7)_4$ have been deposited as ZrO_2 films on glass, alumina, stainless, or carbon steel [194, 195]. These precursors of alkoxides are relatively expensive, highly moisture sensitive, flammable, and irritant. Avoiding such technicalities, a complex $[Zr(C_2O_4)_4]^{4-}$ solution is being used to deposit ZrO_2 films [196]. Below 600 °C, c-ZrO_2 is the only phase. A partial c → m-ZrO_2 conversion encounters over 700–1000 °C. This process has several advantages, e.g., relatively simple apparatus, simultaneous deposition and densification processing, and easily available precursors – not limited to volatile compounds unlike in the chemical vapor deposition.

Precipitation Method

This is a common method for separating inorganic species out of a salt or complex dissolved in a solvent such as water. A metastable compound forms on such species that separates from the solution and ultimately precipitates in insoluble product. This occurs under specific conditions of concentration, pH, and temperature. In case if it is required, two or more kinds of cationic species can be precipitated in situ, in a so-called coprecipitation method. Desired salts are dissolved in a common solution in requisite contents. Adding NH_4OH, NaOH, ethylenediamine, oxalic acid, citric acid, etc. induce the precipitation. A specimen of mixed metal hydroxides or oxyhydroxides precipitates on controlled reaction with NH_4OH. Calcining recovered precipitates (after washing out by-product species in distilled water) results in a ceramic powder.

For example, Hong et al. [197] prepared Al_2O_3–ZrO_2 composites via a reaction of chlorides with NH_4OH. On calcining, the coprecipitate transforms to γ-Al_2O_3 with amorphous ZrO_2 over 400–700 °C. c-/t-ZrO_2 develops along with γ → δ → θ-Al_2O_3 transformations in steps followed by a α-Al_2O_3 and t-/m-ZrO_2 solid solution at 800–1500 °C. t-ZrO_2 forms with a small amount of δ-Al_2O_3 when heating a coprecipitate of hydroxides, with Al_2O_3 up to 40 mol%, at 600–1000 °C in air [198]. Vasylkiv et al. [199, 200] developed Y^{3+}-stabilized ZrO_2 by a coprecipitated oxalate. Gutzov et al. [201] observed incipient m-ZrO_2 in 1 mol% Eu_2O_3:t-ZrO_2 at 350–1000 °C. A single t-ZrO_2 phase appears only on as large Eu_2O_3 as 9.4 mol% [202]. Ishida et al. [107] explored a new method using hydrazine and reported c-ZrO_2 solid solution with Al_2O_3 up to 40 mol% at 900 °C or lower temperatures. An independent Al_2O_3 grows at 1200 °C [107].

This widely used method of preparing most of ceramics offers a high yield in small period of the processing adequate for mass-scale production in industries. Sometimes, part of some precipitates dissolve back in the mother solution. As a result, only selective ceramics can be prepared and those on merely limited compositions. It is hard to keep homogeneous dispersion in a multicomponent precursor.

Microemulsion Precipitation Method

This method of forming ceramic particles such as ZrO_2 is used so that (i) a hydroxide formation and water removal are done under controlled conditions and (ii) the ceramic particles primarily formed are stabilized against aggregation by a specific combination of surface-active agents. For example, Fiona et al. [203] used water-in-oil emulsions (in preparing dense t-ZrO_2) which stand pretty stable against phase inversion with increasing temperature and different compositions of dispersed phase. Formation of zirconium hydroxide is induced by mixing two emulsions: one containing $ZrCl_4$ and the other the precipitation reagent urotropine, hexamethylenetetramine (HMTA), which decomposes into ammonia and formaldehyde near 60 °C. Surface-active compounds Arkopal 40, nonylphenoltetraethyleneglycolether, didodecyldimethylammonium bromide (DiDAB), and poly(octadecyl methacrylate) (PODMA) as a 25 % solution in toluene were used. Two emulsions were made by dispersing 1.0 g Arkopal 40 (nonionic surfactant), 1.0 g DiDAB (ionic surfactant), 60.0 g decane oil, and 40.0 g 0.02 M $ZrCl_4$ solution in water, or 0.025 M HMTA, using a specific dispersion device. After mixing two emulsions were kept at 60 °C for 15 min to decompose HMTA, leading to form zirconium hydroxide inside the emulsion droplets. Before removing the water by azeotropic distillation, 40.0 mg PODMA solution was added in stabilizing the particles. Other details of precipitation of the emulsions are well-described in an article by Sager et al. [204].

Forced Co-hydrolysis Method In this method, hydrolysis of dispersed metal cations is promoted in a hydrolyzing medium with in situ chemical reduction on part of the metal ion species used thereby of a precursor. For example, Ram [111] prepared a polymer precursor with dispersed Zr^{4+} and Cr^{3+}/Cr^{4+} species in a specific ratio by reaction of (i) $ZrO(OH)_2 \cdot xH_2O$, (ii) $(NH_4)_2Cr_2O_7$, (iii) sucrose, and (iv) PVA in water. A freshly prepared $ZrO(OH)_2 \cdot xH_2O$ by hydrolyzing $ZrO(Cl)_2 \cdot 8H_2O$ with NH_4OH in water is used. The hydrolysis reaction follows (Eq. 15.16) as

$$ZrOCl_2 + (2 - x)H_2O \rightarrow Zr(OH)_2 \cdot xH_2O + 2H^+ + 2Cl^-. \quad (15.16)$$

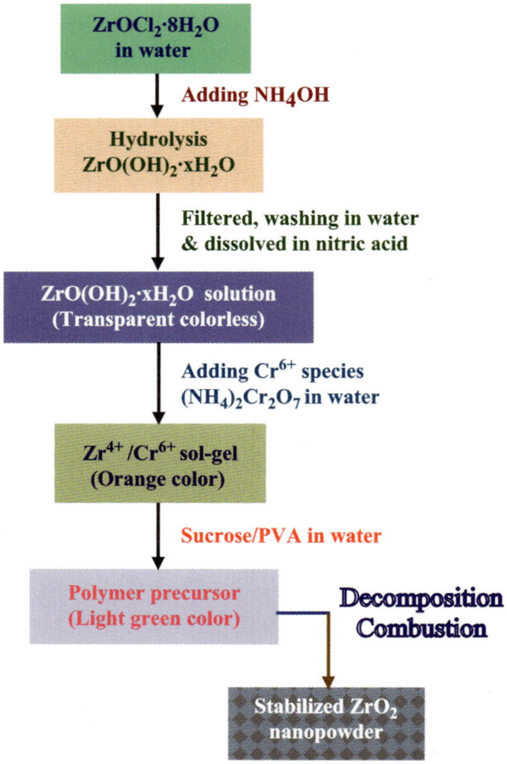

Fig. 15.10 Schematic diagram of forming a polymer precursor with metal cations in an aqueous sucrose-PVA solution and its decomposition to ZrO_2 nanopowder

According to it, both the $2H^+$ and $2Cl^-$ ion concentrations are as large as twice the initial concentration, which was ~0.5 mol dm^{-3} in $ZrO(Cl)_2 \cdot 8H_2O$ solution. This reaction proceeds only if the species react. NH_4OH thus has been added dropwise in order to probe the influence of the $2H^+$ and $2Cl^-$ ions produced by the hydrolysis on the formation processes of $ZrO(OH)_2 \cdot xH_2O$ in an amorphous structure. It reacts with HCl and forms NH_4Cl, which is soluble in water and thus goes to the solution. Average ionic strength of the solution with $ZrO(OH)_2 \cdot xH_2O$ sol or gel thus way does not increase much to govern a controlled reaction in cold water at room temperature. As the $ZrO(OH)_2$ molecules appear, they recombine one another via oxobridging in an amorphous network structure in association with H_2O molecules.

As summarized in Fig. 15.10, a recovered $ZrO(OH)_2 \cdot xH_2O$ after washing in water was pulverized and dissolved in nitric acid in a 0.5 mol dm^{-3} solution. Cr^{6+} species were added in a predetermined ratio through an aqueous $(NH_4)_2Cr_2O_7$ solution in 0.5 mol dm^{-3} concentration with a continuous magnetic stirring. Addition of sucrose and PVA (m_w 1,25,000) by 50–70 % in a batch of 25 g sample yields a polymer precursor solution – a transparent light green color. A molar ratio 13:1 is maintained in sucrose and PVA. It prevents unwanted phase separation in metal cations. The obtained solution is dried as a precursor mass (a dark black characteristic color) at 50–80 °C in air. That is pyrolyzed into a fluffy powder (a light brown to deep blue color depending upon the Cr^{3+}/Cr^{4+} contents) by heating on a hot plate at

250 °C in air. A further calcination at 600–900 °C yields a stabilized c-/t-ZrO_2 of a nanopowder. Samples with Cr^{3+}/Cr^{4+} contents up to 30 at.% thus have been obtained from an amorphous precursor.

Thermal Stability

Let us examine occurrence of successive phases of stabilized ZrO_2 in the presence of chromium ions following a metal ion–polymer complex precursor prepared from a forced hydrolysis described in above forced co-hydrolysis method. To obtain small oxide particles of few nanometers, the precursor is first heated onto a hot plate at low temperature such as 250 °C in air. At this temperature, insufficient to induce its spontaneous combustion, it slowly disintegrates by a dynamic reconstructive organization with polymer molecules in small groups or particles in a self-controlled manner. As such the ZrO_2 composite particles so obtained have thin surface layers of decomposed polymer molecules (mostly carbon in specific allotropes which include graphene) in a dark blackish color.

As demonstrated with a model in Fig. 15.11, each ZrO_2 particle consists of a core–shell structure. We believe that there is a compositional gradient on the ionic species from the interior core to the shell of a specific pattern, which is highly constrained. It helps to design and tailor thermodynamic stability, functional properties, and phase transformation of such systems of composite particles. At higher temperatures, part of the surface layer (shell) reacts with oxygen in air and goes away as oxidized gas species, leaving off modified ZrO_2 composite particles free from agglomerates with light brownish to bluish visible color as per the dispersion of the Cr^{3+}/Cr^{4+} species in a hybrid composite structure. A continuous release of part of the material in a gas from unstable surfaces (keep the particles separated apart) inhibits the stable ceramic particles to recombine and grow at a strictly controlled manner at 900 °C or lower temperatures used here.

A reconstructive molecular decomposition of porous precursor sols (Fig. 15.11a) at controlled combustion is highly effective in shaping the product ZrO_2 composite particles. This is not possible with an adiabatic combustion process. For instance,

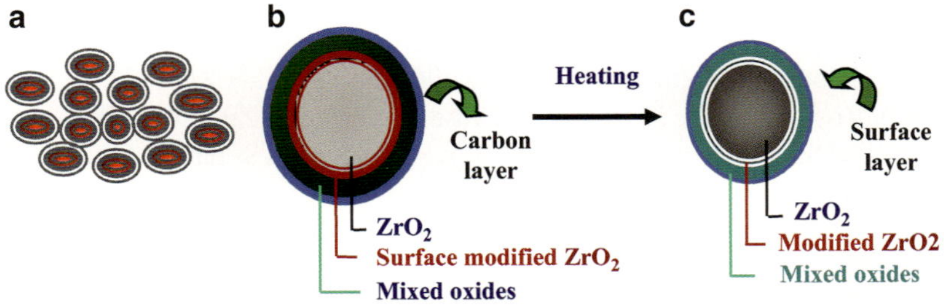

Fig. 15.11 Model structures demonstrating phase transformation of (**a**) porous precursor particles to stabilized ZrO_2 of core–shell particles upon thermal heating a metal ion–polymer complex precursor in air; (**b**) before and (**c**) after leaching out a parent carbon surface layer

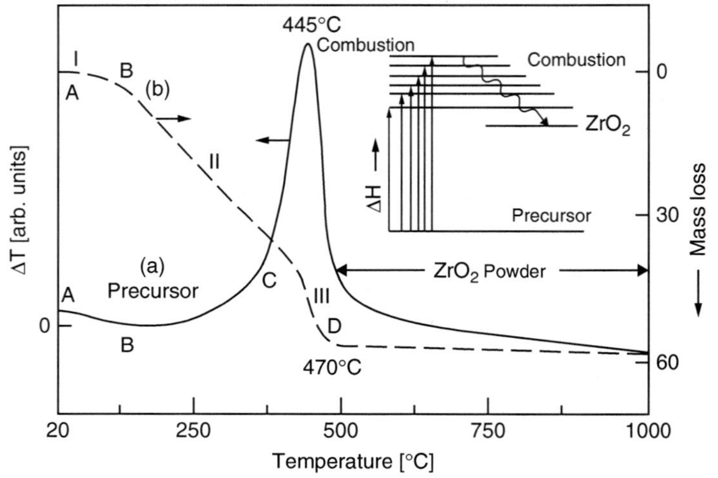

Fig. 15.12 (**a**) DTA and (**b**) TG curves in a polymer precursor (with 10 % Cr^{3+} and Cr^{4+} additives) dried over a water bath at 80 °C. The data are collected by heating the sample at 5 °C min^{-1} in air. A schematic diagram in thermal excitation of the precursor by absorption of enthalpy ΔH followed by dissociation and combustion is given in the *inset* [111]

combustion of a precursor of metal salts dispersed in oxalic dihydrazide yields only a mixture of yttria-stabilized c-/t-ZrO_2 in 1–5 μm particles [205]. No particles of so controlled size appear. In combusting a polymer precursor, the organic part (a reductant) initiates and drives a redox reaction in oxidizing the salt species. A large amount of heat and gas evolve rapidly in a very short period of time. The local heating leads the precursor to disintegrate instantaneously into smaller particles. The gas which relieves readily there from promotes precursor disintegration, but it lasts over a very short period of a few seconds only, and by the time most of the gas escapes, unable to take active part in inhibiting recombination of particles even in a fast cooling after the combustion.

Figure 15.12 shows (a) DTA and (b) TG curves in thermal decomposition and combustion in a continuous heating of 25 mg of a polymer precursor (10 at.% Cr^{3+}/Cr^{4+} content and dried at 80 °C over a water bath) at 5 °C/min in air. A molecular decomposition with a predominant combustion occurs with an exothermic peak in combustion at T_{PC} = 445 °C in the DTA with a total mass loss ΔM = 59.0 % in the TG curve. A total 22.6 % mass loss lies in $ZrO(OH)_2 \cdot H_2O \rightarrow ZrO_2 + 2H_2O$ molecular decomposition while 12.8 % in the case of anhydrite $ZrO(OH)_2$ or 22.6 % in anhydrite $Zr(OH)_4$. A hydrous zirconia, $ZrO_2 \cdot xH_2O$, x ~1.9, has 21.5 % mass loss [206]. Thus the additional value in this polymer precursor imparts primarily from the polymer matrix with the structural H_2O molecules. As marked over the TG curve (Fig. 15.12a), the mass loss in the polymer precursor proceeds in three successive steps I, II, and III between the points A, B, C, and D. Signal I, which involves a very small ΔM ~2 % value and lasts to ~150 °C, indicates desorption of part of H_2O and/or interstitial gases adsorbed during the processing. Generation of CO, CO_2, and NH_3 gases are common in processing of a polymer precursor with metal cations in an acidic

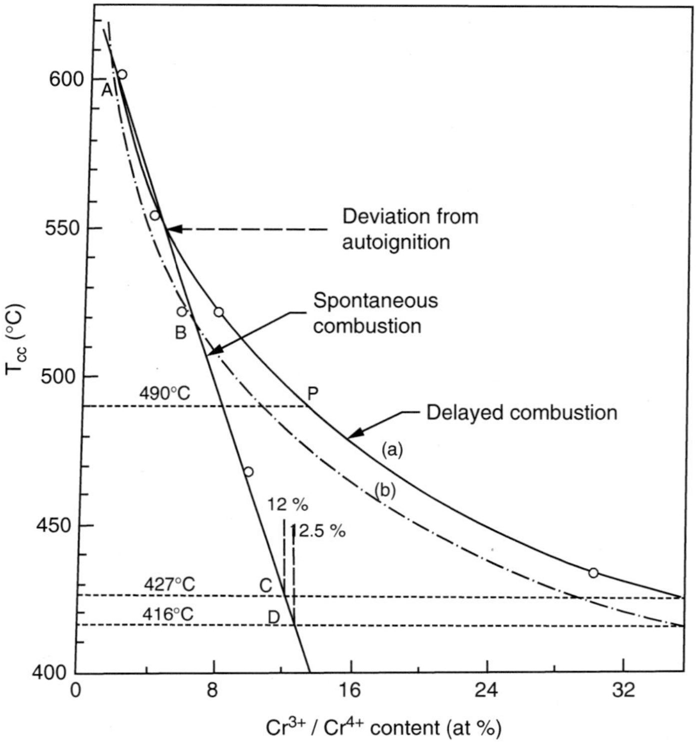

Fig. 15.13 (**a**) A plot of complete combustion (constant weight) temperature T_{CC} in TG with a function of Cr^{3+}/Cr^{4+} content in the polymer precursor. Curve (**b**) is the simulation to that in (**a**). An extrapolation of a linear plot over the early data points of ABCD intersects their tangents at points C and D at (**a**) 12.0 % and (**b**) 12.5 % of Cr^{3+}/Cr^{4+} contents [111]

medium in this example. Signal II has ΔM ~38 % and extends to 430 °C, while signal III, ΔM ~19 %, lasts to point D at T_{CC} ~ 470 °C. No further loss in mass appears in continuing the heating over higher temperatures above 470 °C in air.

In principle, an endothermic process of desorption or decomposition involves absorption of heat (or enthalpy ΔH) by excitation of the system through a series of energy levels (as shown in the inset in Fig. 15.12) in a molecular rearrangement to the limit of its thermodynamic stability from which it ultimately occurs. This is very much reflected in by desorption of the H_2O and interstitial gases in the endothermic signal I in Fig. 15.12. Also the molecular decomposition of the polymer precursor at early temperatures in signal II begins with an altogether endothermic heat output in the DTA. A monotonically increasing rate of heat output leads to a spontaneous combustion of the sample in a prominent exothermic signal III at subsequent temperatures.

As portrayed in Fig. 15.13, the value is decreasing in T_{CC} with an increasing value of Cr^{3+}/Cr^{4+} content (δ). Extrapolation of the curve to $\delta \rightarrow 0$ yields a maximal value $T^m_{CC} = 662$ °C in the virgin polymer with Zr^{4+} species. The present value is as large as 52 % of those of 385 °C and 437 °C, respectively, observed in the precursor with

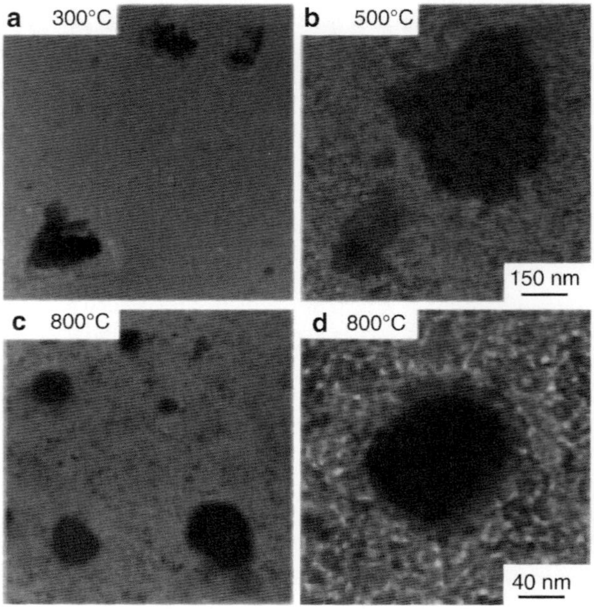

Fig. 15.14 TEM images in 10 % Ag–ZrO_2 annealed at (**a**) 300 °C and (**b**) 500 °C in air, (**c**) 800 °C in H_2 (all at the same magnification), with (**d**) part of a (**c**) close-up [209]

$\delta \sim 30$ %. The effect of δ on diminishing the T_{CC} value seems to be the most prevalent at a specific value of it around 13 %. In the T_{PC} versus δ curve (Fig. 15.4, [111]), it thus reflects in a peak in T_{PC} at 407 °C at this specific value. As marked by point P, the T_{CC} versus δ curve in Fig. 15.13 changes significantly in its slope at around this value ($T_{CC} \sim 490$ °C). In an inorganic precursor with hydrazine $(NH_2)_2 \cdot H_2O$ [207], Cr^{3+} species are rather shown to improve the temperature of the crystallization T_C into ZrO_2 nanoparticles. A large $T_C = 477$ °C value (which can be treated as our $T^m{}_{PC}$ value) thus appears at $\delta = 5$ % and that shifts further to 615 or 713 °C at $\delta = 20$ or 30 % [208].

A simulation to the experimental T_{PC} data points with a function of δ, i.e.,

$$T_{PC} = f(\delta) \tag{15.17}$$

with

$$f(\delta) = T_{PC}(0) - \alpha\delta - \beta \tag{15.18}$$

represents the results, qualitatively, at δ in the 0–0.30 range (Eqs. 15.17 and 15.18). Here, β is a parameter which governs an induced combustion in refined polymer molecules according to their thermal history developed at lower temperatures in heating as a function of temperature at a given rate. Obviously, it is operative only over a specific range of temperature (i.e., 415 to 440 °C according to the observed data [111]) over which the polymer refines in smaller units in support of the Cr^{3+}/Cr^{4+} additives in process to the spontaneous combustion. If we ignore it, for a time being, the relation (15.17) simplifies to the equation of a straight line (Eq. 15.19),

$$T_{PC} = c - \alpha\delta \tag{15.19}$$

with the constant $c = T_{PC}$ (0), the value of T_{PC} at $\delta = 0$. It reproduces the results (shown by the solid line in Fig. 15.13) with $c = 465$ °C and $\alpha = -5.3$ °C per unit value of δ. In a realistic case, β is a function of temperature in the region in question such that it passes through a maximum in superposition with the straight line function to accord with the observed cure (marked by the dots). It can be ascribed in terms of an exponential function as often used to describe a first-order phase transformation [208].

Similarly, an empirical exponential function (Eq. 15.20),

$$\mathrm{T}_{CC} = \gamma T^{\mathrm{m}}_{\mathrm{CC}}\exp - \left[\frac{\mathrm{F_{Cr}}}{\mathrm{F_{Zr}}}\right]^{\mathrm{n}} + f\left(T^{\mathrm{i}}_{\mathrm{CC}}\right) \tag{15.20}$$

with F_{Cr} and F_{Zr} as the effective fractions of the Cr^{3+}/Cr^{4+} and Zr^{4+} cations in the polymer structure, may be used to ascribe the δ-dependent T_{CC} in this example. It reproduces (Eq. 15.21)

$$T_{CC} = \gamma T^{m}_{CC} = T^{m}_{CC} \tag{15.21}$$

at $F_{\mathrm{Cr}} = 0$ with $\gamma = 1$ the ideal value in the polymer before the Cr^{3+}/Cr^{4+} additives. The second term, $f(T^{\mathrm{i}}_{\mathrm{CC}})$, in Eq. 15.20 describes a manifestation in the instantaneous T_{CC} value (which is denoted as T^{i}_{CC}) by the internal heating which initiates at a specific F_{Cr} and F_{Zr} combination. It is a complex function of T^{i}_{CC} which varies from point to point according to the heat produced during the process and the average heat capacity. Assuming it to be negligibly small relative to the first term, the above relation simplifies (Eq. 15.22) as

$$T_{CC} = \gamma T^{m}_{CC}\exp - \left[\frac{F_{Cr}}{F_{Zr}}\right]^{n}. \tag{15.22}$$

Substituting the values for $F_{\mathrm{Cr}} = \delta$ and $F_{\mathrm{Zr}} = 1 - \delta$, in the assumption that all of the Cr^{3+}/Cr^{4+} and Zr^{4+} metal cations in the specimen formed a polymer structure with the PVA and sucrose polymer molecules, relation (15.22) represents

$$T_{CC} = \gamma T^{m}_{CC}\exp - \left[\frac{\delta}{1-\delta}\right]^{n}. \tag{15.23}$$

It describes well the T_{CC} versus Cr^{3+}/Cr^{4+} content plot in the solid curve in Fig. 15.13. A best fit (the dashed curve b) to the experimental data is obtained with an empirical 0.95 value (i.e., close to the ideal value 1) for the correlation constant γ with a value for the exponent $n = ½$. A small difference in the two curves dictates the $f(T^{i}_{CC})$ contribution, especially on Cr^{3+}/Cr^{4+} contents above 10 at.%. It determines a so-called delayed combustion of the precursor at manifested temperatures. This factor is dominating in the case of an inorganic precursor with hydrazine so that the T_{CC} value increases with Cr^{3+}/Cr^{4+} content in an opposite trend. An extrapolation of a linear plot of the initial data points of ABCD intersects the tangents to the two curves a and b at points C and D at 12.0 % and 12.5 % of Cr^{3+}/Cr^{4+} contents, respectively. This straight line ABCD represents the Cr^{3+}/Cr^{4+} limited dynamics of T_{CC} in the spontaneous combustion of the precursor. A deviation from the experimental curve (a) starts in autoignition as early as ~5 % Cr^{3+}/Cr^{4+}.

The above model variations of the T_{CC} and T_{PC} values with the Cr^{3+}/Cr^{4+} contents demonstrate the fact that the metal cations impart the polymer structure of precursor and in turn influence the kinetics of (i) its combustion process and (ii) formation of stabilized ZrO_2 by reaction of its decomposed species of metal cations during the combustion. The Cr^{3+}/Cr^{4+} species behave as an internal catalyst in facilitating combustion at moderate temperature. This is a reason that they function as a stabilizer in forming stabilized ZrO_2 in small particles at moderate temperature insufficient to induce their growth further by a surface diffusion-controlled recombination reaction.

Gonella et al. [209] observed a distinct core–shell structure in 10 mol% silver-modified c-/t-ZrO_2 after annealing at 500–800 °C in air, H_2, or N_2 gas. TEM images in Fig. 15.14 shows marked variation in annealed samples at 300 °C, 500 °C, or 800 °C for 30 min in air or H_2 gas. Selected area electron diffraction (SAED) patterns in Fig. 15.15 characterize the phase evolution. A closer inspection of these images

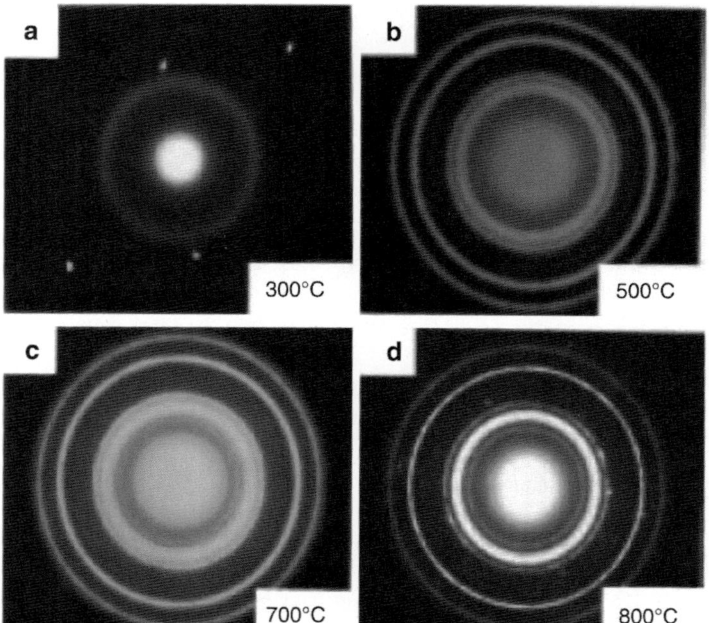

Fig. 15.15 SAED patterns of 10 % Ag–ZrO_2 annealed at (**a**) 300 °C and (**b**) 500 °C in air, (**c**) 700 °C in N_2, and (**d**) 800 °C in H_2 [209]

clearly reveals a shell (few nanometers thick) surrounding the Ag particles (Fig. 15.14d). A sample annealed at 300 °C in air is amorphous with two diffraction halos of positions matching with the values in the (111) and (220) reflections of c-ZrO_2, with some diffraction spots in part from fcc-Ag being precipitated in rather large crystallites (100–200 nm clusters are visible in TEM images in Fig. 15.14a). A weak (011) diffraction ring develops on an m-ZrO_2 phase transition especially on annealing at temperature as early as 700 °C in N_2 or 800 °C in H_2.

Absorption and Emission

Polymeric Liquids and Nanogels

Colored inorganic and/or organic polymeric liquids and nanogels consisting of zirconium ions with selective doping of metal nanoparticles, molecules, or ionic species (of transition and/or rare-earth metals) in a dielectric host, such as water, exhibit useful optical absorption and emission bands for several applications, including color pigments, microbarcodes, nanoink, optical devices, camouflage, pharmacy, biological sensors, catalysts, etc. For example, Fig. 15.16 compares absorption spectra obtained in the 200–800 nm range on Cr^{3+}/Cr^{4+}:ZrO(OH)·xH_2O gels in distilled water. In the presence of PVA polymer molecules (serve as a reductant) in water, the initial ionic state Cr^{6+} (added through CrO_3 in water) in part of the reactants (used to obtain these hydrogels) gets converted into Cr^{4+}/Cr^{3+} states,

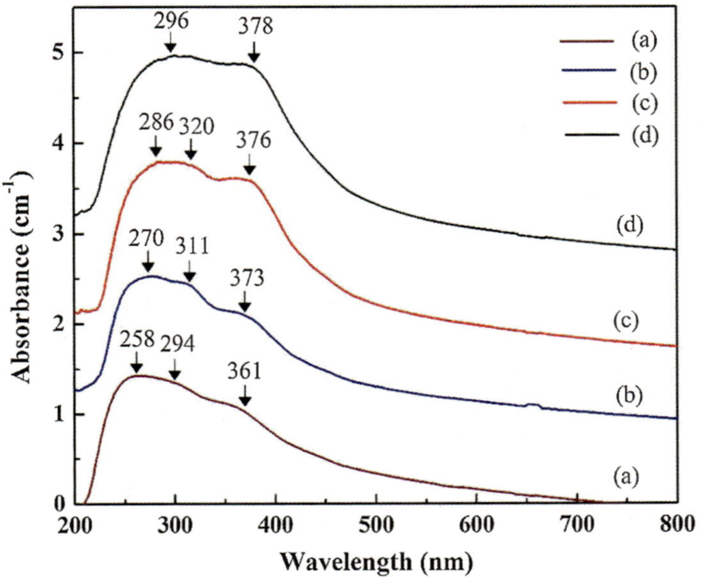

Fig. 15.16 Optical absorption spectra in (**a**) 2 mol%, (**b**) 5 mol%, (**c**) 10 mol%, and (**d**) 15 mol% Cr^{3+}/Cr^{4+}:ZrO(OH)·xH_2O gels

Table 15.3 Absorption bands in the various $Cr^{3+}/Cr^{4+}:ZrO(OH)_2{\cdot}xH_2O$ hydrogel samples

		Absorption bands (nm)			
S. No.	Samples	1	2	3	
(a)	2 mol $Cr^{4+}:ZrO(OH)_2$ gel	258 (1.4)	294 (1.3)	351 (1.1)	4–5
(b)	5 mol $Cr^{4+}:ZrO(OH)_2$ gel	270 (2.5)	311 (2.5)	373 (2.3)	4–5
(c)	10 mol $Cr^{4+}:ZrO(OH)_2$ gel	286 (4.0)	320 (3.9)	376 (3.7)	4–5
(d)	15 mol $Cr^{4+}:ZrO(OH)_2$ gel	286 (5.0)	378 (4.8)		4–5

The absorbance values are given in the parentheses

resulting in a modified electronic absorption spectrum. That occurs in groups over 250–400 nm intrinsic of the 3d–3d electronic transitions in the Cr^{3+} ($3d^3$) and Cr^{4+} ($3d^2$) species. Both the bandgroups in this example arise from Cr^{4+} ($3d^2$) in the 250–320 nm $^3T_1 \rightarrow {}^3T_2$ (P) and 340–400 nm $^3T_1 \rightarrow {}^3T_1$ (G) of two transitions. No such absorption spectrum appears in this region in a bare hydrogel gel on Zr^{4+}($4d^0$) species, which absorb at still lower wavelengths. A doublet bandgroup is occurring over 220–230 nm or still lower range in a pure ZrO_2 (hydrated) or polymorphic zirconias according to the optical bandgap (Table 15.1).

The bands from Cr^{3+} ($3d^3$) are observed in relatively weak intensity in bandgroups at 300–340 nm ($^4A_{2g} \rightarrow {}^4T_{1g}$ transition). It is observed that on increasing the Cr^{3+}/Cr^{4+} concentration from 2 to 15 mol%, the bands position shift over longer wavelengths by as much as 30 nm in spectra (a) and (d). All these transitions are spin-allowed so that they reflect in sufficiently large intensity. Loss of part absorption intensity has been observed on interaction between the optical species in these gels so that the value is not increasing proportionately to their contents. Individual bands observed in selective gels are given in Table 15.3. Result infers $Cr^{3+} \rightarrow Cr^{4+}$ transformation in growing c-/t-ZrO_2 nanoparticles. It is important in designing Cr^{4+}/Cr^{3+}-modified $ZrO(OH)_2{\cdot}xH_2O$ gels and derived c-/t-ZrO_2 of small particulates and hybrid composites.

Figure 15.17 shows emission spectra in selective $Cr^{3+}/Cr^{4+}:ZrO(OH)_2{\cdot}xH_2O$ hydrogels in which the Cr^{3+}/Cr^{4+}content is varied from 2 to 15 mol%. The four spectra are measured under identical conditions on exciting at 250 nm from a xenon lamp. The emission bands occur over extended 330–700 nm region of the absorption bands in Fig. 15.16. The prominent bandgroup 430–500 nm is characteristic of the Cr^{4+} (d^2) species ($^3T_{2g} \leftarrow {}^3T_{1g}$ transition). Another weak band of Cr^{4+} species is arising around 340 nm. The characteristic weak band for Cr^{3+} species ($^4A_{2g} \leftarrow {}^2E_{1g}$ transition) lies in the 660–670 nm region. Both of these low intensity bands disappear on increasing the chromium species. In general, the emission bands shift over higher wavelengths similar to the absorption bands. No emission appears from bare Zr^{4+} species in the parent gel.

In another example of optical polymeric liquids, obtained by a chemical route as demonstrated in Fig. 15.10 following the work of Ram [111], an addition of sucrose followed by PVA (dissolved in water) to a transparent mixed solution (say A) of $Zr(OH)_4 \cdot xH_2O$ and $(NH_4)_2Cr_2O_7$ in a specific ratio results in a specimen of dispersed metal cations in a polymer matrix of sucrose and PVA polymer molecules. The sucrose reacts with $(NH_4)_2Cr_2O_7$ in the solution and reduces its oxidation state

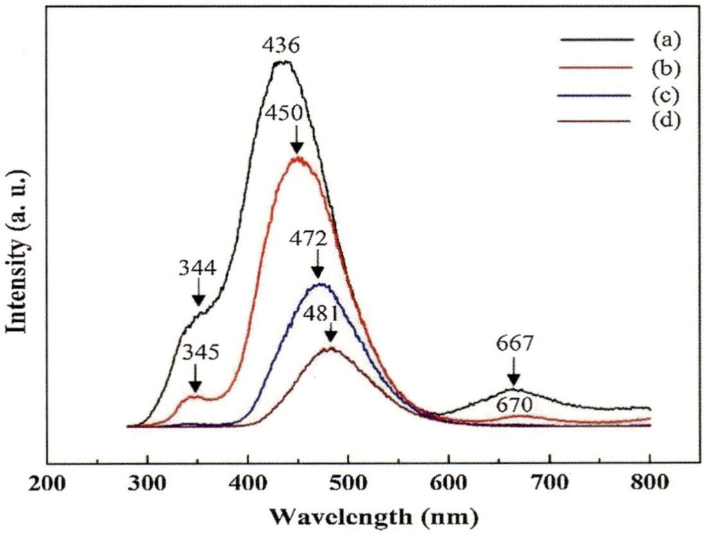

Fig. 15.17 Photoluminescence (on irradiating at 250 nm from a xenon lamp) in Cr^{3+}/Cr^{4+}:ZrO $(OH)_2$ gels; (a) 2 mol%, (b) 5 mol%, (c) 10 mol%, and (d) 15 mol% Cr^{3+}/Cr^{4+} species

of chromium $Cr^{6+} \rightarrow Cr^{4+}/Cr^{3+}$ with a change of its initial color from orange to a light green one. The light green equilibrium color becomes stable at a stable $Cr^{4+} \leftrightarrow Cr^{3+}$ configuration in the polymer as per the experimental conditions. A similar change in color also appears on a reaction of PVA with solution A but that reverts in a characteristic light orange color within a period of 5–10 h of the reaction in cold water at room temperature in open air.

Phenomenologically, a characteristic orange color in $(NH_4)_2Cr_2O_7$ in water occurs in a ligand–metal ($L \rightarrow Cr^{6+}$) charge transfer band, otherwise the $Cr^{6+}(3d^0)$ species has no optical absorption or emission spectrum in this region. Here, the ligand L is formed by O^{2-} anions surrounding the Cr^{6+} cations. The original Cr^{6+} oxidation state is thus reduced to Cr^{4+} or Cr^{3+} on dispersing in sucrose or PVA polymer molecules in water. This is analyzed by a chemical test as follows. In a 20 ml of $(NH_4)_2Cr_2O_7$ solution (with PVA or sucrose) in water is added 0.1 g of KI followed by 5.0 ml CH_3COOH in a beaker. The sample is covered with a watch glass, to avoid its oxidation with air, and stirred for a few min to let the iodide oxidize by the Cr^{4+} species in the sample as per the $I^- + Cr^{4+} \rightarrow I_2 + Cr^{3+}$ chemical reaction. The I_2 gas turns color of the sample from light green into violet, which changes in an intense blue color if added with starch. The blue color disappears instantaneously on adding sodium thiosulfate, reduces iodine gas to iodide, $I_2 + S_2O_3^{2-} \rightarrow I^- + S_4O_6^{2-}$, again. The intense blue color in the so-called charge transfer band disappears as the I_2–starch complex breaks down.

Now let us comment on the above chemical transformations in terms of the optical absorption spectra of the optical solutions achieved after selective chemical reactions of $(NH_4)_2Cr_2O_7$ with PVA and sucrose molecules in water. For example,

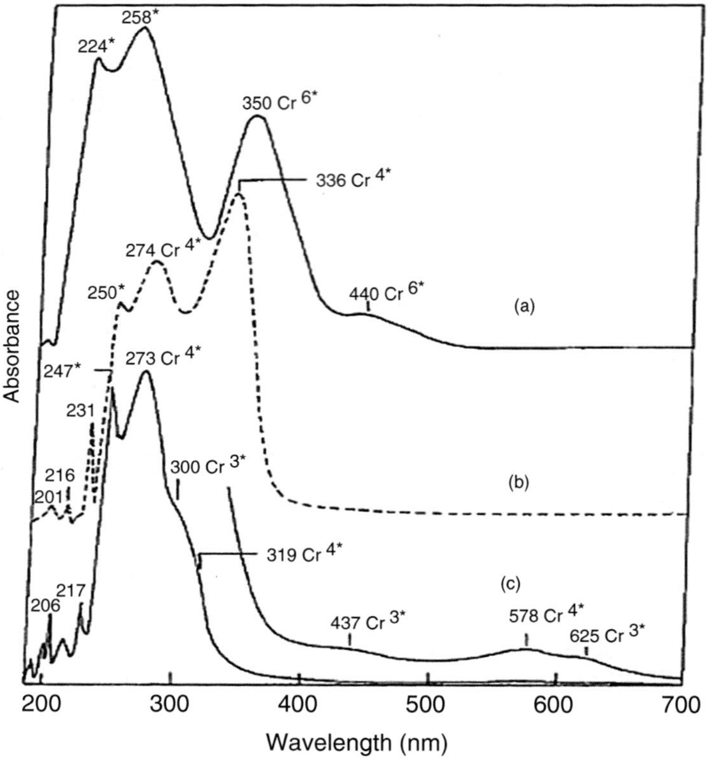

Fig. 15.18 Electronic absorption spectra in (**a**) Cr^{6+} in 0.0002 M $(NH_4)_2Cr_2O_7$ in water and that of transformed into Cr^{4+} and Cr^{3+} after adding 2 vol.% of (**b**) sucrose or (**c**) PVA in 10 % solution in water. *Charge transfer bands [111]

the electronic absorption spectra of Cr^{6+} species in $(NH_4)_2Cr_2O_7$ before (a) and after adding (b) sucrose or (c) PVA (Fig. 15.18) in the acidic solution in water confirm the transformation of the initial Cr^{6+} species to (b) Cr^{4+} or (c) Cr^{4+}/Cr^{3+} species. As summarized in Table 15.4, two absorption bands at 440 and 350 nm in spectrum (a) arise in the ${}^1A_{1g} \rightarrow {}^3A_{2g}$ and ${}^1A_{1g} \rightarrow {}^1T_{2g}$ electronic transitions of $Cr^{6+}(3d^0)$ species.[210,211] Other two bands, which appear at much lower wavelengths (or higher energy) at 258 and 224 nm, arise from the $L \rightarrow Cr^{6+}$ charge transfer bands which occur through a ligand with the O^{2-} anions in an oxygen polyhedron with the cation Cr^{6+} at the center. As expected on the $Cr^{6+} \rightarrow Cr^{4+}$ chemical reaction with the organic species in the solution, two distinct bands due to the resulting $Cr^{4+}(3d^2)$ species are developed at 336 nm (${}^3T_1 \rightarrow {}^3T_1$) and 274 nm (${}^3T_1 \rightarrow {}^3T_2$) in spectrum (b), and those are shifted at 319 nm and 273 in spectrum (c). A third characteristic band of the Cr^{4+} species (${}^3T_1 \rightarrow {}^3A_2$) has developed at 578 nm in spectrum (c).

Furthermore, in Fig. 15.18, the absorption bands from the Cr^{3+} $(3d^3)$ species in spectrum (c) are observed at 625 nm (${}^4A_{2g} \rightarrow {}^2E_g$), 437 nm (${}^4A_{2g} \rightarrow {}^4T_{2g}$), and 300 nm (${}^4A_{2g} \rightarrow {}^4T_{1g}$) in relatively poor intensities. All the electronic transitions,

Table 15.4 The electronic absorption bands in modified $Cr^{6+} \rightarrow Cr^{4+}/Cr^{3+}$ oxidation states in $(NH_4)_2Cr_2O_7$ dissolved in sucrose and PVA polymer in water

Band positions (nm)			
$(NH_4)_2Cr_2O_7$	$(NH_4)_2Cr_2O_7$ in sucrose	$(NH_4)_2Cr_2O_7$ in PVA	Transitions
		625 (vw)	$^2A_{2g} \rightarrow {}^2E_g\ (Cr^{3+})$
		578 (vw)	$^3T_1 \rightarrow {}^3A_2\ (Cr^{4+})$
440 (vw)	–	–	$^1A_{1g} \rightarrow {}^1T_{2g}\ (Cr^{6+})$
	–	437 (vw)	$^4A_{2g} \rightarrow {}^2T_{2g}\ (Cr^{3+})$
350 (ms)	–	–	$^1A_{1g} \rightarrow {}^3A_{2g}\ (Cr^{6+})$
	336 (vs)	319 (w)	$^3T_1 \rightarrow {}^3T_1\ (Cr^{4+})$
	–	300 (ms)	$^4A_{2g} \rightarrow {}^4T_{1g}\ (Cr^{3+})$
	274 (s)	273 (vs)	$^3T_1 \rightarrow {}^3T_2\ (Cr^{4+})$
258 (vs)	250 (ms)	247 (vs)	
224 (s)	–	–	Charge transfer bands
	231 (vw)	–	
	216 (vw)	217 (vw)	Vibronic band
	201 (vw)	206 (vw)	

Relative band intensities are given in the parentheses: vs, very strong; s, strong; ms, medium strong; w, weak; and vw, very weak intensity values

except the $^4A_{2g} \rightarrow {}^2E_g$ in Cr^{3+} species, in the above examples of optical liquids are spin-allowed and thus appear with pretty large band intensities. Rather easier 3d–3d excitation in three free 3d-electrons in Cr^{3+} species occurs in the bands over relatively lower energy in comparison to those in the Cr^{4+} species with a lesser number of only two free 3d-electrons. The sharp (and weak) absorption bands observed between 250 and 200 nm in the absorption spectra (b) and (c) in Fig. 15.18 are possibly vibronic bands, which co-occur in strong electron–phonon coupling in the local oxygen polyhedron with the chromium ions.

All the electronic transitions, except the $^4A_{2g} \rightarrow {}^2E_g$ in the Cr^{3+} species, in these examples of the optical liquids with chromium ions in polymer molecules in water are spin-allowed and thus appear with reasonably large band intensities. Rather easier 3d–3d electron excitation in three free 3d-electrons in Cr^{3+} species occurs over relatively lower energy of the bands in comparison to those in the Cr^{4+} species with a lesser number of only two free 3d-electrons. The sharp (and weak) bands observed between 250 and 200 nm in absorption spectra (b) and (c) in Fig. 15.18 are possibly vibronic bands.

It should be mentioned that sucrose promotes PVA molecules cross-linking in form of a copolymer with dispersed metal cations in water. The sucrose plays a multifunctional role. At first, it forms an inorganic–organic complex with metal cations by coordinating through hydroxyl groups in small groups of a kind of micelles. The micelle circumvents selective precipitation of encapsulated cations within it while evaporating part of the water in the solution in obtaining a dried polymeric precursor mass. Sucrose, being in excess to the cations, behaves as a strong chelating agent and ensures an atomistical distribution of cations in a polymer structure. As soon as added to the solution, the sucrose gets hydrolyzed to fructose and glucose, which ultimately oxidizes to gluconic acid or a polyhydroxyl acid as follows (Eq. 15.24)

$$\begin{array}{c} C_{12}H_{22}O_{11}(\text{sucrose}) + H_2O \\ \downarrow \\ C_6H_{12}O_6(\text{glucose}) + C_6H_{12}O_6(\text{fructose}) \\ \downarrow \\ C_6H_{12}O_7(\text{gluconic acid}). \end{array} \tag{15.24}$$

Gluconic acid, with a carboxylic acid group in one end of the carbon chain and five linear hydroxyl groups, easily forms a metal ion complex with chromium and zirconium ions. It forms a branched polymer with PVA in water [111, 212]. A dried sample obtained with sucrose in the presence of PVA in a small 10 % amount results in a desirably crushable, fluffy powder. In addition to dispersing and bonding the metal cations in a specific network structure, a polymer matrix serves as an internal fuel to decompose and burn out organic part of the sample spontaneously into a refined powder in Cr^{3+}-/Cr^{4+}-stabilized ZrO_2 of composite nanoparticles. The ceramic particles nucleate and grow in support with a template (or electrode) from the organic species during this combustion process in air. A modified carbon skeleton retains on early stages of the burning in reaction with air.

ZrO_2 and Composites

Monolithic ZrO_2 Particles

Cao et al. [69] studied absorption and emission spectra in monolithic t-ZrO_2 prepared in form of nanowires with 80 nm average diameter and length of over 10 μm at room temperature. A typical absorption spectrum measured in the 200–800 nm range from a sample which was dispersed in a NaOH solution is shown in Fig. 15.19. It consists of a peak at 232.5 nm (5.33 eV in photon energy) in the electronic transition from the valence band (VB) to the conduction band (CB). This is caused by an

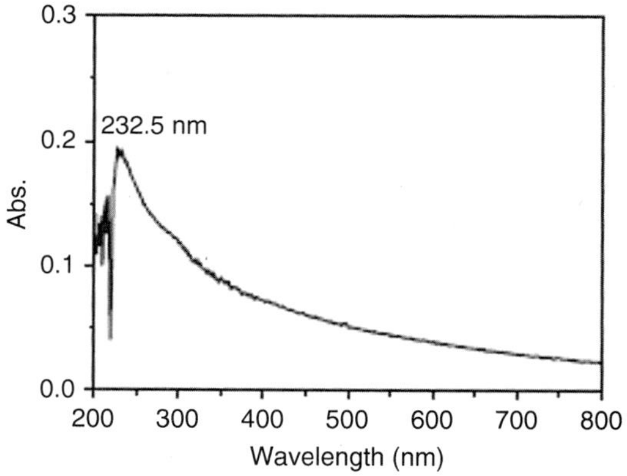

Fig. 15.19 Absorption spectrum of t-ZrO_2 nanowires in NaOH solution [69]

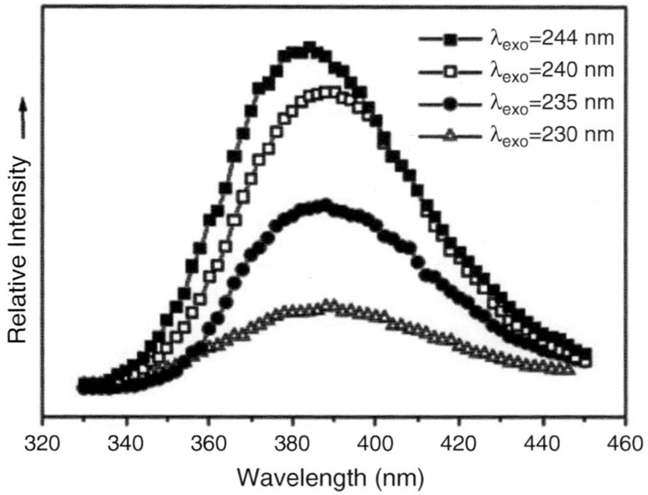

Fig. 15.20 PL emission spectra from ZrO_2 nanowires measured by exciting at selective wavelengths 230, 235, 240, or 244 nm by a xenon laser [69]

$O^{2-}(2p) \rightarrow Zr^{4+}$ (A_{1g}) charge- transfer transition in the excitation of electrons from VB (mainly has O^{2-} 2p character) to the CB (mainly having Zr^{4+} $4d^0$ character). As expected, no d–d transition is observed in the longer wavelength region because the configuration of Zr^{4+} ions is $4d^0$. The photon energy 5.33 eV is comparable to the optical bandgap 5.2 eV observed in bulk ZrO_2 [213]. Note that ZrO_2 is a direct bandgap insulator with two direct band-to-band transitions at 5.20 and 5.79 eV [213]. As a result, it serves as an active and typical photon absorber and photocatalyst among wide bandgap metal oxides.

Figure 15.20 shows the PL emission spectra (in the 330–450 nm range) of the t-ZrO_2 nanowires in a NaOH solution measured at room temperature by exciting at selective wavelengths 230, 235, 240, and 244 nm from a xenon laser as the excitation source. Varying the excitations on these four wavelengths produces the emission peaks at 388, 388, 390, and 384 nm, respectively. The excitations with light of 230 and 235 nm, i.e., close to the absorption band of 232.5 nm in Fig. 15.19, produce the emission peaks of the same value 388 nm. This is evidence of the "normal photo-physical behavior" and rules out the possibility of this emission band arising from any impurity.

The PL emission band in ZrO_2 nanowires possibly arises due to the oxygen vacancies which can induce the formation of new energy levels in the bandgap. Formation of an oxygen vacancy in such oxides can be described as follows (Eq. 15.25):

$$ZrO_2 \leftrightarrow ZrO_{2\text{-}n} + \frac{n}{2} O_2(g), \qquad (15.25)$$

which can undergo several modifications such as (Eqs. 15.26, 15.27, and 15.28)

$$O_o^x \leftrightarrow \frac{1}{2} O_2(g) + V_o, \quad (15.26)$$

$$V_o \leftrightarrow V_o^{\bullet} + e^-, \text{ and} \quad (15.27)$$

$$V_o^{\bullet} \leftrightarrow V_o^{\bullet\bullet} + e^- \quad (15.28)$$

where V_o represents a neutral oxygen vacancy, $V_o^{\bullet}$ a singly ionized oxygen vacancy, $V_o^{\bullet\bullet}$ a doubly ionized oxygen vacancy, O_o^X an oxide anion in a regular lattice site, and e^- an electron in the CB in the Kröger's notation [214]. In general, the emission spectrum in this example can be divided into two major categories, i.e., the near-band-edge (NBE) emission and deep-level (DL) emission [69, 215]. The position and intensity of the emission band depend upon the kinds of the optical species and their relative concentrations. Especially in nanowires, high shape anisotropy and quantum confinement are two major factors which render intense emission of ultraviolet light at room temperature in ZrO_2 and other similar wide bandgap oxides, e.g., ZnO [216, 217], Al_2O_3 [218, 219], or SiO_2 [220, 221]. Primarily, singly ionized oxygen vacancies or interstitials are responsible for exhibiting an intense light emission in the ultraviolet region. The transition involves a radiative recombination of a photo-generated hole with an electron occupying such sites in the lattice. In c-/t-ZrO_2 nanoparticles, Emeline et al. [73] observed a similar PL emission band of 387 nm as of those described above in t-ZrO_2 nanowires, which are believed to be better suitable for devising building blocks in optical CDs or other nanodevices.

Hybrid ZrO_2–Polymer Nanocomposites

Dey and De [82] studied optical and dielectric spectra in hybrid nanocomposites of t-ZrO_2 nanoparticles reinforced polypyrrole (PPY) at room temperature. An optimized t-ZrO_2 content 34 wt% leads the dielectric constant to be improved dramatically, i.e., as large value as 12,000 (at 0.7 kHz) relative to a value 570 in pure PPY. Such conducting polymer composites have several applications, e.g., solid electrolytes, microwave devices, and microelectronics. They offer architectural flexibility useful to shape the final product according to the requirement of the particular device. Figure 15.21 compares plot of $(\alpha h\nu)^2$ versus photon energy $(h\nu)$ for (a) ZrO_2 nanoparticles and (b) a derived ZrO_2–66 wt% PPY nanocomposite. The absorption spectra of (a) the bare ZrO_2 nanoparticles and (b) dilute and (c) colloidal dispersion of two ZrO_2–PPY nanocomposites (66 and 88.6 wt% PPY) are included in the inset in this figure. In virgin ZrO_2, a strong absorption starts at around 3.8 eV, and the bandgap corresponding to the band edge structure gives an activation energy 4.0 eV, which stands relatively smaller than the E_g-value 5.2 eV reported above in bulk ZrO_2. The absorption band below 5 eV implies some extrinsic states, such as surface trap or defect states (possible due to Zr^{3+} species and/or oxygen vacancies) formed within the bandgap. The bonding and antibonding states of polaron and bipolaron of PPY show absorption peak over 2–2.5 eV [222]. Hence, there occurs a great overlap of optical transition between PPY and extrinsic states of t-ZrO_2.

A simple $\alpha = 2.303 \times 10^3 A\rho/Cl$ relation [223] is used to determine the absorption coefficient α in the three samples, where A is the absorbance, ρ is the density,

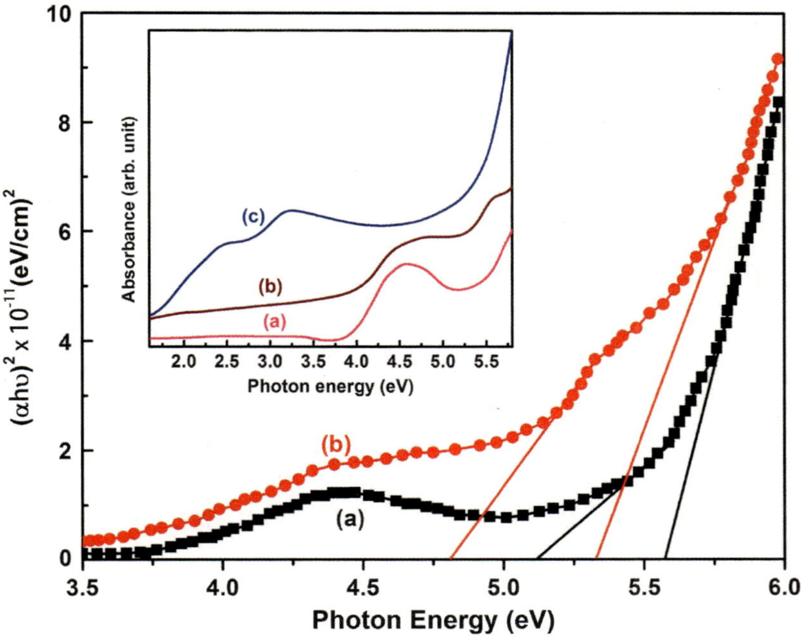

Fig. 15.21 Plot of $(\alpha h\nu)^2$ versus photon energy ($h\nu$) for (**a**) ZrO_2 nanoparticles and (**b**) ZrO_2–66 wt% PPY nanocomposite. The *inset* shows the absorption spectra of (**a**) ZrO_2, (**b**) ZrO_2 dispersed in 66 wt% PPY, and (c) 88.6 wt% PPY [82]

C is the concentration in g l^{-1}, and l is the optical path length of the sample. The α-value near the absorption edge of a semiconductor sample is described as (Eq. 15.29) [224]

$$\alpha h\nu = A\left(h\nu - E_g\right)^m \tag{15.29}$$

where $h\nu$ is the photon energy and m is the nature of the transition, which has different values such as 1/2, 2, 3/2, or 3 for allowed direct, allowed indirect, forbidden direct, or forbidden indirect transition, respectively. For allowed direct transitions, one can plot $(\alpha h\nu)^2$ against $h\nu$ as presented in Fig. 15.21 and extrapolate the linear portion of it to $\alpha = 0$ value to obtain the corresponding E_g-value. In this case, $m = 1/2$ and so the interband transition is allowed direct. The estimated E_g-values for pure ZrO_2 are 5.6 and 5.1 eV corresponding to two direct band-to-band transitions. The first one decreases to 4.2 eV and the second one to 4.0 eV with increasing PPY content. Combined experimental [70, 225] and theoretical [226] observations reveal that the valence and conduction bands of ZrO_2 are separated by 5–6 eV. The valence band consists of O^{2-} 2p states and the conduction band is primarily of $4d^0$-Zr^{4+} states. The absorption bands arise due to the transition from occupied O^{2-} 2p bands to unoccupied $4d^0$-Zr^{4+} bands of t-ZrO_2.

In another example of organic–inorganic ZrO_2 hybrid composites, Marusaki et al. [227] studied optical absorption and emission in spin-coated films of (a) poly

Fig. 15.22 Optical absorption spectra of (**a**) P (DHS-co-AAEM) and (**b**) PDHS–ZrO_2 hybrid thin films at different temperatures [227]

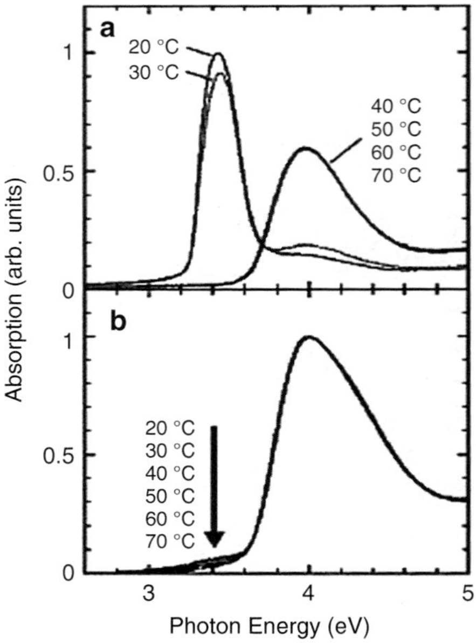

(di-*n*-hexylsilane) (PDHS) dispersed in a zirconia matrix. Thermochromism is suppressed usefully in the composite films. Poly(di-*n*-alkylsilanes) are known for applied electronic and structural properties. Many of such polymers have an order–disorder phase transition (known as thermochromism) slightly above room temperature that is accompanied by changes in the optical properties. In PDHS, the optical absorption shifts from 317 nm in the high temperature disorder phase to 374 nm in the low temperature ordered phase. The polysilane films used here were prepared from the copolymers [(P(DHS-co-AAEM)] using $Zr(O\text{-}n\text{-}C_4H_9)_4$ in a sol–gel method. A polysilane copolymer has been synthesized by photopolymerization of 2-(acetoacetoxy)ethylmethacrylate (AAEM). Figure 15.22 compares absorption spectra of (a) P(DHS-*co*-AAEM) and (b) PDHS–ZrO_2 films at selective temperatures 20 °C, 30 °C, 40 °C, 50 °C, 60 °C, and 70 °C. The absorption peak in P (DHS-*co*-AAEM) of 3.4 eV at 20 °C shifts to 4.0 eV upon heating above 40 °C, indicating that the sample undergoes an order–disorder phase transition. The Si–Si main chain conformation of P(DHS-*co*-AAEM) changes from "transoid" to "disorder" at 30–40 °C. This is because of the presence of AAEM blocks and thermochromism [228, 229]. No such spectral shift occurs in PDHS–ZrO_2 films (Fig. 15.22b) with increasing temperature.

Evidently, the thermochromism of PDHS is completely suppressed in the ZrO_2 matrix in these ZrO_2–PDHS hybrid composite films. The Si–Si main chain conformation of PDHS–ZrO_2 hybrid has the "disorder" form because the hybrids were baked at 150 °C, i.e., the main chain conformation is fixed in at this temperature. A similar suppression of the order–disorder phase transition has been also observed in liquid crystals containing dispersed nanoparticles of silica of spherical shapes [230].

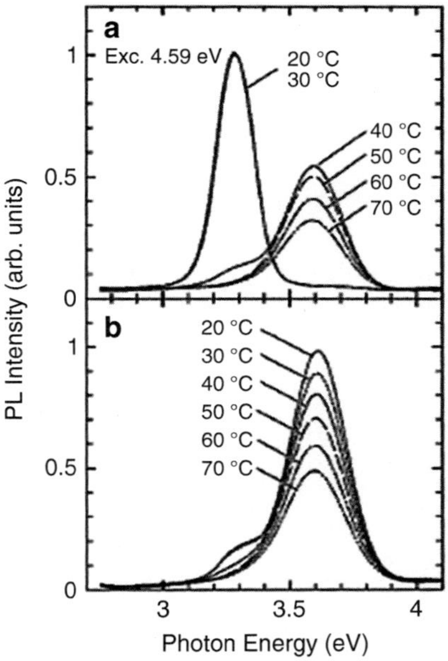

Fig. 15.23 PL spectra of (**a**) P(DHS-co-AAEM) and (**b**) PDHS–ZrO_2 hybrid thin films at different temperatures [227]

Figure 15.23 shows PL spectra of (a) P(DHS-co-AAEM) and (b) PDHS–ZrO_2 hybrid composite thin films measured at selective 20 °C, 30 °C, 40 °C, 50 °C, 60 °C, and 70 °C temperatures by exciting at 270 nm. In P(DHS-co-AAEM) thin films, the PL peak position shifts from 3.3 eV at room temperature to as large value as 3.6 eV on raising the temperature to 40 °C or larger, while in PDHS–ZrO_2 hybrid thin films, the PL peak position stands essentially the same over the entire temperature range 20–70 °C. In both the samples, the PL intensity is decreasing in example of thermal quenching of PL emission in charge carriers. As described in a reference book "*Survey of semiconductor physics: Electrons and other particles in bulk semiconductors*" by Böer [229], this is quite a common phenomenon in semiconductors. The same authors had also studied PDHS–SiO_2 and PDHS–TiO_2 hybrid composites of similar thin films but those do not have suppression of the thermochromism of PDHS. Furthermore, refractive index n of PDHS–ZrO_2 films decreases linearly (instead of undergoing an abrupt change at the order disorder phase transition in PDHS) with increasing temperature, showing a value -4×10^{-4} K^{-1} for dn/dT, i.e., the temperature coefficient of n or thermo-optic (TO) coefficient. Such high TO value makes such composite films candidate material for optical devices such as TO switches, cross-connects, and optical deflectors. The optical devices play a key role in communication network because they control the optical path. Currently used materials of TO switches are silica and optical polymers such as poly(methyl methacrylate).

Metal-Doped ZrO_2 Nanocomposites (Cermets)

Metals, particularly noble metals (copper, silver, gold) are very useful in modifying electronic, optical, and other properties of high-k (dielectric constant) metal oxides

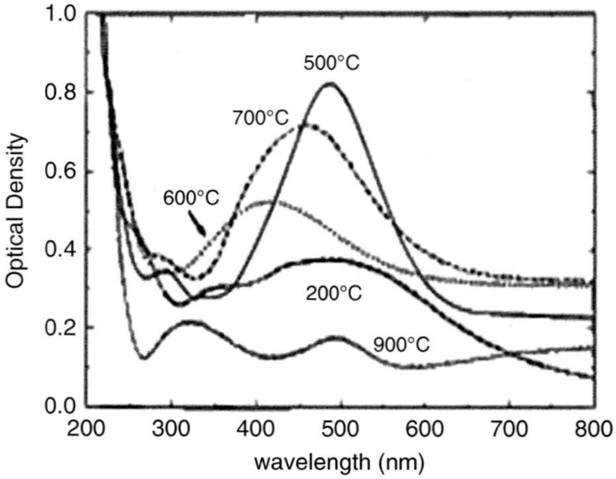

Fig. 15.24 Optical absorption spectra of ZrO_2:Ag nanoparticles annealed at 200 °C, 600 °C, 700 °C, 800 °C, and 900 °C in H_2 for 1 h [209]

such as SiO_2, Al_2O_3, Ta_2O_5, TiO_2, HfO_2, ZrO_2, etc., for several applications. High electrical conductivity in the bulk and unusual optical properties that depend on size and shape in small particles are important in engineering novel properties by selective inclusions in such metal oxides in form of a hybrid composite of cermets. As mentioned above in Sect. 1.2, a cermet of engineering properties can be designed in several ways including (i) dispersion of small metal particles in a ceramic host, (ii) coating ceramic particles using a metal, (iii) dispersion of ceramic particles in a metallic host, and (iv) coating small metal (or alloy) particles using a metal oxide. In all these four kinds of cermets, a core–shell structure governs the hybrid composites properties. Confining electron distribution in metal of a nanostructured region results in surface plasmon resonance (SPR) bands which fall in ultraviolet, visible, or near-infrared region as per (i) the shape, surface–interface and size of the basic nano-structure, and (ii) the crystal field imposed by the surrounding. In cermets, high-k metal oxide sensitively tailors the electronic energy levels in the counterpart metal nanostructure and vice versa [230–234]. Moreover, the metal species invite new color centers and/or modify the color centers in the oxide species engineering especially useful optical properties for applications.

For example, following the work of Gonella et al. [209], Fig. 15.24 shows optical absorption spectra obtained for a 10 mol% silver-modified c-/t-ZrO_2 of nanoparticles which were annealed at 200 °C, 600 °C, 700 °C, 800 °C, and 900 °C in H_2 gas for 1 h. Annealing ensures silver metal particles occurring with stabilized c-/t-ZrO_2 phase. Two bands are observed in the as-grown sample (annealed at 200 °C) at 270 and 482 nm. The first band ascribes to the ceramic matrix c-/t-ZrO_2 in agreement with the work of Jana et al. [173] and Morell et al. [235] on Y^{3+}-stabilized zirconias, while the second band which is characteristically broad arises from the silver nanoparticles. When the annealing temperature reaches 600 °C, the silver band reveals a blue shift to 419 nm. In the Mie theory [236], such a shift can be explained only assuming a reduction to about $n = 1.8$ of the refractive index of the host with respect to the pure ZrO_2 value of $n \sim 2$.

Palpant et al. [237] suggested two possible mechanisms for a change in the *n*-value in a hybrid silver–ZrO_2 composite. The first is related to the nature of the metal–oxide interface. The different chemical properties of silver and ZrO_2 imply the possibility of defects around the silver particles that develop as a locally enhanced porosity between silver particles and matrix. Thus, a dielectric screening of the host is reduced in this local effect, describing a blue shift of the overall band on a phenomenological ground equivalent to reducing the *n*-value of the matrix. The second mechanism is related to the structural evolution of silver-modified ZrO_2 particles at selective temperatures. As described above in Sect. 4, above 300 °C, an amorphous-to-crystalline transition takes place with subsequent c/t-ZrO_2 phase transitions, which involve significant volume changes [209, 238]. Different structural arrangements develop in varied annealing conditions, leading to varied dielectric and optical properties.

It has been found that an effectively high annealing temperature anneals out possible structural defects, restoring the theoretically expected dielectric behavior of the silver–ZrO_2 interface. Indeed, when the annealing temperature is raised at 700 °C in H_2 gas, the absorption band in Fig. 15.24 increases in intensity and red-shifts again, with as large value as 487 nm (458 nm after 700 °C) achieved after annealing at 800 °C, i.e., precisely the Mie SPR band in silver nanoparticles in ZrO_2. On increasing the temperature further, bulk silver particles grow readily, showing diminished intensity of absorption band near 510 nm. Argon and other annealing atmospheres are not so effective in precipitating silver nanoparticles out of zirconia with such varied optical properties. In another example, De et al. [239] studied optical properties in silver, copper, and silver–copper nanoclusters in a silica matrix after annealing (over 500–1000 °C) in air, argon, or a mixture of N_2 and H_2. Ag nanoparticles have strong SPR bands over 300–550 nm. Cermets with metal particles embedded in glasses have enhanced nonlinear optical properties, with large values of the intensity-dependent refractive index, which is related to the real part of the third-order susceptibility (useful for optoelectronics).

ZrO_2 cermets, with ferromagnetic metal nanoparticles such as nickel, cobalt, iron, etc., can offer new properties for magnet technology, magnetoresistance, robotics, and also solid oxide fuel cells. The metallic part extends to design a mechanically strong tool or device with desirably improved ductility, toughness, and heat transfer [240–242]. Such kinds of superior mechanical and magnetic properties also develop in ferromagnetic cermets consisting of Al_2O_3 and other oxide ceramics. Unfortunately, not many reports are available on optical properties in such kinds of cermets. Occurrence of magnetic and optical properties with good heat transfer in a single hybrid composite material of cermets of specific shapes opens a window for new materials and applications.

Rare-Earth Oxide-Modified ZrO_2 Composites

There are several reasons to believe that ZrO_2 serves as an excellent optical host for engineering highly intense light-emitting nanocomposite materials (subject to suitable irradiation on ultraviolet, visible, and near-infrared regions) as follows. It is chemically highly inert and stands to be highly stable against any unwanted photochemical change in ambient atmosphere. A high value of refractive index $n = 2.02$ and low phonon energy under 800 cm^{-1} control part of the energy loss in the electron → phonon energy transfer and other nonradiative processes [36, 243, 244]. In accordance with energy law, a large separation between emitting and terminal energy levels inhibits probability of nonradiative decay of the excited optical species. Lower host phonon energy has a greater number of phonons connecting the emitting level with the next lower level. Suppressed nonradiative energy loss on large numbers of phonons needed to gap the energy facilitates the radiative light emission. Using a heavy host lattice of low energy phonons overcomes the problem of the electron → phonon energy transfer. Moreover, ZrO_2 provides an ideal dielectric medium that promotes the light emission of highly luminescent material.

Furthermore, ZrO_2 allows desirable foreign chemical doping using rare-earth ions and other similar optical species essential for incorporating the light emitters of selective wavelengths in form of an efficient luminophor. As a promising dispersive medium, it disperses such optical species uniformly at length scale as small as single species and that offers a monodispersed design of the optical centers within the host ZrO_2 with their ideal single microscopic electronic energy levels. It occurs with a specific molecular designing with selective polyhedral units ZrO_r in part of the matrix ($r \geq 4$, as per average Zr^{4+} coordination number which varies in amorphous and crystalline ZrO_2 phases). As a result, the local host structure stipulates the surface interface as well as the effective shapes and topologies in the optical centers and those modulate very effectively the dielectric and optical properties out of a hybrid composite structure.

Within the recent past few years, limited studies (in comparison to those using a ceramic host of silica, alumina, or other oxides) have been carried out concerning the photoluminescence emission in doped ZrO_2 with selective rare-earth ions [244–250]. For example, Savoini et al. [245] studied Pr^{3+}-doped and Pr^{3+}-stabilized c-ZrO_2. Other studies include luminescence of Eu^{3+}, Tb^{3+}, and Sm^{3+} in amorphous and crystallized t-/m-ZrO_2 [244, 246–249], Er^{3+}:t-ZrO_2 [250], or Mn^{2+}: t-ZrO_2 [251]. In all these cases, the optical and other spectroscopic properties highly depend on the crystalline structure of the host ZrO_2. Gosh and Patra [243] studied photoluminescence emission (550–700 nm) in Eu^{3+}-doped and Eu^{3+}-coated ZrO_2 nanocrystals (42 nm diameter). The volume fraction of t-ZrO_2 phase increases considerably from 28.08 to 91.56 % on the surface coating with 1.0 mol% Eu_2rO_3, with a core–shell structure. The intensity of the prominent band at 611–617 nm ($^5D_0 \rightarrow {}^7F_2$ transition) of Eu^{3+} species has been found to be sensitive to the environment in both kinds of these samples. Eu^{3+}-doped ZrO_2 exhibits larger intensity in this band relative to a coated sample.

It is quite phenomenological to argue that, in a Eu^{3+}-coated t-ZrO_2 of nanocrystals in the above example, a strong shell (Eu^{3+}) → core (ZrO_2) energy transfer renders smaller intensity of radiative emission in the Eu^{3+} species ($^5D_0 \rightarrow {}^7F_J$, J = 1, 2 transitions). Decay time of doped ZrO_2 increases upon

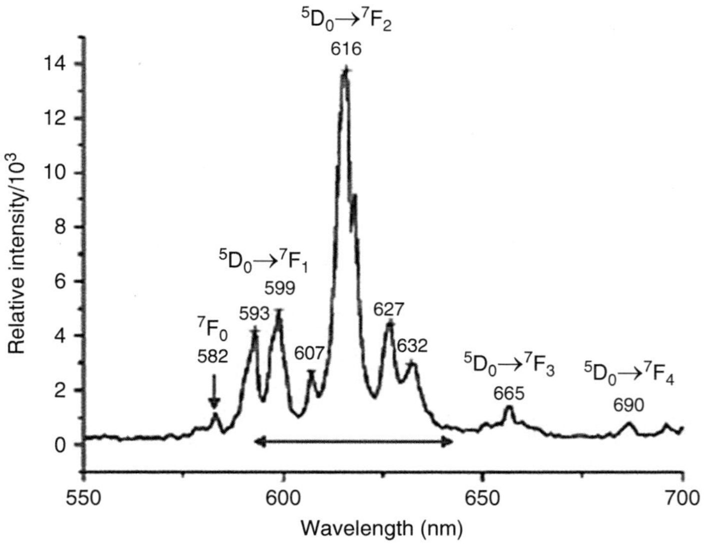

Fig. 15.25 Emission spectrum (recorded after exciting at 395 nm) in $^5D_0 \rightarrow {}^7F_J$ (J = 0 → 4) transitions of Eu^{3+} ions in t-/m-ZrO_2 nanocrystals [254]

increasing Eu^{3+} content (0.05–1.0 mol% Eu_2rO_3) and/or the temperature (900–1100 °C in air for 1 h) of heating [243]. Average decay times were found to be 770 and 488 μs for the 1.0 mol% Eu_2rO_3-doped and Eu_2rO_3-coated ZrO_2 (heated at 1100 °C). The local site symmetry of the optical species plays most important role in engineering the radiative and nonradiative relaxation processes in a hybrid composite of core–shell nanoparticles on limited shell thickness.

As an illustration, Fig. 15.25 shows a typical emission spectrum measured from a 2 mol% Eu_2O_3-doped t-/m-ZrO_2 of nanocrystals (~30 nm diameter) after exciting at 395 nm. The sample has been prepared by a combustion method from $Zr(NO_3)_4 \cdot 5\ H_2O$ and Eu_2O_3 (dissolved in nitric acid) with urea as a fuel at 400 °C in air. As marked over the bands, sharp emission bands are observed in orange–red and red regions of the spectrum in three major bandgroups in the $^5D_0 \rightarrow {}^7F_0$, $^5D_0 \rightarrow {}^7F_1$, and $^5D_0 \rightarrow {}^7F_2$ transitions of the Eu^{3+} ions. As expected, the $^5D_0 \rightarrow {}^7F_0$ transition appears in a single band with a poor intensity at 580 nm, i.e., a forbidden transition in all respects of the selection rules in 4f–4f intraband transitions in rare-earth ions. This is an electric dipole forbidden, magnetic dipole forbidden, as well as an electric quadrupole forbidden transition [252]. The $^5D_0 \rightarrow {}^7F_1$ transition (magnetic dipole transition) consists of two distinct bands of 593 and 599 nm. Five bands at 607, 616, 619, 627, and 632 nm constitute the most intense transition $^5D_0 \rightarrow {}^7F_2$ (forced electric dipole transition) of the spectrum. The Stark splitting of the ground state level $^7F_2$ into "2 J + 1" = 5 distinct sublevels as per the J = 2 value demonstrates that the Eu^{3+} ions occupy the possible lowest local site symmetry in the lattice c-/t-ZrO_2. A lower number of J-splitting represents higher site symmetry of the Eu^{3+} ions

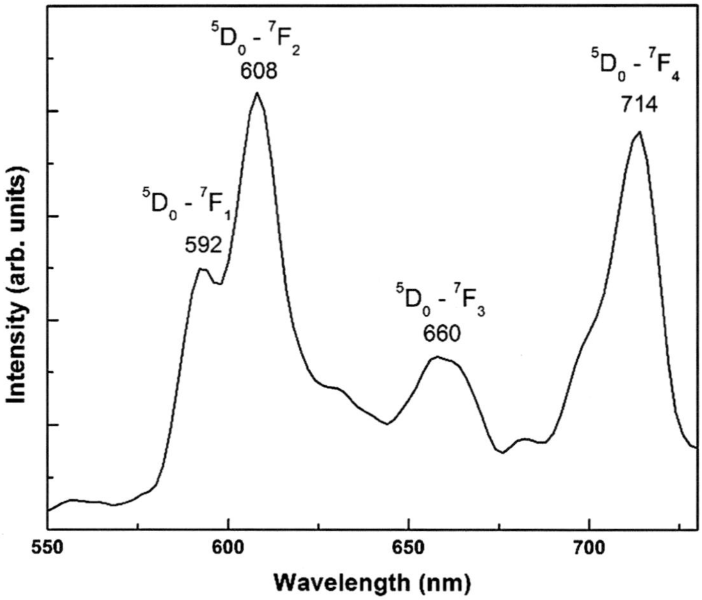

Fig. 15.26 Surface-enhanced $^5D_0 \rightarrow {}^7F_J$ ($J = 0 \rightarrow 4$) photoluminescence emission (recorded on exciting at 465.8 nm from an Ar + laser) in a Eu^{3+}:t-ZrO_2 nanoparticles (heated at 650 °C for 2 h in air) following an oxalate precursor [244]

in the host lattice. In ZrO_2, the site symmetries for Zr^{4+} ions are C_{2h}, D_{4h}, and O_h in the m-, t-, and c- phases, respectively [244, 253]. Other weak bandgroups in Fig. 15.25 arise near 665 nm 690 nm in the $^5D_0 \rightarrow {}^7F_J$ ($J = 3, 4$) in the Eu^{3+} ions.

Mondal and Ram studied similar samples of 2 mol% Eu^{3+}-doped t-ZrO_2 of nanocrystals (average size varies from 16 to 49 nm as per the final annealing temperature 650–1200 °C) after dehydrating an oxalate gel at 200 °C followed by 2 h of heating at 650 °C, 1000 °C, and 1200 °C in air [244]. As per the X-ray diffraction, a modified t-ZrO_2 of nanocrystals persists of a single-phase compound after annealing at these temperatures. The t-ZrO_2 phase, whose X-ray diffraction hardly differs from c-ZrO_2, has confirmed unambiguously in terms of the Raman spectrum that consists of six Raman active modes (A_{1g}, $2B_{1g}$, $3E_g$) of vibration – only a single vibration F_{2g} (triply degenerate) merges on increasing the symmetry in c-ZrO_2 [69, 255, 256]. These six bands observed of 150, 265, 318, 468, 609, and 640 cm^{-1} are very similar to those of pure t-ZrO_2 [255], except slightly increased values on the lattice expansion with modified oxygen vacancies upon the Eu^{3+} species present. As shown in Fig. 15.26, an unusually enhancement of the photoluminescence emission of Eu^{3+} species is observed in the $^5D_0 \rightarrow {}^7F_J$ ($J = 0 \rightarrow 4$) bands over 550–730 nm as follows.

The four bandgroups observed in Fig. 15.26 in Eu^{3+}:t-ZrO_2 arise in the (i) 560–600 nm, (ii) 600–645 nm, (iii) 645–680 nm, and (iv) 680–730 nm regions in the four well-known $^5D_0 \rightarrow {}^7F_1$, $^5D_0 \rightarrow {}^7F_2$, $^5D_0 \rightarrow {}^7F_3$, and $^5D_0 \rightarrow {}^7F_4$ transitions, respectively, in the Eu^{3+} ($4f^6$) species. The relative peak intensities in the four bands are

67, 100, 51, and 93 %. These bands differ markedly from those in Fig. 15.25 in a similar sample of average chemical composition but prepared by using a different method. In general, the bands are broadening considerably in a magnified intensity. Both the electric dipole forbidden and magnetic dipole forbidden $^5D_0 \rightarrow {}^7F_0$ transitions are not visible in this example. The $^5D_0 \rightarrow {}^7F_1$ bandgroup (near 592 nm) retains its features of moderate intensity characteristic of the magnetic dipole transition. The $^5D_0 \rightarrow {}^7F_3$ bandgroup with average position of 658 nm arises out of a forbidden transition by the electric dipole as well as the magnetic dipole moments. It is the weakest band in intensity of the present spectrum as expected in the parent selection rules in the 4f–4f transitions. Nevertheless, also this much intensity stands pretty larger if comparing to the value in this band in Fig. 15.25. The $D_0 \rightarrow {}^7F_2$ bandgroup over 600–645 nm follows the general trend of the most intense bandgroup of the spectrum. It represents a hypersensitive band in a kind of a forced electric dipole transition in a specific structure of hybrid nanocomposite particles.

In Fig. 15.26 in Eu^{3+}:t-ZrO_2 nanoparticles, the $^5D_0 \rightarrow {}^7F_4$ bandgroup over 680–730 nm appears in the second most intense bandgroup of the spectrum. Like the $^5D_0 \rightarrow {}^7F_2$ bandgroup, this is another forced electric dipole transition in Eu^{3+} species that is highly sensitive of the surface topology of core–shell Eu^{3+}:t-ZrO_2 nanoparticles. No so enhanced intensity occurs in either Eu_2O_3-doped or Eu_2O_3-coated ZrO_2 of nanocrystals when prepared through a sol–gel method using zirconium alkoxide [243]. The above results altogether confer the fact that part of the Eu^{3+} species occupies very low symmetry sites without an inversion center. This is very much feasible in core–shell Eu^{3+}:t-ZrO_2 nanoparticles in which majority of the Eu^{3+} species constitute the shell which suffers from a high level of constrain. X-ray diffraction data reveals a significant $Eu^{3+} \rightarrow Zr^{4+}$ substitution in the Eu^{3+}:t-ZrO_2 lattice, and it causes significant microstrain in the lattice [244]. All these factors induce a surface-enhanced light emission in Eu^{3+} species in a kind of a forced $^5D_0 \rightarrow {}^7F_J$ ($J = 0 \rightarrow 4$) transition. Microstrain is a major source of enhanced electron–phonon coupling[30,35,36] and that is believed to result in promoted light emission intensity in the forced electric dipole transitions, especially in core–shell nanoparticles.

In rare-earth ions, the electron–phonon coupling induces intermixing of the 4f energy levels that modifies optical transition probabilities in selective transitions in the rare-earth-doped materials [245, 257]. Zr^{4+} ion has an average sevenfold coordination with O^{2-} ions in m-ZrO_2 (C_s Zr^{4+}-site symmetry) and eightfold coordination in t-/c-ZrO_2 phase (nearly D_{4d} Zr^{4+}-site symmetry).[12,13,16] An analysis of site symmetry in Eu^{3+} species as a probe thus allows analyzing site symmetries for the Zr^{4+} species in ZrO_2 as a host in various polymorphous and/or in various environments. A low Eu^{3+}-site symmetry according to the light emission infers a partial $Eu^{3+} \rightarrow Zr^{4+}$ substitution in Eu^{3+}: ZrO_2 in the low symmetry sites. We can use the Judd–Ofelt parameter Ω_2 to express the intensity in the $^5D_0 \rightarrow {}^7F_2$ band (a hypersensitive transition) in terms of the ratio of its intensity I_{21} with respect to the $^5D_0 \rightarrow {}^7F_1$ band, i.e., a rather local environment insensitive transition. The ratio between the electric dipole and the magnetic dipole ($e/m \cong I_{21}$) in these bands is a measure of the site symmetry in the Eu^{3+} species. As large I_{21} value as 1.71 is observed in Eu^{3+}:t-ZrO_2 nanoparticles in Fig. 15.26. Gosh and Patra [243] observed as large Ω_2 -value as 2.41 in 1 mol% Eu^{3+}-doped t-/m-ZrO_2 (heated at 1100 °C for

1 h), describing a markedly increased covalent Eu^{3+}–O^{2-} character when m-ZrO_2 is present. A still larger Ω_2-value 2.81 is shown in a porous 2 mol% Eu^{3+}:t-/mZrO_2 powder (Fig. 15.25). The pores seem to increase the Eu^{3+}–O^{2-} covalent character further. Similar spectral features are observed in Eu^{3+}-doped silicate glasses [258].

In this conjuncture, an ionic doping in ZrO_2 or a similar ceramic host is supposed to reduce the e/m value. This is what has been shown when co-doping (stabilizes the t-/c-ZrO_2 phases) ZrO_2 using 5 and 8 mol% Ca^{2+} with 2 mol% Eu^{3+}. The e/m value 2.81 (before Ca^{2+} addition) has dropped to 1.65 and 1.62 at 5 and 8 mol% Ca^{2+}, respectively [254]. Sequentially, a value 3.19 is observed in a melt glass 1 mol% Eu^{3+}:30BaO-70P_2O_5 which consists of a covalent network interbridging Eu^{3+}–O^{2-} through the P^{5+}–O^{2-} bonds [259]. The $^5D_0 \rightarrow {}^7F_1$ and $^5D_0 \rightarrow {}^7F_2$ bandgroups have three and five distinct bands, respectively, in the $^7F_1$ and $^7F_2$ Stark splittings. The Eu^{3+} species in the different sites have the same type of ligand structure, but are not located in the different phases.

Reisfeld et al. [179] observed strong light emission on incorporating Eu_2O_3 and dibenzoylmethane chelate Eu(DBM)$_3$ into zirconia xerogel glasses. The DBM chelate was the most efficient in luminescence intensity. The intensity was even stronger in hybrid xerogels prepared out of a nonaqueous precursor. Aromatic and heteroaromatic β-diketones are good sensitizers for the light emission in rare-earth ions (especially Eu^{3+} and Tb^{3+}) upon ultraviolet excitation. These authors studied light emission in ZrO_2, Eu^{3+}-doped ZrO_2, and Eu(DBM)$_3$-doped ZrO_2 films. A dramatic enhancement of the signal, a doublet of two distinct bands of 590 nm ($^5D_0 \rightarrow {}^7F_1$) and 618 nm ($^5D_0 \rightarrow {}^7F_2$), occurs in the complex. A combination of organic and inorganic compounds in a hybrid xerogel designs hydrophobic domains in which H_2O molecules are expelled out to a major extent. This is well demonstrated with a value of intensity further enhanced roughly by an order of magnitude in Eu(DBM)$_3$ in ZrO_2-3-glycidoxypropyl-trimethoxysilan films.

Nanoporous ZrO_2 with Optical Inclusions

A nanoporous ZrO_2 is expected to present an efficient optical host of current interest with a new kind of local environments on rare-earth ions or other possible optical species used as a dopant in form of a hybrid nanocomposite in comparison to a nonporous amorphous or polymorphic ZrO_2. Small pores invite kinds of structural defects in selective sites of the host that facilitate desired doping in selective sites according to their local structure, stability, and chemistry. They can be treated a co-dopant or a sensitizer in tuning the dynamics of the optical parameters. Such effects can be observed very sensitively in terms of the optical signals such as light emission from the dopants, the doping modified nanoporous composite structure, or a combined effect of these two components. Considering copious reports on porous powders and films using an Al_2O_3 or silica host, not many reports are available on zirconia system. As described above (Figs. 15.19 and 15.20), a porous ZrO_2 absorbs ultraviolet light in a strong band of 233 nm and emits strongly in a broad band over 340–450 nm (due to ionized oxygen vacancies) [71], useful to promote light emission from selective impurity levels after proper doping, chemical modification, or thermal annealing in selective atmospheres.

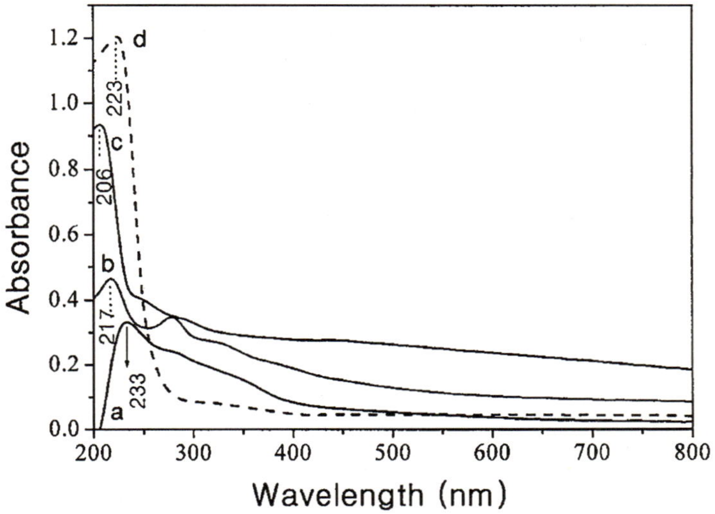

Fig. 15.27 Absorption spectra in (**a**) as-prepared u-ZrO_2, (**b**) phosphoric acid-treated u-ZrO_2 (heated at 500 °C in air), (**c**) ZrP_2O_7, and (**d**) ZrO_2 nanocrystals [260]

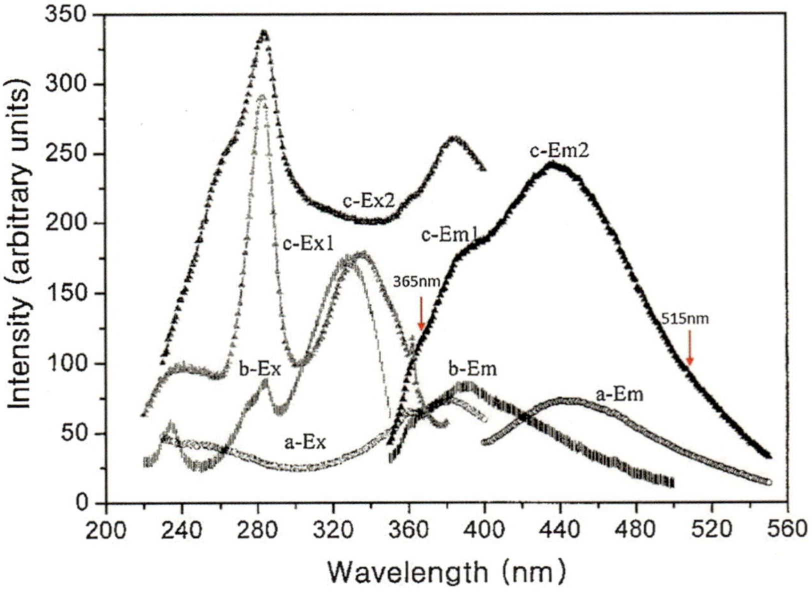

Fig. 15.28 PL emissions (−Em) and excitations (−Ex) spectra from (**a**) u–ZrO_2, (b) ZrP_2O_7, and (**b**) phosphoric acid-treated u-ZrO_2 (heated at 500 °C in air). (**c**) The excitation spectrum c-Ex1 was recorded for 390 nm emission, and the excitation spectrum c-Ex2 was recorded for 440 nm emission [260]

For example, Fig. 15.27 shows absorption spectra of (a) an as-prepared mesoporous ZrO_2 (u-ZrO_2) from a combustion method, (b) phosphoric acid-treated u-ZrO_2 after heating in flowing air at 500 °C for 6 h to remove the surfactant, (c) ZrP_2O_7, and (d) ZrO_2 nanocrystals. As demonstrated by Chen et al. [260], the acid-treated sample consists of a specific phosphate surface layer. In u-ZrO_2, the absorption band has red-shifted to 233 nm with considerably reduced intensity relative to a reference of ZrO_2 nanocrystals at 223 nm. Marked modification arises of the band edge absorption in u-ZrO_2 on forming a mesoporous structure, which accounts in several factors such as the interfacial chemical condition, ambient circumstance around ZrO_2, a change in oxidation state, and/or a coordination number in the surface zirconium ions. A phosphoric acid-treated u-ZrO_2 (heated) shows a distinctive blue shift and stronger absorption band at 217 nm. A further smaller value 206 nm of absorption band stands in ZrP_2O_7, supporting the fact that a distinctive blue shift is arising in this example from the incorporation of phosphate groups. Further, strong band edge absorptions from the onset absorption near 410 nm involve more absorption states in defect energy bands in the acid-treated u-ZrO_2 sample. This could be a reason for strong light emission in this sample (Fig. 15.28).

Figure 15.28 shows excitation and emission spectra from (a) u-ZrO_2, (b) ZrP_2O_7, and (c) phosphoric acid-treated u-ZrO_2 (heated at 500 °C in air). Sample (c) exhibits two emission bands at 390 nm 440 nm on exciting at selective wavelengths. The

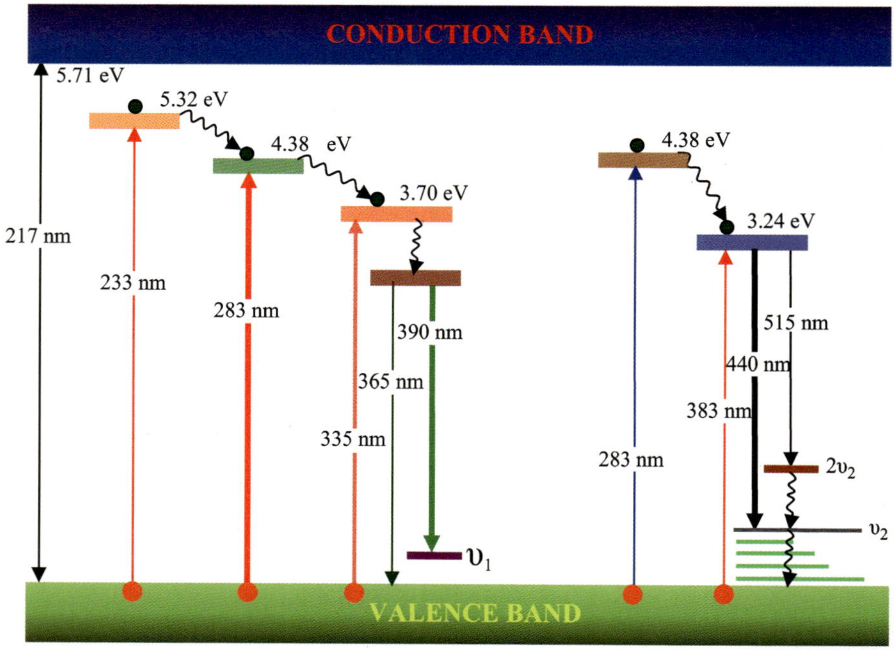

Fig. 15.29 The energy levels in observed transitions in absorption, emission, and excitation spectra in mesoporous u-ZrO_2 treated with phosphoric acid and heated at 500 °C in air. The phonon levels account for the multiple structure in the 390 and 440 nm emission bandgroups

390 nm emission could be exited at 233 nm, 283 nm, and 335 nm (spectrum c-Ex1), while the 440 nm emission was mainly excited at 283 and 383 nm (spectrum c-Ex2), suggesting the presence of two kinds of optical centers. The 390 nm emission in ZrP_2O_7 and 440 nm emission in u-ZrO_2 reproduce the two distinctive emission bands of this sample (c). Also the excitation bands of these two samples are matching to those of the acid-treated u-ZrO_2. Both the intrinsic optical defects from the oxo-phosphors in a model tetrapodal zirconium framework and the oxygen vacancies in a mesoporous u-ZrO_2 itself impart the light emission in a recombined material. The stronger light emission shown over those in u-ZrO_2 and ZrP_2O_7 is possible in the presence of many absorption states or kinds of defects, owing to the pore structure with a high surface area according to the absorption bands. The probability of the transition between energy levels is greatly enhanced as a result of their interactions in a hybrid mesoporous composite.

The two emission bandgroups of 390 and 440 nm in sample (c) in Fig. 15.29 have different decay times 0.935 ns (also 2.565 ns) and 2.010 ns, respectively, characterizing two different kinds of the emission centers. A proposed energy level diagram in Fig. 15.29 describes possible transitions in such centers lying in between the conduction band (at 217 nm according to the absorption band) and the valence band. Two distinct decay rates measured in the 390 nm emission describes that it involves two bands, which are clearly seen in Fig. 15.29c at 390 and 365 nm. A phonon spectrum clearly spans over longer wavelengths to the 440 nm primary emission. We believe that two distinct intermediate levels ν_1 (1750 cm^{-1}) and ν_2 (3380 cm^{-1}) mediate the emission in two distinct bandgroups. The level ν_1 stands very close to the valence band, and level ν_2 nearly doubles the ν_1 -value in a strongly correlated electronic band structure with the e-p coupling.

Thermoluminescence

Thermoluminescence (TL) is associated with the properties of materials to glow when heating at elevated temperature. The TL signal is obtained through the following steps: a sample of insulator or semiconductor is excited by radiation, X-rays, or ultraviolet light; when the sample is heated, luminescence (after glow) is emitted. Life times τ_1 of the glow distribute in the range from minutes to 4.6×10^9 years, and the applications are radiation dosimetry, age determination, geology, etc. [261, 262]. The impurities and defects, i.e., the traps of localized energy levels in the bandgap, are the source of the TL signal. Some traps are deep located at a considerable distance from the valance or conduction band. Such levels play the role of traps for charge carriers. For electron, at such level to be able to recombine with a hole, it should first be excited to the conduction band. The process can be stimulated by an electric field, light, or simply heating the sample. A light emission coming out of a pre-excited sample stimulated by heating is TL. It reveals the nature of the defects produced in such materials by radiation in general of all wavelengths. High-energy irradiation such as X-rays or γ-rays (which add such traps according to the sample) promotes and tunes the final TL signal and its dynamics.

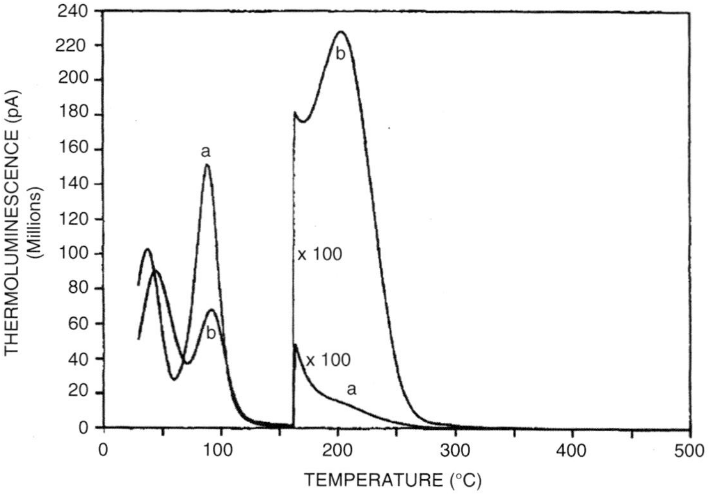

Fig. 15.30 TL spectra recorded after 302 nm irradiation at 12,000 pJ cm^{-2} exposure; (**a**) 1 mol% ZnO:ZrO_2 and (**b**) pure ZrO_2 sintered pellets at 1100 °C [268]

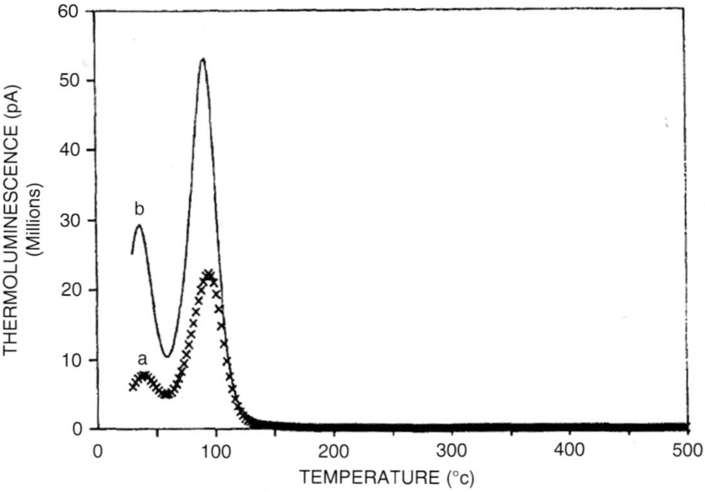

Fig. 15.31 TL spectra recorded after 302 nm irradiation at 12,000 pJ cm^{-2} exposure; (**a**) 1 mol% ZnO:ZrO_2 and (**b**) pure ZrO_2 sintered pellets at 1300 °C [268]

Ultraviolet irradiation-induced TL in ZrO_2 has been known for many years [263, 264]. The Curie model [265, 266] describes trap centers exhibiting TL peaks at 135, 190, 200, and 285 K in TiO_2-doped ZrO_2. A new trapping level of activation energy of order 1.29 eV appears in TL in rare-earth oxide-doped ZrO_2 after X-ray irradiation [267]. Let us describe TL spectra with a typical example of ZnO-doped ZrO_2 after the work of Hsieh and Su [268]. Figures 15.30 and 15.31 compare TL recorded from (a) 1 mol% ZnO:ZrO_2 and (b) pure ZrO_2 (sintered at 1100 °C and 1300 °C for 7 h)

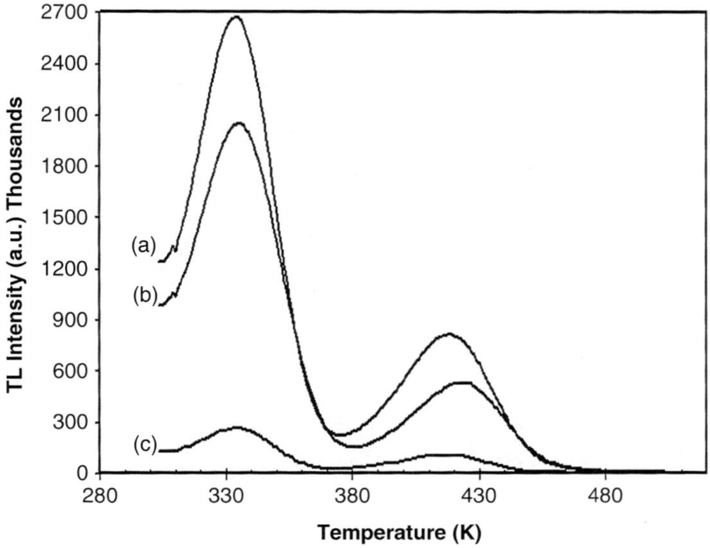

Fig. 15.32 TL spectra induced by ultraviolet irradiation in m-ZrO_2; (**a**) pure, (**b**) with added Ag nanoparticles, (**c**) with impregnated Ag (exposition time: 30 s) [271]

by irradiating at 302 nm at 12,000 pJ cm^{-2} exposure. In 1100 °C sintering, pure ZrO_2 has two peaks at 45 °C and 90 °C, which are shifted to 40 °C and 85 °C on doping. Intensity is increased in the 85 °C peak roughly by three times on the doping. High temperature peak at 210 °C (relatively weak compared to the low temperature peaks) has an order of lowered value of the intensity in pure ZrO_2. On increasing sintering temperature at 1300 °C, only two peaks persist, i.e., 40 °C and 85 °C before doping and 45 °C and 90 °C after doping. Increasing ZnO content, such as 20 mol%, results in a minor shift in the TL peaks near 85 °C over higher temperatures with a further enhanced intensity, with no significant change in the 210 °C peak.

Figure 15.32 shows TL spectra induced by ultraviolet irradiation in sol–gel-synthesized m-ZrO_2: (a) pure, (b) with 1 wt % silver nanoparticles (heated at 1000 °C in air), and (c) with 1 wt% impregnated silver (via a solution having AgI and m-ZrO_2 and then dried at 300 °C in air). Two peaks, which arise at 334 K and 417 K in m-ZrO_2 before and after silver impregnation, shift to 336 and 423 K in m-ZrO_2 with added Ag nanoparticles. Commercial ZrO_2 also has two peaks at 343 and 403 K [268, 269], while an Er_2O_3-doped t-ZrO_2 has modified peaks at 363 and 483 K [270]. Since the wavelength of ultraviolet radiation does not alter the position of the TL peaks and the quantity of radiation (dose) does influence the signal intensity, these spectral changes arise primarily in variation of sample crystallite size (also morphology) and TL reading condition (heating rate).

Figure 15.33 portrays the fading characteristics of the integrated TL signal of the m-ZrO_2 measured after ultraviolet irradiating for 30s at different storage times (a) 0.25, (b) 0.50, (c) 1.0, and (d) 24 h. A strong decrease (~20 %) occurs in

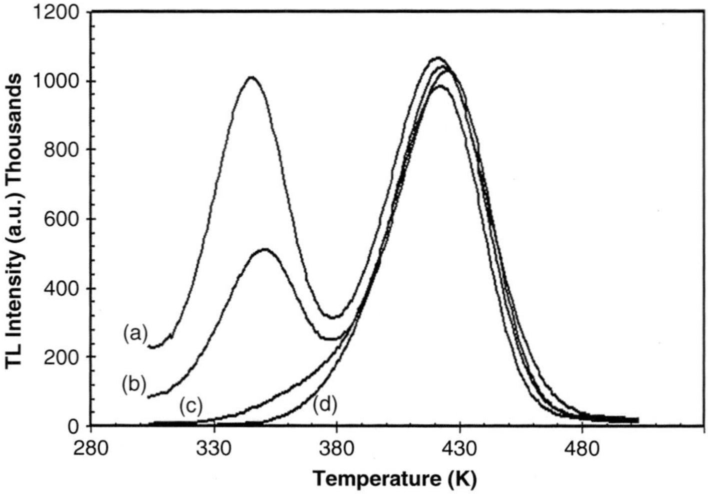

Fig. 15.33 TL fading characteristics of pure m-ZrO_2 after ultraviolet irradiation for 30 s, with storage time (**a**) 0.25, (**b**) 0.50, (**c**) 1.0 and (**d**) 24 h [271]

Table 15.5 TL glow curve areas of pure and doped m-ZrO_2 exposed to γ-radiation

Dose (Gy)	ZrO_2	ZrO_2:Ag(s)[a]	ZrO_2:Ag(I)[b]
0.5	73.81 ± 5.47	68.89 ± 8.35	4.70 ± 0.50
1	152.60 ± 14.52	138.14 ± 10.74	9.08 ± 0.91
2	300.16 ± 10.92	326.81 ± 27.17	18.40 ± 1.77
3	598.53 ± 64.09	583.28 ± 49.62	33.14 ± 2.58
5	1188.72 ± 82.17	1066.50 ± 94.17	68.44 ± 6.26
7.5	2248.27 ± 158.12	1871.33 ± 130.17	133.55 ± 7.35
10	3125.78 ± 261.91	2537.44 ± 277.57	199.41 ± 12.53
15	4995.94 ± 197.99	3855.77 ± 320.44	305.76 ± 27.83
20	6617.45 ± 595.79	4854.10 ± 430.34	398.98 ± 34.16
30	8484.17 ± 436.51	6750.30 ± 422.28	548.36 ± 23.78
50	13960.74 ± 810.35	10709.69 ± 903.87	967.42 ± 91.95

[a]Monoclinic ZrO_2 with added Ag nanoparticles
[b]Monoclinic ZrO_2 with impregnated Ag nanoparticles

integrated TL signal during the first hour after irradiation (Fig. 15.33c), showing stable features afterwards (Fig. 15.33d). The strong fading observed during the first hour is mainly associated to the fade of the low temperature peak at 334 K in pure m-ZrO_2. A similar behavior has been observed in Ag-doped m-ZrO_2. Table 15.5 summarizes the data for the TL intensities studied on the three kinds of the m-ZrO_2 samples exposed to γ-radiation for 30 s. The curves were obtained by plotting the integrated TL intensity as a function of the γ-radiation in the 0.5 50 Gy range. The integration of the TL signal was carried out 15 min after sample irradiation in order to standardizing the fading effects.

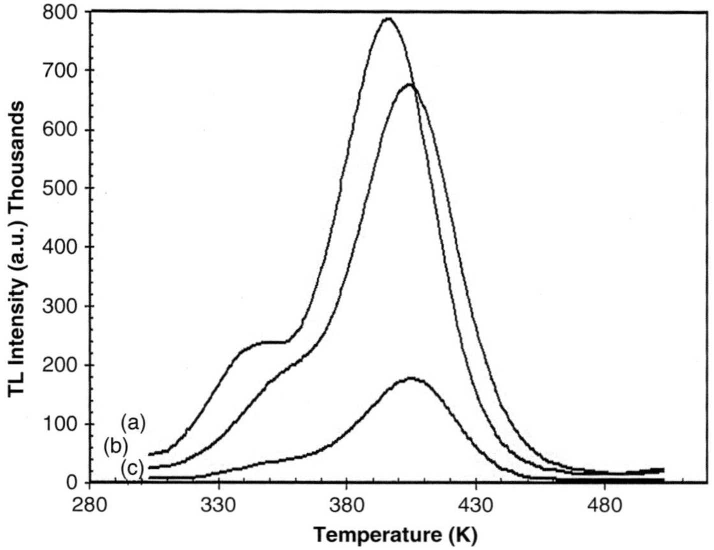

Fig. 15.34 TL response induced by γ-radiation in m-ZrO_2; (**a**) pure, (**b**) with added Ag nanoparticles, (**c**) with impregnated Ag (dose 30 Gy) [271]

Figure 15.34 shows the TL signals obtained from m-ZrO_2: (a) pure, (b) with 1 wt% Ag nanoparticles, and (c) with 1 wt% impregnated silver after 15 min γ-irradiation with a dose of 30 Gy. As shown with ultraviolet irradiation, two TL peaks occurring at 342 K and 397 K in m-ZrO_2 before and after silver impregnation are shifted to 349 K and 404 K in ZrO_2 with added Ag nanoparticles. The intensity patterns in these samples stand in the same order as those shown above with ultraviolet irradiation (Fig. 15.32). It is important to analyze the role that Ag nanoparticles in m-ZrO_2 are playing on the TL signal. When Ag nanoparticles are added during synthesizing m-ZrO_2, the TL intensity diminishes slightly; however, there is a noticeable shift toward higher temperature values (Fig. 15.34b). On the other hand, when silver is impregnated on m-ZrO_2 surfaces, the TL intensity suffers a considerable decrease despite no change in the peak values (Fig. 15.34c). Thin films of 5 mol% Eu^{3+}-doped t-ZrO_2 exhibit two TL peaks at 120 °C and 290 °C [272]. The first peak, which is prominent, corresponds to the main peak shown in m-ZrO_2.

Applications

Growing demand for ZrO_2 and hybrid nanocomposites is based on their unique properties, which make them indispensable to many high-performance applications such as in thermal and chemical barrier coating, buffer layer for high temperature superconducting films, ball heads for hip replacements, and automobiles. Doped ZrO_2, using selective oxides of rare earths, alkaline earths, or transition metals, offers several other multifunctional properties, which include a high refractive index, a high degree of luminescence, catalytic activities, superior mechanical toughness, and high wear resistance. Specific applications include structural tools, medical tools, and turbine blades, dielectric formulations in electronic chips such as multilayer ceramic capacitors for almost all kinds of printed circuit boards and electronic devices. In view of optical properties, the efforts of developing rare-earth and/or transition metal oxide-based ZrO_2 of ceramics, glasses, and hybrid nanocomposites stem from unusually self-confined vibronic structure (strong e-p coupling) with support of low energy phonons in a special kind of materials of insulators or wideband semiconductors. Novel materials include mesoporous composites wherein the pores in spite of generating new optical centers mediate designing a discrete structure of the optical (or also the spin) centers of self-minimized energy loss in the concentration quenching, the cross-relaxations, and other possible processes. Specific applications include optical systems and devices, optoelectronics, optical waveguides, optical data storage systems, optical communications, optical displays, and series of optical and/or biological sensors.

In brief, polymorphic ZrO_2 and hybrid composites offer widespread high-tech applications in numerous disciplines as follows:

(i) Monolithic ZrO_2 polymorph
- Transparent windows for near ultraviolet, visible, and near-infrared regions
- Ultraviolet filters $\leq$ 250 nm
- Catalysis for producing heavy alcohols
- Thermoluminescence
- Biomaterials and medial tools
- Gas sensors

(ii) Stabilized ZrO_2 and composites
- Ultraviolet filters
- High temperature solid fuel cells
- Catalysis
- Optical materials and lasers
- Camouflage and color pigments
- Micro barcodes
- Dosimeter
- Memory devices
- Ferroics and applications
- Structural tools, medical tools, and biomaterials
- Porous membranes for hot gas filters
- Quantum cutting
- Down energy conversion
- Automobile industries
- Substrates
- Solar energy conversion
- Optoelectronics
- Power gates

- Thermoluminescence
- Robotics
- Bone replacements (Al_2O_3- and MgO-based lightweight composites)

Concluding Remarks

In recent years of development, rare-earth ions-doped polymorphic ZrO_2 as light-emitting materials are found to be highly commensurate for excellent optical and other properties and applications under varied experimental conditions of temperature and pressure. Oxide glasses and polymer–ceramic composites, which offer extended shape formability, are especially important for designing and fabricating varieties of products of desired shapes and sizes of thin films, quantum dots, sheets, plates, rings, rods, thin wires, or coils for specific engineering applications. A core–shell nanostructure supports superplasticity which promotes useful mechanical, optical, electronic, and other desired properties of functionalized materials of hybrid nanocomposites.

References

1. Wilson AH (1954) The theory of metals. Cambridge Press, New York
2. Coutts TJ (1974) Electrical conduction in thin metal films. Elsevier, New York
3. Ashcroft NW, Mermin ND (1976) Solid state physics. WB Saunders, Philadelphia
4. Colombo P (2006) Conventional and novel processing methods for cellular ceramics. Philos Trans R Soc A 364:109
5. Farah AA, Hall N, Morin S, Pietro WJ (2006) Poly(ε-caprolactone)-block-polystyrene metallopolymers via sequential ROP and ATRP condition with in situ generated ruthenium catalyst. Polymer 47:4282
6. Motaung DE, Malgas GF, Christopher JA, Mavundla SE, Clive JO, Knoesen D (2009) The influence of thermal annealing on the morphology and structural properties of a conjugated polymer in blends with an organic acceptor material. J Mater Sci 44:3192
7. Lucas JP, Moody NR, Robinson SL, Hanrock J, Hwang RQ (1995) Determining fracture toughness of vitreous silica glass. Scripta Mater 35:743
8. Tomozawa M (2001) Silicon-based materials and devices. In: Nalwa HS (ed) Hand book of nanostructured materials and nanotechnology. Academic, New York, p 127
9. Břuning R, Levelut C, Faivre A, LeParc R, Simon JP, Bley F, Hazemann JL (2005) Characterization of the glass transition in vitreous silica by temperature scanning small-angle X-ray scattering. Europhys Lett 70:211
10. Wang J, Lin QH, Zhou RQ, Xu BK (2002) Humidity sensor based on composite materials of nano-Bati O_3 and polymer RMX. Sens Actuators B 81:248
11. Li N, Wang J, Li M (2003) Electrochemistry at carbon nanotube electrodes. Rev Anal Chem 22:19
12. Cho HJ, Choi GM (2008) Effect of milling methods on performance of Ni-Y_2O_3-stabilized ZrO_2 anode for solid oxide fuel cell. J Power Sources 176:96
13. Wohlfarth EP (1982) Ferromagnetic materials. North Holland, New York
14. O'Handley RC (2000) Modern magnetic materials principles and applications. Wiley, New York

15. Zhua XH, Li J, Zheng DN (2007) Frequency and temperature dependence of tunable dielectric properties of $Ba(Zr_{0.2}Ti_{0.8})O_3$ thin films grown on (001) MgO. Appl Phys Lett 90:142913
16. De Groot RA, Mueller FM, Van Engen PG, Buschow KHJ (1983) New class of materials: half metallic ferromagnets. Phys Rev Lett 50:2024
17. Buschow KHJ (1991) Handbook of magnetic materials. North-Holland, Amsterdam
18. Garter M, Scofield J, Birkhahn R, Steckl AJ (1999) Visible and infrared rare-earth-activated electroluminescence from indium tin oxide Schottky diodes to GaN:Er on Si. Appl Phys Lett 74:182
19. Huh C, Lee JM, Kim DJ, Park SJ (2002) Improvement in light-output efficiency of InGaN/GaN multiple-quantum well light-emitting diodes by current blocking layer. J Appl Phys 92:2248
20. Xu HY, Liu YC, Liu YX, Xu CS, Shao CL, Mu R (2005) Ultraviolet electroluminescence from p-GaN/i-ZnO/*n*-ZnO heterojunction light-emitting diodes. Appl Phys B 80:871
21. Lakshminarasimhan N, Varadaraju UV (2005) White-light generation in $Sr_2SiO_4:Eu^{2+}$, Ce^{3+} under near-UV excitation. J Electrochem Soc 152:H152
22. Liu Y, Xu CN, Matsui H, Imamura T, Watanabe T (2000) Preparation and luminescence of rare-earth-activated Y_2SiO_5 thin films by metallorganic decomposition. J Lumin 87–89:1297
23. Heber J, Mühlig C, Triebel W, Danz N, Thielsch R, Kaiser N (2002) Deep UV laser induced luminescence in oxide thin films. Appl Phys A 75:637
24. Sreseli OM, Goryachev DN, Belyakov LV, Vul SP, Zakharova IB, Alekseeva EA (2004) Effect of a fullerene coating on the photoluminescence of porous silicon. Semiconductors 38:120
25. Louisa C, Rouxa S, Ledouxa G, Dujardina C, Tillementa O, Cheng BL, Perriat P (2006) Luminescence enhancement by energy transfer in core-shell structures. Chem Phys Lett 429:157
26. Yang P, Zhao D, Margolese DI, Chmelka BF, Stucky GD (1998) Generalized syntheses of large-pore mesoporous metal oxides with semicrystalline frameworks. Nature 396:152
27. Cai W, Zhang Y, Jia J, Zhang L (1998) Semiconducting optical properties of silver/silica mesoporous composite. Appl Phys Lett 73:2709
28. Murakami S, Herren M, Morita M (1998) Excitation energy transfer and low-temperature anomaly in luminescence of RE^{31} codoped PLZT ceramics. J Lumin 76&77:460
29. Ishizaka T, Nozaki R, Kurokawa Y (2002) Luminescence properties of Tb^{3+} and Eu^{3+} -doped alumina films prepared by sol-gel method under various conditions and sensitized luminescence. J Phys Chem Solids 63:613
30. Mohanty P, Ram S (2003) Enhanced photoemission in dispersed Eu_2O_3 nanoparticles in amorphous Al_2O_3. J Mater Chem 13:3021
31. Ram S, Kundu TK (2004) Synthesis and unusual electron paramagnetic resonance spectrum of metastable nanoclusters of ZnO semiconductor crystallites. J Nanosci Nanotech 4:1076
32. Bang J, Yang H, Holloway PH (2005) Enhanced luminescence of $SiO_2:Eu^{3+}$ by energy transfer from ZnO nanoparticles. J Chem Phys 123:084709
33. Kawai C (2005) Porous silicon nitride ceramic phosphors. J Mater Sci 40:2591
34. Balakrishnan S, Gun'ko YK, Perova TS, Rafferty A, Astrova EV, Moore RA (2005) Porous silicon – rare earth doped xerogel and glass composites. Phys Status Solidi (a) 202:1693
35. Mohanty P, Ram S (2006) Light emission associated with the $^5D_0 \rightarrow {}^7F_3$ forbidden transition in Eu^{3+} cations dispersed in an Eu^{3+}: Al_2O_3 mesoporous structure. Philos Mag Lett 86:375
36. Meltzer RS, Hong KS (2000) Electron-phonon interactions in insulating nanoparticles: Eu_2O_3. Phys Rev B 61:3396
37. Reimanis IE (1997) A review of issues in the fracture of interfacial ceramics and ceramic composites. Mater Sci Eng A 237:159
38. Ajayan PM, Schadler LS, Braun PV (2003) Nanocomposite science and technology. Wiley-VCH, Germany
39. Tjong SC (2006) Structural and mechanical properties of polymer nanocomposites. Mater Sci Eng R 53:73

40. Nalwa HS (2000) Handbook of nanostructured materials and nanotechnology. Academic, New York
41. Thakoor AP, Raj B, Pandya DK, Chopra KL (1981) Metal chalcogenide-oxide composite coatings prepared by spray pyrolysis. Thin Solid Films 83:231
42. Gai PL, Harmer MA (2002) Surface atomic defect structures and growth of gold nanorods. Nano Lett 2:771
43. Shenhar R, Rotello VM (2002) Nanoparticles: scaffolds and building blocks. Acc Chem Res 36:549
44. Kamat PV (2002) Photoinduced transformations in semiconductor-metal nanocomposite assemblies. Pure Appl Chem 74:1693
45. Chen MMY, Katz A (2002) Steady-state fluorescence-based investigation of the interaction between protected thiols and gold. Langmuir 18:2413
46. Gao Y, Jiang P, Liu DF, Yuan HJ, Yan XQ, Zhou ZP, Wang JX, Song L, Liu LF, Zhou WY, Wang G, Wang CY, Xie SS (2003) Synthesis, characterization and self-assembly of silver nanowires. Chem Phys Lett 380:146
47. Thomas KG, Kamat PV (2003) Chromophore-functionalized gold nanoparticles. Acc Chem Res 36:888
48. Edelstein AS, Cammaratra RC (eds) (1996) Nanomaterials: synthesis, properties and applications. Taylor & Francis Group, New York
49. Xu R, Wang D, Zhang J, Li Y (2006) Shape-dependent catalytic activity of silver nanoparticles for the oxidation of styrene. Chem Asian J 1:888
50. Hansma PK, Turner PJ, Ruoff RF (2007) Optimized adhesives for strong, lightweight, damage-resistant, nanocomposite materials new insights from natural materials. Nanotechnology 18:44026
51. Brydson RM, Hammond C (2000) Genetic methodologies for nanotechnology: classification and fabrication. In: Kelsall RW, Hamley IW, Geoghegan M (eds) Nanoscale science and technology. Wiley, Germany, p 1
52. Sōmiya S, Akiba T (1999) A high potential material-zirconia. Bull Mater Sci 22:207
53. Li Y, Meng GW, Zhang LD, Phillip F (2000) Ordered semiconductor ZnO nanowires Arrays and their photoluminescence properties. Appl Phys Lett 76:2011
54. Huang MH, Mao S, Feick H, Yan H, Wu Y, Weber E, Russo R, Yang P (2001) Room-temperature ultraviolet nanowire nanolasers. Science 292:1897
55. Huang MH, Wu Y, Feick H, Tran N, Weber E, Yang P (2001) Catalytic growth of zinc oxide nanowires by vapor transport. Adv Mater Res 13:113
56. Kong YC, Yu DP, Zhang B, Fang W, Feng SQ (2001) Ultraviolet-emitting ZnO nanowires synthesized by a physical vapor deposition approach. Appl Phys Lett 78:407
57. Sands D, Brunson KM, Cheung CC, Thomas CB (1988) $ZnS_{1-x}O_x$: Mn and ZnS_{1-x} Se_x: Mn as thin film electroluminescent materials. Semicond Sci Technol 3:816
58. Blasse G, Grabmaier BC (1994) Luminescent materials. Springer, Heidelberg
59. Justel T, Krupa J, Wiechert DU (2001) VUV spectroscopy of luminescent materials for plasma display panels and Xe discharge lamps. J Lumin 93:179
60. Minami T (2003) Oxide thin-film electroluminescent devices and materials. Sol State Elect 47:2237
61. Wilk GD, Wallance RM, Anthony JM (2001) High-k gate dielectrics current status and materials properties considerations. J Appl Phys 89:5243
62. Zhang JY, Boyd IW (2002) Structural and electrical properties of tantalum oxide films grown by photo-assisted pulsed laser deposition. Appl Surf Sci 186:40
63. Kadoshima M, Hiratani M, Shimamoto Y, Torii K, Miki H, Kimyra S, Nabatame T (2003) Rutile-type TiO_2 thin film for high-k gate insulator. Thin Solid Films 424:224
64. Fang Q, Zhang JY, Wang ZM, Wu JX, Osullivan BJ, Hurley PK, Leedham TL, Davies H, Audier MA, Jimenez C, Senateur JP, Boyd IW (2003) Characterisation of HfO_2 deposited by photo-induced chemical vapour deposition. Thin Solid Films 427:391
65. Forst CJ, Ashman CR, Schwarz K, Blochl PE (2004) The interface between silicon and a high-k oxide. Nature 427:53

66. Hubbard KJ, Schlom DG (1996) Thermodynamic stability of binary oxides in contact with silicon. J Mater Res 11:2757
67. Puthenkovilakam R, Chang JP (2004) Valence band structure and band alignment at the ZrO_2/Si interface. Appl Phys Lett 84:1353
68. Robertson J (2000) Band offsets of wide-band-gap oxides and implications for future electronic devices. Vac J Sci Technol B 18:1785
69. Cao H, Xianqing Q, Luo B, Zhang Y, Tan R, Zhao M, Zhu Q (2004) Synthesis and room temperature ultraviolet photoluminescence properties of zirconia nanowires. Adv Funct Mater 14:243
70. French RH, Glass SJ, Ohuchi FS, Xu YN, Ching WY (1994) Experimental and theoretical determination of the electronic structure and optical properties of three phases of ZrO_2. Phys Rev B 49:5133
71. Aita CR, Hoppe EE, Sorbello RS (2003) Fundamental optical absorption edge of undoped tetragonal zirconium dioxide. Appl Phys Lett 82:677
72. Kralik B, Chang EK, Louie SG (1998) Structural properties and quasiparticle band structure of zirconia. Phys Rev B 57:7027
73. Emeline A, Kataeva GV, Litke AS, Rudakova AV, Ryabchuk VK, Serpone N (1998) Spectroscopic and photoluminescence studies of a wide band gap insulating material: powdered and colloidal ZrO_2 sols. ACS Langmuir 14:5011
74. Miyazaki S (2001) Photoemission study of energy-band alignments and gap-state density distributions for high-k gate dielectrics. J Vac Sci Technol B 19:2212
75. Zhu LQ, Fang Q, He G, Liu M, Zhang LD (2005) Microstructure and optical properties of ultra-thin zirconia films prepared by nitrogen-assisted reactive magnetron sputtering. Nanotechnology 16:2865
76. Liang L, Sheng Y, Xu Y, Wu D, Sun Y (2007) Optical properties of sol-gel derived ZrO_2-TiO_2 composite films. Thin Solid Films 515:7765
77. Yawen Z, Shu J, Liao C, Yan CH (2002) Microstructures and optical properties of nanocrystalline rare earth stabilized zirconia thin films deposited by a simple sol-gel method. Mater Lett 56:1030
78. Kosacki I, Petrovsky V, Anderson HU (1999) Band gap energy in nanocrystalline ZrO_2:16%Y thin films. Appl Phys Lett 74:341
79. Hartridge A, Krishna MG, Bhattacharya AK (2001) Temperature and ionic size dependence on the structure and optical properties of nanocrystalline lanthanide doped zirconia thin films. Thin Solid Films 384:254
80. Zhu LQ, Fang Q, He G, Liu M, Zhang LD (2006) Interfacial and optical properties of ZrO_2/Si by reactive magnetron sputtering. Mater Lett 60:888
81. Zhou Y, Kojima N, Sasaki K (2008) Growth and dielectric properties of tetragonal ZrO_2 films by limited reaction sputtering. J Phys D Appl Phys 41:175414
82. Dey A, De SK (2006) Impedance and dielectric spectra in zirconia-polypyrrole hybrid nano composites. J Phys D Appl Phys 39:4077
83. Smith DK, Newkirk W (1965) The crystal structure of baddeleyite (monoclinic ZrO_2) and its relation to the polymorphism of ZrO_2. Acta Cryst 18:983–991
84. McCullough JD, Trueblood KN (1959) The crystal structure of baddeleyite (monoclinic ZrO_2). Acta Cryst 12:507
85. Ruff O, Ebert F (1929) Die formen des zirkondioxyds. Z Anorg Allg Chem 180:19
86. Lynch CT, Vahldiek FW, Rabinson LB (1961) Monoclinic-tetragonal transition of zirconia. J Am Ceram Soc 44:147
87. Intrater J, Herwitt S (1961) High temperature, high vacuum, diffractometer attachment. Rev Sci Instrum 32:905
88. Balmer ML, Lange FF, Levi CG (1994) Metastable phase selection and partitioning for $Zr_{(1-x)}Al_xO_{(2-x/2)}$ materials synthesized with liquid precursors. J Am Ceram Soc 77:2069
89. Moreau S, Gervais M, Douy A (1997) Formation of metastable solid solution in the ZrO_2 rich part of the system ZrO_2-Al_2O_3. Solid State Ion 101–103:625

90. Mondal A, Ram S (2003) Controlled phase transformations in Al^{3+} stabilized ZrO_2 nanoparticles via forced hydrolysis of metal cations in water. Mater Lett 53:1696
91. Hu Michael ZC, Hunt RD, Payzant EA, Hubbard CR (1999) Nanocrystallization and phase transformation in monodispersed ultrafine zirconia particles from various homogeneous precipitation. Methods J Am Ceram Soc 82:2313
92. Koji M, Michiharu O (2001) Formation mechanism of hydrous zirconia particles produced by the hydrolysis of $ZrOCl_2$ solutions: III, kinetics study for the nucleation and crystal-growth processes of primary particles. J Am Ceram Soc 84:2303
93. Hannink-Richard HJ, Kelly PM, Muddle BC (2000) Transformation toughening in zirconia-containing ceramics. J Am Ceram Soc 83:461
94. Suyama R, Ashida T, Kume S (1985) Synthesis of the orthorhombic phase of ZrO_2. J Am Ceram Soc 68:C314
95. Adams DM, Leonard S, Russell DR, Cernik RJ (1991) X-ray diffraction study of hafnia under high pressure using synchrotron radiation. J Phys Chem Solids 52:1181
96. Dewhurst JK, Lowther JE (1998) Relative stability, structure, and elastic properties of several phases of pure zirconia. Phys Rev B 57:741
97. Lowther JE, Dewhurst JK, Leger JM, Haines J (1999) Relative stability of ZrO_2 and HfO_2 structural phases. Phys Rev B 60:14485
98. Fabris S, Paxton AT, Finnis MW (2001) Free energy and molecular dynamics calculations for the cubic-tetragonal phase transition in zirconia. Phys Rev B 63:94101
99. X-ray Powder Diffraction File JCPDS-ICDD (Joint Committee on Powder Diffraction Standards-International Centre for Diffraction Data, Swarthmore, PA) (1999) (a) 27–0997, c-ZrO_2, (b) 79–1771, and 81–1327, t-ZrO_2, (c) 13–0307, m-ZrO_2 and (d) 33–1483, 37–1413, 79–1796, and 83–0810, o-ZrO_2
100. X-ray Powder Diffraction File JCPDS-ICDD (Joint Committee on Powder Diffraction Standards-International Centre for Diffraction Data, Swarthmore, PA) (2002) (a) 27–0997, c-ZrO_2, (b) 24–1164, t-ZrO_2, (c) 13–307, m-ZrO_2, and (d) 41–0017, 79–1796, 83–0810 and 37–1413, o-ZrO_2.
101. Mondal A, Ram S (2003) Formation of a new polymorph of ZrO_2 with orthorhombic crystal structure contained in a mesoporous structure. Chem Phy Lett 382:297
102. Fabris S, Paxton AT, Finnis MW (2000) Relative energetics and structural properties of zirconia using a self-consistent tight-binding model. Phys Rev B 61:6617
103. Ryshkewitch E (1960) Oxide ceramics: physical chemistry and technology. Academic, New York, p 3
104. Garvie RC, Nicholson PS (1972) Structure and thermomechanical properties of partially stabilized zirconia in the CaO-ZrO_2 system. J Am Ceram Soc 55:152
105. Garvie RC, Hannink RH, Pascoe RT (1975) Ceramic steel? Nature 258:703
106. Kelly PM, Rose LRF (2002) The martensitic transformation in ceramics-its role in transformation toughening. Prog Mater Sci 47:463
107. Ishida K, Hirot K, Yamaguchi O, Kume H, Inamura S, Miyamoto H (1994) Formation of zirconia solid solutions containing alumina prepared by new preparation method. J Am Ceram Soc 77:1391
108. Tsukada T, Venigalla S, Morrone AA, Adair JH (1999) Low-temperature hydrothermal synthesis of yttrium-doped zirconia powders. J Am Ceram Soc 82:1169
109. Xia B, Duan L, Xie Y (2000) ZrO_2 nanopowders prepared by low-temperature vapor-phase hydrolysis. J Am Ceram Soc 83:1077
110. Raghavan S, Wang H, Porter WD, Dinwiddie RB, Mayo MJ (2001) Thermal properties of zirconia co-doped with trivalent and pentavalent oxides. Acta Mater 49:169
111. Ram S (2003) Synthesis and structural and optical properties of metastable ZrO_2 nanoparticles with intergranular Cr^{3+}/Cr^{4+} doping and grain surface modification. J Mater Sci 38:643
112. Lee JH, Mori T, Li JG, Ikegami T, Takenouchi S (2001) The influence of alumina addition and its distribution upon grain boundary conduction in 15 mol% calcia-stabilized zirconia. Ceram Intern 27:269

113. Cormack AN, Parker SC (1990) Some observations on the role of dopants in phase transitions in zirconia from atomistic simulations. J Am Ceram Soc 73:3220
114. Schelling PK, Phillpot SR, Wolf D (2001) Mechanism of the cubic-to-tetragonal phase transition in zirconia and yttria-stabilized zirconia by molecular-dynamics simulation. J Am Ceram Soc 84:1609
115. Sōmiya S, Bradt RC (eds) (1987) Fundamental structural ceramics. Terra Scientific, Tokyo, p 464
116. Ruehle M (1997) Microscopy of structural ceramics. Adv Mater 9:195
117. Mondal A, Ram S (2003) Al^{3+}-stabilized c-ZrO_2 nanoparticles at low temperature by forced hydrolysis of dispersed metal cations in water. Solid State Ion 160:169
118. Garvie RC (1965) The occurrence of metastable tetragonal zirconia as a crystallite size effect. J Phys Chem 69:1238
119. Holmes H, Fuller EJ, Gamage R (1972) Heats of immersion in the zirconium oxide-water system. J Phys Chem 76:1497
120. Livey DT, Murray P (1956) Surface energies of solid oxides and carbides. J Am Ceram Soc 39:363
121. Bailey JE, Lewis D, Librant ZM, Porter LJ (1965) Phase transformations in milled zirconia. Trans J Br Ceram Soc 71:25
122. Murase Y, Kato E (1983) Tetragonal-monoclinic phase transformation of ZrO_2. J Am Ceram Soc 66:196
123. Christiansen A, Carter EA (1998) First-principles study of the surfaces of zirconia. Phys Rev B 58:8050
124. Chraska T, King AH, Berndt CC (2000) On the size-dependent phase transformation in nanoparticulate zirconia. Mater Sci Eng A 286:169
125. Livage J, Doi K, Mazieres C (1968) Nature and thermal evolution of amorphous hydrated zirconium oxide. J Am Ceram Soc 51:349
126. Tani E, Yoshimura M, Sōmiya S (1983) Revised phase diagram of the system ZrO_2-CeO_2 below 1400°C. J Am Ceram Soc 66:506
127. Osendi MI, Moya JS, Serna CJ, Soria J (1985) Metastability of tetragonal zirconia powders. J Am Ceram Soc 68:135
128. Mitshuhashi T, Ichiara M, Tatsuke V (1974) Characterization and stabilization of metastable tetragonal ZrO_2. J Am Ceram Soc 57:97
129. Skandan G, Halm H, Roddy M, Cannon WR (1994) Ultrafine-grained dense monoclinic and tetragonal zirconia. J Am Ceram Soc 77:1706
130. Winterer M, Nitsche R, Redfern SAT, Schmahl WW, Hahn H (1995) Phase stability in nanostructured and coarse grained zirconia at high pressures. Nanostruct Mater 5:679
131. Nitsche R, Rodewald M, Skandan G, Fuess H, Halm H (1996) HRTEM study of nano crystalline zirconia powders. Nanostruct Mater 7:535
132. Khan MS, Islam MS, Bates DR (1998) Cation doping and oxygen diffusion in zirconia a combined atomistic simulation and molecular dynamics study. J Mater Chem 8:2299
133. Srinivasan R, Rice L, Davis BH (1990) Effect of pH on crystal phase of ZrO_2 from solution and calcined at 600°C. J Am Ceram Soc 68:135
134. Kuwabara A, Katamura J, Ikuhara Y, Sakuma T (2002) Influence of interaction between neighboring oxygen ions on phase stability in cubic zirconia. J Am Ceram Soc 85:2557
135. Cohen I, Schaner BE (1963) A metallographic and x-ray study of the UO_2-ZrO_2 system. J Nucl Mater 9:18
136. Li P, Chen IW, Penner-Hahn JE (1994) Effect of dopants on zirconia stabilization-An X-ray absorption study: II, trivalent dopants. J Am Ceram Soc 77:1281
137. Hellmanm JR, Stinican VS (1983) Stable and metastable phase relations in the system ZrO_2-CaO. J Am Ceram Soc 66:260
138. Murase Y, Kato E, Diamon K (1986) Stability of ZrO_2 phases in ultrafine ZrO_2-Al_2O_3 mixtures. J Am Ceram Soc 69:83

139. Scott HG (1975) Phase relationships in the zirconia-yttria system. J Mater Sci 10:1527
140. Ho SM (1982) On the structural chemistry of zirconium oxide. Mater Sci Eng 54:23
141. Li P, Chen IW, Penner-Hahn JE (1994) Effect of dopants on zirconia stabilization-An X-ray absorption study: I, trivalent dopants. J Am Ceram Soc 77:118
142. Itoh T (1985) Particle and crystallite sizes of ZrO_2 powder obtained by the calcination of hydrous zirconia. J Mater Sci Lett 4:431
143. Cotton FA, Wilkinson G (1972) Advanced inorganic chemistry. Wiley-Inter science, New York
144. Kauffman AJ, Dilling ED (1955) The metallurgy of zirconium. McGraw-Hill, New York, p 69
145. Mohammed NA, Daher AM (2002) Preparation of high-purity zirconia from Egyptian zircon: an anion-exchange purification process. Hydrometallurgy 65:103
146. Yoshimura M (1988) Phase stability of zirconia. J Am Ceram Soc Bull 67:1950
147. Dodd AC, Mc Cormick PG (2002) Synthesis of nanocrystalline ZrO_2 powders by mechanochemical reaction of $ZrCl_4$ with LiOH. J Eur Ceram Soc 22:1823
148. Tsuzuki T, McCormick PG (2001) Synthesis of ultrafine ceria powders by mechanochemical processing. J Am Ceram Soc 84:1453
149. Michel D, Faudot E, Gaffet E, Mazerolles L (1993) Stabilized zirconias prepared by mechanical alloying. J Am Ceram Soc 76:2884
150. Rana S, Ram S, Roy SK (2002) Processing of Cr^{3+}/Cr^{4+} stabilized c-ZrO_2 nanoparticles through thermomechanical attrition. Mater Manuf Process 17:529
151. Johnson DW (1982) In: Chin GY (ed) Advances in powder technology. American Society for Metals, Metal Park, p 23
152. Skadan G, Hahn H, Roddy M, Cannon WR (1994) Ultrafine-grained dense monoclinic and tetragonal zirconia. J Am Ceram Soc 77:1706
153. Biringer R (1989) Nanocrystalline materials. Mater Sci Eng A 117:33
154. Okazaki K, Maiwa H, Ichinose N (1994) Preparation of (Pb, La)TiO_3 thin films by multiple cathode sputtering. In: Bhalla AS, Nair KM, Lloyd IK, Yanagida H, Payne DA (eds) Ceramic transactions, ferroic materials: design, preparation, and characteristics, vol 43. The American Ceramic Society, Ohio, pp 15–26
155. Frantti J, Lantto VJ (1994) Characterization of $Pb_{0.97}Nd_{0.02}(Zr_{0.55}Ti_{0.45})O_3$ thin films prepared by pulsed laser ablation. J Appl Phys 76:2139
156. Malinosfsky WW, Babbit RK (1961) Fine-grained ferrites. I. Nickel ferrite. J Appl Phys 32:2375
157. Nagashima M, Nakayama T, Yamanaka S, Fujikane M, Hayashi Y, Sekino T, Kusunose T, Niihara K (2003) Fabrication of metastable ZrO_{2-x} single nano-sized particles. Mater Lett 57:4023
158. Bondioli F, Ferrari AM, Leonelli C, Siligardi C, Pellacani GC (2001) Microwave-hydrothermal synthesis of nanocrystalline zirconia powders. J Am Ceram Soc 84:2728
159. Noh HJ, Seo DS, Kim H, Lee JK (2003) Synthesis and crystallization of anisotropic shaped ZrO_2 nanocrystalline powders by hydrothermal process. Mater Lett 57:2425
160. Yoshimura M, Sōmiya S (1999) Hydrothermal synthesis of crystallized nano-particles of rare earth-doped zirconia and hafnia Mater. Chem Phys 61:1
161. Tsukada T, Venigalla S, Morrone AA, Adair JH (1999) Low temperature hydrothermal synthesis of yttrium-doped zirconia powders. J Am Ceram Soc 82:1169
162. Morgan PDE (1984) Synthesis of 6-nm ultrafine monoclinic zirconia. J Am Ceram Soc 67: C204
163. Komarneni S, Roy R, Li QH (1992) Microwave hydrothermal synthesis of ceramic powder. Mater Res Bull 2:1393
164. Komarneni S, D'Arrigo MC, Leonelli C, Pellacani GC, Katsuki H (1998) Microwave-hydrothermal synthesis of nanophase ferrites. J Am Ceram Soc 88:3041
165. Sōmiya S, Akiba T (1999) A high potential material zirconia. Bull Mater Sci 22:207
166. Sōmiya S, Yoshimura M, Nakai Z, Nakai K, Hishinuma K, Kumaki T (1986) Ceramic microstructural development of hydrothermal powder and ceramics microstructure. In: Pask JA, Evans AG (eds) Ceramic microstructures: role of interfaces. Plenum, New York, p 465

167. Sōmiya S, Akiba T (1999) Hydrothermal zirconia powders: a bibliography. J Eur Ceram Soc 19:81
168. Dell'Agli G, Mascolo G (2000) Hydrothermal synthesis of ZrO_2-Y_2O_3 solid solutions at low temperature. J Eur Ceram Soc 20:139
169. Xie Y (1999) Preparation of ultrafine zirconia particles. J Am Ceram Soc 82:768
170. Dominguez JM, Hernandez JL, Sandoval G (2000) Surface and catalytic properties of Al_2O_3-ZrO_2 solid solutions prepared by sol–gel methods. Appl Catal A Gen 197:119
171. Yoshimura M, Oh ST, Sando M, Niihara K (1999) Crystallization and microstructural characterization of ZrO_2 (3 mol% Y_2 O_2) nano-sized powders with various Al_2O_3 contents. J Alloys Comp 290:284
172. Córdova-Martínez W, De la Rosa-Cruz E, Díaz-Torres LA, Salas P, Montoya A, Avendaño M, Rodríguez RA, Barbosa-García O (2002) Nanocrystalline tetragonal zirconium oxide stabilization at low temperatures by using rare earth ions: Sm^{3+} and Tb^{3+}. Opt Mater 20:263
173. Jana S, Biswas PK (1997) Characterization of oxygen deficiency and trivalent zirconium in sol-gel derived zirconia films. Mater Lett 30:53
174. Harizanov O, Ivanova T, Harizanova A (2001) Study of sol–gel TiO_2 and TiO_2 – MnO obtained from a peptized solution. Mater Lett 49:165
175. Ehrhart G, Capeon B, Robbe O, Boy P, Turrell S, Bouazaoui M (2006) Structural and optical properties of *n*-propoxide sol-gel derived ZrO_2 thin films. Thin Solid Films 496:227
176. Hipolite MG, Galcony C, Frutis MAA, Nieto JA (2001) Synthesis and characterization of luminescent ZrO_2: Mn, Cl powders. Appl Phys Lett 79:4369
177. Liang L, Xu Y, Wu D, Sun Y (2009) A simple sol-gel route to ZrO_2 films with high optical performances. Mater Chem Phys 114:252
178. Reisfeld R, Zelner M, Patra A (2000) Fluorescence study of zirconia films doped by Eu^{3+}, Tb^{3+} and Sm^{3+} and their comparison with silica films. J Alloys Comp 300–301:147
179. Reisfeld R, Saraidarov T, Pietraszkiewick M, Lis S (2001) Luminescence of europium(III) compounds in zirconia xerogels. Chem Phys Lett 349:266
180. Zhang Y, Jin S, Liao C, Yan CH (2002) Microstructures and optical properties of nanocrystalline rare earth stabilized zirconia thin films deposited by a simple sol-gel method. Mater Lett 56:1030
181. Pechini MA (1967) Method of preparing +2 valent metal yttrium and rare earth ferrites. U. S. Patent US 3438723 A
182. Kikkawa S, Kijima A, Hirota K, Yamaguchi O (2002) Soft solution preparation methods in a ZrO_2-Al_2O_3 binary system. Solid State Ion 151:359
183. Tai LW, Lessing PA (1992) Modified resin-intermediate processing of perovskite powders: Part I. Optimization of polymeric precursors. J Mater Res 7:502
184. Quinelato AL, Longo E, Perazolli LA, Varela J (2000) Effect of ceria content on the sintering of ZrO_2 based ceramics synthesized from a polymeric precursor. J Euro Ceram Soc 20:1077
185. Laberty-Robert C, Ansart F, Deloget C, Gaudon M, Rousset A (2001) Powder synthesis of nanocrystalline ZrO_2-8% Y_2O_3 via a polymerization route. Mater Res Bull 36:2083
186. Muccillo ENS, Rocha RA, Muccillo R (2002) Preparation of Gd_2O_3-doped ZrO_2 by polymeric precursor techniques. Mater Lett 53:353
187. Zhang Y, Li A, Yan Z, Xu G, Liao C, Yan C (2003) $(ZrO_2)_{0.85}(REO_{1.5})_{0.15}$ (RE = Sc, Y) solid solutions prepared via three Pechini-type gel routes: sol-gel formation and calcination behaviors. J Solid State Chem 171:434
188. Bhaduri S, Bhaduri SB (1997) Enhanced low temperature toughness of Al_2O_3-ZrO_2 nano/nanocomposites. Nanostruct Mater 8:755
189. Juáre RE, Lamas DG, Lascalea GE, Walsöe de Reca NE (2000) Synthesis of nanocrystalline zirconia powders for TZP ceramics by a nitrate – citrate combustion route. J Eur Ceram Soc 20:133
190. Vallet-Regí M, Nicolopoulos S, Román J, Martíneza JL, González-Calbetb JM (1997) Structural characterization of ZrO_2 nanoparticles obtained by aerosol pyrolysis. J Mater Chem 7:1017

191. Cho SY, Kim IT, Kim DY, Park SJ, Kim BK, Lee JH (1997) Effects of H_2O_2 on the morphology of ZrO_2, powder prepared by ultrasonic spray pyrolysis. Mater Lett 32:271
192. Xiaming D, Qing-feng L, Yuying T (1993) Study of phase formation in spray pyrolysis of ZrO_2 and ZrO_2-; Y_2O_3 powders. J Am Ceram Soc 76:760
193. Jayaram V, Mishra RS, Majumdar B, Lesher C, Mukherjee A (1998) Dense nanometric ZrO_2-Al_2O_3 from spray-pyrolysed powders. Coll Surf A: Physicochem Eng Aspects 133:25
194. Stelzer NHJ, Schoonman J (1996) Synthesis of terbia-doped yttria-stabilized zirconia thin films by electrostatic spray deposition (ESD). J Mater Synth Process 4:429
195. Ruiz H, Vesteghem H, Di Giampaolo AR, Lira J (1997) Zirconia coatings by spray pyrolysis. Surf Coat Technol 89:77
196. Peshev P, Stambolova I, Vassilev S, Stefanov P, Blaskov V, Starbova K, Starbov N (2003) Spray pyrolysis deposition of nanostructured zirconia thin films. Mater Sci Eng B 97:106
197. Hong JS, De la Torre SD, Miyamoto K, Miyamoto H, Gao L (1998) Crystallization of Al_2O_3/ZrO_2 solid solution powders prepared by coprecipitation. Mater Lett 37:6
198. Kikkawa S, Kijima A, Hirota K, Yamamoto O (2002) Crystal structure of zirconia prepared with alumina by coprecipitation. J Am Ceram Soc 85:721
199. Vasylkiv O, Sakka Y (2000) Nonisothermal synthesis of yttria-stabilized zirconia nanopowder through oxalate processing: I, characteristics of Y-Zr oxalate synthesis and its decomposition. J Am Ceram Soc 83:2196
200. Huang C, Tang Z, Zhang Z (2001) Differences between zirconium hydroxide ($Zr(OH)_4$. nH_2O) and hydrous zirconia ($ZrO_2.nH_2O$). J Am Ceram Soc 84:1637
201. Gutzov S, Ponahlo J, Lengauer CL, Beran A (1994) Phase characterization of precipitated zirconia. J Am Ceram Soc 77:1649
202. Moore DA, Ferguson IF (1982) Zirconia-stabilized cubic Europia. J Am Ceram Soc 65:414
203. Fiona CMW, Sager FCW, Sibelt GMN, Henk V (2001) Dense nanostructured t-ZrO_2 coatings at low temperatures via modified emulsion precipitation. Adv Mater 13:514
204. Sager W, Eicke HF, Sun W (1993) Precipitation of nanometre-sized uniform ceramic particles in emulsions. Colloid Surf A 79:199
205. Venkatachari KR, Huang D, Ostrander SP, Schulze WA, Stangle GC (1995) A combustion synthesis process for synthesizing nanocrystalline zirconia powders. J Mater Res 10:748
206. Huang C, Tang Z, Zhang Z (2001) Differences between zirconium hydroxide ($Zr(OH)_4 \cdot nH_2O$) and hydrous zirconia ($ZrO_2.nH_2O$). J Am Ceram Soc 84:1637
207. Hirano S, Yoshinaka M, Hirota K, Yamaguchi O (1996) Formation, characterization, and hot isostatic pressing of Cr_2O_3-doped ZrO_2 (0.3mol%Y_2O_3) prepared by hydrazine method. J Am Ceram Soc 79:171
208. Ram S (1994) Kinetics of the desorption of interstitial hydrogen in stable $Nd_2Fe_{14}BH_x$, $x > 5$. Phys Rev B 49:9632
209. Gonella F, Mattei G, Mazzoldi P, Battaglin G, Quaranta A, De G, Montecchi M (1999) Structural and optical properties of silver-doped zirconia and mixed zirconia-silica matrices obtained by sol-gel processing. Chem Mater 11:814
210. Shrirer DF, Atkins PW, Langford CH (1990) Inorganic chemistry. Wiley, New York, p 434
211. Ram S, Kanik R, Shukla BS (1992) Optical absorption and EPR studies of borate glasses with $PbCrO_4$ and Pb_2CrO_5 microcrystals. J Mater Sci 27:511
212. Ray JC, Pramanik P, Ram S (2001) Formation of Cr^{3+} stabilized ZrO_2 nanocrystals in a single cubic metastable phase by a novel chemical route with a sucrose-polyvinyl alcohol polymer matrix. Mater Lett 48:281
213. Emeline AV, Serpone N (2001) Relaxation dynamics of processes in colloidal zirconia nanosols. Dependence on excitation energy and temperature. Chem Phys Lett 345:105
214. Kroger FA (1964) Chemistry of imperfect crystals. North-Holland, Amesterdum
215. Bagnall DM, Chen TF, Shen MY, Zhu Z, Goto T, Yao T (1998) Room temperature excitonic stimulated emission from zinc oxide epilayers grown by plasma-assisted MBE. J Cryst Growth 184–185:605
216. Nicoll FH (1966) Ultraviolet Zno laser pumped by an electron beam. Appl Phys Lett 9:13

217. Hvam JM (1971) Temperature induced wavelength shift of electron beam pumped lasers from Cdse, Cds, and ZnO. Phys Rev B 4:4459
218. Romero-Salazar C, Perez-Rodrıguez F (2003) Elliptic flux-line-cutting critical-state model. Appl Phys Lett 83:5226
219. Robertson J (2000) Band offsets of wide-band-gap oxides and implications for future electronic devices. J Vac Sci Technol B 18:1785
220. Zhu LQ, Fang Q, He G, Liu M, Zhang LD (2006) Effect of annealing on optical properties and band alignments of ZrO_2/Si(100) by nitrogen-assisted reactive sputtering. Phys D: Appl Phys 39:5285
221. David C, Mclntyre PC (2004) Film and interface layer properties of ultraviolet-ozone oxidized hafnia and zirconia gate dielectrics on silicon substrates. Appl Phys Lett 85:4699
222. JBredas JL, Scott JC, Yakushi K, Street GB (1984) Polarons and bipolarons in polypyrrole: evolution of band structure and optical spectrum upon doping. Phys Rev B 30:1023
223. Serpone N, Lawless D, Khairutdinov R (1995) Size effects on the photophysical properties of colloidal anatase TiO_2 particles: size quantization versus direct transitions in this indirect semiconductor? J Phys Chem 99:16646
224. Pankove JI (1971) Optical processes in semiconductors. Prentice Hall, New Jersey
225. Kwok CK, Aita CR (1989) Near band gap optical behavior of sputter deposited α and α + β-ZrO_2 films. J Appl Phys 66:2756
226. Kralik B, Chang EK, Louie SG (1998) Structural properties and quasiparticle band structure of zirconia. Phys Rev B 57:7027
227. Marusaki M, Naito H, Matsuura Y, Matsukawa K (2006) Optical properties of poly (di-*n*-hexylsilane)-zirconia hybrid thin films: suppression of thermochromism and large thermo-optic coefficients. Appl Phys Lett 86:191907
228. Bellini T, Clark NA, Muzny CD, Wu L, Garland CW, Schaefer DW, Oliver BJ (1992) Phase behavior of liquid crystal 8CB in a silica aerogel. Phys Rev Lett 69:788
229. Böer KW (1990) Survey of semiconductor physics: electrons and other particles in bulk semiconductors. Van Nostrand Reinhold, New York, p 1034
230. Kerker M (1969) The scattering of light and other electromagnetic radiation. Academic, New York
231. Bohren CF, Hoffman DR (1983) Adsorption and scattering of light by small particles. Wiley, New York
232. Krieberg U, Vollmer M (1995) Optical properties of metal clusters. Springer, Berlin
233. Link S, El-Sayed MA (2000) Shape and size dependence of radiative, non-radiative and photothermal properties of gold nanocrystals. Int Rev Phys Chem 19:409
234. Cao G (2007) Nanostructures and nanomaterials: synthesis, properties and applications. Imperials College Press, London
235. Morell G, Katiyar RS, Torres D, Paje SE, Llopis J (1997) Raman scattering study of thermally reduced stabilized cubic zirconia. J Appl Phys 81:2830
236. Mie G (1908) Beitréige zur optik triiber medien speziell kolloidaler metallosungen. Ann Phys 25:377
237. Palpant B, Prevel B, Lerme J, Cottancin E, Pellarin M, Treilleux M, Perez A, Vialle JL, Broyer M (1998) Optical properties of gold clusters in the size range 2–4 nm. Phys Rev B 57:1963
238. Uchikoshi T, Sakka Y, Ozawa K, Hiraga K (1998) Preparation of fine-grained monoclinic zirconia ceramics by colloid processing. J Mater Res 13:840
239. De G, Gusso M, Tapfer L, Catalano M, Gonella F, Mattei G, Mazzoldi P, Battaglin GJ (1996) Annealing behavior of Cu, Ag and Ag-Cu nanoclusters doped silica by sol-gel processing. Appl Phys 80:6734
240. Hao L, Lawrence J (2004) CO_2 laser induced microstructure features in magnesia partially stablised zirconia bioceramic and effects thereof on the wettability characteristics. Mater Sci Eng A 364:171
241. Hirvonen A, Nowaka R, Yamamoto Y, Sekino T, Niihara K (2006) Fabrication, structure, mechanical and thermal properties of zirconia-based ceramic nanocomposites. J Eur Ceram Soc 26:1497

242. Chaim R (1992) Microstructure and bending strength in the ternary (Mg, Ca)-partially-stabilized zirconia. J Am Ceram Soc 75:694–699
243. Ghosh P, Patra A (2006) Role of surface coating in ZrO_2/Eu^{3+} nanocrystals. Langmuir 22:6321
244. Mondal A, Ram S (2008) Enhanced phase stability and photoluminescence of Eu^{3+} modified t-ZrO_2 nanoparticles. J Am Ceram Soc 91:329
245. Savoini B, Santiuste JEM, Conzalez R (1997) Optical characterization of Pr^{3+}doped yttria-stabilized zirconia single crystals. Phys Rev B 56:5856
246. Martĭnez WC, De la Rosa-Cruz E, Diaz-Torres LA, Salas P, Montoya A, Avendano M, Rodriguez RA, Garcia OB (2002) Nanocrystalline tetragonal zirconium oxide stabilization at low temperatures by using rare earth ions: Sm^{3+} and Tb^{3+}. Opt Mater 20:263
247. Pereyra-Perea E, Estrada-Yanes MR, Garcia M (1998) Preliminary studies on luminescent terbium-doped thin films prepared by the sol-gel process. J Phys D Appl Phys 31:L7
248. Reisfeld R, Eyal M (1987) Energy transfer between inorganic ions in amorphous solids. Acta Phys Polonica A71:799
249. Reisfeld R, Zelner M, Patra A (2000) Fluorescence study of zirconia films doped by Eu^{3+}, Tb^{3+} and Sm^{3+} and their comparison with silica films. J Alloy Compd 300–301:147
250. Patra A, Friend CS, Kapoor R, Prasad PN (2003) Effect of crystal nature on up conversion luminescence in Er^{3+}: ZrO_2 nano crystals. Appl Phys Lett 83:284
251. Garcia-Hipolito M, Falcony C, Aguilar-Frutis MA, Azorin-Nieto J (2001) Synthesis and characterization of luminescent ZrO_2:Mn, Cl powders. Appl Phys Lett 79:4369
252. Gschneid KA Jr, Eyring LR (1979) Handbook on the physics and chemistry on rare-earths. North-Holland, Amsterdam
253. Lopez EF, Escribano VS, Panizza M, Carnasciali MM, Busca G (2001) Vibrational and electronic spectroscopic properties of zirconia powders. J Mater Chem 11:1891
254. Wang SF, Gu F, Kai M, Yang ZS, Zhou GJ, Zhang HP, Zhou YY, Wang SM (2006) Structure evolution and photoluminescence properties of $ZrO_2:Eu^{3+}$ nanocrystals. Opt Mater 28:1222
255. Rignanese GM, Detraux F, Gonze X (2001) First-principles study of dynamical and dielectric properties of tetragonal zirconia. Phys Rev B 64:134301–134307
256. Merle T, Guinebretiere R, Mirgorodsky A, Quintard P (2002) Polarized Raman spectra of tetragonal pure ZrO_2 measured on epitaxial films. Phys Rev B 65:144302
257. Macfarlane RM, Shelby RM (1987) Coherent transient and hole burning spectroscopy of rare earth ions in solids. In: Kaplyanskii AA, Macfarlane RM (eds) Spectroscopy of crystals containing rare-earth ions. North-Holland, Amsterdam, p 51
258. You H, Nogami M (2004) Optical properties and local structure of Eu^{3+} ions in sol-gel TiO_2-SiO_2 glasses. J Phys Chem B108:12003
259. Nogami M, Umehara N, Hayakawa T (1998) Effect of hydroxyl bonds on persistent spectral hole burning in Eu^{3+}-doped BaO-P_2O_5 glasses. Phys Rev B 58:6166
260. Chen HR, Shi JL, Yong Y, Li YS, Yan DS, Shi CS (2002) Violet-blue photoluminescent properties of mesoporous zirconia modified with phosphoric acid. Appl Phys Lett 81:2761
261. McKeever SWS (1985) Thermoluminance of solids. Cambridge University Press, Cambridge
262. Miyoshi T, Makidera Y, Kawamura T, Kashima S, Matsuo N, Kaneda T (2002) Thermoluminescence of light and X-ray irradiated semiconductor-doped glasses. J Appl Phys 41:5262
263. Harrison DE, Melamed NT, Subbarao EC (1963) A new family of self-activated phosphors. J Electrochem Soc 110:23
264. Bettinali C, Ferraresso G, Manconi JW (1969) Thermoluminescence of ZrO_2. J Chem Phys 50:3957
265. Iacconi P, Lapraz D, Caruba R (1978) Traps and emission centres in thermoluminescent ZrO_2. Phys Status Solidi (a) 50:275
266. Curie D (1957) Modèles pour les divers types de pièges dans le sulfure de zinc phosphorescent. libération thermique et optique des électrons piégés. J Phys Radium 18:214
267. Arsenev PA, Bagdasarov KS, Niklas A, Ryazantsev AD (1980) X-ray and thermostimulated luminescence of 0.9 ZrO_2-0.1 Y_2O_3 single crystals. Phys Status Solidi (a) 62:395

268. Hsieh WC, Su CS (1994) Thermoluminescence in ZrO_2 with Impurity of ZnO induced by UV radiation. Appl Phys A 58:459
269. Salas P, De la Rosa-Cruz E, Diaz-Torres LA, Castano VM, Melendrez R, Barboza-Flores RM (2003) Monoclinic ZrO_2 as a broad spectral response thermoluminescence UV dosemeter. Radiat Meas 37:187
270. Hsieh WC, Su CS (1994) UV induced thermoluminescence in ZrO_2 doped by Er_2O_3. J Phys D Appl Phys 27:1763
271. Sanchéz GV, Anaya DM, Gutierrez-Wing C, Hernandez RP, Martinez PRG, Chavez CA (2007) Ag nanoparticle effects on the thermoluminescent properties of monoclinic ZrO_2 exposed to ultraviolet and gamma radiation. Nanotechnology 18:265703
272. Azorín J, Rivera T, Martínez E, García M (1998) Thermoluminescence of Eu-doped ZrO_2 thin films exposed to ultraviolet and visible light. Radiat Meas 29:315

$xAgI\text{-}(1\text{-}x)MPO_3$ [M = Ag, Li) Superionic Composite Glasses and Their Current Issues

16

D. P. Singh, L. Sowntharya, K. Shahi, and Kamal K. Kar

Contents

D.P. Singh
Solid State Ionics Laboratory, Materials Science Programme, Indian Institute of Technology Kanpur, Kanpur, Uttar Pradesh, India
e-mail: dpphysics@gmail.com

L. Sowntharya
Advanced Nanoengineering Materials Laboratory, Materials Science Programme, Indian Institute of Technology Kanpur, Kanpur, Uttar Pradesh, India
e-mail: sowsindhu@gmail.com

K. Shahi
Department of Physics, Indian Institute of Technology Kanpur, Kanpur, Uttar Pradesh, India
e-mail: kshahi@iitk.ac.in

K.K. Kar (✉)
Advanced Nanoengineering Materials Laboratory, Materials Science Programme, Indian Institute of Technology Kanpur, Kanpur, Uttar Pradesh, India

Advanced Nanoengineering Materials Laboratory, Department of Mechanical Engineering, Indian Institute of Technology Kanpur, Kanpur, Uttar Pradesh, India
e-mail: kamalkk@iitk.ac.in

K.K. Kar (ed.), *Composite Materials*,
DOI 10.1007/978-3-662-49514-8_16

Abstract
Superionic conductors are special class of materials, which conduct electricity via movement of ions. They are popular due to their various applications and for many physical phenomena, which are still to be understood fully. Phosphate glass is a good candidate in this category, and it is necessary to explore its practical application and rich science of super ionic conductors as they are easier to prepare and contain nontoxic elements.

Keywords
Superionic conductors • Solid electrolytes • AC conductivity

Superionic Conductors

Superionic conductors (SICs), fast ionic conductors (FICs), or solid electrolytes are materials in which the ionic conductivity is very high and in some cases it can be comparable to that of molten salts and ionic solutions. It is found that the high ionic conductivity in these materials is related to the high degree of disorder. For example, the extraordinarily high ionic conductivity (~1.3 S cm^{-1}) reported in α-phase of silver iodide (α-AgI) (which is stable above 150 °C, until melting point) is comparable to its liquid phase, whereas most of ionic solids have an ionic conductivities of the order of ~10^{-8} S cm^{-1} at moderate temperatures (e.g., sodium chloride (NaCl) at 200 °C) [1]. Compound α-AgI is the first super ionic conductor ever discovered. The stable phase below 150 °C is β-AgI, which has hexagonal wurtzite structure (space group, $P6_3mc$). There is another low-temperature phase, i.e., γ-AgI, which is stable below 135 °C and has cubic symmetry having a zinc blend structure (space group, F-43 m) [1]. Both of these phases transform to a bcc structure (α-AgI, space group, Im3m) at 150 °C. This phase change is accompanied by a considerable volume contraction (5.4 %). In this high conducting phase, iodide ions form a bcc sub-lattice, while the two silver ions of the unit cell are distributed over 42 equipoints with equal probability. It is assumed that the silver ions flow like a liquid because of a very high number of vacant sites. The structure of α-AgI is shown in Fig. 16.1 [2, 3]. Out of 42 sites available to 2 Ag^+ ions, 6 are octahedral, 12 are tetrahedral, and 24 are trigonal bipyramidal sites. The phase transition in AgI is close to the melting of cationic sub-lattice, and the entropy change during the transition is comparable to that of melting solid [4]. Soon after the discovery of α-AgI, superionic phases of many other materials have been reported like α-silver sulfide (α-Ag_2S), α-silver selenide (α-Ag_2Se), α-silver telluride (α-Ag_2Te), etc. [5].

Apart from the superionic α-phases of AgI, there are other materials also, which show very high ionic conductivity. Based on the structure, crystalline superionic conductors may be classified into three categories [7]:

(i) **Type-I:** In this category, these are solid electrolytes, which show superionic conduction after a first-order (order → disorder) phase transition. At phase transition, a rearrangement of immobile ion sub-lattice and complete disorder

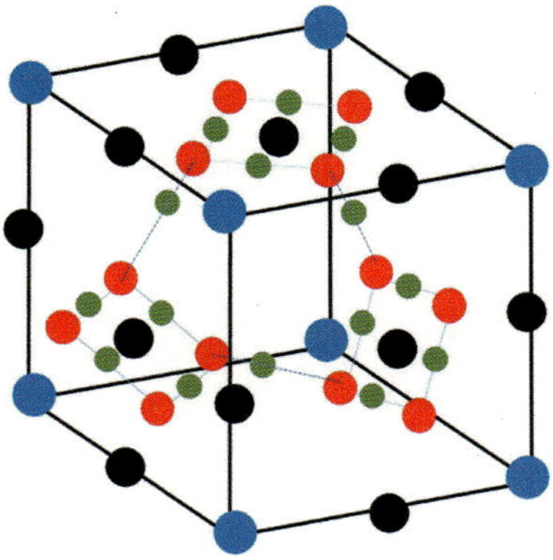

Fig. 16.1 Crystal structure of α-AgI. There are 6 octahedral sites, 12 tetrahedral sites, and 24 trigonal sites, which are available to two Ag^+ ions [2, 3, 6]

Table 16.1 Some crystalline fast ion conductors with their room temperature (unless stated) conductivities

System	σ_{max} (S cm^{-1})	System	σ_{max} (S cm^{-1})
α-AgI [1]	1.31 (148 °C)	$Ag_6I_4WO_4$ [12]	4×10^{-2}
$RbAg_4I_5$ [8, 9]	0.27	Li_3N [13]	1.2×10^{-3}
Ag_3SI [10]	10^{-2}	α-Li_2SO_4 [14]	0.86 (550 °C)
Ag_2HgI_4 [11]	1.5×10^{-3} (60 °C)	$Li_{10}GeP_2S_{12}$ [15]	1.2×10^{-2}

of mobile ion sub-lattice take place. Examples: silver iodide (AgI), silver sulfide (Ag_2S), copper sulfide (CuS), silver selenide (Ag_2Se), silver mercury (II) iodide (Ag_2HgI_4), lithium sulfate (Li_2SO_4), silver telluride (Ag_2Te), etc. [7].

(ii) **Type-II:** In this case, the immobile sub-lattice remains the same in both phases, but a gradual mobile sub-lattice disorder takes place at the phase transition. Examples: lead fluoride (PbF_2), calcium fluoride (CaF_2), sodium oxide (Na_2O), potassium sulfide (K_2S), etc. [7].

(iii) **Type-III:** Here, no phase transition takes place, though high ionic conductivity is noticed. Due to their favorable structure of cationic transport, these materials show a high ionic conductivity. Examples: sodium beta alumina (Na-β-Al_2O_3), potassium beta alumina (K-β-Al_2O_3), lithium nitride (Li_3N), etc. [7]. A list of some crystalline fast ionic conductors is given in Table 16.1. The room temperature conductivity is also given in the same Table 16.1. Figure 16.2 shows the electrical conductivity as a function of inverse temperature for some selected crystalline superionic conductors. It also includes concentrated sulfuric acid (H_2SO_4) for comparison. While searching for the new superionic conductors, one aims for the top right-hand corner of the diagram (i.e., high conductivity at low temperatures). It is clear from the diagram that most of the crystalline superionic conductors except rubidium silver iodide ($RbAg_4I_5$) (which has $\sigma \sim 10^{-1}$ S cm^{-1} at room temperature) have very low room temperature conductivity. The early work in the field of superionic conductors has been aimed to stabilize the high-temperature phase at room temperature, which is responsible for such high conductivity. The highly disordered structure is the central feature of the crystalline superionic

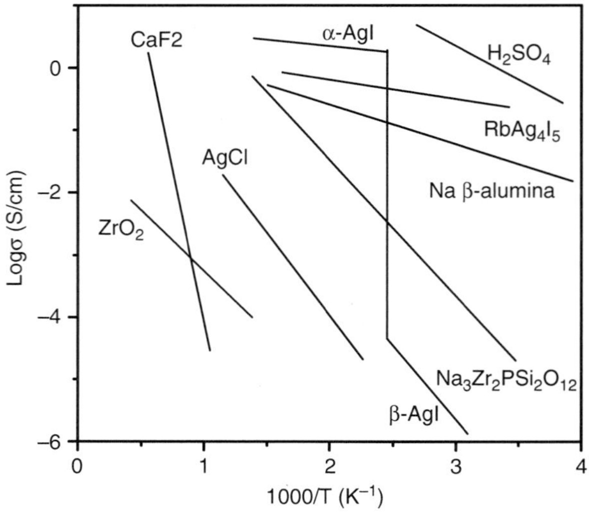

Fig. 16.2 Electrical conductivity of various superionic conductors as a function of temperatures. H_2SO_4 is included for comparison (Figure redrawn from *Solid State Chemistry and Its Applications* [17])

conductors (especially type-I superionic conductors). So the obvious choice is to stabilize these α-phases in amorphous/glassy matrix. Since α-AgI has the highest ionic conductivity till date in any known solid electrolyte, most work has been concentrated on AgI-based conducting glasses. In many glass matrices, doping of AgI has been studied earlier, and room temperature conductivity of ~10^{-2} S cm^{-1} has been reported in glasses containing high amounts of AgI [16]. The AgI does not show crystalline nature in many glass matrices and gives an X-ray diffraction (XRD) pattern similar to glass.

Examples of Glassy Superionic Conductors

A list of some fast ion-conducting glasses is given in Table 16.2 along with their room temperature conductivities.

Phosphate Glasses

Among glass-based solid electrolytes, phosphate glasses have been of interest among researchers because of their many unique properties. They have high transmission in ultraviolet region and are suitable material for high power lasers [33]. These glasses are biocompatible also. When these glasses are doped with suitable inorganic salts, they become fast ion conductors at room temperature with conductivities up to ~10^{-2} S cm^{-1}. Materials with such high conductivity have found several applications in energy storage devices, solid state batteries, sensors, etc. [34–37]. AgI-$AgPO_3$ glasses of high ionic conductivity have found applications in nanotechnology, such as growing silver nanowire within glass [38], resistive switching devices based on nanocrystalline AgI-$AgPO_3$ [39], AgI nanowire-based

Table 16.2 Selected glassy fast ion conductors with their room temperature conductivities

System	σ_{max} (S cm^{-1})	System	σ_{max} (S cm^{-1})
$60Li_2S$-$40SiS_2$ [18]	5.1×10^{-4}	$55Ag_2S$-$45GeS_2$ [19]	2.0×10^{-3}
$50Li_2S$-$50GeS_2$ [19]	2.0×10^{-4}	52.3AgI-48.7 (Ag_2S-GeS_2) [19]	1.0×10^{-2}
$40Li_2Se$-$60GeSe_2$ [20]	1.8×10^{-4}	70AgI-$10Ag_2Se$-20-P_2Se_5 [19]	2.0×10^{-2}
$14SiS_2$-$9P_2S_5$-$47Li_2S$-30LiI [21]	2.1×10^{-3}	80AgI-$20Ag_3AsO_4$ [27]	1.4×10^{-2}
$37Li_2S$-$18P_2S_5$-45LiI [22]	1.0×10^{-3}	80AgI-$20Ag_2CrO_4$ [28]	1.5×10^{-2}
67.67AgI-$25Ag_2O$-$8.33P_2O_5$ [23]	2.0×10^{-2}	75AgI-$25Ag_2SeO_4$ [28]	3.0×10^{-2}
70AgI-$15Ag_2O$-$15B_2O_3$ [24]	9.7×10^{-3}	Ag_2S-Sb_2S_3 [29]	5.6×10^{-2}
75AgI-$25Ag_2MoO_4$ [25]	1.0×10^{-2}	50AgI-$50LiPO_3$ [30]	7.4×10^{-4}
67.67AgI-$25Ag_2O$-$8.33V_2O_5$ [26]	5.3×10^{-2}	$19CdI_2$-$81AgPO_3$ [31]	1.0×10^{-2}
90.1AgI-$9.9Ag_3BO_3$ [16]	1.0×10^{-1}	57.5AgI-$42.5AgPO_3$ [32]	1.5×10^{-2}

nano-devices [40], electrochemical nanoimprinting [41], etc. Phosphate glasses with Cu^{2+} ions can display semiconducting properties, and with trivalent rare earth ions, these glasses have luminescent and magnetic properties [42].

Synthesis of Phosphate Glass

The conventional technique for preparing the ion-conducting glasses is the melt quenching. In this method, appropriate amounts of chemicals are first melted in a crucible at high temperatures, and then the molten salt and glass mixture is quenched by pouring it into liquid nitrogen, by pressing it in between two metal plates, or by pouring between fast rotating twin rollers. Various quenching techniques have different quenching rates. The obtained flakes of the glasses are crushed using mortar and pestle and a fine powder is made. Figure 16.3 shows the twin roller assembly used in sample preparations. The powdered samples are pressed into pellets of thicknesses 1–1.5 mm and diameter 10 mm for electrical measurements by using a hydraulic press. Electrodes are made on the surfaces of the pellets by applying a mixture of electrolyte and silver powder (1:1 ratio) during pellet preparation. This method of electrode preparation removes the electrode polarization substantially.

The processing temperature for preparing silver metaphosphate and lithium metaphosphate is given in Table 16.3.

The typical heating rate is kept at 100 °C/h. The reactions for $AgPO_3$ and $LiPO_3$ glasses are shown below (Eqs. 16.1 and 16.2).

For $AgPO_3$:

$$AgNO_3 + NH_4H_2PO_4 \rightarrow AgPO_3 + 3H_2O + \frac{1}{2}N_2 + NO \tag{16.1}$$

For $LiPO_3$:

$$NH_4H_2PO_4 + \frac{1}{2}Li_2CO_3 \rightarrow LiPO_3 + \frac{1}{2}H_2O + \frac{1}{2}CO_2 + NH_3 \tag{16.2}$$

Fig. 16.3 Twin-roller assembly

Table 16.3 Processing temperature for different compositions

System	Processing temperature (°C)
$AgPO_3$	At 300 for 2 h and 5 h at 500
xAgI-(1-x)$AgPO_3$ (x = 0.1–0.5)	At 300 for 2 h and 5 h at 700
$LiPO_3$	At 300 for 2 h and 5 h at 750
xAgI-(1-x)$LiPO_3$ (x = 0.05–0.5)	At 300 for 2 h and 5 h at 750

The evolved gases and water are removed during the final stage of heat processing. In the case of composite glasses, appropriate amount of salt (AgI) is added in the initial mixture, and powder is mixed thoroughly before keeping in the furnace. The molten glass-salt mixture is quenched by using twin roller as explained earlier.

It is observed that a large amount of silver iodide (up to 76 mol%) can be amorphized in $AgPO_3$ glass matrix [43]. Above this composition, extra silver iodide is precipitated in the form of β-AgI crystallites. However, in $LiPO_3$ glass, a relatively less amount (<20 mol%) of silver iodide can be amorphized, and above this composition, excess silver iodide is precipitated in the form of γ-AgI nano-crystallites [30]. The maximum room temperature conductivities achieved in $AgPO_3$ and $LiPO_3$ systems upon AgI addition are 10^{-2} S cm^{-1} and 10^{-3} S cm^{-1}, respectively, which are obtained by adding 50 mol% of silver iodide to the glass matrices.

Structural and Thermal Characterizations

The structure of phosphate glasses (P_2O_5) is consisted of PO_4 tetrahedra connected by corners, which makes a three-dimensional network by connecting three of their four corners. The fourth corner is the terminal/non-bridging double bonded oxygen. The introduction of network modifiers, such as Ag_2O, Li_2O, etc., to the base glass results in the breaking of some of the bonds and in creating new terminal oxygen

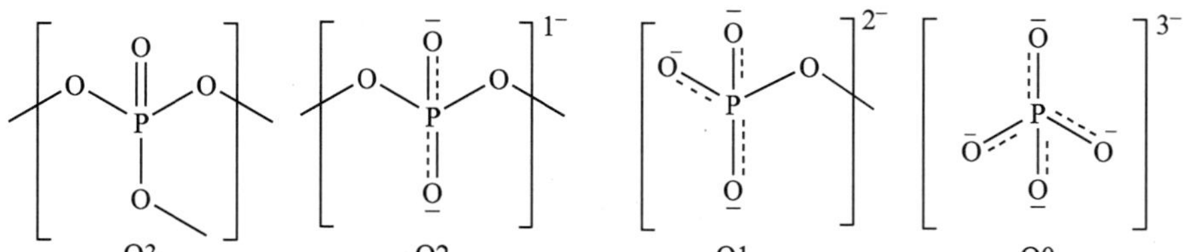

Fig. 16.4 Different tetrahedral phosphorous sites that can be formed in phosphate glasses [51]

atoms. This terminal oxygen can form ionic bonds with the metallic elements of the added modifier. Depending upon the number of bridging oxygen atoms, the phosphate tetrahedral is termed as Q^i, where i represents the number of bridging oxygen atoms and may have a value of 0, 1, 2, or 4. A structure built up by Q^3-type tetrahedra denotes a fully polymerized structure, while a structure formed by Q^2-type units gives a two-dimensional structure made of chains and rings. Q^1 means two tetrahedra connected by a corner, whereas Q^0 implies isolated tetrahedra. The different Q^i units are shown in Fig. 16.4. The structure of $AgPO_3$ and $LiPO_3$ glasses ideally consists of infinite long-chain corner-shared Q^2-type PO_4 tetrahedra, and the metal ions are associated with two non-bridging oxygen atoms. Several studies on the glass structure of doped and undoped phosphate glasses have been done by using infrared and Raman spectroscopy techniques [33, 42, 44–63].

In the IR spectra of phosphate glasses (Fig. 16.5), different bands appear at certain wave numbers, which are due to the presence of corresponding vibrational modes in the polymeric phosphate chains. The main features of IR spectra of phosphate glasses are as follows: asymmetric vibrations due to terminal oxygens $\nu_{as}(P\text{-}O_t)$ in the long polymeric chains P-O-P at 1225–1260 cm^{-1}. Asymmetric ionic stretching band $\nu_{as}(PO_3^{2-})$ is observed around 1055–1105 cm^{-1}. In the range of 900–920 cm^{-1}, a band appears due to asymmetric vibrations $\nu_{as}(P\text{-}O_{br})$ of bridging oxygens in the long chains. Two bands due to asymmetric vibrations $\nu_{as}(P\text{-}O_{br})$ of bridging oxygen atoms in the rings of different sizes are seen in the range 463–473 cm^{-1} (bending mode) and 515–535 cm^{-1} (bending mode), respectively. Apart from these, some symmetric vibrations of oxygen atoms in the ring and chain structures are also observed in the IR spectra of phosphate glasses. In the range of 717–731 cm^{-1} and 765–773 cm^{-1}, two bands are observed, which are due to symmetric vibrations of bridging oxygen atoms in the rings of different sizes. A band in the range 370–390 cm^{-1} is also noticed, which is due to the symmetric bending mode of bridging oxygen atoms in large rings. These peak assignments are taken from references mentioned earlier in this section [44, 55, 59].

The Raman spectra of phosphate glasses (Fig. 16.6 for Raman spectra of $LiPO_3$ glass) are dominated by two bands at 1140 and 680 cm^{-1}, which are due to symmetric vibrations of terminal and bridging oxygen atoms in the long polymeric chains, respectively. The asymmetric counterparts of these bands are also seen in the IR spectra around 1230 and 900 cm^{-1}, respectively. There are other bands due to ring structures, which are observed in the Raman spectra of phosphate glasses and

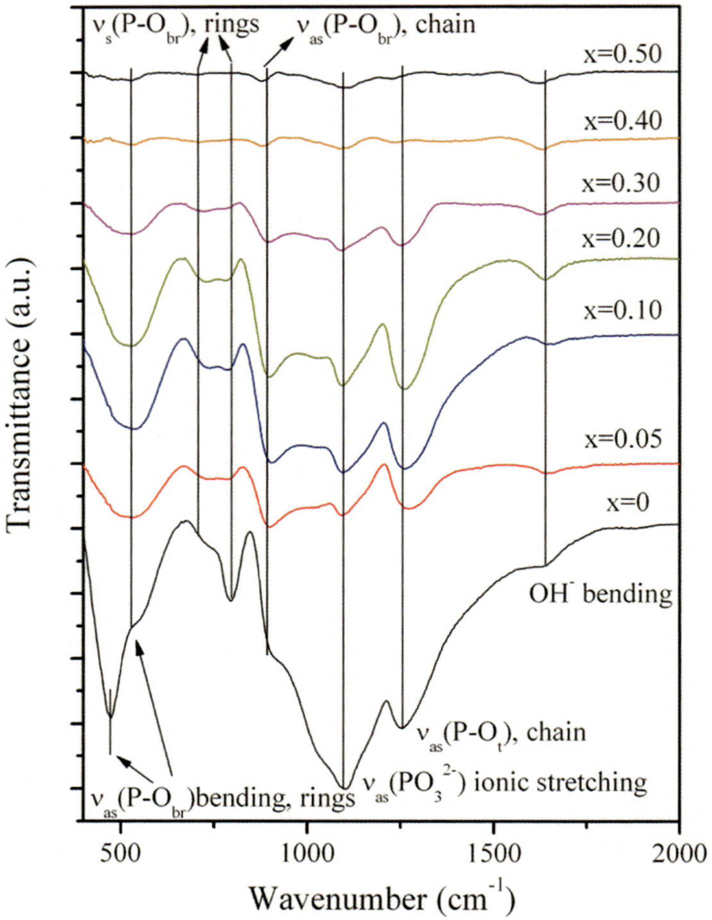

Fig. 16.5 FTIR spectra of xAgI-(1-x)$LiPO_3$ composite glass electrolytes [44, 55, 59]

are much weaker in scattering strengths. The IR and Raman studies of silver metaphosphate glasses have suggested that the glass structure largely consists of polymeric chains with small fraction of rings, and the AgI added to this glass decouples the glass network, and at higher AgI, long chains are converted to rings [58, 59, 64]. It has also been found that as the amount of AgI increases in the base glass, the Raman mode at 1140 cm^{-1} redshifts as can be seen from Fig. 16.7. This shows that the network steadily softens and the interchain spacing increases with an increase of AgI. A unique feature of Raman spectra of phosphate glasses is the observation of so-called Boson mode near 29 cm^{-1}. The origin of this peak is not clear. In the metaphosphate glasses, it has been found that different glass modifiers affect the rigidity of the glass network [42]. The prominent Raman modes in $LiPO_3$ are found to be blue-shifted in comparison to $AgPO_3$. This suggests a more rigid structure of $LiPO_3$ glass in comparison to $AgPO_3$ glass. Again this is due to the

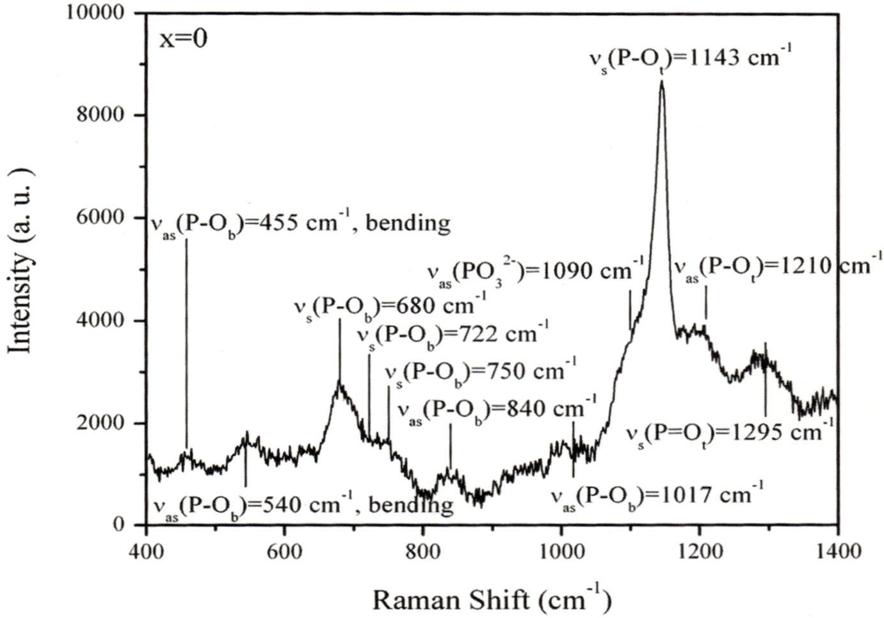

Fig. 16.6 Raman spectra of LiPO$_3$ glass [58, 59]

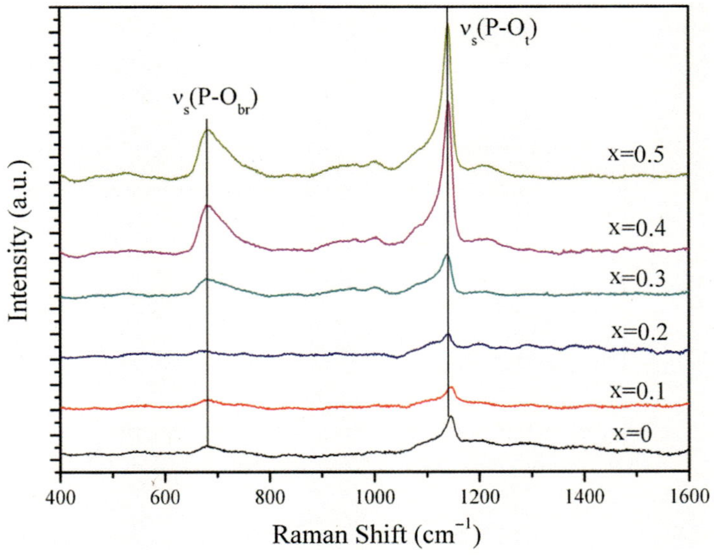

Fig. 16.7 Raman spectra of xAgI-(1-x)AgPO$_3$ glass electrolytes [65]

strong ionic character of Li^+, which makes bond with O^{2-}. Ag^+ has sufficient covalent contribution. As the two bands in Raman spectra represent the vibrations of bridging and terminal oxygen in the polymeric chain, the relative intensity $\frac{I(P-O_{br})}{I(P-O_t)}$ of these peaks can provide the information about the decrease or increase of covalence character in the chains [65]. In the Raman spectra of glassy materials, symmetric vibrations are seen more prominently than asymmetric vibrations. A shift of 1143 cm^{-1} band toward left (to lower wave numbers) is easily observed for $x > 0.1$ which is an indication of softening of glass network upon AgI addition.

A number of structural studies on AgI-doped phosphate glasses have been done using X-ray diffraction [66–68], inelastic and quasi-elastic neutron scattering [69–79], extended X-ray absorption fine structure (EXAFS) [71, 80] and NMR [81], etc. The conductivity of glasses generally increases upon doping with suitable salts. The increase in conductivity of $AgPO_3$ upon addition of AgI has been attributed to the formation of AgI microdomains or clusters in the glass matrix. The structure of AgI microdomains is considered to be similar to α-AgI, which has very high ionic conductivity, due to its highly disordered and open structure. The presence of AgI microdomains in AgI-$AgPO_3$ glass electrolytes is supported by X-ray and neutron scattering, which show that the AgI does not alter the short range order of the glass network [66]. Moreover, AgI does not enter into the chain structure of the glass network. The atomic distances of P-O, P-P, and O-O bonds obtained by X-ray diffraction do not change with composition [67]. This is consistent with results obtained by neutron scattering [82] and Monte Carlo simulation [83]. In the X-ray and neutron scattering of these glasses, a sharp diffraction peak is observed at $Q \sim 0.6–0.8$ Å. Here, $Q = \frac{4\pi \sin\theta}{\lambda}$ and θ are the half scattering angle between incident and diffracted beams and λ is the wavelength of X-ray. This peak is known as first sharp diffraction peak (FSDP) (Fig. 16.8) and is attributed to the medium-range order (<10 Å) present in the glass [67] or to the AgI clusters in the glass [75]. However, some objections are raised on the explanation based on AgI

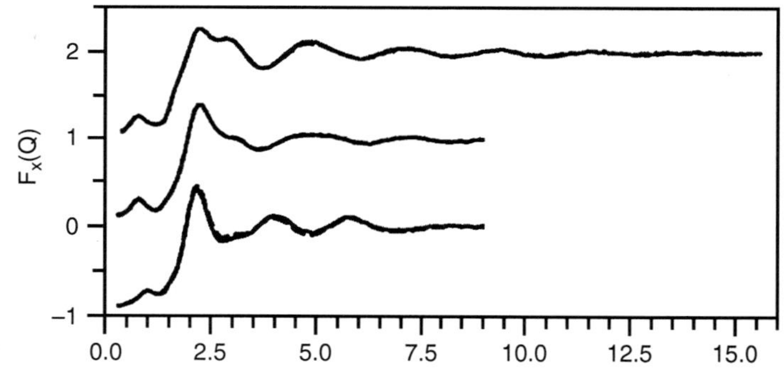

Fig. 16.8 First sharp diffraction (FSDP) peak for xAgI-(1-x)$AgPO_3$ glass electrolytes, which is due to medium-range order. This peak shifts to lower Q value with increasing AgI content indicating more open structure with increasing x (see text for details). The lower cure is for $x = 0$, middle $x = 0.2$, and upper $x = 0.5$. The figure is taken from literature [71]

clusters, and alternative explanations are given by others [71]. It is proposed by Wicks et al. that the polymer chain density gives rise to FSDP. In xAgI-(1-x)AgPO$_3$ system, the intensity of FSDP increases, and its position shifts to lower value of Q with an increase of x [70, 75]. The shift of FSDP at lower value of Q for high x is attributed to chain density fluctuation as the chains are pushed apart by the addition of AgI to the glass matrix. Thus, it is advised that the AgI addition leads to more open structure in the expanded glass network, which helps in achieving the high silver ion conductivity. The Debye-Waller temperature factor of these glasses also increases with an increase of x, just like conductivity showing a connection of conductivity enhancement with composition [68]. In a recent letter, a correlation has been reported between the conductivity enhancement and I-Ag distance in AgI-doped glasses by using EXAFS technique [80]. This technique is important to study the local environment of an X-ray absorbing atom. They have argued that a pathway volume is formed in which majority of silver ions, which contribute to ion conduction, have mixed oxide and iodide coordination [84]. The pathway volume is similar to free volume, but they have said that the conductivity enhancement in the glass depends on the spatial distribution of surrounding ions. They have also reported the local structure using EXAFS around iodine atom in a number of AgI-doped glasses. They have found that the glasses with longer I-Ag distances show high ionic conductivity, which is independent of their host glass matrix. Moreover, progressive AgI addition to the glass leads to an increase I-Ag distance from 2.77 to 2.85 Å, where short I-Ag distance glasses possess high activation energy and low conductivity, and glasses with long I-Ag distance possess low activation energy and high conductivity. This progressive increase in I-Ag distance leads to increase in pathway volume for ionic conduction. A correlation between medium-range order, conductivity enhancement, free volume, and FSDP wave number can also be obtained [78, 85]. The length scale $d = 2\pi/Q$ characterizes the medium-range order. By doping the glass with AgI, the characteristic length d increases (due to the decrease in corresponding FSDP wave number). This increase of d (intermediate-range order) termed as the glass network expansion or increase of free volume [71] adds to the enhancement of conductivity.

Electrical Characterization

There are several techniques from which motion of ions in the solid can be probed indirectly. AC conductivity is one such technique used to indirectly probe the ion dynamics. A review of various dynamic processes in glasses and experimental techniques has been given by Angell [86]. The AC conductivity of the system is measured by applying an AC electrical signal with varying frequency and measuring the system response by using impedance analyzer/LCR meter. The time-dependent mean square displacement of ions is directly related to the AC conductivity via Fourier transform as shown in the following Eq. 16.3 [87]:

$$\left\langle r^2(t)\right\rangle = \frac{12k_B \mathrm{TH_R}}{\mathrm{Nq^2\pi}} \int_0^t dt' \int_0^\infty \frac{\sigma'(\nu)}{\nu} \sin\left(2\pi\nu t'\right) \mathrm{d}\nu \tag{16.3}$$

In this equation, $\langle r^2(t)\rangle$ is the time-dependent mean square displacement of mobile ions, k_B is the Boltzmann constant, T is the temperature, N is the mobile ion density, q is the ionic charge, $\sigma'(\nu)$ is the real part of AC conductivity, and ν represents the frequency of applied AC signal voltage. $\mathrm{H_R}$ is the Haven ratio, which takes into account the strength of ion-ion correlations. $\mathrm{H_R}$ is also defined as the ratio of tracer diffusion coefficient to conductivity diffusion coefficient ($\mathrm{H_R} = \mathrm{D^*/D_\sigma}$). For low ion concentrations, $\mathrm{H_R}$ is close to unity and for high ion concentrations it can be as low as 0.3. In the limit of high frequencies, it approaches to unity, independent of ion concentrations [87].

The ionic transport and frequency-dependent conductivity of conducting glasses are still not understood fully. In comparison to crystalline materials, the ion transport process in glasses is more complicated and difficult to understand due to the disordered structure of the glass. The mechanism of ion transport is affected by glass structure. Apart from the glass structure, ion transport is also affected by the correlations between the surrounding mobile ions. When some dopant or network modifier is added, an increase of mobile ion concentration and changes in the glass network are expected. The correlated motion of ions gives the dispersive conductivity [88]. When an ion leaves a preoccupied site in the glass network, it leaves an uncompensated charge attached to that site. After the positive ion has moved, this site will keep the memory of that ion as long as there is negatively charge attached to it. After the ion goes, the environment around this site relaxes back to the initial state. This process decreases the backward movement of the ion after the jump, and this motion is considered as correlated motion and gives strong dispersion in the AC conductivity. This correlated motion of ions is very common to ionic conductors and gives "universal dynamic response" (UDR) or "first universality." The real part of the complex AC conductivity, σ' (c.f. $\sigma^*(\nu) = \sigma'(\nu) + i\sigma''(\nu)$) in the UDR regime, has a power law dependence on frequency. This behavior is also known as Jonscher power law (JPL) behavior [89]. The form of this equation is represented by the following Eq. 16.4, known as Almond-West formulism [90]:

$$\sigma'(\nu) = \sigma_{dc}\left[1 + \left(\frac{\nu}{\nu^*}\right)^n\right] \tag{16.4}$$

In this Eq. 16.4, σ_{dc} represents the DC conductivity. The quantity ν^* represents a characteristic frequency mark, i.e., the onset of AC conduction. The AC conduction is assumed to begin when $\sigma'(\nu^*) = 2\sigma_{dc}$. The frequency ν^* separates the long-time (or low-frequency) dynamics from the short-time (high-frequency) dynamics. Long-time dynamics is diffusive and is due to the random motion of ions over long ranges; whereas, short-time dynamics is sub-diffusive which is due to the correlated motion of ions over the short distances. Frequency index, n, has weak temperature dependence and is usually less than 1 (n < 1). The conductivity spectra of ion-conducting glasses are shown in Fig. 16.9 for sodium tellurium oxide ($0.1\mathrm{Na_2O}$-$0.9\mathrm{TeO_2}$) and

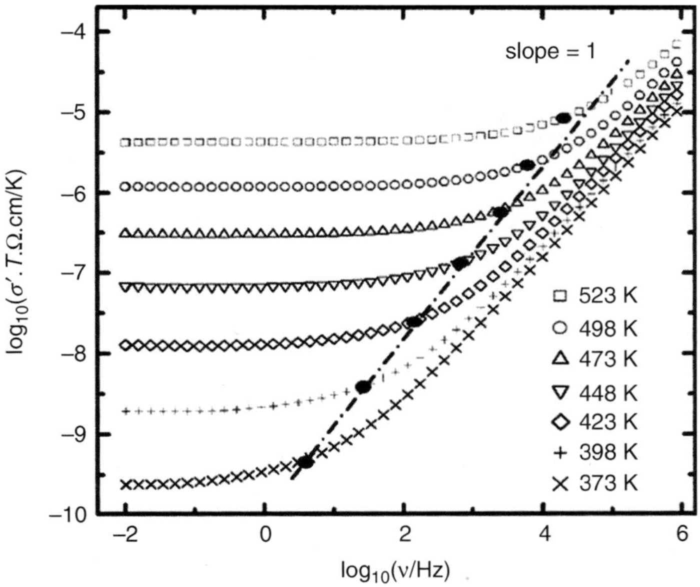

Fig. 16.9 Conductivity spectra of $0.1Na_2O$-$0.9TeO_2$ glass at different temperatures. The *filled circles* on the plot, acquired by imposing the condition $\sigma'(\nu = \nu^*) = 2\sigma_{dc}$, represent the onset of dispersive (AC) conduction [91]

Fig. 16.10 for 0.2AgI-0.8LiPO$_3$ glasses at various temperatures. The transition points separating the dispersive behavior (AC) from DC conducting (plateau) part are also shown in the former. For this system, these points do not fall on the line of slope 1 as seen in the figure (i.e., DC conductivity is not linearly proportional to the hopping frequency). An equivalent form of Eq. 16.4 is written as

$$\sigma'(\omega) = \sigma_{dc} + A\omega^n \tag{16.5}$$

In Eq. 16.5, A is the temperature-dependent constant, $\omega = 2\pi\nu$ and $n < 1$. A comparison of Eqs. 16.4 and 16.5 gives $A = \frac{\sigma_{dc}}{(\nu^*)^n}$.

The dispersive (AC) conductivity is less temperature dependent than DC conductivity. DC conductivity has the following relation with the mobile ion density N and mobility μ:

$$\sigma_{dc} = Nq\mu \tag{16.6}$$

By the use of Nernst-Einstein relation $D = \mu kT/q$, and $D = \frac{a^2\Gamma}{6}$ for random ion hopping, Eq. 16.6 can be written as follows:

$$\sigma_{dc} = \frac{Nq^2a^2\Gamma}{6k_BT} \tag{16.7}$$

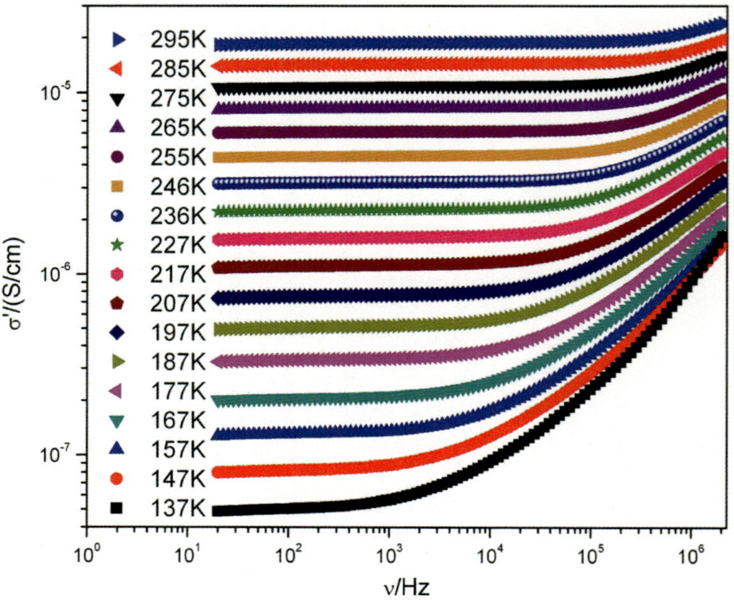

Fig. 16.10 AC conductivity of 0.2AgI-(0.8) $LiPO_3$ glass at various temperatures. This type of behavior is known as universal dynamic response behavior as it is observed in a vast variety of materials including glass, semiconductors, polymers, etc. [30]

where N is the mobile ion density, q is the charge on each ion, a is the jump distance, T is absolute temperature, and Γ is the hopping rate for random ion hopping.

The hopping rate has exponential temperature dependence as

$$\Gamma = \Gamma_0 \exp(-\Delta E_h / k_B T) \tag{16.8}$$

DC conductivity is also exponentially dependent on temperature through hopping rate,

$$\sigma_{dc} T = \sigma_0 \exp(-\Delta E_{dc} / k_B T) \tag{16.9}$$

The activation energies for hopping rate (ΔE_h) and DC conductivity (ΔE_{dc}) are nearly equal when carrier concentration is independent of temperature [92] and they are different when carrier concentration is temperature dependent. The hopping distance, a, and mobile ion concentration, N, in fast ion conductors are nearly independent of temperature in a wide range of temperature [93]. Most glasses including phosphate glasses obey Arrhenius temperature dependence of DC conductivity as shown in Fig. 16.11 for xAgI-(1-x) $LiPO_3$ glasses. Some glasses such as alkali borate [94] and alkali silicate glasses [95] have shown deviations from the Arrhenius Eq. 16.9 at high temperatures (but well below their glass transition temperature). In these glasses, structural changes are noticed, which give rise to

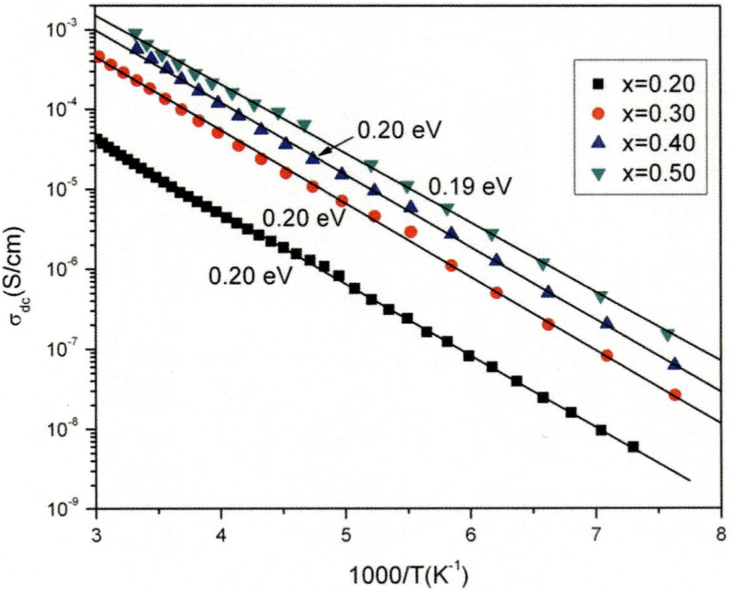

Fig. 16.11 Arrhenius temperature dependence of DC conductivity in xAgI-(1-x) $LiPO_3$ glasses. Most of the glass electrolytes show this behavior [30]

non-Arrhenius DC conductivity. These structural changes are sensitive to composition, temperature, and thermal history. A vacancy like transport mechanism is suggested in these glasses. The non-Arrhenius temperature dependence of DC conductivity is shown in Fig. 16.12 for alkali silicate glasses.

The mobile ion concentration (or mobile ion density), which is seen in Eq. 16.7, is not a well-defined quantity, since in the very long-time limit every ion can be thought of as being mobile. So to define mobile ion density, the time consideration is necessary. One natural choice is the time scale corresponding to the crossover frequency, ν^* which separates DC conductivity from dispersive AC conductivity [96]. This crossover frequency has been identified as hopping rate (or hopping frequency) in the literature [97]. With this assumption Eq. 16.7 can be written as

$$\sigma_{dc} = \frac{n_{mob} q^2 a^2 \nu^*}{6 k_B T} \tag{16.10}$$

Equation 16.10 defines the mobile ion density, n_{mob} as the number of ions, which has moved from its initial position in the time of τ $(=1/\nu^*)$. Solid-state NMR techniques such as motional narrowing experiments provide the information about the mobile ion density [98]. If mobile ion density n_{mob} is comparable to the total ion density, then the electrolyte is called as strong electrolyte [28]. In such a case, the doping salt dissociates completely in the glass matrix and most of the ions take part in the conduction process. This is the case of many fast ion-conducting glasses. If the n_{mob}

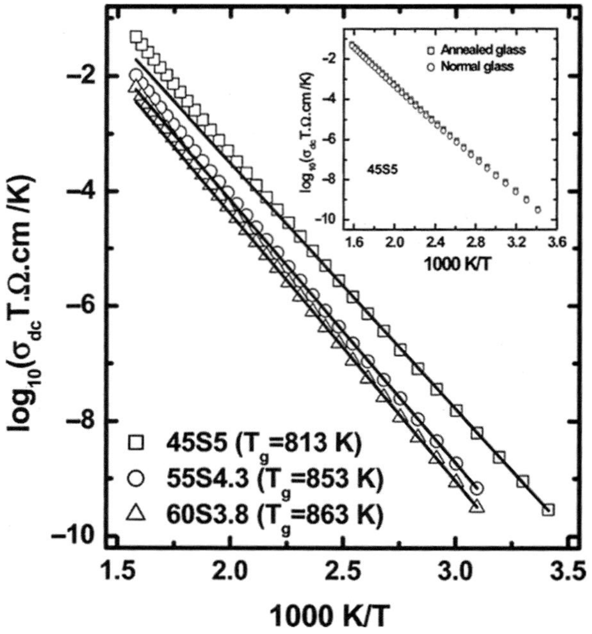

Fig. 16.12 Arrhenius plots of DC conductivity of alkali silicate glasses. The departure from Arrhenius behavior is observed at higher temperatures (see text for details) [95]

is very small in comparison to total ion density, then it is a weak electrolyte [99]. In the weak electrolytes, ion transport proceeds via vacancy mechanism [100]. Here (normally in crystalline materials), the number of vacancies is strongly temperature dependent, whereas in glasses the vacant sites are determined by the history of glass formation, and their number in glasses is significantly higher than crystals [101]. Because of higher vacancy concentration, interactions between them take place, and in turn ion transport in glasses becomes very complex. In deriving the Eqs. 16.7 and 16.10, ion-ion correlations have been ignored and it is assumed that all ions make equal jump distances. Also similar jump rates have been assumed for all ions. Equation 16.10 gives a good estimate of mobile ion density by the measured DC conductivity. When there is a distribution of jump rates (as in the case of random barrier model), the mobile ion density is less than that given by Eq. 16.10. To a good approximation, Eq. 16.10 gives an upper limit to the mobile ion density [96]. If ion-ion correlations are considered, Eq. 16.10 would become

$$\sigma_{dc} = \frac{n_{mob} q^2 a^2 \nu^*}{6 k_B T H_R} \tag{16.11}$$

where H_R is the time scale-dependent Haven ratio. The value of H_R lies in between 0.3 and 1.

The transformation of Eq. 16.4 gives the following time dependencies for $\langle r^2(t) \rangle$ in the long- and short-time limits

$$\langle r^2(t)\rangle \approx \begin{cases} \dfrac{a^2 t}{\tau}, t > \tau \rightarrow dc \\ a^2\left(\frac{t}{\tau}\right)^{1-n}, t < \tau \rightarrow ac \end{cases} \tag{16.12}$$

where $\tau = 1/(\nu^*)$. Equation 16.12 states that mean square displacement $\langle r^2(t)\rangle$ is a linear function with time for $t > \tau$. In this regime, the ion motion can be modeled by the classical random walk model of diffusion. This motion is known as the diffusive motion. For the time intervals $t < \tau$, the mean square displacement of ions is nonlinear with time function. In this regime, the correlated motion of ions takes place up to a relatively short distances and the motion is called as sub-diffusive to distinguish it from diffusive motion. The correlated motion in the sub-diffusive regime is best understood as a motion in which the ions perform several "unsuccessful" back-and-forth jumps [88] before any "successful" jump takes place. Equation 16.12 clearly states that for $t > \tau$, DC conduction dominates, whereas for $t < \tau$, AC conduction dominates. But there can still be AC conduction for $t > \tau$ (at low frequencies). This equation wrongly predicts a continuously increasing dielectric constant with decreasing the frequency instead of a low-frequency plateau. If the AC contribution to root mean square displacement is also considered for $t > \tau$, this error is removed. For $t > \tau$, the AC contribution to mean square displacement is given by $\langle r^2(t)\rangle_{ac} \approx a^2$ [102]. Experimentally observed length scale that separates diffusive transport from sub-diffusive transport is a $\approx$ 2Å. The ion movements over the long ranges (>2Å) are considered as diffusive, and the ion movements over the short ranges (<2Å) are considered as sub-diffusive [102].

In the time domain, Eq. 16.5 is equivalent to a stretched potential $\phi(t)$ known as Kohlrausch-Williams-Watts (KWW) potential [103]

$$\phi(t) = \exp(-t/\tau)^{1-n} \tag{16.13}$$

where $\phi(t)$ represents the decay of the electric field within the material after a fixed polarization that is applied at $t = 0$ and τ is the relaxation time. The ionic conductivity may be obtained by taking the Fourier transform of Eq. 16.13.

Current Issues

Ion conduction phenomenon in glasses, polymers, crystals, etc. plays a highly significant role in various technologies. The application of ion-conducting materials has been realized in many devices such as solid oxide fuel cells, batteries, supercapacitors, oxygen separation membranes, electrochemical windows, etc. So these applications related to the ion-conducting materials motivate the researchers to work in this area. There is a permanent quest for the materials having high ionic conductivity. The level of ionic conductivity achievable in materials has reached at saturation level, and the main impeding factor toward the development of new materials with high ionic conductivity is the lack of complete understanding of ion conduction phenomenon. Proper understanding of the ion conduction in solids has been a fundamental problem since many years. Several attempts have been made by researchers to understand the phenomena of ion conduction in disordered solids. Currently, researchers are much interested in the frequency dependence of low-temperature AC conductivity of disordered ionic solids. At low temperatures, apart from UDR or "first universality," the AC conductivity of disordered ionic solids obeys a different frequency response, which was first noticed by Nowick et al. [104, 105]. In this case, the AC conductivity varies almost linearly with frequency, or equivalently dielectric loss ($\varepsilon'' = \sigma'/2\pi\varepsilon_0\nu$) becomes independent of frequency. This behavior is termed as "nearly constant loss" (NCL) behavior or "second universality." Figure 16.7 shows the NCL behavior in sodium phosphate ($NaPO_3$) glass at different temperatures. Generally, the NCL is observed at low temperatures and/or high frequencies when the DC conduction ceases, as NCL conductivity is very small and could not be observed when there is an existence of sufficient amount of DC conduction. In the NCL regime, the conductivity has no or very little temperature dependence. The whole conductivity spectra of glasses and crystalline materials could be approximated by superposition of JPL behavior and NCL behavior as written below:

$$\sigma'(\nu) = \sigma_{dc}\left[1 + \left(\frac{\nu}{\nu^*}\right)^n\right] + A\nu \tag{16.14}$$

where the first term is the usual UDR behavior, the last term represents the NCL behavior, and A is a constant. As stated earlier, the origin of AC conductivity in the first universality is the correlated hopping motion of mobile ions [88]. However, the origin of NCL is still unclear and is much debated. One view is that the NCL arises because of the relaxations of low energy excitations over an energy barrier described by asymmetric double-well potential (ADWP) [106]. In a more recent view, it is thought to arise because of the collective localized movements of mobile ions having Coulomb interactions between them [107]. Another view is that the NCL is caused by vibrational motion of ions in strong anharmonic potential or from hopping in cage potential defined by neighboring ions [96]. However, this view focuses on very high-frequency NCL in the terahertz range, where AC conductivity joins vibrational absorption [108]. Yet another behavior of AC conductivity has been reported less frequently, which is known as superlinear power law (SPL) [109, 110]. In this case, the frequency index, n, in the power law-dependent conductivity can even exceed 1. This behavior appears in the same temperature and frequency window as NCL. Researchers are still not sure whether the NCL and SPL have same physical origins or different. A lot of work has been done on NCL behavior in disordered materials, some of it is discussed in brief as follows. There are few reports in which origins of Jonscher and NCL processes are considered to be same [111, 112]. This group has suggested with supports from random barrier model and analysis of experimental results that the entire conductivity spectra is governed by hopping dynamics of ions and any demarcation between JPL and NCL behaviors is artificial (Figs. 16.13 and 16.14).

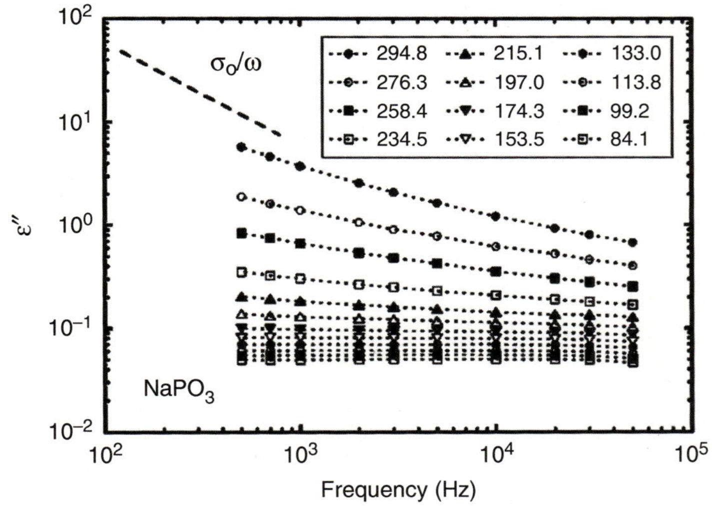

Fig. 16.13 Dielectric losses as a function of frequency for $NaPO_3$ glass at 12 temperatures (in Kelvin). It is clear, that lower curves have no frequency and temperature dependence. The *dashed line* (with slope −1) on the *top left* corner shows the approaching DC behavior at low frequencies [113]

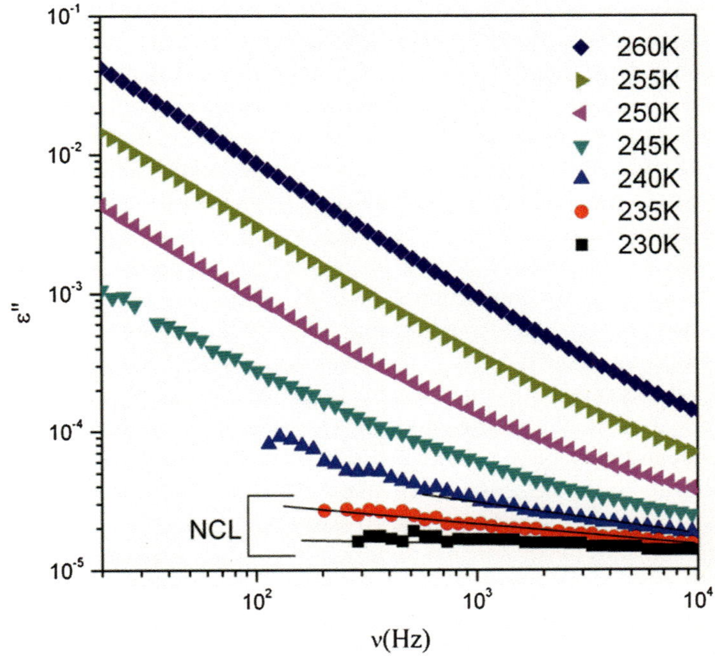

Fig. 16.14 NCL behavior in $0.1AgI\text{-}0.9LiPO_3$ composite glass electrolyte at low temperatures. The dielectric loss has near-zero temperature dependence below 235 K [30]

At low temperatures, the Jonscher behavior changes smoothly to NCL behavior without any change of conduction mechanism. However, this view could not get much support due to other's criticism [114]. Nowick et al. have studied the NCL in sodium silicate glasses [105, 115]. They have given a description of NCL by a two level system, which is based on asymmetric double-well potential. In this model the conductivity at low temperatures is given by hopping and tunneling processes.

Sidebottom has studied the AC conductivity behavior of several metaphosphate glasses [113, 116]. They have noticed that the frequency index, n, of the power law conductivity depends on the cation constriction (i.e., cation size relative to chain separation). They found that n decreases when increasing the constriction of the cation. This decrease results from the reduction in the coordination of the cation's local conduction space, which is caused by increased constriction. In other words, it can be said that the value of n depends on the dimensionality of conduction space. In silver- and alkali halide-doped glasses, n is found to be less than 2/3, which means cations in these systems have conduction space of lower dimensionality. The value of $n = 2/3$ is related to the 3D conductors. But the 2D conductors such as Na β-alumina give a value of n around 0.58 and 1D conductors such as hollandite give a value of n to be about 0.3. The AgI-free $AgPO_3$ glass gives a value of n ~0.6 suggestive of 2D ion dynamics. When AgI is added in the $AgPO_3$ glass matrix, n increases to 0.67, implying 3D ion dynamics. Lower index in pure $AgPO_3$ glass indicates a network of lower dimensionality than 3D. Here the PO_3 chains collapse and reduce the number of pathways in the vicinity. However, in contrast to $AgPO_3$ glass, pure alkali phosphate glasses such as $LiPO_3$ and $NaPO_3$ give 3D ion dynamics. Here, the size of the cation comes into picture; the size of silver cation is much larger than lithium or sodium cation. Thus, in alkali phosphate glasses, the collapse of chains is insufficient, which behave as network of dimensionality lower than 3D. Moreover, when the alkali ion size is higher such as for Cs, the network behaves like a lower dimensionality and gives a smaller n. Thus, expansion of glass network and cation size both determine the observed value of frequency exponent, n.

Ngai and coworkers have done a lot of work on AC conductivity and NCL behavior in the fast ion conductors [117–124]. They have argued that the NCL contribution and cooperative hopping contribution to the total conductivity are not additive processes; instead, NCL crosses over to cooperative ion hopping contribution with a decrease of temperature [120]. Therefore, the use of Eq. 16.14 to fit the conductivity data should be discontinued. (However, there are reports, which consider both contributions to be additive [125].) To understand the ion dynamics and NCL behavior of ionic conductors, they have proposed a coupling model [126–128]. They have shown that the crossover from Jonscher behavior to NCL behavior is activated by an activation energy, which is much less than the activation energy of DC conduction. They have found a mixed alkali-type effect in NCL too. A partial replacement of alkali ions reduces the NCL effect. However, this effect is smaller than the corresponding decrease in DC conductivity. They have analyzed the cation mass dependence of NCL in a number of alkali triborate glasses and found

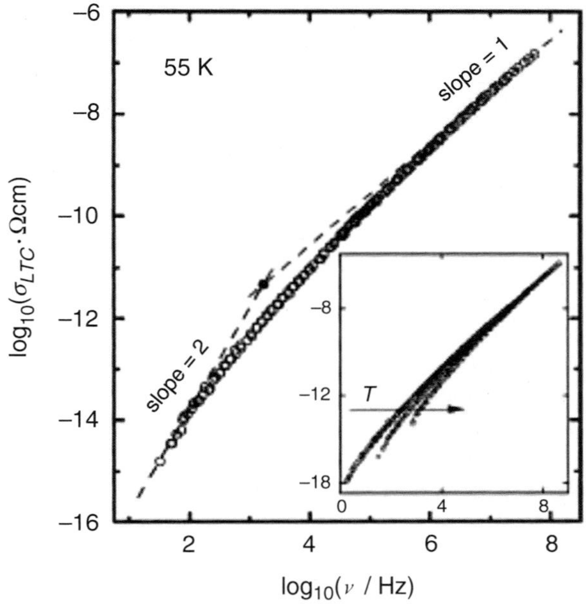

Fig. 16.15 Low-temperature AC conductivity σ_{LTC} in $0.10Na_2O$-$0.9B_2O_3$ glass showing 2-to-1 slope change in log-log plot of conductivity versus frequency. The *inset* shows that slope 2 region has temperature dependence, whereas slope 1 region does not have [107]

that the magnitude of NCL decreases systematically with an increase of alkali ion mass. Recently, NCL has been reported in disordered pyrochlore-type oxide ion conductor [129]. Apart from many glassy materials, NCL has been observed in polycrystalline materials, viz., lithium aluminum inosilicate ($LiAlSi_2O_6$) [130], Na β-alumina [131], and $Li_xLa_{1-x}TiO_3$ [132] too. This shows that the structural disorder is not a prerequisite for the observation of NCL; however, the structural disorder may enhance the magnitude of NCL as the results suggest. Recently, NCL has also been detected in Mn-doped bismuth magnesium titanate-lead titanate ($Bi(Mg_{1/2}\ Ti_{1/2})O_3$-$PbTiO_3$) disordered ferroelectric ceramics also [133].

Significant works on AC conductivity and NCL in ion-conducting glasses have also been carried out by Funke and his coworkers [88, 107, 125, 134–155]. They have developed a mismatch generated relaxation for the accommodation and transport of ION (MIGRATION) concept, which literally means that a mismatch is introduced by the initial hop of the ion, resulting in relaxation (rearrangement) of the neighborhood; this relaxation of the neighborhood finally accommodates the ion at its new position. As soon as the accommodation of the ion at the new location is done, an elementary step of macroscopic transport is completed by the ion. This model uses rate equations to derive time correlation functions and conductivity spectra. A formal description of their model can be found in many of their papers including most recent papers mentioned earlier. Their model predicts a continuous increase of frequency exponent, n, with frequency having a limiting value of 1. Moreover, this model yields a finite value of conductivity at very high frequencies. By MIGRATION concept, they have predicted that the slope of log σ versus log ν

plot first changes from two to one and then one to zero at high frequencies as reported in an earlier study done by using Monte Carlo simulation [156]. The NCL region lies within the two crossover points. All of the three regions have been observed in sodium borate ($0.3Na_2O$-$0.7B_2O_3$) glass [107, 155]. Both type of conductivity behaviors, viz., frequency squared (corresponding to slope 2) and linear (corresponding to slope 1), have same origins (localized movement of coulomb interaction of ions); however, squared dependence is observed at much lower temperatures than linear. Moreover, squared dependent conductivity (on frequency) is temperature activated, whereas the linear frequency-dependent conductivity is temperature independent. The frequency dependence of low-temperature AC conductivity (σ_{LTC}) of 10 M percent sodium borate ($0.10Na_2O$-$0.9B_2O_3$) is shown in Fig. 16.15.

The current understanding of the conductivity in the disordered ionic conductors is that the DC conductivity is caused by the long-range hops of mobile ions, whereas Jonscher-type power law conductivity results from the correlated ionic hops. The NCL effects are caused by the collective phenomena of many charge carriers with Coulomb interaction among them. These interacting charge carriers are supposed to move locally in time-dependent single-particle double-well potentials.

Conclusions

In conclusion, glassy superionic conductors are of significant importance for electrochemical charge storage, and energy conversion devices. Among others, silver halide doped phosphate glasses exhibit high ionic conductivities, which, combined with their easy processing, makes them suitable for practical applications. Highly amorphous nature of the glasses is suitable for better ionic conduction. Significant research works have been carried out to develop high ionic conducting solid electrolytes and to understand the ion conduction mechanism. However, complete understanding of the ion conduction has to be achieved for further development of solid electrolytes with improved conductivity.

Acknowledgment The authors acknowledge the financial support provided by the Council of Scientific and Industrial Research (CSIR), India, for carrying out this work.

References

1. Tubandt C (1921) Electrical conduction in solid crystallized compounds. II. Transport and migration of the ions in simple solid electrolytes. Z Anorg Allgem Chem 115:105
2. Strock LW (1934) Crystal structure of high-temperature silver iodide α-AgI. Z Phys Chem B 25:411
3. Strock LW (1935) Additions and corrections to: crystal structure of the high-temperature silver iodide α-AgI. Z Phys Chem B 31:132
4. O'Keeffe M, Hyde BG (1976) The solid electrolyte transition and melting in salts. Philos Mag 33:219

5. Tubandt C (1932) Handbuch der Experimental Physik. Akad. Verlagsgesellschaft, Leipzig, vol 12(1), p 383
6. Boolchand P, Bresser WJ (2001) Mobile silver ions and glass formation in solid electrolytes. Nature 410(6832):1070
7. Boyce JB, Huberman BA (1979) Superionic conductors: transitions, structures, dynamics. Phys Rep 51(4):189
8. Bradley JN, Greene PD (1976) Solids with high ionic conductivity in group 1 halide systems. Trans Faraday Soc 63:424
9. Owens BB, Argue GR (1967) High-conductivity solid electrolytes: MAg_4I_5. Science 157:308
10. Takahashi T, Yamamoto O (1964) Solid electrolyte cell III. The conductivity of solid electrolytes. 3. The conductivity of Ag_3SI. Denki Kgaku 32(8):610
11. Suchow L, Pond GR (1953) Electrical conductivity of Ag_2HgI_4, Cu_2HgI_4 and their eutectoid. J Am Chem Soc 75(21):5242
12. Shahi K, Chandra S (1975) Electrical conductivity and thermoelectric power of $Ag_6I_4WO_4$ solid electrolyte. Phys Status Solidi A 28(2):653
13. Huggins RA (1977) Recent results on lithium ion conductors. Electrochim Acta 22:773
14. Kvist A, Lunden A (1965) Electrical conductivity of solid and molten Li_2SO_4 Z Naturf 20A:235
15. Kamaya N et al. (2011) A lithium superionic conductor. Nat Mater 10(9):682
16. Tatsumisago M, Shinkuma Y, Minami T (1991) Stabilization of superionic a-AgI at room temperature in a glass matrix. Nature 354(21):217
17. West AR (2003) Solid state chemistry and its applications. Wiley, New Delhi
18. Pradel A, Ribes M (1986) Electrical properties of lithium conductive silicon sulfide glasses prepared by twin roller quenching. Solid State Ionics 18–19, Part 1(0): 351 (1986)
19. Julien C, Nazri GA (1994) Solid state batteries: materials design and optimization. Kluwer Academic Publications, Boston
20. Kikkawa S, Miyai T, Koizumi M (1988) New lithium ionic conductor, Li-Ge-Se glasses. Solid State Ionics 28–30(Part 1):743
21. Kennedy JH, Zhang Z (1988) Improved stability for the SiS_2-P_2S_5-Li_2S-LiI glass system. Solid State Ionics 28–30(Part 1):726
22. Malugani JP, Robert G (1980) Preparation and electrical properties of the 0,37Li_2S-0,18P_2S_5-0,45LiI glass. Solid State Ionics 1(5-6):519
23. Minami T, Imazawa K, Tanaka M (1977) Superionic conducting glasses: glass formation and conductivity in the AgI-Ag_2O-P_2O_5 system. J Electrochem Soc 124(11):1659
24. Magistris A, Chiodelli G, Schiraldi A (1979) Formation of high conductivity glasses in the system AgI – Ag_2O – B_2O_3. Electrochim Acta 24(2):203
25. Minami T and Tanaka M (1980) Structure and ionic transport of superionic conducting glasses in the system AgI – Ag_2O – MoO_3. J Non-Cryst Solids 38–39, Part 1(0): 289
26. Hariharan K, Kaushik R (1987) The superionic AgI-Ag_2O-V_2O_5 system: electrical conductivity studies on glass and polycrystalline forms. J Mater Sci 22(9):3335
27. Grant RJ et al. (1978) Optimized ionic conductivity in glass:vitreous silver arsenate iodide ($Ag_7I_4AsO_4$) electrolytes. J Phys Chem 82(26):2838
28. Tuller HL, Button DP, Uhlmann DR (1980) Fast ion transport in oxide glasses. J Non-Cryst Solids 40(1–3):93
29. Sun HW et al. (1987) Physical and chemical characterization and ionic conductivity of rapidly quenched glasses in the Sb_2S_3-Ag_2S-AgI system. J Solid State Chem 70(1):141
30. Singh DP, Shahi K, Kar KK (2013) Scaling behavior and nearly constant loss effect in AgI-$LiPO_3$ composite glasses. Solid State Ionics 231:102
31. Malugani JP et al. (1978) Electrical conductivity and Raman diffusion spectra of mixed glasses silver metaphosphate-MI2 (M = Cd, Hg, Pb): Correlation between conductivity and structure. Mater Res Bull 13(10):1009
32. Malugani JP et al. (1978) Ion conductivity in silver metaphosphate-AgX glasses (X = I, Br, Cl). Mater Res Bull 13(5):427

33. Nocun M (2004) Structural studies of phosphate glasses with high ionic conductivity. J Non-Cryst Solids 333(1):90
34. Borsa F et al. (1992) Relaxation and fluctuations in glassy fast-ion conductors: wide-frequency-range NMR and conductivity measurements. Phys Rev B 46(2):795
35. Chowdari BVR et al. (1995) Electrical and structural studies of lithium fluorophosphate glasses. Solid State Ionics 76(3–4):189
36. Winter R, Siegmund K, Heitjans P (1997) Nuclear magnetic and conductivity relaxations by Li diffusion in glassy and crystalline $LiAlSi_4O_{10}$. J Non-Cryst Solids 212(2–3):215
37. Bardi U, Caporali S, Tolstogouzov A (2009) Study on sublimation of solid electrolyte $(AgI)_{05}$-$(AgPO_3)_{05}$ with Knudsen effusion mass spectrometry. Rapid Commun Mass Spectrom 23(1):147
38. Mukherjee PK et al. (2007) Giant dielectric permittivity in aligned silver nanowires grown within (AgI)($AgPO_3$) glasses. J Phys Chem C 111(10):3914
39. Guo HX et al. (2007) Resistive switching devices based on nanocrystalline solid electrolyte $(AgI)_{05}(AgPO_3)_{05}$. Appl Phys Lett 91(24):243513
40. Liang C et al. (2007) Anomalous phase transition and ionic conductivity of AgI nanowire grown using porous alumina template. J Appl Phys 102(12):124308
41. Hsu KH et al. (2007) Electrochemical nanoimprinting with solid-state superionic stamps. Nano Lett 7(2):446
42. Matic A, Borjesson L (1998) Structure and dynamics of phosphate glasses. Philos Mag B 77 (2):357
43. Tatsumisago M et al. (1992) Formation of frozen a-AgI in superionic glass matrices at ambient temperature by rapid quenching. Solid State Ionics 50(3–4):273
44. Bartholomew RF (1972) Structure and properties of silver phosphate glasses – infrared and visible spectra. J Non-Cryst Solids 7(3):221
45. Galeener FL, Mikkelsen JC (1979) The Raman spectra and structure of pure vitreous P_2O_5. Solid State Commun 30(8):505
46. Malugani J-P, Mercier R (1984) Vibrational properties of and short range order in superionic glasses $AgPO_3$-AgX (X = I, Br, Cl). Solid State Ionics 13(4):293
47. Condrate RA (1986) Vibrational spectra of structural units in glass. J Non-Cryst Solids 84 (1–3):26
48. Fontana A, Rocca F, Tomasi A (1990) Light scattering in AgI containing superionic glasses. J Non-Cryst Solids 123(1–3):230
49. Benassi P et al. (1991) Light scattering in superionic glasses $(AgI)_x(AgPO_3)_{1-x}$: Brillouin and Raman scattering. Phys Rev B 43(2):1756
50. Koo J, Bae BS, Na HK (1997) Raman spectroscopy of copper phosphate glasses. J Non-Cryst Solids 212(2–3):173
51. Hudgens JJ et al. (1998) Raman spectroscopy study of the structure of lithium and sodium ultraphosphate glasses. J Non-Cryst Solids 223(1–2):21
52. Efimov AM (1999) Vibrational spectra, related properties and structure of inorganic glasses. J Non-Cryst Solids 253(1–3):95
53. Swenson J et al. (2000) Experimental insight into the mixed mobile ion effect in glasses. Solid State Ionics 136–137:1055
54. Uchino T, Yoko T (2000) Structure and vibrational properties of alkali phosphate glasses from ab initio molecular orbital calculations. J Non-Cryst Solids 263–264:180
55. Rossi F, Fontana A, Righetti L (2002) Temperature behaviour of quasielastic scattering in silver phosphate glass. Philos Mag B 82(3):323
56. Metwalli E et al. (2004) Properties and structure of copper ultraphosphate glasses. J Non-Cryst Solids 344(3):128
57. Doweidar H et al. (2007) Infrared spectra of Fe_2O_3-PbO-P_2O_5 glasses. Vib Spectrosc 37:91 (2005)
58. Boolchand P, et al. Raman scattering as a probe of intermediate phases in glassy networks. J Raman Spectrosco 38, 660

59. Novita DI, Boolchand P (2007) Synthesis and structural characterization of dry $AgPO_3$ glass by Raman scattering, infrared reflectance and modulated differential scanning calorimetry. Phys Rev B 76(18):184205
60. Padmaja G, Kistaiah P (2009) Infrared and Raman spectroscopic studies on alkali borate glasses: evidence of mixed alkali effect. J Phys Chem A 113(11):2397
61. Money BK, Hariharan K (2010) Phase dependent heterogeneous dynamics of Li^+ ion in $LiPO_3$ based systems. Integr Ferroelectr 120:75
62. Santagneli SH et al. (2011) Structural investigations of tungsten silver phosphate glasses by solid state NMR, vibrational and X-ray absorption near edge spectroscopies. J Non-Cryst Solids 357(10):2126
63. Kalampounias AG (2012) Short-time vibrational dynamics of metaphosphate glasses. J Phys Chem Solids 73(2):148
64. Novita DI et al. (2007) Fast-ion conduction and flexibility of glassy networks. Phys Rev Lett 98(19):195501
65. Konidakis I, Varsamis CP, Kamitsos EI (2011) Effect of synthesis method on the structure and properties of $AgPO_3$-based glasses. J Non-Cryst Solids 357(14):2684
66. Musinu A, Piccaluga G, Pinna G (1988) Short range order in AgI–$AgPO_3$ glasses by X-ray diffraction. J Chem Phys 89(2):1074
67. Takahashi H, Matsubara E, Waseda Y (1994) Structural study of superionic conducting glasses Agl-$AgPO_3$ by X-ray diffraction. J Mater Sci 29(9):2536
68. Thazin A et al. (2004) Effective Debye-Waller temperature parameter in superionic conducting Agl-$AgPO_3$ glasses. Solid State Ionics 175(1–4):675
69. Dianoux AJ, et al. (1991) Neutron scattering by superionic conductor glasses. J Non-Cryst Solids 131–133, Part 2(0): 973
70. Rousselot C et al. (1991) Characterization of intermediate-range order in superionic $AgPO_3$-AgX (X = I, Br, Cl) glasses by neutron diffraction. Solid State Ionics 44(3–4):151
71. Wicks JD et al. (1995) Structure and ionic conduction in $(AgI)_x(AgPO_3)_{1-x}$ glasses. Phys Rev Lett 74(5):726
72. Matic A et al. (1999) Ionic motion of silver in super-ionic glasses. Physica B 266(1–2):69
73. Kartini E, Collins MF (2000) Nature of the precipitate in $(AgI)_{07}(AgPO_3)_{03}$ glass. Physica B 276–278:467
74. Kartini E et al. (2000) Neutron scattering from the superionic glasses $(AgI)_x(AgPO_3)_{1-x}$, through the glass transition. Solid State Ionics 138(1–2):115
75. Kartini E et al. (2000) Anomalous temperature dependence of the first diffraction peak in the superionic glass $(AgI)_x(AgPO_3)_{1-x}$. Phys Rev B 61(2):1036
76. Kartini E et al. (2004) Anion effect on the structure of Ag_2S-$AgPO_3$ superionic glasses. Solid State Ionics 167(1–2):65
77. Nakamura M, et al. (2006) Low energy vibrational excitations characteristic of superionic glass. Physica B 385–386, Part 1(0): 552
78. Aniya M (2008) Medium range structure and power law conductivity dispersion in superionic glasses. J Non-Cryst Solids 354(2–9):365
79. Gunawan M, Kartini E, Putra EGR (2008) Small angle neutron scattering experiments on solid electrolyte $(AgI)_x(AgPO_3)_{1-x}$. J Solid State Electrochem 12(7–8):903
80. Sanson A et al. (2008) Correlation between I-Ag distance and ionic conductivity in AgI fast-ion-conducting glasses. Phys Rev Lett 101(15):155901
81. Tomasi C et al. (2001) Electric, thermodynamic and NMR evidence of anomalies in (x)AgI (1-x)$AgPO_3$ glasses. J Non-Cryst Solids 293–295:785
82. Tachez M et al. (1987) Structure determination of $AgPO_3$ and $(AgPO_3)_{05}(AgI)_{05}$ glasses by neutron diffraction and small angle neutron scattering. Solid State Ionics 25(4):263
83. Borjesson L, McGreew RL, Howels WS (1992) Fractal aspects of superionic glasses from reverse Monte Carlo simulations. Philos Mag B 65(2):261

84. Adams S, Swenson J (2000) Determining ionic conductivity from structural models of fast ionic conductors. Phys Rev Lett 84(18):4144
85. Aniya M (2000) Average electronegativity, medium-range-order and ionic conductivity in superionic glasses. Solid State Ionics 136–137:1085
86. Angell CA (1990) Dynamic processes in ionic glasses. Chem Rev 90(3):523
87. Roling B, Martiny C, Bruckner S (2001) Ion transport in glass: influence of glassy structure on spatial extent of nonrandom ion hopping. Phys Rev B 63(21):214203
88. Funke K (1993) Jump relaxation in solid electrolytes. Prog Solid State Chem 22:111
89. Jonscher AK (1977) The 'universal' dynamic response. Nature 267:673
90. Almond DP, West AR (1983) Anomalous conductivity prefactors in fast ion conductors. Nature 306(5942):456
91. Murugavel S, Roling B (2004) Ionic transport in glassy networks with high electronic polarizabilities: conductivity spectroscopic results indicating a vacancy-type transport mechanism. J Phys Chem B 108(8):2564
92. Almond DP, Duncan GK, West AR (1983) The determination of hopping rates and carrier concentrations in ionic conductors by a new analysis of ac conductivity. Solid State Ionics 8 (2):159
93. Cutroni M et al. (2002) Ionic conduction and dynamical regimes in silver phosphate glasses. J Non-Cryst Solids 307–310:963
94. Murugavel S, Roling B (2007) Ion transport mechanism in borate glasses: influence of network structure on non-Arrhenius conductivity. Phys Rev B 76(18):180202
95. Murugavel S et al. (2010) Ion transport mechanism in glasses: non-arrhenius conductivity and nonuniversal features. J Phys Chem B 114(42):13381
96. Dyre JC et al. (2009) Fundamental questions relating to the ion conduction in disordered solids. Rep Prog Phys 72:046501
97. Almond DP and West AR (1983) Mobile ion concentrations in solid electrolytes from an analysis of ac conductivity. Solid State Ionics 9–10, Part 1(0): 277
98. Rivera A, Sanz J (2004) Lithium dynamics in the fast ionic conductor $Li_{018}La_{061}TiO_3$ probed by ^{7}Li NMR spectroscopy. Phys Rev B 70(9):094301
99. Ingram MD, Moynihan CT and Lesikar AV (1980) Ionic conductivity and the weak electrolyte theory of glass. J Non-Cryst Solids 38–39, Part 1(0) 371
100. Bohnke O, Bohnke C, Fourquet JL (1996) Mechanism of ionic conduction and electrochemical intercalation of lithium into the perovskite lanthanum lithium titanate. Solid State Ionics 91(1–2):21
101. Dyre JC (2003) Is there a native band gap in ion conducting glasses? J Non-Cryst Solids 324 (1–2):192
102. Sidebottom DL (2009) Colloquium: understanding ion motion in disordered solids from impedance spectroscopy scaling. Rev Mod Phys 81(3):999
103. Williams G, Watts DC (1970) Non-symmetrical dielectric relaxation behaviour arising from a simple empirical decay function. Trans Faraday Soc 66:80
104. Lee WK, Lie JF, Nowick AS (1991) Limiting behavior of ac conductivity in ionically conducting crystals and glasses: a new universality. Phys Rev Lett 67(12):1559
105. Nowick AS, Lim BS, Vaysleyb AB (1994) Nature of the ac conductivity of ionically conducting crystals and glasses. J Non-Cryst Solids 172–174(Part 2):1243
106. Lu X, Jain H (1994) Low temperature AC conductivity of oxide glasses. J Phys Chem Solids 55(12):1433
107. Laughman DM, Banhatti RD, Funke K (2010) New nearly constant loss feature detected in glass at low temperatures. Phys Chem Chem Phys 12(42):14102
108. Burns A et al (1989) Dielectric spectra of ionic conducting oxide glasses to 2 GHz. Phys Chem Glasses 30:264
109. Elliott SR (1978) On the super-linear frequency dependent conductivity of amorphous semiconductors. Solid State Commun 28(11):939

110. Tiwari JP, Shahi K (2007) Super-linear frequency dependence of ac conductivity of disordered $Ag_2S–Sb_2S_3$ at cryogenic temperatures. Philos Mag 87(29):4475
111. Roling B, Martiny C, Murugavel S (2001) Ionic conduction in glass: new information on the interrelation between the Jonscher behavior and the nearly constant-loss behavior from broadband conductivity spectra. Phys Rev Lett 87(8):085901
112. Murugavel S, Roling B (2003) Nearly constant dielectric loss of glasses containing different mobile alkali ions. J Non-Cryst Solids 330(1–3):122
113. Sidebottom DL (2005) Constriction effect in the nearly constant loss of alkali metaphosphate glasses. Phys Rev B 71(13):134206
114. Leon C et al. (2002) Comment on ionic conduction in glass: new information on the interrelation between the 'Jonscher behavior' and the 'nearly constant-loss behavior' from broadband conductivity spectra. Phys Rev Lett 89(7):079601
115. Nowick AS, Vaysleyb AB, Liu W (1998) Identification of distinctive regimes of behaviour in the ac electrical response of glasses. Solid State Ionics 105(1–4):121
116. Sidebottom DL (2000) Influence of cation constriction on the ac conductivity dispersion in metaphosphate glasses. Phys Rev B 61(21):14507
117. Leon C, Lunkenheimer P, Ngai KL (2001) Test of universal scaling of ac conductivity in ionic conductors. Phys Rev B 64(18):184304
118. Leon C et al. (2001) Origin of constant loss in ionic conductors. Phys Rev Lett 86(7):1279
119. Leon C et al. (2002) Crossover of near-constant loss to ion hopping relaxation in ionically conducting materials: experimental evidences and theoretical interpretation. J Non-Cryst Solids 305(1–3):88
120. Ngai KL, Casaline R (2002) Near-constant loss and the loss from cooperative ion hopping in ionic conductors are not additive contributions. Phys Rev B 66(13):132205
121. Ngai KL, Lenon C (2002) Cage decay, near constant loss and crossover to cooperative ion motion in ionic conductors: Insight from experimental data. Phys Rev B 66(6):064308
122. Ngai KL, Wang Y, Moynihan CT (2002) The mixed alkali effect revisited: the importance of ion-ion interactions. J Non-Cryst Solids 307–310:999
123. Rivera A et al. (2002) Crossover from ionic hopping to nearly constant loss in the fast ionic conductor $Li_{018}La_{061}TiO_3$. Phys Rev B 65(22):224302
124. Rivera A et al. (2002) Cation mass dependence of the nearly constant dielectric loss in alkali triborate glasses. Phys Rev Lett 88(12):125902
125. Banhatti RD et al. (2011) Nearly constant loss effect in sodium borate and silver metaphosphate glasses: new insights. Solid State Ionics 192(1):70
126. Ngai KL et al. (2003) A combined molecular dynamics simulation, experimental and coupling model study of the ion dynamics in glassy ionic conductors. J Phys Condens Mat 15 (16):S1607
127. Rivera A et al. (2003) Temperature dependence of the near constant loss in ionic conductors: a coupling model approach. J Phys Condens Mat 15(16):S1633
128. Leon C, Ngai KL, Rivera A (2004) Correlation between ion hopping conductivity and near constant loss in ionic conductors. Phys Rev B 69(13):134303
129. Diaz-Guillen MR et al. (2010) Crossover to nearly constant loss in ac conductivity of highly disordered pyrochlore-type ionic conductors. Phys Rev B 82(17):174304
130. Rizos AK et al. (2001) Near constant loss in glassy and crystalline $LiAlSi_2O_6$ from conductivity relaxation measurements. J Chem Phys 114(2):931
131. Ngai KL, Strom U (1988) High-frequency dielectric loss of Na b-alumina: evidence for relaxation crossover. Phys Rev B 38(15):10350
132. Leon C et al. (1998) Universal scaling of the conductivity relaxation in crystalline ionic conductors. Phys Rev B 57(1):41
133. Ke S et al. (2010) Crossover from a nearly constant loss to a superlinear power-law behavior in Mn-doped $Bi(Mg_{1/2}Ti_{1/2})O_3–PbTiO_3$ ferroelectrics. J Appl Phys 107(8):084112

134. Funke K et al. (1981) Diffusion dynamics in AgI-type solid electrolytes. Solid State Ionics 3–4:45
135. Funke K (1982) A simple non-hopping model for the self-diffusion of the silver ions in the network of <100> channels in a-Ag_2Se. Solid State Ionics 6(1):93
136. Funke K et al. (1983) On the microwave conductivity of a-AgI part I: the effect of sample preparation. Solid State Ionics 11(3):247
137. Funke K, Schneider H (1984) Ionic conductivity of a-$RbAg_4I_5$ up to far-infrared frequencies. Solid State Ionics 13(4):335
138. Funke K and Hoppe R (1990) Jump-relaxation model yields Kohlrausch-Williams-Watts behaviour. Solid State Ionics 40–41, Part 1(0): 200
139. Funke K, et al. (1990) Hopping and non-hopping localized ionic motion: b-AgI and b-Ag_3SI. Solid State Ionics 40–41, Part 1(0): 275
140. Funke K, et al. (1992) Jump relaxation in $RbAg_4I_5$ by dynamic conductivity and quasielastic neutron scattering. Solid State Ionics 53–56, Part 2(0): 947
141. Funke K et al. (1996) Ionic and polaronic glassy conductors: conductivity spectra and implications for ionic hopping in glass. Solid State Ionics 85(1–4):293
142. Funke K, et al. (1996) On the dynamics of frenkel defect formation and ionic hopping in AgCl, AgBr and b-AgI. Solid State Ionics 86–88, Part 1(0): 141
143. Funke K (1997) Ion transport in fast ion conductors – spectra and models. Solid State Ionics 94 (1–4):27
144. Funke K, Roling B, Lange M (1998) Dynamics of mobile ions in crystals, glasses and melts. Solid State Ionics 105(1–4):195
145. Funke K, Wilmer D (2000) Concept of mismatch and relaxation derived from conductivity spectra of solid electrolytes. Solid State Ionics 136–137:1329
146. Funke K (2002) Ionic motion in materials with disordered structures. Phys Chem Chem Phys 4 (14):vii–14
147. Funke K et al. (2002) Ionic motion in materials with disordered structures: conductivity spectra and the concept of mismatch and relaxation. Phys Chem Chem Phys 4(14):3155
148. Funke K et al. (2002) Dynamics of mobile ions in crystals, glasses and melts, described by the concept of mismatch and relaxation. Solid State Ionics 154–155:65
149. Banhatti RD, Fuke K (2004) Dielectric function and localized diffusion in ion conducting glasses. Solid State Ionics 175(1–4):661
150. Funke K, Banhatti RD (2004) Modelling frequency-dependent conductivities and permittivities in the framework of the migration concept. Solid State Ionics 169(1–4):1
151. Funke K, Ross I, Banhatti RD (2004) Nearly constant loss behavior in g-$RbAg_4I_5$: microwave conductivity plateau identified. Solid State Ionics 175(1–4):819
152. Funke K, Banhatti RD, Cramer C (2005) Correlated ionic hopping processes in crystalline and glassy electrolytes resulting in migration-type and nearly-constant-loss-type conductivities. Phys Chem Chem Phys 7(1):157
153. Funke K, Banhatti RD (2006) Ionic motion in materials with disordered structures. Solid State Ionics 177(19–25):1551
154. Funke K, Singh P, Banhatti RD (2007) Conductivity dispersion in supercooled calcium potassium nitrate: caged ionic motion viewed as part of standard behaviour. Phys Chem Chem Phys 9(41):5582
155. Laughman DM, Banhatti RD, Funke K (2009) Nearly constant loss effects in borate glasses. Phys Chem Chem Phys 11(17):3158
156. Dieterich W, Philipp M (2002) Non-debye relaxations in disordered ionic solids. Chem Phys 284(1–2):439

Carbon Nanotube-/Graphene-Reinforced Ceramic Composites 17

Raghunandan Sharma and Kamal K. Kar

Contents

Abstract

Carbon nanomaterials such as carbon nanotubes (CNTs) and graphene are employed to improve the fracture toughness of ceramics. The presence of

R. Sharma
Advanced Nanoengineering Materials Laboratory, Materials Science Programme, Indian Institute of Technology Kanpur, Kanpur, Uttar Pradesh, India
e-mail: raghumsp@iitk.ac.in

K.K. Kar (✉)
Advanced Nanoengineering Materials Laboratory, Materials Science Programme, Indian Institute of Technology Kanpur, Kanpur, Uttar Pradesh, India

Advanced Nanoengineering Materials Laboratory, Department of Mechanical Engineering, Indian Institute of Technology Kanpur, Kanpur, Uttar Pradesh, India
e-mail: kamalkk@iitk.ac.in

K.K. Kar (ed.), *Composite Materials*,
DOI 10.1007/978-3-662-49514-8_17

exceptionally strong carbon nanomaterials in small fractions in a ceramic matrix restricts the crack propagation mechanism significantly, which leads to enhanced fracture toughness. A review of CNTs as well as graphene-reinforced ceramics has been presented. Effects of various factors such as dispersion and volume fraction of reinforcing phase, type of ceramic matrix, processing conditions, etc. on the performance of composites have been reviewed in detail. The reinforcement mechanism has also been highlighted, and an overview of emerging applications of the ceramic composites has been presented.

Keywords
Reinforcement • Fracture toughness • Crack propagation • Dispersion • Alumina • Zirconia • Sintering

Introduction

Carbon nanotubes (CNTs) are nanosized cylindrical tubes consisting of one or more graphitic carbon layers rolled to form a seamless cylinder with a diameter ranging from ~1 nm to a few tens of nanometers and a length of few microns [1, 2]. These graphitic carbon layers have sp^2-hybridized carbon–carbon (C–C) bonds and an unhybridized p_z electron. The high strength of the C–C bonds provides exceptional mechanical strength and modulus to CNTs. Again, the loosely bound p_z electrons are delocalized to form an electron cloud, which is responsible for the high electronic conductivity of graphitic carbons.

The cylindrical geometry and strong C–C bonding of CNTs provide them unique mechanical, thermal, and electronic properties [3–8]. Both theoretical and experimental studies predict the exceptional mechanical properties of CNTs. Numerous studies based on the theoretical models predict the high strength for the single as well as multiwalled CNTs (MWCNTs) [6, 7]. The mechanical properties of CNTs measured by various methods reveal high values of Young's modulus ranging from 0.4 to ~4.15 TPa [3, 5, 9, 10]. Similarly, the shear modulus of singlewalled carbon nanotubes (SWCNTs) has been reported to be 0.5 TPa [5]. Another study by Demczyk et al. has revealed the tensile strength of individual CNTs to be 0.15 TPa [10]. Apart from excellent mechanical properties, CNTs have unique electronic and thermal properties. The electronic conductivity of SWCNTs is determined by their rolling direction and diameter, i.e., the chirality [11, 12]. SWCNTs can be metallic or semiconducting in nature [13]. MWCNTs, however, are mostly metallic in nature due to the presence of multiple layers of varying chiralities. Ebbesen et al. have reported the electrical conductivities of individual metallic nanotubes to be as high as 1.2×10^9 S/m [8]. CNTs have excellent thermal conductivity along the nanotube axis due to the "ballistic transport" phenomena. Pop et al. have reported the thermal conductivity of individual SWCNTs at room temperature to be close to 3500 Wm^{-1}

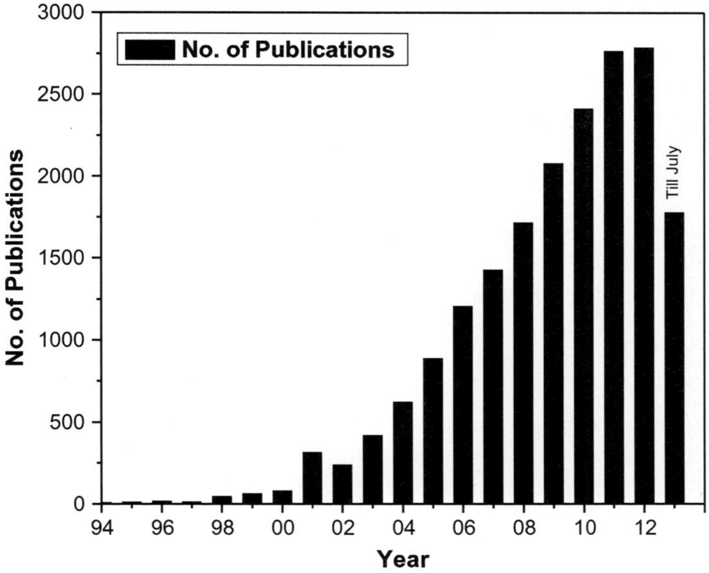

Fig. 17.1 A number of articles published from 1994 to July 2013 on "carbon nanotube composites." The data shown here is obtained from SciFinder search using "carbon nanotube composites" search string anywhere in the publication [25]

K^{-1}, almost 10 times higher than that of copper [14]. Moreover, CNTs show high thermal stability up to 2000 °C in vacuum [15].

Similar to CNTs, graphene is the other allotrope of carbon with similar hexagonal arrangement of sp^2-hybridized carbon. Graphene exhibits excellent mechanical, thermal, and electronic properties, which are comparable or superior to that of CNTs. The shear modulus of graphene has been reported to be 0.2 to 0.5 TPa using various techniques [16–19]. The thermal conductivity of graphene has been found in between 3080–5150 $Wm^{-1}\,K^{-1}$ by Ghosh et al. [20]. The electron mobility of graphene at room temperature has been predicted to be as high as temperature mobility to 200,000 $cm^2\,V^{-1}\,s^{-1}$ [21].

Since the unveiling discovery of their exceptional properties, CNTs have immersed promising materials as advanced composite reinforcement [22]. During last two decades, their application as reinforcing material for various matrices such as polymers, ceramics, metals, etc. has been explored significantly [23, 24]. Various studies, both from fundamental and technical points of view, have been carried out around the globe. The number of research articles on the carbon nanotube (CNT)-reinforced composites published yearly in the last two decades is shown in Fig. 17.1 [25]. This reveals the increasing scientific interest, possibilities, and the challenges in the field.

CNTs are used to reinforce various classes of materials such as polymers, metals, and ceramics. Numerous techniques are employed to synthesize these composites.

For a particular matrix, the process parameters need to be optimized, separately. Despite considerable efforts, CNT-reinforced composites face a number of technical issues, such as poor dispersion of CNTs in the matrix and low CNT–matrix interaction, resulting in poor improvements of the properties. The CNT dispersion and CNT–matrix interaction are governed by the properties of matrix, processing conditions and properties of CNTs (length, diameter, purity, surface activity, presence of defects, etc.). Effects of these parameters on the composite properties have been studied comprehensively [3, 22, 26, 27].

Ceramic materials are inorganic materials such as oxides, carbides, nitrides and borides. Owing to their high stiffness and strength, excellent chemical inertness, low thermal expansion and high thermal stability, ceramics are used in applications ranging from household pottery to aerospace and space shuttles. Being nonmetallic compounds with ionic bonds, ceramics are brittle in nature and have low toughness [28]. Moreover, the fracture of ceramic materials is attributed by the presence of micro cracks leading to the large scattering in mechanical properties. The brittle nature and lack of toughness reduces their mechanical reliability [29]. The reliability and toughness of ceramics can be improved by the incorporation of a reinforcing material having high toughness in the ceramic matrix [28]. Both fibers as well as particulate reinforcements can be used to improve these properties [30]. Here, we focus on the ceramic composites reinforced by CNTs and graphene. Alumina (Al_2O_3), zirconia (ZrO_2), silicon carbide (SiC), silicon nitrite (Si_3N_4 and SiN), and their derivatives are the frequently referred matrices in the CNT-/graphene-reinforced ceramic composites.

Synthesis of Carbon Nanotube-/Graphene-Reinforced Ceramic Composites

Ceramic composites can be synthesized through various techniques either from ceramic precursors or by pre-synthesized ceramic particles (Fig. 17.2). Chemical vapor infiltration (CVI), directed metal oxidation and pyrolysis of polymer

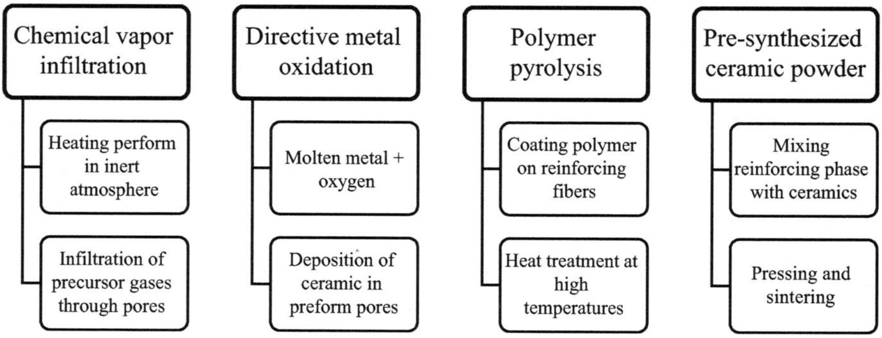

Fig. 17.2 Techniques to synthesize ceramic composites

precursors are the common techniques to synthesize these composites using ceramic precursors [31]. In these processes, ceramic matrix is incorporated in the preform consisting of reinforcing phase. The ceramic matrix is synthesized by infiltration of gaseous ceramic precursors through a porous reinforcing preform kept inside a chemical vapor deposition (CVD) chamber at high temperatures. In directed metal oxidation, molten metal such as aluminum is oxidized to form the ceramic material, which is directed to fill in the porous preform of reinforcing phase. In the third method, ceramic precursor polymers are pyrolyzed at high temperatures to form ceramic fibers/particles [31–39]. Another route to prepare ceramic composites is to blend pre-synthesized ceramic particles with reinforcing phase. Ceramic particles such as metal oxides, carbides, nitrides and their derivatives are mixed with the reinforcing materials followed by sintering at high temperatures [31]. This technique is frequently used to synthesize CNT-reinforced composites.

Carbon Nanotube-Reinforced Ceramic Composites

CNT-reinforced ceramic composites are synthesized mostly by mixing CNTs with ceramic matrix followed by sintering at high temperatures. During mixing, CNT should be uniformly coated with ceramics and should form a homogeneous composition without forming agglomerations or CNT-rich areas. Sintering is performed to increase the density of the ceramic matrix. Figure 17.3 shows a schematic of CNT coated with ceramic particles before and after sintering. Other aforementioned ceramic composite synthesis methods along with a few special methods have also been reported.

For effective incorporation of CNTs in ceramic matrix, there are two important factors to be considered [40]. These are uniform dispersion of CNT and excellent CNT–matrix interaction. The basic steps of fabricating CNT-reinforced ceramic composites are synthesis of CNT, formation of CNT–ceramic dispersion, and sintering to attain high density. The following sections describe these steps with experimental variations.

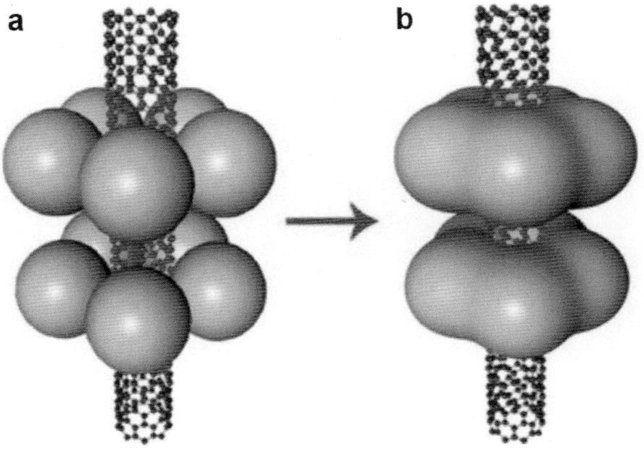

Fig. 17.3 Implantation of ceramic particles on CNTs. Particles coated on CNT by heterocoagulation method (**a**) before and (**b**) after sintering (Reprinted with permission from [40])

Synthesis of Carbon Nanotubes

Numerous techniques such as arc discharge, laser ablation, and chemical vapor deposition (CVD) are used to synthesize CNTs [41, 42]. In arc discharge and laser ablation techniques, atomic carbon is produced by vaporization of graphite either by an electric arc between two electrodes or by laser irradiation [43–47]. The atomic carbon gets deposited on the reactor walls in the form of soot, which consists mostly of CNTs and amorphous carbon [41, 43, 48]. In CVD synthesis, CNTs are grown by depositing carbon atoms on the transition metal (mostly Fe, Co, Ni, and their alloys) catalyst nanoparticles at high temperatures (700–1000 °C). The catalyst particles can either be coated on a substrate or synthesized in situ by using organometallic precursors. A carbon precursor such as acetylene or methane is pyrolyzed in the presence of the catalyst particles to produce CNTs [22, 49–52]. For applications such as electronic devices, where uniformity and alignment of CNTs is more important, anodized alumina templates can be used to form an extremely regular array of CNTs embedded in the thin film of alumina matrix [53].

Dispersion of Carbon Nanotubes in Ceramic Matrix

To disperse CNTs uniformly in the ceramic matrix, various mixing methods have been used. Ultrasonication, molecular level mixing, sol–gel synthesis, hydrothermal synthesis, and ball milling are the commonly used techniques. Apart from this, CNTs can be directly grown on ceramic nanoparticles by CVD. The dispersion of CNT in ceramic matrix is limited by the CNT weight fraction. Incorporation of higher weight fraction of CNT beyond certain limit (depending on the dispersion process) leads to CNT agglomeration, which ultimately reduces the strength and toughness of composites. Scanning electron microscopy (SEM) images of a uniformly dispersed and an agglomerated CNT-reinforced alumina matrix are shown in Fig. 17.4 [54].

Ultrasonication: Ultrasonication can be used to disperse both CNTs and ceramic particles. Generally, separate dispersions of CNTs and ceramic particles are prepared using various solvents and other additives. Ultrasonication of CNTs in organic or aqueous solutions is the most frequently used method [55–59]. Similarly, ceramic

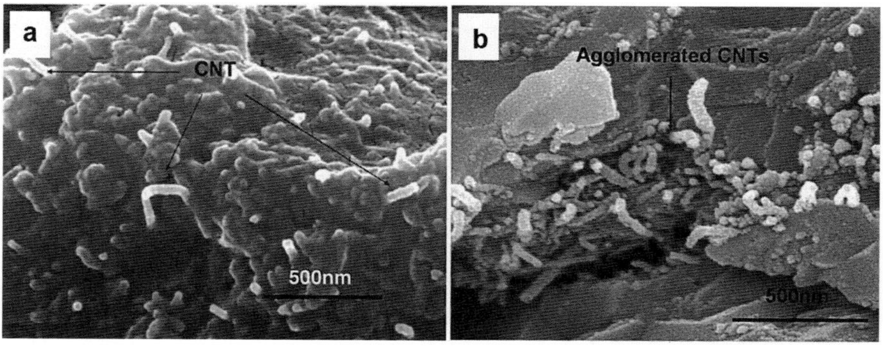

Fig. 17.4 SEM images of (**a**) uniformly dispersed and (**b**) agglomerated CNT–alumina (Reprinted with permission from [54])

particles are also dispersed by mechanical stirring or ultrasonication in organic solvents such as polyethylene glycol (PEG). Surfactants such as polyethylenimine, dodecyl sodium sulfate (SDS), dodecyl maranil, etc. are used to disperse CNTs [56–59].

Colloidal: In colloidal route, Sun et al. have used polyethylenimine as surfactant to disperse CNTs and polyacrylic acid to disperse alumina nanoparticles. Both dispersions are mixed to coat alumina on CNT. The alumina-coated CNTs are further mixed with concentrated alumina dispersion in ethanol to form a CNT–alumina mixture having ~1 wt% CNT [57, 58]. Fan et al. have used such a process shown in Fig. 17.5 to synthesize CNT-reinforced alumina composites. To disperse CNT with alumina, the CNT dispersion is dripped in the alumina dispersion with ultrasonic agitation. Surfactants such as SDS are used to improve the CNT dispersion. The surfactant makes the CNT surface electronegative, while the alumina surface becomes electropositive in PEG. Due to the electrostatic attraction between two species, alumina particles are deposited on CNT surface. This process, where two different species in a suspension carrying opposite charge at certain pH are attracted toward each other to coagulate, is termed as heterocoagulation. Other surfactants may be used to make CNT surface electronegative or electropositive [60].

Molecular level mixing: In another study, Cha et al. have mixed CNTs with alumina by molecular level mixing to improve the interfacial bonding and homogeneity of dispersion. In this process, CNTs act as nucleation sites for the ceramic nanoparticles. CNTs are functionalized by acid treatment and oxidation before dispersing in water by ultrasonication for 24 h. Alumina precursor such as aluminum nitrate is added by ultrasonicating the dispersion for 24 h. After drying at 100 °C, the powder is oxidized at 350 °C to obtain the CNT–alumina mixture for further sintering to form the composite [54]. Formation of chemical bonds between CNT and alumina has been observed in the process. Wang et al. have used similar process to prepare SWCNT–SiC composites. To mix ceramic (SiC) with CNTs at molecular level, acid-treated SWCNTs are mixed with chlorotrimethylsilane (SiC precursor) ultrasonically. The mixture is treated with microwave at a pressure of 125 psi for 10 min to produce SWCNT–SiC composites [61]. SiC nanoparticles are produced and coated on CNTs during the microwave treatment. Other ceramic composites can be synthesized as well by using appropriate precursor/solvents [62].

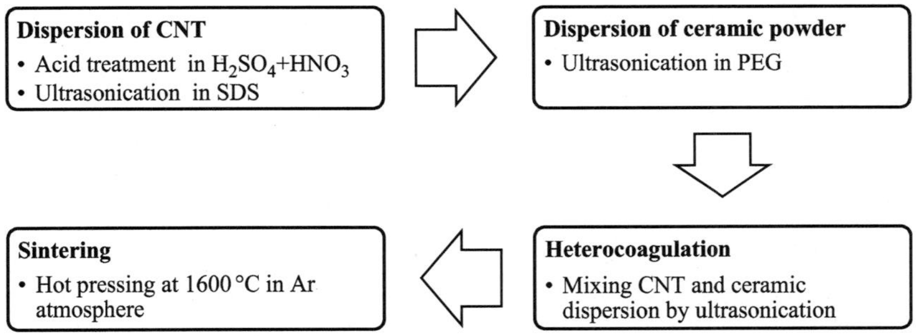

Fig. 17.5 Flow chart of CNT-reinforced ceramic composite synthesis by ultrasonic dispersion and heterocoagulation

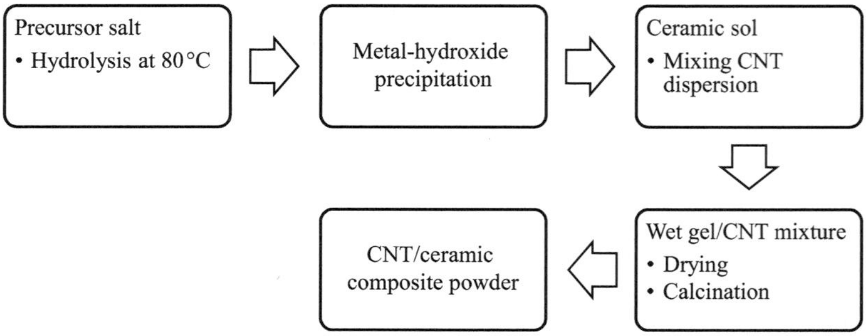

Fig. 17.6 Flow chart of CNT-reinforced ceramic composite synthesis by sol–gel process

Sol–gel: Sol–gel synthesis of ceramic particles on the CNT surface is another route to prepare CNT-reinforced ceramic composites. A ceramic precursor sol is mixed with CNT suspension prepared by ultrasonication or mechanical stirring. On gelation, dispersed CNTs are entrapped in the gel network. The gel is further dried at 200–300 °C and calcined at 600 °C to form CNT–ceramic composite [63]. Mo et al. have synthesized CNT-reinforced alumina composites using this process. In the first step, CNTs are dispersed in ethanol. The alumina precursor sol is prepared by hydrolysis of an aluminum salt. CNT dispersion in ethanol is mixed dropwise in the sol during the gel formation and dried at 350 °C to obtain CNT-dispersed alumina. The flow chart of the process is shown in Fig. 17.6 [64].

Hydrothermal: Hydrothermal synthesis of ceramic nanoparticles on the CNT surface is another method to attain molecular level mixing. In hydrothermal process, precursor hydroxides are crystallized into ceramic oxides at high temperatures and high pressures. As reported by Lupo et al. [65], ceramic precursors (hydroxides) are mixed in an aqueous dispersion of CNT and held at 200 °C for 8 h to form a thin coating of ZrO_2 on CNTs. Lu et al. have used comparatively lower temperature (90 °C) to prepare MWCNT–zirconia core–shell structural nanocomposites with MWCNT core in zirconia shell [66].

Ball milling: Apart from solution-based processes, ball milling also has been used to disperse carbon nanofibers (CNFs)/CNTs with ceramic matrix. A mixture of ceramic nanoparticles and reinforcing material in ethanol solution is ball-milled for several hours to ensure uniform dispersion [67–70]. The process is often supported by ultrasonication, where CNTs are ultrasonicated in ethanol before mixing with ceramic for ball milling [71–73]. Long duration ball milling, however, may cause damage to the CNT surface with reduction in the average CNT length as well.

Direct coating of CNT on ceramic powder: All of the above processes use pre-synthesized CNTs to disperse in ceramic matrix. In a completely dissimilar process, CNTs can be grown on the ceramic nanoparticles by CVD. Using this process, ceramic particles are coated uniformly with CNTs. Here, catalytic decomposition of carbon precursors on ceramic surface is employed to deposit CNTs [74, 75].

Sintering of CNT-Dispersed Ceramic Powder

Sintering at high temperatures is performed to attain high density. In conventional sintering, the CNT-dispersed ceramic powders are hot pressed at high temperatures for longer durations [40]. Sun et al. have used a pressure of 30 MPa along with a temperature of 1300 to 1500 °C for 1 h to obtain CNT-reinforced alumina composites [57]. However, during the conventional sintering process, application of high temperature for longer duration (few hours) is required. This may damage or destroy the CNTs and effectively reduce their impact on the reinforcement [76]. Hence, low temperature or small duration sintering is desired. Spark plasma sintering (SPS) is employed for this purpose, where a smaller time is required to sinter the material due to the very fast sintering process. A pulsed electric current is used to heat the graphite mold, which can attain a heating rate of ~300 °C/min as the current directly passes through the mold as well as the sample. A schematic diagram of SPS is shown in Fig. 17.7. The high heating rate makes it possible to complete the sintering in short time and, hence, reduces the chances of CNT wall damage.

Parameters such as applied pressure, sintering time, temperature, and heating rate affect the final densification and microstructure [77]. The optimum values of these parameters vary with the ceramic system to be sintered. Spark plasma sintering has been performed at 1300 °C and 50 MPa pressure for 5 min by Sun et al. for CNT–alumina composites [58]. Even lower temperature sintering of CNT-nanocrystalline alumina at 1150 °C using SPS is reported by Zhan et al. [78]. SPS at higher temperatures, such as 1650 °C at a lower applied pressure (25 MPa), has also been reported for the similar ceramic system [64]. For CNT/yttria (Y_2O_3)-stabilized–zirconia (ZrO_2) composites, Shi et al. and Mazahari et al. have used comparatively lower sintering temperatures (1250–1350 °C) [79, 80].

Application of pressure may be unidirectional or isostatic. Sintering by a hot isostatic press (HIP), which produces more uniform sintering, is also reported [68]. Again, the pressure applied during sintering affects the final density, which in turn determines the mechanical properties. Balazsi et al. have studied the effect of pressure during hot isostatic press sintering of CNT-reinforced silicon nitride (Si_3N_4) composites. The density has been found to increase with increasing the applied

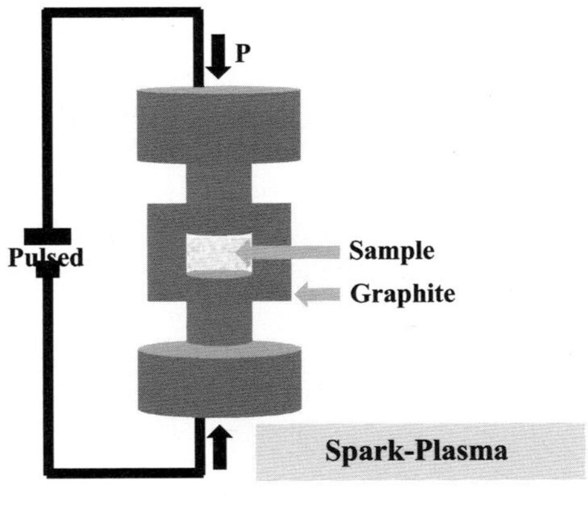

Fig. 17.7 Schematic of spark plasma sintering

pressure during sintering. Tensile strength and bending strength show similar increasing trends [68].

The sintering route affects the ceramic properties significantly. A study by Dusza et al. has revealed higher specific density, hardness, and toughness of the SPS sintered zirconia–carbon nanofiber (CNF) composites as compared to the hot-pressed composites with similar composition [81]. Again, the sintering and densification is limited by the weight fraction of CNTs. For high CNT weight fraction, a poor densification takes place as the sintering process is hindered by the presence of CNTs. Moreover, the grain growth is also impeded by their presence. Fine microstructure has been observed in the presence of CNFs in zirconia matrix [81]. However, SPS sintering is found to be more effective in terms of densification.

Graphene-Reinforced Ceramic Composites

The process to synthesize graphene-reinforced ceramic composites is similar to that of CNT-reinforced ceramic composites. Pre-synthesized graphene is dispersed in ceramic matrix followed by sintering to obtain the composite.

Synthesis of Graphene

Graphene can be synthesized by various processes such as exfoliation of graphite, chemical vapor deposition, and chemical synthesis [82–84]. Reduction of graphite oxide (GO) at high temperatures in an inert atmosphere is one of the frequently used methods to synthesize graphene. The earliest method to synthesize GO is the Hummers' method, where graphite is treated by $KMnO_4$ and $NaNO_3$ in a H_2SO_4 solution [85]. With increasing scientific interest to synthesize graphene, numerous modified versions of Hummers' method have been reported. For ceramic matrix reinforcement, Walker et al. have prepared GO by oxidizing graphite flakes by reaction with an acid solution consisting of H_2SO_4, HNO_3, HCl, and $KClO_3$ [86]. The detailed synthesis of graphene oxide and graphene can be found elsewhere [87].

Dispersion of Graphene in Ceramic Matrix

The first step to synthesize the graphene-reinforced ceramic composites is the uniform dispersion of graphene in the ceramic matrix. In this step, similar to CNT–ceramic composites, the dispersion can be attained by various routes. Both graphene and GO can be dispersed with ceramic matrix. Wang et al. have used ultrasonication and mechanical stirring to prepare the GO–alumina dispersion. Separate dispersions of alumina and GO are prepared by ultrasonication. The GO dispersion is added in the alumina dispersion gradually under continuous mechanical stirring. Finally, the GO–alumina dispersion is reduced by using a reducing agent such as hydrazine monohydrate to form graphene–alumina dispersion [88]. Kun et al. and Tapaszto et al. have used attrition milling to disperse multilayered graphene in silicon nitride matrix [89, 90]. Aqueous colloidal processing has been used by Walker et al. to disperse graphene platelets in silicon nitride (Si_3N_4) matrix [86].

Sintering of Graphene-Dispersed Ceramic Powder

After dispersing the graphene with ceramic matrix, sintering is employed to attain high density. The process is similar to that of the CNT-reinforced ceramic composites, where the use of both conventional and SPS has been reported. Sintering of graphene-reinforced composites by hot isostatic pressing (HIP) requires relatively higher temperatures (~1700 °C), longer sintering times (~3 h), and pressure of ~20 MPa [89–91]. On the other hand, SPS can be used at comparatively lower temperatures (~1300 °C) with shorter sintering times [88].

Properties of Carbon Nanotube-/Graphene-Reinforced Ceramic Composites

As stated previously, the ceramic materials have a lack of toughness and exhibit brittle fracture with scattered strength values. Incorporation of carbon nanomaterials (CNT, CNF, etc.) in the ceramic matrix can improve toughness by absorbing energy during fracture. The presence of CNT hinders the crack propagation by exerting forces during the cracks propagating around them. Again, the energy required to break or pull out of CNTs during crack growth increases fracture toughness significantly [92]. The energy absorbed in the later process is determined by the strength of CNTs and CNT–matrix interaction. Numerous studies have been carried out to enhance the toughness, hardness, and electronic conductivity of ceramic materials by incorporation of CNTs. The mechanical properties are measured either by indentation or crack length measurements [93]. This section provides a review of CNT-reinforced ceramic composites of a few important ceramic systems.

Carbon Nanotube/Alumina Composites

Reinforcement of alumina matrix using CNTs has been studied by numerous groups with a variation of CNT weight fraction, synthesis method, and sintering parameters. Mechanical properties reported by some of the studies dedicated to CNT-reinforced alumina composites are summarized in Table 17.1. Depending on CNT weight fraction and processing parameters, the properties show a large variation. However, incorporation of CNTs in a small weight fraction improves the toughness and bending strength significantly. Yamamoto et al. have observed an increase of 25 % and 27 % in fracture toughness and bending strength, respectively [94]. On the other hand, Fan et al. have shown that the incorporation of a small weight fraction (1 wt%) of SWCNTs can enhance the fracture toughness of the alumina matrix by a factor of 2 (~102 % enhancement). Similarly, the bending strength shows an improvement of ~19 % [40]. Moreover, by using a lower sintering temperature, Zhang et al. have reported a threefold improvement in the toughness of SWCNT-reinforced alumina [78].

Table 17.1 Effect of various parameters on the mechanical properties of CNT–alumina composites

Reinforcing material	% loading (wt%)	Bending strength (MPa)	Toughness (MPa m^{-2})	Hardness (GPa)	References
SWCNT	0.5	402	5.12	–	[40]
SWCNT	1.0	423	6.4	–	[40]
MWCNT	0.01	–	3.05	15.49	[55]
SWCNT	0.01	–	3.73	14.6	[55]
MWCNT	–	536	4.0		[57]
CNT	0.1	–	4.9	–	[58]
MWCNT	4 (vol.%)	370	6.9	18.2	[59]
MWCNT	10 (vol.%)	331	6.1	15.3	[59]
CNF	1.0	301	2.5	18.5	[67]
SWCNT	–	–	9.7	–	[78]
MWCNT	0.9	689.6	5.9	16.8	[94]

With increasing the fractional weight/volume of CNTs above a certain limit, the mechanical properties tend to decrease [59]. This may be attributed to the increased chances of CNT agglomeration and formation of localized clusters having boundaries prone to crack propagation. It is also important to note that the presence of CNTs hinders the sintering process as well as the grain growth. It has been observed that increasing the CNT fraction beyond 4 wt% prevents the improvement. This, despite the homogeneity of dispersion, results in poor mechanical properties [55]. A smaller grain size with lower specific density has been reported by Ahmad et al. with increasing MWCNT volume fraction from 4 to 10 % (Table 17.1) [59]. A smaller weight fraction (<1 wt%) has been found to be most suitable for these composites.

Properties of the reinforcing CNTs such as number of walls, length, and purity also affect the composite properties. Apparently, this is due to the change in the mechanical properties of reinforcing phase. Also, the CNT–matrix interaction is affected by the surface structure of CNTs, which in turn affects the properties of composites. In a comparison between SWCNT- and MWCNT-reinforced alumina composites, the SWCNT composites exhibit significantly high fracture toughness [55]. With reference to Table 17.1, reinforcement by CNFs shows the lowest value of fracture toughness [67].

The fracture toughness and strength of CNT-based composites is affected largely by the processing temperature. Sun et al. have studied the effect of temperature on 1 wt% CNT-reinforced alumina composites. As shown in Fig. 17.8, the fracture toughness of composites decreases with increasing the temperature, whereas for pure alumina, it remains almost constant. The bending strength of the composites as well as pure alumina decreases with increasing the sintering temperature [57]. This can be attributed to the fact that at high sintering temperature, the outer walls of CNTs face partial damage leading to degradation of their mechanical strength. Lowering the sintering temperature increases the effectiveness of reinforcement as the CNT walls remain intact.

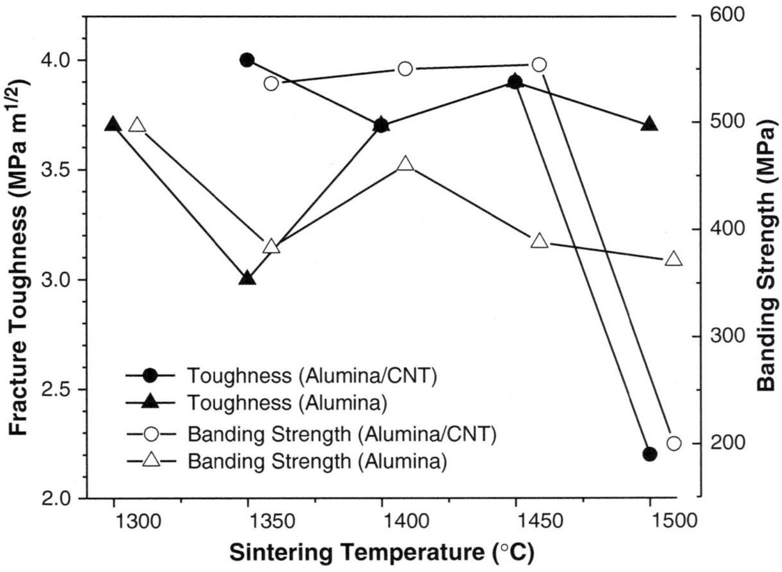

Fig. 17.8 Effect of sintering temperature on mechanical properties of 1 wt% MWCNT-reinforced alumina composites (Data retrieved from [57])

Carbon Nanotube/Zirconia Composites

Zirconia is another important ceramic material used in applications such as refractories, thermal insulation, abrasives, etc. Stabilized zirconia is an important electroceramic material due to its high ionic conductivity. For example, yttria-stabilized zirconia (YSZ), being a good conductor of O^{2-} ions, is used in applications such as sensors and solid electrolyte membranes for oxygen separation. Owing to its inertness and stability at high temperatures, YSZ is utilized as electrolyte membranes in solid oxide fuel cells (SOFCs), where it provides a conduction path to O^{2-} ions in the harsh chemical conditions (presence of molten salts at high temperature: ~800 °C).

Incorporation of CNT in zirconia matrix alters their mechanical and electronic properties. However, the contradictory results have been reported for the mechanical properties of CNT-incorporated zirconia composites. Duszova et al. have studied the electrical and mechanical properties of CNT-reinforced (~1 wt%) zirconia composites synthesized by hot pressing. Compared to pure hot-pressed zirconia, the fracture toughness and hardness of CNT-reinforced zirconia have been found to be decreased by 10 and 30 %, respectively [76]. The electronic conductivity, however, increases significantly by a 13 orders of magnitude. On the other hand, Mazaheri et al. and Dusza et al. have shown a significant improvement in toughness and a marginal improvement in hardness (Fig. 17.9). Moreover, the toughness increases with increasing CNT weight faction and reaches toward saturation, whereas the hardness

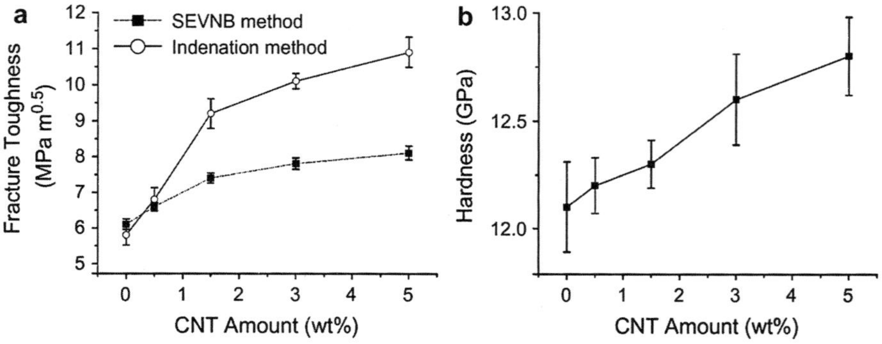

Fig. 17.9 Fracture toughness and hardness of CNT–zirconia composites with CNT content (Reprinted with permission from [80])

increases linearly [80, 81]. As revealed by scanning electron micrograph, the crack bridging has been attributed to be the toughness enhancement mechanism.

Zirconia nanoparticles can be coated on CNTs by chemical routes. Such nanoparticle-coated CNTs have been used to reinforce alumina matrix. Coating of zirconia nanoparticles on CNT provides additional toughness enhancement by phase transformation, micro-cracking, and internal stresses caused by different thermal expansion coefficients of alumina/zirconia [63]. Comparisons of toughness and bending strength of pure alumina, CNT-reinforced alumina (Al_2O_3/CNT), zirconia-mixed alumina (Al_2O_3 + ZrO_2), and zirconia nanoparticle-coated CNT (ZrO_2/CNT)-reinforced alumina (Al_2O_3 + ZrO_2/CNT) have been shown in Fig. 17.10. The Al_2O_3 + ZrO_2/CNT composites show the largest improvements in toughness (~102 %) and bending strength (~67 %) compared to the pure alumina matrix.

A study of electric properties of CNT-reinforced zirconia composites shows a significant improvement in electric conductivities. This improvement is attributed to the high electronic conductivity of CNTs. In the uniformly dispersed CNT–ceramic matrix, CNTs provide conducting channels to the electrons. For lower CNT concentrations, where the CNT network in matrix is discontinuous, the conduction takes place through tunneling effect [95]. MWCNT-reinforced YSZ composites synthesized by SPS sintering by Shi et al. exhibit a significant enhancement in DC conductivity. As shown in Fig. 17.11, with a variation of MWCNT weight fraction from 3 to 8 wt%, the conductivity at 300 K increases by a factor of ~165 [79, 95].

The DC conductivity (σ_{dc}) of the composites follows the scaling law as $\sigma_{dc} \propto (p - p_c)^t$ with a percolation concentration (p_c) of 0.017 (1.7 %), a DC transport exponent (t) of 3.3, and (p) being the conducting component volume fraction. Based on the study of temperature variation of conductivity, authors have suggested that at temperatures higher than 35 K, the conduction takes place through tunneling of electrons through the CNT–zirconia separations. On the other hand, at lower temperatures, the conduction is dominated by the hoping mechanism [95].

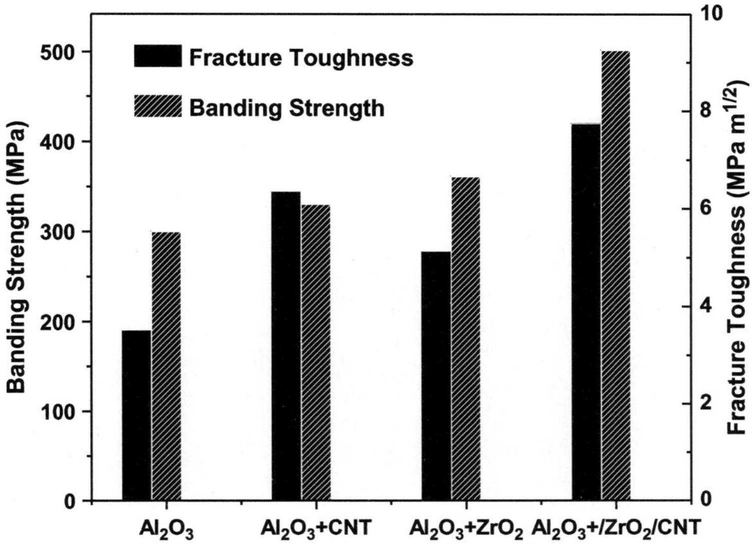

Fig. 17.10 Fracture toughness and bending strength of Al_2O_3, Al_2O_3/CNTs, Al_2O_3 + ZrO_2, and Al_2O_3 + ZrO_2/CNT (Data retrieved from [63])

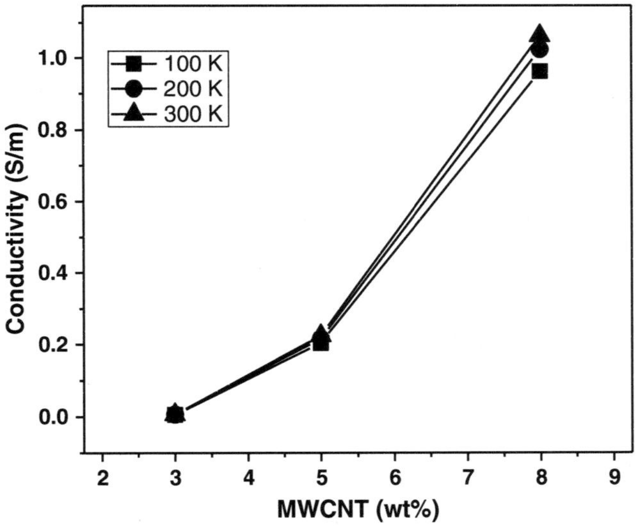

Fig. 17.11 Conductivity of MWCNT-reinforced zirconia composite as a function of MWCNT weight fraction at different temperatures (Data retrieved from [95])

In conclusion, CNT-reinforced zirconia shows high electric conductivity as well improved mechanical properties. A proper dispersion and CNT–matrix interaction is necessary to obtain strengthening effect. For applications such as solid electrolyte membrane in SOFCs, the increased electric conductivity of CNT-reinforced YSZ is not favorable, as the SOFC electrolyte membrane should not conduct electrons.

Carbon Nanotube-/Silicon-Derived Ceramic Composites

Silicon carbide, silicon nitride, and their derivatives such as SiCN are the common silicon-derived ceramics. Owing to their unique properties such as high strength, hardness, thermal conductivity, inertness, low thermal expansion coefficient, etc., these are used in numerous applications including abrasives, cutting tools, bulletproof jackets, automobiles, electronic devices, power plants, space applications, and many other applications. Similar to alumina and zirconia, silicon-derived ceramic matrices such as Si_3N_4 [69, 70, 73, 96], SiCN [62], SiC [97–99], etc. have also been explored as the ceramic matrices for CNT reinforcement. For silicon nitride matrix, a small amount of sintering aids (Al_2O_3, Y_2O_3, and ZrO_2) is frequently used to facilitate the sintering process [69, 70, 73, 96]. Similar to other ceramic systems, sintering of these ceramic composites in the presence of CNT provides finer microstructure. As revealed by Wing et al., the presence of MWCNTs (1 wt%), however, does not affect the sintering process as the density is similar to the pure matrix.

The fracture toughness shows a significant improvement with a value of 6.6 MPa $m^{1/2}$ for CNT–SiN composite as compared to that of 4.8 MPa $m^{1/2}$ for SiN matrix. However, a marginal decrease in the hardness (15.0 GPa for CNT–SiN composites compared with 15.7 GPa for SiN matrix) after CNT incorporation has been observed [70]. CNT–SiN composites exhibit a fracture mechanism dominated by crack deflection and crack bridging, where the toughness enhancement is attributed to the CNT pullout [70].

CNT-reinforced SiC composites can be synthesized by various processes such as polymer precursor route and chemical vapor infiltration (CVI) [97, 98]. Wang et al. have prepared MWCNT-reinforced SiC composites by pyrolysis of CNT–polycarbosilane blend. For 0.5 wt% loading of MWCNTs, Young's modulus exhibits an enhancement of 93.6 %, whereas the tensile strength improves by 38.5 % [97]. CNT-reinforced SiC composites show high-temperature stability. Composites fabricated by CVI can withstand oxidation at temperatures as high as 1600 °C [98].

Graphene/Ceramic Composites

Graphene, a two-dimensional honeycomb structure of sp^2-hybridized carbon, owns superior mechanical [19, 100, 101], thermal [20], and electronic properties [21, 102]. Incorporation of graphene in a matrix can alter its properties significantly. Graphene-based composites have attained significant scientific interest in the last decade [20]. Significant enhancements in the mechanical as well as electric properties have been reported for the graphene-reinforced ceramic systems such as Si_3N_4 and alumina [86, 89–91].

Graphene-reinforced composites show a large enhancement in the mechanical properties. Walker et al. have reported an increment of ~235 % in fracture toughness on the incorporation of a small (1.5 vol.%) fraction of graphene in an Si_3N_4 matrix. Apart from mechanical property enhancement, owing to its high electronic conductivity [102] and highest known intrinsic mobility (2×10^5 cm^2/Vs [21]), graphene-reinforced ceramic composites exhibit a high electronic conductivity. Graphene nanosheet (GNS)-reinforced alumina composites studied by Fan et al. exhibit an electronic conductivity of 5709 S/m for a GNS volume fraction of 0.15, which is 170 % higher than the best ever reported conductivity for CNT–alumina composites [103]. Similar results have been reported by Wang et al. by using 2 wt% GNS incorporated in alumina matrix with an electronic conductivity of 172 S/m and a toughness of 5.21 MPa $m^{1/2}$ [88]. Compared to pure alumina, it exhibits the enhancements of 53 % and 13 orders of magnitude for toughness and electric conductivity, respectively [88].

Reinforcement

The extent of reinforcement of a matrix depends not only on the mechanical properties of the reinforcing phase but also on the interfacial bonding between these two phases. Depending on the interfacial interaction, composite failure may take place by various mechanisms. For lower interfacial strength, the failure takes place by the detachment of the interface, leading to inferior mechanical properties. On the other hand, breaking of the reinforcing fibers takes place for strong interfacial bonding.

Carbon Nanotube/Ceramic Composites

In CNT-reinforced ceramics, the matrix fracture is explained by two mechanisms, namely, CNT pullout and CNT breaking. These are governed by the CNT–matrix interfacial bonding strength. For the lower bonding strength, failure occurs by complete interface debonding leading to CNT pullout, whereas for higher bonding strength, CNT breaking takes place. A mixed mode failure is also possible. A detailed study of the CNT failure mechanism in a CNT-reinforced (0.9 wt%) alumina matrix composite has been reported by Yamamoto et al. [104]. The CNT–alumina composites are prepared from a dispersion of aluminum hydroxide ($Al(OH)_3$) and CNTs in ethanol. The mixture is heated at 600 °C in argon to form CNT-dispersed alumina. Further, the composite is made by sintering the CNT–alumina by spark plasma sintering. The sintering parameters, namely, time, temperature, and pressure are kept to be 20 min, 1500 °C, and 20 MPa, respectively. To study the CNT failure mechanism, single-CNT pullout experiments are performed by using a nanomanipulator system attached with the scanning electron microscope (SEM) [104]. As shown in Fig. 17.12, the composite sample (platinum coated) and the atomic-force microscope (AFM) cantilever tip are mounted on two different linear motion stages in such a way that the CNTs protruding from the fractured surface are facing the cantilever tip. A single CNT can be clamped with AFM tip by deposition of a carbonaceous material by local electron beam-induced deposition (EBID) [105].

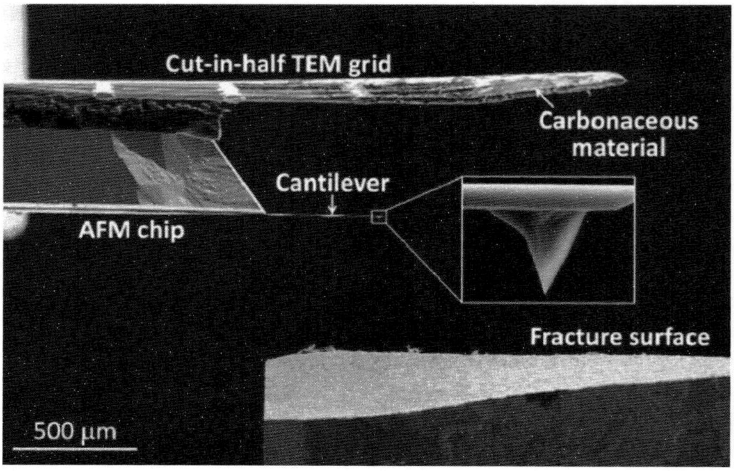

Fig. 17.12 Experimental setup for CNT pullout test (Reprinted with permission from [104])

The pullout experiments for several CNTs show that rather than CNT pullout, the CNT failure occurs by complete or sword in sheath fracture. The absence of CNT pullout from the matrix is attributed by the large CNT–matrix interaction. Further, the transmission electron microscope (TEM) images of the CNTs after pullout test show a reduction in diameter near the failure point. This reduction in diameter suggests a failure by breaking of the outer graphitic layers of multiwalled CNT followed by pulling away the inner walls. Hence, the outer layer removal from the CNT reduces its diameter. SEM images of CNT failure mechanism based on this observation are shown in Fig. 17.13.

The fracture mechanism reveals that during the crack opening, CNT takes the tensile load and a partial debonding of CNT–matrix interface takes place. With increasing the tensile load, i.e., crack opening, outer layers of the CNT are detached. With increasing the displacement (load), the inner core pulls away leaving the fragmented CNT inside the matrix. In this process, CNTs make a bridge between the two fractured surfaces preventing the crack propagation (Fig. 17.14). This suggests that the toughness enhancement is caused by the energy absorbed by CNT during fragmentation and partial debonding and not by the CNT pullout. Hence, the mechanical strength of CNTs plays an important role to enhance the fracture toughness of the ceramic composites. However, in the case of singlewalled CNTs, the above mechanism, where a few layers are fragmented and remained in the matrix, is not possible.

In another study, Xia et al. have used anodized alumina template gown CNT film with a thickness of 90 μm to study the failure mechanism. During indention test, the crack formed by an indentation parallel to the CNT axis (indenter on the top of the alumina template) is deflected by the presence of CNTs (shown by white arrow in

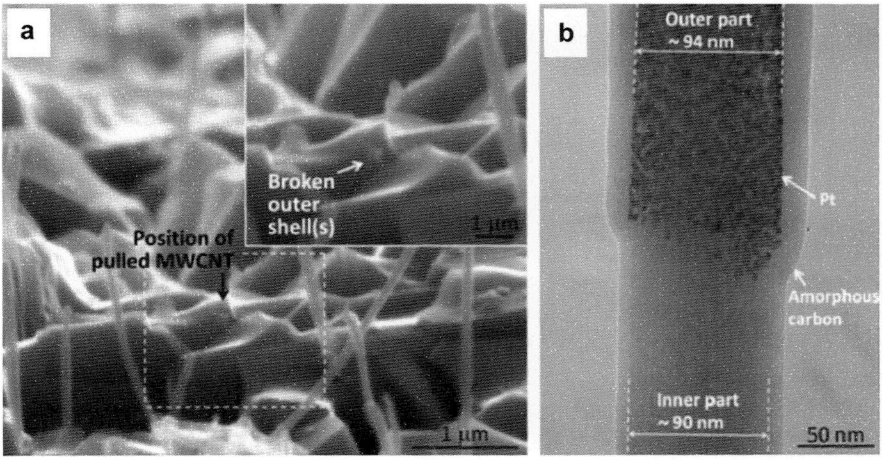

Fig. 17.13 Fracture mechanism studied using AFM (**a**) SEM image shows the fractured CNT outer layers remaining in the matrix (shown by *arrow*) (**b**) TEM image of fractured CNT showing a reduction in the diameter at the lower part attributed to the detachment of upper walls in the matrix (Reprinted with permission from [104])

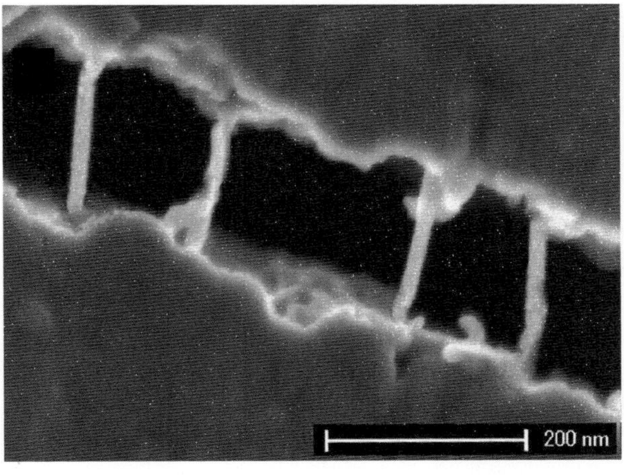

Fig. 17.14 Crack bridging by CNTs in CNT-reinforced YSZ matrix composites (Reprinted with permission from [80])

Fig. 17.15a). The crack deflection due to the presence of CNTs increases the crack length. More energy is absorbed before the failure takes place. On the other hand, the fractured surface for an indentation perpendicular to CNTs (indenter on the side face of the alumina template) is shown in Fig. 17.15b. Similar to the previous example, the failure occurs by interface-debonding and crack-bridging mechanism [53].

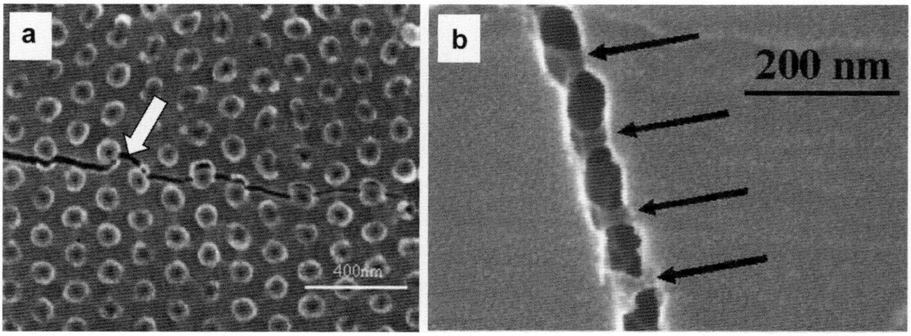

Fig. 17.15 (**a**) Crack deflection by CNTs in a CNT-reinforced alumina matrix (anodized alumina template) for crack parallel to CNT axis and (**b**) crack bridging in the same sample for a sideways indentation (crack perpendicular to CNT axis) (Reprinted with permission from [53])

Graphene/Ceramic Composites

The toughness is improved as the graphene layers hinder the crack propagation by means of crack bridging, crack deflection, and crack branching [91]. These layers anchor the matrix grains and form a wall along the grain boundaries (Fig. 17.16). This prevents the crack propagation in the plane. The crack must propagate in three dimensions requiring more energy, which results in higher toughness.

Emerging Applications of Carbon Nanotube/Graphene Ceramic Composites

Ceramic composites, owing to their characteristic properties, such as inertness, thermal stability, and high strength, are being used conventionally in various applications. Advanced applications such as space, aviation, defense, power generation, and biomedicine require ceramic materials with specific properties. This has opened up new emerging fields for these materials. As, for example, in high-performance aircraft engines, materials with low density, high strength, and thermal stability should be used to improve weight to thrust ratio. The turbine inlet temperature should be increased to attain high performance, which requires materials with high thermal stability and creep resistance [106]. Turbine blades, subjected to high temperatures and stress variations, are composed of silicon carbide. On the other hand, solid oxide fuel cells use YSZ as electrolyte material [107, 108]. Having high-temperature stability, ceramic composites also may be used in nuclear reactors as a blanket. However, more research is required to improve the properties such as radiation stability and creep resistance, which have not been found highly satisfactory for these applications [109].

Hydroxyapatite (HAP) is generally used in orthopedic transplants such as hip and dental implants due to its biocompatibility. However, the lower tensile strength and toughness of HAP makes it unsuitable for implants requiring load-bearing capacity.

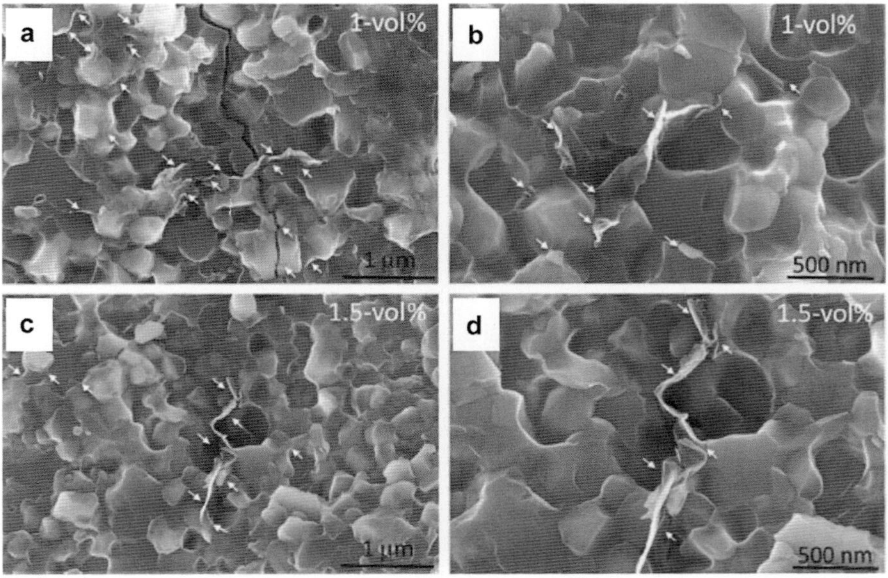

Fig. 17.16 SEM images of graphene platelet (GPL)-reinforced Si_3N_4 matrix composites with 1 vol.% (**a**, **b**) and 1.5 vol.% (**c**, **d**) GPL at different resolutions. Arrows show the presence of GPL in the matrix incorporated along the grain boundary (Reprinted with permission from [86])

CNTs and graphene are being explored as a prospective reinforcing material for HAP matrix to improve its toughness and strength [110]. The CNT-/graphene-reinforced HAP can be used to prepare implants or to provide biocompatible coating on the metallic implants. However, more studies have to be carried out to establish the biocompatibility and toxicity of CNTs/graphene and to obtain good mechanical properties by means of proper CNT/graphene–HAP interaction.

Concluding Remarks

CNT-/graphene-reinforced ceramic composites are promising materials with high toughness. They form an important class of prospective materials for applications such as defense, power generation, aviation, space, biomaterials, etc.

The fracture toughness and other properties of ceramics can be improved by the incorporation of carbon nanomaterials such as CNTs, CNFs, and graphene. Composites of ceramic matrices such as alumina, zirconia, silicon carbide, silicon nitride, etc. have been explored frequently for their mechanical properties. Special attention has been made to the fracture toughness of the composites. The improvement of toughness is governed by the dispersion of CNT/graphene in the matrix as well as CNT/graphene–matrix interaction. Uniform dispersion is possible only for small weight fractions of CNT/graphene, and, hence, the appropriate CNT/graphene weight fraction is limited up to ~4 wt%, beyond which, properties decrease due to the agglomeration. Process parameters such as dispersion method, nature and purity of CNTs/graphene, sintering temperature, and nature of matrix affect the performance of composites. Depending on the processing conditions, the toughness values show a large scattering. For commercial applications, optimization of the parameters for each type of ceramic matrix is required.

Especially, graphene is an emerging reinforcing material for these composites. Apart from mechanical properties, graphene-reinforced composites promise very high electronic and thermal conductivities compared to CNT-reinforced composites. Again, large-scale synthesis of few layered graphene, uniform dispersion, and graphene–ceramic interaction are the key factors governing their successful applications.

Acknowledgment The authors acknowledge the financial support provided by Indian Space Research Organization, India, for carrying out this research work.

References

1. Iijima S, Ichihashi T (1993) Single-shell carbon nanotubes of 1-nm diameter. Nature 363:603
2. Bethune DS, Klang CH, de Vries MS, Gorman G, Savoy R, Vazquez J, Beyers R (1993) Cobalt-catalysed growth of carbon nanotubes with single-atomic-layer walls. Nature 363:605
3. Salvetat JP, Bonard JM, Thomson NH, Kulik AJ, Forró L, Benoit W, Zuppiroli L (1999) Mechanical properties of carbon nanotubes. Appl Phys A 69:255
4. Despres JF, Daguerre E, Lafdi K (1995) Flexibility of graphene layers in carbon nanotubes. Carbon 33:87
5. Lu JP (1997) Elastic properties of carbon nanotubes and nanoropes. Phys Rev Lett 79:1297
6. Kudin KN, Scuseria GE, Yakobson BI (2001) C_2F, BN, and C nanoshell elasticity from ab initio computations. Phys Rev B 64:235406
7. Li C, Chou T-W (2003) A structural mechanics approach for the analysis of carbon nanotubes. Int J Solids Struct 40:2487
8. Ebbesen TW, Lezec HJ, Hiura H, Bennett JW, Ghaemi HF, Thio T (1996) Electrical conductivity of individual carbon nanotubes. Nature 382:54
9. Treacy MMJ, Ebbesen TW, Gibson JM (1996) Exceptionally high Young's modulus observed for individual carbon nanotubes. Nature 381:678
10. Demczyk BG, Wang YM, Cumings J, Hetman M, Han W, Zettl A, Ritchie RO (2002) Direct mechanical measurement of the tensile strength and elastic modulus of multiwalled carbon nanotubes. Mater Sci Eng A Struct Mater Prop Microstruct Process 334:173
11. Saito R, Fujita M, Dresselhaus G, Dresselhaus MS (1992) Electronic structure of chiral graphene tubules. Appl Phys Lett 60:2204
12. Saito R (1998) Physical properties of carbon nanotubes. Imperial College Press, London
13. Bandaru PR (2007) Electrical properties and applications of carbon nanotube structures. J Nanosci Nanotechnol 7:4
14. Pop E, Mann D, Wang Q, Goodson K, Dai H (2005) Thermal conductance of an individual single-wall carbon nanotube above room temperature. Nano Lett 6:96

15. Kim YA, Muramatsu H, Hayashi T, Endo M, Terrones M, Dresselhaus MS (2004) Thermal stability and structural changes of double-walled carbon nanotubes by heat treatment. Chem Phys Lett 398:87
16. Zakharchenko KV, Katsnelson MI, Fasolino A (2009) Finite temperature lattice properties of graphene beyond the quasiharmonic approximation. Phys Rev Lett 102:046808
17. Sakhaee-Pour A (2009) Elastic properties of single-layered graphene sheet. Solid State Commun 149:91
18. Tsai J-L, Tu J-F (2010) Characterizing mechanical properties of graphite using molecular dynamics simulation. Mater Des 31:194
19. Min K, Aluru NR (2011) Mechanical properties of graphene under shear deformation. Appl Phys Lett 98:013113
20. Ghosh S, Calizo I, Teweldebrhan D, Pokatilov EP, Nika DL, Balandin AA, Bao W, Miao F, Lau CN (2008) Extremely high thermal conductivity of graphene: prospects for thermal management applications in nanoelectronic circuits. Appl Phys Lett 92:151911
21. Chen J-H, Jang C, Xiao S, Ishigami M, Fuhrer MS (2008) Intrinsic and extrinsic performance limits of graphene devices on SiO_2. Nat Nanotechnol 3:206
22. Kar KK, Hodzic A (2011) Carbon nanotube based nanocomposites: recent development. Research Publishing, Singapore
23. Ajayan P, Zhou O (2001) Applications of carbon nanotubes. In: Dresselhaus MS, Dresselhaus G, Avouris P (eds) Carbon nanotubes; synthesis, structure, properties, and applications. Springer, Berlin, p 391
24. Baughman RH, Zakhidov AA, de Heer WA (2002) Carbon nanotubes – the route toward applications. Science 297:787
25. (2007) SciFinder web version. Chemical Abstracts Service (CAS); Columbus
26. Luo ZP, Koo JH (2008) Quantitative study of the dispersion degree in carbon nanofiber/polymer and carbon nanotube/polymer nanocomposites. Mater Lett 62:3493
27. Ruoff RS, Lorents DC (1995) Mechanical and thermal properties of carbon nanotubes. Carbon 33:925
28. Chawla K (1998) Ceramic matrix composites. Composite Materials. Springer, New York, p. 212
29. Munz D (1988) Failure criteria in ceramic materials. J Nucl Mater 155–157(Part 1):77
30. Donald IW, McMillan PW (1976) Ceramic-matrix composites. J Mater Sci 11:949
31. US Department of Defense (2002) Composite Materials Handbook, Vol. 5: Ceramic Matrix Composites, MIL-HDBK-17-5, USA
32. Kita K, Narisawa M, Nakahira A, Mabuchi H, Sugimoto M, Yoshikawa M (2010) Synthesis and properties of ceramic fibers from polycarbosilane/polymethylphenylsiloxane polymer blends. J Mater Sci 45:3397
33. Narisawa M, Nishioka M, Okamura K, Oka K, Dohmaru T (2006) SiC Ceramic Fibers Synthesized from Polycarbosilane-Polymethylsilane Polymer Blends. In: Kohyama A, Singh M, Lin HT, Katoh Y (eds) Advanced SiC/SiC Ceramic Composites: Developments and Applications in Energy Systems. The American Ceramic Society, Westerville, p. 173
34. Yajima S, Okamura K, Hayashi J, Omori M (1976) Synthesis of continuous SiC fibers with high tensile strength. J Am Ceram Soc 59:324
35. Bernard S, Weinmann M, Gerstel P, Miele P, Aldinger F (2005) Boron-modified polysilazane as a novel single-source precursor for SiBCN ceramic fibers: synthesis, melt-spinning, curing and ceramic conversion. J Mater Chem 15:289
36. Bai HW, Wen G, Huang XX, Han ZX, Zhong B, Hu ZX, Zhang XD (2011) Synthesis and structural characterization of SiBOC ceramic fibers derived from single-source polyborosiloxane. J Eur Ceram Soc 31:931
37. Penn BG, Ledbetter FE, Clemons JM, Daniels JG (1982) Preparation of silicon carbide–silicon nitride fibers by the controlled pyrolysis of polycarbosilazane precursors. J Appl Polym Sci 27:3751

38. Kimura Y, Kubo Y, Hayashi N (1994) High-performance boron-nitride fibers from poly (borazine) preceramics. Compos Sci Technol 51:173
39. Fazen PJ, Carroll PJ, McGhie AR, Sneddon LG (1995) Synthesis, properties, and ceramic conversion reactions of polyborazylene. A high-yield polymeric precursor to boron nitride. Chem Mater 7:1942
40. Fan J-P, Zhuang D-M, Zhao D-Q, Zhang G, Wu M-S, Wei F, Fan Z-J (2006) Toughening and reinforcing alumina matrix composite with single-wall carbon nanotubes. Appl Phys Lett 89:121910
41. Suzuki T, Guo Y, Inoue S, Zhao X, Ohkohchi M, Ando Y (2006) Multiwalled carbon nanotubes mass-produced by dc arc discharge in He–H_2 gas mixture. J Nanopart Res 8:279
42. Journet C, Bernier P (1998) Production of carbon nanotubes. Appl Phys A 67:1
43. Ando Y, Iijima S (1993) Preparation of carbon nanotubes by arc-discharge evaporation. Jpn J Appl Phys Lett 32(Part 2):107
44. Iijima S (1991) Helical microtubules of graphitic carbon. Nature 354:56
45. Maser WK, Muñoz E, Benito AM, Martínez MT, de la Fuente GF, Maniette Y, Anglaret E, Sauvajol JL (1998) Production of high-density single-walled nanotube material by a simple laser-ablation method. Chem Phys Lett 292:587
46. Thess A, Lee R, Nikolaev P, Dai H, Petit P, Robert J, Xu C, Lee YH, Kim SG, Rinzler AG, Colbert DT, Scuseria GE, Tomanek D, Fischer JE, Smalley RE (1996) Crystalline ropes of metallic carbon nanotubes. Science 273:483
47. Kokai F, Takahashi K, Yudasaka M, Yamada R, Ichihashi T, Iijima S (1999) Growth dynamics of single-wall carbon nanotubes synthesized by CO_2 Laser vaporization. J Phys Chem B 103:4346
48. Ebbesen TW, Ajayan PM (1992) Large-scale synthesis of carbon nanotubes. Nature 358:220
49. Colomer JF, Stephan C, Lefrant S, Van Tendeloo G, Willems I, Kónya Z, Fonseca A, Laurent C, B. Nagy J (2000) Large-scale synthesis of single-wall carbon nanotubes by catalytic chemical vapor deposition (CCVD) method. Chem Phys Lett 317:83
50. Cheng HM, Li F, Su G, Pan HY, He LL, Sun X, Dresselhaus MS (1998) Large-scale and low-cost synthesis of single-walled carbon nanotubes by the catalytic pyrolysis of hydrocarbons. Appl Phys Lett 72:3282
51. Mukhopadhyay K, Koshio A, Sugai T, Tanaka N, Shinohara H, Konya Z, Nagy JB (1999) Bulk production of quasi-aligned carbon nanotube bundles by the catalytic chemical vapour deposition (CCVD) method. Chem Phys Lett 303:117
52. Li WZ, Wen JG, Ren ZF (2002) Effect of temperature on growth and structure of carbon nanotubes by chemical vapor deposition. Appl Phys A 74:397
53. Xia Z, Riester L, Curtin WA, Li H, Sheldon BW, Liang J, Chang B, Xu JM (2004) Direct observation of toughening mechanisms in carbon nanotube ceramic matrix composites. Acta Mater 52:931
54. Cha SI, Kim KT, Lee KH, Mo CB, Hong SH (2005) Strengthening and toughening of carbon nanotube reinforced alumina nanocomposite fabricated by molecular level mixing process. Scr Mater 53:793
55. Bocanegra-Bernal MH, Echeberria J, Ollo J, Garcia-Reyes A, DomÃnguez-Rios C, Reyes-Rojas A, Aguilar-Elguezabal A (2011) A comparison of the effects of multi-wall and single-wall carbon nanotube additions on the properties of zirconia toughened alumina composites. Carbon 49:1599
56. Strano MS, Moore VC, Miller MK, Allen MJ, Haroz EH, Kittrell C, Hauge RH, Smalley RE (2003) The role of surfactant adsorption during ultrasonication in the dispersion of single-walled carbon nanotubes. J Nanosci Nanotechnol 3:81
57. Sun J, Gao L, Jin X (2005) Reinforcement of alumina matrix with multi-walled carbon nanotubes. Ceram Int 31:893
58. Sun J, Gao L, Li W (2002) Colloidal processing of carbon nanotube/alumina composites. Chem Mater 14:5169
59. Ahmad I, Cao H, Chen H, Zhao H, Kennedy A, Zhu YQ (2010) Carbon nanotube toughened aluminium oxide nanocomposite. J Eur Ceram Soc 30:865

60. Garmendia N, Santacruz I, Moreno R, Obieta I (2009) Slip casting of nanozirconia/MWCNT composites using a heterocoagulation process. J Eur Ceram Soc 29:1939
61. Wang Y, Iqbal Z, Mitra S (2006) Rapid, low temperature microwave synthesis of novel carbon nanotube-silicon carbide composite. Carbon 44:2804
62. Katsuda Y, Gerstel P, Narayanan J, Bill J, Aldinger F (2006) Reinforcement of precursor-derived Si-C-N ceramics with carbon nanotubes. J Eur Ceram Soc 26:3399
63. Zhu Y-F, Shi L, Liang J, Hui D, K-t L (2008) Synthesis of zirconia nanoparticles on carbon nanotubes and their potential for enhancing the fracture toughness of alumina ceramics. Composites Part B 39:1136
64. Mo CB, Cha SI, Kim KT, Lee KH, Hong SH (2005) Fabrication of carbon nanotube reinforced alumina matrix nanocomposite by sol–gel process. Mater Sci Eng A 395:124
65. Lupo F, Kamalakaran R, Scheu C, Grobert N, RÃ¼hle M (2004) Microstructural investigations on zirconium oxide-carbon nanotube composites synthesized by hydrothermal crystallization. Carbon 42:1995
66. Lu J, Zang JB, Shan SX, Huang H, Wang YH (2008) Synthesis and characterization of core-shell structural MWNT-zirconia nanocomposites. Nano Lett 8:4070
67. Maensiri S, Laokul P, Klinkaewnarong J, Amornkitbamrung V (2007) Carbon nanofiber-reinforced alumina nanocomposites: fabrication and mechanical properties. Mater Sci Eng A 447:44
68. Balazsi C, Konya Z, Weber F, BL P, Arato P (2003) Preparation and characterization of carbon nanotube reinforced silicon nitride composites. Mater Sci Eng C 23:1133
69. Balazsi C, Weber F, Kover Z, Shen Z, Konya Z, Kasztovszky Z, Vertesy Z, Biro LP, Kiricsi I, Arato P (2006) Application of carbon nanotubes to silicon nitride matrix reinforcements. Curr Appl Phys 6:124
70. Pasupuleti S, Peddetti R, Santhanam S, Jen K-P, Wing ZN, Hecht M, Halloran JP (2008) Toughening behavior in a carbon nanotube reinforced silicon nitride composite. Mater Sci Eng A 491:224
71. Zhan G-D, Kuntz JD, Garay JE, Mukherjee AK (2003) Electrical properties of nanoceramics reinforced with ropes of single-walled carbon nanotubes. Appl Phys Lett 83:1228
72. Ye F, Liu L, Wang Y, Zhou Y, Peng B, Meng Q (2006) Preparation and mechanical properties of carbon nanotube reinforced barium aluminosilicate glass-ceramic composites. Scr Mater 55:911
73. Balazsi C, Fenyi B, Hegman N, Kover Z, Weber F, Vertesy Z, Konya Z, Kiricsi I, Biro LP, Arato P (2006) Development of CNT/Si_3N_4 composites with improved mechanical and electrical properties. Compos Part B 37:418
74. Veedu VP, Cao A, Li X, Ma K, Soldano C, Kar S, Ajayan PM, Ghasemi-Nejhad MN (2006) Multifunctional composites using reinforced laminae with carbon-nanotube forests. Nat Mater 5:457
75. Peigney A, Laurent C, Flahaut E, Rousset A (2000) Carbon nanotubes in novel ceramic matrix nanocomposites. Ceram Int 26:677
76. Duszova A, Dusza J, Tomasek K, Blugan G, Kuebler J (2008) Microstructure and properties of carbon nanotube/zirconia composite. J Eur Ceram Soc 28:1023
77. Shen Z, Johnsson M, Zhao Z, Nygren M (2002) Spark plasma sintering of alumina. J Am Ceram Soc 85:1921
78. Zhan G-D, Kuntz JD, Wan J, Mukherjee AK (2003) Single-wall carbon nanotubes as attractive toughening agents in alumina-based nanocomposites. Nat Mater 2:38
79. Shi S-L, Liang J (2006) Effect of multiwall carbon nanotubes on electrical and dielectric properties of yttria-stabilized zirconia ceramic. J Am Ceram Soc 89:3533
80. Mazaheri M, Mari D, Hesabi ZR, Schaller R, Fantozzi G (2011) Multi-walled carbon nanotube/nanostructured zirconia composites: Outstanding mechanical properties in a wide range of temperature. Compos Sci Technol 71:939
81. Jn D, Blugan G, Morgiel J, Kuebler J, Inam F, Peijs T, Reece MJ, Puchy V (2009) Hot pressed and spark plasma sintered zirconia/carbon nanofiber composites. J Eur Ceram Soc 29:3177

82. Avouris P, Dimitrakopoulos C (2012) Graphene: synthesis and applications. Mater Today 15:86
83. Stankovich S, Dikin DA, Piner RD, Kohlhaas KA, Kleinhammes A, Jia Y, Wu Y, Nguyen ST, Ruoff RS (2007) Synthesis of graphene-based nanosheets via chemical reduction of exfoliated graphite oxide. Carbon 45:1558
84. Choi W, Lahiri I, Seelaboyina R, Kang YS (2010) Synthesis of graphene and its applications: a review. Crit Rev Solid State Mater Sci 35:52
85. Marcano DC, Kosynkin DV, Berlin JM, Sinitskii A, Sun Z, Slesarev A, Alemany LB, Lu W, Tour JM (2010) Improved synthesis of graphene oxide. ACS Nano 4:4806
86. Walker LS, Marotto VR, Rafiee MA, Koratkar N, Corral EL (2011) Toughening in graphene ceramic composites. ACS Nano 5:3182
87. Zhu Y, Murali S, Cai W, Li X, Suk JW, Potts JR, Ruoff RS (2010) Graphene and graphene oxide: synthesis, properties, and applications. Adv Mater 22:3906
88. Wang K, Wang Y, Fan Z, Yan J, Wei T (2011) Preparation of graphene nanosheet/alumina composites by spark plasma sintering. Mater Res Bull 46:315
89. Kun P, Tapaszto O, Weber F, Balazsi C (2012) Determination of structural and mechanical properties of multilayer graphene added silicon nitride-based composites. Ceram Int 38:211
90. Tapaszto O, Tapaszto L, Marko M, Kern F, Gadow R, Balazsi C (2011) Dispersion patterns of graphene and carbon nanotubes in ceramic matrix composites. Chem Phys Lett 511:340
91. Kvetkova L, Duszova A, Hvizdoa P, Duszs J, Kun P, Balazsi C (2012) Fracture toughness and toughening mechanisms in graphene platelet reinforced Si_3N_4 composites. Scr Mater 66:793
92. Curtin WA, Sheldon BW (2004) CNT-reinforced ceramics and metals. Mater Today 7:44
93. Sheldon BW, Curtin WA (2004) Nanoceramic composites: tough to test. Nat Mater 3:505
94. Yamamoto G, Omori M, Hashida T, Kimura H (2008) A novel structure for carbon nanotube reinforced alumina composites with improved mechanical properties. Nanotechnology 19:315708
95. Shi S-L, Liang J (2007) Electronic transport properties of multiwall carbon nanotubes/yttria-stabilized zirconia composites. J Appl Phys 101:023708
96. Balazsi C, Shen Z, Konya Z, Kasztovszky Z, Weber F, Vertesy Z, Biro LP, Kiricsi I, Arato P (2005) Processing of carbon nanotube reinforced silicon nitride composites by spark plasma sintering. Compos Sci Technol 65:727
97. Wang H-Z, Li X-D, Ma J, Li G-Y, Hu T-J (2012) Multi-walled carbon nanotube-reinforced silicon carbide fibers prepared by polymer-derived ceramic route. Compos Part A 43:317
98. Gu Z, Yang Y, Li K, Tao X, Eres G, Howe JY, Zhang L, Li X, Pan Z (2011) Aligned carbon nanotube-reinforced silicon carbide composites produced by chemical vapor infiltration. Carbon 49:2475
99. Patterson E, Backman D, Cloud G, Carlson T, Marsh C, Kriven W, Stynoski P, Welch C. Processing, microstructure, and properties of carbon nanotube reinforced silicon carbide. Composite Materials and Joining Technologies for Composites, Volume 7: Springer New York 2013, p. 147.
100. Lee C, Wei X, Kysar JW, Hone J (2008) Measurement of the elastic properties and intrinsic strength of monolayer graphene. Science 321:385
101. Faccio R, Denis PA, Pardo H, Goyenola C, MombrÃ ÃW (2009) Mechanical properties of graphene nanoribbons. J Phys Condens Matter 21:285304
102. El-Kady MF, Strong V, Dubin S, Kaner RB (2012) Laser scribing of high-performance and flexible graphene-based electrochemical capacitors. Science 335:1326
103. Fan Y, Wang L, Li J, Li J, Sun S, Chen F, Chen L, Jiang W (2010) Preparation and electrical properties of graphene nanosheet/Al_2O_3 composites. Carbon 48:1743
104. Yamamoto G, Shirasu K, Hashida T, Takagi T, Suk JW, An J, Piner RD, Ruoff RS (2011) Nanotube fracture during the failure of carbon nanotube/alumina composites. Carbon 49:3709
105. Koops HWP, Weiel R, Kern DP, Baum TH (1988) High-resolution electron-beam induced deposition. J Vac Sci Technol B 6:477
106. Ohnabe H, Masaki S, Onozuka M, Miyahara K, Sasa T (1999) Potential application of ceramic matrix composites to aero-engine components. Composites Part A 30:489

107. Fergus JW (2011) Solid oxide fuel cells, In: Liu RS, Zhang L, Sun X, Liu Y, Zhang J (eds) Electrochemical technologies for energy storage and conversion. Vol. 1 & 2 Wiley-VCH, Verlag GmbH & Co. KGaA, Weinheim, Germany, p. 671
108. Courtin E, Boy P, Cm R, Bianchi L, Bruneton E, Poirot N, Laberty-Robert C, Cm S (2012) Optimized sol–gel routes to synthesize yttria-stabilized zirconia thin films as solid electrolytes for solid oxide fuel cells. Chem Mater 24:4540
109. Colombo P, Riccardi B, Donato A, Scarinci G (2000) Joining of SiC/SiCf ceramic matrix composites for fusion reactor blanket applications. J Nucl Mater 278:127
110. White AA, Best SM, Kinloch IA (2007) Hydroxyapatite–carbon nanotube composites for biomedical applications: a review. Int J Appl Ceram Technol 4:1

Bamboo Fiber-Based Polymer Composites 18

Mahuya Das

Contents

Abstract

Bamboo is a potential natural fiber with its stiffness, rigidity, high availability, and hydrophilic nature. This hydrophilicity renders improper interfacial adhesion during polymer matrix-based composite fabrication. This adhesion can be improved by suitable surface treatment of bamboo fiber which is an important research topic to many scientists and covered to some extent in this chapter.

Keywords

Bamboo • Fiber • Composites • Natural • Interfacial adhesion • Mechanical properties

Introduction

There is a growing interest in natural fiber-based composites, mainly due to their high specific modulus [1–4], light weight [4], low cost, and resistance to deforestation in addition to the other usual advantages. Moreover, natural fibers possess fairly

M. Das (✉)
JIS College of Engineering, Kalyani, Nadia, West Bengal, India
e-mail: d_mahuya@yahoo.com

K.K. Kar (ed.), *Composite Materials*,
DOI 10.1007/978-3-662-49514-8_18

reactive cell walls that allow modification on the surface. Other potential advantages over man-made fibers include reduced tool wear, safer handling, and working conditions. In addition to these technical advantages, there are also the significant environmental advantages of using fibers which come from continually renewable resources and which utilize atmospheric CO_2 [4].

Bamboo, like other natural fiber, is hygroscopic and exhibits a tendency to be in moisture equilibrium with the relative humidity of the surrounding atmosphere, either by taking up moisture from or giving out moisture to the atmosphere [4]. They are very much prone to swell/warp and shrink when exposed to moist and hot weather conditions, respectively [5]. However, for application like composites, this aspect is detrimental so far as its dimensional stability is concerned. Natural fibers absorb moisture as the cell wall polymers contain hydroxyl or other oxygenated groups that attract moisture through H-bonding [6]. Taking the advantage of plenty of reactive group's present, modification of cell wall using proper surface modifier can be done to increase the scope of utilization of the natural fibers as reinforcement [7–10].

Importance of Bamboo as Reinforcement

All over the world, bamboo exists in about 75 genera and 1250 species. Monsoon is the best season for bamboo thriving to attain their maximum growth and grow into shrubs in temperate regions and at high altitudes. It is a long fleshy plant which technically comes under grass family but the appearance is never like grass. With respect to the species variety and distribution of bamboo, India is the second largest country in the world only after China. There are 136 genetic species of bamboo including eleven exotic species in India. Among all these 136 species, 58 species belong to 10 genera which are found in the Northeast. In many countries of the world like the USA, Japan, China, India, etc., bamboo is being grown as plantation besides natural occurrence. It is soft toward the center and hard toward its periphery. Bamboo is mostly grown in tropical countries and is naturally occurring composites. Bamboos are largely used for the purpose like housing, forestry, agroforestry, agricultural activities, utensils, and weapons. It is mainly planted in Asian countries and constitutes about 65 % of the total bamboo resources found in the world [11].

Bamboo shows the mechanical properties which are analogous to that of wood. Bamboo shows better mechanical properties as compared to fibers such as sisal, banana, vakka, etc. Bamboo can be used in a different form to synthesize a composite product. These can be either in a form of a long strip, whole bamboo, sections, and short bamboo fibers. The selection of their fiber kind depends upon the property to be imparted in the composite [12, 13]. There are some basic differences between wood and bamboo which is mainly due to the differences in chemical component. In bamboo, instead of rays or knots as in the case of woods, there are nodes. Absence of rays or knots leads to far more evenly distributed stresses throughout the length of bamboo. Bamboo is hollow internally producing tubelike part in between the nodes, and the outer wall may be thick or thin. Consequently, it is more difficult to join bamboo than pieces of wood with nailing, whereas due to variation in chemical extractives, bamboo can be glued very well [14]. Bamboo possesses some favorable properties which are contributed by its macroscopic property like diameter, thickness, internodal length, and the fiber distribution [15]. The main disadvantages of bamboo are its moisture affinity, low mildew resistance, and nature of flowering.

Anatomy of Bamboo

Throughout the length bamboo has some nodes and the place between the nodes is known as internodes. The internodes are hollow internally up to the nodal plane and the cavity is called a lacuna. Internodes has culm wall of varying thickness surrounding the lacuna. Only a few species possess solid internodes. A solid bamboo without cavity or lacuna in internodes is sometimes called a male bamboo. With increasing height of the bamboo itself, the internodes will show a continuous decrease in length and also in diameter as the bamboo culms generally taper from base to tip. The fibers are characterized by their slender form – long, tapered at both ends, and sometimes, forked [16–18]. Their length increases inward to a maximum in the outer third of the culm wall and decreasing again toward the inner wall. The fiber length to width ratio varies across the culm wall from 70:1 to 150:1. Owing to the shorter fibers around the nodes, the values there are lesser than those at the internodes. Figure 18.1 represents cross-section of a bamboo culm.

All natural fibers are lingo-cellulosic material consisting of cellulose microfibrils embedded in the amorphous matrix of lignin and hemicellulose. Cellulose fibers consist of several microfibrils which are distributed microscopically along the length of the fiber. In turn individual fibril composed of a thin primary wall surrounding by a thicker secondary layer leading to a polylamellate complex structure. The angle

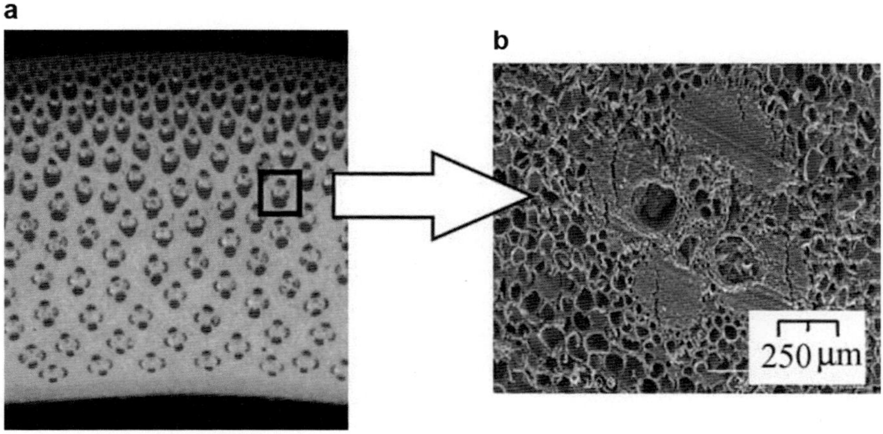

Fig. 18.1 Cross section of bamboo [19]

between the fiber axis and the alignment of microfibrils is termed as the microfibrillar angle. This microfibrillar angle and cellulose content of the natural fiber will determine the mechanical behavior of the fiber itself [16].

The fine structure of bamboo fibers is of special interest because of its significance for culm strength. Detailed investigations have been made by different worker [19–21]. Tono and Ono [22] and Parameswaran and Liese [23–25], as well as contributions by Fujii et al. [26], Kawase et al. [27], and Murphy and Alvin [28], have revealed the numerous layers which constitute the cell wall. Different results have been reported about the number of lamellae, with a maximum of 18. This can be partly attributed to the position of the vascular bundles, the location of the fibers within, and the maturity state of the fibers. In most of the earlier observations, the effect of aging was hardly considered. Often younger culms were investigated with lesser lamellae because of the technical difficulties encountered when sectioning the extremely hard tissues of older ones. Preston and Singh [29] employed X-ray analysis to study wall structures, while Tono and Ono [22] used swelling reactions resulting from acid treatment. Bamboo is also a lingo-cellulosic composite material like the other natural fiber consisting of cellulose and hemicellulose and lignin as amorphous matrix material. Ray et al. studied the variation of mechanical strength with microstructure of bamboo in their study with a piece of bamboo [30]. It was observed that not only the quantity but also the fiber quality varies from outer to inner most peripheries, and it causes variation of tensile strength. The hemicellulose and lignin matrix materials have found to be removed satisfactorily with 10 % NaOH solution without affecting the fibers very much, while 20 % or stronger alkali solution deteriorates the strength property of fibers. Low et al. carried out investigation of Australian bamboo using synchrotron radiation diffraction and Vickers indentation and concluded that mechanical properties of that bamboo was a strong variant of its age [31]. Result analysis exhibited that the older counterpart has a lower strength, elastic stiffness, and fracture toughness than the young one [31]. Furthermore, bamboo expands along the interfacial damage zone and exhibited indentation creep which makes hardness of bamboo both load dependent and time dependent, respectively. Fiber debonding, crack deflection, and crack bridging in bamboo involve high-energy dissipation, leading to very high toughness [19].

Bamboo Polymer Composites

Bamboo, due to its moisture affinity, changes its dimension during its service life when exposed to moist and hot weather conditions. However, with respect to the dimensional stability and also the good interfacial adhesion in composite application, this aspect is detrimental. Natural fibers absorb moisture through H-bonding taking the advantage of plenty of reactive functional group's (hydroxyl or other oxygenated groups) present in the constituent component. Natural fibers are hydrophilic in nature, and to increase the scope of utilization as reinforcement modification, using proper surface modifier can be done [32–37]. Mechanical property of a polymer composite reinforced by fiber is primarily controlled by three factors [31]:

(a) strength and modulus of the fiber, (b) strength and chemical stability of the resin, and (c) effectiveness of interfacial adhesion between resin and fiber while transferring stress across the interface [4]. Low volume fraction of fiber can exhibit reinforcing action if and only if good interfacial bonding exists between the fiber and the matrix. The strength of interfacial bond also depends upon the modification of fiber and chemical structure of the resin. There are many physical and chemical methods of different efficiencies for the adhesion between fiber and matrix which can be used to optimize the interface. Examples of physical methods are stretching [38], calendering [39, 40], thermo-treatment [41], corona and cold plasma treatment [42], etc. When fiber and matrix are totally incompatible (hydrophilic fiber and hydrophobic polymer), then chemical method is employed to get proper compatibility with the help of any other material, which itself has compatibility with both fiber and matrix [43]. The example of chemical methods employed are mercerization [38, 44–47], change of surface tension, impregnation of fibers, chemical coupling, graft copolymerization, treatment with compounds which contain methanol groups, treatment with isocyanates, triazine coupling agents, and organosilanes as coupling agent.

Among all these methods, mercerization is one of the most suitable low-cost treatments for natural fibers by which the noncellulosic substances will be extracted, rendering increased wetting ability of fiber [48, 49]. The main objective of alkali treatment is to produce fibers with improved wetting and spreading characteristics and the consequent changes in the dimension, fine structure, morphology, and mechanical properties [50]. Das et al. have carried out detailed study regarding effect of mercerization of bamboo fiber in either strip or dust form on physical, mechanical, and thermal properties of bamboo fiber itself and thermoset polymer (e.g., novolac, resol, and polyester resin) composite based on alkali-treated bamboo fibers [50–58].

Treatment of bamboo strips with alkali solutions of different concentrations, e.g., 5, 10, 15, 20, 25, and 50 %, for one hour leads to swelling of bamboo strips which introduces a change in average density of ~15 %. The weight loss value increases with increase in alkali solution concentration with a maximum of 21.94 % at 50 % alkali treatment. The cellulose chains present in bamboo fiber react with sodium hydroxide forming the stable product, sodium cellulosate or soda cellulose, by replacement of the hydrogen ion present in the –OH groups of cellulose with Na+ ions. By subsequent neutralization with dilute H_2SO_4 and followed by washing with distilled water, the sodium ion present in the soda cellulose compound is completely replaced further by hydrogen ion regaining the cellulose structure but with a lattice transformation from cellulose-I to cellulose-II results [48]. The experimental conditions will govern the extent of transformation between cellulose-I to cellulose-II. During this crystalline lattice transformation, a fraction of the swollen cellulose frequently does not gain the crystalline structure further resulting in amorphous cellulose, which can be detected with X-ray diffraction and tabulated in Table 18.1. It was also revealed that with increasing concentration of alkali solution percentage of bamboo sample ranging from 0 % to 50 %, amount of cellulose-II increases and maximum amount of cellulose-II is obtained with 50 % alkali solution for bamboo

Table 18.1 d-spacing and percent amorphousness data of alkali-treated bamboo sample in both strips and dust form

	d-spacing		Amorphous (%)		Cellulose-I (%)		Cellulose-II (%)	
Sample	Strip	Dust	Strip	Dust	Strip	Dust	Strip	Dust
B-0	4.019	4	54.4	56.4	45.5	43.5	0	0
B-10	4.013	3.97	49.9	54.1	35.4	38.2	14.6	7.6
B-15	4.01	4	48.2	55.1	34.0	31.0	17.8	14.8
B-20	4.026	4.03	52.5	58.4	29.7	8.5	17.8	33.0
B-50	4.074	4.01	82.1	81.4	1.9	1.7	15.8	16.8

strip samples [48]. The same phenomenon has also been observed with the dust sample but at an earlier concentration owing to higher interaction between alkali solution and cellulose molecule provided with larger surface area of dust samples showing maximum conversion to cellulose-II for the sample treated 20 % alkali solution with the glucosidic bond cleavage of cellulose. The explanation may lie possibly in the breakdown of the cellulose crystal and the recrystallization or reorientation of the cellulose molecules present in bamboo fiber from which the cementing material, hemicelluloses, has been depleted with the action of caustic solution. The recrystallization or reorientation of the cellulose molecules results from the formation of new H-bonds among a part of the cellulose chains leading to a closer packing leading to increased average density and mechanical properties. The cellulose chains are somehow strained due to the presence of the matrix material hemicellulose/lignin matrix and that initial strain is released by the removal of hemicellulose upon the action of caustic solution [49]. This is also evident from the trend of d-spacing (Table 18.1) values of lattice as obtained from X-ray analysis. For bamboo sample treated with 50 % alkali solution, the crystallinity is due to cellulose-II structure.

Analysis of the mechanical properties of the alkali-treated bamboo strips showed a steady increase with increasing concentration of caustic soda, exhibiting a comparable increment with 15 % and 20 % alkali solution treatment followed by a gradual fall. This fall of mechanical properties after 20 % alkali treatment can be well explained on the basis of the molecular degradation of cellulose chain itself in the presence of strong alkali solution [50]. The composite structure of bamboo fiber destructed thoroughly by the removal of the cementing material with alkali treatment, leading to splitting of the fiber into a finer element, fibrils. Fibrils are the smaller fibers which all together forms the composite fiber bundle. Fibrillation leads to a decrease in the spiral angle leading to an increase in molecular orientation. It was also reported by Das et al. that alkali treatment with increasing concentration of the fibers is continuously more and more fibrillated as evident from polarizing light microscope [48].

Differential scanning calorimetric (DSC) study of those treated fiber showed that the moisture desorption peak (and the enthalpy values associated with it) has shifted rightward up to 15 % alkali treatment concentration, and beyond that concentration further reduced [50]. In case of all samples for alkali treatments, up to 15 % single broad exotherm is a characteristic of all the DSC plots, and beyond that concentration, two comparatively smaller exothermic peaks appeared probably as an identification of the presence of both cellulose-I and cellulose-II. Dynamic mechanical thermal analysis (DMTA) of bamboo strip samples revealed that the storage modulus (E') value of 15 % alkali-treated samples at room temperature has been increased by ~400 % over that of untreated bamboo strips, and the rate of fall in the modulus over the temperature range of 140–180 °C is also maximum for alkali-treated samples [51]. The primary loss modulus (E'') peak at 111.8 °C for untreated bamboo samples has been shifted to a higher region in case of alkali-treated samples [51]. Untreated sample provides maximum damping parameter (tan δ) due to the presence of higher amount of matrix material.

From the thermogravimetric analysis plot, it can be concluded that the B-15 and B-20 samples have undergone three-step degradation, whereas parent bamboo degrades in two steps in the temperature range of 50–150 °C and 150–426.11 °C. The weight loss during the first step of degradation for the parent bamboo sample is 6.65 % which is due to moisture loss. This value has been reduced to 1.75 and 1.25 for B-15 and B-20 samples, respectively, for the first degradation step commencing at 100 °C for both samples. A continuous decreasing trend was observed for the amount of moisture adsorbed with increasing treatment concentration of alkali, and the moisture liberation temperature is higher for alkali-treated samples. The explanation lies in the increase in crystallinity and change in morphology of bamboo fibers incorporated with the alkali treatment. Therefore the ease of adsorbed moisture release upon heating will be reduced as moisture is strongly entrapped with tightly packed structure leading to a higher finishing temperature. Leaching of increasing amount hemicelluloses with higher alkali concentration is mainly responsible for the decreased amount of adsorbed water as they are also active sites for moisture absorption. It can be then concluded from the data of table that treated samples have higher thermal stability than the untreated among which B-15 shows the maximum. The explanation lies in the fact that mercerization reduced the alkali-sensitive material in considerable amount forming more stable lignin-cellulose complex than the native one, thereby increasing the residual weight [59]. This process involves increased rate of formation of free radicals that are stabilized by condensed carbon ring formation in the char. The activation energy also helps to conclude about the thermal stability of fibers. Increase in the activation energy will lead to more thermal stability of the fiber (Table 18.2).

There are many other works based on bamboo cellulose modification. The species of Bamboo, *Bambusa balcooa* was coated with polyethylene glycol-based polyurethane (PU) and its semi-interpenetrating polymer network (SIPN) with polystyrene (PS) after initial alkali treatment by which the tensile behavior and chemical resistance of the fiber improved [60]. Ray et al. also observed similar type of removal of the adhered matrix with 10 % NaOH solution with a little effect on fibers, while 20 % or stronger alkali leads to reduction in the strength of fiber [30]. Green composites were developed with a starch-based resin and alkali-treated short bamboo fiber, and in this case the reinforcing action of bamboo fibers was not observed with a small aspect ratio of 20 but was filler [61].

Table 18.2 Results for TG analysis of untreated and treated bamboo fiber

Sample designation	Residual weight (%) at 600 °C	Activation Energy Eact J/K
B-U (untreated bamboo fiber)	40	23.22
B-10 (10 % alkali-treated fiber)	46.45	25.4
B-15 (15 % alkali-treated fiber)	56.4	31.91
B-20 (20 % alkali-treated fiber)	56.25	29.07

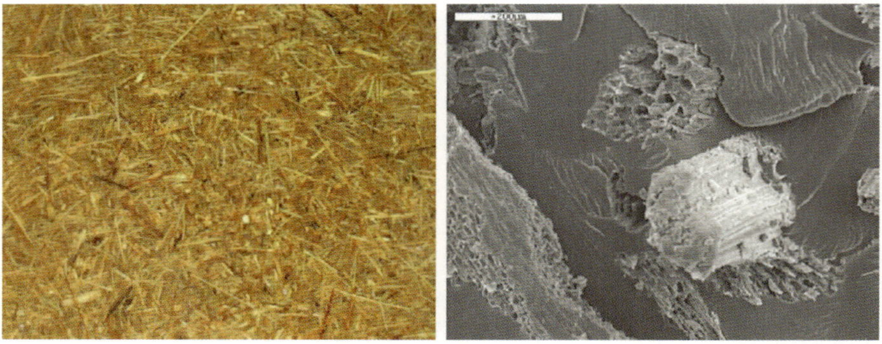

Fig. 18.2 Bamboo fiber preform for VIM technique and SEM of fracture surface of epoxy-based composite made with that fiber preform [62]

Natural fiber-based polymeric composites have been developed using both thermoset and thermoplastic polymer. Thermoset polymers like polyester, epoxies, and phenolics are most commonly used. An et al. developed bamboo fiber epoxy composite by vacuum injection molding technique (VIM) with good and more uniform flexural and compressive modulus than tensile strength (Fig. 18.2) [62].

Das et al. further studied the polymer composite based on this alkali-treated bamboo fiber and novolac, resol, and polyester resin widely [52–58]. They have reported that all types of composite specimens based on bamboo strips and novolac, resol, or polyester resin were subjected to mechanical property study, thermal property study, dynamic mechanical study, and weathering property study, and the results showed that the alkali-treated bamboo fiber composites are superior with all aspect. These findings were confirmed by the strong interfacial interaction between –OH groups of cellulose and methylol in case of novolac or resol resin with the formation of aryl alkyl ether as evident from infrared spectroscopy. The alkali solution concentrations which are used for bamboo fiber treatment is the main controlling factor regarding the dynamic mechanical and thermal properties evaluated by means of dynamic mechanical, differential scanning calorimetry, and thermogravimetric analysis. Twenty percent alkali-treated fiber composites exhibited maximum storage modulus (E′) value following an increasing trend of storage modulus with increasing alkali treatment concentration. Due to removal of alkali-sensitive binding material, there will be proper wetting of fiber with the resin leading to the increasing interaction thereby restricting the chain movement at an elevated temperature. The number of free cellulosic –OH will continuously increase with increasing alkali treatment resulting to more restricted chain mobility. The result is therefore increased glass transition temperature of composites compared to that of the resin as evident from loss modulus peak. Due to proper wetting of fiber by resin, the amount of matrix which can be penetrated into the crevices by volume is more sufficient to absorb the vibrational energy of molecules properly leading to increasing tan δ value of composite with increasing concentration alkali treatment. It was concluded by Das et al. that 20 % alkali-treated fiber composites exhibited the best dynamic mechanical properties, and the alkali-treated fiber composites attains better thermal.

Thermoplastic polymers like polyethylene, polystyrene, polypropylene, etc. are nowadays getting attention as matrices [63–65]. These polymers are hydrophobic in nature, whereas natural fibers are hydrophilic imparting poor adhesion between them. Consequently, the reinforcing action of the fibers will vary widely according to the nature of the matrix. Therefore after overall evaluation of entire sets of composite, it is considered as essential to compare the properties of different composites developed on the basis of interaction with resin. There are different works based on bamboo fiber and different polymers which are made by employing different surface treatment method to obtain improved properties which are noted below.

Epoxy resin composite reinforced with varying content of short bamboo fiber has been evaluated with respect to weight reduction, density, and void content [66]. The density of these composites was found to be decreasing with fiber content. In another work it was reported that 30 wt% of bamboo fiber-mixed epoxy resin is giving optimum mechanical properties compared to 20 % and 40 % loading of fiber [67]. The addition of bamboo fibers has improved mechanical properties of epoxy resin and increased water absorption of the material. Rao et al. also fabricated hybrid composite based on bamboo/glass fibers and epoxy resin using treated and untreated bamboo fiber. They have studied effect of alkali treatment of the bamboo fibers on frictional coefficient and impact strength; dielectric strength and chemical resistance were also studied. These properties were improved with treated bamboo fibers because of the removal of amorphous hemicellulose with alkali treatment imparting improved crystallinity of the bamboo fibers.

Kumar et al. have studied the effect of bamboo fiber loading on mechanical properties of epoxy resin composites reinforced with bamboo fiber and concluded that the tensile strength and flexural strength of bamboo/epoxy composite increase to the certain level of fiber loading with a maximum at 25 wt% of fiber loading and then start decreasing on further fiber loading [68]. It also shows that the microsurface hardness of bamboo/epoxy composite improves till 25 wt% of fiber loading, and after that the hardness is nearly constant.

In fabrication of bamboo fiber-reinforced polypropylene composites, maleic anhydride-grafted polypropylene has been selected as the most suitable compatibilizer [63–65]. The weight percent of maleic anhydride in MAPP may vary from 0.5 % to higher in different cases [65]. It was found that in specific cases of 24 %, the mechanical properties of the composite increased significantly such as tensile strength of 32–36 MPa and a tensile modulus of 5–6 GPa. The developed material has more than three times higher than that of the commercial product with light weight, low cost, and good water resistance. Incorporation of steam-exploded fiber improved the tensile strength of BFRP further. Along with MAPP other surface treatment may be there [63], Mi et al. [64] tried maleation of PP prior to inclusion of the bamboo fibers in the PP matrix to improve the interfacial bonding between the bamboo fibers and the PP matrix. Comparison of bamboo fiber-reinforced polypropylene composites (BFRP) and bamboo/glass fiber-reinforced polypropylene hybrid composite (BGRP) on the basis of their hygrothermal aging and their fatigue behavior under cyclic tensile load was made by Thwe et al. [68]. They have suggested that BGRP has superior fatigue resistance to BFRP at all load levels tested. When the samples were allowed to age for 6 months at 25 °C, tensile strength and elastic modulus of both the samples have shown moderate reduction; however, there were considerable reductions when the aging condition was changed to 75 °C for 3 months. Hybridization of small amount of glass fibers with bamboo fiber renders enhanced durability of bamboo fiber-reinforced polypropylene [69].

Bamboo fibers have been tried as reinforcing agent in cementitious matrix (Li et al.) to develop two types of laminates: (a) double-layer reformed bamboo (cross-ply) laminate and (b) fiber-reinforced cementitious (FRC) plate-reformed bamboo laminate. Reformed bamboo and extruded FRC plates were used to fabricate the second laminate [70–75]. The advantage of reformed bamboo is its ductility and toughness and high tensile strength, while for the FRC plate, it is higher compressive strength. Evaluation of composite with respect to flexural, impact, and compression after impact (CAI) tests has revealed high flexural strength (up to 96 MPa), excellent impact resistance, good ductility, and outstanding toughness for the newly developed FRC plate-reformed bamboo laminate having a good potential in engineering applications.

Y. F. Shih et al. fabricated epoxy resin composites based on chemically modified bamboo fibers obtained by treating wastewater bamboo by coupling agents [76]. These novel composites were evaluated with respect to morphologies, mechanical properties, and heat resistance. The results indicated improved thermal resistance and mechanical properties, and the bamboo fiber was mostly cellulose-I type even after the treatment. Better compatibility in between modified fibers and the polymer matrix was evident from the morphology analysis with application of coupling agents [77]. The effects of a silane coupling agent and filler loading on curing characteristics and mechanical properties of bamboo fiber-filled natural rubber composites were studied by Ismail et al., and they have observed an improvement in mechanical properties with the addition of silane compound as coupling agent [78, 79]. Bamboo/epoxy and bamboo/polyester composites were obtained with improved water resistance and mechanical properties when surface modification of bamboo fibers was made with polyesteramide polyol [78, 80]. Fiberboards using bamboo and wood fibers as the raw materials exhibited improve in the MOR,

MOE retention ratio, and linear expansion of boards after boiling with increasing mixing ratio of bamboo fiber [78]. On the other hand, incorporation of bamboo fiber with partial replacement of spruce had no effect on quality and properties of MDF panels [80]. Comparing with wood fiberboard, it was observed that the mixed fiberboard has more or less the same MOE but lower MOR. The wood fiberboard is also superior to bamboo fiber with respect to water absorption and thickness swelling. Acetylation of bamboo fiber incorporates more hydrophobicity in fiber reducing equilibrium moisture content compared to untreated fiber [81]. Natural fibers are becoming a competitive option as reinforcement of polymeric composite materials due to their bio-based character, good specific mechanical properties, low cost, and inexhaustible supply [82]. In this study, the tensile properties for single bamboo (*Guadua angustifolia*) fibers at four different span lengths after machine compliance correction show that the specific E-modulus is comparable with glass fiber and the specific strength is only 10 % lower. Strength values of around 800 MPa were obtained by using a novel green mechanical extraction process. Untreated unidirectional bamboo/epoxy composites show good flexural strength and also, alkali treatment at low concentrations benefits the stiffness of the composite. Flexural tests with both fiber orientations were performed with UD bamboo fiber thermoplastic composites using polypropylene and maleic anhydride-grafted polypropylene. Material performance was reasonable, but further work is necessary to improve fiber-matrix interfacial strength.

Takagi et al. evaluated the effect of alkali-treated bamboo fiber length on mechanical properties of the "green" composites fabricated using a starch-based resin as matrix with short bamboo fiber as reinforcing agent [61]. They concluded that bamboo fibers were acting as filler rather than reinforcement having small aspect ratio of 20. In another work of Takagi et al., they have applied hot-pressing technique to steam-exploded bamboo stripe and bamboo fibers to make biomass-based composites only from bamboo without resin. Their mechanical properties depend on the pressing temperature and lower than that of synthetic resin-based composite [83].

Lee et al. esterified bamboo fiber with maleic anhydride in the presence of dicumyl peroxide as a radical initiator and developed eco-composite from polylactic acid (PLA) using those bamboo fibers (BF) by employing mechanochemical compositing [84]. It was confirmed by scanning electron microscope and polarizing microscope observation that interfacial properties between BF and PLA were improved after the addition of BF-e-MA. Lee et al. tried lysine-based diisocyanate (LDI) as a coupling agent to fabricate biocomposite from poly(lactic acid) (PLA), poly(butylene succinate) (PBS), and bamboo fiber (BF) [85]. The composites were characterized with respect to tensile properties, water resistance, and interfacial adhesion of both PLA/BF and PBS/BF, and an improvement was observed by the addition of LDI. LDI leads to cross-linking between polymer matrix and BF, rendering restricted thermal flow. Okubo et al. used commercially available microfibrillated cellulose obtained from wood pulp as an enhancer, to improve mechanical properties of polylactic acid-based bamboo fiber composite [86]. The composite showed significant improvement with respect to three-point bending strength and mode-I fracture toughness due to application of the enhancer. The

Fig. 18.3 Scanning electron micrograph of steam-exploded bamboo fiber bundle surface [19]

development of composites for ecological purposes using bamboo fibers and their basic mechanical property evaluation were also made by Okubo et al. [19]. Bamboo fibers were extracted with the help of the steam explosion technique. Bamboo fibers (bundles) itself exhibited a sufficient specific strength comparable to that of conventional glass fibers. They have made PP-based composites using steam-exploded bamboo fiber and concluded about 15 % and 30 % increment in tensile strength and modulus, respectively, due to well impregnation and reduced number of voids as exhibited by the SEM photograph in Fig. 18.3.

Despande et al. employed a combination of chemical and mechanical methods for the extraction of bamboo fibers [87]. The fibers were separated mechanically by applying conventional methods of compression molding technique (CMT) and roller mill technique (RMT). Fibers from both the techniques were evaluated with respect to mechanical properties of the fibers and were used to make unidirectional composites of polyester resin [87]. All types of composite samples exhibited high values of tensile strength where the failure mode is predominantly fiber-matrix interfacial cracking. The mechanical properties of biodegradable polymer composite with carbonized bamboo fibers were evaluated by Matsui et al. [88]. Poly(butylene succinate) (PBS) was used as the biodegradable plastic matrix while the condition of carbonization was varied. By increasing fiber content, tensile modulus was confirmed to increase. In particular, the tensile modulus of composite filled with semi-carbonized bamboo displayed higher values than the uncarbonized bamboo fiber composite. The values of tensile strength decreased according to the increase of fiber content; however, the carbonized bamboo fiber composites experienced less decrease than the uncarbonized ones. The surface resistivity of carbonized bamboo fiber composites was lower than that of bamboo fibers and also decreased with the increase in fiber content in each case. Thermal degradation of bamboo fiber reinforcement was also concluded as a two-stage process by Rajulu et al. [89]. This fact

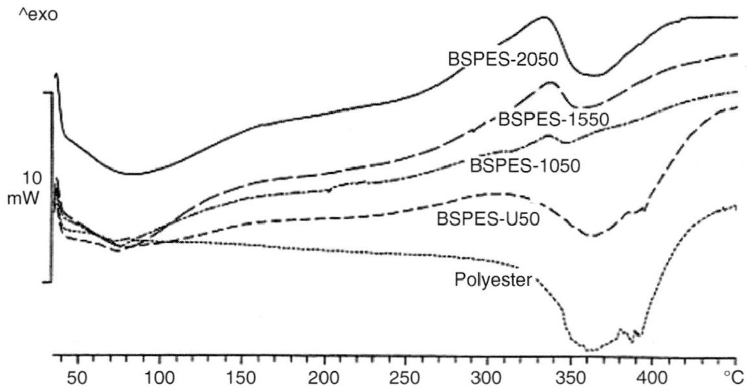

Fig. 18.4 DSC plot of bamboo strips reinforced polyester resin composites

is also evident with Das et al. when they work with mercerized bamboo fiber composites based on phenolic resin and polyester resin [56, 57].

The DSC curves of bamboo/polyester composites using untreated and different alkali-treated fiber and also of polyester resin are presented in Fig. 18.4. It is interesting to observe that the decomposition pattern for all the composite samples is more or less similar because of the presence of a high weight percent of fiber in the composites that makes it behave predominantly like the fiber and not like the resin matrix [10]. An endothermic peak below 100 °C for composite samples indicates moisture loss which is absent for polyester resin due to removal of increasing amount of alkali-sensitive material with increasing strength of mercerization, leading to more and more amount of free cellulosic –OH groups available for bonding with resin. When the interaction is strong enough, moisture desorption became restricted and the enthalpy associated with it will increase. The first exothermic peak is due to cellulose decomposition and has been shifted to higher region with increased enthalpy for all the treated composites indicating more thermal stability [54]. The successive endothermic peak, responsible for resin decomposition, has been shifted toward lower temperature with an increased enthalpy for all the treated fiber composites (Table 18.3). The shifting of both the peak is possibly due to the formation of H-bonding between the cellulosic –OH of bamboo fiber and polyester resin. This is further supported by the increase in decomposition temperature of α-cellulose in case of composites reinforced with alkali-treated fibers. Similar types of DSC tracing were also exhibited with phenolic resin composites [55].

Comparing the DSC data from Table 18.3 for 20 % alkali-treated fiber composite with different resin, it is observed that the powdered novolac resin is providing the best thermal property.

Figure 18.5 represents the plots obtained from thermogravimetric analysis of parent bamboo, polyester resin, and all composite samples. The first peak intended for moisture loss is at 69.5 °C for BSPES-U and shifted to 74.5, 82.25, and 88.4 °C, respectively, for BSPES-10, BSPES-15, and BSPES-20. The trend of releasing moisture at a higher temperature with increased enthalpy value in case of the treated

Table 18.3 DSC data for 20 % alkali-treated fiber composite with different resin

Sample designation	Peak temp (°C)	Nature of peak	ΔH (J/g)
B-20 (20 % alkali-treated sample)	64.6	Endo	8.94
	345.3	Exo	19.5
	392.5	Exo	4.2
BSNC-20 (20 % alkali-treated fiber and novolac resin)	90.31	Endo	−57.55
	257.65	Endo	−5.45
	334.18	Exo	19.08
BSRC-20 (20 % alkali-treated fiber and resol resin)	82.31	Endo	−149.81
	251.56	Endo	−11.45
	328.24	Exo	30.19
	379.21	Exo	3.26
BSPES-20 (20 % alkali-treated fiber polyester resin)	84.31	Endo	−47.55
	332.69	Exo	60.01
	363.16	Endo	−64.08

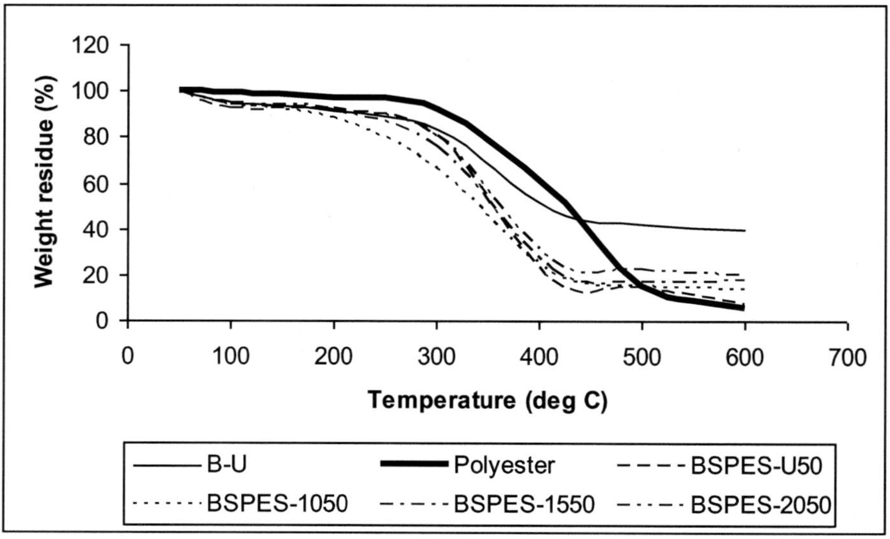

Fig. 18.5 TGA plot of untreated bamboo strips and polyester resin

composites could be due to the improved spreading of the resin over the fibrillated fibers resulting in a stronger bond at the interface. The next transition was observed at around 255 °C in case of composite samples which might be for degradation of hemicellulose [55]. The increasing amount of hemicelluloses leaching with increasing alkali treatment concentration is responsible for continuously decreasing energy and complete absence after 15 % alkali treatment. The maximum temperature for the third transition, which could be due to cellulose decomposition in the fiber part of the composites, is 384.5 °C for BSPES-U and has been shifted to 319.5 °C, 317.5 °C, and 316.65 °C, respectively, for BSPES-10, BSPES-15, and BSPES-20 indicating

Table 18.4 TGA data for 20 % alkali-treated fiber composite with different resin

Sample designation	No. of transition	Transition temperature (°C) T_i	T_m	T_f	Degradation at corresponding transition (%)	Residual weight (%) at 600 °C
Untreated bamboo fiber	1	50	91.03	150	7.65	40
	2	150	342.89	426.11	52.35	
BSNC-20 (20 % alkali-treated fiber)	1	50	86.05	189.48	6.5	36.62
	2	189.48	252.63	300.00	18.13	
	3	300	315.78	344.74	18.75	
	4	344.24	425.56	560.05	16.25	
BSRC-20 (20 % alkali-treated fiber)	1	50	92	190	4.5	47.37
	2	190	264	316	13.13	
	3	316	404	436	28.75	
	4	436	536	600	6.25	
BSPES-20 (20 % alkali-treated fiber)	1	50	88.4	233.3	8.75	20
	2	233.3	316.65	344.4	44.25	
	3	344.4	378.5	600	27	

increasing interaction with resin. A continuous increase in weight loss is observed. The treated fiber composites are showing higher thermal stability indicated by higher amount of residual weight at 600 °C for treated one than the untreated fiber composite. Comparing the TGA data from Table 18.4 for 20 % alkali-treated fiber composite with different resin, it is clear that the resol resin is providing the best thermal property.

Concluding Remarks

The potential applicability of bamboo fiber in polymeric composites has getting more and more attention. Due to a great number of attractive properties, availability, eco-friendly nature of bamboo, construction, automotive, furniture, and electronics, pharmaceuticals, cosmetics, and biomedical applications are also being considered [90]. Mechanical properties like as high strength and stiffness, the surface reactivity (with numerous hydroxyl groups), and the specific surface treatment may tailor composite material made with these bamboos for our growing need.

References

1. Van de Velde K, Kiekens P (1999) Wettability of natural fibres used as reinforcement for composites. Angew Makromo Chem 272:87–93
2. Eichhorn SJ, Baillie CA, Zafeiropoulos N, Mwaikambo LY, Ansell MP, Dufresne A, Entwistle KM, Herrera-Franco PJ, Escamilla GC, Groom L, Hughes M, Hill C, Rials TG, Wild PM (2001) Review: current international research into cellulosic fibres and composites. J Mater Sci 36:2107–2131
3. Gassan J, Vs G, Bledzki AK (2000) About the surface characteristics of natural fibres. Macromol Mater Eng 283:132–139

4. Bledzki AK, Gassan J (1999) Composites reinforced with cellulose based fibres. Prog Polym Sci 24:221–274
5. http://www.tifac.org.in/
6. Md R, Hamdan S, Ahmed AS, Islam MS, Talib ZA, Abdullah WFW, Mat MSC (2011) Thermogravimetric analysis and dynamic Young's modulus measurement of N, N-dimethylacetamide-impregnated wood polymer composites. J Vinyl Addit Technol 17:177–183
7. Guho Roy TK, Chatterjee SK, Adhikari D, Mukherjee AK (1988) Studies on the bleaching jute. J Textile Inst 79:108–125
8. Dash BN, Rana AK, Mishra SC, Misra HK, Nayak SK, Tripathy SS (1999) Novel low-cost jute-polyester composites, Part-I: processing, mechanical properties and SEM analysis. Polym Compos 20:62–71
9. Rao RMVGK, Balasubramanian N, Chand M (1981) Moisture absorption phenomenon in permeable fiber polymer composites. J Appl Polym Sci 26:4069–4079
10. Dash BN, Rana AK, Misra HK, Nayak SK, Tripathy SS (2000) Novel low-cost jute–polyester composites. III. Weathering and thermal behavior. J Appl Polym Sci 78:1671–1679
11. Bisanda ETN (2000) The effect of alkali treatment on the adhesion characteristics of sisal fibres. Appl Comp Mater 7:331–339
12. Chand N, Rohatgi PK (1986) Adhesion of sisal fibre-polyester system. Polym Commun 27:157–160
13. Prasad SV, Pavithran C, Rohatgi PK (1983) Alkali treatment of coir fibers for coir–polyester composites. J Mater Sci 18:1443–1454
14. http://sites01.lsu.edu/wp/graduateschool/thesis-and-dissertation-library/
15. Amada S, Ichikawa Y, Munekata T, Nagase V, Shimizu H (1997) Fiber texture and mechanical graded structure of bamboo. Compos Part B 28:13–20
16. Abd Latif M (1993) Effects of age and height on the machining properties of Malaysian bamboo. J Trop Forest Sci 5:528–535
17. Abd Latif M (1995) PhD thesis. Faculty Forestry, UPM. 281
18. Abdul Latif M, Khoo KC, Jamaluddin K, Jalil AA (1994) Fiber morphology and chemical properties of *Gigantochloa scortechinii*. J Trop Forest Sci 64:397–407
19. Okubo K, Fujii T, Yamashita N (2004) Development of bamboo-based polymer composites and their mechanical properties. Compos Part A Appl Sci Manuf 35:377–383
20. Janssen JJA (1995) Building with bamboo, 2nd edn. Intermediate Technology Publication Limited, London, p 65
21. Abdul Latif M, Khoo KC, NorAzah MA (1991) Proceedings of the 4th international bamboo workshop, Chiangmai, Thailand, 27–30 November, International Development Research Centre, Ottawa; Forestry Research Support Programme for Asia and the Pacific, Bangkok, 1994, p 227
22. Tono T, Ono K (1962) The layered structure and its morphological transformation by acid treatment. J Jpn Wood Res Soc 8:245–249
23. Parameswaran N, Liese W (1976) On the fine structure of bamboo fibres. Wood Sci Technol 10:231–246
24. Parameswaran N, Liese W (1980) Ultrastructural aspects of bamboo cells. Cellulose Chem Technol 14:587–609
25. Parameswaran N, Liese W (1981) 6–17 September. In: Higuchi T (eds) Bamboo production and utilization. Proceedings of the XVII IUFRO World Congress, Kyoto, Japan, Kyoto University, Kyoto, Japan, 1981, p 178
26. Fujii T (1985) Cell-wall structure of the culm of Azumanezaza (*Pleioblastus chino* Max.). Mokuzai Gakhaishi 31:865–872
27. Kawase K, Sato K, Imagawa H, Ujie M (1986) Studies on utilization of sasa-bamboos as forest. 4. Pulping of young culms and histological change of cell structure of the culms in growing process resources. Res Bull College Forest Hokkaido Univ Jpn 43:73–97
28. Murphy RJ, Alvin KL (1992) Variation in fibre wall structure of bamboo. IAWA Bull 13:403–410
29. Preston RD, Singh K (1950) The fine structures of bamboo fibres. J Exp Bot 1:214–230

30. Ray AK, Mondal S, Das SK, Ramachandrarao P (2005) Bamboo – a functionally graded composite-correlation between microstructure and mechanical strength. J Mater Sci 40:5249–5253
31. Low IM, Che ZY, Latella BA (2006) Mapping the structure, composition and mechanical properties of bamboo. J Mater Res 8:1969–1976
32. Ray D, Sarkar BK, Basak RK, Rana AK (2002) Study of thermal behaviour of alkali treated jute fibres. J Appl Polym Sci 85:2594–2599
33. Ray D, Sarkar BK, Rana AK, Bose NR (2000) The mechanical properties of vinylester resin matrix composites reinforced with alkali-treated jute fibres. Compos Part A 32:119–127
34. Ray D, Sarkar BK, Bose NR (2002) Impact fatigue behaviour of vinylester resin matrix composites reinforced with alkali treated jute fibres. Compos Part A 33:233–241
35. Ray D, Sarkar BK, Das S, Rana AK (2002) Dynamic mechanical and thermal analysis of vinyl ester-resin-matrix composites reinforced with untreated and alkali-treated jute fibres. Comp Sci Technol 62:911–917
36. Wang B, Panigrahi S, Tabil L, Crerar W (2004a) Effects of chemical treatments on mechanical and physical properties of flax fiber-reinforced rotationally molded composites. In: 2004 ASAE/CSAE annual international meeting, Paper Number 046083, Ottawa, Ontario, August 1–4. ASAE, St. Joseph
37. Gassan J, Mildner I, Bledzki AK (1999) Influence of fiber structure modification on the mechanical properties of flax fiber–epoxy composites. Compos Mater 35:435–440
38. Zeronian SH, Kawabata H, Alger KW (1990) Factors affecting the tensile properties of nonmercerized and mercerized cotton fibers. J Textile Res 60:179–183
39. Semsarzadeh MA (1986) Fiber matrix interactions in jute reinforced polyester resin. Polym Comp 7:23–25
40. Semsarzadeh MA, Lotfali AR, Mirazadeh H (1984) Jute reinforced polyester structures. Polym Comp 5:141–142
41. Ray PK, Chakravarty AC, Bandyopadhya SB (1976) Fine structure and mechanical properties of jute differently dried after retting. J Appl Polym Sci 20:1765–1767
42. Goa S, Zeng Y (1993) Surface modification of ultrahigh molecular weight polyethylene fibers by plasma treatment. I. Improving surface adhesion. J Appl Polym Sci 47:2065–2071
43. Mittal KL (1992) Silanes and other coupling agents. VSP: Utrecht, Netherlands 5:295–304
44. Nevell TP, Zeronian SH (1985) Cellulose chemistry and its application, vol 25. Halsted/Wiley, New York, p 552, 87–88
45. Ugbolue SCO (1990) Structure-property relationships in textile fibres. Textile Inst 20:1–43
46. Kroschwitz JI (1990) Polymers: fibers and textiles. A compendium. Wiley, New York, pp 634–669
47. Safonov VV (1991) Treatment of textile materials. Legprombitizdat, Moscow
48. Das M, Chakraborty D (2006) Influence of alkali treatment on the fine structure and morphology bamboo fibers. J Appl Polym Sci 102:5050–5056
49. Varma DS, Varma M, Varma IK (1984) Coir fibers. Part 1: effect of physical and chemical treatment on properties. Textile Res Inst 54:827–832
50. Das M, Chakraborty D (2008) Evaluation of improvement of physical and mechanical properties of bamboo fibers due to alkali treatment. J Appl Polym Sci 107(1):522–527
51. Das M, Chakraborty D (2006) Influence of mercerization on dynamic mechanical properties of bamboo – a natural lignocellulosic composite. Ind Eng Chem Res 45:6489–6492
52. Das M, Pal A, Chakrabarty D (2006) Effects of mercerization of the bamboo strips on the mechanical properties of uni-directional bamboo-novolac composites. J Appl Polym Sci 100:238–244
53. Das M, Chakraborty D (2007) Role of mercerization of the bamboo strips on the impact properties and morphology of uni-directional bamboo strips-novolac composites. Polym Comp 28:57–60
54. Das M, Chakraborty D (2009) Processing of the uni-directional powdered phenolic resin–bamboo fiber composites and resulting dynamic mechanical properties. J Reinf Plast Comps 28:1339–1348

55. Das M, Chakraborty D (2009) Thermo-gravimetric and weathering study of novolac resin composites: reinforced with mercerized bamboo fiber. Polym Comp 30:1408–1416
56. Das M, Chakraborty D (2008) Study of impact properties and morphology of uni-directional bamboo strips-polyester composites: Effect of mercerization. Comp Interf 15:829–839
57. Das M, Chakraborty D (2009) The effect of alkalization and fibre loading on the mechanical properties of bamboo fibre composites: part 2 – resol matrix. J Appl Polym Sci 112:447–453
58. Das M, Chakraborty D (2009) The effect of alkalization and fibre loading on the mechanical properties of bamboo fibre composites: part 1 – polyester resin matrix. J Appl Polym Sci 112:489–495
59. Saha SC, Ray PK, Pandey SN, Goswamy K (1991) IR and X-ray diffraction study of raw and chemically treated pineapple leaf fiber (PALF). J Polym Sci 42:2767–2772
60. Siddaramaiah KH (2005) Study of chemical and tensile properties of PU and PU/PS coated bamboo fibers. Polym Plast Technol Eng 44:1369–1377
61. Takagi H, Ichihara Y (2004) Effect of fiber length on mechanical properties of "Green" composites using a starch-based resin and short bamboo fibers. J JSME Int Ser A 47:551–555
62. An NH, Hoa TP, Chuong B (2007) Preparation of bamboo fiber polymer composite by vacuum injection molding technique and some properties of composite material. J Chem 45:214–220
63. Tung NH, Amamoto HY, Matsuoka T, Fujii T (2004) Effect of surface treatment on interfacial strength between bamboo fiber and PP resin. J JSME Int Ser A 47:561–565
64. Mi YL, Chen XY, Guo QP (1997) Bamboo fiber-reinforced polypropylene composites: crystallization and interfacial morphology. J Appl Polym Sci 64:1267–1273
65. Chen XY, Guo QP, Mi YL (1998) Bamboo fiber-reinforced polypropylene composites: a study of the mechanical properties. J Appl Polym Sci 69:1891–1899
66. Rajulu AV, Baksh SA, Reddy GR, Chary KN (1998) Chemical resistance and tensile properties of short bamboo fiber reinforced epoxy composites. J Reinf Plast Compos 17:1507–1511
67. Banga H, Singh VK, Choudhary SK (2015) Fabrication and study of mechanical properties of bamboo fibre reinforced bio-composites. Innov Syst Design Eng 6:84–98
68. Thwe MM, Liao K (2003) Durability of bamboo–glass fibre reinforced polymer matrix hybrid composite. Comp Sci Technol 63:375–387
69. Thwe MM, Liao K (2002) Effects of environmental aging on the mechanical properties of bamboo–glass fiber reinforced polymer matrix hybrid composites. Comp Part A 33:43–52
70. Li Z, Liu C, Yu T (2002) Laminate of reformed bamboo and extruded fiber-reinforced cementitious plate. J Mater Civil Eng 14:359–365
71. Yao W, Li Z (2003) Flexural behavior of bamboo fiber reinforced mortar laminates. Cem Concr Compos 33:15–19
72. Ghavami K (2005) Bamboo as reinforcement in structural concrete elements. Cem Concr Compos 27:637–649
73. Ghavami K, Filho RDT, Barbosa NP (1999) Behaviour of composite soil reinforced with natural fibres. Cem Concr Compos 21:39–48
74. Coutts RSP, Ni Y (1995) Autoclaved bamboo pulp fibre reinforced cement. Cem Concr Compos 17:99–106
75. Mansur MA, Aziz MA (1983) Study of bamboo-mesh reinforced cement composites. Int Cem Compos Lightw Concr 5:165–171
76. Ismail H, Shuhelmy S, Edyham MR (2002) The effects of a silane coupling agent on curing characteristics and mechanical properties of bamboo fibre filled natural rubber composites. J Eur Polym 38:39–47
77. Ismail H, Edyham MR, Wirjosentono B (2002) Bamboo fibre filled natural rubber composites: the effects of filler loading and bonding agent. Polym Test 21:39–144
78. Saxena M, Gowri VS (2003) Studies on bamboo polymer composites with polyester amide polyol as interfacial agent. Polym Comp 24:428–436
79. Zhang M, Kawai S, Sasaki H, Yamawaki T, Yoshida Y, Kashihara M (1995) Manufacture and properties of bamboo/wood composite fiberboard. Mokuzai Gakkaishi 41:903

80. Van AJ, Geyter DJD, Stevens SM (2000) Bamboo as a raw material for wood processing in Europe. Paper presented at the IUFRO meeting in Kuala Lumpur, Malaysia, August 2000
81. Rowell RM, Norimoto M (1987) Acetylation of bamboo fiber. Mokuzai Gakkaishi 33:907–910
82. Trujillo E, Osorio L, Van Vuure AW, Ivens J, Verpoest I (2010) Characterization of polymer composite materials. Based on bamboo fibres. 14th European conference on composite materials, 7–10 June 2010, Budapest. Paper ID: 344-ECCM14, pp 1–10
83. Takagi H, Mori H, Mori T, Sujito S, Pandey JK (2009) Development of high strength biomass composites made from bamboo. In: 18th International conference on composite materials, July 27th, 2009–July 31st, 2009, Edinburgh, Scotland
84. Lee SH, Ohkita T, Kitagawa K (2004) Eco-composite from poly(lactic acid) and bamboo fiber. Holzforschung 58:529–536
85. Lee SH, Wang S (2006) Biodegradable polymers/bamboo fiber biocomposite with bio-based coupling agent. Compos Part A 37:80–91
86. Okubo K, Fujii T, Yamashita N (2005) Improvement of interfacial adhesion in bamboo polymer composite enhanced with micro-fibrillated cellulose. JSME Int J Ser A Solid Mech Mater Eng 48:199–204
87. Despande AP, Rao MB, Rao CL (2000) Extraction of bamboo fibres and their use as reinforcement in polymeric composites. J Appl Polym Sci 76:83–92
88. Matsui H, Kitagawa K, Semba T, Okumura H, Ishiaku US, Hamada H (2004) Antec Conf Proc 62:1636
89. Rajulu AV, Devi RR, Devi LG (2005) Thermal degradation parameters of bamboo fiber reinforcement. J Reinf Plast Comp 24:1407–1411
90. Kalia S, Dufresne A, Bibin MC, Kaith BS, Averous L, Njuguna J, Nassiopouloset E (2011) Cellulose-based bio- and nanocomposites: a review. Int J Polym Sci 2011:1–35

Superhydrophobic and Superoleophobic Surfaces in Composite Materials

19

Oindrila Manna, Sarit K. Das, Raghunandan Sharma, and
Kamal K. Kar

Contents

O. Manna • S.K. Das (✉)
Department of Mechanical Engineering, Indian Institute of Technology Madras, Chennai, Tamil Nadu, India
e-mail: oindrilamanna@gmail.com; skdas@iitm.ac.in

R. Sharma
Advanced Nanoengineering Materials Laboratory, Materials Science Programme, Indian Institute of Technology Kanpur, Kanpur, Uttar Pradesh, India
e-mail: raghumsp@iitk.ac.in

K.K. Kar
Advanced Nanoengineering Materials Laboratory, Materials Science Programme, Indian Institute of Technology Kanpur, Kanpur, Uttar Pradesh, India

Advanced Nanoengineering Materials Laboratory, Department of Mechanical Engineering, Indian Institute of Technology Kanpur, Kanpur, Uttar Pradesh, India
e-mail: kamalkk@iitk.ac.in

K.K. Kar (ed.), *Composite Materials*,
DOI 10.1007/978-3-662-49514-8_19

Abstract

Liquid-repelling surfaces have received tremendous attention owing to their unique self-cleaning, anti-icing, anti-sticking, and antireflective properties. Various natural surfaces, owing to their specific surface structure, exhibit high water/oil repellency. Study of such naturally occurring superhydrophobic/superoleophobic surfaces has led to the understanding that the wettability of surfaces depends on their surface energy and surface structure. Surfaces with multilevel roughness exhibit high contact angles due to formation of air–liquid interfaces. The present chapter reviews the basic physical understanding and the structure–property correlations of the liquid-repelling surfaces. The chapter further explores the structures of various naturally occurring liquid-repellent surfaces and reports the recent progress achieved toward the development of artificial liquid-repellent surfaces by mimicking the natural ones. Finally, various techniques being employed to fabricate such surfaces have been described.

Keywords

Wetting • Hydrophobic • Oleophobic • Lotus effect • Surface morphology

Introduction

Mother Nature, the habitat of all living and inanimate objects on Earth, has evolved over 3.8 giga-years [1]. All living beings that exist on Earth are the end result of millenniums of experiments in nature's laboratory, scientifically delineated as evolution. Even though nature continues to evolve even today, the present status of our existence is the end (if not the best) result that nature has attained until now. Hence, it may be rationally assumed that nature has experienced all possible dilemma and hurdles and found optimal solutions to them in the course of time, under the given circumstances, somewhere along the way. In substance, nature may inspire solutions to several problems that we face now or may encounter in the future, only if we are clever enough to decipher the code and terminology of nature.

Wettability of surfaces has many applications: In nature, wettability is important for plants and animals for survival, self-cleaning, and predatory purposes or simply for keeping dry. In the industry, wettability plays an important role in water purification, fluid transport, food processing, power generation, transportation, architecture and infrastructure maintenance, corrosion prevention, etc. In addition, wettability is a phenomenon of great importance in the field of fluid and thermal sciences. Hence, it is important to understand the physics behind wettability and to find ways applying this knowledge for our development.

Physics of Wetting

Wettability may be defined as the dampening of a surface in the presence of a liquid due to intermolecular interactions between them. The degree of wettability is determined by the strength of adhesion (attractive force between unlike molecules) and cohesion (attractive force between like molecules) between the solid and liquid. The physical parameters that govern wettability are surface roughness and surface energy.

Contact Angle and Hysteresis

The prevalent way of characterizing wettability of a surface is the static contact angle (denoted by θ) for a drop of liquid deposited on the surface. It is the angle between the baseline of solid–liquid interface and tangent at the liquid–air interface, measured from the liquid phase (as shown in Fig. 19.1a). It is theoretically equated by Young's equation (as derived later in Eq. 19.3). It depends on several factors such as surface energy, surface roughness, and the cleanliness of the surface [2]. It is interesting to note that a liquid drop placed on a surface may show a range of θ values, varying by a few degrees. This is because θ also depends upon the way a drop is placed on the surface. However, a surface is characterized by the maximum and minimum θ values only. The maximum value of θ is termed as advancing contact angle and the minimum as receding contact angle. All other intermediate points are metastable.

Contact angle of a liquid on a solid varies as $0^\circ \leq \theta_c \leq 180^\circ$. Liquids are said to wet the surface if $\theta_c < 90^\circ$ and not if $\theta_c > 90^\circ$. Surfaces are considered superhydrophilic if $\theta_c < 10^\circ$ and superhydrophobic if $\theta_c > 150^\circ$, in the case of water. The same applies for oils. It should be mentioned that the terms hydrophobic/hydrophilic are restricted to water only, while oleophobic/oleophilic refers to oils, organic liquids, and petroleum products. Terms ambiphobic/ambiphilic refer to

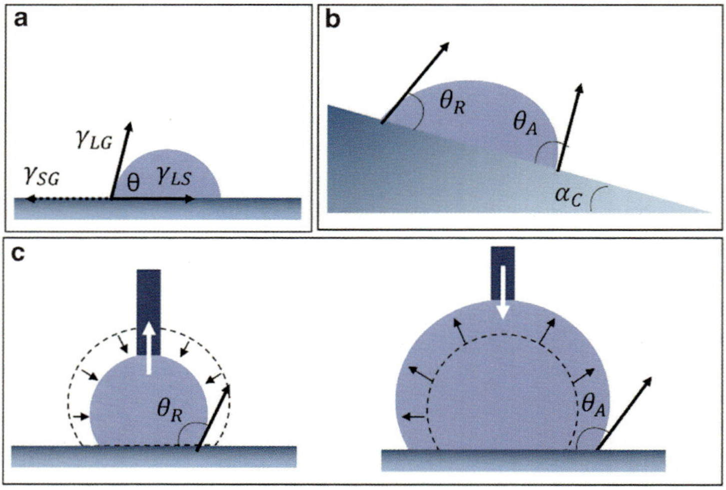

Fig. 19.1 Schematic showing (**a**) Young's contact angle and advancing and receding angle by (**b**) tilting method [3] and (**c**) droplet injection method [4] (Reprinted with permission from [3, 4])

everything within the extremes of water and oils. Surfaces can hence be classified into different degrees of wettability depending upon the nature of liquid and its contact angle.

The angle to which a surface is tilted for a drop to roll/slide is termed as the critical tilt angle. A surface can be characterized by the contact angles of a drop on a tilted surface. At the critical tilt angle, the leading edge of drop gives the advancing angle (θ_A) while the rear the receding angle (θ_R), as shown in Fig. 19.1b. It should be mentioned here that θ_A and θ_R can be measured in two ways, one of which is the tilting base method mentioned above. The second way of measuring θ_A and θ_R is by injecting a small volume of liquid into a previously deposited drop in a controlled manner and recording the increase in its contact angle as function of volume (this is θ_A). This is subsequently followed by decreasing the volume of the drop by sucking the liquid out of the drop and recording the decreasing contact angle as a function of volume (this is θ_R), as shown in Fig. 19.1c.

If $\theta_A > \theta_R$, then their difference is termed as the contact angle hysteresis (CAH). This is a measure of the irreversibility of the wetting/non-wetting process, due to dissipation of energy during movement of the solid–liquid interface line along the surface. CAH indicates the deviation of real surfaces from ideal, in terms of smoothness, rigidity, and chemical and mechanical heterogeneity. These include surface roughness, morphology/microporosity of the solid, reorientation of moieties at the surface, transfer of molecules from solid to liquid by surface diffusion, and evaporation. CAH determines the motion of a drop as at low CAH values, liquid drops tend to roll, while the drops slide at higher CAH values. However, drops may show combined rolling and sliding at low CAH and are favorable for self-cleaning surfaces. Real surfaces have nonzero CAH and surfaces with CAH $< 10^{\circ}$ show self-cleaning properties.

Surface Tension and Surface Free Energy

The units of surface free energy (J/m^2) and surface tensions (N/m) are equal. Thermodynamically, surface tension, surface tension force, surface energy density, and surface free energy are unique entities that share the similar units like pressure and stress or work and energy. However, in the study of wettability, these entities can be assumed equivalent provided temperature and pressure are kept constant and there is no adsorption at the interfaces [5]. Surface free energy and surface tension are the results of molecular interactions in matter. In order to lower their potential energy, molecules in bulk form bonds with their immediate neighbors. However, molecules at the surface are imbalanced due to lack of neighbors on all sides and are at higher energy states. This is why matters with free surface, such as liquids, have a tendency to convolute within themselves and attain a shape that offers the least surface area, a sphere. A non-wetting (low-energy) surface is at a lower energy state when dry, and hence, a liquid drop on such a surface assumes a spherical shape to decrease its solid–liquid interface as well as maintain the least liquid–air surface area. The vice versa is true for wettable (high-energy) surfaces where the drop assumes a flatter shape balanced by larger solid–liquid and lesser air–liquid interface. Roughness can also enhance the effect of surface energy. This is why rough low-energy surfaces show high liquid contact angles, while rough high-energy surfaces show low liquid contact angles. Intuitively, the surface tension of the liquid is also an important factor of wetting. Liquids with lower surface tension such as oils spread more easily since their energy requirement is lesser. Hence, surfaces that are even superhydrophobic may not repel oils. On the other hand, surfaces, which are oil repelling by virtue of surface energy, are most certainly water repelling, provided the surface does not contain chemically hydrophilic moieties.

Capillary Length, Contact Line, and Spreadability

The shape of a liquid drop is also a function of its length scale, which determined if it is under the influence of surface forces or gravity. As the size of a drop reduces, the dominant force changes from gravity to surface tension. If the magnitude of forces acting on a waterdrop is plotted (as shown in Fig. 19.2a), we observe a crossover at ~2.73 mm. This is the length scale of water and is termed as capillary length. Capillary length is defined as the characteristic length applicable for interfaces between two liquids, which are subject to these forces. It is denoted by l_c and equals $\sqrt{\frac{\gamma_{lv}}{\rho g}}$. Drops having size smaller than this are shaped by surface tension and unaffected by gravity. Larger drops are de-shaped by gravity and egg shaped [6].

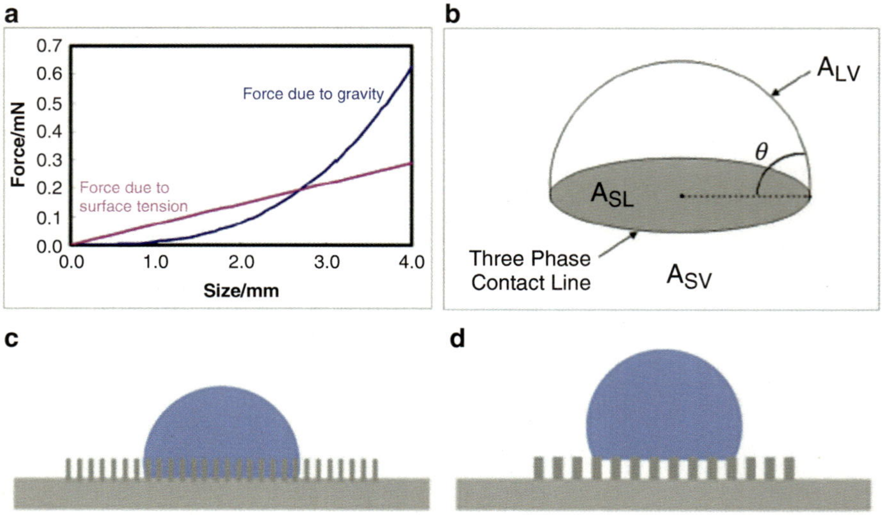

Fig. 19.2 (**a**) Variation of surface and gravity forces in increasing length scale (for water) [6], (**b**) schematic showing 3D contact line of droplet [5], (**c**) Wenzel state, and (**d**) Cassie state (Reprinted with permission from [5, 6])

The three-phase contact line is the interface along which the solid, liquid, and gas (air) coexist (Fig. 19.2b). It is a 3D spread of the contact angle, determined by the interplay of interfacial surface tensions γ_{sl}, γ_{lv}, and γ_{sv}. Each of these interfacial tensions has energy costs since at the boundaries, the chemical bonds of a phase with its own atoms and molecules are missing.

Droplets resting on a surface can be pulled into a film or remain as droplets depending on the balance of the interfacial surface tensions in play. This condition is governed by spreadability of the surface (S), given by Eq. 19.1 [7, 8]:

$$S = \gamma sv - (\gamma sl + \gamma lv) \tag{19.1}$$

S is a measure of the energy cost per unit area for a dry surface to be wetted by a liquid film. If $S > 0$, wetting is favored; else, the wetting is finite and the liquid forms a definite θ_c. Hydro-/oleophilic materials hence have $S > 0$ and vice versa.

Theoretical Wetting Models

Young's Equation: Wettability was first scrupulously investigated in 1805 by Young while studying wetting on ideal surfaces, that is, surfaces that possess neither physical nor chemical heterogeneities. A liquid drop touches such an ideal surface in the shape of a disk, with the solid–liquid interface enclosed within the solid–liquid–air interface. Young substantiated that the shape of a liquid drop is governed by minimization of total Gibbs free energy (Young [9]). This is equal to the summation of solid–liquid, liquid–vapor, and solid–vapor interfacial free energies (denoted by γ_{sl}, γ_{lv}, and γ_{sv}, respectively). Let us consider a liquid drop on such an ideal solid surface. Let the solid–liquid interface area be increased by dA. Hence, the interface energies increase by $\gamma_{sl}\, dA$ and $\gamma_{lv} \cos\theta\, dA$ while decrease by $-\gamma_{sv}\, dA$. It may be noted that liquid–vapor interfacial energy γ_{lv} depends on the shape of the drop, which is accounted for by dA cos θ. Hence, the net change in interfacial energy is given by Eq. 19.2 [10]:

$$dE = \gamma_{sl} dA - \gamma_{sv} dA + \gamma_{lv} \cos\theta dA \tag{19.2}$$

Now for the drop to come to rest and assume a stable contact angle θ_c, Gibbs free energy has to be minimized. Differentiating Eq. 19.2 with respect to A and equating it to zero (at minima), the classical Young's equation (Eq. 19.3), which correlates the surface energies with the equilibrium contact angle, θ_Y (also shown in Fig. 19.1a), is obtained [10]:

$$\gamma_{sl} - \gamma_{sv} + \gamma_{lv} \cos\theta_Y = 0 \tag{19.3}$$

If $\theta_c < \theta_Y$ for a liquid drop on a solid surface, the non-wetted region becomes unstable. Therefore, the energy contribution due to solid–liquid interfacial area increases with corresponding decrease in solid–vapor interfacial area by increasing the radius of the drop till $\theta_c = \theta_Y$. Similarly, if $\theta_c > \theta_Y$, the wetted region becomes unstable. Thus, energy contribution due to solid–liquid contact decreases with corresponding increase in solid–vapor interfacial area by decreasing the radius of the drop till $\theta_c = \theta_Y$. This definition is however limited to ideal surfaces only.

Wenzel Model: Wenzel reported wetting of real surfaces that differ from real surface in terms of surface heterogeneities [11]. It was suggested that liquid drops on real surfaces are impaled into the morphology of its surface roughness (Fig. 19.2c). Wenzel proposed a modified Young's equation that accounted for real surfaces in terms of a "roughness factor" and provided a relation between surface energy, roughness, and contact angle (Eq. 19.4) [11]:

$$\cos\theta_W = r \cos\theta_Y \tag{19.4}$$

Here, θ_W is the Wenzel/actual contact angle on rough surfaces, θ_Y is the Young's/equilibrium contact angle, and r is the ratio of water–solid contact area to the projected water–solid area. For smooth surfaces, $r = 1$, and hence $\theta_W = \theta_Y$. On the other hand, for rough surfaces, $r > 1$, and hence $\theta_W > \theta_Y$ when $\theta_Y > 90°$ (hydrophobic surfaces) and $\theta_W < \theta_Y$ when $\theta_Y < 90°$ (hydrophilic surfaces). Hence, when roughness of a hydrophobic surface is increased ($r > 1$), it becomes more hydrophobic. Conversely, hydrophilic surfaces become more hydrophilic when roughness is increased. However, the Wenzel model is limited to homogeneous wetting only.

Cassie–Baxter Model: Cassie and Baxter attempted to explain the wetting phenomenon observed in textiles and bird/duck feathers, which could not be explained by the Wenzel model [12]. They proposed that liquid drops do not always get impaled but may also stay suspended on the apex of the surface morphology. This is known as heterogeneous wetting as the droplet is supported at places by the solid and cushioned by entrapped air pockets in the morphology, elsewhere (as shown in Fig. 19.2d). Their model explained the tendency of the waterdrops to minimize its liquid–solid as well as liquid–air interface allowing it to attain a spherical shape. The mathematical model (Eq. 19.5) was derived by incorporating a modified factor for surface roughness in the Young's equation [7]:

$$f \cos \theta_C = f(\cos \theta_Y + 1) - 1 \tag{19.5}$$

Here, θ_C is the Cassie–Baxter/actual contact angle, and f is the ratio of water–solid contact area to projected water area. Hence, as roughness increases, f decreases, while θ_C increases. The Cassie–Baxter and Wenzel models are limited to understanding static wetting only. However, there are a number of static phenomena that cannot be explained by these models such as shape of a drop when its size is much larger than that of roughness scale, transition between homogeneous and heterogeneous wetting and its causes, etc. Most importantly, these models cannot explain the dynamics of wetting.

Other Models: McCarthy et al. showed the Cassie–Baxter model is not well rounded as it fails to take into account the motion of contact line (when a drop is moving on a surface) [13]. They went on to explain how the structure of the three-phase contact line is important for understanding dynamic wetting (deliberated in terms of advancing and receding contact angles and contact angle hysteresis). However, the nature of motion of the three-phase contact line is not yet well understood.

To summarize the above, Young's equation forms the basis of wetting theory for ideal surfaces. The Wenzel model takes into consideration the roughness of real surfaces and proposes a model for homogeneous wetting. Later, the Cassie–Baxter model improves our understanding of heterogeneous wetting states. However, these theories are not able to explain statics as well as dynamics satisfactorily. Models attempting to explain dynamic wetting are being developed.

Naturally Occurring Superhydrophobic Surfaces

In nature, there are several examples of terrestrial and aquatic plants/animals that exhibit water repellence mostly for drying, self-cleaning, or predatory purposes. Leaves are important to plant since they are the primary site for photosynthesis and transpiration. It is important for plants to maintain clean and dry leaf surfaces for these processes. The most famous example of wettability in nature is the mysteriously clean lotus leaf, whose reputation of non-contamination has often elevated it to symbols of pious purity. Barthlott et al. examined with the scanning electron

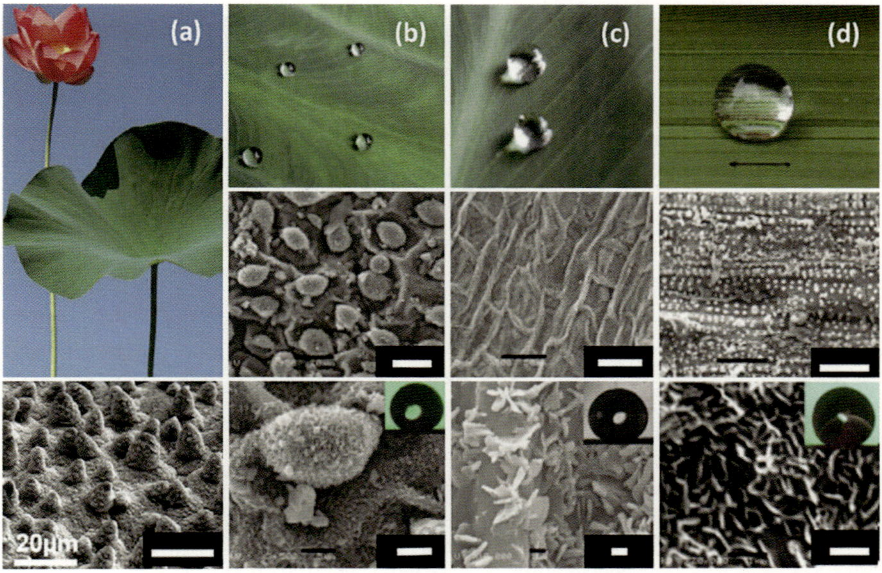

Fig. 19.3 Images of plant leaves and their SEM images at different magnifications. (**a**) Lotus leaf [17], (**b**) taro leaf [5], (**c**) Indian canna [5], and (**d**) rice leaf [5] (Reprinted with permission from [5, 17])

microscope (SEM) and finally explained the mysterious properties of the lotus (*Nelumbo nucifera*) leaves [14]. Further research revealed the surfaces were a composite of microscaled cuticles and nanoscaled wax crystals. These scales combined to form a hierarchical topography, which, along with the hydrophobic nature of the wax, led to water-repelling properties. Water droplets on the leaf surface cannot infiltrate the morphology, but sit on the apex of the surface roughness while air in the cavities cushioned them. The drop is therefore suspended in the Cassie state. The surface roughness also causes solid contaminants like dust to settle on its apex and, when the leaves sway in breeze or because of ripples, sets the waterdrops rolling. The rolling drops then pick up any contaminants in its way, cleaning up the surface. Water is reported to have ~162° contact angle on lotus leaves. This self-cleaning ability of lotus leaves is often termed as "lotus effect." A similar self-cleaning effect is observed in taro (*Colocasia esculenta*) leaves, which contain elliptical surface protrusions (diameter ~10 μm) covered in nanoscaled pins that impart a hierarchical roughness. Water is reported to have ~159° contact angle [15]. The Indian canna (*Canna generalis* Bailey) and the rice leaves also show similar water repellence. This is due to their microbinary–nanobinary morphology [16]. Some images of hydrophobic and superhydrophobic plant leaves and their surface morphology (using SEM) have been shown in Fig. 19.3.

It may hence be noted that non-wetting of surfaces is dependent on the hierarchical morphology of surfaces. However, this morphology has many variations in

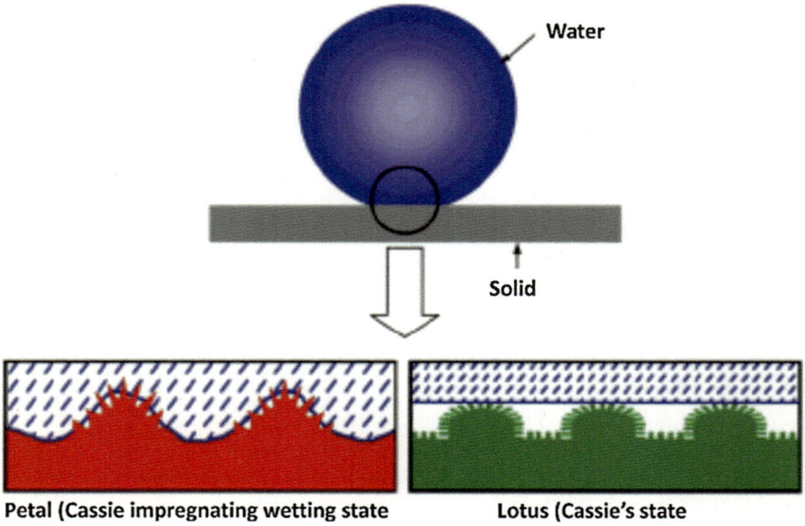

Fig. 19.4 Schematic showing the difference in morphology for petal effect and lotus effect (Reprinted with permission from [19])

nature. For instance, among the leaves already described, the rice plant leaves are known to have critical tilt angles ranging between 4° and 12°, because of anisotropic morphology [16]. The lotus leaves are however reported to show 2° critical tilt angles independent of direction by virtue of their isotropic morphology [18].

The hierarchical morphology may also show different kinds of wetting that do not conform to high contact angles and low critical tilt angles. Petals of flowers, such as the red rose (rosea Rehd.), show high contact angles as well as pinning, such that the waterdrops refuse to roll off even when held vertically. The surface of the petal is reported to be covered in bumps (larger than that of lotus leaves), which is infiltrated with water when deposited on it. The bumps are however covered in finer structures that give the petals its hierarchical structure, which cannot be infiltrated by water (showed in Fig. 19.4). Hence, a modified Cassie state is reached, which gives the petals this unique wetting property [19].

Nonetheless, hierarchical structure and surface roughness leading to water-repelling properties are not simply restricted to plants but also found in animals (also shown in Fig. 19.5). For example, an interesting phenomenon is observed among the insects, water striders, or pond skaters (*Gerris lacustris*) on the surface of open water bodies. The striders appear to defy the laws of physics by not only running across water but also standing on it. Gao et al. reported that the strider's feet are covered with oriented hair and fine nanogroves, covered in cuticle wax [20]. This special architecture allows the striders to trap air bubbles between its feet and water surface and provide enough buoyancy to support their weight. Another interesting feat of nature is the water collection ability of the silk webs of the cribellate spiders [21]. The structure of silk fibers consists of puffs composed of randomly oriented

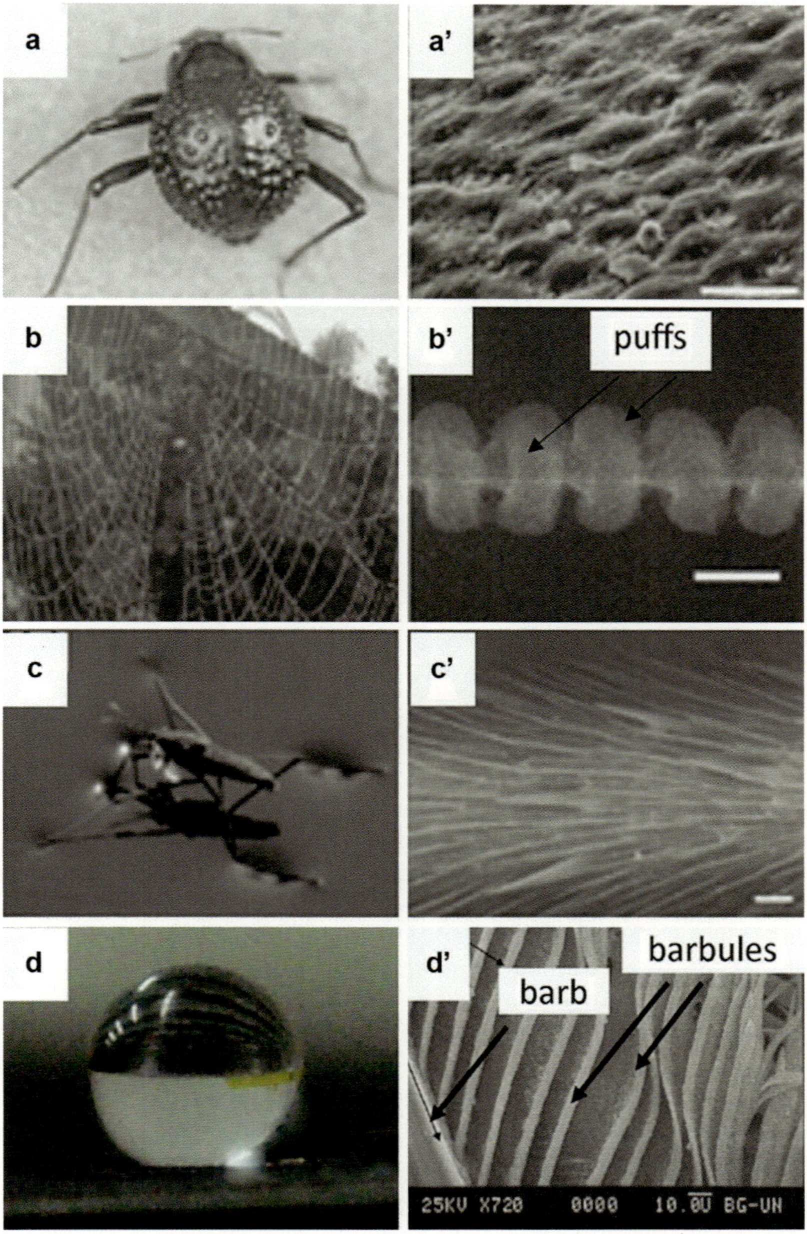

Fig. 19.5 Examples of a few liquid-repelling natural surfaces from nature. (**a**) Namib Desert beetle and (**a'**) SEM image of (**a**) showing macro-scaled bumps [22]. (**b**) Silk webs of the cribellate spiders and (**b'**) their low-magnification SEM image [21]. (**c**) Pond skater on water surface and (**c'**) SEM image of their foot [20, 25]. (**d**) Water droplet on pigeon feather and (**d'**) SEM image showing barb and barbule structures of pigeon feather [24] (Reprinted with permission from [20–22, 24, 25])

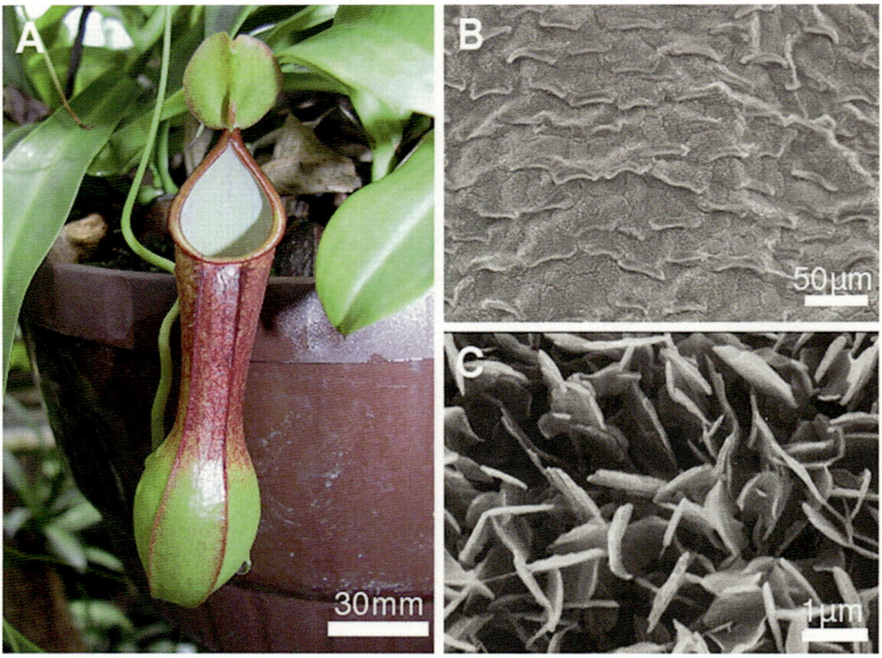

Fig. 19.6 Surface of pitcher plants: (**a**) as seen by the naked eye and SEM images of (**b**) the conductive zone from an air-dried pitcher and (**c**) the outermost parts of the epicuticular wax platelets [27] (Reprinted with permission from [27])

nanofibrils, which are hydrophilic in nature (Fig. 19.5c'). Dry spider webs are reported to undergo changes in their structure when dampened by moist air. The process starts with tiny drops of water condensing on the web fibers, structured similar to the beads on a string. As the beads come in contact with water, they shrink to form a directional slope. The beads also aid the formatin of Laplace pressure differences. As the drops grow larger, the combined driving forces overcome its resisting hysteresis and allow it to coalesce and flow in the direction of the gradient. A similar phenomenon is observed among the desert beetles in Namib Desert, which are known to drink water that condenses on their backs [22]. Among them, the tenebrionid beetle (*Stenocara* sp.) is known to tilt its body against the wind, which aids water droplets in the air to condense on its fused wings. This is covered with macro-scaled bumps whose apexes are hydrophilic and troughs covered with microscaled roughness that makes them hydrophobic. The water droplets start condensing at the apex, and after sufficient growth, they encounter the hydrophobic parts, which aid their flow. Once the droplets are large enough to overcome the hysteresis, it flows into the beetle's mouth parts. Wetting in case of bird feathers is slightly more complex with wettability being an integral effect of the texture of the wings and the oils generated by the glands [23]. Pigeon feathers, having a textured structure consisting of network formed by barbs and barbules (Fig. 19.5e), exhibit Cassie–Baxter wetting with high contact angles [24].

However, surface morphology is not only restricted to hydrophobicity in nature but may also include hydrophilicity. A particularly intriguing wetting phenomenon is the insect capturing mechanism of carnivorous plants using wettability. These plants are known to exploit hydrophilicity since they grow in the tropical humid belts, for survival. Two common mechanisms of prey capture among such plants are sticky and slippery surfaces. For example, the carnivorous plants sundew (*Drosera*) and butterworts (*Pinguicula*) are known to secrete sticky enzymes and adhesives, which attract preys by giving the illusion of water droplets or wet surfaces. Once the insect touches the sticky surface, they get stuck in the viscous enzymes and are digested by the plant [2]. On the other hand, the genus *Nepenthes* of pitcher plants, which are native of the tropics, use bright colors or nectar to attract prey into their slippery cupped leaves, referred to as pitchers. The rims of these pitchers become slippery when wetted by rain, condensation, or nectar. The insects perching on them tend to slip into the cavity aided by the plant secretions, waxy surfaces, and surface morphology, which slopes inward [27]. This also prevents the insects from flying or crawling out of the pitchers, where they are gradually digested by the enzymes or other means. Figure 19.6 shows scanning electron microscopy images of the surface of pitcher plants at different magnifications. Hence, it may be inferred that water-repelling or water-attracting properties in nature are, predominantly, the result of congenially hierarchical surface morphology in micron- and nanoscale.

Controlling Surface Energy of Liquid-Repelling Surfaces

Fluorocarbons

Polymers with fluorine side chains are an attractive class of chemicals for liquid repellence because of their extremely low surface energies. This, in combination with surface roughness, has often been studied and reported in literature. One of the most commonly used fluorinated polymers for this purpose is polytetrafluoroethylene or Teflon. Zhang et al. reported the superhydrophobic properties of Teflon films when stretched to induce void spaces in the otherwise continuous film in the form of tears, holes, and protruding fibrous crystals [28]. This generated the necessary surface morphology for transitioning the hydrophobic Teflon to superhydrophobic. The study reported the increase in water contact angle on the stretched films, from 118° to 165° when the films were stretched to almost twice the original. Shiu et al. reported treating Teflon films with oxygen plasma to induce roughness resulting in water contact angles of 168° [29].

Since fluorinated polymers are not readily soluble in most solvents, they are often linked or blended with other materials to tailor non-wetting properties. Yabu et al. created a porous superhydrophobic membrane by casting a fluorinated block polymer solution under controlled evaporation and humid conditions [30]. The study showed the effect of pore size of the honeycombed film in terms of water repellence and transparency. Xu et al. reported polypyrrole films, which, by virtue of inherent electrical conductivity, can transition between superhydrophobic and superhydrophilic [31]. The substrate was created by a combination of electropolymerization and chemical polymerization to introduce roughness on the polypyrrole film surface doped with perfluorooctane sulfonate (PFOS). The transition in wettability occurred because the oxidation state at the surface changed depending on the applied electrochemical potential.

Silicones

Besides polymers with fluorinated side chains, others with repeating units of siloxanes, commonly known as silicones, also possess very low surface energies and are used to achieve non-wetting surfaces. A commonly used silicone is the polydimethylsiloxane (PDMS), and its use is prevalent in existing literature [32].

Khorasani et al. reported treating PDMS with CO_2 pulsed laser to generate surface porosity and achieving water contact angles of almost 175° [33]. The study reasoned that the superhydrophobicity was an effect of both surface porosity and chain ordering inherent in PDMS. Sun et al. also reported the use of actual lotus leaves to form its negative templates using PDMS and using this casting to reproduce the original morphology of the lotus leaf by using DMS again [34]. Although the replica showed comparable water repellence as the lotus leaf, it was expected to be better since the surface energy of the paraffinic wax on the leaves is 30–32 mN/m, while that of PDMS is ~20 mN/m.

Apart from the direct use of PDMS, the use of its block copolymers has also been reported. For example, Ma et al. reported superhydrophobic electrospun fiber mats using poly(styrene-b-dimethylsiloxane) (PS-PDMS) block copolymer blended with polystyrene (PS) [35]. The study reports water contact angles of ~163° due to a combination of low surface energy of PDMS and surface features arising from the polymer fibers (diameter between 150 and 400 nm). The advantage of the hydrophobic fiber mats lies in their flexibility and porosity allowing them to be of use in the textile industry. Some surfaces and fibers developed in the described studies have been shown in Fig. 19.7.

Other Organic Materials

Along with fluorocarbons and silicones, there are a few other examples of nature, which possess non-wetting properties such as paraffinic waxes. Lu and coworkers showed superhydrophobic properties of the porous surface polyethylene (PE) achieved by controlled crystallization by introducing a non-solvent (cyclohexanone) into the PE solution in xylene [36]. This aided the formation of surface floral patterns in nanoscale and endowed the surfaces with water contact angles of ~173°. Also Jiang et al. showed similar hydrophobic properties of polystyrene (PS) mats obtained by electrospinning its solution in dimethylformamide (DMF), resulting in a fabric of microfibers–nanofibers [37].

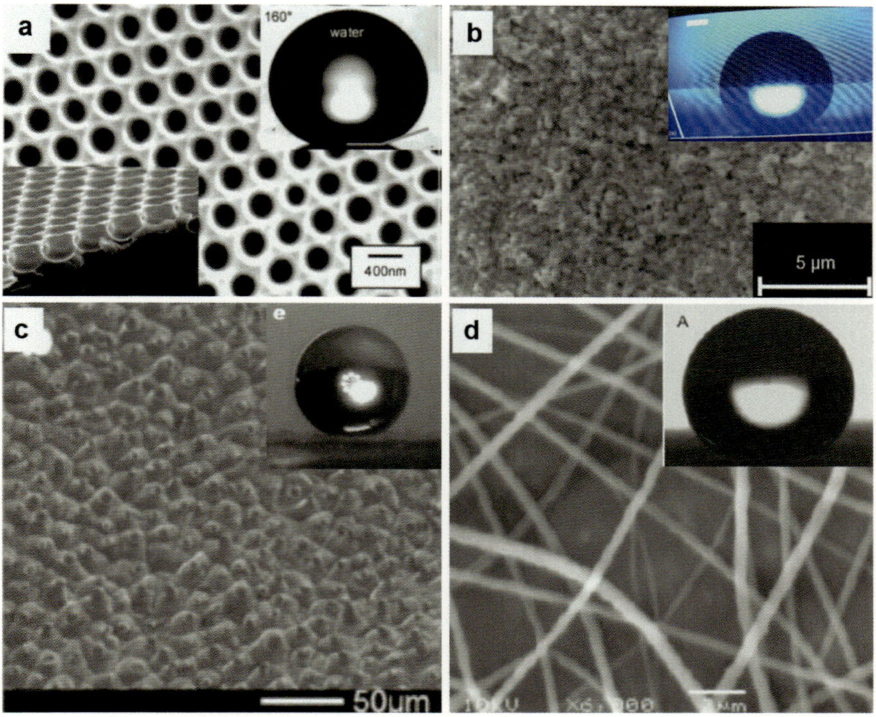

Fig. 19.7 Images of morphologies of surfaces coated with low surface energy materials and fabricated using diverse techniques. (**a**) Honeycomb-patterned fluorinated polymer films [30], (**b**) PDMS surface treated with CO_2 pulsed laser [33], (**c**) artificial lotus leaf structure of PDMS [34], and (**d**) superhydrophobic electrospun fiber mats of PS-PDMS [35]. *Inset* images show the contact angles for water (Reprinted with permission from [30, 33–35])

Other examples of non-wetting organic materials include polyamide [38], polycarbonate [39], and alkylketene dimer [40]. It may be noted here that most of these materials are accompanied with necessary surface morphological features, to achieve the desired degree of non-wetting.

Inorganic Materials

Besides organic materials, there are a few examples of inorganic materials possessing non-wetting properties. For example, Feng et al. demonstrated the use of ZnO nanorods for tunable wetting properties [41]. ZnO when in the (001) plane possesses very low surface energy which in combination with surface roughness leads to superhydrophobic properties. However, when this surface is exposed to UV radiation, generation of electron–hole pairs leads to adsorption of hydroxyl at the surface, making it superhydrophilic. Yang et al. again demonstrated the use of ZnO agave-like nanowire structures on the surface leading to superhydrophobic properties [42]. Another inorganic material, which demonstrates tunable wettability, is TiO_2 or titania. Feng et al. demonstrated the superhydrophobic properties of glass coated with nanoscaled TiO_2 rods created by a low-temperature hydrothermal approach exploiting the tendency of TiO_2 to grow in a particular plane orientation (110) [43]. The study also demonstrated the transition of the surface from superhydrophobic to superhydrophilic when exposed to UV radiation and reversal after being kept in the dark for ~2 weeks.

Composites

Apart from individual organic/inorganic substances, coatings of the composites consisting of two or more phases have been widely investigated to form superhydrophobic surfaces. As is known widely, the naturally occurring superhydrophobic surfaces have multiscale surface roughness, which, along with the low surface energies of the constituent materials, is responsible for their non-wetting properties. The surface roughness and hence the hydrophobicity of a surface can be modified effectively by applying coatings of composites such as inorganic/polymeric micro-/nanosized particles dispersed in a low surface energy polymeric matrix. The particles provide desired surface roughness, while the polymeric matrix, apart from providing low surface energy, acts as a binder to the micro-/nanosized particles as well. Apart from this, the superhydrophobic coatings need to be stable under harsh environmental conditions such as variable temperatures, pH values, etc., depending on their applications. A better combination of non-wetting properties and durability can be attained by the use of the composite coatings.

The superhydrophobic composite coatings are generally formed by incorporation of nanosized inclusions in the polymeric matrices. Micro-/nanosized fillers of various types, such as polymers, ceramics (silica, TiO_2, ZnO, etc.), CNTs, graphene, etc., can be used to develop superhydrophobic surfaces. For example, Zhang et al. reported simultaneous superhydrophobic and superoleophilic natures of a composite consisting of porous polyurethane (PU) film combined with polystyrene (PS) microspheres [44]. Similarly, silica nanoparticles have been used widely to modify the surface structure [45, 46]. Fabrication of surfaces having multilevel surface roughness, which could mimic naturally occurring superhydrophobic surfaces such as lotus leaves, has been attained by using such composite coatings. Cui et al. prepared multiscale structures on microstructured epoxy paint surface by coating nanosilica/epoxy. The prepared composite surface exhibited high contact angle as high as 167.8°, along with excellent stability in neutral and basic aqueous solutions [45]. Similarly, superhydrophobic membranes having multiscale roughness were fabricated by using TiO_2 nanoparticles to modify the roughness of microporous PVDF membranes [47].

Besides polymeric and ceramic fillers, carbon nanomaterials, such as CNTs and graphene, being mechanically robust, chemically inert, and hydrophobic in nature, may be used to develop highly durable superhydrophobic composite coatings. Lau et al. reported the superhydrophobic nature of aligned CNT surface coated with

PTFE, attributed to the inherent nanoscaled roughness of the CNT surface and the hydrophobic nature of PTFE [48]. Similarly, Wang et al. reported such superhydrophobic coatings of composites consisting of randomly oriented CNTs and a fluoroplastic such as perfluoroalkoxy alkane (PFA) resin, prepared by spray coating a mixture of CNT dispersion and resin [49]. Again, hierarchical structures mimicking natural hydrophobic surfaces have also been fabricated by using CNT microstructures. For example, Jung et al. reported hierarchical CNT/epoxy resin composite structures by spray coating of the CNT/epoxy resin on a microstructured substrate of epoxy. The superhydrophobic CNT composite structures exhibited high mechanical stability and a high contact angle of 170° [50]. The hierarchical structure was found to have superior stability compared to nanostructures made of lotus wax.

Graphene, owing to its excellent mechanical properties and chemical inertness, has also found significant attention toward its application in superhydrophobic surfaces having high chemical inertness and stability. Zha et al. fabricated PVDF/graphene porous composites with a low (~1 wt%) loading of graphene. The contact angle for PVDF/graphene composite in water was measured to be ~152° [51]. The hydrophobicity may be enhanced further by functionalization of graphene with various hydrophobic functional groups. Other strategies to attain high contact angles and superior stabilities may include the coatings of polymeric matrix composites consisting of multiple types of particulate fillers. For example, Asmatulu et al. have reported superhydrophobic electrospun PS and PVC composite fibers containing TiO_2 nanoparticles and graphene nanoflakes and reported a contact angle of ~178° in water [52].

Special Design Requirements for Superoleophobic Surfaces

A lot of progress has been made in the design and fabrication of hydrophobic or superhydrophobic materials. The same principles and methods may be used to make oleophobic or superoleophobic materials [53].

Overhang Surfaces

Existing literature shows transition of hydrophilic material to superhydrophobic by introducing micron- or nanosized textures with overhang structures [54, 55]. The overhang structures prevent water from surface morphology by virtue of capillary forces, hence suspending the water droplets in a metastable Cassie state (also schematically represented in Fig. 19.8a). The same principle has been reported to be applicable for fabrication of superoleophobic surfaces. Cao et al. investigated this by depositing silica and polystyrene nanoparticles in layer-by-layer (LBL) deposition method [56]. Subsequently, the polystyrene was removed by calcination leaving behind a bed of silica particles with intermittent void spaces that formed the necessary overhang structures. The surfaces were reported to show water contact angles above 150° and repelled hexadecane (surface tension ~27 mN/m) with

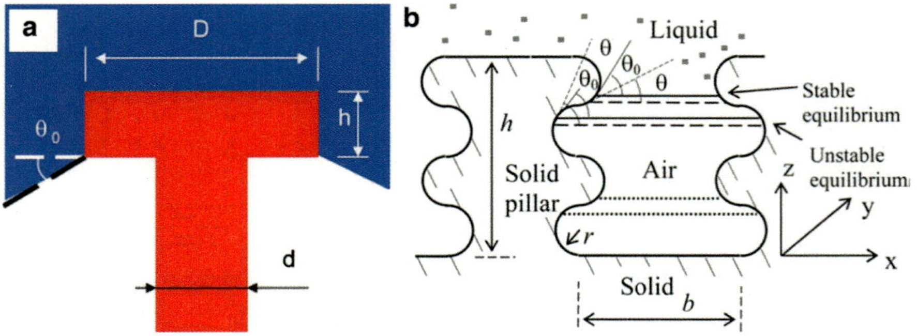

Fig. 19.8 Schematic of (**a**) overhang structures [65] and (**b**) reentrant structure [61] necessary for oil-repelling surfaces (Reprinted with permission from [61, 65])

contact angles of ~90°. Kumar et al. also showed the overhang structures aiding superamphiphobic properties [57]. The study showed silicon surfaces coated with thin layer of hydrophobic material etched using SF_6 and oxygen gas in an ion etching technique to form T-shaped nanorods or nanograss. These structures prevent liquids including oils from entering the surface pores aided by capillary forces and air pockets, hence making the surfaces amphiphobic. Cao et al. reported fabrication of porous silicon films using a gold-assisted electroless etching process that lead to a hierarchical morphology of micron-sized asperities imposed on nanoscaled pores [58]. This type of overhang structures was reasoned prevented oil and water from wetting the surfaces. Aulin et al. also reported overhang structures resulting from micron-sized pores with nanoscaled pillars or wires fabricated by plasma etching process [59]. This surface when coated with fluorinated trichlorosilicones was able to repel both water and oils, but is otherwise oleophilic. The study also demonstrated fabrication of porous cellulose films using a structured silicon surface and coating it with monolayer of the same.

It may be inferred that overhang structures can prevent low surface tension liquids such as oils from penetrating the surface texture. This along with the cushioning of droplets by the air–solid composite can suspend droplets in metastable Cassie state, hence making the surface apparently non-wettable. However, quantitative evaluation of the stability of these metastable states along with the dynamic behavior of droplets needs further investigation.

Reentrant Surfaces

Hydrophobicity due to surface roughness is intrinsic and involves the effect of its manifold scale ranges. Liquid repellence is characterized by the ability of a surface to achieve high contact angle and low contact angle hysteresis, and the predominant factor for this is formation of a composite solid–liquid–air interface in the form of air pockets in the valleys between the roughness apexes. The lesser the interface is between solid and liquid, the lesser is the adhesion between droplet and surface leading to lower contact angle hysteresis [60, 61]. This composite interface may however be metastable and irreversibly transform into a homogeneous solid–liquid interface (also schematically represented in Fig. 19.8b). To prevent this, recent literature contains suggestions of reentrant surface structures consisting of semicircular ridges and troughs that prevent the liquid from entering all the crevices of the surface.

Nosonovsky reasoned that since the interface depends on both the size of droplet and length scales of surface roughness, multiscale roughness is a necessary condition of stability [61]. The study also demonstrated the improved stability for surfaces covered with convex structures that pinned the interface in place and prevented liquid from entering the crevices, hence stabilizing its suspension. Tuteja et al. emphasized the importance of reentrant surface features necessary for stable interfaces and fabrication of superoleophobic surfaces, in addition to low surface energy and roughness [62]. The study reported the synthesis of types of fluorinated polyhedral oligomeric silsesquioxane (fluorooctyl and fluorodecyl POSS) which have very low surface energy, blended with polymethyl methacrylate (PMMA) for the substrate PMMA-POSS. The reentrant surfaces were created by electrospinning and compared with spin-coated flat surfaces to show improvement. Joly et al. explained the principle of reentrant surfaces in terms of surface free energy and numerically modeled the same [63]. It may hence be inferred that reentrant surfaces aid formation of large static contact angles and low sliding angles, particularly aided by their top curvatures [64].

Naturally Occurring Superoleophobic Surfaces

It is known that surface tension of oils is usually lower than that of water. It is difficult to come across surfaces in nature that have sufficiently low surface energies or intricate morphology capable of repelling oils. Developments of superoleophobic materials require a combination of surface energy and unique surface morphology termed as reentrant surfaces. Reentrant surfaces are structures whose morphologies are such that the cross section at the inlet is small compared to its body and are commonly found in the form of arrays of T- or mushroom like structures. These surfaces are capable of trapping air inside their constricted entry that is capable of forming a negative Laplace pressure that forms a convex liquid–air interface instead of concave, hence restricting infiltration by liquids. The usual method of achieving surfaces having sufficiently low surface energies is to coat them with artificial chemicals such as fluorosilanes and silicones.

There exist a few examples of natural materials exhibiting superoleophobicity, though their functionality being limited to the environment, direction, etc. Some of such examples have been shown in Fig. 19.9. Surprisingly again, of the most available examples in nature is the lotus leaf, bottom surface of which exhibits superoleophobic properties. It was reported that the underside of lotus leaf is covered

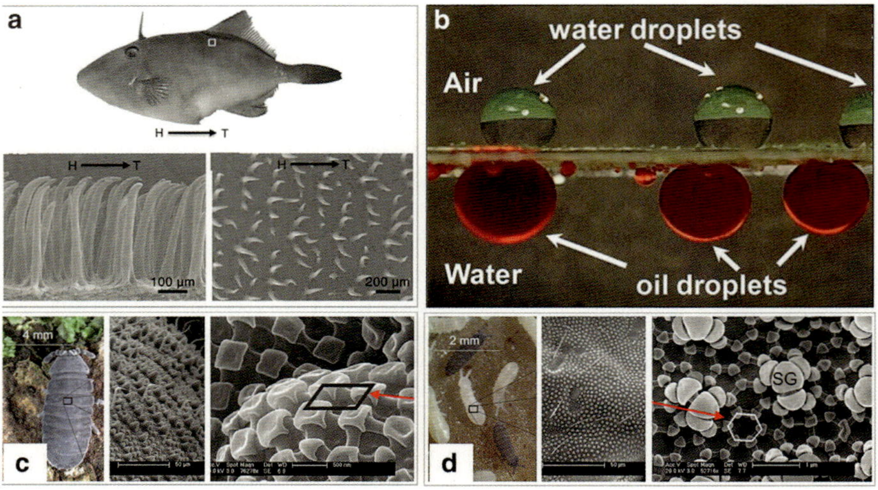

Fig. 19.9 Images of oil-repelling surfaces in nature. (**a**) Filefish [72], (**b**) both sides of lotus leaf (with the underside showing oil repellence in water) [66], (**c**) *Tetrodontophora bielanensis* springtail with its regular comblike rhombic patterns and (**d**) *Orthonychiurus stachianus* springtails with their regular comblike hexagonal patterns [70] (Reprinted with permission from [66, 70, 72])

with micro-papillae and nanogroves (similar to red rose petals) as opposed to the micro-papillae – nanoprojections and wax crystal morphology of the top surface [66]. Another example of underwater superoleophobicity is fishes. The skin of filefish (*Navodon septentrionalis*) has a unique morphology consisting of hooklike spines growing in a particular direction that allow oils to slip from the fish's head to tail. The skin is however ineffective in any other orientation, allowing the fish to survive in man-made calamities like oil spills by keeping its head and gills clean [67, 68]. Unfortunately, surfaces which are oleophobic in water are not so in air, and there are very few natural examples of the latter. For example, a few parts of leafhopper (Hemiptera, Cicadellidae) are covered with bronchosomes, the highly structured particles made of polar protein molecules [69]. These particles, having a hollow constitution (200–700 nm in diameter) of honeycomb structure, are found to be loosely attached to the insect bodies and are erodible, giving the ants a chance to escape when captured by predators that employ adhesive tactics. Another interesting example of oleophobicity is the springtail worms (some species of Collembola) exploiting complex surface morphology. These insects, which live in the soil and breathe through their skin, are found to be covered in regular geometric comblike patterns (rhombic or hexagonal, as shown in Fig. 19.9c, d). These structures form a negative overhang that is characteristic of reentrant surfaces [70]. Some of these naturally oil-repelling examples in nature have been shown in Fig. 19.9.

It may again be noted that not all the specialized surface morphologies result in specialized wettability, as in the case of shark skin. The shark's body is covered in toothlike scales (or denticles) that are ribbed with longitudinal grooves. These groves are aligned such that water flows along them, allowing them to reduce any vortex associated with the shark's movement. This enables sharks to experience reduced drag and hence move faster [71].

Other Synthetic Liquid-Repelling Surfaces

As established so far, Wenzel and Cassie–Baxter are mutually exclusive wetting states. It has been investigated that the former may make a transition into the latter with increase in surface roughness. The point of transition is analytically [60] proved by equating the Wenzel and Cassie–Baxter equations as $\theta_W = \theta_C$. Therefore, $\cos\theta_Y = \frac{f-1}{r-f}$ [3]. It may be noted that since $r > 1$ and $f < 1$, by simple estimation, $\theta_Y > 90^{\circ}$. This condition, however, implies that transition from Wenzel to Cassie may take place only if the material is non-wetting for the liquid, i.e., $\theta_Y > 90^{\circ}$. It also demonstrates the difficulty of developing oleophobic surfaces since materials with $\theta_Y > 90^{\circ}$ for oils and other low surface tension liquids are rarely natural. It is interesting to note that Tuteja et al. suggested that this handicap be overcome by the inclusion of "reentrant" surfaces in addition to surface roughness and surface energy, as necessary parameters for developing oleophobic surfaces [62].

Experimentally, there have been many studies establishing that higher roughness increases the water contact angle for non-wetting surfaces and decreases for wetting surfaces. Yost et al. established that smooth copper surface which has a contact angle of ~15–20° with Sn–Pb decreases with increase in roughness of the surface [73]. Shibuichi et al. established that with increasing roughness, wettable liquid contact angle decreases, while with decreasing roughness, non-wettable liquid contact angle increases, by studying the wettability of various mixtures of water and 1,4-dioxane on alkylketene dimer (AKD) substrate [74]. Semal et al. studied the dynamic spreading of a sessile drop of squalane on a heterogeneous substrate to establish that increase in microroughness decreases the rate of dynamic spreading [75]. Similarly, Burton et al. established that for patterned surfaces, with increase in roughness, contact angle decreased for hydrophilic surfaces and increased for hydrophobic surfaces [76]. Koch et al. investigated the influence of micro-, nano-, and hierarchical roughness on wetting parameters (contact angle, hysteresis, tilt angle, etc.) and established that increasing roughness indeed increases hydrophobicity of surfaces [77].

Similar to water, lower surface tension liquids such as oils have been investigated for factors that modify its wettability. Numerous studies have established the effect of surface chemistry or energy in this matter. However, Tuteja et al. established the importance of "reentrant" surface curvatures in addition to surface chemistry and roughness that are essential for successful oleophobic surfaces [62]. Liu et al. showed that hydrophilic oleophilic surfaces may be altered to hydrophilic oleophobic surfaces underwater [26].

Fabrication Technology of Liquid-Repelling Surfaces

Liquid-repelling surfaces in nature are mostly characterized by micro- or nanoroughness and intricate morphology. Another important feature is multiscaled roughness often observed as nanoprojections of micron-scaled morphological features. Reproducing such features on artificial surfaces requires modern fabrication technologies. Some of such fabrication techniques and final surfaces have been described in the following section.

Imprinting Methods

One way of achieving superhydrophobic surface is conventional imprinting techniques (involving a master and replica) such as lithography and templating. Such methods can be independent or used in combination with others to minimize the complexity of the total fabrication process. For instance, a substrate with micron-scaled morphology can be created using conventional lithography and modified with nanopores or pillars by using plasma treatment. The final substrate may even be used as a master for future replicas.

Lithography: The basic principle of lithography involves copying information from the master to the replica, which may be an exact or negative reproduction depending on the desired outcome. Lithographic techniques can be broadly classified into subcategories depending on the method employed, nature of substrate, power source, etc. A few popular categories include optical or photolithography, X-ray lithography, electron beam lithography, soft lithography, and nanolithography.

The most common lithography technique is photolithography, usually done on silicon wafers. It involves copying the geometric patterns of a "photomask" onto a photosensitive substrate using light energy (such as ultraviolet, X-ray, or electron beams) as a function of time and space. In case scattered light is replaced with UV, X-ray, or electron beams, the lithographic process is named accordingly. Conventionally, the process starts with cleaning the substrate to remove organic and inorganic impurities before coating it with a photosensitive material termed as "photoresist." Photoresist can be used to form positive or negative replicas. Positive replicas are formed if the resist is exposed to light at areas where the underlying substrate needs to be removed. These areas become more soluble and are washed away easily when the sample is cleaned with developer solution. In case of negative replicas, the resist polymerizes and becomes insoluble in the light-exposed areas. Hence, a negative impression is created when the substrate is cleaned with the developer solution. The substrate is finally treated to harden the photoresist and improves its adhesion to the base [78–80].

Soft lithography requires mechanically soft or elastomeric materials for substrate such as polydimethylsiloxane (PDMS). It can be classified into further categories of patterning, replica molding, and embossing. In the case of patterning, the protruding parts of the elastomeric "stamp" are inked with the material to be transferred and pressed onto the substrate. The substrate can then be soft stamped and can then be peeled off leaving the substrate inked in the desired pattern. The second category of replica molding involves transferring a pattern from the master by allowing a liquefied melt to solidify when in contact with the mold. The replica can then be peeled off. Soft lithography can be further categorized as microtransfer molding, micromolding via capillaries, and UV molding. In the case of the former, the mold is usually made of PDMS so that it can be peeled off and the liquefied melt usually is cured thermally. In the case of micromolding via capillaries, the mold containing capillary channels is placed facedown on the substrate and the liquefied melt placed at the channel ends. Once the channel is completely flooded, the melt is cured and the master removed. In the case of UV molding, the master is usually of hard material. It is filled with the liquefied melt that cures when exposed to UV radiation and covered with a transparent substrate, before curing. The top substrate and master are removed at the end of the process. The devices used in soft lithography consist of electromeric stamps, molds, masks, etc. [81]. Other lithographic techniques include nanoimprint lithography [82], electron beam lithography [83], and colloidal lithography [84].

As already mentioned, one lithographic process by itself is generally incapable of developing the desired surface morphology all by itself. This is because different techniques have different accuracies and advantages at different length scales and on different substrates. This is best emphasized when fabricating dual-scaled or hierarchical roughness on solids. For example, Jeong et al. reported using a UV-assisted capillary micromolding soft lithography method for fabricating hydrophobic surfaces [85]. The study reported the creation of two different molds made of polyurethane acrylate (PUA) and PDMS. The microtextured PUA mold was created by pouring the polymer melt onto a silicon master (which was fabricated by photolithography or electron beam lithography) and covered with a film of optically transparent polyethylene terephthalate (PET), before curing by UV radiation. The mold was subsequently peeled off the master. The nanotextured PDMS mold was created by casting it on a similar photolithography-created silicon master and solidified using a chemical curing agent. The actual hierarchical structure was created by a two-step process. The microtextured mold was placed on a silicon or polyethylene terephthalate (PET) substrate spin coated with a PUA resin which responds to UV. After the melt infiltrated the mold's cavities, it was partially cured by exposure to UV. The specific requirements for the curing process allow the base of the melt to be cured, but only partially cure the melt within the cavities. The mold is then removed and replaced with the other, which is nanotextured at a low pressure. The unsolidified resin is shaped by the nanosized cavities and wholly cured by adequate UV exposure. The surface morphology of the final surfaces, as revealed by SEM imaging, has been shown in Fig. 19.10. There are numerous other examples of lithographic techniques used in combination to fabricate surfaces of desired morphologies [18].

Templating: By principle, templating is similar to the manufacturing process of molding. It involves fabricating a template master, using this to mold the replica, and separating it on completion. Often the original prototypes used for making the templates include parts of living organisms with desirable surface features. Gecko feet [86], reptile skin [87], plant leaves [88], etc. have been widely investigated as

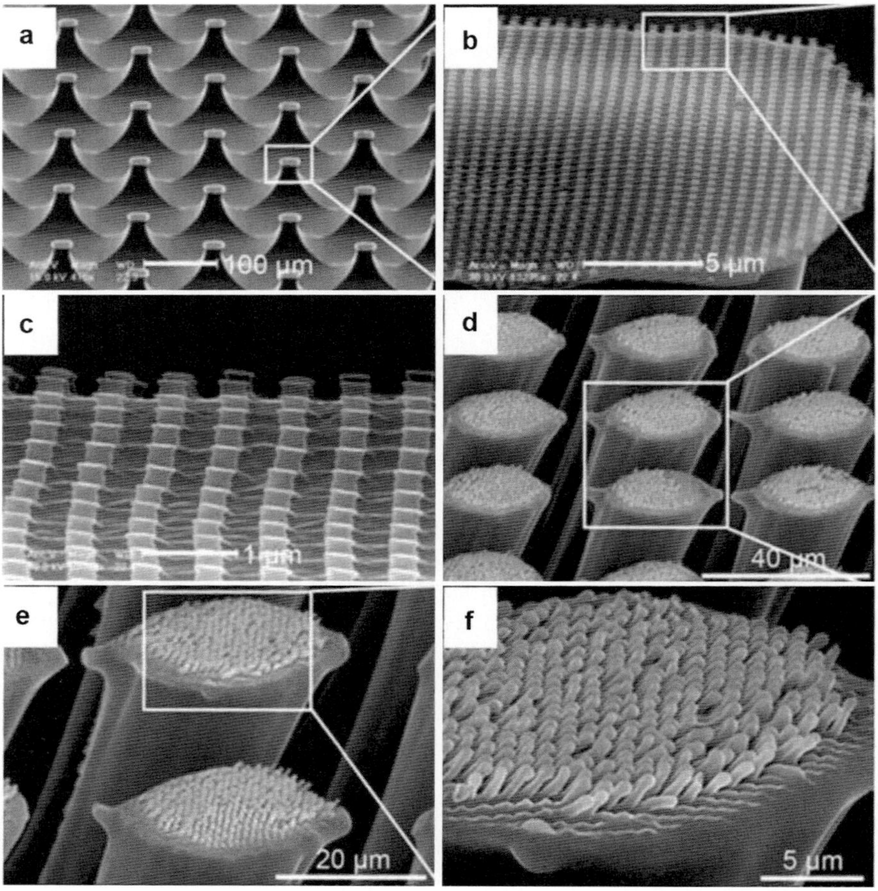

Fig. 19.10 SEM images of the superhydrophobic surfaces developed by using UV-assisted capillary micromolding soft lithography (Reprinted with permission from [54])

templates to fabricate superhydrophobic surfaces. Cho et al. attempted to fabricate superhydrophobic surfaces by mimicking the hairy features on gecko's feet [86]. The study reported the use of nanotextured anodic aluminum oxide membranes as the template because of the ease of tailoring the pillar sizes, diameters, spacing, etc. by changing the parameters of the anodization process. These membranes were created by anodizing polished aluminum surfaces in H_3PO_4 to remove the resident protective oxide layer. This resulted in a textured pure aluminum surface, which, when anodized again, produced a coating of anodic aluminum oxide (AAO), uniform to the surface resulting in cylindrical pillars whose centers coincided with that of the concaves. This served as the template in the study and was spin coated with the precursor polymer solution of hard PDMS (h-PDMS). The h-PDMS was covered with vinyl-terminated glass slide to facilitate easy removal of the PDMS after curing.

The final PDMS replica was reported to show water contact angles of above 150°. In addition, Yuan et al. reported the use of taro leaves for templating superhydrophobic surfaces [88]. The study reported coating a taro leaf with a solution of PDMS and a catalyzer. This was cured and carefully peeled off to be used as the negative template. The template was then cast with polystyrene (PS) dissolved in tetrahydrofuran and allowed to dry in air. The h-PDMS template was then peeled off the solidified PS replica, which was reported to show water contact angles of ~158° and sliding angles of 3°.

It may be inferred that templating is a simple fabrication process that can be used to replicate regular surface patterns. Depending on the quality of the template, it can also be used to create a large number of replicas and reproduce micro- and nanostructures and even hierarchical morphologies depending on the casting material provided caution is exercised. The technique however might not be able to comply surfaces with excessively complex features since they can cause damage during peeling off.

Plasma Treatment: Plasma treatment is often used as an additional surface patterning technique before [89, 90] or after [91, 92] the actual fabrication process. It may even be used intermittently during other processes [93].

Berendsen et al. reported superhydrophobic surfaces using thermoplastic polymers embossing submicron features using a heated nickel stamp, followed by surface coating with fluorocarbon compounds [89]. The nickel stamp was fabricated by laser interference lithography followed by electroplating to contain the desired submicron features. The stamp was then pressed onto a film of polystyrene (PS) and heated to emboss the pattern. The fabrication was completed after the PS film then underwent plasma deposition of a hydrophobic fluorocarbon. The films were reported to show water contact angles of 167°. Tsougeni et al. reported the use of plasma processing technique to fabricate polymeric microfluidics using polymethyl methacrylate (PMMA) and poly(ether ether ketone) (PEEK) [92]. The fabrication was initiated by spin coating the PMMA or PEEK sheets with a photoresist film (either a thin inorganic Si-containing photosensitive PDMS or organic–organic polymer, ORMOCER, or instead an independent thick layer of organic photoresist) which was thermally cured. This was exposed to UV through a photomask and allowed to wash and develop in solutions of methyl isobutyl ketone, isopropyl alcohol, and tetramethylammonium hydroxide, to get the positive impressions of the mask. This was followed by deep O_2 plasma etching to develop different selectivities for the residual photoresist and the bare polymer substrate because of different etching rates. The developed channels are reported to be superhydrophilic. The study also reported that parts of the channel could be made superhydrophobic using C_4F_8 fluorocarbon plasma deposition method and stencil masks.

Chemical Deposition

Chemical deposition of thin films on selective substrates is another popular fabrication method. Some common deposition techniques include chemical vapor

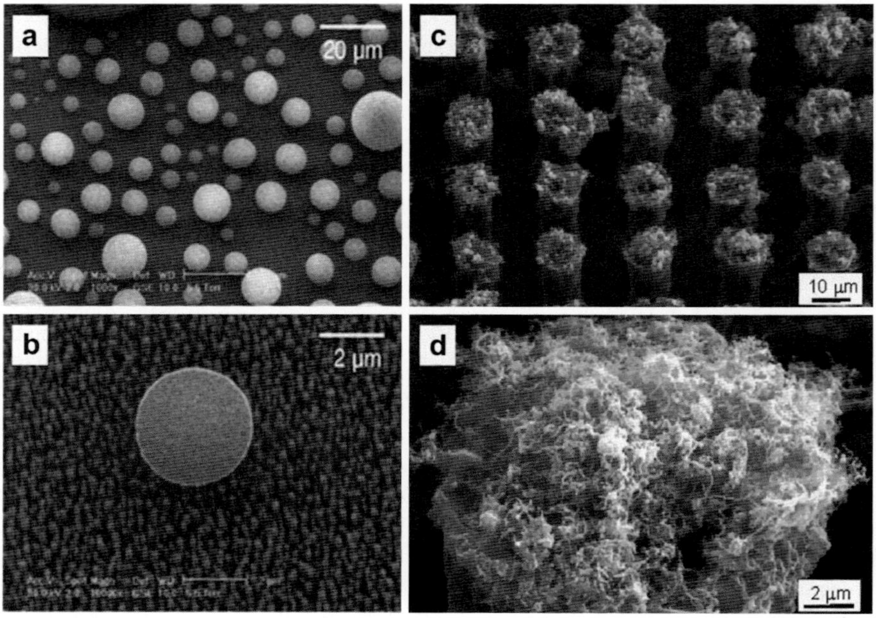

Fig. 19.11 (**a**, **b**) SEM images of micron-sized water droplets on aligned CNT (PECVD synthesized)/PTFE composite (tilt angle 15°) [48]. (**c**, **d**) SEM images of hierarchical CNT structures fabricated by spray coating of CCVD-synthesized CNTs on epoxy resin-patterned substrate (tilt angle 45°) [50] (Reprinted with permission from [48, 50])

deposition (CVD), electrochemical deposition, and LBL deposition, as explained in the following sections.

CVD-Based Surface Treatment: The CVD process involves exposing a reactive substrate surface to a gaseous precursor in order to allow the deposition of a film or layer of the latter. There may be chemical reactions involved during this process.

Recently, a plasma-enhanced chemical vapor deposition (PECVD) has garnered a lot of attention from fabrication of superhydrophobic surfaces as reported by Borras et al. [94]. The study reported Ag/TiO_2 core–shell nanofibers prepared using low-temperature PECVD method. The fibers were fabricated on an inner nanocrystalline silver thread, coated with a TiO_2 layer by plasma deposited using a titanium tetraisopropoxide (TTIP) precursor and O_2 or Ar + O_2 plasma. The developed fibers were reported to reach water contact angles of almost 180°. The water contact angle was however reported to be a function of both the width of the fibers and their concentration. The fibers were also reported to become superhydrophilic when irradiated with UV light because of change in the crystalline state of TiO_2 along with size of the individual TiO_2 domains covering the fibers. Similarly, vertically aligned CNTs grown on a nickel-coated silicon substrate by PECVD using DC plasma discharge of acetylene and ammonia have been used to form superhydrophobic composite by applying a coating of PTFE on the top surface of

CNTs [48]. Figure 19.11a exhibits the SEM images of micron-sized water droplets (tilt angle 15°) on the superhydrophobic aligned CNT/PTFE composite.

Besides PECVD, catalyst-assisted chemical vapor deposition (CCVD) is another recently developed chemical deposition method. Jung et al. reported multiwalled carbon nanotubes (CNTs) which are known to show high water contact angles and low tilt angles [50]. The study reports the use of iron catalysts to initiate the growth of nanotubes with natural gas as the carbon source and Ar/H_2 used as the buffer gas. The multiwalled CNT was then sonicated in acetone to form a dispersion and sprayed onto an epoxy resin film. This was heated to elevated temperatures such that the epoxy that covered the fibers could melt and move into the crevices while exposing the CNT at the surface. The study was inspired by the hierarchical structure observed in lotus leaves and superhydrophobic water contact angles, good mechanical durability, and resistance to wear and friction (shown in Fig. 19.11b).

CVD-based methods were frequently used to produce superhydrophobic surfaces depending on their applicability to material and ability to maintain or enhance the mechanical strength of the formed structure. They are also reported to be able to create complex surface structures up to the nanoscale and are hence very useful. However, their drawback lies in the extensive operation procedures involved and the difficulty of controlling the exact details of deposition at individual spot on the surface.

Layer-by-Layer (LBL) Deposition: LBL deposition is a fabrication technique for coating substrates with thin films. Early literature reports numerous studies of multilayer assembly [95, 96], but it is only in the last two decades that further developments have been made [97]. The LBL method is a periodic process, which involves the formation of layer by adsorption of charged material onto a substrate. Subsequent layer is formed by adsorption of an oppositely charged material on top of the previous layer. The result is a single bilayer whose thickness is on the nanoscale. This deposition can then be repeated periodically to attain a desired thickness [98]. The films in certain cases can even sustain themselves without the original substrate to form membranes [99] or freestanding films [100]. The LBL technique can be practiced outside the domain of electrostatic interactions as well, such as the molecular interactions of covalent or hydrogen bonding. It is applicable for a wide range of materials such as polymers including proteins, lipids, and nucleic acids [101].

Wu et al. reported the fabrication of thin film Bragg stacks using an aqueous layer-by-layer assembly method [102]. The process was initiated by dipping glass slides or silicon wafers in a solution of polyelectrolyte for adsorption, serving as the charged base or substrate. The substrate was then alternately dipped in poly (diallyldimethylammonium chloride) (PDAC), which served as the cationic solution, and poly(sodium 4-styrenesulfonate) (SPS), which served as the anionic solution, and the process included intermediate wash steps. After four bilayers of PDAC and SPS were formed, the substrate was alternatively dipped in a dispersion of TiO_2 nanoparticles and SPS to form another 30 bilayers. This was followed by heating to elevated temperatures to melt the polymer. Further alternate dipping in a dispersion of SiO_2 nanoparticles and SPS was carried out to generate another

20 bilayers. This was again heated to elevated temperature to melt the polymer. In this, TiO_2 and SiO_2 nanoparticles were deposited in layers, and the process is repeated according to the desired thickness or effect. The objective of the study was to study the color and structural characteristics of the nanoporous Bragg stacks. The study also reported the stacks to be superhydrophilic, and they were studied for self-cleaning properties.

The widespread use of LBL assembly in various fields requiring different process requirements and substrates has led to the development of a variety of LBL techniques such as dipping, centrifugation, spinning, high gravity, spraying, electrodeposition, and magnetic assembly, among others. The assemble method is however important for determining process properties as well as physicochemical properties of the films like thickness, homogeneity, etc. [99].

Colloidal Assembly and Aggregation

Colloidal assembly is a fabrication process that depends on the formation of assemblies of monodispersed particles or colloidal particles, aided by chemical bonding or van der Waals forces. If the sizes of the colloidal particles are in different length scales, even hierarchical structures can be formed. The process has been reported to be initiated by dip- or spin coating the substrate with a solution of the particles [103, 104] and allowing the solvent to dry and the particulates to aggregate due to capillary forces [105]. Aggregations may also be due to internal interactions of the suspended particulates such as molecular and van der Waals forces. Zhu et al. reported superhydrophobic dandelion-like 3D microstructure that was fabricated by the self-assembly of 1D nanofibers of polyaniline (PANI) in the presence of perfluorosebacic acid (PFSEA) [106]. The study reported that the PANI 1D nanofiber self-assembly was driven by the combined interactions of hydrogen bonding, π–π stacking, and hydrophobic interactions. Xue et al. reported superhydrophobic surfaces using epoxy-functionalized cotton, which was customized, coated with amino- and epoxy-functionalized silica nanoparticles in order to achieve a dual-sized hierarchical structure [107]. The study reported the covalent bonding of the silica particles with the cotton fibers and the outer surface of the fibers rich with epoxy groups that could be used for grafting. The cotton was made hydrophobic by grafting its rough surface using stearic acid or 1H, 1H, 2H, 2H-perfluorodecyltrichlorosilane (PFTDS) or a combination of the two.

Hence, it may be inferred that colloidal assembly techniques are effective in modifying the surface energy as well as modifying surface roughness to change the wettability of a surface. However, to do so, extensive knowledge of the chemistry involved is vital. Nonetheless, the colloidal assembling technique is relatively more economical and efficient when compared to other traditional fabrication methods [108].

Phase Separation: Colloidal aggregation is known to be used in combination with phase separation method, which involves the separation of the solid phase from its unstable mixture by changing the surrounding conditions of temperature, pressure, etc. for fabricating non-wetting surfaces. The length scale of the formed surface structure and roughness may be in macroscopic, microscopic, and even nanoscopic scale [109, 110]. This method is also often used to assist other methods such as solgel [111], plasma treatment [112], electrospinning [34], and self-aggregation, to make superhydrophobic surfaces. Note that the phase separation processes involved in these various methods are more or less connected to colloidal polymerization.

Electrospinning and Electrospraying

Electrospinning is a simple fabrication process that can form continuous polymer fibers in the micron- as well as nanoscale [113], which can be assembled to form surfaces with inherent surface roughness [62]. The process of electrospraying although similar can fabricate films ranging between beads and fibers. Conventionally, electrospinning is associated with fibers while electrospraying to beads [114].

Ding et al. reported the fabrication of superhydrophobic surfaces using electrospinning [115]. The study showed electrospinning an aqueous solution of polyvinyl alcohol (PVA) and zinc acetate on a grounded and rotating aluminum foil-wrapped cylindrical object. The resulting fibrous composite film was then calcinated. Similarly, a pure PVA nanofibrous film was electrospun, while a pure zinc oxide nanofibrous film was created using the solgel method described later. The process ended with calcinating all three films and modifying their wettability by coating with fluoroalkylsilanes (FAS) in hexane. The study found that while the pure PVA film remained superhydrophilic in spite of the silane modification, the composite and pure ZnO films showed enhanced hydrophilicity of 132° and 165°, respectively. It is interesting to note that the ZnO film without the silane coating was reported to show water contact angles of 0°, since it shows the effect of surface energy on wettability of a surface. Electrospinning techniques have also been reported to enhance the mechanical strength of the films when two or more materials are blended, similar to composites [116].

Burkarter et al. showed the superhydrophobic properties of electrosprayed films of polytetrafluoroethylene (PTFE) fabricated using an aqueous solution of the same [117]. The coating was done on conducting fluorine-doped tin oxide (FTO)-coated glass slides as substrates. The study reported that in order to aid the solvent, the substrate was heated to 150 °C using a hot plate to aid the water evaporation upon coming in contact with the substrate. Typical deposition times ranged from 30 s to 20 min in the particular configuration reported here. The as-deposited samples are hydrophilic and wet easily, and the coating starts to float on the substrate upon wetting. We found out experimentally that the heat treatment of the sprayed coatings in air at 265 °C already removes the wetting agents, and we used this temperature in the results reported below. The developed film was reported to show contact angles of ~160° and sliding angle of ~2°.

Miscellaneous Methods

There are many other fabrication techniques such as solgel method [118–126], ATRP [81, 82], polymerization [94, 107, 127–131], pyrolysis [132], spraying [133, 134], etc.

Application and Uses

It is evident that fundamental aspects of superhydrophobic surfaces, their fabrication and applications, have been studied extensively in the recent years. Although the study of non-wetting surfaces originated from the lotus leaf and development in the initial years was limited to biomimicry, the achievements of recent years have far succeeded them. The customized textured low surface energy material-coated surfaces have been reported to not only repel water and oils but also show antibacterial, self-cleaning, anti-icing, and antireflective properties. The following section summarizes a few such studies highlighting these applications.

Antibacterial

Tomsic et al. reported modified cellulose fibers that in addition to being oleophobic and hydrophobic were also antibacterial, as studied for *Escherichia coli* and *Staphylococcus aureus* [135]. The fibers were modified using a fluoroalkyl functional waterborne siloxane, a nanosized silver, and a reactive organic–inorganic binder, by solgel process. In a similar study, Vilcnik et al. reported the antibacterial properties of superhydrophobic–oleophobic cotton coated with a siloxane solgel hybrid [136]. The study showed a 100 % reduction in bacterial growth of the *E. coli* bacteria against unwashed cotton fabric. Jin et al. also showed similar antibacterial properties of the commercial filter paper modified by the coating of TiO_2-containing composite polymer thin films through chemical etching process [137].

Antireflection

Yan et al. reported antireflective hydro-/oleophobic films employing a simple dipcoating method on fused silica substrates [138]. Fluoroalkylsilanes (FAS) were used as precursor, and by base-catalyzed hydrolysis and condensation of tetraethoxysilane, FAS-modified SiO_2 films were created. This single-layered SiO_2 film dipcoated on either side of the fused silica substrate was reported to exhibit high optical transmittance of 99.5 %, and laser damage threshold had an average value of 22.6 J/cm^2 at 351 nm, when a pulsed laser (1 ns duration) was used. Zhang et al. also fabricated superoleophobic coatings, using silicone nanofilaments through a grow-from approach [139]. They reported the silicone nanofilaments showed an increase in transmittance from 91.2 % (bare glass slide) to about 94 % at 600 nm. They also reported that after modification, the transmittance could be decreased, while its transparency was reasonably retained. The study reasoned that transparency could be attributed to uniform growth of nanofilaments, which reduced light scattering.

Corrosion Resistance

Bulk metallic glass (BMG) is a class of metallic alloys with distinct amorphous structure that provides it with exceptional high yield strength and superior elastic strain limit. However, BMGs have drawbacks in applications because of their corrosive nature. For example, the BMGs containing Ca and Li show poor resistance to oxidation and corrosion [53]. To overcome this, Zhao et al. modified CaLi-based bulk metallic glass by surface etching with water followed by modification using fluoroalkylsilanes [140]. The modified BMG surfaces were reported to show high resistance to corrosion as well as superhydrophobicity and superoleophobicity with contact angles higher than 150°. In addition, Xiong et al. fabricated superamphiphobic cover slips using silica particles coated with a diblock copolymer [141]. The particles were also reported to show resistance to etching when soaked in an aqueous basic solution for extended times, with no changes in contact angles.

Device

As already reported, the legs of water striders are covered with multiscaled structures of oriented needle-shaped microsetae and helical nanogrooves. This structure stabilizes the water–air interface and allows the water striders to walk on the surface of water. Jin et al. reported superhydrophobic and superoleophobic nanocellulose aerogels inspired by these insects [142]. The study showed that fibrillar networks and aggregates consisting of structures of different length scales can support weights of up to three times of their own by exploiting the surface tension acting at different length scales. They suggested that while the macroscopic scale is effective along the perimeter, the microscopic scales are responsible in the interior of the carrier for ensuring buoyancy. The study claimed the wide scope applicability of these aerogel carriers for miniature sensors and devices that are required to float on liquid surfaces, because of qualities like buoyancy; water, oil, and dirt repellence; gas permeability; and flexibility. Jin et al. proposed a design that ensured continuous self-propulsion in soft floating devices using vapor-induced Marangoni effect [143]. The study claimed that velocity of the device could be controlled by the choice of fuels having different vapor pressures.

Self-Cleaning

The lotus leaf is renowned for its self-cleaning abilities by virtue of its high water contact angles and low sliding angles. This has frequently been the inspiration for biomimicry of a large number of artificially developed self-cleaning superhydrophobic materials. The necessary conditions for a surface to be self-cleaning include large static contact angle and small sliding angle. Kim et al. reported a superhydrophobic and superoleophobic surface fabricated by wet etching of silicon wafers followed by forming a fluorinated self-assembled monolayer [144]. This resulting surface was reported to be textured with nanostructures and nanoholes. Self-cleaning properties of the surface were studied by contaminating the surfaces with aluminum hydroxide using glycerine drops. The surfaces were also reported to show superamphiphilic properties when exposed to UV light.

Des et al. reported developing a coating procedure for conductive polymer composite films composed of hollow core carbon nanofibers (CNFs) and fluoroacrylic copolymers [145]. The films could be sprayed on smooth or microtextured surfaces and allowed to dry to form conformal coatings. The coatings were studied to show oil and water contact angles up to 164°s and could be varied by changing the concentrations of the CNF and polymer. The coatings were also reported to show self-cleaning properties. Another study conducted by Yuan et al. [146] showed a bioinspired approach in fabricating ultrathin silica nanowire structures on substrates. The study showed that silica nanostructures could be governed by controlling both the concentration of the polymer and the substrate surface properties, among other experimental parameters. The surfaces were then modified with fluorocarbons to show high liquid repellence against complex liquids like commercial ink-jet ink, milk, etc. The study also showed that the surfaces when contaminated with these complex fluids could be easily cleaned by water flow.

Heat Resistance

There are also studies which show thermally and chemically durability in addition to being hydrophobic or/and oleophobic. Uyanik et al. reported such surfaces could be fabricated using fluoro- or ceramic powders such as SiO_2, SiC, and Al_2O_3 instead of Teflon, by spin coating of aluminum substrates [147]. The study reported the synthesis of polyimides, high thermal-resistant heteroaromatic polymers, which were added with fluor-oligomers to obtain the coatings of desired properties. The final films were reported to show desired thermal, abrasion, scratch, and chemical-resistant properties in addition to liquid repellence.

Oil and Water Collection

The water-collecting abilities of Namib Desert beetles inspired studies for water collection in air atmosphere and oil collection in aqueous environment [53]. Studies report that while 1H, 1H, 2H, 2H-perfluorodecyltrichlorosilane (FTS)-derived organosilane-coated surfaces show superamphiphobicity in air, they become superoleophilic underwater. This knowledge enables the collection oil droplets in water such as during oil drilling or even oil spills. Liu et al. reported that when an FTS-derived glass tube was placed underwater, oil droplets were found to gather its end and coalesce together, which could subsequently be extracted from the water surface.

Oil/Water Separation

Oil/water separation is another important application, and there are numerous studies which have investigated this potential of textured non-wetting surfaces. The use of electric field is one such method, which can be used to exploit the polar nature of water opposed to that of nonpolar oils. Kwon et al. reported a membrane-based operation that achieved this application using gravity-driven separation of oil in oil as well as oil in water emulsions and reported separation efficiencies above 99.9 % [148].

Others

There are numerous other applications to textured superamphiphobic surfaces besides the ones mentioned above such as drag reduction, energy efficiency, anti-icing, cell screening, etc. These surfaces show a wide potential in numerous fields ranging from fluid transportation, fuel economy, infrastructure maintenance, food industry, health-care industry, etc.

Conclusions

Superhydrophobic/superoleophobic surfaces, being important to several advanced applications, are of significant scientific and technologicalinterest. Study of naturally occurring superhydrophobic/superoleophobicsubstances suggests the surfaces with a combination of the low surface energy and the multiple length-scale roughness to be suitable for such applications. Development of artificial liquid-repellent surfaces by mimicking the natural ones has achieved great success. The surface energy of such artificial surfaces can be controlled by application of various organic/inorganic substances containing liquid-repellent functional groups. Composites, on the other hand, can be used to provide both the desired surface roughness and the low surface energy by the combinations of two or more suitable phases. Hence, application of composites, combined with numerous fabrication methods to synthesize surfaces with desired surface morphologies,offerssignificant potential towards the application-oriented development of artificial liquid-repelling surfaces.

References

1. Gordon JE (1991) The new science of strong materials: or why you don't fall through the floor. Penguin, London

2. Bhushan B, Jung YC (2011) Natural and biomimetic artificial surfaces for superhydrophobicity, self-cleaning, low adhesion, and drag reduction. Prog Mater Sci 56(1):1–108
3. Guo H-Y, Li Q, Zhao H-P, Zhou K, Feng X-Q (2015) Functional map of biological and biomimetic materials with hierarchical surface structures. RSC Adv 5:66901
4. Korhonen JT, Huhtamäki T, Ikkala O, Ras RHA (2013) Reliable measurement of the receding contact angle. Langmuir 29:3858
5. Yan YY, Gao N, Barthlott W (2011) Mimicking natural superhydrophobic surfaces and grasping the wetting process: a review on recent progress in preparing superhydrophobic surfaces. Adv Colloid Interface Sci 169(2):80–105
6. Shirtcliffe NJ, McHale G, Atherton S, Newton MI (2010) An introduction to superhydrophobicity. Adv Colloid Interface Sci 161(1):124–138
7. Dussan EB (1979) On the spreading of liquids on solid surfaces: static and dynamic contact lines. Annu Rev Fluid Mech 11(1):371–400
8. De GPG (1985) Wetting: statics and dynamics. Rev Mod Phys 57(3):827
9. Young T (1805) An essay on the cohesion of fluids. Philos Trans R Soc Lond 85:65–87
10. Nosonovsky M, Bhushan B (2008) Multiscale dissipative mechanisms and hierarchical surfaces: friction, superhydrophobicity, and biomimetics. Springer Science & Business Media, Heidelberg
11. Wenzel RN (1936) Resistance of solid surfaces to wetting by water. Ind Eng Chem 28 (8):988–994
12. Cassie ABD, Baxter S (1944) Wettability of porous surfaces. Trans Faraday Soc 40:546–551
13. Öner D, McCarthy TJ (2000) Ultrahydrophobic surfaces. Effects of topography length scales on wettability. Langmuir 16(20):7777–7782
14. Barthlott W, Neinhuis C (1997) Purity of the sacred lotus, or escape from contamination in biological surfaces. Planta 202(1):1–8
15. Guo Z, Liu W, Su BL (2011) Superhydrophobic surfaces: from natural to biomimetic to functional. J Colloid Interface Sci 353(2):335–355
16. Guo Z, Liu W (2007) Biomimic from the superhydrophobic plant leaves in nature: binary structure and unitary structure. Plant Sci 172(6):1103–1112
17. Koch K, Bhushan B, Barthlott W (2009) Multifunctional surface structures of plants: an inspiration for biomimetics. Prog Mater Sci 54:137
18. Lee SM, Kwon TH (2007) Effects of intrinsic hydrophobicity on wettability of polymer replicas of a superhydrophobic lotus leaf. J Micromech Microeng 17(4):687
19. Feng L, Zhang Y, Xi J, Zhu Y, Wang N, Xia F, Jiang L (2008) Petal effect: a superhydrophobic state with high adhesive force. Langmuir 24(8):4114–4119
20. Gao X, Jiang L (2004) Biophysics: water-repellent legs of water striders. Nature 32(7013):36
21. Zheng Y, Bai H, Huang Z, Tian X, Nie FQ, Zhao Y, Jiang L (2010) Directional water collection on wetted spider silk. Nature 463(7281):640–643
22. Parker AR, Lawrence CR (2001) Water capture by a desert beetle. Nature 414(6859):33–34
23. Genzer J, Marmur A (2008) Biological and synthetic self-cleaning surfaces. MRS bull 33 (08):742–746
24. Bormashenko E, Bormashenko Y, Stein T, Whyman G, Bormashenko E (2007) Why do pigeon feathers repel water? Hydrophobicity of pennae, Cassie–Baxter wetting hypothesis and Cassie–Wenzel capillarity-induced wetting transition. J Colloid Interface Sci 311:212–216
25. Bhushan B (2009) Biomimetics: lessons from nature – an overview. Philos Trans R Soc Lond Ser A 367(1893):1445–1486
26. Liu M, Wang S, Wei Z, Song Y, Jiang L (2009) Bioinspired design of a superoleophobic and low adhesive water/solid interface. Adv Mater 21(6):665–669
27. Scholz I, Bückins M, Dolge L, Erlinghagen T, Weth A, Hischen F, Mayer J, Hoffmann S, Riederer M, Riedel M, Baumgartner W (2010) Slippery surfaces of pitcher plants: Nepenthes wax crystals minimize insect attachment via microscopic surface roughness. J Exp Biol 213:1115
28. Zhang J, Li J, Han Y (2004) Superhydrophobic PTFE surfaces by extension. Macromol Rapid Commun 25(11):1105–1108

29. Shiu JY, Kuo CW and Chen P (2004) Fabrication of tunable superhydrophobic surfaces. Proc. SPIE 5648, Smart Materials III, pp 325–332
30. Yabu H, Shimomura M (2005) Single-step fabrication of transparent superhydrophobic porous polymer films. Chem Mater 17(21):5231–5234
31. Xu L, Chen W, Mulchandani A, Yan Y (2005) Reversible conversion of conducting polymer films from superhydrophobic to superhydrophilic. Angew Chem Int Ed 44(37):6009–6012
32. Ma M, Hill RM (2006) Superhydrophobic surfaces. Curr Opin Colloid Interface Sci 11 (4):193–202
33. Khorasani MT, Mirzadeh H, Kermani Z (2005) Wettability of porous polydimethylsiloxane surface: morphology study. Appl Surf Sci 242(3):339–345
34. Sun M, Luo C, Xu L, Ji H, Ouyang Q, Yu D, Chen Y (2005) Artificial lotus leaf by nanocasting. Langmuir 21(19):8978–8981
35. Ma M, Hill RM, Lowery JL, Fridrikh SV, Rutledge GC (2005) Electrospun poly (styrene-block-dimethylsiloxane) block copolymer fibers exhibiting superhydrophobicity. Langmuir 21 (12):5549–5554
36. Lu X, Zhang C, Han Y (2004) Low-density polyethylene superhydrophobic surface by control of its crystallization behavior. Macromol Rapid Commun 25(18):1606–1610
37. Jiang L, Zhao Y, Zhai J (2004) A lotus-leaf-like superhydrophobic surface: a porous microsphere/nanofiber composite film prepared by electrohydrodynamics. Angew Chem 116 (33):4438–4441
38. Zhang J, Lu X, Huang W, Han Y (2005) Reversible superhydrophobicity to superhydrophilicity transition by extending and unloading an elastic polyamide film. Macromol Rapid Commun 26:477–480
39. Zhao N, Xu J, Xie QD, Weng LH, Guo XL, Zhang XL et al (2005) Fabrication of biomimetic superhydrophobic coating with a micro-nano-binary structure. Macromol Rapid Commun 26:1075–1080
40. Mohammadi R, Wassink J, Amirfazli A (2004) Effect of surfactants on wetting of superhydrophobic surfaces. Langmuir 20:9657–9662
41. Feng X, Feng L, Jin M, Zhai J, Jiang L, Zhu D (2004) Reversible super-hydrophobicity to super-hydrophilicity transition of aligned ZnO nanorod films. J Am Chem Soc 126(1):62–63
42. Yang YH, Li ZY, Wang B, Wang CX, Chen DH, Yang GW (2005) Self-assembled ZnO agave-like nanowires and anomalous superhydrophobicity. J Phys Condens Matter 17(35):5441
43. Feng X, Zhai J, Jiang L (2005) The fabrication and switchable superhydrophobicity of TiO_2 nanorod films. Angew Chem Int Ed 44(32):5115–5118
44. Zhang J, Huang W, Han Y (2006) A composite polymer film with both superhydrophobicity and superoleophilicity. Macromol Rapid Commun 27:804
45. Cui Z, Yin L, Wang Q, Ding J, Chen Q (2009) A facile dip-coating process for preparing highly durable superhydrophobic surface with multi-scale structures on paint films. J Colloid Interface Sci 337:531
46. Milionis A, Ruffilli R, Bayer IS (2014) Superhydrophobic nanocomposites from biodegradable thermoplastic starch composites (Mater-Bi®): hydrophobic nano-silica and lycopodium spores. RSC Adv 4:34395
47. Razmjou A, Arifin E, Dong G, Mansouri J, Chen V (2012) Superhydrophobic modification of TiO_2 nanocomposite PVDF membranes for applications in membrane distillation. J Membr Sci 415–416:850
48. Lau KKS, Bico J, Teo KBK, Chhowalla M, Amaratunga GAJ, Milne WI, McKinley GH, Gleason KK (2003) Superhydrophobic carbon nanotube forests. Nano Lett 3:1701
49. Wang K, Hu N-X, Xu G, Qi Y (2011) Stable superhydrophobic composite coatings made from an aqueous dispersion of carbon nanotubes and a fluoropolymer. Carbon 49:1769
50. Jung YC, Bhushan B (2009) Mechanically durable carbon nanotube – composite hierarchical structures with superhydrophobicity, self-cleaning, and low-drag. ACS Nano 3:4155
51. D-a Z, Mei S, Wang Z, Li H, Shi Z, Jin Z (2011) Superhydrophobic polyvinylidene fluoride/graphene porous materials. Carbon 49:5166

52. Asmatulu R, Ceylan M, Nuraje N (2011) Study of superhydrophobic electrospun nanocomposite fibers for energy systems. Langmuir 27:504
53. Liu K, Tian Y, Jiang L (2013) Bio-inspired superoleophobic and smart materials: design, fabrication, and application. Prog Mater Sci 58(4):503–564
54. Cao LL, Hu HH, Gao D (2007) Design and fabrication of micro-textures for inducing a superhydrophobic behaviour on hydrophilic materials. Langmuir 23:4310–4314
55. Herminghaus S (2000) Roughness-induced non-wetting. Europhys Lett 52:165–170
56. Cao L, Gao D (2010) Transparent superhydrophobic and highly oleophobic coatings. Faraday Discuss 146:57–65
57. Kumar RTR, Mogensen KB, Bøggild P (2010) Simple approach to superamphiphobic overhanging silicon nanostructures. J Phys Chem C 114(7):2936–2940
58. Cao L, Price TP, Weiss M, Gao D (2008) Super water-and oil-repellent surfaces on intrinsically hydrophilic and oleophilic porous silicon films. Langmuir 24(5):1640–1643
59. Aulin C, Yun SH, Wågberg L, Lindström T (2009) Design of highly oleophobic cellulose surfaces from structured silicon templates. ACS Appl Mater Interfaces 1(11):2443–2452
60. Tuteja A, Choi WJ, McKinley GH, Cohen RE, Rubner MF (2008) Design parameters for superhydrophobicity and superoleophobicity. MRS Bull 33:752–758
61. Nosonovsky M (2007) Multiscale roughness and stability of superhydrophobic biomimetic interfaces. Langmuir 23(6):3157–3161
62. Tuteja A, Choi W, Ma M, Mabry JM, Mazzella SA, Rutledge GC, Cohen RE (2007) Designing superoleophobic surfaces. Science 318(5856):1618–1622
63. Joly L, Biben T (2009) Wetting and friction on superoleophobic surfaces. Soft Matter 5 (13):2549–2557
64. Zhao H, Law KY, Sambhy V (2011) Fabrication, surface properties, and origin of superoleophobicity for a model textured surface. Langmuir 27(10):5927–5935
65. Ahuja A, Taylor JA, Lifton V, Sidorenko AA, Salamon TR, Lobaton EJ, Kolodner P, Krupenkin TN (2008) Nanonails: a simple geometrical approach to electrically tunable superlyophobic surfaces. Langmuir 24:9
66. Cheng Q, Li M, Zheng Y, Su B, Wang S, Jiang L (2011) Janus interface materials: superhydrophobic air/solid interface and superoleophobic water/solid interface inspired by a lotus leaf. Soft Matter 7(13):5948–5951
67. Bhushan B, Jung YC, Koch K (2009) Self-cleaning efficiency of artificial superhydrophobic surfaces. Langmuir 25:3240–3248
68. Darmanin T, Guittard F (2015) Superhydrophobic and superoleophobic properties in nature. Mater Today 18(5):273–285
69. Rakitov R, Gorb SN (2013) Brochosomal coats turn leafhopper (Insecta, Hemiptera, Cicadellidae) integument to superhydrophobic state. Proc R Soc Lond Ser B 280(1752):20122391
70. Helbig R, Nickerl J, Neinhuis C, Werner C (2011) Smart skin patterns protect springtails. PLOS One 6(9):e25105
71. Bhushan B (2011) Biomimetics inspired surfaces for drag reduction and oleophobicity/philicity. Beilstein J Nanotechnol 2(1):66–84
72. Cai Y, Lin L, Xue Z, Liu M, Wang S, Jiang L (2014) Filefish-inspired surface design for anisotropic underwater oleophobicity. Adv Funct Mater 24:809
73. Yost FG, Michael FR, Eisenmann ET (1995) Extensive wetting due to roughness. Acta Metall Mater 43(1):299–305
74. Shibuichi S, Onda T, Satoh N, Tsujii K (1996) Super water-repellent surfaces resulting from fractal structure. J Phys Chem 100(50):19512–19517
75. Semal S, Blake TD, Geskin V, De Ruijter MJ, Castelein G, De Coninck J (1999) Influence of surface roughness on wetting dynamics. Langmuir 15(25):8765–8770
76. Burton Z, Bhushan B (2005) Hydrophobicity, adhesion, and friction properties of nanopatterned polymers and scale dependence for micro- and nanoelectromechanical systems. Nano Lett 5(8):1607–1613

77. Koch K, Bhushan B, Jung YC, Barthlott W (2009) Fabrication of artificial Lotus leaves and significance of hierarchical structure for superhydrophobicity and low adhesion. Soft Matter 5 (7):1386–1393
78. Kwon Y, Patankar N, Choi J, Lee J (2009) Design of surface hierarchy for extreme hydrophobicity. Langmuir 25(11):6129–6136
79. Wong TS, Huang APH, Ho CM (2009) Wetting behaviors of individual nanostructures. Langmuir 25(12):6599–6603
80. Fürstner R, Barthlott W, Neinhuis C, Walzel P (2005) Wetting and self-cleaning properties of artificial superhydrophobic surfaces. Langmuir 21(3):956–961
81. Xia Y, Whitesides GM (1998) Soft lithography. Annu Rev Mater Sci 28(1):153–184
82. Choi YW, Han JE, Lee S, Sohn D (2009) Preparation of a superhydrophobic film with UV imprinting technology. Macromol Res 17(10):821–824
83. Wang JZ, Zheng ZH, Li HW, Huck WTS, Sirringhaus H (2004) Dewetting of conducting polymer inkjet droplets on patterned surfaces. Nat Mater 3(3):171–176
84. Zhang X, Zhang J, Ren Z, Li X, Zhang X, Zhu D, Yang B (2009) Morphology and wettability control of silicon cone arrays using colloidal lithography. Langmuir 25(13):7375–7382
85. Jeong HE, Kwak MK, Park CI, Suh KY (2009) Wettability of nanoengineered dual-roughness surfaces fabricated by UV-assisted capillary force lithography. J Colloid Interface Sci 339 (1):202–207
86. Sato O, Kubo S, Gu ZZ (2008) Structural color films with lotus effects, superhydrophilicity, and tunable stop-bands. Acc Chem Res 42(1):1–10
87. Cho WK, Choi IS (2008) Fabrication of hairy polymeric films inspired by geckos: wetting and high adhesion properties. Adv Funct Mater 18(7):1089–1096
88. Yuan Z, Chen H, Tang J, Gong H, Liu Y, Wang Z, Chen X (2007) A novel preparation of polystyrene film with a superhydrophobic surface using a template method. J Phys D: Appl Phys 40(11):3485
89. Berendsen CW, Škereň M, Najdek D, Černý F (2009) Superhydrophobic surface structures in thermoplastic polymers by interference lithography and thermal imprinting. Appl Surf Sci 255 (23):9305–9310
90. Winkleman A, Gotesman G, Yoffe A, Naaman R (2008) Immobilizing a drop of water: fabricating highly hydrophobic surfaces that pin water droplets. Nano Lett 8(4):1241–1245
91. Manca M, Cortese B, Viola I, Aricò AS, Cingolani R, Gigli G (2008) Influence of chemistry and topology effects on superhydrophobic CF4-plasma-treated poly (dimethylsiloxane) (PDMS). Langmuir 24(5):1833–1843
92. Tsougeni K, Papageorgiou D, Tserepi A, Gogolides E (2010) "Smart" polymeric microfluidics fabricated by plasma processing: controlled wetting, capillary filling and hydrophobic valving. Lab Chip 10(4):462–469
93. Lee SM, Jung ID, Ko JS (2008) The effect of the surface wettability of nanoprotrusions formed on network-type microstructures. J Micromech Microeng 18(12):125007
94. Borras A, Barranco A, González-Elipe AR (2008) Reversible superhydrophobic to superhydrophilic conversion of Ag@ TiO_2 composite nanofiber surfaces. Langmuir 24 (15):8021–8026
95. Kirkland JJ (1965) Porous thin-layer modified glass bead supports for gas liquid chromatography. Anal Chem 37(12):1458–1461
96. Iler RK (1966) Multilayers of colloidal particles. J Colloid Interface Sci 21(6):569–594
97. Decher G (1997) Fuzzy nanoassemblies: toward layered polymeric multicomposites. Science 277(5330):1232–1237
98. Richardson JJ, Björnmalm M and Caruso F (2015) Technology-driven layer-by-layer assembly of nanofilms. Science 348(6233):aaa2491 (1–11)
99. Ott P, Trenkenschuh K, Gensel J, Fery A, Laschewsky A (2010) Free-standing membranes via covalent cross-linking of polyelectrolyte multilayers with complementary reactivity. Langmuir 26(23):18182–18188

100. Donath E, Sukhorukov GB, Caruso F, Davis SA, Möhwald H (1998) Novel hollow polymer shells by colloid-templated assembly of polyelectrolytes. Angew Chem Int Ed 37 (16):2201–2205
101. Zhang X, Chen H, Zhang H (2007) Layer-by-layer assembly: from conventional to unconventional methods. Chem Commun 14:1395–1405
102. Wu Z, Lee D, Rubner MF, Cohen RE (2007) Structural color in porous, superhydrophilic, and self-cleaning SiO_2/TiO_2 Bragg stacks. Small 3(8):1445–1451
103. Zhao Y, Li M, Lu Q, Shi Z (2008) Superhydrophobic polyimide films with a hierarchical topography: combined replica molding and layer-by-layer assembly. Langmuir 24 (21):12651–12657
104. Hsieh CT, Chen WY, Wu FL, Shen YS (2008) Fabrication and superhydrophobic behavior of fluorinated silica nanosphere arrays. J Adhes Sci Technol 22(3-4):265–275
105. Min WL, Jiang P, Jiang B (2008) Large-scale assembly of colloidal nanoparticles and fabrication of periodic sub wavelength structures. Nanotechnology 19(47):475604
106. Zhu Y, Li J, Wan M, Jiang L (2008) Superhydrophobic 3D microstructures assembled from 1D nanofibers of polyaniline. Macromol Rapid Commun 29(3):239–243
107. Xue CH, Jia ST, Zhang J, Tian LQ (2009) Superhydrophobic surfaces on cotton textiles by complex coating of silica nanoparticles and hydrophobization. Thin Solid Films 517 (16):4593–4598
108. Yan YY, Gao N, Barthlott W (2011) Mimicking natural superhydrophobic surfaces and grasping the wetting process: a review on recent progress in preparing superhydrophobic surfaces. Adv Colloid Interface Sci 169(2):80–105
109. Chen H, Yuan Z, Zhang J, Liu Y, Li K, Zhao D, Tang J (2009) Preparation, characterization and wettability of porous superhydrophobic poly (vinyl chloride) surface. J Porous Mater 16 (4):447–451
110. Li X, Chen G, Ma Y, Feng L, Zhao H, Jiang L, Wang F (2006) Preparation of a superhydrophobic poly (vinyl chloride) surface via solvent–nonsolvent coating. Polymer 47 (2):506–509
111. Nakajima A, Abe K, Hashimoto K, Watanabe T (2000) Preparation of hard super-hydrophobic films with visible light transmission. Thin Solid Films 376(1):140–143
112. Song W, Veiga DD, Custódio CA, Mano JF (2009) Bioinspired degradable substrates with extreme wettability properties. Adv Mater 21(18):1830–1834
113. Zhu M, Zuo W, Yu H, Yang W, Chen Y (2006) Superhydrophobic surface directly created by electrospinning based on hydrophilic material. J Mater Sci 41(12):3793–3797
114. Zheng J, He A, Li J, Xu J, Han CC (2006) Studies on the controlled morphology and wettability of polystyrene surfaces by electrospinning or electrospraying. Polymer 47 (20):7095–7102
115. Ding B, Ogawa T, Kim J, Fujimoto K, Shiratori S (2008) Fabrication of a super-hydrophobic nanofibrous zinc oxide film surface by electrospinning. Thin Solid Films 516(9):2495–2501
116. Li X, Ding B, Lin J, Yu J, Sun G (2009) Enhanced mechanical properties of superhydrophobic microfibrous polystyrene mats via polyamide 6 nanofibers. J Phys Chem C 113 (47):20452–20457
117. Burkarter E, Saul CK, Thomazi F, Cruz NC, Roman LS, Schreiner WH (2007) Superhydrophobic electrosprayed PTFE. Surf Coat Technol 202(1):194–198
118. Lee D, Rubner MF, Cohen RE (2006) All-nanoparticle thin-film coatings. Nano Lett 6 (10):2305–2312
119. Bravo J, Zhai L, Wu Z, Cohen RE, Rubner MF (2007) Transparent superhydrophobic films based on silica nanoparticles. Langmuir 23(13):7293–7298
120. Cebeci FÇ, Wu Z, Zhai L, Cohen RE, Rubner MF (2006) Nanoporosity-driven superhydrophilicity: a means to create multifunctional antifogging coatings. Langmuir 22(6):2856–2862
121. Zhai L, Cebeci FC, Cohen RE, Rubner MF (2004) Stable superhydrophobic coatings from polyelectrolyte multilayers. Nano Lett 4(7):1349–1353

122. Chunder A, Etcheverry K, Londe G, Cho HJ, Zhai L (2009) Conformal switchable superhydrophobic/hydrophilic surfaces for microscale flow control. Colloids Surf A 333 (1):187–193
123. Lvov Y, Ariga K, Onda M, Ichinose I, Kunitake T (1997) Alternate assembly of ordered multilayers of SiO_2 and other nanoparticles and polyions. Langmuir 13(23):6195–6203
124. Ling XY, Phang IY, Vancso GJ, Huskens J, Reinhoudt DN (2009) Stable and transparent superhydrophobic nanoparticle films. Langmuir 25(5):3260–3263
125. Manca M, Cannavale A, De Marco L, Arico AS, Cingolani R, Gigli G (2009) Durable superhydrophobic and antireflective surfaces by trimethylsilanized silica nanoparticles-based sol–gel processing. Langmuir 25(11):6357–6362
126. Liu Y, Chen X, Xin JH (2006) Super-hydrophobic surfaces from a simple coating method: a bionic nanoengineering approach. Nanotechnology 17(13):3259
127. Milella A, Di Mundo R, Palumbo F, Favia P, Fracassi F, d'Agostino R (2009) Plasma nanostructuring of polymers: different routes to superhydrophobicity. Plasma Processes Polym 6(6-7):460–466
128. Han ZJ, Tay BK, Shakerzadeh M, Ostrikov K (2009) Superhydrophobic amorphous carbon/ carbon nanotube nanocomposites. Appl Phys Lett 94(22):223106
129. Ishizaki T, Saito N, InoueY BM, Takai O (2007) Fabrication and characterization of ultra-water-repellent alumina–silica composite films. J Phys D: Appl Phys 40(1):192
130. Chen W, McCarthy TJ (1997) Layer-by-layer deposition: a tool for polymer surface modification. Macromolecules 30(1):78–86
131. Soeno T, Inokuchi K, Shiratori S (2004) Ultra-water-repellent surface: fabrication of complicated structure of SiO_2 nanoparticles by electrostatic self-assembled films. Appl Surf Sci 237 (1):539–543
132. Lee D, Rubner MF, Cohen RE (2006) All-nanoparticle thin-film coatings. Nano Lett 6 (10):2305–2312
133. Lai Y, Lin Z, Huang J, Sun L, Chen Z, Lin C (2010) Controllable construction of ZnO/TiO_2 patterning nanostructures by superhydrophilic/superhydrophobic templates. New J Chem 34 (1):44–51
134. Amigoni S, Taffin de Givenchy E, Dufay M, Guittard F (2009) Covalent layer-by-layer assembled superhydrophobic organic–inorganic hybrid films. Langmuir 25(18):11073–11077
135. Tomšič B, Simončič B, Orel B, Černe L, Tavčer PF, Zorko M, Kovač J (2008) Sol–gel coating of cellulose fibers with antimicrobial and repellent properties. J Sol-Gel Sci Technol 47 (1):44–57
136. Vilcnik A, Jerman I, Šurca Vuk A, Kozelj M, Orel B, Tomšič B, Kovac J (2009) Structural properties and antibacterial effects of hydrophobic and oleophobic sol–gel coatings for cotton fabrics. Langmuir 25(10):5869–5880
137. Jin C, Jiang Y, Niu T, Huang J (2012) Cellulose-based material with amphiphobicity to inhibit bacterial adhesion by surface modification. J Mater Chem 22(25):12562–12567
138. Yan L, Lv H, Wang C, Yuan X (2011) Hydro-oleophobic silica antireflective films with high laser-damage threshold. Opt Laser Technol 43(1):232–236
139. Zhang J, Seeger S (2011) Superoleophobic coatings with ultralow sliding angles based on silicone nanofilaments. Angew Chem Int Ed 50(29):6652–6656
140. Zhao K, Liu KS, Li JF, Wang WH, Jiang L (2009) Superamphiphobic CaLi-based bulk metallic glasses. Scr Mater 60(4):225–227
141. Xiong D, Liu G, Hong L, Duncan ES (2011) Superamphiphobic diblock copolymer coatings. Chem Mater 23(19):4357–4366
142. Jin H, Kettunen M, Laiho A, Pynnönen H, Paltakari J, Marmur A, Ras RH (2011) Superhydrophobic and superoleophobic nanocellulose aerogel membranes as bioinspired cargo carriers on water and oil. Langmuir 27(5):1930–1934
143. Jin H, Marmur A, Ikkala O, Ras RH (2012) Vapour-driven Marangoni propulsion: continuous, prolonged and tunable motion. Chem Sci 3(8):2526–2529

144. Kim TI, Tahk D, Lee HH (2009) Wettability-controllable super water-and moderately oil-repellent surface fabricated by wet chemical etching. Langmuir 25(11):6576–6579
145. Das A, Schutzius TM, Bayer IS, Megaridis CM (2012) Superoleophobic and conductive carbon nanofiber/fluoropolymer composite films. Carbon 50(3):1346–1354
146. Yuan JJ, Jin RH (2011) Direct generation of silica nanowire-based thin film on various substrates with tunable surface nanostructure and extreme repellence toward complex liquids. Langmuir 27(15):9588–9596
147. Uyanik M, Arpac E, Schmidt H, Akarsu M, Sayilkan F, Sayilkan H (2006) Heat-resistant hydrophobic–oleophobic coatings. J Appl Polym Sci 100(3):2386–2392
148. Kwon G, Kota A, Li Y, Sohani A, Mabry JM, Tuteja A (2012) On-demand separation of oil-water mixtures. Adv Mater 24(27):3666–3671

Printed by Printforce, the Netherlands